SETS OF FINITE PERIMETER AND GEOMETRIC VARIATIONAL PROBLEMS

The marriage of analytic power to geometric intuition drives many of today's mathematical advances, yet books that build the connection from an elementary level remain scarce. This engaging introduction to geometric measure theory bridges analysis and geometry, taking readers from basic theory to some of the most celebrated results in modern analysis.

The theory of sets of finite perimeter provides a simple and effective framework. Topics covered include existence, regularity, analysis of singularities, characterization, and symmetry results for minimizers in geometric variational problems, starting from the basics about Hausdorff measures in Euclidean spaces, and ending with complete proofs of the regularity of area-minimizing hypersurfaces up to singular sets of codimension (at least) 8.

Explanatory pictures, detailed proofs, exercises, and remarks providing heuristic motivation and summarizing difficult arguments make this graduate-level textbook suitable for self-study and also a useful reference for researchers. Readers require only undergraduate analysis and basic measure theory.

Francesco Maggi is an Associate Professor at the Università degli Studi di Firenze, Italy.

CAMBRIDGE STUDIES IN ADVANCED MATHEMATICS

All the titles listed below can be obtained from good booksellers or from Cambridge University Press. For a complete series listing visit: http://www.cambridge.org/mathematics.

Already published

94 B. Conrad *Modular forms and the Ramanujan conjecture*
95 M. Schechter *An introduction to nonlinear analysis*
96 R. Carter *Lie algebras of finite and affine type*
97 H. L. Montgomery & R. C. Vaughan *Multiplicative number theory, I*
98 I. Chavel *Riemannian geometry (2nd Edition)*
99 D. Goldfeld *Automorphic forms and L-functions for the group GL(n,R)*
100 M. B. Marcus & J. Rosen *Markov processes, Gaussian processes, and local times*
101 P. Gille & T. Szamuely *Central simple algebras and Galois cohomology*
102 J. Bertoin *Random fragmentation and coagulation processes*
103 E. Frenkel *Langlands correspondence for loop groups*
104 A. Ambrosetti & A. Malchiodi *Nonlinear analysis and semilinear elliptic problems*
105 T. Tao & V. H. Vu *Additive combinatorics*
106 E. B. Davies *Linear operators and their spectra*
107 K. Kodaira *Complex analysis*
108 T. Ceccherini-Silberstein, F. Scarabotti & F. Tolli *Harmonic analysis on finite groups*
109 H. Geiges *An introduction to contact topology*
110 J. Faraut *Analysis on Lie groups: An introduction*
111 E. Park *Complex topological K-theory*
112 D. W. Stroock *Partial differential equations for probabilists*
113 A. Kirillov, Jr *An introduction to Lie groups and Lie algebras*
114 F. Gesztesy *et al. Soliton equations and their algebro-geometric solutions, II*
115 E. de Faria & W. de Melo *Mathematical tools for one-dimensional dynamics*
116 D. Applebaum *Lévy processes and stochastic calculus (2nd Edition)*
117 T. Szamuely *Galois groups and fundamental groups*
118 G. W. Anderson, A. Guionnet & O. Zeitouni *An introduction to random matrices*
119 C. Perez-Garcia & W. H. Schikhof *Locally convex spaces over non-Archimedean valued fields*
120 P. K. Friz & N. B. Victoir *Multidimensional stochastic processes as rough paths*
121 T. Ceccherini-Silberstein, F. Scarabotti & F. Tolli *Representation theory of the symmetric groups*
122 S. Kalikow & R. McCutcheon *An outline of ergodic theory*
123 G. F. Lawler & V. Limic *Random walk: A modern introduction*
124 K. Lux & H. Pahlings *Representations of groups*
125 K. S. Kedlaya *p-adic differential equations*
126 R. Beals & R. Wong *Special functions*
127 E. de Faria & W. de Melo *Mathematical aspects of quantum field theory*
128 A. Terras *Zeta functions of graphs*
129 D. Goldfeld & J. Hundley *Automorphic representations and L-functions for the general linear group, I*
130 D. Goldfeld & J. Hundley *Automorphic representations and L-functions for the general linear group, II*
131 D. A. Craven *The theory of fusion systems*
132 J. Väänänen *Models and games*
133 G. Malle & D. Testerman *Linear algebraic groups and finite groups of Lie type*
134 P. Li *Geometric analysis*
135 F. Maggi *Sets of finite perimeter and geometric variational problems*
136 M. P. Brodmann & R. Y. Sharp *Local cohomology (2nd Edition)*
137 C. Muscalu & W. Schlag *Classical and multilinear harmonic analysis, I*
138 C. Muscalu & W. Schlag *Classical and multilinear harmonic analysis, II*
139 B. Helffer *Spectral theory and its applications*

Sets of Finite Perimeter and Geometric Variational Problems

An Introduction to Geometric Measure Theory

FRANCESCO MAGGI
Università degli Studi di Firenze, Italy

CAMBRIDGE
UNIVERSITY PRESS

CAMBRIDGE
UNIVERSITY PRESS

University Printing House, Cambridge CB2 8BS, United Kingdom

Cambridge University Press is part of the University of Cambridge.

It furthers the University's mission by disseminating knowledge in the pursuit of education, learning and research at the highest international levels of excellence.

www.cambridge.org
Information on this title: www.cambridge.org/9781107021037

First published 2012

A catalogue record for this publication is available from the British Library

Library of Congress Cataloguing in Publication data
Maggi, Francesco, 1978–
Sets of finite perimeter and geometric variational problems : an introduction to geometric measure theory / Francesco Maggi, Universita degli Studi di Firenze, Italy.
pages cm – (Cambridge studies in advanced mathematics ; 135)
Includes bibliographical references and index.
ISBN 978-1-107-02103-7
1. Geometric measure theory. I. Title.
QA312.M278 2012
515′.42 – dc23 2012018822

ISBN 978-1-107-02103-7 Hardback

To Chiara

Contents

Preface

Everyone talks about rock these days;
the problem is they forget about the roll.
Keith Richards

The theory of sets of finite perimeter provides, in the broader framework of Geometric Measure Theory (hereafter referred to as GMT), a particularly well-suited framework for studying the existence, symmetry, regularity, and structure of singularities of minimizers in those geometric variational problems in which surface area is minimized under a volume constraint. Isoperimetric-type problems constitute one of the oldest and more attractive areas of the Calculus of Variations, with a long and beautiful history, and a large number of still open problems and current research. The first aim of this book is to provide a pedagogical introduction to this subject, ranging from the foundations of the theory, to some of the most deep and beautiful results in the field, thus providing a complete background for research activity. We shall cover topics like the Euclidean isoperimetric problem, the description of geometric properties of equilibrium shapes for liquid drops and crystals, the regularity up to a singular set of codimension at least 8 for area minimizing boundaries, and, probably for the first time in book form, the theory of minimizing clusters developed (in a more sophisticated framework) by Almgren in his AMS Memoir [Alm76].

Ideas and techniques from GMT are of crucial importance also in the study of other variational problems (both of parametric and non-parametric character), as well as of partial differential equations. The secondary aim of this book is to provide a multi-leveled introduction to these tools and methods, by adopting an expository style which consists of both heuristic explanations and fully detailed technical arguments. In my opinion, among the various parts of GMT, the theory of sets of finite perimeter is the best suited for this aim. Compared to the theories of currents and varifolds, it uses a lighter notation and, virtually, no preliminary notions from Algebraic or Differential Geometry. At the same time, concerning, for example, key topics like partial regularity properties of minimizers and the analysis of their singularities, the deeper structure of many fundamental arguments can be fully appreciated in this simplified framework. Of course this line of thought has not to be pushed too far. But it is my conviction that a careful reader of this book will be able to enter other parts of GMT with relative ease, or to apply the characteristic tools of GMT in the study of problems arising in other areas of Mathematics.

The book is divided into four parts, which in turn are opened by rather detailed synopses. Depending on their personal backgrounds, different readers may like to use the book in different ways. As we shall explain in a moment, a short "crash-course" is available for complete beginners.

Part I contains the basic theory of Radon measures, Hausdorff measures, and rectifiable sets, and provides the background material for the rest of the book. I am not a big fan of "preliminary chapters", as they often miss a storyline, and quickly become boring. I have thus tried to develop Part I as independent, self-contained, and easily accessible reading. In any case, following the above mentioned "crash-course" makes it possible to see some action taking place without having to work through the entire set of preliminaries.

Part II opens with the basic theory of sets of finite perimeter, which is presented, essentially, as it appears in the original papers by De Giorgi [DG54, DG55, DG58]. In particular, we avoid the use of functions of bounded variation, hoping to better stimulate the development of a geometric intuition of the theory. We also present the original proof of De Giorgi's structure theorem, relying on Whitney's extension theorem, and avoiding the notion of rectifiable set. Later on, in the central portion of Part II, we make the theory of rectifiable sets from Part I enter into the game. We thus provide another justification of De Giorgi's structure theorem, and develop some crucial cut-and-paste competitors' building techniques, first and second variation formulae, and slicing formulae for boundaries. The methods and ideas introduced in this part are finally applied to study variational problems concerning confined liquid drops and anisotropic surface energies.

Part III deals with the regularity theory for local perimeter minimizers, as well as with the analysis of their singularities. In fact, we shall deal with the more general notion of (Λ, r_0)-perimeter minimizer, thus providing regularity results for several Plateau-type problems and isoperimetric-type problems. Finally, Part IV provides an introduction to the theory of minimizing clusters. These last two parts are definitely more advanced, and contain the deeper ideas and finer arguments presented in this book. Although their natural audience will unavoidably be made of more expert readers, I have tried to keep in these parts the same pedagogical point of view adopted elsewhere.

As I said, a "crash-course" on the theory of sets of finite perimeter, of about 130 pages, is available for beginners. The course starts with a revision of the basic theory of Radon measures, temporarily excluding differentiation theory (Chapters 1–4), plus some simple facts concerning weak gradients from Section 7.2. The notion of distributional perimeter is then introduced and used to prove the existence of minimizers in several variational problems, culminating with the solution of the Euclidean isoperimetric problem (Chapters 12–14). Finally, the differentiation theory for Radon measures is developed (Chapter 5), and then applied to clarify the geometric structure of sets of finite perimeter through the study of reduced boundaries (Chapter 15).

Each part is closed by a set of notes and remarks, mainly, but not only, of bibliographical character. The bibliographical remarks, in particular, are not meant to provide a complete picture of the huge literature on the problems considered in this book, and are limited to some suggestions for further reading. In a similar way, we now mention some monographs related to our subject.

Concerning Radon measures and rectifiable sets, further readings of exceptional value are Falconer [Fal86], Mattila [Mat95], and De Lellis [DL08].

For the classical approach to sets of finite perimeter in the context of functions of bounded variation, we refer readers to Giusti [Giu84], Evans and Gariepy [EG92], and Ambrosio, Fusco, and Pallara [AFP00].

The partial regularity theory of Part III does not follow De Giorgi's original approach [DG60], but it is rather modeled after the work of authors like Almgren, Allard, Bombieri, Federer, Schoen, Simon, etc. in the study of area minimizing currents and stationary varifolds. The resulting proofs only rely on direct comparison arguments and on geometrically viewable constructions, and should provide several useful reference points for studying more advanced regularity theories. Accounts and extensions of De Giorgi's original approach can be found in the monographs by Giusti [Giu84] and Massari and Miranda [MM84], as well as in Tamanini's beautiful lecture notes [Tam84].

Readers willing to enter into other parts of GMT have several choices. The introductory books by Almgren [Alm66] and Morgan [Mor09] provide initial insight and motivation. Suggested readings are then Simon [Sim83], Krantz and Parks [KP08], and Giaquinta, Modica, and Souček [GMS98a, GMS98b], as well as, of course, the historical paper by Federer and Fleming [FF60]. Concerning the regularity theory for minimizing currents, the paper by Duzaar and Steffen [DS02] is a valuable source for both its clarity and its completeness. Finally (and although, since its appearance, various crucial parts of the theory have found alternative, simpler justifications, and several major achievements have been obtained), Federer's legendary book [Fed69] remains the ultimate reference for many topics in GMT.

I wish to acknowledge the support received from several friends and colleagues in the realization of this project. This book originates from the lecture notes of a course that I held at the University of Duisburg-Essen in the Spring of 2005, under the advice of Sergio Conti. The successful use of these unpublished notes in undergraduate seminar courses by Peter Hornung and Stefan Müller convinced me to start the revision and expansion of their content. The work with Nicola Fusco and Aldo Pratelli on the stability of the Euclidean isoperimetric inequality [FMP08] greatly influenced the point of view on sets of finite perimeter adopted in this book, which has also been crucially shaped (particularly in connection with the regularity theory of Part III) by several, endless, mathematical discussions with Alessio Figalli. Alessio has also lectured at the University of Texas at Austin on a draft of the first three parts, supporting me with hundreds of comments. Another important contribution came from Guido De Philippis, who read the entire book *twice*, giving me much careful criticism and many useful suggestions. I was lucky to have the opportunity of discussing with Gian Paolo Leonardi various aspects of the theory of minimizing clusters presented in Part IV. Comments and errata were provided to me by Luigi Ambrosio (his lecture notes [Amb97] have been a major source of inspiration), Marco Cicalese, Matteo Focardi, Nicola Fusco, Frank Morgan, Matteo Novaga, Giovanni Pisante and Berardo Ruffini. Finally, I wish to thank Giovanni Alberti, Almut Burchard, Eric Carlen, Camillo de Lellis, Michele Miranda, Massimiliano Morini, and Emanuele Nunzio Spadaro for having expressed to me their encouragement and interest in this project.

I have the feeling that while I was busy trying to talk about the rock without forgetting about the roll, some errors and misprints made their way to the printed page. I will keep an errata list on my webpage.

This work was supported by the European Research Council through the Advanced Grant n. 226234 and the Starting Grant n. 258685, and was completed during my visit to the Department of Mathematics and the Institute for Computational Engineering and Sciences of the University of Texas at Austin. My thanks to the people working therein for the kind hospitality they have shown to me and my family.

Francesco Maggi

Notation

Notation 1 We work in the **n-dimensional Euclidean space** $\mathbb{R}^n$, that is the n-fold cartesian product of the space of real numbers $\mathbb{R}$. Therefore $x = (x_1, ..., x_n)$ is the generic element of $\mathbb{R}^n$, and $\{e_i\}_{i=1}^n$ is the **canonical orthonormal basis** of $\mathbb{R}^n$. We associate with $x \in \mathbb{R}^n \setminus \{0\}$ the one-dimensional linear subspace $\langle x \rangle$ of $\mathbb{R}^n$, $\langle x \rangle = \{tx : t \in \mathbb{R}\}$, called the **space spanned** by x. We endow $\mathbb{R}^n$ with the **Euclidean scalar product** $x \cdot y = \sum_{i=1}^n x_i y_i$. Given a linear subspace H of $\mathbb{R}^n$, we denote by $\dim(H)$ its dimension. If $\dim(H) = k$, then the **orthogonal space** to H in $\mathbb{R}^n$ is the $(n-k)$-dimensional linear space defined by

$$H^\perp = \left\{y \in \mathbb{R}^n : \text{if } x \in H \text{ then } y \cdot x = 0\right\},$$

and we set $x^\perp = \langle x \rangle^\perp$ for $x \neq 0$. The **Minkowski sum** of $E, F \subset \mathbb{R}^n$ is defined as

$$E + F = \left\{x + y : x \in E,\ y \in F\right\},$$

with $x + F = \{x\} + F$ if $x \in \mathbb{R}^n$. A **k-dimensional plane** π in $\mathbb{R}^n$ is a set of the form $\pi = x + H$ where $x \in \mathbb{R}^n$ and H is a k-dimensional space in $\mathbb{R}^n$. When $k = 1$ we simply say that π is a **line** in $\mathbb{R}^n$. Given $E \subset \mathbb{R}^n$ and $\lambda > 0$ we set

$$\lambda E = \left\{\lambda x : x \in E\right\}.$$

Defining the **Euclidean norm** $|x| = (\sum_{i=1}^n x_i^2)^{1/2}$, the **Euclidean open ball** in $\mathbb{R}^n$ of center x and radius $r > 0$ is

$$B(x, r) = \left\{y \in \mathbb{R}^n : |y - x| < r\right\}.$$

When $x = 0$ we set $B(0, r) = B_r$ and $B_1 = B$, so that $B(x, r) = x + B_r = x + rB$. We also set $S^{n-1} = \partial B = \{x \in \mathbb{R}^n : |x| = 1\}$ for the unit sphere in $\mathbb{R}^n$. Given $E, F \subset \mathbb{R}^n$, the **diameter of** E and the **distance between** E **and** F are

$$\operatorname{diam}(E) = \sup\left\{|x - y| : x, y \in E\right\},$$

$$\operatorname{dist}(E, F) = \inf\left\{|x - y| : x \in E,\ y \in F\right\}.$$

The **interior**, **closure**, and **topological boundary** (in the Euclidean topology) of $E \subset \mathbb{R}^n$ are denoted as usual as $\mathring{E}$, $\overline{E}$, and ∂E respectively. We write $E \subset\subset A$ and say E is **compactly contained** in A if $\overline{E} \subset A$.

Notation 2 A family $\mathcal{F}$ of subsets of $\mathbb{R}^n$ is **disjoint** if $F_1, F_2 \in \mathcal{F}$, $F_1 \neq F_2$ implies $F_1 \cap F_2 = \emptyset$; it is **countable** if there exists a surjective function $f\colon \mathbb{N} \to \mathcal{F}$; it is a **covering** of $E \subset \mathbb{R}^n$ if $E \subset \bigcup_{F \in \mathcal{F}} F$. A **partition** of E is a disjoint covering of E which is composed of subsets of E.

Notation 3 (Linear functions) We denote by $\mathbb{R}^m \otimes \mathbb{R}^n$ the vector space of linear maps from $\mathbb{R}^n$ to $\mathbb{R}^m$. If $T \in \mathbb{R}^m \otimes \mathbb{R}^n$, then $T(\mathbb{R}^n)$, the **image** of T, is a linear subspace of $\mathbb{R}^m$, and $\mathrm{Ker}\, T = \{T = 0\}$, the **kernel** of T, is a linear subspace of $\mathbb{R}^n$. The dimension of $T(\mathbb{R}^n)$ is called the **rank** of T, and T has **full rank** if $\dim(T(\mathbb{R}^n)) = m$. On $\mathbb{R}^m \otimes \mathbb{R}^n$ we define the **operator norm**,

$$\|T\| = \sup\left\{|Tx| : x \in \mathbb{R}^n, |x| < 1\right\}, \qquad T \in \mathbb{R}^m \otimes \mathbb{R}^n .$$

We notice that $\|T\| = \mathrm{Lip}(T)$, the Lipschitz constant of T on $\mathbb{R}^n$; see Chapter 7. If $T \in \mathbb{R}^m \otimes \mathbb{R}^n$, then we define a linear map $T^* \in \mathbb{R}^n \otimes \mathbb{R}^m$, called the **adjoint** of T, through the identity

$$(Tx) \cdot y = x \cdot (T^* y), \qquad \forall x \in \mathbb{R}^n, y \in \mathbb{R}^m .$$

Given $v \in \mathbb{R}^n$ and $w \in \mathbb{R}^m$, we define a linear map $w \otimes v$ from $\mathbb{R}^n$ to $\mathbb{R}^m$, setting

$$(w \otimes v)x = (v \cdot x)w, \qquad x \in \mathbb{R}^n .$$

When $v \neq 0$ and $w \neq 0$ we say that $w \otimes v$ is a **rank-one map**, as we clearly have

$$(w \otimes v)(\mathbb{R}^n) = \langle w \rangle , \qquad \mathrm{Ker}\,(w \otimes v) = v^\perp .$$

We also notice the useful relations

$$(w \otimes v)^* = v \otimes w, \qquad \|w \otimes v\| = |v|\,|w| .$$

Rank-one maps induce a canonical identification of $\mathbb{R}^m \otimes \mathbb{R}^n$ with the space $\mathbb{R}^{m \times n}$ of $m \times n$ matrices $(a_{i,j})$ $(1 \le i \le n,\ 1 \le j \le m)$, having m rows and n columns. Indeed, if $V = \{v_j\}_{j=1}^n$ and $W = \{w_i\}_{i=1}^m$ are orthonormal bases of $\mathbb{R}^n$ and $\mathbb{R}^m$ respectively, then, by definition of $w_i \otimes v_j$, we find that

$$T = \sum_{j=1}^{n} \sum_{i=1}^{m} \left(w_i \cdot (T v_j)\right) w_i \otimes v_j .$$

Correspondingly, we associate T with the $m \times n$ matrix $(T_{i,j})$ with (i, j)th entry given by $T_{i,j} = w_i \cdot (T v_j)$. When $n = m$, this identification allows us to define the notions of **determinant** and **trace** of a matrix for a linear map, by setting

$$\det T = \det(T_{i,j}), \qquad \mathrm{trace}\, T = \mathrm{trace}(T_{i,j}) .$$

The functions $\det : \mathbb{R}^n \otimes \mathbb{R}^n \to \mathbb{R}$ and $\mathrm{trace} : \mathbb{R}^n \otimes \mathbb{R}^n \to \mathbb{R}$ are then independent of the choice of V underlying the identification of $\mathbb{R}^m \otimes \mathbb{R}^n$ with the space $\mathbb{R}^{m\times n}$, and inherit their usual properties. For example, we have

$$\det(TS) = \det(T)\det(S), \qquad \forall T, S \in \mathbb{R}^n \otimes \mathbb{R}^n,$$

and $\det(\mathrm{Id}_n) = 1$, where of course $\mathrm{Id}_n x = x$ ($x \in \mathbb{R}^n$). If we denote by $\mathbf{GL}(n)$ the set of **invertible linear functions** $T \in \mathbb{R}^n \otimes \mathbb{R}^n$, then

$$\mathbf{GL}(n) = \left\{ T \in \mathbb{R}^n \otimes \mathbb{R}^n : \det T \neq 0 \right\}.$$

In particular, if $n \geq 2$ then $\det(w \otimes v) = 0$ for every $v, w \in \mathbb{R}^n$. The trace defines a linear function on $\mathbb{R}^n \otimes \mathbb{R}^n$ with $\mathrm{trace}(\mathrm{Id}_n) = n$ and, for every $T, S \in \mathbb{R}^n \otimes \mathbb{R}^n$,

$$\mathrm{trace}(T^*) = \mathrm{trace}(T), \qquad \mathrm{trace}(TS) = \mathrm{trace}(ST).$$

It is also useful to recall that for every $v, w \in \mathbb{R}^n$ we have

$$\mathrm{trace}(w \otimes v) = v \cdot w.$$

The trace operator can also be used to define a scalar product on $\mathbb{R}^m \otimes \mathbb{R}^n$:

$$T : S = \mathrm{trace}(S^* T) = \mathrm{trace}(T^* S), \qquad T, S \in \mathbb{R}^m \otimes \mathbb{R}^n.$$

The norm corresponding to this scalar product (which does not coincide with the operator norm) is defined as

$$|T| = \sqrt{\mathrm{trace}(T^* T)}, \qquad T \in \mathbb{R}^m \otimes \mathbb{R}^n.$$

Notation 4 (Standard product decomposition of $\mathbb{R}^n$ into $\mathbb{R}^k \times \mathbb{R}^{n-k}$) When we need to decompose $\mathbb{R}^n$ as the cartesian product $\mathbb{R}^k \times \mathbb{R}^{n-k}$, $1 \leq k \leq n-1$, we denote by $\mathbf{p} : \mathbb{R}^n \to \mathbb{R}^k \times \{0\} = \mathbb{R}^k$ and $\mathbf{q} : \mathbb{R}^n \to \{0\} \times \mathbb{R}^{n-k} = \mathbb{R}^{n-k}$ the horizontal and vertical projections, so that $x = (\mathbf{p}x, \mathbf{q}x)$, $x \in \mathbb{R}^n$. We then introduce the cylinder of center $x \in \mathbb{R}^n$ and radius $r > 0$,

$$\mathbf{C}(x, r) = \left\{ y \in \mathbb{R}^n : |\mathbf{p}(y - x)| < r, |\mathbf{q}(y - x)| < r \right\},$$

and the k-dimensional ball of center $z \in \mathbb{R}^k$ and radius $r > 0$,

$$\mathbf{D}(z, r) = \left\{ w \in \mathbb{R}^k : |z - w| < r \right\}.$$

Moreover, we always abbreviate

$$\mathbf{C}(0, r) = \mathbf{C}_r, \qquad \mathbf{C}_1 = \mathbf{C}, \qquad \mathbf{D}(0, r) = \mathbf{D}_r, \qquad \mathbf{D}_1 = \mathbf{D}.$$

When $k = n - 1$, we alternatively set $\mathbf{p}x = x'$ and $\mathbf{q}x = x_n$, so that $x = (x', x_n)$. Correspondingly we denote the gradient operator in $\mathbb{R}^n$ and in $\mathbb{R}^{n-1}$, respectively, by ∇ and $\nabla' = (\partial_1, \ldots, \partial_{n-1})$. If $u : \mathbb{R}^n \to \mathbb{R}$ has gradient $\nabla u(x) \in \mathbb{R}^n$ at $x \in \mathbb{R}^n$, then we set $\nabla u(x) = (\nabla' u(x), \partial_n u(x))$.

PART ONE

Radon measures on $\mathbb{R}^n$

Synopsis

In this part we discuss the basic theory of Radon measures on $\mathbb{R}^n$. Roughly speaking, if $\mathcal{P}(\mathbb{R}^n)$ denotes the set of the parts of $\mathbb{R}^n$, then a Radon measure μ on $\mathbb{R}^n$ is a function $\mu\colon \mathcal{P}(\mathbb{R}^n) \to [0, \infty]$, which is countably additive (at least) on the family of Borel sets of $\mathbb{R}^n$, takes finite values on bounded sets, and is completely identified by its values on open sets. The Lebesgue measure on $\mathbb{R}^n$ and the Dirac measure δ_x at $x \in \mathbb{R}^n$ are well-known examples of Radon measures on $\mathbb{R}^n$. Moreover, any locally summable function on $\mathbb{R}^n$, as well as any k-dimensional surface in $\mathbb{R}^n$, $1 \le k \le n-1$, can be naturally identified with a Radon measure on $\mathbb{R}^n$. There are good reasons to look at such familiar objects from this particular point of view. Indeed, the natural notion of convergence for sequences of Radon measures satisfies very flexible compactness properties. As a consequence, the theory of Radon measures provides a unified framework for dealing with the various convergence and compactness phenomena that one faces in the study of geometric variational problems. For example, a sequence of continuous functions on $\mathbb{R}^n$ that (as a sequence of Radon measures) is converging to a surface in $\mathbb{R}^n$ is something that cannot be handled with the notions of convergence usually considered on spaces of continuous functions or on Lebesgue spaces. Similarly, the existence of a tangent plane to a surface at one of its points can be understood as the convergence of the (Radon measures naturally associated with) re-scaled and translated copies of the surface to the (Radon measure naturally associated with the) tangent plane itself. This peculiar point of view opens the door for a geometrically meaningful (and analytically powerful) extension of the notion of differentiability to the wide class of objects, the family of rectifiable sets, that one must consider in solving geometric variational problems.

Part I is divided into two main portions. The first one (Chapters 1–6) is devoted to the more abstract aspects of the theory. In Chapters 1–4, we introduce the main definitions, present the most basic examples, and prove the fundamental representation and compactness theorems about Radon measures. (These results already suffice to give an understanding of the basic theory of sets of finite perimeter as presented in the first three chapters of Part II.) Differentiation theory, and its applications, are discussed in Chapters 5–6. In the second portion of Part I (Chapters 7–11), we consider Radon measures from a more geometric viewpoint, focusing on the interaction between Euclidean geometry and Measure Theory, and covering topics such as Lipschitz functions, Hausdorff measures, area formulae, rectifiable sets, and measure-theoretic differentiability. These are prerequisites to more advanced parts of the theory of sets of finite perimeter, and can be safely postponed until really needed. We now examine more closely each chapter.

In Chapters 1–2 we introduce the notions of Borel and Radon measure. This is done in the context of *outer measures*, rather than in the classical context of standard measures defined on σ-algebras. We simultaneously develop both the basic properties relating Borel and Radon measures to the Euclidean topology of $\mathbb{R}^n$ and the basic examples of the theory that are obtained by combining the definitions of Lebesgue and Hausdorff measures with the operations of restriction to a set and push-forward through a function.

In Chapter 3 we look more closely at Hausdorff measures. We establish their most basic properties and introduce the notion of Hausdorff dimension. Next, we show equivalence between the Lebesgue measure on $\mathbb{R}^n$ and the n-dimensional Hausdorff measure on $\mathbb{R}^n$, and we study the relation between the elementary notion of length of a curve, based on the existence of a parametrization, and the notion induced by one-dimensional Hausdorff measures.

In Chapter 4 we further develop the general theory of Radon measures. In particular, the deep link between Radon measures and continuous functions with compact support is presented, leading to the definition of *vector-valued* Radon measures, of weak-star convergence of Radon measures, and to the proof of the fundamental Riesz's representation theorem: every bounded linear functional on $C^0_c(\mathbb{R}^n;\mathbb{R}^m)$ is representable as integration with respect to an $\mathbb{R}^m$-valued Radon measure on $\mathbb{R}^n$. This last result, in turn, is the key to the weak-star compactness criterion for sequences of Radon measures.

Chapters 5–6 present differentiation theory and its applications. The goal is to compare two Radon measures ν and μ by looking, as $r \to 0^+$, at the ratios

$$\frac{\nu(B(x,r))}{\mu(B(x,r))},$$

which are defined at those x where μ is supported (i.e., $\mu(B(x,r)) > 0$ for every $r > 0$). The Besicovitch–Lebesgue differentiation theorem ensures that, for μ-a.e. x in the support of μ, these ratios converge to a finite limit $u(x)$, and that restriction of ν to the support of μ equals integration of u with respect to μ. Differentiation theory plays a crucial role in proving the validity of classical (or generalized) differentiability properties in many situations.

In Chapter 7 we study the basic properties of Lipschitz functions, proving Rademacher's theorem about the almost everywhere classical differentiability of Lipschitz functions, and Kirszbraun's theorem concerning the optimal extension problem for vector-valued Lipschitz maps.

Chapter 8 presents the area formula, which relates the Hausdorff measure of a set in $\mathbb{R}^n$ with that of its Lipschitz images into any $\mathbb{R}^m$ with $m \geq n$. As a consequence, the classical notion of area of a k-dimensional surface M in $\mathbb{R}^n$ is seen to coincide with the k-dimensional Hausdorff measure of M. Some applications of the area formula are presented in Chapter 9, where, in particular, the classical Gauss–Green theorem is proved.

In Chapter 10 we introduce one of the most important notions of Geometric Measure Theory, that of a k-dimensional rectifiable set in $\mathbb{R}^n$ $(1 \leq k \leq n-1)$. This is a very broad generalization of the concept of k-dimensional C^1-surface, allowing for complex singularities but, at the same time, retaining tangential differentiability properties, at least in a measure-theoretic sense. A crucial result is the following: if the k-dimensional blow-ups of a Radon measure μ converge to k-dimensional linear spaces (seen as Radon measures), then it turns out that μ itself is the restriction of the k-dimensional Hausdorff measure to a k-dimensional rectifiable set.

In Chapter 11, we introduce the notion of tangential differentiability of a Lipschitz function with respect to a rectifiable set, extend the area formula to this context, and prove the divergence theorem on C^2-surfaces with boundary.

1

Outer measures

Denote by $\mathcal{P}(\mathbb{R}^n)$ the set of all subsets of $\mathbb{R}^n$. An **outer measure** μ on $\mathbb{R}^n$ is a set function on $\mathbb{R}^n$ with values in $[0, \infty]$, $\mu \colon \mathcal{P}(\mathbb{R}^n) \to [0, \infty]$, with $\mu(\emptyset) = 0$ and

$$E \subset \bigcup_{h \in \mathbb{N}} E_h \quad \Rightarrow \quad \mu(E) \le \sum_{h \in \mathbb{N}} \mu(E_h) .$$

This property, called **σ-subadditivity**, implies the **monotonicity** of μ,

$$E \subset F \quad \Rightarrow \quad \mu(E) \le \mu(F) .$$

1.1 Examples of outer measures

Simple familiar examples of outer measures are the Dirac measure and the counting measure. The **Dirac measure** δ_x at $x \in \mathbb{R}^n$ is defined on $E \subset \mathbb{R}^n$ as

$$\delta_x(E) = \begin{cases} 1, & x \in E, \\ 0, & x \notin E, \end{cases}$$

while the **counting measure** $\#$ of E is

$$\#(E) = \begin{cases} \text{number of elements of } E, & \text{if } E \text{ is finite}, \\ +\infty, & \text{if } E \text{ is infinite}. \end{cases}$$

The two most important examples of outer measures are Lebesgue and Hausdorff measures.

Lebsegue measure: The **Lebesgue measure** of a set $E \subset \mathbb{R}^n$ is defined as

$$\mathcal{L}^n(E) = \inf_{\mathcal{F}} \sum_{Q \in \mathcal{F}} r(Q)^n ,$$

where $\mathcal{F}$ is a countable covering of E by cubes with sides parallel to the coordinate axes, and $r(Q)$ denotes the side length of Q (the cubes Q are not assumed to be open, nor closed). The Lebesgue measure $\mathcal{L}^n(E)$ is interpreted as the n-dimensional volume of E. Usually, we write

$$\mathcal{L}^n(E) = |E|,$$

and refer to $|E|$ as the *volume* of E. Clearly, $\mathcal{L}^n$ is an outer measure. Moreover, it is translation-invariant, that is $|x + E| = |E|$ for every $x \in \mathbb{R}^n$, and satisfies the scaling law $|\lambda E| = \lambda^n |E|$, $\lambda > 0$. If $B = \{x \in \mathbb{R}^n : |x| < 1\}$ is the Euclidean unit ball of $\mathbb{R}^n$, then we set $\omega_n = |B|$. It is easily seen that $\omega_1 = 2$.

Hausdorff measure: Let $n, k \in \mathbb{N}$, with $n \geq 2$ and $1 \leq k \leq n - 1$. A bounded open set $A \subset \mathbb{R}^k$ and a function $f \in C^1(\mathbb{R}^k; \mathbb{R}^n)$ define a **k-dimensional parametrized surface** $f(A)$ in $\mathbb{R}^n$ provided f is injective on A with $Jf(x) > 0$ for every $x \in A$. Here $Jf(x)$ denotes the **Jacobian** of f at x, namely

$$Jf(x) = \sqrt{\det(\nabla f(x)^* \nabla f(x))},$$

where, if $k = 1$, this means that $Jf(x) = |f'(x)|$. The condition $Jf(x) > 0$ ensures that $\nabla f(x)(\mathbb{R}^k)$ is a k-dimensional subspace of $\mathbb{R}^n$. The k-dimensional area of $f(A)$ is then classically defined as

$$k\text{-dimensional area of } f(A) = \int_A Jf(x)\,dx. \tag{1.1}$$

In the study of geometric variational problems we need to extend this definition of k-dimensional area to more general sets than k-dimensional C^1-images. Hausdorff measures are introduced to this end. To avoid the use of parametrizations the definition is based on a covering procedure, as in the construction of the Lebesgue measure. Given $n, k \in \mathbb{N}$, $\delta > 0$, the **k-dimensional Hausdorff measure of step** δ of a set $E \subset \mathbb{R}^n$ is defined as

$$\mathcal{H}^k_\delta(E) = \inf_{\mathcal{F}} \sum_{F \in \mathcal{F}} \omega_k \left(\frac{\operatorname{diam}(F)}{2} \right)^k, \tag{1.2}$$

where $\mathcal{F}$ is a countable covering of E by sets $F \subset \mathbb{R}^n$ such that $\operatorname{diam}(F) < \delta$; see Figure 1.1. The **$k$-dimensional Hausdorff measure** of E is then

$$\mathcal{H}^k(E) = \sup_{\delta \in (0,\infty]} \mathcal{H}^k_\delta(E) = \lim_{\delta \to 0^+} \mathcal{H}^k_\delta(E). \tag{1.3}$$

It is trivial to see that, for every $\delta \in (0, \infty]$, $\mathcal{H}^k_\delta$ is an outer measure. As an immediate consequence, $\mathcal{H}^k$ is an outer measure too. In a similar way one proves that $\mathcal{H}^k$ is translation-invariant and that it satisfies the scaling law $\mathcal{H}^k(\lambda E) = \lambda^k \mathcal{H}^k(E)$, $\lambda > 0$. The fact that $\mathcal{H}^k(f(A))$ agrees with the classical notion of area on a k-dimensional parametrized surface $f(A)$ as defined in (1.1) is the content of the important *area formula*, discussed in Chapter 8.

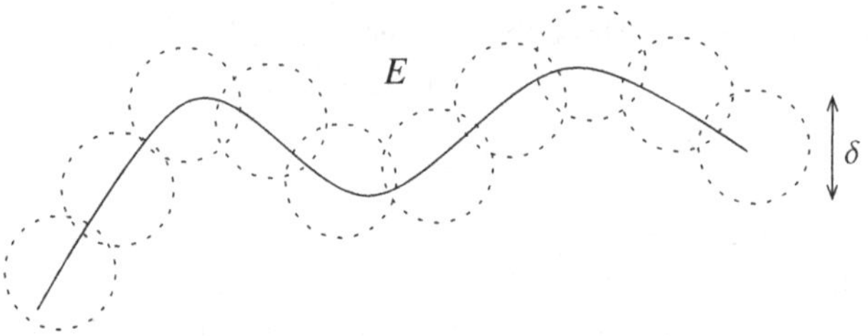

Figure 1.1 When computing $\mathcal{H}^k_\delta(E)$ one sums up, corresponding to each element F of a covering $\mathcal{F}$ of E, the k-dimensional measure of a k-dimensional ball of diameter $\mathrm{diam}(F)$. The minimization process used to compute $\mathcal{H}^k_\delta(E)$ does not detect any "deviation from straightness" of E taking place at a scale smaller than δ; see also Remark 1.1. Hence, one takes the limit $\delta \to 0^+$.

Remark 1.1 The idea behind the definition of Hausdorff measures is readily understood by considering the following statements concerning the case $k = 1$, $n = 2$ (see Chapter 3 for proofs).

(i) If E is a segment, then, for every $\delta > 0$, $\mathcal{H}^1_\delta(E)$ and $\mathcal{H}^1(E)$ coincide with the Euclidean length of E. If E is a polygonal curve composed of finitely many segments of length at least d, then, for every $\delta \in (0, d)$, $\mathcal{H}^1_\delta(E)$ and $\mathcal{H}^1(E)$ both agree with the Euclidean length of E.

(ii) If E is a curve of diameter d and $\delta \geq d$, then $\mathcal{H}^1_\delta(E) \leq d$ (use the covering $\mathcal{F} = \{E\}$ of E in (1.2)), while, of course, the length of E can be arbitrarily large. It is only in the limit $\delta \to 0^+$ that $\mathcal{H}^1_\delta(E)$ approaches the length of E; see Section 3.2.

(iii) If E is countable (hence, zero-dimensional), then $\mathcal{H}^1(E) = 0$.

(iv) If E is an open set of $\mathbb{R}^2$ (i.e., a two-dimensional set), then $\mathcal{H}^1(E) = \infty$.

Remark 1.2 Given $s \in [0, \infty)$, the s-dimensional Hausdorff measures $\mathcal{H}^s_\delta$ and $\mathcal{H}^s$ are defined by simply replacing k with s in (1.2) and (1.3). The normalization constant ω_k is replaced by

$$\omega_s = \frac{\pi^{s/2}}{\Gamma(1 + s/2)}, \qquad s \geq 0,$$

where $\Gamma\colon (0, \infty) \to [1, \infty)$ is the Euler Gamma function

$$\Gamma(s) = \int_0^\infty t^{s-1} e^{-t} \mathrm{d}t, \qquad s > 0.$$

This is consistent as $\omega_k = \pi^{k/2}\Gamma(1+k/2)^{-1}$ for $k \in \mathbb{N}$, $k \geq 1$. Once again $\mathcal{H}^s_\delta$ and $\mathcal{H}^s$ are translation-invariant outer measures, with $\mathcal{H}^s(\lambda E) = \lambda^s \mathcal{H}^s(E)$, $\lambda > 0$.

Exercise 1.3 Clearly, in the definition of $\mathcal{H}^s_\delta(E)$, we may equivalently consider coverings of E by subsets of E. Similarly,

(i) we may use coverings of E by closed convex sets intersecting E: indeed, the diameter of a set is the same as the diameter of its closed convex hull, and, if a set in $\mathcal{F}$ does not intersect E, it is convenient to discard it;
(ii) we may use coverings of E by open sets intersecting E: indeed, for every $F \subset \mathbb{R}^n$ and $\varepsilon > 0$, the **ε-neighborhood of** F,

$$I_\varepsilon(F) = \left\{x \in \mathbb{R}^n : \operatorname{dist}(x, F) < \varepsilon\right\}, \tag{1.4}$$

is open, contains F, and is such that $\operatorname{diam}(F_\varepsilon) \le \operatorname{diam}(F) + 2\varepsilon$.

1.2 Measurable sets and σ-additivity

Given a family $\mathcal{F}$ of subsets of $\mathbb{R}^n$, we say that the outer measure μ on $\mathbb{R}^n$ is **σ-additive** on $\mathcal{F}$, provided

$$\mu\Big(\bigcup_{h\in\mathbb{N}} E_h\Big) = \sum_{h\in\mathbb{N}} \mu(E_h),$$

for every disjoint sequence $\{E_h\}_{h\in\mathbb{N}} \subset \mathcal{F}$ (i.e., $E_h \cap E_k = \emptyset$ if $h \ne k$). Accordingly to our naive intuition about the notion of measure, we would expect any reasonable measure to be σ-additive on $\mathcal{P}(\mathbb{R}^n)$. However, this fails even in the case of the Lebesgue measure $\mathcal{L}^1$ on $\mathbb{R}$. To show this, let us consider the classical **Vitali's example**. Define an equivalence relation $\approx$ on $(0, 1)$, so that $x \approx y$ if and only if $x - y$ is rational. By the axiom of choice, there exists a set $E \subset (0, 1)$ containing exactly one element from each of the equivalence classes defined by $\approx$ on $(0, 1)$. If $\{x_h\}_{h\in\mathbb{N}} = \mathbb{Q} \cap (0, 1)$, then the sequence of sets

$$E_h = \Big(x_h + \big(E \cap (0, 1 - x_h)\big)\Big) \cup \Big((x_h - 1) + \big(E \cap (1 - x_h, 1)\big)\Big)$$

is, by construction of E, disjoint. By the translation invariance of $\mathcal{L}^1$,

$$|E_h| = |E \cap (0, 1 - x_h)| + |E \cap (1 - x_h, 1)| = |E|,$$

with $(0, 1) = \bigcup_{h\in\mathbb{N}} E_h$. The σ-additivity of $\mathcal{L}^1$ on $\{E_h\}_{h\in\mathbb{N}}$ would then imply

$$1 = |(0, 1)| = \sum_{h\in\mathbb{N}} |E|,$$

against $|E| \in [0, \infty]$. Hence, $\mathcal{L}^1$ is not σ-additive on $\mathcal{P}(\mathbb{R})$. As we are going to prove in Section 2.1, $\mathcal{L}^1$ is, however, σ-additive on a large family of subsets of $\mathbb{R}^n$. A first step towards this kind of result is the following theorem, which provides, given outer measure μ, a natural domain of σ-additivity for μ.

Theorem 1.4 (Carathéodory's theorem) *If μ is an outer measure on $\mathbb{R}^n$, and $\mathcal{M}(\mu)$ is the family of those $E \subset \mathbb{R}^n$ such that*

$$\mu(F) = \mu(E \cap F) + \mu(F \setminus E), \qquad \forall F \subset \mathbb{R}^n, \tag{1.5}$$

then $\mathcal{M}(\mu)$ is a σ-algebra, and μ is a measure on $\mathcal{M}(\mu)$.

Remark 1.5 We recall that $\mathcal{M} \subset \mathcal{P}(\mathbb{R}^n)$ is a **σ-algebra** on $\mathbb{R}^n$ if $E \in \mathcal{M}$ implies $\mathbb{R}^n \setminus E \in \mathcal{M}$, $\{E_h\}_{h\in\mathbb{N}} \subset \mathcal{M}$ implies $\bigcup_{h\in\mathbb{N}} E_h \in \mathcal{M}$, and $\mathbb{R}^n \in \mathcal{M}$. If $\mathcal{M}$ is a σ-algebra, then a set function $\mu \colon \mathcal{M} \to [0, \infty]$ is a **measure** on $\mathcal{M}$ if $\mu(\emptyset) = 0$ and μ is σ-additive on $\mathcal{M}$.

Remark 1.6 A set E belongs to $\mathcal{M}(\mu)$ if it can be used to divide *any test set* $F \subset \mathbb{R}^n$ into two parts on which μ is additive. Notice that, by σ-subadditivity of μ, $E \in \mathcal{M}(\mu)$ if and only if

$$\mu(F) \geq \mu(F \setminus E) + \mu(F \cap E), \qquad \forall F \subset \mathbb{R}^n \text{ s.t. } \mu(F) < \infty. \tag{1.6}$$

Elements of $\mathcal{M}(\mu)$ are called **μ-measurable sets**.

Proof *Step one:* We prove that $\mathcal{M}(\mu)$ is a σ-algebra. Clearly, $\emptyset \in \mathcal{M}(\mu)$ and $E \in \mathcal{M}(\mu)$ implies $\mathbb{R}^n \setminus E \in \mathcal{M}(\mu)$. We now let $\{E_h\}_{h\in\mathbb{N}} \subset \mathcal{M}(\mu)$, set $E = \bigcup_{h\in\mathbb{N}} E_h$, and prove that $E \in \mathcal{M}(\mu)$. Given $F \subset \mathbb{R}^n$, as $E_0 \in \mathcal{M}(\mu)$, we have

$$\mu(F) = \mu(F \setminus E_0) + \mu(F \cap E_0).$$

As $E_1 \in \mathcal{M}(\mu)$ we also have

$$\begin{aligned}\mu(F \setminus E_0) &= \mu\Big((F \setminus E_0) \setminus E_1\Big) + \mu\Big((F \setminus E_0) \cap E_1\Big)\\ &= \mu\Big(F \setminus (E_0 \cup E_1)\Big) + \mu\Big((F \setminus E_0) \cap E_1\Big),\end{aligned}$$

and thus $\mu(F) = \mu(F \setminus (E_0 \cup E_1)) + \mu((F \setminus E_0) \cap E_1) + \mu(F \cap E_0)$. By induction,

$$\mu(F) = \mu\Big(F \setminus \bigcup_{h=0}^{k} E_h\Big) + \sum_{h=0}^{k} \mu\Big(\Big(F \setminus \bigcup_{j=0}^{h-1} E_j\Big) \cap E_h\Big), \tag{1.7}$$

for every $k \in \mathbb{N}$, $k \geq 1$. Since $F \setminus E \subset F \setminus \bigcup_{h=0}^{k} E_h$, by monotonicity we find

$$\mu(F) \geq \mu(F \setminus E) + \sum_{h=0}^{k} \mu\Big(\Big(F \setminus \bigcup_{j=0}^{h-1} E_j\Big) \cap E_h\Big).$$

Letting first $k \to \infty$, and then using σ-subadditivity, we find $E \in \mathcal{M}(\mu)$, as

$$\mu(F) \geq \mu(F \setminus E) + \sum_{h \in \mathbb{N}} \mu\Big(\Big(F \setminus \bigcup_{j=0}^{h-1} E_j\Big) \cap E_h\Big) \tag{1.8}$$

$$\geq \mu(F \setminus E) + \mu\Big(\bigcup_{h \in \mathbb{N}}\Big(F \setminus \bigcup_{j=0}^{h-1} E_j\Big) \cap E_h\Big) = \mu(F \setminus E) + \mu(F \cap E).$$

Step two: We show that μ is σ-additive on $\mathcal{M}(\mu)$. Let $\{E_h\}_{h \in \mathbb{N}}$ be a disjoint sequence in $\mathcal{M}(\mu)$. Setting $F = E = \bigcup_{h \in \mathbb{N}} E_h$ in (1.8), we find that $\mu(E) \geq \sum_{h \in \mathbb{N}} \mu(E_h)$. As μ is σ-subadditive, we conclude the proof of the theorem. □

Exercise 1.7 If μ and ν are outer measures on $\mathbb{R}^n$, then $\mu + \nu$ is an outer measure on $\mathbb{R}^n$, with $\mathcal{M}(\mu) \cap \mathcal{M}(\nu) \subset \mathcal{M}(\mu + \nu)$.

Exercise 1.8 If μ is an outer measure on $\mathbb{R}^n$ and $\{E_h\}_{h \in \mathbb{N}} \subset \mathcal{M}(\mu)$, then

$$E_h \subset E_{h+1}, \quad \forall h \in \mathbb{N} \quad \Rightarrow \quad \mu\Big(\bigcup_{h \in \mathbb{N}} E_h\Big) = \lim_{h \to \infty} \mu(E_h),$$

$$\begin{cases} E_{h+1} \subset E_h, & \forall h \in \mathbb{N} \\ \mu(E_1) < \infty, \end{cases} \quad \Rightarrow \quad \mu\Big(\bigcap_{h \in \mathbb{N}} E_h\Big) = \lim_{h \to \infty} \mu(E_h).$$

1.3 Measure Theory and integration

By Theorem 1.4, every outer measure on $\mathbb{R}^n$ can be seen as a measure on a σ-algebra on $\mathbb{R}^n$. In this way, various classical results from Measure Theory are immediately recovered in the context of outer measures. For the sake of clarity, in this chapter we gather those definitions and statements that will be used in the rest of the book. Let μ be a measure on the σ-algebra $\mathcal{M}$ on $\mathbb{R}^n$ (if μ is an outer measure on $\mathbb{R}^n$, then we take by convention $\mathcal{M} = \mathcal{M}(\mu)$). A function $u \colon E \to [-\infty, \infty]$ is a **μ-measurable function on $\mathbb{R}^n$** if its domain E covers μ-almost all of $\mathbb{R}^n$, that is $\mu(\mathbb{R}^n \setminus E) = 0$, and if, for every $t \in \mathbb{R}$, the super-level sets

$$\{u > t\} = \Big\{x \in E : u(x) > t\Big\}$$

belong to $\mathcal{M}$. We say that u is a **μ-simple function on $\mathbb{R}^n$** if u is μ-measurable and the image of u is countable. For a non-negative, μ-simple function u, the **integral of u with respect to μ** is defined in $[0, \infty]$ as the series

$$\int_{\mathbb{R}^n} u \, d\mu = \sum_{t \in u(\mathbb{R}^n)} t\, \mu\Big(\{u = t\}\Big),$$

with the convention that $0 \cdot \infty = 0$. When u is μ-simple, and either $\int_{\mathbb{R}^n} u^+ \, d\mu$ or $\int_{\mathbb{R}^n} u^- \, d\mu$ is finite (here, $u^+ = \max\{u, 0\}$, $u^- = \max\{-u, 0\}$), we say that u is a **μ-integrable simple function**, and set

$$\int_{\mathbb{R}^n} u \, d\mu = \int_{\mathbb{R}^n} u^+ \, d\mu - \int_{\mathbb{R}^n} u^- \, d\mu \, .$$

The **upper and lower integrals with respect to** μ of a function u whose domain covers μ-almost all of $\mathbb{R}^n$, and which takes values in $[-\infty, \infty]$, are

$$\int_{\mathbb{R}^n}^* u \, d\mu = \inf\left\{ \int_{\mathbb{R}^n} v \colon v \geq u \ \mu\text{-a.e. on } \mathbb{R}^n \right\},$$
$$\int_{*\mathbb{R}^n} u \, d\mu = \sup\left\{ \int_{\mathbb{R}^n} v \colon v \leq u \ \mu\text{-a.e. on } \mathbb{R}^n \right\},$$

where v ranges over the family of μ-integrable simple functions on $\mathbb{R}^n$. If u is μ-measurable and its upper and lower integrals coincide, then we say that u is a **μ-integrable function**, and this common value is called the **integral of** u **with respect to** μ, denoted by $\int_{\mathbb{R}^n} u \, d\mu$. The following example suggests that μ-integrable functions define a large subfamily of μ-measurable functions.

Example 1.9 If u is μ-measurable on $\mathbb{R}^n$ and $u \geq 0$ μ-a.e. on $\mathbb{R}^n$, then u is μ-integrable. Indeed, if $\mu(\{u = \infty\}) > 0$, then for every $t > 0$ we have

$$\int_{*\mathbb{R}^n} u \, d\mu \geq t \, \mu\big(\{u = \infty\}\big),$$

so that, in particular, u is μ-integrable with $\int_{\mathbb{R}^n} u \, d\mu = \infty$. If, instead, $u(x) < \infty$ for μ-a.e. $x \in \mathbb{R}^n$, then given $t > 1$ we may construct a partition $\{E_h\}_{h \in \mathbb{Z}}$ of μ-almost all of $\mathbb{R}^n$ by setting $E_h = \{t^h \leq u < t^{h+1}\}$, $h \in \mathbb{Z}$. By looking at the μ-simple functions $\sum_{h \in \mathbb{Z}} t^h 1_{E_h}$ and $\sum_{h \in \mathbb{Z}} t^{h+1} 1_{E_h}$, we thus conclude that

$$\int_{\mathbb{R}^n}^* u \, d\mu \leq t \int_{*\mathbb{R}^n} u \, d\mu \, , \qquad \forall t > 1 \, .$$

Finally, u is a **locally μ-summable function**, or $u \in L^1_{\mathrm{loc}}(\mathbb{R}^n, \mu)$, if it is μ-measurable and $\int_K |u| \, d\mu < \infty$ for every compact set $K \subset \mathbb{R}^n$; it is **μ-summable**, $u \in L^1(\mathbb{R}^n, \mu)$, if $\int_{\mathbb{R}^n} |u| \, d\mu < \infty$. The L^p-spaces $L^p(\mathbb{R}^n, \mu)$ and $L^p_{\mathrm{loc}}(\mathbb{R}^n, \mu)$, $1 < p \leq \infty$, are defined as usual. We shall also set for brevity $L^p(\mathbb{R}^n) = L^p(\mathbb{R}^n, \mathcal{L}^n)$.

Theorem (Monotone convergence theorem) *If $\{u_h\}_{h \in \mathbb{N}}$ is a sequence of μ-measurable functions $u_h \colon \mathbb{R}^n \to [0, \infty]$ such that $u_h \leq u_{h+1}$ μ-a.e. on $\mathbb{R}^n$, then*

$$\lim_{h \to \infty} \int_{\mathbb{R}^n} u_h \, d\mu = \int_{\mathbb{R}^n} \sup_{h \in \mathbb{N}} u_h \, d\mu \, .$$

If, instead, $u_h \ge u_{h+1}$ μ*-a.e. on* $\mathbb{R}^n$*, and* $u_1 \in L^1(\mathbb{R}^n; \mu)$*, then*

$$\lim_{h\to\infty} \int_{\mathbb{R}^n} u_h \, d\mu = \int_{\mathbb{R}^n} \inf_{h\in\mathbb{N}} u_h \, d\mu \,.$$

Theorem (Fatou's lemma) *If* $\{u_h\}_{h\in\mathbb{N}}$ *is a sequence of* μ*-measurable functions* $u_h : \mathbb{R}^n \to [0, \infty]$*, then*

$$\int_{\mathbb{R}^n} \liminf_{h\to\infty} u_h \, d\mu \le \liminf_{h\to\infty} \int_{\mathbb{R}^n} u_h \, d\mu \,.$$

Theorem (Dominated convergence theorem) *If* $\{u_h\}_{h\in\mathbb{N}}$ *is a sequence of* μ*-measurable functions with pointwise limit* u *defined* μ*-a.e. on* $\mathbb{R}^n$*, and if there exists* $v \in L^1(\mathbb{R}^n, \mu)$ *such that* $|u_h| \le v$ μ*-a.e. on* $\mathbb{R}^n$*, then*

$$\int_{\mathbb{R}^n} u \, d\mu = \lim_{h\to\infty} \int_{\mathbb{R}^n} u_h \, d\mu \,.$$

Example 1.10 (Integral measure) If μ is an outer measure on $\mathbb{R}^n$ and $u : \mathbb{R}^n \to [0, \infty]$ is μ-measurable, then a set function $u\,\mu : \mathcal{M}(\mu) \to [0, \infty]$ is defined as

$$u\,\mu(E) = \int_{\mathbb{R}^n} 1_E \, u \, d\mu = \int_E u \, d\mu \,, \qquad E \in \mathcal{M}(\mu) \,.$$

Notice that $u\,\mu$ is defined on $\mathcal{M}(\mu)$ (and, possibly, not on the whole $\mathcal{P}(\mathbb{R}^n)$) because we need $1_E \, u$ to be μ-measurable. It follows from the above theorems that $u\,\mu$ is a measure on the σ-algebra $\mathcal{M}(\mu)$. If $\mu(E) < \infty$, then we shall set

$$⨍_E u \, d\mu = \frac{1}{\mu(E)} \int_E u \, d\mu \,.$$

Theorem (Egoroff's theorem) *If* $\{u_h\}_{h\in\mathbb{N}}$ *is a sequence of* μ*-measurable functions with pointwise limit* u*, then, for every* $\varepsilon > 0$ *and for every* $E \in \mathcal{M}(\mu)$ *with* $\mu(E) < \infty$*, there exists* $F \in \mathcal{M}(\mu)$ *such that*

$$\mu(E \setminus F) \le \varepsilon \qquad \text{and} \qquad u_h \to u \text{ uniformly on } F \,.$$

A fundamental result from Measure Theory is, of course, Fubini's theorem, which can be stated in the language of outer measures as follows. Let μ be an outer measure on $\mathbb{R}^n$ and let ν be an outer measure on $\mathbb{R}^m$. An outer measure $\mu \times \nu : \mathcal{P}(\mathbb{R}^n \times \mathbb{R}^m) \to [0, \infty]$ is defined at $G \subset \mathbb{R}^n \times \mathbb{R}^m$ by setting

$$\mu \times \nu(G) = \inf_{\mathcal{F}} \sum_{E\times F\in\mathcal{F}} \mu(E)\nu(F) \,,$$

where $\mathcal{F}$ is a countable covering of G by sets of the form $E \times F$, where $E \in \mathcal{M}(\mu)$ and $F \in \mathcal{M}(\nu)$. We call $\mu \times \nu$ the **product measure** of μ and ν. To every $x \in \mathbb{R}^n$ there corresponds a **vertical section** $G_x \subset \mathbb{R}^m$ of $G \subset \mathbb{R}^n \times \mathbb{R}^m$,

$$G_x = \left\{ y \in \mathbb{R}^m : (x, y) \in G \right\} .$$

The standard proof of Fubini's theorem can then be adapted in the present context to prove the following result (see [EG92, Section 1.4], [Fed69, 2.6.2]).

Theorem (Fubini's theorem) *Let μ be an outer measure on $\mathbb{R}^n$ and let ν be an outer measure on $\mathbb{R}^m$.*

(i) If $E \in \mathcal{M}(\mu)$ and $F \in \mathcal{M}(\nu)$, then $E \times F \in \mathcal{M}(\mu \times \nu)$ and

$$\mu \times \nu(E \times F) = \mu(E)\nu(F).$$

(ii) For every $G \subset \mathbb{R}^n \times \mathbb{R}^m$ there exists $H \in \mathcal{M}(\mu \times \nu)$ such that

$$G \subset H, \qquad \mu \times \nu(G) = \mu \times \nu(H).$$

(iii) If $G \subset \mathbb{R}^n \times \mathbb{R}^m$ is σ-finite with respect to $\mu \times \nu$, then $G_x \in \mathcal{M}(\nu)$ for μ-a.e. $x \in \mathbb{R}^n$ and, moreover,

$$x \in \mathbb{R}^n \mapsto \nu(G_x) \text{ is } \mu\text{-measurable on } \mathbb{R}^n,$$
$$\mu \times \nu(G) = \int_{\mathbb{R}^n} \nu(G_x)\, d\mu(x).$$

(iv) If $u \in L^1(\mathbb{R}^n \times \mathbb{R}^m, \mu \times \nu)$, then

$$x \in \mathbb{R}^n \mapsto \int_{\mathbb{R}^m} u(x,y) d\nu(y) \text{ belongs to } L^1(\mathbb{R}^n, \mu),$$
$$\int_{\mathbb{R}^n \times \mathbb{R}^m} u\, d(\mu \times \nu) = \int_{\mathbb{R}^n} d\mu(x) \int_{\mathbb{R}^m} u(x,y) d\nu(y).$$

Remark 1.11 Note that statement (ii) provides a strong "regularity" property of product measures (compare with Section 2.2). Concerning statement (iii), recall that G is σ-finite with respect to $\mu \times \nu$, if there exists a countable covering of G by $\mu \times \nu$-measurable sets with finite $\mu \times \nu$-measure. This is a necessary assumption for the validity of (iii). Consider, for example, the outer measures on $\mathbb{R}$ given by $\mu = \#$ and $\nu = \mathcal{L}^1$, and let $G = \{(x,y) \in \mathbb{R}^2 : x = y \in [0,1]\}$. Clearly, G is not σ-finite with respect to $\mu \times \nu$, and $\mu \times \nu(G) = \infty$. Moreover, $G_x = \{y \in \mathbb{R} : (x,y) \in G\} = \{x\}$, so that $\nu(G_x) = 0$, for every $x \in [0,1]$, and $G^y = \{x \in \mathbb{R} : (x,y) \in G\} = \{y\}$, so that $\mu(G^y) = 1$, for every $y \in [0,1]$. Thus,

$$\mu \times \nu(G) = \infty, \qquad \int_{\mathbb{R}} \nu(G_x)\, d\mu(x) = 0, \qquad \int_{\mathbb{R}} \mu(G^y)\, d\nu(y) = 1.$$

Exercise 1.12 If $\varphi \in C^1_c(\mathbb{R}^n)$, then

$$\int_{\mathbb{R}^n} \nabla\varphi(x) dx = 0.$$

Hint: Combine Fubini's theorem and the fundamental theorem of Calculus.

Exercise 1.13 (Layer-cake formula) If $u \in L^p(\mathbb{R}^n, \mu)$, $p \in [1, \infty)$, $u \geq 0$, and $\{u > t\} = \{x \in \mathbb{R}^n : u(x) > t\}$, then

$$\int_{\mathbb{R}^n} |u|^p \, d\mu = p \int_0^\infty t^{p-1} \mu\left(\{|u| > t\}\right) dt . \tag{1.9}$$

Hint: Apply Fubini's theorem to $\mu \times \mathcal{L}^1$ and $f(x,t) = pt^{p-1} 1_{(0,|u(x)|)}(t)$.

Exercise 1.14 If $\{u_k\}_{k=1}^N \subset L^1(\mathbb{R}^n; \mu)$ and $\mu(\mathbb{R}^n) < \infty$, then

$$\mu\left(\bigcap_{k=1}^N \left\{x \in \mathbb{R}^n : u_k(x) \leq \frac{2N}{\mu(\mathbb{R}^n)} \int_{\mathbb{R}^n} u_k \, d\mu\right\}\right) \geq \frac{\mu(\mathbb{R}^n)}{2} .$$

Exercise 1.15 The theory of outer measures discussed in this chapter can be repeated *verbatim* if we replace the ambient space $\mathbb{R}^n$ with a generic set X.

2

Borel and Radon measures

We now introduce the notion of Radon measure on $\mathbb{R}^n$. Section 2.1 introduces the related notion of Borel measure, with a useful characterization result due to Carathéodory. In Section 2.2 we define Borel regular measures, while in Section 2.3 we prove two fundamental approximation theorems. The definition of Radon measure appears in Section 2.4, with a brief discussion about basic examples, operations, and useful facts.

2.1 Borel measures and Carathéodory's criterion

Given an outer measure μ, Theorem 1.4 provides a σ-algebra $\mathcal{M}(\mu)$ on which μ is σ-additive. Nevertheless, $\mathcal{M}(\mu)$ could be trivial, that is, it could be equal to $\{\emptyset, \mathbb{R}^n\}$. The following theorem furnishes a valuable criterion for $\mathcal{M}(\mu)$ to contain the family $\mathcal{B}(\mathbb{R}^n)$ of the **Borel sets** of $\mathbb{R}^n$. Let us recall that $\mathcal{B}(\mathbb{R}^n)$ is defined as the σ-algebra generated by the open sets of $\mathbb{R}^n$, that is the intersection of all the σ-algebras containing the family of open sets (it is easily seen that this intersection defines a σ-algebra). By definition, a **Borel measure** on $\mathbb{R}^n$ is an outer measure μ on $\mathbb{R}^n$ such that $\mathcal{B}(\mathbb{R}^n) \subset \mathcal{M}(\mu)$. The following theorem provides an useful characterization of Borel measures on $\mathbb{R}^n$.

Theorem 2.1 (Carathéodory's criterion) *If μ is an outer measure on $\mathbb{R}^n$, then μ is a Borel measure on $\mathbb{R}^n$ if and only if*

$$\mu(E_1 \cup E_2) = \mu(E_1) + \mu(E_2), \tag{2.1}$$

for every $E_1, E_2 \subset \mathbb{R}^n$ such that $\operatorname{dist}(E_1, E_2) > 0$.

Example 2.2 By Theorem 2.1, the Lebesgue measure is a Borel measure. Indeed, let us prove that $|E_1 \cup E_2| \geq |E_1| + |E_2|$ for every $E_1, E_2 \subset \mathbb{R}^n$ with

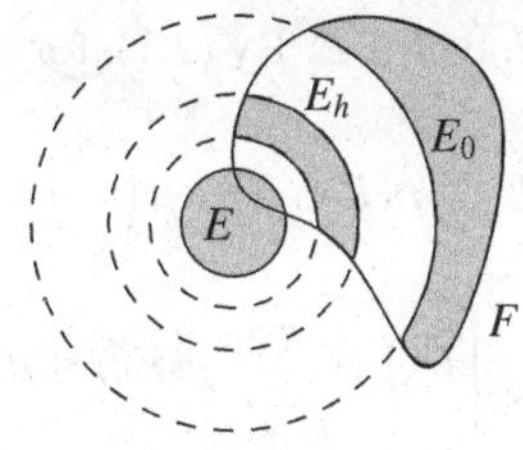

Figure 2.1 Construction of the sets E_h.

$d = \mathrm{dist}(E_1, E_2) > 0$. Let $\mathcal{F}$ be a countable family of disjoint cubes (with sides parallel to the coordinate axes) such that

$$E_1 \cup E_2 \subset \bigcup_{Q \in \mathcal{F}} Q .$$

Since $\mathcal{L}^n$ is additive on finite disjoint unions of cubes with sides parallel to the coordinate axes, up to further division of each $Q \in \mathcal{F}$ into finitely many sub-cubes, we may also assume that $\mathrm{diam}(Q) < d$ for every $Q \in \mathcal{F}$. If $\mathcal{F}_h = \{Q \in \mathcal{F} : Q \cap E_h \neq \emptyset\}$, then $\mathcal{F}_1 \cap \mathcal{F}_2 = \emptyset$ and $E_h \subset \bigcup_{Q \in \mathcal{F}_h} Q$ for $h = 1, 2$, so that

$$\sum_{Q \in \mathcal{F}} r(Q)^n \geq \sum_{Q \in \mathcal{F}_1} r(Q)^n + \sum_{Q \in \mathcal{F}_2} r(Q)^n \geq |E_1| + |E_2| .$$

By the arbitrariness of $\mathcal{F}$ we conclude that $|E_1 \cup E_2| \geq |E_1| + |E_2|$, as required.

Exercise 2.3 Let $n \geq 1$, $s \in (0, n)$, and $\delta \in (0, \infty]$. Show that $\mathcal{H}^s$ is a Borel measure on $\mathbb{R}^n$, and that $\mathcal{H}^s_\delta$ is *never* a Borel measure on $\mathbb{R}^n$.

Proof of Theorem 2.1 *Step one:* Let μ be a Borel measure and let $E_1, E_2 \subset \mathbb{R}^n$, with $\mathrm{dist}(E_1, E_2) > 0$. Since $\overline{E_1}$ is a Borel set, it is μ-measurable. Thus, by testing the μ-measurability of $\overline{E_1}$ on the set $E_1 \cup E_2$, while taking into account that $\overline{E_1}$ and E_2 are disjoint, we deduce (2.1), that is

$$\mu(E_1 \cup E_2) = \mu\Big((E_1 \cup E_2) \cap \overline{E_1}\Big) + \mu\Big((E_1 \cup E_2) \setminus \overline{E_1}\Big) = \mu(E_1) + \mu(E_2) .$$

Step two: We now assume the validity of (2.1). By Remark 1.6, to prove that μ is a Borel measure it suffices to show that if $E \subset \mathbb{R}^n$ is closed, then

$$\mu(F) \geq \mu(E \cap F) + \mu(F \setminus E) , \tag{2.2}$$

for every $F \subset \mathbb{R}^n$ with $\mu(F) < \infty$. Since E is closed, setting (for $h \geq 1$)

$$E_h = \Big\{x \in F : \frac{1}{h+1} \leq \mathrm{dist}(x, E) < \frac{1}{h}\Big\}, \qquad E_0 = \Big\{x \in F : \mathrm{dist}(x, E) \geq 1\Big\},$$

we define a partition $\{E_h\}_{h\in\mathbb{N}}$ of $F \setminus E$. By (2.1), for every $N \in \mathbb{N}$,

$$\mu(F \cap E) + \mu(F \setminus E) \le \mu(F \cap E) + \mu\Big(\bigcup_{h=0}^{N} E_h\Big) + \sum_{h=N+1}^{\infty} \mu(E_h)$$

$$= \mu\Big((F \cap E) \cup \bigcup_{h=0}^{N} E_h\Big) + \sum_{h=N+1}^{\infty} \mu(E_h) \le \mu(F) + \sum_{h=N+1}^{\infty} \mu(E_h). \quad (2.3)$$

Again by (2.1), we have

$$\sum_{h=0}^{N} \mu(E_{2h}) = \mu\Big(\bigcup_{h=0}^{N} E_{2h}\Big) \le \mu(F) < \infty,$$

$$\sum_{h=0}^{N} \mu(E_{2h+1}) = \mu\Big(\bigcup_{h=0}^{N} E_{2h+1}\Big) \le \mu(F) < \infty,$$

so that $\sum_{h\in\mathbb{N}} \mu(E_h) < \infty$. We let $N \to \infty$ into (2.3) to prove (2.2). □

2.2 Borel regular measures

The σ-algebra of the μ-measurable sets with respect to a Borel measure μ may strictly contain the family of Borel sets. Therefore, knowing a Borel measure on the family of Borel sets may return only a partial description of the measure itself. We say that a Borel measure μ is **regular** if for every $F \subset \mathbb{R}^n$ there exists a Borel set E such that

$$F \subset E, \qquad \mu(E) = \mu(F).$$

Thus, a regular Borel measure is completely determined by its values on Borel sets. It turns out that the most relevant examples of Borel measures are regular.

Example 2.4 The Lebesgue measure $\mathcal{L}^n$ and the Hausdorff measure $\mathcal{H}^s$ ($s > 0$) are Borel regular measures on $\mathbb{R}^n$. Let us prove the Borel regularity of $\mathcal{H}^s$ (the proof is analogous in the case of $\mathcal{L}^n$). By Exercise 1.3, given $E \subset \mathbb{R}^n$ and $k \in \mathbb{N}$ we can find a countable covering $\mathcal{F}_k$ of E by closed sets with diameter bounded by k^{-1} such that

$$\omega_s \sum_{F\in\mathcal{F}_k} \Big(\frac{\operatorname{diam}(F)}{2}\Big)^s \le \mathcal{H}^s_{1/k}(E) + \frac{1}{k}.$$

Since elements of $\mathcal{F}_k$ are closed, the set $G = \bigcap_{k\in\mathbb{N}} \bigcup_{F\in\mathcal{F}_k} F$ is a Borel set with $E \subset G$. Since $\mathcal{F}_k$ is a competitor in the definition of $\mathcal{H}^s_{1/k}(G)$,

$$\mathcal{H}^s_{1/k}(G) \le \omega_s \sum_{F\in\mathcal{F}_k} \Big(\frac{\operatorname{diam}(F)}{2}\Big)^s \le \mathcal{H}^s_{1/k}(E) + \frac{1}{k},$$

for every $k \in \mathbb{N}$. We let $k \to \infty$ to find that $\mathcal{H}^s(G) \le \mathcal{H}^s(E)$, as required.

Remark 2.5 We show the equivalence of the Lebesgue measure $\mathcal{L}^n$ on $\mathbb{R}^n$ and the n-fold product $(\mathcal{L}^1)^n$ of the Lebesgue measure $\mathcal{L}^1$ on $\mathbb{R}$, that is

$$\mathcal{L}^n = \mathcal{L}^1 \times \cdots \times \mathcal{L}^1 \qquad (n \text{ times}).$$

Indeed, define a set function $\mu\colon \mathcal{P}(\mathbb{R}^n) \to \mathbb{R}$, $\mu(E) = \mathcal{L}^n(E) - (\mathcal{L}^1)^n(E)$, $E \subset \mathbb{R}^n$. First, we prove that μ is non-negative: indeed, if $\mathcal{F}$ is a countable covering of $E \subset \mathbb{R}^n$ by cubes with sides parallel to the coordinate axes, then

$$\sum_{Q\in\mathcal{F}} r(Q)^n = \sum_{Q\in\mathcal{F}} (\mathcal{L}^1)^n(Q) \ge (\mathcal{L}^1)^n\Big(\bigcup_{Q\in\mathcal{F}} Q\Big) \ge (\mathcal{L}^1)^n(E).$$

By the arbitrariness of $\mathcal{F}$, we find $\mu(E) \ge 0$ for every $E \subset \mathbb{R}^n$. At the same time, $\mu(Q) \le 0$ for every cube Q parallel to the coordinate axes, since, by definition, $\mathcal{L}^n(Q) \le r(Q)^n = (\mathcal{L}^1)^n(Q)$. Now, by Example 2.2 and Example 2.4, $\mathcal{L}^n$ is a Borel regular measure. Similarly, we see that $(\mathcal{L}^1)^n$ is a Borel regular measure. Hence, $\mathcal{L}^n$ and $(\mathcal{L}^1)^n$ are σ-additive on $\mathcal{B}(\mathbb{R}^n)$, and we easily see that μ is σ-additive on $\mathcal{B}(\mathbb{R}^n)$ (in particular, μ is monotone on $\mathcal{B}(\mathbb{R}^n)$). By looking at a countable disjoint partition of $\mathbb{R}^n$ by cubes with sides parallel to the coordinate axes, we thus conclude that $\mu(\mathbb{R}^n) = 0$. By monotonicity, $\mu = 0$ on $\mathcal{B}(\mathbb{R}^n)$. By Borel regularity, $\mu = 0$ on $\mathcal{P}(\mathbb{R}^n)$.

Exercise 2.6 If μ and ν are Borel regular measures on $\mathbb{R}^n$ such that $\mu = \nu$ on $\mathcal{B}(\mathbb{R}^n)$, then $\mu = \nu$ on $\mathcal{P}(\mathbb{R}^n)$.

Exercise 2.7 If μ and ν are Borel regular measures on $\mathbb{R}^n$, then $\mu + \nu$ is a Borel regular measure on $\mathbb{R}^n$.

2.3 Approximation theorems for Borel measures

A Borel measure is characterized on $\mathcal{B}(\mathbb{R}^n)$ by its behavior on compact sets (Theorem 2.8), and open sets may be used in place of compact sets if a local finiteness condition holds true (Theorem 2.10). These two important approximation theorems prepare the ground for the definition of Radon measure.

Theorem 2.8 (Inner approximation by compact sets) *If μ is a Borel measure on $\mathbb{R}^n$, and E is a Borel set in $\mathbb{R}^n$ with $\mu(E) < \infty$, then for every $\varepsilon > 0$ there exists a compact set $K \subset E$ such that $\mu(E \setminus K) < \varepsilon$. In particular,*

$$\mu(E) = \sup\big\{\mu(K) : K \subset E,\ \ K \text{ is compact}\big\}.$$

Remark 2.9 Consider the Borel measure $\mu = \sum_{h\in\mathbb{N}} \delta_{1/h}$ on $\mathbb{R}$. If $E = (0,1)$, then $\mu(E) = \infty$ and $\mu(E \setminus K) = \infty$ for every compact set $K \subset E$.

Proof of Theorem 2.8 *Step one:* We prove that for every $\varepsilon > 0$ there exists a closed set $C \subset E$ such that $\mu(E \setminus C) < \varepsilon$. To this end, we introduce a finite Borel measure ν on $\mathbb{R}^n$, defined as

$$\nu(F) = \mu(E \cap F), \qquad \forall F \subset \mathbb{R}^n,$$

and consider the family of sets

$$\mathcal{G} = \Big\{F \subset \mathcal{M}(\mu) : \forall \varepsilon > 0, \exists C \subset F, C \text{ closed}, \nu(F \setminus C) < \varepsilon\Big\},$$
$$\mathcal{F} = \Big\{F \in \mathcal{G} : \mathbb{R}^n \setminus F \in \mathcal{G}\Big\}.$$

We claim that, if $\{F_h\}_{h\in\mathbb{N}} \subset \mathcal{G}$, then $\bigcup_{h\in\mathbb{N}} F_h \in \mathcal{G}$ and $\bigcap_{h\in\mathbb{N}} F_h \in \mathcal{G}$. The claim concludes the proof of step one. First, by the claim, $\mathcal{F}$ is a σ-algebra; second, since $\mathcal{G}$ contains the closed sets, and since every open set of $\mathbb{R}^n$ is a countable union of closed sets, by the claim $\mathcal{G}$ contains the open sets. Hence, by construction, $\mathcal{F}$ contains the open sets, and so $\mathcal{B}(\mathbb{R}^n) \subset \mathcal{F} \subset \mathcal{G}$. In particular, $E \in \mathcal{G}$, and step one is proved. We now prove the claim. For every $h \in \mathbb{N}$, let $C_h \subset F_h$ be closed with $\nu(F_h \setminus C_h) < \varepsilon/2^h$. Then $C = \bigcap_{h\in\mathbb{N}} C_h$ is closed, with

$$\nu\Big(\bigcap_{h\in\mathbb{N}} F_h \setminus C\Big) = \nu\Big(\bigcap_{h\in\mathbb{N}} F_h \setminus \bigcap_{h\in\mathbb{N}} C_h\Big) \le \nu\Big(\bigcup_{h\in\mathbb{N}} (F_h \setminus C_h)\Big)$$
$$\le \sum_{h\in\mathbb{N}} \nu(F_h \setminus C_h) < \varepsilon,$$

and $\bigcap_{h\in\mathbb{N}} F_h \in \mathcal{F}$. At the same time, since $\nu(\mathbb{R}^n) = \mu(E) < \infty$,

$$\lim_{N\to\infty} \nu\Big(\bigcup_{h\in\mathbb{N}} F_h \setminus \bigcup_{h=1}^{N} C_h\Big) = \nu\Big(\bigcup_{h\in\mathbb{N}} F_h \setminus \bigcup_{h\in\mathbb{N}} C_h\Big) \le \nu\Big(\bigcup_{h\in\mathbb{N}} (F_h \setminus C_h)\Big)$$
$$\le \sum_{h\in\mathbb{N}} \nu(F_h \setminus C_h) < \varepsilon.$$

For N large enough and $C' = \bigcup_{h=1}^{N} C_h$, we have $\nu\left(\bigcup_{h\in\mathbb{N}} F_h \setminus C'\right) < \varepsilon$. Since C' is closed, $\bigcup_{h\in\mathbb{N}} F_h \in \mathcal{F}$.

Step two: We prove that for every $\varepsilon > 0$ we can find a compact set $K \subset E$ with $\mu(E \setminus K) < \varepsilon$. By step one, for every $\varepsilon > 0$ we can find a closed set $F \subset E$ such that $\mu(E \setminus F) < \varepsilon/2$. Now $K_h = \overline{B_h} \cap F$ is compact for every h. Moreover, as $\mu(F) \le \mu(E) < \infty$, we have $\lim_{h\to\infty} \mu(F \setminus K_h) = 0$, and we deduce $\mu(F \setminus K) < \varepsilon/2$ if we choose $K = K_h$ for h big enough. In conclusion $\mu(E \setminus K) \le \mu(E \setminus F) + \mu(F \setminus K) < \varepsilon$. □

An outer measure μ on $\mathbb{R}^n$ is **locally finite** if $\mu(K) < \infty$ for every compact set $K \subset \mathbb{R}^n$. For example, the Lebesgue measure is locally finite. A locally finite Borel measure admits outer approximation by open sets. This property fails on generic Borel measures: if $\mu = \mathcal{H}^1$ and E is a bounded segment in $\mathbb{R}^2$, then $\mathcal{H}^1(E) < \infty$ but $\mathcal{H}^1(A) = \infty$ whenever $A \subset \mathbb{R}^2$ is open (see Chapter 3).

Theorem 2.10 (Outer approximation by open sets) *If μ is a locally finite Borel measure on $\mathbb{R}^n$, and E is a Borel set, then*

$$\mu(E) = \inf\{\mu(A) : E \subset A,\ A \textit{ is open}\}, \tag{2.4}$$

$$= \sup\{\mu(K) : K \subset E,\ K \textit{ is compact}\}. \tag{2.5}$$

Proof *Step one:* In proving (2.4) we can directly assume $\mu(E) < \infty$. Indeed, given $\varepsilon > 0$, let us construct an open set A with $E \subset A$ and $\mu(A \setminus E) < \varepsilon$. Since $B_h \setminus E$ is a Borel set, and, by local finiteness, $\mu(B_h \setminus E) < \infty$, by Theorem 2.8 for every $h \in \mathbb{N}$ we can find $C_h \subset \mathbb{R}^n$ closed, with $C_h \subset B_h \setminus E$ and

$$\mu\Big(\big(B_h \setminus E\big) \setminus C_h\Big) = \mu\Big(B_h \setminus \big(E \cup C_h\big)\Big) < \frac{\varepsilon}{2^h}.$$

Thus $A_h = B_h \setminus C_h$ is open, with $B_h \cap E \subset A_h$, and such that

$$\mu\Big(A_h \setminus \big(B_h \cap E\big)\Big) = \mu\Big(B_h \setminus \big(E \cup C_h\big)\Big) < \frac{\varepsilon}{2^h}.$$

If we set $A = \bigcup_{h\in\mathbb{N}} A_h$, then A is open, it contains $E = \bigcup_{h\in\mathbb{N}}(E \cap B_h)$, and

$$\mu(A \setminus E) = \mu\Big(\bigcup_{h\in\mathbb{N}} A_h \setminus \bigcup_{h\in\mathbb{N}}(E \cap B_h)\Big) \le \sum_{h\in\mathbb{N}} \mu\big(A_h \setminus (E \cap B_h)\big) < \varepsilon.$$

Step two: We prove (2.5). By local finiteness of μ, $E \cap B_R$ ($R > 0$) is a Borel set with $\mu(E \cap B_R) < \infty$ and $\mu(E \cap B_R) \to \mu(E)$ as $R \to \infty$. By Theorem 2.8, for every $R > 0$ there exists a compact set $K_R \subset E \cap B_R$ such that $\mu(E \cap B_R) \ge \mu(K_R) \ge \mu(E \cap B_R) - R^{-1}$. Thus $\mu(K_R) \to \mu(E)$ as $R \to \infty$. □

2.4 Radon measures. Restriction, support, and push-forward

The results of Section 2.3 motivate the following crucial definition. An outer measure μ is a **Radon measure** on $\mathbb{R}^n$ if it is a locally finite, Borel regular measure on $\mathbb{R}^n$. By Theorem 2.10, if μ is a Radon measure on $\mathbb{R}^n$, then

$$\mu(E) = \inf\{\mu(A) : E \subset A,\ A \text{ is open}\} \tag{2.6}$$

$$= \sup\{\mu(K) : K \subset E,\ K \text{ is compact}\}, \tag{2.7}$$

for every Borel set $E \subset \mathbb{R}^n$. Thus, by Borel regularity, a Radon measure μ is characterized on $\mathcal{M}(\mu)$ by its behavior on compact (or open) sets.

Example 2.11 By Example 2.4, the Lebesgue measure, which is trivially locally finite, is a Radon measure. If $s \in [0, n)$, then $\mathcal{H}^s$ is not locally finite (as $\mathcal{H}^s(A) = \infty$ for every open set A; see Chapter 3), and thus it is not a Radon measure. However, if E is a Borel set with $\mathcal{H}^s(E) < \infty$, then the restriction $\mathcal{H}^s \llcorner E$ of $\mathcal{H}^s$ to E is a Radon measure on $\mathbb{R}^n$; see Proposition 2.13 below.

We now notice that (2.6) and (2.7) have in fact a wider range of validity.

Proposition 2.12 *If μ is a Radon measure, then (2.6) holds true for every $E \subset \mathbb{R}^n$, while (2.7) holds true for every $E \in \mathcal{M}(\mu)$.*

Proof *Step one:* We prove (2.6) for $E \subset \mathbb{R}^n$. By Borel regularity, there exists a Borel set F with $E \subset F$ and $\mu(E) = \mu(F)$. By (2.6) (applied to F),

$$\begin{aligned}\mu(E) = \mu(F) &= \inf\left\{\mu(A) : F \subset A,\ A \text{ is open}\right\}\\ &\geq \inf\left\{\mu(A) : E \subset A,\ A \text{ is open}\right\} \geq \mu(E).\end{aligned}$$

Step two: We prove (2.7) for $E \in \mathcal{M}(\mu)$. As (2.7) holds true on closed set in $\mathbb{R}^n$, it is enough to prove that

$$\mu(E) = \sup\left\{\mu(C) : C \subset E,\ C \text{ is closed}\right\}. \tag{2.8}$$

By Proposition 2.13 below, $\nu = \mu \llcorner E$ is a Radon measure on $\mathbb{R}^n$. By step one,

$$\begin{aligned}0 = \nu(\mathbb{R}^n \setminus E) &= \inf\left\{\nu(A) : \mathbb{R}^n \setminus E \subset A,\ A \text{ is open}\right\}\\ &= \inf\left\{\mu(E \setminus C) : C \subset E, C \text{ is closed}\right\}.\end{aligned} \tag{2.9}$$

If $C \subset E$ is closed, then $\mu(E) = \mu(C) + \mu(E \setminus C)$, and (2.9) implies (2.8). □

We now introduce some basic operations on Radon measures. We begin with restriction, which has already appeared in Example 2.11 and in the proof of Proposition 2.12. Given an outer measure μ on $\mathbb{R}^n$, and $E \subset \mathbb{R}^n$, the **restriction of μ to E** is the outer measure $\mu \llcorner E$ defined as

$$\mu \llcorner E(F) = \mu(E \cap F), \qquad F \subset \mathbb{R}^n.$$

We have $\mathcal{M}(\mu) \subset \mathcal{M}(\mu \llcorner E)$, and the following useful proposition holds true.

Proposition 2.13 (Restriction of Borel regular measures) *If μ is a Borel regular measure on $\mathbb{R}^n$, and $E \in \mathcal{M}(\mu)$ is such that $\mu \llcorner E$ is locally finite, then $\mu \llcorner E$ is a Radon measure on $\mathbb{R}^n$.*

Proof It is easily seen that $\mu \llcorner E$ is a locally finite Borel measure. We are left to check that $\mu \llcorner E$ is Borel regular. Let $F \subset \mathbb{R}^n$. By Borel regularity of μ there exist Borel sets G and H in $\mathbb{R}^n$ such that

$$E \subset G\,, \quad \mu(E) = \mu(G)\,, \qquad F \cap G \subset H\,, \quad \mu(F \cap G) = \mu(H)\,.$$

Then F is contained in the Borel set $H \cup (\mathbb{R}^n \setminus G)$, and since $\mu(G \setminus E) = 0$,

$$\begin{aligned}(\mu \llcorner E)\big(H \cup (\mathbb{R}^n \setminus G)\big) &\le \mu\Big(G \cap \big(H \cup (\mathbb{R}^n \setminus G)\big)\Big) \le \mu(H) \\ &= \mu(F \cap G) = \mu(F \cap E) = (\mu \llcorner E)(F)\,. \end{aligned}$$ □

An outer measure μ on $\mathbb{R}^n$ is **concentrated on** $E \subset \mathbb{R}^n$ if $\mu(\mathbb{R}^n \setminus E) = 0$. The intersection of the closed sets E such that μ is concentrated on E is denoted by $\operatorname{spt}\mu$, and called the **support of** μ. In particular,

$$\mathbb{R}^n \setminus \operatorname{spt}\mu = \Big\{x \in \mathbb{R}^n : \mu(B(x,r)) = 0 \text{ for some } r > 0\Big\}\,.$$

Note that μ may be concentrated on a set strictly contained in its support. Let $\{x_h\}_{h\in\mathbb{N}}$ be a sequence in $\mathbb{R}^n$ and let $\mu = \sum_{h\in\mathbb{N}} 2^{-h}\delta_{x_h}$. Then μ is a Radon measure (with $\mu(\mathbb{R}^n) = 1$), μ is concentrated on the elements of the sequence but $\operatorname{spt}\mu$ contains every accumulation point of the sequence.

Finally, we introduce the operation of push-forward of a measure. Given a function $f\colon \mathbb{R}^n \to \mathbb{R}^m$ and an outer measure μ on $\mathbb{R}^n$, the **push-forward of** μ **through** f is the outer measure $f_\#\mu$ on $\mathbb{R}^m$ defined by the formula

$$f_\#\mu(E) = \mu\big(f^{-1}(E)\big)\,, \qquad E \subset \mathbb{R}^m\,.$$

Sometimes $f_\#\mu$ is also called the **image measure of** μ **through** f. If we think for example that $f_\#(\delta_x) = \delta_{f(x)}$, this terminology suggests an efficient way to visualize the push-forward operation (more complex examples are discussed in Exercise 8.12). Let us recall that $f\colon \mathbb{R}^n \to \mathbb{R}^m$ is **proper** if $f^{-1}(K) \subset \mathbb{R}^n$ is compact whenever $K \subset \mathbb{R}^m$ compact. A continuous and proper function is **closed**, that is, it maps closed sets to closed sets.

Proposition 2.14 (Push-forward of a Radon measure) *If μ is a Radon measure on $\mathbb{R}^n$, and $f\colon \mathbb{R}^n \to \mathbb{R}^m$ is continuous and proper, then $f_\#\mu$ is a Radon measure on $\mathbb{R}^m$,* $\operatorname{spt}(f_\#\mu) = f(\operatorname{spt}\mu)$*, and for every Borel measurable function $u\colon \mathbb{R}^m \to [0,\infty]$ we have*

$$\int_{\mathbb{R}^m} u \, \mathrm{d}(f_\#\mu) = \int_{\mathbb{R}^n} (u \circ f)\, \mathrm{d}\mu\,. \tag{2.10}$$

Proof As μ is locally finite on $\mathbb{R}^n$, and since f is proper, we easily see that $f_\#\mu$ is locally finite on $\mathbb{R}^m$. If $E \subset \mathbb{R}^m$ with $f^{-1}(E) \in \mathcal{M}(\mu)$, then (clearly) $E \in \mathcal{M}(f_\#\mu)$. As f is continuous (and thus Borel measurable), we thus find that $f_\#\mu$ is a Borel measure on $\mathbb{R}^m$. We now prove that $f_\#\mu$ is Borel regular. Consider $E \subset \mathbb{R}^m$ and let A_h be an open set in $\mathbb{R}^n$ such that $f^{-1}(E) \subset A_h$ and $\mu(A_h) \le \mu(f^{-1}(E)) + h^{-1}$, $h \in \mathbb{N}$. Since f is closed, if we let

$$F_h = \mathbb{R}^m \setminus f(\operatorname{spt}\mu \setminus A_h), \qquad h \in \mathbb{N},$$

then each F_h is open, with $E \subset F_h$ and

$$f_\#\mu\,(F_h) \le \mu\,(A_h) \le f_\#\mu(E) + h^{-1}.$$

Hence, $F = \bigcap_{h\in\mathbb{N}} F_h$ is a Borel set, $E \subset F$ and $f_\#\mu(E) = f_\#\mu(F)$. This proves $f_\#\mu$ is Borel regular. If μ is concentrated on a set E, then $f_\#\mu$ is concentrated on $f(E)$, so that $\operatorname{spt}(f_\#\mu) \subset f(\operatorname{spt}\mu)$. Conversely, if $y \in f(\operatorname{spt}\mu)$ then $y = f(x)$ with $x \in \mathbb{R}^n$ and $\mu(B(x,r)) > 0$ for every $r > 0$. By continuity of f, for every $R > 0$ there exists $r > 0$ such that $f(B(x,r)) \subset B(y,R)$; therefore $f_\#\mu(B(y,R)) = \mu(f^{-1}(B(x,R))) \ge \mu(B(x,r)) > 0$ and $y \in \operatorname{spt}(f_\#\mu)$. Finally, (2.10) is trivial on non-negative simple Borel functions. The general case is inferred by approximation. □

We close this chapter with an alternative approximation result to (2.7) and (2.6), and with a proposition about "foliations" by Borel sets which is going to be used repeatedly in the rest of the book.

Proposition 2.15 *If μ is a Radon measure on $\mathbb{R}^n$, E is a bounded set with $\mu(\partial E) = 0$, and $\varepsilon > 0$, then there exists A open and K compact such that*

$$\overline{A} \subset E \subset \mathring{K}, \qquad \mu(K \setminus A) < \varepsilon.$$

Proof For $s, t > 0$ let us consider the sets

$$A_t = \left\{x \in \mathring{E} : \operatorname{dist}(x, \partial E) > t\right\}, \qquad K_s = \left\{x \in \mathbb{R}^n : \operatorname{dist}(x, E) \le s\right\}.$$

Since A_t is open, with $\overline{A_t} \subset E$ and $\mathring{E} = \bigcup_{t>0} A_t$, we have

$$\lim_{t\to 0^+} \mu(A_t) = \mu(\mathring{E}) = \mu(E).$$

Thus, if we take $A = A_t$ for t small enough, then $\mu(E \setminus A) < \varepsilon/2$. Similarly, K_s is compact, with $E \subset \mathring{K}_s$ and $\bigcap_{s>0} K_s = \overline{E}$. Hence

$$\lim_{s\to 0^+} \mu(K_s) = \mu(\overline{E}) = \mu(E),$$

and if we let $K = K_s$ for s small enough, we find $\mu(K \setminus E) \le \varepsilon/2$. □

Proposition 2.16 (Foliations by Borel sets) *If $\{E_t\}_{t\in I}$ is a disjoint family of Borel sets in $\mathbb{R}^n$, indexed over some set I, and μ is a Radon measure on $\mathbb{R}^n$, then $\mu(E_t) > 0$ for at most countably many $t \in I$.*

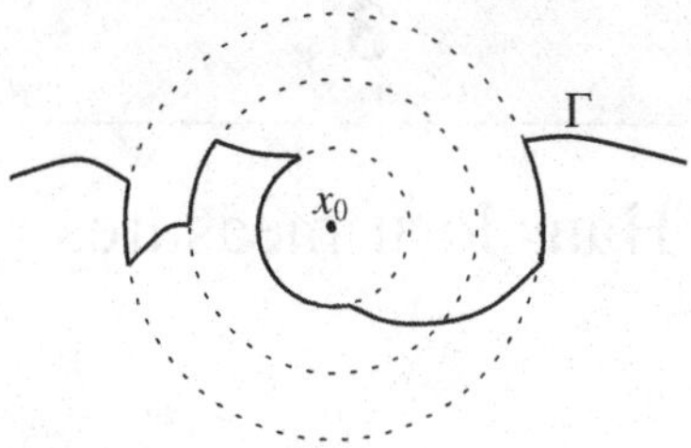

Figure 2.2 As an application of Proposition 2.16, a curve of locally finite length can contain at most countably many circular arcs of positive length.

Proof If $I_k = \{t \in I : \mu(E_t \cap B_k) > k^{-1}\}$, then $\{t \in I : \mu(E_t) > 0\} = \bigcup_{k \in \mathbb{N}} I_k$. But I_k is finite, with $\#(I_k) \le k\,\mu(B_k)$: indeed, if $J \subset I_k$ is finite, then

$$\mu(B_k) \ge \mu\Big(\bigcup_{t\in I} E_t \cap B_k\Big) \ge \mu\Big(\bigcup_{t\in J} E_t \cap B_k\Big) = \sum_{t\in J} \mu(E_t \cap B_k) \ge \frac{\#(J)}{k}. \qquad \square$$

Example 2.17 If $\Gamma \subset \mathbb{R}^n$ is a curve, then $\mu = \mathcal{H}^1 \llcorner \Gamma$ is a Radon measure (see Section 3.2). Now consider the disjoint family of Borel sets $\{\partial B(x_0, r)\}_{r>0}$ defined by all the spheres centered at $x_0 \in \mathbb{R}^n$. By Proposition 2.16,

$$\mathcal{H}^1\big(\Gamma \cap \partial B(x_0, r)\big) > 0,$$

for at most countably many radii $r > 0$; see Figure 2.2.

Exercise 2.18 (Radon measures on open sets, I) Let μ be a regular Borel measure on $\mathbb{R}^n$ and let A be an open set in $\mathbb{R}^n$. We say that μ is a Radon measure on A if μ is locally finite on A, that is, if $\mu(K) < \infty$ for every compact set $K \subset A$. If μ is a Radon measure on $\mathbb{R}^n$, then $\mu \llcorner A$ is a Radon measure on A. However, given a Radon measure on A, there could be no way to "extend" it as a Radon measure on $\mathbb{R}^n$. For example, think of $\mu = \sum_{h\in\mathbb{N}} \delta_{1/h}$, that is a Radon measure on $(0, 1)$ with $\mu((0, 1)) = \infty$. Adapt the theory of this chapter to the case of Radon measures on open sets in $\mathbb{R}^n$.

3

Hausdorff measures

Hausdorff measures provide an important source of examples of Radon measures. For this reason, before further developing the theory of Radon measures, in this chapter we take a closer look at Hausdorff measures. In Section 3.1 we introduce and motivate the notion of Hausdorff dimension. In Section 3.2 we relate $\mathcal{H}^1$ to the classical notion of length of a curve. Finally, in Section 3.3 we show equivalence of the Lebesgue measure and $\mathcal{H}^n$ on $\mathbb{R}^n$.

3.1 Hausdorff measures and the notion of dimension

We begin our discussion by introducing a measure-theoretic notion of dimension. Precisely, given $E \subset \mathbb{R}^n$ we define the **Hausdorff dimension** of E as

$$\dim(E) = \inf\left\{s \in [0,\infty) : \mathcal{H}^s(E) = 0\right\}.$$

Its use as a notion of dimension is justified by the following statements.

(i) If $E \subset \mathbb{R}^n$ then $\dim(E) \in [0,n]$. Moreover $\mathcal{H}^s(E) = \infty$ for every $s < \dim(E)$ and $\mathcal{H}^s(E) \in (0,\infty)$ implies $s = \dim(E)$ (the converse is not necessarily true: it may happen that $\mathcal{H}^s(E) \in \{0,+\infty\}$ for $s = \dim(E)$).

(ii) $\mathcal{H}^0$ is the counting measure.

(iii) If E is a curve, then $\mathcal{H}^1(E)$ coincides with the classical length of E.

(iv) If $1 \le k \le n-1$, $k \in \mathbb{N}$, and E is a k-dimensional C^1-surface, then $\mathcal{H}^k(E)$ coincides with the classical k-dimensional area of E.

(v) If $E \subset \mathbb{R}^n$, then $\mathcal{H}^n(E) = \mathcal{L}^n(E)$.

(vi) If $s > n$, then $\mathcal{H}^s = 0$.

(vii) If A is an open set in $\mathbb{R}^n$, then $\dim(A) = n$.

(viii) For every $s \in [0,n]$ there exists a compact set K such that $\dim(K) = s$.

We now prove properties (i), (ii), and (vi). Properties (iii) and (v) are proved in Sections 3.2 and 3.3, respectively. Property (vii) follows from (i) and (v), since $|(0,1)^n| = 1$. Property (iv) is a consequence of the area formula; see Chapter 8. For property (viii), see [Hut81] and [Fal86].

Proposition 3.1 *If $s > n$, then $\mathcal{H}^s = 0$.*

Proof Let $Q = (0,1)^n$. Since $\lambda^s \mathcal{H}^s(Q) = \mathcal{H}^s(\lambda Q) \to \mathcal{H}^s(\mathbb{R}^n)$ as $\lambda \to \infty$, it suffices to prove $\mathcal{H}^s(Q) = 0$. This follows by letting $k \to \infty$ in the following inequalities, which are obtained by considering a partition of Q by k^n cubes of diameter $k^{-1}\sqrt{n}$:

$$\mathcal{H}^s_{\sqrt{n}/k}(Q) \le \omega_s\, k^n \left(\frac{\sqrt{n}}{2k}\right)^s = \frac{\omega_s\, n^{s/2}}{2^s}\, k^{n-s}. \qquad \square$$

Proposition 3.2 *If $E \subset \mathbb{R}^n$, then $\dim(E) \in [0,n]$, and $\mathcal{H}^s(E) = \infty$ for every $s < \dim(E)$.*

Proof By Proposition 3.1 we always have $\dim(E) \in [0,n]$. We now prove that, if $\mathcal{H}^s(E) < \infty$ for some $s \in [0,n)$, then $\mathcal{H}^t(E) = 0$ for every $t > s$. Indeed, if $\mathcal{F}$ is a countable covering of E by sets of diameter less than δ, then

$$\mathcal{H}^t_\delta(E) \le \omega_t \sum_{F\in\mathcal{F}_\delta} \left(\frac{\operatorname{diam}(F)}{2}\right)^t \le \left(\frac{\delta}{2}\right)^{t-s} \frac{\omega_t}{\omega_s}\, \omega_s \sum_{F\in\mathcal{F}_\delta} \left(\frac{\operatorname{diam}(F)}{2}\right)^s,$$

that is, $\mathcal{H}^t_\delta(E) \le C(t,s)\,\delta^{t-s}\mathcal{H}^s(E)$. We let $\delta \to 0^+$ to find $\mathcal{H}^t(E) = 0$. $\square$

Proposition 3.3 *$\mathcal{H}^0$ is the counting measure.*

Proof If $x \in \mathbb{R}^n$ and $\delta > 0$, then $\mathcal{H}^0_\delta(\{x\}) = \omega_0 = 1$, hence $\mathcal{H}^0(\{x\}) = 1$. Since $\mathcal{H}^0$ is a Borel measure, we find $\mathcal{H}^0(E) = \sum_{x\in E} \mathcal{H}^0(\{x\}) = \#(E)$ whenever E is finite or countable. If now E is infinite, then there exists an infinite countable set $F \subset E$, and thus $\mathcal{H}^0(E) \ge \mathcal{H}^0(F) = \infty$. $\square$

Proposition 3.4 *If $E \subset \mathbb{R}^n$ with $\mathcal{H}^s_\infty(E) = 0$, then $\mathcal{H}^s(E) = 0$.*

Proof The case $s = 0$ is trivial. Let $s > 0$. By $\mathcal{H}^s_\infty(E) = 0$, for every $\varepsilon > 0$ there exists a countable cover $\mathcal{F}$ of E with

$$\omega_s \sum_{F\in\mathcal{F}} \left(\frac{\operatorname{diam}(F)}{2}\right)^s \le \varepsilon \qquad \text{so that} \qquad \sup_{F\in\mathcal{F}} \operatorname{diam}(F) \le 2\left(\frac{\varepsilon}{\omega_s}\right)^{1/s} = \delta(\varepsilon).$$

Thus $\mathcal{H}^s_{\delta(\varepsilon)}(E) \le \varepsilon$ with $\delta(\varepsilon) \to 0$ as $\varepsilon \to 0$. $\square$

We close this introductory section with a simple proposition that provides a first illustration of the deep link between Hausdorff measures and Lipschitz functions. The **Lipschitz constant** $\mathrm{Lip}(f;E)$ of a function $f\colon E \subset \mathbb{R}^n \to \mathbb{R}^m$ is defined as the infimum of the non-negative constants L (if any) such that

$$|f(x) - f(y)| \le L\,|x-y|, \qquad \forall x, y \in E. \tag{3.1}$$

If $\mathrm{Lip}(f;E) < \infty$, then f is a **Lipschitz function on** E. When $E = \mathbb{R}^n$, we simply set $\mathrm{Lip}(f) = \mathrm{Lip}(f;\mathbb{R}^n)$. In Chapter 7 we are going to study Lipschitz functions in great detail. In particular, we are going to show that if f is a Lipschitz function on E, then there exists a Lipschitz function $g\colon \mathbb{R}^n \to \mathbb{R}^m$ such that $\mathrm{Lip}(g) = \mathrm{Lip}(f;E)$. Therefore in the following proposition we can focus directly on the case $E = \mathbb{R}^n$.

Proposition 3.5 *If $f\colon \mathbb{R}^n \to \mathbb{R}^m$ is a Lipschitz function, then*

$$\mathcal{H}^s(f(E)) \le \mathrm{Lip}(f)^s \mathcal{H}^s(E), \tag{3.2}$$

for every $s \in [0,\infty)$ and $E \subset \mathbb{R}^n$. In particular $\dim(f(E)) \le \dim(E)$.

Proof Let $\mathcal{F}$ be a countable covering of E by sets of diameter less than δ. Then $\{f(F) : F \in \mathcal{F}\}$ is a covering of $f(E)$ with

$$\mathrm{diam} f(F) \le \mathrm{Lip}(f)\,\mathrm{diam}(F) \le \mathrm{Lip}(f)\,\delta.$$

Exploiting the arbitrariness of $\mathcal{F}$ in the following inequalities,

$$\mathcal{H}^s_{\mathrm{Lip}(f)\delta}(f(E)) \le \omega_s \sum_{F\in\mathcal{F}} \left(\frac{\mathrm{diam}\, f(F)}{2}\right)^s \le \mathrm{Lip}(f)^s \omega_s \sum_{F\in\mathcal{F}} \left(\frac{\mathrm{diam}(F)}{2}\right)^s,$$

we find $\mathcal{H}^s_{\mathrm{Lip}(f)\delta}(f(E)) \le \mathrm{Lip}(f)^s \mathcal{H}^s_\delta(E)$. We let $\delta \to 0^+$ to prove (3.2). □

Remark 3.6 By Proposition 3.5 we find that Hausdorff measures are decreased under projection over an affine subspace of $\mathbb{R}^n$. Indeed, if H is an affine subspace of $\mathbb{R}^n$ and $f\colon \mathbb{R}^n \to \mathbb{R}^n$ is the projection of $\mathbb{R}^n$ over H, then $\mathrm{Lip}(f) = 1$. The same happens, of course, if we project over a convex set.

Remark 3.7 We say that $f\colon \mathbb{R}^n \to \mathbb{R}^m$ $(1 \le n \le m)$ is an **isometry** if $|f(x) - f(y)| = |x-y|$ for every $x, y \in \mathbb{R}^n$. If $s \ge 0$, $E \subset \mathbb{R}^n$, and f is an isometry, then $\mathcal{H}^s(f(E)) = \mathcal{H}^s(E)$, as we may see either by applying Proposition 3.5 to f and to any extension g of f^{-1} with $\mathrm{Lip}(g) \le 1$, or by the area formula (8.1), see Remark 8.10. In particular, if π is an n-dimensional plane in $\mathbb{R}^m$, then there exists an orthogonal injection (see Section 8.1) $P \in \mathbf{O}(n,m)$ such that $\pi = P(\mathbb{R}^n)$, and thus

$$\mathcal{H}^n \llcorner \pi = P_\# \mathcal{H}^n. \tag{3.3}$$

On the left-hand side, $\mathcal{H}^n$ stands for the n-dimensional Hausdorff measure on $\mathbb{R}^m$, on the right-hand side, it denotes the n-dimensional Hausdorff measure on $\mathbb{R}^n$ (which in turn coincides with $\mathcal{L}^n$; see Theorem 3.10).

3.2 $\mathcal{H}^1$ and the classical notion of length

A set $\Gamma \subset \mathbb{R}^n$ is a **curve** if there exist $a > 0$ and a continuous, injective function $\gamma\colon [0,a] \to \mathbb{R}^n$ such that $\Gamma = \gamma([0,a])$. The function γ is called a **parametrization of** Γ. Given a parametrization $\gamma\colon [0,a] \to \mathbb{R}^n$ and a sub-interval $[b,c]$ of $[0,a]$ we define the **length of** γ **over** $[b,c]$ as

$$\ell(\gamma;[b,c]) = \sup\left\{\sum_{h=1}^{N} |\gamma(t_h) - \gamma(t_{h-1})| \colon b = t_0 < t_{h-1} < t_h < t_N = c,\ N \in \mathbb{N}\right\}.$$

It is easily seen that $\ell(\gamma;[0,a])$ is independent of the parametrization γ of Γ. Therefore, the **length of** Γ is defined as

$$\mathrm{length}(\Gamma) = \ell(\gamma;[0,a]).$$

Whether length(Γ) is finite or not, the following theorem holds true.

Theorem 3.8 *If Γ is a curve, then* $\mathcal{H}^1(\Gamma) = \mathrm{length}(\Gamma)$.

Proof The theorem is proved by Remark 3.7 if Γ is a segment. We now consider a parametrization $\gamma\colon [0,a] \to \mathbb{R}^n$ of Γ and set $\ell = \ell(\gamma;[0,a]) = \mathrm{length}(\Gamma)$. We divide the proof into three steps, and notice that

(i) $\ell(\gamma;[b,c]) \ge |\gamma(b) - \gamma(c)|$, whenever $0 \le b \le c \le a$;
(ii) $\ell(\gamma;[b,c]) = \ell(\gamma;[b,d]) + \ell(\gamma;[d,c])$ whenever $0 \le b \le d \le c \le a$.

Step one: We show that $\mathcal{H}^1(\Gamma) \ge |\gamma(a) - \gamma(0)|$. Since the projection $\mathbf{p}\colon \mathbb{R}^n \to \mathbb{R}^n$ of $\mathbb{R}^n$ onto the line defined by $\gamma(0)$ and $\gamma(a)$ satisfies $\mathrm{Lip}(\mathbf{p}) \le 1$, by Proposition 3.5 we have $\mathcal{H}^1(\mathbf{p}(\Gamma)) \le \mathcal{H}^1(\Gamma)$. At the same time, $\mathbf{p}(\Gamma)$ must contain the segment $[\gamma(0)\gamma(a)]$: otherwise, $\Gamma = \gamma([0,a])$ would be disconnected, against the continuity of γ. Thus $\mathcal{H}^1(\mathbf{p}(\Gamma)) \ge \mathcal{H}^1([\gamma(0)\gamma(a)]) = |\gamma(a) - \gamma(0)|$.

Step two: If $\{t_h\}_{h=0}^N$ is a competitor in the definition of ℓ, then, setting $\Gamma_h = \gamma([t_{h-1}, t_h])$, we have $\Gamma = \bigcup_{h=1}^N \Gamma_h$ and, by the injectivity of γ, $\mathcal{H}^1(\Gamma_h \cap \Gamma_{h+1}) = \mathcal{H}^1(\{\gamma(t_h)\}) = 0$. We thus find $\mathcal{H}^1(\Gamma) \ge \ell$ as, by step one,

$$\mathcal{H}^1(\Gamma) = \sum_{h=1}^{N} \mathcal{H}^1(\Gamma_h) \ge \sum_{h=1}^{N} |\gamma(t_h) - \gamma(t_{h-1})|.$$

Step three: We finally prove that $\mathcal{H}^1(\Gamma) \le \ell$, by constructing a continuous injective function $\gamma^*\colon [0,\ell] \to \mathbb{R}^n$ with $\mathrm{Lip}(\gamma^*) \le 1$ and $\Gamma = \gamma^*([0,\ell])$. Indeed, by Proposition 3.5, the existence of γ^* will imply, as required, that

$$\mathcal{H}^1(\Gamma) = \mathcal{H}^1\big(\gamma^*([0,\ell])\big) \le \mathcal{H}^1\big([0,\ell]\big) = \ell.$$

To construct γ^* (which is just the parametrization by arc length of γ, defined without using derivatives), we define $v\colon [0,a] \to [0,\ell]$ by $v(t) = \ell(\gamma,[0,t])$, $t \in [0,a]$. Then $v(0) = 0$, $v(a) = \ell$ and v is strictly increasing, that is, $v(t) < v(s)$ if $t < s$, as γ is injective. In particular, v is invertible, with a strictly increasing inverse $w\colon [0,\ell] \to [0,a]$. Let then $\gamma^*\colon [0,\ell] \to \mathbb{R}^n$ be defined by $\gamma^*(s) = \gamma(w(s))$, $s \in [0,\ell]$. We easily find that $\mathrm{Lip}(\gamma^*) \le 1$, since, by properties (i) and (ii) above, if $[s_1,s_2] \subset [0,\ell]$, then

$$|\gamma^*(s_1) - \gamma^*(s_2)| \le \ell\big(\gamma^*,[s_1,s_2]\big) = \ell\big(\gamma^*,[0,s_2]\big) - \ell\big(\gamma^*,[0,s_1]\big) = s_2 - s_1\,. \qquad \square$$

Remark 3.9 When Γ admits a C^1-parametrization $\gamma\colon [0,a] \to \mathbb{R}^n$, it is immediately seen that $\mathrm{length}(\Gamma) = \int_0^a |\gamma'(t)|\,dt$. In particular, by Theorem 3.8,

$$\mathcal{H}^1(\Gamma) = \int_0^a |\gamma'(t)|\,dt\,.$$

This is the one-dimensional case of the area formula discussed in Chapter 8.

3.3 $\mathcal{H}^n = \mathcal{L}^n$ and the isodiametric inequality

We show here equivalence of the Lebesgue measure and the n-dimensional Hausdorff measure $\mathcal{H}^n$ on $\mathbb{R}^n$.

Theorem 3.10 *If $E \subset \mathbb{R}^n$, and $\delta \in (0,\infty]$, then $|E| = \mathcal{H}^n(E) = \mathcal{H}^n_\delta(E)$.*

A first tool used in proving Theorem 3.10 is *Vitali's property* of Lebesgue measure: if $A \subset \mathbb{R}^n$ is open and $\delta > 0$, then a countable disjoint family $\mathcal{F}$ of closed balls contained in A with diameter less than δ exists such that

$$\Big|A \setminus \bigcup\big\{\overline{B} : \overline{B} \in \mathcal{F}\big\}\Big| = 0\,. \tag{3.4}$$

Postponing until Section 5.1 the proof of this result, we now introduce the second tool used in proving Theorem 3.10, namely, the *isodiametric inequality*.

Theorem 3.11 (Isodiametric inequality) *Among all sets of fixed diameter, balls have maximum volume. In other words,*

$$|E| \le \omega_n \left(\frac{\mathrm{diam}(E)}{2}\right)^n, \qquad \forall E \subset \mathbb{R}^n\,. \tag{3.5}$$

Proof of Theorem 3.10 *Step one:* We first notice that

$$\omega_n \left(\frac{\sqrt{n}}{2}\right)^n |E| \ge \mathcal{H}^n_\infty(E)\,. \tag{3.6}$$

If the covering $\mathcal{F}$ is a competitor in the definition of $|E|$, and $r(F)$ denotes the side length of the cube $F \in \mathcal{F}$, then $\operatorname{diam}(F) = \sqrt{n}\, r(F)$, so that, in particular,

$$\mathcal{H}^n_\infty(E) \le \omega_n \sum_{F\in\mathcal{F}} \left(\frac{\operatorname{diam}(F)}{2}\right)^n = \omega_n \left(\frac{\sqrt{n}}{2}\right)^n \sum_{F\in\mathcal{F}} r(F)^n .$$

By the arbitrariness of $\mathcal{F}$, we find (3.6).

Step two: We prove $|E| \ge \mathcal{H}^n(E)$. We can assume $|E| < \infty$. Given ε and δ positive, we may consider A open with $E \subset A$ and $|A| \le |E| + \varepsilon$, and, by Vitali's property, a countable disjoint family $\mathcal{F}$ of closed balls contained in A, with diameter less than δ, such that (3.4) holds true. If $F = \bigcup_{\overline{B}\in\mathcal{F}} \overline{B}$, then

$$|E| + \varepsilon \ge |A| = \left|\bigcup_{\overline{B}\in\mathcal{F}} \overline{B}\right| = \sum_{\overline{B}\in\mathcal{F}} |\overline{B}| = \omega_n \sum_{\overline{B}\in\mathcal{F}} \left(\frac{\operatorname{diam}(\overline{B})}{2}\right)^n \ge \mathcal{H}^n_\delta(F) . \qquad (3.7)$$

By (3.6), $\mathcal{H}^n_\infty(A \setminus F) = 0$. Hence, $\mathcal{H}^n_\infty(E \setminus F) = 0$ and, by Proposition 3.4, $\mathcal{H}^n_\delta(E \setminus F) = 0$. By (3.7),

$$\mathcal{H}^n_\delta(E) \le \mathcal{H}^n_\delta(E \cap F) + \mathcal{H}^n_\delta(E \setminus F) \le \mathcal{H}^n_\delta(F) \le |E| + \varepsilon ,$$

and we conclude by letting $\varepsilon, \delta \to 0^+$.

Step three: Given $\delta \in (0, \infty]$, let $\mathcal{F}_\delta$ be a countable cover of E by sets F with $\operatorname{diam}(F) \le \delta$. Hence $|E| \le \mathcal{H}^n_\delta(E) \le \mathcal{H}^n(E)$ since, by Theorem 3.11,

$$\omega_n \sum_{F\in\mathcal{F}_\delta} \left(\frac{\operatorname{diam}(F)}{2}\right)^n \ge \sum_{F\in\mathcal{F}_\delta} |F| \ge \left|\bigcup_{F\in\mathcal{F}_\delta} F\right| \ge |E| . \qquad \square$$

We now prove Theorem 3.11. The assertion is trivial if E is contained in a ball of the same diameter. Although this property may fail when $n \ge 2$ (consider an equilateral triangle), we can always *reduce* to a case where it holds by using *Steiner symmetrization*. We decompose $\mathbb{R}^n$ as $\mathbb{R}^{n-1} \times \mathbb{R}$, with the projections $\mathbf{p}\colon \mathbb{R}^n \to \mathbb{R}^{n-1}$ and $\mathbf{q}\colon \mathbb{R}^n \to \mathbb{R}$, so that $x = (\mathbf{p}x, \mathbf{q}x)$ for $x \in \mathbb{R}^n$ (in particular, $\mathbf{q}x = x_n$). We define the vertical section $E_z \subset \mathbb{R}$ of E as

$$E_z = \left\{t \in \mathbb{R} \colon (z, t) \in E\right\}, \qquad z \in \mathbb{R}^{n-1} .$$

The **Steiner symmetrization E^s of E** is then defined as (see Figure 3.1)

$$E^s = \left\{x \in \mathbb{R}^n : |\mathbf{q}x| \le \frac{\mathcal{L}^1(E_{\mathbf{p}x})}{2}\right\} . \qquad (3.8)$$

By Fubini's theorem, if E is Lebesgue measurable, then E_z is Lebesgue measurable in $\mathbb{R}^{n-1}$, $z \in \mathbb{R}^{n-1} \mapsto \mathcal{L}^1(E_z)$ is Lebesgue measurable, and

$$|E| = \int_{\mathbb{R}^{n-1}} \mathcal{L}^1(E_z)\, dz = \int_{\mathbb{R}^{n-1}} \mathcal{L}^1(E^s_z)\, dz = |E^s| ,$$

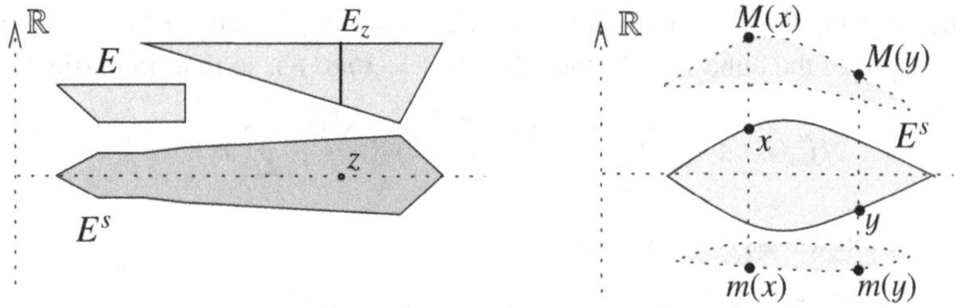

Figure 3.1 Steiner symmetrization (left) and the proof of (3.9).

that is, Lebesgue measure is invariant under Steiner symmetrization. Other relevant geometric quantities, like perimeter (see Section 14.1), decrease under Steiner symmetrization. This is also the case with diameter, that is, we have

$$\mathrm{diam}(E^s) \le \mathrm{diam}(E)\,. \tag{3.9}$$

To show this, given $x \in E^s$, let us consider $m(x), M(x) \in \overline{E}$ such that

$$\mathbf{p}m(x) = \mathbf{p}M(x) = \mathbf{p}x\,, \tag{3.10}$$

$$\mathbf{q}m(x) \le \mathbf{q}z \le \mathbf{q}M(x)\,, \qquad \text{for every } z \in E \text{ with } \mathbf{p}z = \mathbf{p}x. \tag{3.11}$$

Given $x, y \in E^s$, by (3.11) and by construction of E^s (see Figure 3.1)

$$|\,\mathbf{q}x - \mathbf{q}y| \le \max\left\{|\,\mathbf{q}M(x) - \mathbf{q}m(y)|, |\,\mathbf{q}m(x) - \mathbf{q}M(y)|\right\}. \tag{3.12}$$

By the projection constraint (3.10), and since $|z|^2 = (\mathbf{p}z)^2 + (\mathbf{q}z)^2$,

$$|x - y| \le \max\left\{|M(x) - m(y)|, |m(x) - M(y)|\right\} \le \mathrm{diam}(\overline{E}) = \mathrm{diam}(E)\,.$$

Since x and y are arbitrary in E^s we find (3.9).

Proof of Theorem 3.11 Replacing E by $\overline{E}$ will not affect the right-hand side of (3.5) and may only increase the left-hand side. Hence we assume E to be closed and, in particular, Lebesgue measurable. Let F^i denote the Steiner symmetrization of $F \subset \mathbb{R}^n$ with respect to the ith coordinate axis, and set $E_0 = E$, $E_i = (E_{i-1})^i$, $1 \le i \le n$. Since $|E_n| = |E|$ and, by (3.9), $\mathrm{diam}(E_n) \le \mathrm{diam}(E)$, it suffices to prove (3.5) on E_n. By construction, E_n is symmetric under reflection with respect to each coordinate hyperplane, so that $x \in E_n$ if and only if $-x \in E_n$. Thus $E_n \subset B_R$, $R = \mathrm{diam}(E_n)/2$, and (3.5) follows. □

4

Radon measures and continuous functions

Since a Radon measure μ is characterized on $\mathcal{M}(\mu)$ by its behavior on the family of compact/open sets of $\mathbb{R}^n$ (see Chapter 2), integration with respect to μ on $L^1(\mathbb{R}^n, \mu)$ is characterized by its behavior on $C_c^0(\mathbb{R}^n)$, the space of compactly supported continuous functions on $\mathbb{R}^n$ (Section 4.1). At the same time, integration with respect to μ defines a monotone linear functional on $C_c^0(\mathbb{R}^n)$, which is continuous with respect to the natural notion of convergence on $C_c^0(\mathbb{R}^n)$. *Riesz's theorem* shows that this correspondence actually provides an alternative characterization of Radon measures, and naturally leads to the notion of vector-valued Radon measure (Section 4.2). From Riesz's theorem we deduce compactness criteria in *weak-star convergence* for sequences of Radon measures. This natural notion of convergence enjoys good lower semicontinuity properties, and allows us to describe in a unified framework sequences of functions and surfaces undergoing concentration, oscillation, cancellation, and spreading-of-mass phenomena (Section 4.3).

4.1 Lusin's theorem and density of continuous functions

Integration with respect to a Radon measure μ on $L^1(\mathbb{R}^n, \mu)$ is characterized by its behavior on $C_c^0(\mathbb{R}^n)$ as a consequence of *Lusin's theorem.*

Theorem 4.1 (Lusin's theorem) *If μ is a Borel measure on $\mathbb{R}^n$, $u\colon \mathbb{R}^n \to \mathbb{R}$ is a Borel function, and $E \subset \mathbb{R}^n$ a Borel set with $\mu(E) < \infty$, then for every $\varepsilon > 0$ there exists a compact set $K \subset E$ such that u is continuous on K and*

$$\mu(E \setminus K) < \varepsilon. \tag{4.1}$$

Remark 4.2 If $E \subset \mathbb{R}^n$ is a Borel set in $\mathbb{R}^n$, then $u\colon E \to [-\infty, \infty]$ is a **Borel function** provided each $\{x \in E : u(x) > t\}$ ($t \in \mathbb{R}$) is a Borel set. If $\mathbb{R}^n \setminus E$ is non-empty, then, setting $u = 0$ on $\mathbb{R}^n \setminus E$, we extend u to the whole space $\mathbb{R}^n$ as a Borel function. Borel functions usually arise with a domain of definition properly contained in $\mathbb{R}^n$, and are tacitly extended to $\mathbb{R}^n$ as explained.

Proof of Theorem 4.1 For $k \in \mathbb{N}$, let $\{I_h^k\}_{h\in\mathbb{N}}$ be a countable partition of $u(\mathbb{R}^n)$ by (non-empty) Borel sets with $\mathrm{diam}(I_h^k) \le 1/k$, and consider the countable partition of E by Borel sets $\{E_h^k = E \cap u^{-1}(I_h^k)\}_{h\in\mathbb{N}}$. For $k, h \in \mathbb{N}$, we choose $y_h^k \in u(E_h^k)$, and apply Theorem 2.8 to find a compact set K_h^k with

$$K_h^k \subset E_h^k, \qquad \mu(E_h^k \setminus K_h^k) \le \frac{\varepsilon}{2^{h+k}}.$$

Since $\mu(E) < \infty$, we get

$$\lim_{N\to\infty} \mu\Big(E \setminus \bigcup_{h=1}^N K_h^k\Big) = \mu\Big(E \setminus \bigcup_{h\in\mathbb{N}} K_h^k\Big) = \mu\Big(\bigcup_{h\in\mathbb{N}} E_h^k \setminus K_h^k\Big) \le \sum_{h\in\mathbb{N}} \frac{\varepsilon}{2^{h+k}} = \frac{\varepsilon}{2^k}.$$

Thus, for every $k \in \mathbb{N}$, there exists $N(k) \in \mathbb{N}$ such that $\mu(E \setminus D_k) < \varepsilon/2^k$, where $D_k = \bigcup_{h=1}^{N(k)} K_h^k$ is compact. If we define $v_k : D_k \to \mathbb{R}$ by

$$v_k(x) = \sum_{h=1}^{N(k)} y_h^k \, 1_{K_h^k}(x), \qquad x \in D_k. \tag{4.2}$$

then, $\{K_h^k\}_{h\in\mathbb{N}}$ being disjoint, v_k is trivially continuous. Since $\mathrm{diam}(I_h^k) \le 1/k$,

$$|\, v_k(x) - u(x)| \le k^{-1}, \qquad \forall x \in D_k. \tag{4.3}$$

Hence, $v_k \to u$ uniformly on the compact set $D = \bigcap_{k\in\mathbb{N}} D_k$. Since each v_k is continuous on $D \subset D_k$, u is continuous on D. Finally, since $\mu(E \setminus D_k) < \varepsilon/2^k$,

$$\mu(E \setminus D) \le \sum_{k\in\mathbb{N}} \mu(E \setminus D_k) \le \varepsilon. \qquad \square$$

Theorem 4.3 *If μ is a Radon measure on $\mathbb{R}^n$ and $p \in [1, \infty)$, then for every $u \in L^p(\mathbb{R}^n, \mu)$ there exists a sequence $\{u_k\}_{k\in\mathbb{N}} \subset C_c^0(\mathbb{R}^n)$ such that*

$$\lim_{k\to\infty} \int_{\mathbb{R}^n} |u - u_k|^p \, d\mu = 0. \tag{4.4}$$

Proof Given $u \in L^p(\mathbb{R}^n, \mu)$, we easily see that, as $k \to \infty$,

$$u_k = 1_{B(0,k)} \min\big\{k, \max\{-k, u\}\big\} \to u \qquad \text{in } L^p(\mathbb{R}^n, \mu).$$

We may thus assume that $|u| \le M$ and $\mathrm{spt}\, u \subset B_R$, for some fixed $M, R > 0$. Let us now assume that u is a Borel function. Setting $E = B_R$, we define I_h^k, E_h^k, K_h^k, D_k, and y_h^k ($k, h \in \mathbb{N}$) as in the proof of Lusin's theorem, and introduce open sets A_h^k with $E_h^k \subset A_h^k$ and $\mu(A_h^k \setminus E_h^k) \le \varepsilon/2^{h+k}$. Replacing A_h^k with $A_h^k \cap B_R$ preserves these properties; hence we may assume $A_h^k \subset B_R$ as well. We now consider a partition of unity $\{\varphi_h^k\}_{h=1}^{N(k)}$ subordinated to the open covering $\{A_h^k\}_{h=1}^{N(k)}$ of the compact set $D_k = \bigcup_{h=1}^{N(k)} K_h^k$, that is $\varphi_h^k \in C_c^0(A_h^k)$, $0 \le \varphi_h^k \le 1$, and

$$\sum_{h\in\mathbb{N}} \varphi_h^k = 1 \quad \text{on } D_k\,.$$

Correspondingly, we define $u_k\colon \mathbb{R}^n \to \mathbb{R}$ by setting

$$u_k(x) = \sum_{h=1}^{N(k)} y_h^k\, \varphi_h^k(x)\,, \qquad x \in \mathbb{R}^n\,,$$

cf. (4.2). Clearly, $|u_k| \le M$ and $\operatorname{spt} u_k \subset\subset B_R$ for every $k \in \mathbb{N}$. Moreover, u_k is continuous on $\mathbb{R}^n$ with $u_k = v_k$ on D_k, thus, by (4.3), $|u_k - u| \le k^{-1}$ on D_k. In conclusion, (4.4) is proved for u a Borel function by letting $k \to \infty$ in

$$\begin{aligned}\int_{\mathbb{R}^n} |u - u_k|^p \, d\mu &= \int_{D_k} |u - u_k|^p \, d\mu + \int_{B_R\setminus D_k} |u - u_k|^p \, d\mu \\ &\le \frac{\mu(D_k)}{k^p} + (2M)^p\,\mu(B_R \setminus D_k) \le \frac{\mu(B_R)}{k^p} + (2M)^p\,\frac{\varepsilon}{2^k}\,.\end{aligned}$$

If now $u \in L^p(\mathbb{R}^n, \mu)$, then the sets $E_h^k = u^{-1}(I_h^k)$ may not be Borel sets but still belong to $\mathcal{M}(\mu)$. We may thus repeat the above argument by using Proposition 2.12 to find the required compact and open sets K_h^k and A_h^k. □

Remark 4.4 We have in fact proved that if μ is a Radon measure on $\mathbb{R}^n$ and $u \in L^1(\mathbb{R}^n;\mu)$ with $a \le u \le b$ on $\mathbb{R}^n$ and $\operatorname{spt} u \subset B_R$ for some $a, b \in \mathbb{R}$, $R > 0$, then there exists a sequence $\{u_h\}_{h\in\mathbb{N}} \subset C_c^0(B_R)$, with $a \le u_h \le b$ and $\operatorname{spt} u_h \subset B_R$, such that $u_h \to u$ in $L^1(\mathbb{R}^n;\mu)$.

Exercise 4.5 If μ is a Radon measure, $g \in L^1_{\mathrm{loc}}(\mathbb{R}^n;\mu)$ and $\int_{\mathbb{R}^n} u\,g \, d\mu \ge 0$ for every $u \in C_c^0(\mathbb{R}^n)$ with $u \ge 0$, then $g(x) \ge 0$ for μ-a.e. $x \in \mathbb{R}^n$.

4.2 Riesz's theorem and vector-valued Radon measures

If μ is a Radon measure on $\mathbb{R}^n$, then the linear functional $L\colon C_c^0(\mathbb{R}^n) \to \mathbb{R}$,

$$\langle L, \varphi\rangle = \int_{\mathbb{R}^n} \varphi \, d\mu\,, \qquad \varphi \in C_c^0(\mathbb{R}^n)\,,$$

is positive ($\varphi \ge 0$ implies $\langle L, \varphi\rangle \ge 0$) or, equivalently, monotone ($\varphi_1 \le \varphi_2$ implies $\langle L, \varphi_1\rangle \le \langle L, \varphi_2\rangle$). As a consequence, L is continuous with respect to the following notion of convergence on $C_c^0(\mathbb{R}^n)$: $\varphi_h \to \varphi$ in $C_c^0(\mathbb{R}^n)$ if $\varphi_h \to \varphi$ uniformly on $\mathbb{R}^n$ and, for a compact set $K \subset \mathbb{R}^n$,

$$\operatorname{spt}\varphi \cup \bigcup_{h\in\mathbb{N}} \operatorname{spt}\varphi_h \subset K\,.$$

Indeed, $\varphi_h \to \varphi$ in $C_c^0(\mathbb{R}^n)$ implies $\langle L, \varphi_h\rangle \to \langle L, \varphi\rangle$, as we clearly have

$$\sup\Big\{\langle L,\varphi\rangle : \varphi \in C_c^0(\mathbb{R}^n), |\varphi| \le M, \operatorname{spt}\varphi \subset K\Big\} \le M\mu(K) < \infty\,,$$

for every compact set $K \subset \mathbb{R}^n$ and $M > 0$. We are now going to prove that Radon measures can actually be characterized as bounded monotone linear functionals on $C_c^0(\mathbb{R}^n)$, and use this point of view to introduce the important notion of vector-valued Radon measure. Indeed, let us consider a linear functional $L\colon C_c^0(\mathbb{R}^n;\mathbb{R}^m) \to \mathbb{R}$. By linearity, L is continuous with respect to convergence in $C_c^0(\mathbb{R}^n;\mathbb{R}^m)$ if and only if it is **bounded** (see Exercise 4.15), in the sense that, for every compact set $K \subset \mathbb{R}^n$,

$$\sup\Big\{\langle L,\varphi\rangle : \varphi \in C_c^0(\mathbb{R}^n;\mathbb{R}^m), \operatorname{spt}\varphi \subset K, |\varphi| \le 1\Big\} < \infty\,. \tag{4.5}$$

If L is integration with respect to a Radon measure μ on $\mathbb{R}^n$, then L is a linear bounded functional on $C_c^0(\mathbb{R}^n)$, with the additional property of being monotone. We may construct further examples.

Example 4.6 If μ is a Radon measure on $\mathbb{R}^n$ and $f \in L^1_{\rm loc}(\mathbb{R}^n, \mu; \mathbb{R}^m)$, we may define a bounded linear functional $f\mu\colon C_c^0(\mathbb{R}^n;\mathbb{R}^m) \to \mathbb{R}$ setting

$$\langle f\mu, \varphi\rangle = \int_{\mathbb{R}^n} (\varphi\cdot f)\,d\mu\,, \qquad \varphi \in C_c^0(\mathbb{R}^n;\mathbb{R}^m)\,.$$

Riesz's theorem ensures that every bounded linear functional on $C_c^0(\mathbb{R}^n;\mathbb{R}^m)$ can be represented as a product $f\mu$. In particular, the Radon measure μ in this decomposition can be characterized in terms of L as follows. Define the **total variation** $|L|$ of a linear functional L on $C_c^0(\mathbb{R}^n;\mathbb{R}^m)$ as the set function $|L|\colon \mathcal{P}(\mathbb{R}^n) \to [0,\infty]$ such that, for $A \subset \mathbb{R}^n$ open,

$$|L|(A) = \sup\Big\{\langle L,\varphi\rangle : \varphi \in C_c^0(A;\mathbb{R}^m), |\varphi| \le 1\Big\}\,, \tag{4.6}$$

and, for $E \subset \mathbb{R}^n$ arbitrary,

$$|L|(E) = \inf\Big\{|L|(A) : E \subset A \text{ and } A \text{ is open}\Big\}\,. \tag{4.7}$$

Note that (4.7) is consistent with (4.6).

Theorem 4.7 (Riesz's theorem) *If $L\colon C_c^0(\mathbb{R}^n;\mathbb{R}^m) \to \mathbb{R}$ is a bounded linear functional, then its total variation $|L|$ is a Radon measure on $\mathbb{R}^n$ and there* *exists a $|L|$-measurable function $g\colon \mathbb{R}^n \to \mathbb{R}^m$ with $|g| = 1$ $|L|$-a.e. on $\mathbb{R}^n$ and*

$$\langle L, \varphi \rangle = \int_{\mathbb{R}^n} (\varphi \cdot g)\, \mathrm{d}|L|, \qquad \forall \varphi \in C^0_c(\mathbb{R}^n; \mathbb{R}^m), \tag{4.8}$$

that is, $L = g|L|$. Moreover, for every open set $A \subset \mathbb{R}^n$,

$$|L|(A) = \sup\left\{ \int_{\mathbb{R}^n} (\varphi \cdot g)\, \mathrm{d}|L| : \varphi \in C^0_c(A; \mathbb{R}^m), |\varphi| \le 1 \right\}. \tag{4.9}$$

Remark 4.8 When $L = f\mu$ as in Example 4.6, then the total variation $|f\mu|$ and the vector field g in the statement of Riesz's theorem satisfy

$$|f\mu| = |f|\mu, \qquad g = \frac{f}{|f|} \quad |f|\mu\text{-a.e. on } \mathbb{R}^n.$$

Remark 4.9 (Radon measures and monotone linear functionals) If L is a monotone linear functional on $C^0_c(\mathbb{R}^n)$, then L is bounded on $C^0_c(\mathbb{R}^n)$; see Exercise 4.16. By Riesz's theorem, $\langle L, \varphi \rangle = \int_{\mathbb{R}^n} g\, \mathrm{d}|L|$, where $g\colon \mathbb{R}^n \to \mathbb{R}$ is $|L|$-measurable, with $|g| = 1$ $|L|$-a.e. on $\mathbb{R}^n$. Since L is monotone and linear, L is positive. By Exercise 4.5, $g \ge 0$ $|L|$-a.e. on $\mathbb{R}^n$. Hence, $g = 1$ $|L|$-a.e. on $\mathbb{R}^n$, and $\langle L, \varphi \rangle = \int_{\mathbb{R}^n} \varphi \mathrm{d}|L|$ for every $\varphi \in C^0_c(\mathbb{R}^n)$. Note also that if two Radon measures μ_1, μ_2 on $\mathbb{R}^n$ coincide as linear functionals, that is if

$$\int_{\mathbb{R}^n} \varphi\, \mathrm{d}\mu_1 = \int_{\mathbb{R}^n} \varphi\, \mathrm{d}\mu_2, \qquad \forall \varphi \in C^0_c(\mathbb{R}^n),$$

then we have $\mu_1 = \mu_2$. Indeed if K is compact, A is open and $K \subset A$, then there exists $\varphi \in C^0_c(\mathbb{R}^n)$ such that $1_K \le \varphi \le 1_A$. In particular

$$\mu_1(K) = \int_{\mathbb{R}^n} 1_K\, \mathrm{d}\mu_1 \le \int_{\mathbb{R}^n} \varphi\, \mathrm{d}\mu_1 = \int_{\mathbb{R}^n} \varphi\, \mathrm{d}\mu_2 \le \mu_2(A),$$

and by (2.6) and (2.7) we have $\mu_1(E) \le \mu_2(E)$ for every Borel set $E \subset \mathbb{R}^n$. By Borel regularity, $\mu_1 = \mu_2$ on $\mathcal{P}(\mathbb{R}^n)$, so that *Radon measures can be unambiguously identified with monotone linear functionals on $C^0_c(\mathbb{R}^n)$.*

Remark 4.10 (Bounded linear functionals and vector-valued set functions) Let $\mathcal{B}_b(\mathbb{R}^n)$ denote the family of bounded Borel sets of $\mathbb{R}^n$, and $\mathcal{B}(E)$ the family of Borel sets contained in $E \subset \mathbb{R}^n$. If L is a bounded linear functional on $C^0_c(\mathbb{R}^n; \mathbb{R}^m)$, then L induces a $\mathbb{R}^m$-valued set function $\nu\colon \mathcal{B}_b(\mathbb{R}^n) \to \mathbb{R}^m$,

$$\nu(E) = \int_E g\, \mathrm{d}|L|, \qquad E \in \mathcal{B}_b(\mathbb{R}^n), \tag{4.10}$$

that enjoys the σ-additivity property

$$\nu\Big(\bigcup_{h \in \mathbb{N}} E_h \Big) = \sum_{h \in \mathbb{N}} \nu(E_h) \tag{4.11}$$

on every disjoint sequence $\{E_h\}_{h\in\mathbb{N}} \subset \mathcal{B}(K)$, for some K compact in $\mathbb{R}^n$. Thus, *bounded linear functionals on $C^0_c(\mathbb{R}^n;\mathbb{R}^m)$ naturally induce $\mathbb{R}^m$-valued set functions on $\mathbb{R}^n$ that are σ-additive on bounded Borel sets.*

Remark 4.11 (Vector-valued Radon measures) Taking into account Remark 4.9 and Remark 4.10, we define **$\mathbb{R}^m$-valued Radon measures on $\mathbb{R}^n$** as the bounded linear functionals on $C^0_c(\mathbb{R}^n;\mathbb{R}^m)$. When $m = 1$ we speak of **signed Radon measures on $\mathbb{R}^n$**. We shall always adopt the Greek symbols μ, ν, etc. in place of L to denote vector-valued Radon measures, and also set

$$\langle \mu, \varphi \rangle = \int_{\mathbb{R}^n} \varphi \cdot \mathrm{d}\mu \tag{4.12}$$

to denote the value of the $\mathbb{R}^m$-valued Radon measure μ on $\mathbb{R}^n$ at $\varphi \in C^0_c(\mathbb{R}^n;\mathbb{R}^m)$.

Remark 4.12 (Polar decomposition and Jordan decomposition) By Riesz's theorem, every $\mathbb{R}^m$-valued Radon measure μ on $\mathbb{R}^n$ admits a **polar decomposition** $\mu = g\,|\mu|$, so that (4.12) takes the form

$$\langle \mu, \varphi \rangle = \int_{\mathbb{R}^n} (\varphi \cdot g)\mathrm{d}|\mu| .$$

If $g^{(i)}$ denotes the ith component of g, then we define the **ith component of** μ as the signed Radon measure $\mu^{(i)} = g^{(i)}|\mu|$. Note that, clearly,

$$|\mu|(E) \ge |\mu^{(i)}|(E), \qquad \forall E \subset \mathbb{R}^n . \tag{4.13}$$

When μ is a signed Radon measure then $g(x) \in \{1, -1\}$ for $|\mu|$-a.e. $x \in \mathbb{R}^n$ and we define the **positive and negative parts of** μ as

$$\mu^+ = 1_{\{g=1\}}|\mu| = \frac{|\mu| + \mu}{2}, \qquad \mu^- = 1_{\{g=-1\}}|\mu| = \frac{|\mu| - \mu}{2} .$$

Hence, μ^+, μ^- are Radon measures on $\mathbb{R}^n$, the **Jordan decomposition** $\mu = \mu^+ - \mu^-$ holds, and μ^+ and μ^- are mutually singular (see Chapter 5). (Standard) Radon measures coincide with signed Radon measures with zero negative part.

Exercise 4.13 If $f\colon \mathbb{R}^n \to \mathbb{R}^m$ is a bounded Borel vector field, if μ is Radon measure on $\mathbb{R}^n$ with values in $\mathbb{R}^m$ and we set

$$\langle f \cdot \mu, \varphi \rangle = \int_{\mathbb{R}^n} \varphi(x) f(x) \cdot \mathrm{d}\mu(x) , \qquad \varphi \in C^0_c(\mathbb{R}^n) ,$$

then $f \cdot \mu$ is a signed Radon measure with $|f \cdot \mu| = |f|\,|\mu|$.

Exercise 4.14 (Fundamental lemma of the Calculus of Variations) Let A be an open set and ν be a Radon measure on $\mathbb{R}^n$ with values in $\mathbb{R}^m$. If

$$\int_{\mathbb{R}^n} \varphi \cdot \mathrm{d}\nu = 0 , \qquad \forall \varphi \in C^\infty_c(A;\mathbb{R}^m) ,$$

then $|\nu|(A) = 0$. In particular, if $u \in L^1_{\rm loc}(\mathbb{R}^n;\mathbb{R}^m)$ and

$$\int_{\mathbb{R}^n} \big(\varphi(x)\cdot u(x)\big)\,dx = 0\,, \qquad \forall \varphi \in C^\infty_{\rm c}(A;\mathbb{R}^m)\,,$$

then $u = 0$ a.e. on A.

Exercise 4.15 Show that a linear functional $L\colon C^0_{\rm c}(\mathbb{R}^n;\mathbb{R}^m) \to \mathbb{R}$ is continuous if and only if it is bounded. *Hint:* To prove the "only if" part, argue by contradiction: there exist a compact set $K \subset \mathbb{R}^n$ and a sequence $\{\varphi_h\}_{h\in\mathbb{N}} \subset C^0_{\rm c}(\mathbb{R}^n;\mathbb{R}^m)$ such that $|\varphi_h| \le 1$, $\operatorname{spt}\varphi_h \subset K$, and $\langle L,\varphi_h\rangle \ge h$. To conclude, consider $\psi_h = h^{-1/2}\varphi_h$.

Exercise 4.16 If $L\colon C^0_{\rm c}(\mathbb{R}^n) \to \mathbb{R}$ is a monotone linear functional, then L is bounded. *Hint:* Given $K \subset \mathbb{R}^n$ compact, fix $\psi \in C^0_{\rm c}(\mathbb{R}^n)$ such that $\psi \ge 0$ and $\psi = 1$ on K. Then $\langle L,\varphi\rangle \le 2\,\langle L,\psi\rangle$ for every $\varphi \in C^0_{\rm c}(\mathbb{R}^n)$ such that $|\varphi| \le 1$ and $\operatorname{spt}\varphi \subset K$.

4.2.1 Proof of Riesz's Theorem

We now recall, for the sake of completeness, the classical proof of Riesz's theorem. We start by proving that the total variation defines a Radon measure.

Lemma 4.17 *If L is a bounded linear functional on $C^0_{\rm c}(\mathbb{R}^n;\mathbb{R}^m)$, then its total variation $|L|$ is a Radon measure on $\mathbb{R}^n$.*

Proof *Step one:* We prove that $|L|$ is an outer measure. Let us first show that

$$|L|(A) \le \sum_{h\in\mathbb{N}} |L|(A_h)\,, \tag{4.14}$$

for $A = \bigcup_{h\in\mathbb{N}} A_h$, A_h open. Indeed, let $\varphi \in C^0_{\rm c}(A;\mathbb{R}^m)$ with $|\varphi| \le 1$. Since $\operatorname{spt}\varphi \subset A$ is compact, there exists $N \in \mathbb{N}$ such that $\operatorname{spt}\varphi \subset \bigcup_{h=1}^N A_h$. We consider the corresponding partition of unity, that is

$$\varphi_h \in C^0_{\rm c}(A_h)\,, \qquad 0 \le \varphi_h \le 1\,, \qquad \sum_{h=1}^N \varphi_h = 1 \quad \text{on } \operatorname{spt}\varphi.$$

Since $\varphi = \sum_{h=1}^N \varphi\varphi_h$ and $\varphi\varphi_h \in C^0_{\rm c}(A_h;\mathbb{R}^m)$ with $|\varphi\varphi_h| \le 1$, we have

$$\langle L,\varphi\rangle = \sum_{h=1}^N \langle L,\varphi\varphi_h\rangle \le \sum_{h=1}^N |L|(A_h) \le \sum_{h\in\mathbb{N}} |L|(A_h)\,,$$

and (4.14) is proved. We now consider $E \subset \bigcup_{h\in\mathbb{N}} E_h$, and prove that

$$|L|(E) \le \sum_{h\in\mathbb{N}} |L|(E_h)\,.$$

Given $\varepsilon > 0$ and $h \in \mathbb{N}$, by definition of $|L|$ we find A_h open with $E_h \subset A_h$ and $|L|(A_h) \le |L|(E_h) + \varepsilon/2^h$. Hence, by (4.14)

$$|L|(E) \le |L|\Big(\bigcup_{h\in\mathbb{N}} A_h\Big) \le \sum_{h\in\mathbb{N}} |L|(A_h) \le \sum_{h\in\mathbb{N}} |L|(E_h) + \varepsilon .$$

Step two: By Theorem 2.1, $|L|$ is a Borel measure if $\mathrm{dist}(E_1, E_2) > 0$ implies

$$|L|(E_1 \cup E_2) \ge |L|(E_1) + |L|(E_2) . \tag{4.15}$$

When E_1, E_2 are open, (4.15) follows from the definition of $|L|$. In the general case, since $0 < \mathrm{dist}(E_1, E_2) = \mathrm{dist}(\overline{E_1}, \overline{E_2})$, there exist open sets A_1, A_2 such that $\overline{E_j} \subset A_j$ and $\mathrm{dist}(A_1, A_2) > 0$. If A is open and $E_1 \cup E_2 \subset A$, then $\mathrm{dist}(A_1 \cap A, A_2 \cap A) > 0$ and $E_j \subset A_j \cap A$, so that (4.15) on open sets implies

$$|L|(A) \ge |L|\big((A_1\cap A)\cup(A_2\cap A)\big) \ge |L|(A_1\cap A)+|L|(A_2\cap A) \ge |L|(E_1)+|L|(E_2) .$$

As A is arbitrary, (4.15) follows. Hence $|L|$ is a Borel measure, locally finite thanks to (4.5). Finally, $|L|$ is Borel regular (thus a Radon measure), since, if $E \subset \mathbb{R}^n$, $|L|(E) < \infty$ and $\{A_h\}_{h\in\mathbb{N}}$ are open sets with $E \subset A_h$ and $|L|(A_h) \to |L|(E)$, then $F = \bigcap_{h\in\mathbb{N}} A_h$ is a Borel set with $E \subset F$ and $|L|(E) = |L|(F)$. □

By the (elementary) Riesz's representation theorem on Hilbert spaces, if μ is a Radon measure on $\mathbb{R}^n$, and $L\colon L^2(\mathbb{R}^n, \mu) \to \mathbb{R}$ is a linear functional with

$$\sup\Big\{\langle L, u\rangle : u \in L^2(\mathbb{R}^n, \mu), \|u\|_{L^2(\mathbb{R}^n,\mu)} = 1\Big\} = C < \infty ,$$

then there exists $v \in L^2(\mathbb{R}^n, \mu)$ such that $\| v\|_{L^2(\mathbb{R}^n,\mu)} = C$ and

$$\langle L, u\rangle = \int_{\mathbb{R}^n} u\, v \, d\mu , \qquad \forall u \in L^2(\mathbb{R}^n, \mu) .$$

Bounded linear functionals on $L^1(\mathbb{R}^n, \mu)$ are then addressed as follows.

Lemma 4.18 (Riesz's representation theorem in L^1) *If μ is a Radon measure on $\mathbb{R}^n$ and $L\colon L^1(\mathbb{R}^n, \mu) \to \mathbb{R}$ is a linear functional such that*

$$\sup\Big\{\langle L, u\rangle : u \in L^1(\mathbb{R}^n, \mu), \|u\|_{L^1(\mathbb{R}^n,\mu)} = 1\Big\} = C < \infty , \tag{4.16}$$

then there exists a function $v \in L^\infty(\mathbb{R}^n, \mu)$ with $\| v\|_{L^\infty(\mathbb{R}^n,\mu)} = C$ and

$$\langle L, u\rangle = \int_{\mathbb{R}^n} u\, v \, d\mu , \qquad \forall u \in L^1(\mathbb{R}^n, \mu) . \tag{4.17}$$

Proof Setting $E_h = B_{h+1} \setminus \overline{B_h}$, $h \in \mathbb{N}$, let $\{t_h\}_{h\in\mathbb{N}} \subset (0, \infty)$ be such that $w = \sum_{h\in\mathbb{N}} t_h 1_{E_h} \in L^2(\mathbb{R}^n, \mu)$. The linear functional $L_0\colon L^2(\mathbb{R}^n, \mu) \to \mathbb{R}$ defined as

$$\langle L_0, u\rangle = \langle L, wu\rangle , \qquad u \in L^2(\mathbb{R}^n, \mu) ,$$

is continuous on $L^2(\mathbb{R}^n,\mu)$, with norm bounded by $C\|w\|_{L^2(\mathbb{R}^n,\mu)}$. By Riesz's representation theorem on $L^2(\mathbb{R}^n,\mu)$, there exists $z \in L^2(\mathbb{R}^n,\mu)$ such that

$$\langle L, w\,u\rangle = \int_{\mathbb{R}^n} u\,z\,\mathrm{d}\mu\,, \qquad \forall u \in L^2(\mathbb{R}^n,\mu)\,. \tag{4.18}$$

Since $w > 0$ on $\mathbb{R}^n$, the μ-measurable function $v = z/w$ has the required properties. Indeed, as w is uniformly positive on compact sets, if $u \in C^0_c(\mathbb{R}^n)$, then $u/w \in L^2(\mathbb{R}^n,\mu)$. By (4.18) we thus find

$$\langle L, u\rangle = \int_{\mathbb{R}^n} u\,v\,\mathrm{d}\mu\,, \qquad \forall u \in C^0_c(\mathbb{R}^n)\,. \tag{4.19}$$

To show that $v \in L^\infty(\mathbb{R}^n,\mu)$ with $\|v\|_{L^\infty(\mathbb{R}^n,\mu)} \le C$, assume on the contrary that

$$\mu\Big(\big\{x \in \mathbb{R}^n : |v(x)| > C\big\}\Big) > 0\,,$$

so that $|v| > C$ on a Borel set F with $0 < \mu(F) < \infty$. Testing (4.18) with

$$u_0 = 1_{F\cap\{v>C\}} - 1_{F\cap\{v<-C\}} \in L^2(\mathbb{R}^n,\mu)\,,$$

we would then find the following contradiction

$$C\int_F w < \int_F |z| = \int_F u_0 z = L(w u_0) \le C\int_{\mathbb{R}^n} w|u_0| = C\int_F w\,.$$

Since $v \in L^\infty(\mathbb{R}^n,\mu)$, (4.19) defines a continuous functional on $L^1(\mathbb{R}^n,\mu)$. Since L is continuous in $L^1(\mathbb{R}^n,\mu)$ and, by Corollary 4.3, $C^0_c(\mathbb{R}^n)$ is dense in $L^1(\mathbb{R}^n,\mu)$, we deduce (4.17) from (4.19). Finally, $\|v\|_{L^\infty(\mathbb{R}^m,\mu)} < C$ would contradict (4.19) and (4.16). □

Proof of Riesz's theorem By Lemma 4.17, $|L|$ is a Radon measure on $\mathbb{R}^n$. Let us now define a functional $M\colon C^0_c(\mathbb{R}^n;[0,\infty)) \to [0,\infty)$ as

$$\langle M,\varphi\rangle = \sup\Big\{\langle L,\psi\rangle : \psi \in C^0_c(\mathbb{R}^n;\mathbb{R}^m)\,, |\psi| \le \varphi\Big\}\,, \qquad \varphi \in C^0_c(\mathbb{R}^n;[0,\infty))\,.$$

In step one we show that M is additive, positively homogeneous of degree one, and monotone on $C^0_c(\mathbb{R}^n;[0,\infty))$. In step two, we show the inequality

$$\langle M,\varphi\rangle \le \int_{\mathbb{R}^n} \varphi\,\mathrm{d}|L|\,, \qquad \forall \varphi \in C^0_c(\mathbb{R}^n;[0,\infty))\,. \tag{4.20}$$

Finally, in step three, we combine (4.20) with Riesz's representation theorem in $L^1(\mathbb{R}^n,|L|)$ in order to conclude the proof.

Step one: We show that, whenever $\varphi_1,\varphi_2 \in C^0_c(\mathbb{R}^n;[0,\infty))$ and $c \ge 0$, we have

$$\langle M,\varphi_1+\varphi_2\rangle = \langle M,\varphi_1\rangle + \langle M,\varphi_2\rangle\,, \tag{4.21}$$

$$\langle M,c\varphi_1\rangle = c\,\langle M,\varphi_1\rangle\,, \tag{4.22}$$

$$\langle M,\varphi_1\rangle \le \langle M,\varphi_2\rangle\,, \qquad (\text{if } \varphi_1 \le \varphi_2)\,. \tag{4.23}$$

The inequality $\geq$ in (4.21), as well as (4.22) and (4.23), is easily proved. Now let $\psi \in C_c^0(\mathbb{R}^n;\mathbb{R}^m)$ be such that $|\psi| \leq \varphi_1 + \varphi_2$, and set

$$\psi_h = \frac{\varphi_h}{\varphi_1+\varphi_2}\,\psi \quad \text{on } \left\{\varphi_1+\varphi_2>0\right\}, \qquad \psi_h = 0 \quad \text{elsewhere},$$

for $h = 1, 2$. Since $\psi_h \in C_c^0(\mathbb{R}^n;\mathbb{R}^m)$ with $|\psi_h| \leq \varphi_h$ and $\psi = \psi_1 + \psi_2$,

$$\langle L,\psi\rangle = \langle L,\psi_1\rangle + \langle L,\psi_2\rangle \leq \langle M,\varphi_1\rangle + \langle M,\varphi_2\rangle\,,$$

and complete the proof of (4.21) by the arbitrariness of ψ.

Step two: Given $\varphi \in C_c^0(\mathbb{R}^n;[0,\infty))$ and $\varepsilon > 0$, let $\{t_h\}_{h=0}^N \subset \mathbb{R}$ be such that

$$t_0 < 0 < t_1 < \cdots < t_{N-1} < \sup_{\mathbb{R}^n}\varphi < t_N\,, \qquad t_{h+1} - t_h \leq \varepsilon\,, \tag{4.24}$$

and consider the partition $\{E_h\}_{h=1}^N$ of $\operatorname{spt}\varphi$ by disjoint Borel sets, defined as

$$E_h = \left\{x \in \operatorname{spt}\varphi\colon t_{h-1} < \varphi(x) \leq t_h\right\}, \qquad 1 \leq h \leq N.$$

Since $|L|$ is a Radon measure, there exist open sets A_h with $E_h \subset A_h$ and

$$|L|(A_h) \leq |L|(E_h) + \frac{\varepsilon}{N}\,, \qquad 1 \leq h \leq N. \tag{4.25}$$

If necessary replacing A_h with $\{x \in A_h : \varphi(x) < t_h + \varepsilon\}$, we can also assume

$$\varphi < t_h + \varepsilon \qquad \text{on } A_h\,. \tag{4.26}$$

Finally, let $\{\zeta_h\}_{h=1}^N$ be a partition of unity subordinated to the open covering $\{A_h\}_{h=1}^N$ of the compact set $\operatorname{spt}\varphi$, namely $\zeta_h \in C_c^0(A_h)$, $0 \leq \zeta_h \leq 1$, and $\sum_{h=1}^N \zeta_h = 1$ on $\operatorname{spt}\varphi$. Since $\varphi = \sum_{h=1}^N \zeta_h\varphi$, by step one and (4.26) we find that

$$\langle M,\varphi\rangle = \sum_{h=1}^N \langle M,\varphi\zeta_h\rangle \leq \sum_{h=1}^N (t_h+\varepsilon)\langle M,\zeta_h\rangle\,. \tag{4.27}$$

If $\psi \in C_c^0(\mathbb{R}^n;\mathbb{R}^m)$ and $|\psi| \leq \zeta_h$, then $\operatorname{spt}\psi \subset A_h$ and $|\psi| \leq 1$. Hence, $\langle M,\zeta_h\rangle \leq |L|(A_h)$ and, by (4.25), we find that

$$\begin{aligned}
\langle M,\varphi\rangle &\leq \sum_{h=1}^N (t_h+\varepsilon)\Big(|L|(E_h) + \frac{\varepsilon}{N}\Big)\\
\text{(by } t_h \leq t_{h-1}+\varepsilon) \quad &\leq \sum_{h=1}^N (t_{h-1}+2\varepsilon)\Big(|L|(E_h) + \frac{\varepsilon}{N}\Big)\\
\text{(by } t_{h-1} \leq \varphi \text{ on } E_h) \quad &\leq \int_{\mathbb{R}^n} \varphi\, d|L| + t_N\varepsilon + 2\varepsilon|L|(\operatorname{spt}\varphi) + 2\varepsilon^2\\
\text{(by } t_N \leq \textstyle\sup_{\mathbb{R}^n}\varphi + \varepsilon) \quad &\leq \int_{\mathbb{R}^n} \varphi\, d|L| + \varepsilon\Big(\sup_{\mathbb{R}^n}\varphi + \varepsilon + 2|L|(\operatorname{spt}\varphi) + 2\varepsilon\Big).
\end{aligned}$$

We let $\varepsilon \to 0^+$ to prove (4.20).

Step three: Given $e \in S^{m-1}$, we define $L_e \colon C_c^0(\mathbb{R}^n) \to \mathbb{R}$ by

$$\langle L_e, \varphi \rangle = \langle L, \varphi\, e \rangle\,, \qquad \varphi \in C_c^0(\mathbb{R}^n)\,.$$

By (4.20), we find that, for every $\varphi \in C_c^0(\mathbb{R}^n)$,

$$\langle L_e, \varphi \rangle \le \sup\Big\{ \langle L, \psi \rangle : \psi \in C_c^0(\mathbb{R}^n;\mathbb{R}^m)\,, |\psi| \le |\varphi| \Big\} = \langle M, |\varphi| \rangle \le \int_{\mathbb{R}^n} |\varphi| \mathrm{d}|L|\,.$$

By Theorem 4.3, we may extend L_e as a linear functional on $L^1(\mathbb{R}^n;|L|)$ such that $|\langle L_e, u\rangle| \le \int_{\mathbb{R}^n} |u| \mathrm{d}|L|$ for every $u \in L^1(\mathbb{R}^n,|L|)$. Thus, by Lemma 4.18, there exists $g_e \in L^\infty(\mathbb{R}^n,|L|)$ such that

$$\langle L, u\, e \rangle = \int_{\mathbb{R}^n} u\, g_e \,\mathrm{d}|L|\,, \qquad \forall u \in L^1(\mathbb{R}^n,|L|).$$

If we set $g\colon \mathbb{R}^n \to \mathbb{R}^m$, $g^{(i)} = g_{e_i}$, then g is bounded and $|L|$-measurable, with

$$\langle L, \varphi \rangle = \sum_{i=1}^m \langle L_{e_i}, \varphi \cdot e_i \rangle = \sum_{i=1}^m \int_{\mathbb{R}^n} (\varphi \cdot e_i) g^{(i)} \,\mathrm{d}|L| = \int_{\mathbb{R}^n} (\varphi \cdot g)\,\mathrm{d}|L|\,,$$

for every $\varphi \in C_c^0(\mathbb{R}^n;\mathbb{R}^m)$. Moreover, $|g(x)| = 1$ for $|L|$-a.e. $x \in \mathbb{R}^n$. Indeed,

$$|L|(A) = \sup\Big\{ \int_{\mathbb{R}^n} (\varphi \cdot g)\,\mathrm{d}|L| : \varphi \in C_c^0(A;\mathbb{R}^m)\,, |\varphi| \le 1 \Big\} \tag{4.28}$$

for every open set $A \subset \mathbb{R}^n$. By (4.28), $|L|(A) \le \int_A |g| \mathrm{d}|L|$ for every bounded open set A. Hence, $|g| > 0$ $|L|$-a.e. on $\mathbb{R}^n$ and $1_{\{|g|>0\}}(g/|g|) \in L^1(A,|L|;\mathbb{R}^m)$. By Theorem 4.3, there exists $\{\varphi_h\}_{h\in\mathbb{N}} \subset C_c^0(A;\mathbb{R}^m)$ such that $|\varphi_h| \le 1$ and $\varphi_h \to 1_{\{|g|>0\}}(g/|g|)$ in $L^1(A,|L|;\mathbb{R}^m)$. Thus, $\varphi_h \cdot g \to |g|$ in $L^1(A,|L|)$, and

$$|L|(A) \ge \int_{\mathbb{R}^n} (\varphi_h \cdot g) \mathrm{d}|L| \to \int_A |g| \mathrm{d}|L| \ge |L|(A)\,,$$

on every open set $A \subset \mathbb{R}^n$. Hence, $|g(x)| = 1$ for $|L|$-a.e. $x \in \mathbb{R}^n$. □

Exercise 4.19 (Radon measures on open sets, II) Given an open set A in $\mathbb{R}^n$, an $\mathbb{R}^m$-valued Radon measure ν on A is a bounded linear functional on $C_c^0(A;\mathbb{R}^m)$. Extend the theory of Section 4.2 to this case.

4.3 Weak-star convergence

Let $\{\mu_h\}_{h\in\mathbb{N}}$ and μ be Radon measures on $\mathbb{R}^n$ with values in $\mathbb{R}^m$. We say that μ_h **weak-star converges** to μ, and write $\mu_h \overset{*}{\rightharpoonup} \mu$, if

$$\int_{\mathbb{R}^n} \varphi \cdot \mathrm{d}\mu = \lim_{h\to\infty} \int_{\mathbb{R}^n} \varphi \cdot \mathrm{d}\mu_h\,, \qquad \forall \varphi \in C_c^0(\mathbb{R}^n;\mathbb{R}^m)\,.$$

Weak-star convergence captures a wide variety of behaviors, as we now illustrate with some standard examples; see also [GMS98a, Chapter 1, Section 2].

Example 4.20 Given $\{x_h\}_{h\in\mathbb{N}} \subset \mathbb{R}^n$, let $\mu_h = \delta_{x_h}$. If $x_h \to x_0 \in \mathbb{R}^n$, then

$$\int_{\mathbb{R}^n} \varphi\, \mathrm{d}\mu_h = \varphi(x_h) \to \varphi(x_0) = \int_{\mathbb{R}^n} \varphi\, \mathrm{d}\mu, \qquad \forall \varphi \in C^0_c(\mathbb{R}^n),$$

i.e., $\mu_h \overset{*}{\rightharpoonup} \mu$ for $\mu = \delta_{x_0}$. If $|x_h| \to \infty$, then we clearly have

$$\int_{\mathbb{R}^n} \varphi\, \mathrm{d}\mu_h = \varphi(x_h) \to 0 = \int_{\mathbb{R}^n} \varphi\, \mathrm{d}\mu, \qquad \forall \varphi \in C^0_c(\mathbb{R}^n),$$

for $\mu = 0$, that is $\mu_h \overset{*}{\rightharpoonup} 0$. Hence, for every sequence $\{x_h\}_{h\in\mathbb{N}} \subset \mathbb{R}^n$ there exists $h(k) \to \infty$ such that either $\mu_{h(k)} \overset{*}{\rightharpoonup} \delta_{x_0}$ ($x_0 \in \mathbb{R}^n$) or $\mu_{h(k)} \overset{*}{\rightharpoonup} 0$.

Example 4.21 (Concentration of mass) The "n-dimensional" measures $\mu_h = h^n \mathcal{L}^n \llcorner (0, h^{-1})^n$ weak-star converge to the "zero-dimensional" measure $\mu = \delta_0$,

$$\int_{\mathbb{R}^n} \varphi\, \mathrm{d}\mu_h = h^n \int_{(0,1/h)^n} \varphi(x)\mathrm{d}x \to \varphi(0) = \int_{\mathbb{R}^n} \varphi\, \mathrm{d}\mu, \qquad \forall \varphi \in C^0_c(\mathbb{R}^n).$$

Example 4.22 (Spreading of mass) An increasingly diffused lower dimensional distribution of mass may weak-star converge to a "higher-dimensional" measure. If we set $\mu_h = \sum_{k=1}^h h^{-1}\, \delta_{k/h}$, then $\mu_h \overset{*}{\rightharpoonup} \mathcal{L}^1 \llcorner (0,1)$, as

$$\int_{\mathbb{R}^n} \varphi\, \mathrm{d}\mu_h = \sum_{k=1}^h \frac{\varphi(k/h)}{h} \to \int_{(0,1)} \varphi(x)\mathrm{d}x, \qquad \forall \varphi \in C^0_c(\mathbb{R}).$$

Example 4.23 (Averaging effects) Oscillations are compatible with weak-star convergence. For example, by the Riemann–Lebesgue lemma,

$$\int_{\mathbb{R}} \sin(hx)\, \varphi(x)\, \mathrm{d}x \to 0, \qquad \forall \varphi \in C^0_c(\mathbb{R}).$$

In particular, $\mu_h \overset{*}{\rightharpoonup} \mathcal{L}^1$ for $\mu_h = f_h\, \mathcal{L}^1$, $f_h(x) = 1 + \sin(hx)$, $x \in \mathbb{R}$.

Example 4.24 A fundamental idea in Geometric Measure Theory is formulating the existence of tangent spaces in terms of weak-star convergence of Radon measures. This idea, developed in Section 10.2, is sketched here with an example. Let Γ be a smooth curve in $\mathbb{R}^n$, that is $\Gamma = \gamma((a,b))$ for $\gamma\colon (a,b) \to \mathbb{R}^n$ smooth and injective. Given $t_0 \in (a,b)$, the tangent space to Γ at $x_0 = \gamma(t_0)$ is the line $\pi = \{s\gamma'(t_0) : s \in \mathbb{R}\}$. Consider now Γ as a Radon measure, looking at $\mu = \mathcal{H}^1 \llcorner \Gamma$, and define the blow-ups $\mu_{x_0,r}$ of μ at x_0, setting

$$\mu_{x_0,r} = \frac{1}{r}\, (\Phi_{x_0,r})_\# (\mathcal{H}^1 \llcorner \Gamma) = \mathcal{H}^1 \llcorner \left(\frac{\Gamma - x_0}{r}\right),$$

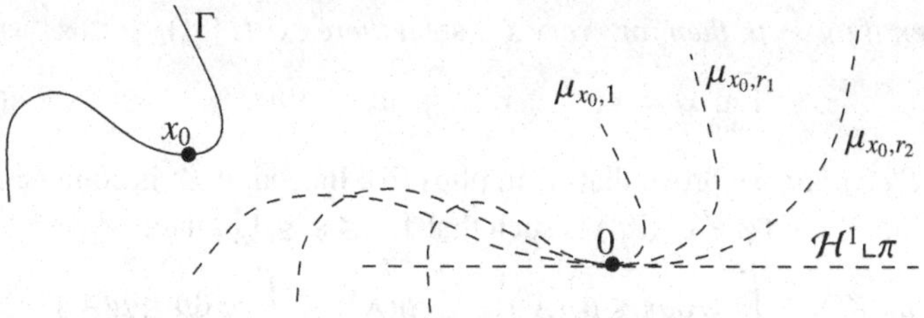

Figure 4.1 The blow-ups at x_0 of the Radon measure $\mathcal{H}^1 \llcorner \Gamma$ weak-star converge to $\mathcal{H}^1 \llcorner \pi$. In the picture, $0 < r_2 < r_1 < 1$.

where $\Phi_{x_0,r}(y) = (y - x_0)/r$, $y \in \mathbb{R}^n$; see Figure 4.1. The fact that π is the tangent space to Γ at x_0 implies that $\mu_{x_0,r} \overset{*}{\rightharpoonup} \mathcal{H}^1 \llcorner \pi$ as $r \to 0^+$. Indeed, if $\varphi \in C^0_c(\mathbb{R}^n)$, then by (2.10) we find that

$$\begin{aligned}\int_{\mathbb{R}^n} \varphi \, d\mu_{x_0,r} &= \frac{1}{r} \int_\Gamma \varphi\Big(\frac{y - x_0}{r}\Big) d\mathcal{H}^1(y) = \frac{1}{r} \int_a^b \varphi\Big(\frac{\gamma(t) - \gamma(t_0)}{r}\Big) |\gamma'(t)| \, dt \\ &= \int_{-(t_0-a)/r}^{(b-t_0)/r} \varphi\Big(\frac{\gamma(t_0 + r\,s) - \gamma(t_0)}{r}\Big) |\gamma'(t_0 + r\,s)| \, ds \\ &\to \int_{\mathbb{R}} \varphi\Big(s\gamma'(t_0)\Big) |\gamma'(t_0)| ds = \int_\pi \varphi \, d\mathcal{H}^1, \qquad \text{as } r \to 0^+.\end{aligned}$$

For a detailed justification of this argument, see the proof of Lemma 10.4.

Example 4.25 Let $\{\mu_h\}_{h\in\mathbb{N}}$ and μ be signed Radon measures with $\mu_h \overset{*}{\rightharpoonup} \mu$. It is not necessarily true that $\mu_h^+ \overset{*}{\rightharpoonup} \mu^+$ or $\mu_h^- \overset{*}{\rightharpoonup} \mu^-$, due to possible cancelations. For example, consider $\mu_h = \delta_{1/h} - \delta_{-1/h}$, or rephrase Example 4.23.

In the following proposition we characterize weak-star convergence of Radon measures in terms of evaluation on sets (note that the strict sign in (4.29) and (4.30) may occur: think of $\mu_h = \delta_{x_h}$ for $x_h \to x_0$).

Proposition 4.26 *If $\{\mu_h\}_{h\in\mathbb{N}}$ and μ are Radon measures on $\mathbb{R}^n$, then the following three statements are equivalent.*

(i) $\mu_h \overset{*}{\rightharpoonup} \mu$.
(ii) If K is compact and A is open, then

$$\mu(K) \ge \limsup_{h\to\infty} \mu_h(K), \tag{4.29}$$

$$\mu(A) \le \liminf_{h\to\infty} \mu_h(A). \tag{4.30}$$

(iii) If E is a bounded Borel set with $\mu(\partial E) = 0$, then

$$\mu(E) = \lim_{h\to\infty} \mu_h(E).$$

Moreover, if $\mu_h \overset{*}{\rightharpoonup} \mu$, *then for every* $x \in \operatorname{spt}\mu$ *there exists* $\{x_h\}_{h\in\mathbb{N}} \subset \mathbb{R}^n$ *with*

$$\lim_{h\to\infty} x_h = x\,, \qquad x_h \in \operatorname{spt}\mu_h\,, \quad \forall h \in \mathbb{N}\,. \tag{4.31}$$

Proof *Step one:* We prove that (i) implies (ii). Indeed, if K' is compact, A' is open, $K' \subset A'$, and $\varphi \in C^0_c(\mathbb{R}^n)$ is such that $1_{K'} \le \varphi \le 1_{A'}$, then

$$\mu_h(K') \le \int_{\mathbb{R}^n} \varphi\,\mathrm{d}\mu_h \le \mu_h(A')\,, \qquad \mu(K') \le \int_{\mathbb{R}^n} \varphi\,\mathrm{d}\mu \le \mu(A')\,.$$

By (i), combining these inequalities we find

$$\limsup_{h\to\infty} \mu_h(K') \le \int_{\mathbb{R}^n} \varphi\,\mathrm{d}\mu \le \mu(A')\,, \tag{4.32}$$

$$\liminf_{h\to\infty} \mu_h(A') \ge \int_{\mathbb{R}^n} \varphi\,\mathrm{d}\mu \ge \mu(K')\,. \tag{4.33}$$

Setting $K' = K$ in (4.32), and by the arbitrariness of A', we deduce (4.29). Setting $A' = A$ in (4.33), and by the arbitrariness of K', we deduce (4.30).

Step two: We prove that (ii) implies (iii). Indeed, $\mu(\mathring{E}) \le \mu(E) \le \mu(\overline{E}) = \mu(\mathring{E}) + \mu(\partial E) = \mu(\mathring{E})$ so that

$$\mu(E) = \mu(\mathring{E}) \le \liminf_{h\to\infty} \mu_h(\mathring{E}) \le \limsup_{h\to\infty} \mu_h(\overline{E}) \le \mu(\overline{E}) = \mu(E)\,.$$

Step three: We prove that (iii) implies (i). Let $\varphi \in C^0_c(\mathbb{R}^n)$, $\varphi \ge 0$. By Proposition 2.16, there exists $I \subset [0,\infty)$ such that $\mu(\{\varphi = t\}) = 0$ for every $t \in I$, where $|\mathbb{R} \setminus I| = 0$. The continuity of φ ensures that $\partial\{\varphi > t\} \subset \{\varphi = t\}$ for every $t \ge 0$. Hence, (iii) implies that

$$\mu\big(\{\varphi > t\}\big) = \lim_{h\to\infty} \mu_h\big(\{\varphi > t\}\big)\,, \qquad \forall t \in I\,.$$

The functions $f_h\colon [0,\infty) \to \mathbb{R}$, defined by $f_h(t) = \mu_h(\{\varphi > t\})$, $t \ge 0$, are decreasing (thus Borel measurable), and have $f\colon [0,\infty) \to \mathbb{R}$, $f(t) = \mu(\{\varphi > t\})$, $t \ge 0$, as their a.e. limit on $(0,\infty)$. Since $|f_h| \le \mu(\operatorname{spt}\varphi)\, 1_{[0,\sup_{\mathbb{R}^n}\varphi]}$ on $(0,\infty)$, by dominated convergence and by the layer-cake formula (1.9),

$$\int_{\mathbb{R}^n} \varphi\,\mathrm{d}\mu = \int_0^\infty \mu\big(\{\varphi > t\}\big)\,\mathrm{d}t = \lim_{h\to\infty} \int_0^\infty \mu_h\big(\{\varphi > t\}\big)\,\mathrm{d}t = \lim_{h\to\infty} \int_{\mathbb{R}^n} \varphi\,\mathrm{d}\mu_h\,.$$

To drop the assumption $\varphi \ge 0$ it suffices to recall that $\varphi = \varphi^+ - \varphi^-$.

Step four: We finally prove (4.31), by showing that for every $\varepsilon > 0$ there exists $h_0 \in \mathbb{N}$ with $\operatorname{spt}\mu_h \cap B(x,\varepsilon) \ne \emptyset$ whenever $h \ge h_0$. By contradiction, there would be $\varepsilon > 0$ and $h(k) \to \infty$ as $k \to \infty$ such that $\operatorname{spt}\mu_{h(k)} \cap B(x,\varepsilon) = \emptyset$ for every $k \in \mathbb{N}$. By (4.30), we would then have

$$\mu(B(x,\varepsilon)) \le \liminf_{k\to\infty} \mu_{h(k)}(B(x,\varepsilon)) = 0\,,$$

against the fact that $x \in \operatorname{spt}\mu$, and thus $\mu(B(x,\varepsilon)) > 0$. □

Exercise 4.27 Let μ be a Radon measure on $\mathbb{R}^n$, let $r > 0$, and define two functions $u, v\colon \mathbb{R}^n \to [0,\infty)$ as $u(x) = \mu(\overline{B}(x,r))$ and $v(x) = \mu(B(x,r))$, $x \in \mathbb{R}^n$. Then u is upper semicontinuous and v is lower semicontinuous. *Hint:* For every $x \in \mathbb{R}^n$ consider the Radon measure $\mu_x = (\tau_x)_\#\mu$ on $\mathbb{R}^n$, where $\tau_x(y) = y - x$. Show that $\mu_x \stackrel{*}{\rightharpoonup} \mu_{x_0}$ whenever $x \to x_0$ and then apply Proposition 4.26.

Remark 4.28 (Limit points of support points and uniform lower bounds) If $\mu_h \stackrel{*}{\rightharpoonup} \mu$, $x_h \in \operatorname{spt}\mu_h$ for every $h \in \mathbb{N}$, and $x_h \to x$, then it is not true, in general, that $x \in \operatorname{spt}\mu$. Consider for example the sequences

$$\mu_h = \Big(1 - \frac{1}{h}\Big)\delta_1 + \frac{1}{h}\delta_{1/h}, \qquad x_h = \frac{1}{h}.$$

This example goes to the heart of the matter, as the implication becomes true as soon as some kind of uniform lower bound on the measure assigned by the μ_h around their support points is assumed. More precisely, let $\{\mu_h\}_{h\in\mathbb{N}}$ be a sequence of Radon measures on $\mathbb{R}^n$, such that, for every $r > 0$,

$$\limsup_{h\to\infty} \inf\Big\{\mu_h(B(x,r)) : x \in \operatorname{spt}\mu_h\Big\} > 0. \tag{4.34}$$

Under this assumption, we claim that, if $\mu_h \stackrel{*}{\rightharpoonup} \mu$, $x_h \to x$, and $x_h \in \operatorname{spt}\mu_h$, then $x \in \operatorname{spt}\mu$. Indeed, let $c(r)$ denote the left-hand side of (4.34). For every $r > 0$, let $h_0 \in \mathbb{N}$ be such that $B(x_h, r) \subset B(x, 2r)$ for every $h \geq h_0$. By (4.29), and if necessary extracting a subsequence so as to exploit (4.34),

$$\mu\big(\overline{B}(x,2r)\big) \geq \limsup_{h\to\infty} \mu_h\big(\overline{B}(x,2r)\big) \geq \limsup_{h\to\infty} \mu_h\big(B(x_h,r)\big) \geq c(r) > 0.$$

By the arbitrariness of r, we find that $x \in \operatorname{spt}\mu$.

We now consider vector-valued Radon measures. Recalling that, by (4.6),

$$|\mu|(A) = \sup\Big\{\int_{\mathbb{R}^n} \varphi \cdot d\mu : \varphi \in C^\infty_c(A;\mathbb{R}^m), |\varphi| \leq 1\Big\}, \tag{4.35}$$

for every $\mathbb{R}^m$-valued Radon measure μ on $\mathbb{R}^n$ and every open set $A \subset \mathbb{R}^n$, the weak-star lower semicontinuity of the total variation is easily proved.

Proposition 4.29 *If μ_h and μ are vector-valued Radon measures with $\mu_h \stackrel{*}{\rightharpoonup} \mu$, then for every open set $A \subset \mathbb{R}^n$ we have*

$$|\mu|(A) \leq \liminf_{h\to\infty} |\mu_h|(A). \tag{4.36}$$

Proof Given $\varphi \in C^0_c(A;\mathbb{R}^m)$ with $|\varphi| \leq 1$, by $\mu_h \stackrel{*}{\rightharpoonup} \mu$ and thanks to (4.35),

$$\int_{\mathbb{R}^n} \varphi \cdot d\mu = \lim_{h\to\infty} \int_{\mathbb{R}^n} \varphi \cdot d\mu_h \leq \liminf_{h\to\infty} |\mu_h|(A).$$

By the arbitrariness of φ, using (4.35) again, we find (4.36). □

Proposition 4.30 *Let $\{\mu_h\}_{h\in\mathbb{N}}$ be $\mathbb{R}^m$-valued Radon measures on $\mathbb{R}^n$.*

(i) If $\mu_h \stackrel{*}{\rightharpoonup} \mu$ and $|\mu_h| \stackrel{*}{\rightharpoonup} \nu$, then for every Borel set $E \subset \mathbb{R}^n$,

$$|\mu|(E) \le \nu(E).$$

Furthermore, if E is a bounded Borel set with $\nu(\partial E) = 0$, then

$$\mu(E) = \lim_{h\to\infty} \mu_h(E).$$

(ii) If $\mu_h \stackrel{*}{\rightharpoonup} \mu$, $|\mu_h|(\mathbb{R}^n) \to |\mu|(\mathbb{R}^n)$, and $|\mu|(\mathbb{R}^n) < \infty$, then $|\mu_h| \stackrel{*}{\rightharpoonup} |\mu|$.

Proof *Step one:* We prove (i). Let A be a bounded open set, and let $A_t = \{x \in A : \mathrm{dist}(x, \partial A) > t\}$. If $\varphi \in C^0_c(A, [0,1])$ is such that $1_{A_t} \le \varphi$, then we have

$$|\mu|(A_t) \le \liminf_{h\to\infty} |\mu_h|(A_t) \le \liminf_{h\to\infty} \int_{\mathbb{R}^n} \varphi \mathrm{d}|\mu_h| = \int_{\mathbb{R}^n} \varphi \,\mathrm{d}\nu \le \nu(A).$$

Thus $|\mu|(A) \le \nu(A)$ for every bounded open set A, and then the assertion follows for every Borel set by approximation. Let us now prove that $\mu_h(E) \to \mu(E)$ whenever $\nu(\partial E) = 0$. Given $\varepsilon > 0$, by Proposition 2.15 we find an open set A and a compact set K such that $\overline{A} \subset E \subset \mathring{K}$ and $\nu(K \setminus A) \le \varepsilon$. Then for every $\varphi \in C^0_c(\mathring{K}; [0,1])$ with $\varphi = 1$ on $\overline{A}$ we find

$$\left|\int_{\mathbb{R}^n} \varphi \mathrm{d}\mu_h - \mu_h(E)\right| \le \int_{\mathbb{R}^n} |\varphi - 1_E| \mathrm{d}|\mu_h| \le |\mu_h|(K \setminus A),$$

$$\left|\int_{\mathbb{R}^n} \varphi \mathrm{d}\mu - \mu(E)\right| \le |\mu|(K \setminus A) \le \nu(K \setminus A),$$

$$\lim_{h\to\infty}\left|\int_{\mathbb{R}^n} \varphi \mathrm{d}\mu_h - \int_{\mathbb{R}^n} \varphi \mathrm{d}\mu\right| = 0.$$

Since $|\mu_h| \stackrel{*}{\rightharpoonup} \nu$ and $K\setminus A$ is compact we have $\limsup_{h\to\infty} |\mu_h|(K\setminus A) \le \nu(K\setminus A)$. Recalling $\nu(K \setminus A) \le \varepsilon$, we thus conclude that

$$\limsup_{h\to\infty} |\mu_h(E) - \mu(E)| \le 2\varepsilon, \qquad \forall \varepsilon > 0.$$

Step two: We prove (ii). By Proposition 4.29, $|\mu|(A) \le \liminf_{h\to\infty} |\mu_h|(A)$ if A is open. By Proposition 4.26, it thus suffices to prove that, if K is compact, then $\limsup_{h\to\infty} |\mu_h|(K) \le |\mu|(K)$. Indeed, let $A = \mathbb{R}^n \setminus K$, and take Proposition 4.29, $|\mu_h|(\mathbb{R}^n) \to |\mu|(\mathbb{R}^n)$, and $|\mu|(\mathbb{R}^n) < \infty$ into account, to find that

$$\begin{aligned}|\mu|(K) = |\mu|(\mathbb{R}^n) - |\mu|(A) &\ge |\mu|(\mathbb{R}^n) - \liminf_{h\to\infty} |\mu_h|(A)\\ &= \lim_{h\to\infty} |\mu_h|(\mathbb{R}^n) + \limsup_{h\to\infty}(-|\mu_h|(A))\\ &= \limsup_{h\to\infty} |\mu_h|(\mathbb{R}^n) - |\mu_h|(A) = \limsup_{h\to\infty} |\mu_h|(K).\end{aligned}$$

□

Exercise 4.31 If $\mu_h \stackrel{*}{\rightharpoonup} \mu$ and $r_k \to \infty$ as $k \to \infty$, then

$$\lim_{h\to\infty} |\mu_h|(B_{r_k}) = |\mu|(B_{r_k}) \qquad \forall k \in \mathbb{N} \qquad \Rightarrow \qquad |\mu_h| \stackrel{*}{\rightharpoonup} |\mu| .$$

Exercise 4.32 (Local weak-star convergence) If $A_0 \subset \mathbb{R}^n$ is open and $\{\mu_h\}_{h\in\mathbb{N}}$, μ are Radon measures on $\mathbb{R}^n$, we say that $\mu_h \stackrel{*}{\rightharpoonup} \mu$ on A_0 if

$$\int_{\mathbb{R}^n} \varphi \, d\mu = \lim_{h\to\infty} \int_{\mathbb{R}^n} \varphi \, d\mu_h , \qquad \forall \varphi \in C^0_c(A_0) .$$

Show that, in this case:

(i) if $A, K \subset A_0$ with A open and K compact, then

$$\mu(A) \le \liminf_{h\to\infty} \mu_h(A) , \qquad \mu(K) \ge \limsup_{h\to\infty} \mu_h(K) ;$$

(ii) if E is a Borel set with $E \subset\subset A_0$ and $\mu(\partial E) = 0$, then

$$\mu(E) = \lim_{h\to\infty} \mu_h(E) ;$$

(iii) for every $x \in A_0 \cap \operatorname{spt}\mu$ there exists $\{x_h\}_{h\in\mathbb{N}} \subset A_0$ such that

$$\lim_{h\to\infty} x_h = x , \qquad x_h \in A_0 \cap \operatorname{spt}\mu_h , \qquad \forall h \in \mathbb{N} . \tag{4.37}$$

4.4 Weak-star compactness criteria

Having in mind the rich variety of phenomena that are compatible with weak-star convergence, it is certainly not surprising that very flexible compactness criteria hold true. To prove these we just have to combine the classical diagonal argument with Riesz's theorem.

Theorem 4.33 (Compactness criterion for Radon measures) *If $\{\mu_h\}_{h\in\mathbb{N}}$ is a sequence of Radon measures on $\mathbb{R}^n$ such that, for every compact set K in $\mathbb{R}^n$,*

$$\sup_{h\in\mathbb{N}} \mu_h(K) < \infty ,$$

then there exist a Radon measure μ on $\mathbb{R}^n$ and a sequence $h(k) \to \infty$ as $k \to \infty$ such that $\mu_{h(k)} \stackrel{*}{\rightharpoonup} \mu$.

Proof Applying a diagonal argument to the sequences $\{\mu_h \llcorner B_k\}_{h\in\mathbb{N}}$ $(k \in \mathbb{N})$, we may directly reduce to considering the case that

$$\sup_{h\in\mathbb{N}} \mu_h(\mathbb{R}^n) < \infty . \tag{4.38}$$

If $\mathcal{F}^+$ is countable and dense (with respect to uniform convergence with equibounded supports) in $C_c^0(\mathbb{R}^n;[0,\infty))$, then $\mathcal{F} = \{u_1 - u_2 : u_1, u_2 \in \mathcal{F}^+\}$ is countable and dense in $C_c^0(\mathbb{R}^n)$. For every $\varphi \in \mathcal{F}$, $\{\int_{\mathbb{R}^n} \varphi \, d\mu_h\}_{h\in\mathbb{N}}$ is bounded, as

$$\sup_{h\in\mathbb{N}} \left| \int_{\mathbb{R}^n} \varphi \, d\mu_h \right| \le \sup_{\mathbb{R}^n} |\varphi| \sup_{h\in\mathbb{R}^n} \mu_h(\mathbb{R}^n) < \infty . \tag{4.39}$$

By a diagonal argument, we find a sequence $h(k) \to \infty$ as $k \to \infty$ such that, for every $\varphi \in \mathcal{F}$, there exists $a(\varphi) \in \mathbb{R}$ with the property that

$$a(\varphi) = \lim_{k\to\infty} \int_{\mathbb{R}^n} \varphi \, d\mu_{h(k)} .$$

We may thus define $L\colon C_c^0(\mathbb{R}^n) \to \mathbb{R}$ by setting

$$\langle L, \varphi \rangle = \lim_{h\to\infty} a(\varphi_h) , \qquad \varphi \in C_c^0(\mathbb{R}^n) ,$$

provided $\{\varphi_h\}_{h\in\mathbb{N}} \subset \mathcal{F}$ with $\varphi_h \to \varphi$ in $C_c^0(\mathbb{R}^n)$. Since, by construction,

$$|\langle L, \varphi \rangle| \le \sup_{\mathbb{R}^n} |\varphi| \sup_{h\in\mathbb{R}^n} \mu_h(\mathbb{R}^n) , \qquad \forall \varphi \in \mathcal{F} ,$$

the definition is well posed. Since $\langle L, \varphi \rangle = a(\varphi)$ for $\varphi \in \mathcal{F}$, L is linear on $\mathcal{F}$ and $\langle L, \varphi \rangle \ge 0$ for $\varphi \in \mathcal{F}^+$. Hence, by density of $\mathcal{F}$ in $C_c^0(\mathbb{R}^n)$ and of $\mathcal{F}^+$ in $C_c^0(\mathbb{R}^n;[0,\infty))$, L defines a monotone, linear functional on $C_c^0(\mathbb{R}^n)$. By Riesz's theorem, Remark 4.9, for every $\varphi \in C_c^0(\mathbb{R}^n)$ we have $\langle L, \varphi \rangle = \int_{\mathbb{R}^n} \varphi \, d\mu$, where μ is a Radon measure on $\mathbb{R}^n$. By (4.35), and thanks to (4.38), $\mu(\mathbb{R}^n) < \infty$. Finally, let us prove that $\mu_h \stackrel{*}{\rightharpoonup} \mu$. Indeed, we have

$$\int_{\mathbb{R}^n} \varphi \, d\mu = \lim_{k\to\infty} \int_{\mathbb{R}^n} \varphi \, d\mu_{h(k)} , \qquad \forall \varphi \in \mathcal{F} . \tag{4.40}$$

Thus for every $\psi \in C_c^0(\mathbb{R}^n)$ we find that

$$\begin{aligned} &\left| \int_{\mathbb{R}^n} \psi d\mu_{h(k)} - \int_{\mathbb{R}^n} \psi d\mu \right| \\ &\le \left| \int_{\mathbb{R}^n} \varphi \, d\mu_{h(k)} - \int_{\mathbb{R}^n} \varphi \, d\mu \right| + \left| \int_{\mathbb{R}^n} (\varphi - \psi) d\mu_{h(k)} \right| + \left| \int_{\mathbb{R}^n} (\psi - \varphi) d\mu \right| \\ &\le \left| \int_{\mathbb{R}^n} \varphi \, d\mu_{h(k)} - \int_{\mathbb{R}^n} \varphi \, d\mu \right| + \sup_{\mathbb{R}^n} |\psi - \varphi| \Big(\mu(\mathbb{R}^n) + \sup_{h\in\mathbb{N}} \mu_h(\mathbb{R}^n) \Big) . \end{aligned}$$

By (4.40), and by density of $\mathcal{F}$ in $C_c^0(\mathbb{R}^n)$, we see that $\mu_h \stackrel{*}{\rightharpoonup} \mu$. □

Corollary 4.34 *If $\{\mu_h\}_{h\in\mathbb{N}}$ are $\mathbb{R}^m$-valued Radon measures on $\mathbb{R}^n$, with*

$$\sup_{h\in\mathbb{N}} |\mu_h|(K) < \infty , \qquad \forall K \subset \mathbb{R}^n \text{ compact} , \tag{4.41}$$

then there exist a $\mathbb{R}^m$-valued Radon measure μ on $\mathbb{R}^n$ and $h(k) \to \infty$ as $k \to \infty$ such that $\mu_{h(k)} \stackrel{*}{\rightharpoonup} \mu$.

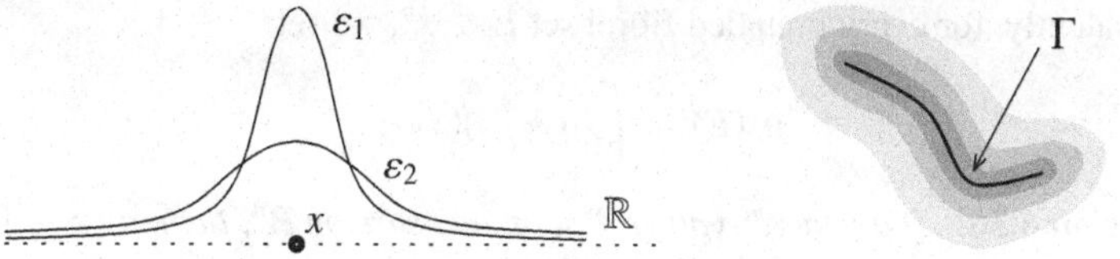

Figure 4.2 On the left, the functions $\mu \star \rho_\varepsilon$ relative to $\varepsilon_1 < \varepsilon_2$ for the measure $\mu = \delta_x$. On the right, a level set representation of $\mu \star \rho_\varepsilon$ for $\mu = \mathcal{H}^1 \llcorner \Gamma$.

Proof Apply Theorem 4.33 to $\{\mu_h^{(i),+}\}_{h\in\mathbb{N}}$ and $\{\mu_h^{(i),-}\}_{h\in\mathbb{N}}$ for $1 \le i \le m$. □

Remark 4.35 It is sometimes useful to recall that, by the Banach–Steinhaus theorem, if $\mu_h \overset{*}{\rightharpoonup} \mu$, then $\sup_{h\in\mathbb{N}} |\mu_h|(K) < \infty$ for every $K \subset \mathbb{R}^n$ compact. In particular, by Theorem 4.33, if $\mu_h \overset{*}{\rightharpoonup} \mu$, then we can always assume $|\mu_{h(k)}| \overset{*}{\rightharpoonup} \lambda$ for a suitable subsequence $\{h(k)\}_{k\in\mathbb{N}}$ and Radon measure λ.

4.5 Regularization of Radon measures

A **regularizing kernel** is a function $\rho \in C_c^\infty(B, [0,\infty))$ with $\int_B \rho = 1$ and $\rho(-x) = \rho(x)$ for every $x \in \mathbb{R}^n$. Given $\varepsilon \in (0,1)$, if we set

$$\rho_\varepsilon(x) = \frac{1}{\varepsilon^n}\rho\left(\frac{x}{\varepsilon}\right), \qquad x \in \mathbb{R}^n, \tag{4.42}$$

then $\rho_\varepsilon \in C_c^\infty(B_\varepsilon, [0,\infty))$ and $\int_{\mathbb{R}^n} \rho_\varepsilon(x)\mathrm{d}x = 1$. Given $u \in L^1_{\mathrm{loc}}(\mathbb{R}^n)$ we define the **ε-regularization** of u as the convolution between u and ρ_ε, that is

$$u_\varepsilon(x) = (u \star \rho_\varepsilon)(x) = \int_{\mathbb{R}^n} \rho_\varepsilon(x-y)u(y)\mathrm{d}y, \qquad x \in \mathbb{R}^n.$$

If $u \in C_c^0(\mathbb{R}^n)$ then, clearly, $u_\varepsilon \to u$ in $C_c^0(\mathbb{R}^n)$. If now μ is an $\mathbb{R}^m$-valued Radon measure on $\mathbb{R}^n$, then we define the functions $(\mu \star \rho_\varepsilon)\colon \mathbb{R}^n \to \mathbb{R}^m$ as

$$(\mu \star \rho_\varepsilon)(x) = \int_{\mathbb{R}^n} \rho_\varepsilon(x-y)\mathrm{d}\mu(y), \qquad x \in \mathbb{R}^n;$$

see Figure 4.2. It is easily seen that $(\mu \star \rho_\varepsilon) \in C^\infty(\mathbb{R}^n, \mathbb{R}^m)$, with

$$\nabla(\mu \star \rho_\varepsilon)(x) = \big(\mu \star (\nabla\rho_\varepsilon)\big)(x) = \int_{\mathbb{R}^n} \nabla\rho_\varepsilon(x-y)\mathrm{d}\mu(y).$$

The **ε-regularization** μ_ε of μ is the $\mathbb{R}^m$-valued Radon measure on $\mathbb{R}^n$,

$$\langle \mu_\varepsilon, \varphi\rangle = \int_{\mathbb{R}^n} \varphi(x)\cdot(\mu \star \rho_\varepsilon)(x)\mathrm{d}x, \qquad \varphi \in C_c^0(\mathbb{R}^n;\mathbb{R}^m).$$

Equivalently, for every bounded Borel set $E \subset \mathbb{R}^n$, we set

$$\mu_\varepsilon(E) = \int_E (\mu \star \rho_\varepsilon)(x)\mathrm{d}x\,.$$

Theorem 4.36 *If μ is a $\mathbb{R}^m$-valued Radon measure on $\mathbb{R}^n$, then, as $\varepsilon \to 0^+$,*

$$\mu_\varepsilon \stackrel{*}{\rightharpoonup} \mu\,, \qquad |\mu_\varepsilon| \stackrel{*}{\rightharpoonup} |\mu|\,.$$

Moreover, if $I_\varepsilon(E) = \{x \in \mathbb{R}^n : \operatorname{dist}(x, E) < \varepsilon\}$, then for every Borel set $E \subset \mathbb{R}^n$

$$|\mu_\varepsilon|(E) \le |\mu|(I_\varepsilon(E))\,. \tag{4.43}$$

Proof Since $\varphi_\varepsilon \to \varphi$ in $C^0_\mathrm{c}(\mathbb{R}^n)$ and, by Fubini's theorem,

$$\int_{\mathbb{R}^n} \varphi \cdot \mathrm{d}\mu_\varepsilon = \int_{\mathbb{R}^n} (\varphi_\varepsilon) \cdot \mathrm{d}\mu\,, \qquad \forall \varphi \in C^0_\mathrm{c}(\mathbb{R}^n; \mathbb{R}^m)\,,$$

we find $\mu_\varepsilon \stackrel{*}{\rightharpoonup} \mu$. Again by Fubini's theorem, if E is a Borel set in $\mathbb{R}^n$, then

$$\begin{aligned}|\mu_\varepsilon|(E) &= \int_E |\mu \star \rho_\varepsilon(x)|\mathrm{d}x \le \int_E \mathrm{d}x \int_{B(x,\varepsilon)} \rho_\varepsilon(x-y)\mathrm{d}|\mu|(y)\\ &\le \int_{I_\varepsilon(E)} \mathrm{d}|\mu|(y) \int_{B(y,\varepsilon)\cap E} \rho_\varepsilon(x-y)\mathrm{d}x \le |\mu|(I_\varepsilon(E))\,,\end{aligned}$$

that is (4.43). Let us now consider a sequence $\{r_k\}_{k\in\mathbb{N}}$ such that $r_k \to +\infty$ and $|\mu|(\partial B_{r_k}) = 0$ for every $k \in \mathbb{N}$ (this choice is possible thanks to Proposition 2.16). We apply (4.43) to $E = B_{r_k}$, and then let $\varepsilon \to 0$, to find that

$$|\mu|(B_{r_k}) \le \liminf_{\varepsilon\to 0} |\mu_\varepsilon|(B_{r_k}) \le \limsup_{\varepsilon\to 0} |\mu|(B_{r_k+\varepsilon}) = |\mu|(\overline{B}_{r_k}) = |\mu|(B_{r_k})\,.$$

Hence, by Exercise 4.31, $|\mu_\varepsilon| \stackrel{*}{\rightharpoonup} |\mu|$ as $\varepsilon \to 0^+$. □

Exercise 4.37 (Radon measures on open sets, III) Extend the theory of this chapter to vector-valued Radon measures on an open set A.

5

Differentiation of Radon measures

The results from Chapters 3 and 8 show that many different objects can be represented as Radon measures. For example, a non-negative function $u \in L^1_{\text{loc}}(\mathbb{R}^n)$ and a k-dimensional C^1-surface M in $\mathbb{R}^n$, $1 \le k \le n-1$, are naturally associated with the Radon measures $\mu = u\,\mathcal{L}^n$ and $\nu = \mathcal{H}^k \llcorner M$. The behavior with respect to $\mathcal{L}^n$ of these two measures is opposite. Indeed, given a Borel set $E \subset \mathbb{R}^n$, $|E| = 0$ implies $\mu(E) = 0$, while, at the same time, $|M| = 0$ and $\nu(M) > 0$. In the first case, μ does not charge null sets for $\mathcal{L}^n$; in the second case, instead, ν and $\mathcal{L}^n$ are somehow orthogonal, as ν is concentrated on a null set for $\mathcal{L}^n$. These considerations motivate the following definitions (which are extended to vector-valued Radon measures in Remark 5.10). We say that ν is **absolutely continuous with respect to** μ, and write $\nu \ll \mu$, if

$$E \in \mathcal{B}(\mathbb{R}^n), \quad \mu(E) = 0 \qquad \Rightarrow \qquad \nu(E) = 0,$$

that is, if ν vanishes on any Borel sets on which μ vanishes; we say that μ and ν are **mutually singular**, and write $\mu \perp \nu$, if, for a Borel set $E \subset \mathbb{R}^n$,

$$\mu(\mathbb{R}^n \setminus E) = \nu(E) = 0,$$

that is, μ is concentrated on E, while ν is concentrated on $\mathbb{R}^n \setminus E$. The two concepts are complementary, since $\mu \perp \nu$ and $\nu \ll \mu$ clearly imply $\nu = 0$.

The Lebesgue–Besicovitch differentiation theorem (Section 5.2) asserts that, given two Radon measures μ and ν on $\mathbb{R}^n$, we can always decompose ν as a sum $\nu_1 + \nu_2$, where $\nu_1 \ll \mu$, and $\nu_2 \perp \mu$; moreover, $\nu_1 = u\,\mathrm{d}\mu$, where $u\colon \mathbb{R}^n \to [0,\infty]$ is a Borel measurable function, characterized at μ-a.e. $x \in \operatorname{spt}\mu$ as the limit

$$u(x) = \lim_{r \to 0^+} \frac{\nu(B(x,r))}{\mu(B(x,r))};$$

at the same time, ν_2 is concentrated on the Borel set

$$Y = \left(\mathbb{R}^n \setminus \operatorname{spt}\mu\right) \cup \left\{x \in \operatorname{spt}\mu\colon \limsup_{r\to 0^+} \frac{\nu(B(x,r))}{\mu(B(x,r))} = \infty\right\}.$$

The proof of this fundamental result is in turn based on Besicovitch's covering theorem, which is discussed in Section 5.1. Finally, in Section 5.3, we apply the Lebesgue–Besicovitch differentiation theorem to prove another important result, the Lebesgue points theorem.

5.1 Besicovitch's covering theorem

We discuss here Besicovitch's covering theorem, one of the most frequently used technical tools in Geometric Measure Theory. To simplify the notation, in this chapter we denote by $\overline{B}$ a generic closed ball of $\mathbb{R}^n$ (and not, as in the rest of the book, the closure of the Euclidean unit ball of $\mathbb{R}^n$), while $\overline{B}(x,r)$ denotes as usual the closed ball of center x and radius r in $\mathbb{R}^n$.

Theorem 5.1 (Besicovitch's covering theorem) *If $n \ge 1$, then there exists a positive constant $\xi(n)$ with the following property. If $\mathcal{F}$ is a family of closed non-degenerate balls of $\mathbb{R}^n$, and either the set C of the centers of the balls in $\mathcal{F}$ is bounded or*

$$\sup\left\{\operatorname{diam}(\overline{B}) : \overline{B} \in \mathcal{F}\right\} < \infty\,, \tag{5.1}$$

then there exist $\mathcal{F}_1, ..., \mathcal{F}_{\xi(n)}$ (possibly empty) subfamilies of $\mathcal{F}$ such that

(i) Each family $\mathcal{F}_i$ is disjoint and at most countable;
(ii) $C \subset \bigcup_{i=1}^{\xi(n)} \bigcup_{\overline{B}\in\mathcal{F}_i} \overline{B}$.

Thus, a ball from $\mathcal{F}_i$ does not intersect other balls from $\mathcal{F}_i$, but could intersect countably many balls from the families $\mathcal{F}_j$ if $j \ne i$. At the same time, every point $x \in C$ intersects *at most* $\xi(n)$ balls in $\mathcal{G} = \bigcup_{i=1}^{\xi(n)} \mathcal{F}_i$. It is this last property which makes Theorem 5.1 so useful in the study of Radon measures, as exemplified by the following statement. (Another important corollary is the validity of Vitali's property for Radon measures; see Corollary 5.5.)

Corollary 5.2 *If μ is an outer measure on $\mathbb{R}^n$, and $\mathcal{F}$ and C are as in Theorem 5.1, then there exists a countable disjoint subfamily $\mathcal{F}'$ of $\mathcal{F}$ with*

$$\mu(C) \le \xi(n) \sum_{\overline{B}\in\mathcal{F}'} \mu(C \cap \overline{B})\,.$$

If, moreover, μ is a Borel measure and C is μ-measurable, then

$$\mu(C) \le \xi(n)\,\mu\Big(C \cap \bigcup\big\{\overline{B} : \overline{B} \in \mathcal{F}'\big\}\Big).$$

Proof of Corollary 5.2 If $\mathcal{G} = \bigcup_{i=1}^{\xi(n)} \mathcal{F}_i$, then, by Theorem 5.1 (ii),

$$\mu(C) \le \sum_{i=1}^{\xi(n)} \sum_{\overline{B} \in \mathcal{F}_i} \mu(C \cap \overline{B}).$$

Set $\mathcal{F}' = \mathcal{F}_i$, for i maximizing $\sum_{\overline{B} \in \mathcal{F}_h} \mu(C \cap \overline{B})$ over $h \in \{1, ..., \xi(n)\}$. □

When C is not bounded and (5.1) is not true, a counterexample to Theorem 5.1 is given by the family of closed non-degenerate balls in $\mathbb{R}^n$,

$$\mathcal{F} = \big\{\overline{B}(k\,e, k + \varepsilon) : k \in \mathbb{N}\big\},$$

where $\varepsilon > 0$ and $e \in S^{n-1}$ are fixed. The set of centers $C = \{k\,e\}_{k \in \mathbb{N}}$ is unbounded and $\sup\{\mathrm{diam}(\overline{B}) : \overline{B} \in \mathcal{F}\} = \infty$. If a subfamily $\mathcal{G}$ of $\mathcal{F}$ covers C, then $\mathcal{G}$ is necessarily infinite, and thus every ball in $\mathcal{G}$ contains the origin. Hence, it is impossible to extract a Besicovitch-type covering from $\mathcal{F}$. When $n \ge 2$ we may try to modify this example to make C bounded, by placing the centers of the balls on different directions. This attempt meets the following obstruction that, in turn, plays a role in the proof of Theorem 5.1.

Lemma 5.3 *If $\delta \in (0, 1)$ and $n \ge 1$, then there exists a constant $C(n, \delta)$ such that $\#(I) \le C(n, \delta)$ whenever $\{x_\alpha\}_{\alpha \in I} \subset S^{n-1}$, with $|x_{\alpha_1} - x_{\alpha_2}| \ge \delta$ for $\alpha_1, \alpha_2 \in I$.*

Proof By contradiction, and by compactness of S^{n-1}. □

The following lemma is motivated by the construction of the family $\mathcal{G}$ in step one of the proof of Theorem 5.1; see also Figure 5.1.

Lemma 5.4 *For every $n \ge 1$ there exists a positive constant $\eta(n)$ with the following property. If $\{\overline{B}_k = \overline{B}(x_k, r_k)\}_{k=1}^{N+1}$ is a finite family of closed non-degenerate balls with*

$$|x_k - x_h| > r_h\,, \qquad r_k \le \frac{3}{2}\, r_h\,,$$

whenever $1 \le h < k \le N + 1$, then

$$\#\Big\{k : \overline{B}(x_k, r_k) \cap \overline{B}(x_{N+1}, r_{N+1}) \neq \emptyset\Big\} \le \eta(n).$$

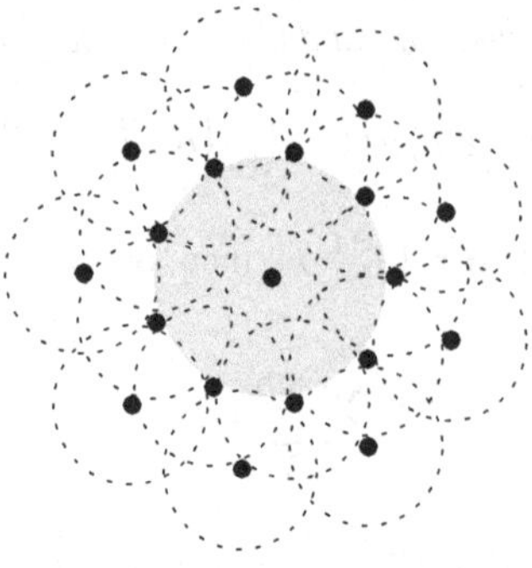

Figure 5.1 In Lemma 5.4, the radius of each $\overline{B}_h$ is, at least, two-thirds the radius of any subsequent ball from the family, that is, $r_h > (2/3)r_k$ whenever $k > h$: in particular, there is a uniform lower bound on the various radii in terms of the radius of the last ball of the family, $\overline{B}_{N+1}$, namely $r_h \ge (2/3)r_{N+1}$. Moreover, the center of each $\overline{B}_k$ does not belong to any previous ball from the family, that is, $|x_k - x_h| > r_h$ whenever $k > h$. Under these conditions, the maximum number of balls from the family that can intersect the last ball $\overline{B}_{N+1}$ depends on the dimension of the ambient space only. In the picture we have tried to maximize the number of balls intersecting a given $\overline{B}_{N+1}$ (depicted in gray), having first minimized the radius of each $\overline{B}_k$, that is, having taken $r_k = (2/3)r_{N+1}$. Note that we face the obstruction of Lemma 5.3.

Proof Up to a translation, we may set $x_{N+1} = 0$ and $r_{N+1} = r > 0$, and reduce to considering a family of closed non-degenerate balls $\{\overline{B}(x_k, r_k)\}_{k=1}^N$, with

$$|x_k - x_h| > r_h\,, \qquad r_k \le \frac{3}{2} r_h\,, \qquad 1 \le h < k \le N\,, \tag{5.2}$$

$$|x_k| > r_k\,, \qquad r \le \frac{3}{2} r_k\,, \qquad 1 \le k \le N\,. \tag{5.3}$$

We want to estimate the cardinality of the set

$$I = \left\{k : \overline{B}(x_k, r_k) \cap \overline{B}(0, r) \ne \emptyset\right\}.$$

To this end, we note that

$$|x_k| \le r_k + r\,, \qquad \forall k \in I\,, \tag{5.4}$$

we introduce a parameter $t > 1$ to be fixed in a moment, and we consider the partition $\{I_1, I_2\}$ of I defined by

$$I_1 = \left\{k \in I\colon r_k \ge t\,r\right\}, \qquad I_2 = \left\{k \in I\colon r_k < t\,r\right\}.$$

Estimate for I_1: If t is large enough, then there exists $\theta \in (0,1)$ such that

$$\frac{x_k}{|x_k|} \cdot \frac{x_h}{|x_h|} \le \theta\,, \qquad \forall h, k \in I_1\,, h \ne k\,. \tag{5.5}$$

From (5.5) we shall deduce that

$$\left|\frac{x_k}{|x_k|}-\frac{x_h}{|x_h|}\right|\ge\sqrt{2(1-\theta)}>0\,,\qquad\forall h,k\in I_1\,,h\ne k\,,$$

and thus conclude by Lemma 5.3 that $\#(I_1)\le C(n,\delta)$, where $\delta=\sqrt{2(1-\theta)}$. To prove (5.5), we notice that, by (5.2), (5.3), and (5.4), and for $k>h$,

$$\begin{aligned}\frac{x_k}{|x_k|}\cdot\frac{x_h}{|x_h|}&=\frac{|x_k|^2+|x_h|^2-|x_k-x_h|^2}{2|x_k||x_h|}\\&\le\frac{|x_k|^2+(r+r_h)^2-r_h^2}{2|x_k|r_h}=\frac{|x_k|}{2r_h}+\frac{r^2}{2|x_k|r_h}+\frac{r}{|x_k|}\,.\end{aligned}$$

Again by (5.2), (5.3), and (5.4), if $h,k\in I_1$, $k>h$, then we find

$$\begin{aligned}\frac{|x_k|}{2r_h}&\le\frac{r+r_k}{2r_h}\le\frac{1}{2t}+\frac{3}{4}\,,\\\frac{r^2}{2|x_k|r_h}&\le\frac{r^2}{2r_kr_h}\le\frac{1}{2t^2}\,,\\\frac{r}{|x_k|}&\le\frac{r}{r_k}\le\frac{1}{t}\,.\end{aligned}$$

Hence, (5.5) follows by choosing t large enough (not depending on n).

Estimate for I_2: First, we remark that $\{\overline{B}(x_k,r_k/3)\}_{k=1}^{N+1}$ is disjoint. Indeed, if $1\le h<k\le N+1$ and $x\in\overline{B}(x_k,r_k/3)\cap\overline{B}(x_h,r_h/3)$, then by (5.2)

$$r_h<|x_k-x_h|\le|x_k-x|+|x_h-x|\le\frac{r_k}{3}+\frac{r_h}{3}\le\frac{1}{3}\left(\frac{3}{2}r_h\right)+\frac{r_h}{3}=\frac{5}{6}r_h\,,$$

which is a contradiction. Next, we remark that, for every $k\in I_2$,

$$\overline{B}\left(x_k,\frac{r}{9}\right)\subset\overline{B}\left(x_k,\frac{r_k}{3}\right)\subset\overline{B}\big(0,\gamma(t)r\big)\,,\tag{5.6}$$

where $\gamma(t)=1+(4/3)t$. The first inclusion in (5.6) is trivial from (5.3). The second inclusion in (5.6) follows because, if $x\in\overline{B}(x_k,r_k/3)$, then by the triangle inequality, by $r_k<t\,r$, and by (5.4), we get

$$|x|\le\frac{r_k}{3}+|x_k|\le\frac{t}{3}r+r+r_k\le\left(\frac{t}{3}+1+t\right)r=\gamma(t)\,r\,.$$

The family $\{\overline{B}(x_k,r/9):k\in I_2\}$ is disjoint by the first inclusion in (5.6). Hence, by the second inclusion in (5.6), we find $\#(I_2)\le 9^n\gamma(t)^n$, as

$$\#(I_2)\,\omega_n\left(\frac{r}{9}\right)^n=\left|\bigcup_{k\in I_2}B\left(x_k,\frac{r}{9}\right)\right|\le\left|\bigcup_{k\in I_2}\overline{B}\left(x_k,\frac{r_k}{3}\right)\right|\le\left|\overline{B}\big(0,\gamma(t)r\big)\right|=\omega_n\gamma(t)^nr^n\,.$$

□

Proof of Theorem 5.1 *Step one:* Let C be bounded. In this case, we may freely assume that (5.1) holds too. We define a countable subfamily $\mathcal{G} = \{\overline{B}_k\}_{k=1}^M$ ($M \in \mathbb{N} \cup \{\infty\}$) of $\mathcal{F}$ as follows. By (5.1), there exists $\overline{B}_1 \in \mathcal{F}$ with

$$\operatorname{diam}(\overline{B}_1) \geq \frac{2}{3} \sup\left\{\operatorname{diam}(\overline{B}) : \overline{B} \in \mathcal{F}\right\}.$$

We inductively define $\overline{B}_k$, $k \geq 2$, to be any ball from $\mathcal{F}$ whose center does not lie in $\bigcup_{h=1}^{k-1} \overline{B}_h$, and such that

$$\operatorname{diam}(\overline{B}_k) \geq \frac{2}{3} \sup\left\{\operatorname{diam}(\overline{B}) : \overline{B} \in \mathcal{F}, \text{ the center of } \overline{B} \text{ is not in } \textstyle\bigcup_{h=1}^{k-1} \overline{B}_h\right\}.$$

If this procedure stops after k steps, then we set $M = k$; otherwise, we set $M = \infty$. By construction, we have that

$$|x_k - x_h| > r_h, \qquad r_k \leq \frac{3}{2} r_h, \tag{5.7}$$

whenever $1 \leq h < k < M$. Hence, by Lemma 5.4, we find that

$$\#\left\{k : 1 \leq k < N, \overline{B}_k \cap \overline{B}_N \neq \emptyset\right\} \leq \eta(n), \tag{5.8}$$

whenever $2 \leq N < M$. We are now left to show that: (a) C is covered by $\mathcal{G}$; (b) $\mathcal{G}$ can be further divided into $\eta(n) + 1$ subfamilies $\mathcal{F}_i$, where each $\mathcal{F}_i$ is disjoint.

If $M < \infty$, then (a) is trivial. If $M = \infty$, then, arguing by contradiction, there exists $x \in C \setminus \bigcup_{k=1}^{\infty} \overline{B}_k$. As $x \in C$, $\overline{B}(x, r) \in \mathcal{F}$ for some $r > 0$. By construction, $r_k \geq (2/3)r$ for every k. Thus $\{r_k\}_{k=1}^{\infty}$ is bounded from below, and, by the first inequality in (5.7), the bounded sequence $\{x_k\}_{k=1}^{\infty} \subset C$ does not admit any converging subsequence. Thus (a) is proved.

We prove (b) by constructing subfamilies $\{\mathcal{F}_i\}_{i=1}^{\eta(n)+1}$ of $\mathcal{G}$ (with $\mathcal{G} = \bigcup_{i=1}^{\eta(n)+1} \mathcal{F}_i$ and each $\mathcal{F}_i$ is disjoint) via an inductive procedure which stops in (at most) $\eta(n) + 1$ steps due to (5.8). We put the ball $\overline{B}_k \in \mathcal{G}$ in the family $\mathcal{F}_i$ with the *lowest* index i such that $\overline{B}_k$ is disjoint from every ball we have already put in $\mathcal{F}_i$ (in the previous steps of the induction). By construction, each $\mathcal{F}_i$ is disjoint, and the union of the $\mathcal{F}_i$ is $\mathcal{G}$. If the family $\mathcal{F}_{\eta(n)+2}$ were created, there would be $\overline{B}_N \in \mathcal{G}$, $N \geq \eta(n) + 2$, such that $\overline{B}_N$ intersects at least one ball from each family $\mathcal{F}_i$, for $1 \leq i \leq \eta(n) + 1$. But then, contradicting (5.8),

$$\#\left\{k : 1 \leq k < N, \overline{B}_k \cap \overline{B}_N \neq \emptyset\right\} \geq \eta(n) + 1.$$

Step two: Letting $R = \sup\{\operatorname{diam}(\overline{B}) : \overline{B} \in \mathcal{F}\}$, we now prove the theorem under the assumption (5.1), that is, $R < \infty$. For every $k \geq 1$, let

$$C_k = \left\{x \in C : 3R(k-1) \leq |x| < 3Rk\right\},$$
$$\mathcal{G}_k = \left\{\overline{B} \in \mathcal{F} : \text{the center of } \overline{B} \text{ belongs to } C_k\right\}.$$

By step one, for every $k \in \mathbb{N}$, there exist $\{\mathcal{G}_{k,i}\}_{i=1}^{\eta(n)+1}$ subfamilies of $\mathcal{G}_k$, with $\mathcal{G}_{k,i}$ countable and disjoint, and with $C_k \subset \bigcup_{i=1}^{\eta(n)+1} \mathcal{G}_{k,i}$. We rearrange these families into $\xi(n) = 2\,\eta(n) + 2$ subfamilies $\mathcal{F}_i$ of $\mathcal{F}$ by setting

$$\mathcal{F}_i = \bigcup_{k \text{ is odd}} \mathcal{G}_{k,i}\,, \qquad \mathcal{F}_{i+\eta(n)+1} = \bigcup_{k \text{ is even}} \mathcal{G}_{k,i}\,,$$

whenever $1 \le i \le \eta(n) + 1$. Clearly, each $\mathcal{F}_i$ is countable. Since $\mathcal{G}_k \cap \mathcal{G}_h = \emptyset$ if $|k-h| \ge 2$, we find that, by construction, each $\mathcal{F}_i$ is disjoint. Finally,

$$C = \bigcup_{k\in\mathbb{N}} C_k \subset \bigcup_{k\in\mathbb{N}} \bigcup_{i=1}^{\eta(n)+1} \bigcup_{\overline{B}\in\mathcal{G}_{k,i}} \overline{B} = \bigcup_{i=1}^{\xi(n)} \bigcup_{\overline{B}\in\mathcal{F}_i} \overline{B}\,. \qquad \square$$

Corollary 5.5 (Vitali's property) *If μ is a Radon measure on $\mathbb{R}^n$, $\mathcal{F}$ is a family of closed non-degenerate balls whose set of centers C is bounded and μ-measurable, and, for every $x \in C$,*

$$\inf\left\{\operatorname{diam}(\overline{B}) : \overline{B} \in \mathcal{F},\ \overline{B} \text{ has center in } x\right\} = 0, \tag{5.9}$$

then there exists a countable disjoint subfamily $\mathcal{G}$ of $\mathcal{F}$ such that

$$\mu\Big(C \setminus \bigcup \{\overline{B} : \overline{B} \in \mathcal{G}\}\Big) = 0\,.$$

Proof By Corollary 5.2, there exists $\mathcal{F}_i \subset \mathcal{F}$, countable and disjoint, with

$$\frac{\mu(C)}{\xi(n)} \le \mu\Big(C \cap \bigcup \{\overline{B} : \overline{B} \in \mathcal{F}_i\}\Big).$$

As $\mu(C) < \infty$ we can select a *finite* subfamily $\mathcal{F}_i' \subset \mathcal{F}_i$ with the property that

$$\frac{\mu(C)}{2\xi(n)} \le \mu\Big(C \cap \bigcup \{\overline{B} : \overline{B} \in \mathcal{F}_i'\}\Big).$$

As a consequence, if we set $\theta = 1 - (2\xi(n))^{-1}$, then

$$\mu\Big(C \setminus \bigcup \{\overline{B} : \overline{B} \in \mathcal{F}_i'\}\Big) \le \theta\,\mu(C)\,.$$

If now $\mathcal{F}^*$ is the family of balls from $\mathcal{F}$ disjoint from every ball in $\mathcal{F}_i'$, and

$$C^* = C \setminus \bigcup \{\overline{B} : \overline{B} \in \mathcal{F}_i'\},$$

then, by (5.9), and since $\mathcal{F}_i'$ is a finite family of closed balls, we see that C^* is the set of the centers of the balls in $\mathcal{F}^*$, and that $\mathcal{F}^*$ and C^* satisfy the same assumptions relating $\mathcal{F}$ and C. By iterating this argument, we find $N_h \to \infty$, $C_h \subset C$, and a countable family of disjoint closed balls $\{\overline{B}_j\}_{j=1}^{\infty} \subset \mathcal{F}$ with

$$C_h = C \setminus \bigcup_{j=1}^{N_h} \overline{B}_j\,, \qquad \mu(C_h) \le \theta^h \mu(C)\,.$$

Since $\mu(C) < \infty$, by monotone convergence we find that $\mu(C \setminus \bigcup_{j=1}^{\infty} \overline{B}_j) = 0$. $\square$

Remark 5.6 Besicovitch's covering theorem and Corollary 5.2 remain valid if we replace closed balls by open balls. The use of closed balls becomes necessary in the proof of Vitali's property; see [AFP00, Example 2.20].

5.2 Lebesgue–Besicovitch differentiation theorem

Let μ and ν be Radon measures on $\mathbb{R}^n$. The **upper μ-density** and the **lower μ-density** of ν are the functions $D_\mu^+\nu\colon \operatorname{spt}\mu \to [0,\infty]$ and $D_\mu^-\nu\colon \operatorname{spt}\mu \to [0,\infty]$, defined as, respectively,

$$D_\mu^+\nu(x) = \limsup_{r\to 0^+} \frac{\nu(\overline{B}(x,r))}{\mu(\overline{B}(x,r))}, \quad D_\mu^-\nu(x) = \liminf_{r\to 0^+} \frac{\nu(\overline{B}(x,r))}{\mu(\overline{B}(x,r))}, \qquad x \in \operatorname{spt}\mu\,.$$

If the two limits exist and are finite, then we denote by $D_\mu\nu(x)$ their common value, and call it the **μ-density** of ν at x. We have thus defined a function

$$D_\mu\nu\colon \left\{x \in \operatorname{spt}\mu : D_\mu^+\nu(x) = D_\mu^-\nu(x)\right\} \to [0,\infty]\,.$$

Remark 5.7 By Exercise 4.27, and since $\operatorname{spt}\mu$ is a closed set, $D_\mu^+\nu$ and $D_\mu^-\nu$ are Borel functions, which, by Remark 4.2, we may consider as defined on the whole of $\mathbb{R}^n$. With the same *caveat*, $D_\mu\nu$ is a Borel function on $\mathbb{R}^n$. By Proposition 2.16, for every $x \in \mathbb{R}^n$ there exist at most countably many values of $r > 0$ such that either $\mu(\partial B(x,r)) > 0$ or $\nu(\partial B(x,r)) > 0$. As a consequence, if $D_\mu\nu$ is defined at x, then it satisfies

$$D_\mu\nu(x) = \lim_{r\to 0^+} \frac{\nu(B(x,r))}{\mu(B(x,r))}\,. \tag{5.10}$$

In other words, in evaluating $D_\mu\nu$, we may indifferently use open or closed balls. The use of closed balls in the definition of $D_\mu^+\nu$ and $D_\mu^-\nu$ is instead necessary in order to apply Vitali's property in the proof of the following theorem.

Theorem 5.8 (Lebesgue–Besicovitch differentiation theorem) *If μ and ν are Radon measures on $\mathbb{R}^n$, then $D_\mu\nu$ is defined μ-a.e. on $\mathbb{R}^n$, $D_\mu\nu \in L^1_{\mathrm{loc}}(\mathbb{R}^n,\mu)$, and, in fact, $D_\mu\nu$ is Borel measurable on $\mathbb{R}^n$. Furthermore,*

$$\nu = (D_\mu\nu)\,\mu + \nu_\mu^s \qquad \text{on } \mathcal{M}(\mu)\,, \tag{5.11}$$

where the Radon measure ν_μ^s is concentrated on the Borel set

$$\begin{aligned} Y &= \mathbb{R}^n \setminus \left\{x \in \operatorname{spt}\mu : D_\mu^+\nu(x) < \infty\right\} \\ &= \left(\mathbb{R}^n \setminus \operatorname{spt}\mu\right) \cup \left\{x \in \operatorname{spt}\mu : D_\mu^+\nu(x) = \infty\right\}. \end{aligned} \tag{5.12}$$

In particular, $\nu_\mu^s \perp \mu$.

Remark 5.9 The Radon measure ν_μ^s is called the **singular part of** ν **with respect to** μ. If we set $\nu_\mu^a = (D_\mu\nu)\mathrm{d}\mu$, then $\nu_\mu^a \ll \mu$. Thus ν_μ^a is called the **absolutely continuous part of** ν **with respect to** μ. This kind of additive decomposition is unique on $\mathcal{M}(\mu)$.

Proof of Theorem 5.8 We set $\{D_\mu^+\nu \geq t\} = \{x \in \operatorname{spt}\mu : D_\mu^+\nu(x) \geq t\}$, define similarly $\{D_\mu^-\nu \leq t\}$ etc., and divide the argument into four steps.

Step one: We show that we can reduce to proving (5.11) on the family of bounded Borel sets $\mathcal{B}_b(\mathbb{R}^n)$. First, by Example 1.10, $(D_\mu\nu)\,\mu$ is a measure on $\mathcal{M}(\mu)$, so that, by intersecting with balls with increasingly larger radii, we see that it suffices to prove (5.11) on bounded μ-measurable sets. Second, if $E \in \mathcal{M}(\mu)$ is bounded, then by the Borel regularity of μ, there exists $F \in \mathcal{B}_b(\mathbb{R}^n)$ with $E \subset F$ and $\mu(F) = \mu(E)$; moreover, by Proposition 2.13, $\nu - \nu_\mu^s$ is a Radon measure on $\mathbb{R}^n$, and, again by Borel regularity, there exists a bounded Borel set G with $E \subset G$ and $(\nu - \nu_\mu^s)(E) = (\nu - \nu_\mu^s)(G)$; combining these facts with the validity of (5.11) on F and G, we thus conclude that

$$\begin{aligned}(\nu - \nu_\mu^s)(E) = (\nu - \nu_\mu^s)(G) &= \int_G D_\mu\nu\,\mathrm{d}\mu \geq \int_E D_\mu\nu\,\mathrm{d}\mu = \int_F D_\mu\nu\,\mathrm{d}\mu \\ &= (\nu - \nu_\mu^s)(F) \geq (\nu - \nu_\mu^s)(E)\,.\end{aligned}$$

Step two: We prove that, if $t \in (0,\infty)$ and E is a bounded Borel set in $\mathbb{R}^n$, then

$$E \subset \left\{D_\mu^-\nu \leq t\right\} \Rightarrow \nu(E) \leq t\,\mu(E)\,, \tag{5.13}$$
$$E \subset \left\{D_\mu^+\nu \geq t\right\} \Rightarrow \nu(E) \geq t\,\mu(E)\,.$$

It is sufficient to prove (5.13). Let us fix $\varepsilon > 0$ and let A be an open bounded set such that $E \subset A$ and $\mu(A) \leq \varepsilon + \mu(E)$. As $E \subset \{D_\mu^-\nu \leq t\}$, the family of balls

$$\mathcal{F} = \left\{\overline{B}(x,r) : x \in E,\ \overline{B}(x,r) \subset A\,,\ \nu\left(\overline{B}(x,r)\right) \leq (t+\varepsilon)\mu\left(\overline{B}(x,r)\right)\right\}$$

satisfies the assumptions of Corollary 5.5. Hence, there exists a countable disjoint subfamily $\{\overline{B}(x_h,r_h)\}_{h\in\mathbb{N}} \subset \mathcal{F}$ such that $\nu(E \setminus \bigcup_{h\in\mathbb{N}} \overline{B}(x_h,r_h)) = 0$, and

$$\nu(E) = \sum_{h\in\mathbb{N}} \nu\left(\overline{B}(x_h,r_h)\right) \leq (t{+}\varepsilon)\sum_{h\in\mathbb{N}} \mu\left(\overline{B}(x_h,r_h)\right) \leq (t{+}\varepsilon)\,\mu(A) \leq (t{+}\varepsilon)\left(\mu(E){+}\varepsilon\right).$$

Step three: We prove that $D_\mu\nu(x)$ exists and it is finite for μ-a.e. $x \in \mathbb{R}^n$. It is enough to prove that the two sets

$$Z = \left\{D_\mu^+\nu = \infty\right\}, \quad Z_{q,p} = \left\{D_\mu^-\nu < q < p < D_\mu^+\nu\right\}, \quad p,q \in \mathbb{Q}\,,$$

have μ-measure zero. Indeed $Z \subset \{D_\mu^+\nu \geq t\}$ for every $t > 0$, and thus

$$\mu(Z \cap B_R) \leq \frac{\nu(Z \cap B_R)}{t} \leq \frac{\nu(B_R)}{t}\,.$$

Since $\nu(B_R)$ is finite, by letting $t \to \infty$, and then $R \to \infty$, we find that $\mu(Z) = 0$. Concerning $Z_{p,q}$ we notice that, again by step one, for every $R > 0$,

$$\nu\big(Z_{p,q} \cap B_R\big) \le q\,\mu\big(Z_{p,q} \cap B_R\big) \le \frac{q}{p}\,\nu\big(Z_{p,q} \cap B_R\big).$$

Since $(q/p) < 1$, we have $\mu(Z_{p,q} \cap B_R) = 0$, and thus $\mu(Z_{p,q}) = 0$.

Step four: Let us set $\nu = \nu_1 + \nu_2$, where

$$\nu_1 = \nu\llcorner(\mathbb{R}^n \setminus Y), \qquad \nu_2 = \nu\llcorner Y, \qquad Y = \big(\mathbb{R}^n \setminus \operatorname{spt}\mu\big) \cup \big\{D_\mu^+\nu = \infty\big\}.$$

By step three, $\mu(Y) = 0$, thus $\nu_2 \perp \mu$. We are thus left to prove that

$$\nu\Big(E \cap \big\{D_\mu^+\nu < \infty\big\}\Big) = \int_E D_\mu\nu\,d\mu,$$

for every Borel set $E \subset \mathbb{R}^n$. By step two,

$$\nu\Big(\big\{D_\mu^-\nu = 0\big\} \cap B_R\Big) \le \nu\Big(\big\{D_\mu^-\nu \le \varepsilon\big\} \cap B_R\Big) \le \varepsilon\,\mu(B_R),$$

therefore $\nu(\{D_\mu^-\nu = 0\}) = 0$. As $D_\mu\nu$ exists and is finite μ-a.e. on $\operatorname{spt}\mu$, we are thus left to show that

$$\nu(E \cap W) = \int_E D_\mu\nu\,d\mu, \tag{5.14}$$

for every Borel set $E \subset \mathbb{R}^n$, where we have set

$$W = \big\{x \in \operatorname{spt}\mu : D_\mu\nu(x) \text{ exists, } 0 < D_\mu\nu(x) < \infty\big\}.$$

To prove (5.14), we fix $t \in (1, \infty)$ and let

$$E_k = E \cap \big\{x \in W : t^k \le D_\mu\nu(x) < t^{k+1}\big\}, \qquad k \in \mathbb{Z}.$$

As $\{E_k\}_{k\in\mathbb{N}}$ is a sequence of disjoint Borel sets with $E \cap W = \bigcup_{k\in\mathbb{Z}} E_k$, we find

$$\int_E D_\mu\nu\,d\mu = \int_{E\cap W} D_\mu\nu\,d\mu = \sum_{k\in\mathbb{Z}} \int_{E_k} D_\mu\nu d\mu, \qquad \nu(E \cap W) = \sum_{k\in\mathbb{Z}} \nu(E_k).$$

By step two, we have $\nu(E_k) \le t^{k+1}\mu(E_k)$, and thus

$$\begin{aligned}\nu(E \cap W) &= \sum_{k\in\mathbb{Z}} \nu(E_k) \le \sum_{k\in\mathbb{Z}} t^{k+1}\mu(E_k) = t\sum_{k\in\mathbb{Z}} t^k\mu(E_k)\\ &\le t\sum_{k\in\mathbb{Z}} \int_{E_k} D_\mu\nu\,d\mu = t\int_{E\cap W} D_\mu\nu\,d\mu.\end{aligned} \tag{5.15}$$

Again by step two, we have $\nu(E_k) \ge t^k\mu(E_k)$, so that

$$\begin{aligned}\nu(E \cap W) &= \sum_{k\in\mathbb{Z}} \nu(E_k) \ge \sum_{k\in\mathbb{Z}} t^k\mu(E_k) = \frac{1}{t}\sum_{k\in\mathbb{Z}} t^{k+1}\mu(E_k)\\ &\ge \frac{1}{t}\sum_{k\in\mathbb{Z}} \int_{E_k} D_\mu\nu\,d\mu = \frac{1}{t}\int_{E\cap W} D_\mu\nu\,d\mu.\end{aligned} \tag{5.16}$$

We let $t \to 1^+$ in (5.15) and (5.16) to conclude the proof. □

Remark 5.10 A vector-valued Radon measure ν is **absolutely continuous** with respect to the Radon measure μ if $|\nu| \ll \mu$. At the same time, we say that μ and ν are **mutually singular** if $|\nu| \perp \mu$.

Corollary 5.11 *If ν is an $\mathbb{R}^m$-valued Radon measure on $\mathbb{R}^n$, and μ is a Radon measure on $\mathbb{R}^n$, then for μ-a.e. $x \in \mathbb{R}^n$ there exists the limit*

$$D_\mu \nu(x) = \lim_{r \to 0^+} \frac{\nu(B(x,r))}{\mu(B(x,r))} \in \mathbb{R}^m ,$$

which defines a Borel vector field $D_\mu \nu \in L^1_{\mathrm{loc}}(\mathbb{R}^n, \mu; \mathbb{R}^m)$, with the property that

$$\nu = (D_\mu \nu)\, \mu + \nu_\mu^s \qquad \text{on } \mathcal{M}(\mu),$$

where $\nu_\mu^s \perp \mu$.

Proof By the Jordan decomposition, $\nu = \sum_{i=1}^m (\nu_+^{(i)} - \nu_-^{(i)}) e_i$ with $\nu_+^{(i)}$ and $\nu_-^{(i)}$ Radon measures on $\mathbb{R}^n$. We apply Theorem 5.8 to differentiate $\nu_+^{(i)}$ and $\nu_-^{(i)}$ with respect to μ. □

Example 5.12 Let ν be an $\mathbb{R}^m$-valued Radon measure on $\mathbb{R}^n$. By Riesz's theorem (Theorem 4.7), $\nu = g\,|\nu|$, where $g \colon \mathbb{R}^n \to \mathbb{R}^m$ is $|\nu|$-measurable, with $|g| = 1$ $|\nu|$-a.e. on $\mathbb{R}^n$. Since $\nu \ll |\nu|$, by Corollary 5.11, $g = D_{|\nu|}\nu$, $|\nu|$-a.e. on $\mathbb{R}^n$. In particular, $|D_{|\nu|}\nu(x)| = 1$ for $|\nu|$-a.e. $x \in \mathbb{R}^n$.

Exercise 5.13 If ν_1 and ν_2 are mutually singular $\mathbb{R}^m$-valued Radon measures on $\mathbb{R}^n$, then $|\nu_1 + \nu_2| = |\nu_1| + |\nu_2|$. *Hint:* Differentiate $\nu_1 + \nu_2$ with respect to $|\nu_1|$ and to $|\nu_2|$, and use Exercise 4.13.

Exercise 5.14 (Monotone functions) If $m \colon \mathbb{R} \to \mathbb{R}$ is an increasing function, that is $m(s) \le m(t)$ whenever $s \le t$, then $m'(t)$ exists for a.e. $t \in \mathbb{R}$ and moreover

$$\int_{\mathbb{R}} m'(t)\mathrm{d}t \le \lim_{t \to \infty} m(t) - \lim_{t \to -\infty} m(t) .$$

Hint: Define an outer measure μ on $\mathbb{R}$, by setting

$$\mu(E) = \inf \left\{ \sum_{h \in \mathbb{N}} (m(b_h) - m(a_h)) : E \subset \bigcup_{h \in \mathbb{N}} (a_h, b_h) \right\}, \qquad E \subset \mathbb{R},$$

(note that $\mu((a,b)) = m(b^-) - m(a^+)$). Show that μ is a Radon measure on $\mathbb{R}$, and differentiate μ with respect to $\mathcal{L}^1$.

Exercise 5.15 (Weak compactness in L^p spaces, $p > 1$) Let μ be a Radon measure on $\mathbb{R}^n$. If $\{u_h\}_{h \in \mathbb{N}} \subset L^p(\mathbb{R}^n, \mu)$ $(1 < p \le \infty)$ satisfies

$$\sup_{h \in \mathbb{N}} \|u_h\|_{L^p(\mathbb{R}^n, \mu)} < \infty ,$$

then there exist a sequence $h(k) \to \infty$ as $k \to \infty$ and $u \in L^p(\mathbb{R}^n, \mu)$ such that

$$\lim_{k\to\infty} \int_{\mathbb{R}^n} \varphi\, u_{h(k)}\, d\mu = \int_{\mathbb{R}^n} \varphi\, u\, d\mu\,,$$

for every $\varphi \in L^{p'}(\mathbb{R}^n, \mu)$ ($p' = 1$ if $p = \infty$, $p' = p/(p-1)$ if $p \in (1, \infty)$). *Hint:* By Corollary 4.34, there exists a signed Radon measure ν such that $\mu_{h(k)} = u_{h(k)}\, \mu \overset{*}{\rightharpoonup} \nu$. Show that $\nu \ll \mu$ and that, in fact, $\nu = u\,\mu$ for $u \in L^p(\mathbb{R}^n, \mu)$.

5.3 Lebesgue points

By the mean value theorem, if $u \in C^0(\mathbb{R})$ and μ is a Radon measure on $\mathbb{R}^n$, then

$$\lim_{r\to 0^+} \frac{1}{\mu(B(x,r))} \int_{B(x,r)} |u(x) - u|\, d\mu = 0\,, \qquad \forall x \in \mathbb{R}^n\,.$$

If now $u \in L^1_{\mathrm{loc}}(\mathbb{R}^n, \mu)$, then this property still holds at μ-a.e. $x \in \mathbb{R}^n$.

Theorem 5.16 (Lebesgue points theorem) *If μ is a Radon measure on $\mathbb{R}^n$, $p \in [1, \infty)$ and $u \in L^p_{\mathrm{loc}}(\mathbb{R}^n, \mu)$, then for μ-a.e. $x \in \mathbb{R}^n$*

$$\lim_{r\to 0^+} \frac{1}{\mu(B(x,r))} \int_{B(x,r)} |u(x) - u|^p\, d\mu = 0\,. \tag{5.17}$$

In this case, we say that x is a **Lebesgue point** *of u with respect to μ.*

Example 5.17 Given $E \subset \mathbb{R}^n$ and $x \in \mathbb{R}^n$, if the limit

$$\theta_n(E)(x) = \lim_{r\to 0^+} \frac{|E \cap B(x,r)|}{\omega_n\, r^n}$$

exists, it is called the ***n*-dimensional density of *E* at *x***. By the same argument as in Remark 5.7, $\theta_n(E)$ defines a Borel function $\mathbb{R}^n$. If E is a Lebesgue measurable set in $\mathbb{R}^n$, and we apply Theorem 5.16 to the Radon measure $\mu = \mathcal{L}^n \llcorner E$, then we deduce immediately that $\theta_n(E)(x)$ exists for a.e. $x \in \mathbb{R}^n$. In particular,

$$\theta_n(E) = 1 \text{ a.e. on } E\,, \qquad \theta_n(E) = 0 \text{ a.e. on } \mathbb{R}^n \setminus E\,. \tag{5.18}$$

Given $t \in [0, 1]$, the **set of points of density *t* of *E*** is defined as

$$E^{(t)} = \left\{x \in \mathbb{R}^n : \theta_n(E)(x) = t\right\},$$

and it turns out to be a Borel set. *Every Lebesgue measurable set is equivalent to the set of its points of density one*, since, by (5.18),

$$\left|E \Delta E^{(1)}\right| = 0\,, \qquad \left|(\mathbb{R}^n \setminus E) \Delta E^{(0)}\right| = 0\,. \tag{5.19}$$

Proof of Theorem 5.16 We first note that for μ-a.e. $x \in \mathbb{R}^n$

$$\lim_{r\to 0^+} \frac{1}{\mu(B(x,r))} \int_{B(x,r)} u\, d\mu = u(x)\,. \tag{5.20}$$

The signed Radon measure $\nu = u\,\mu$ is absolutely continuous with respect to μ, and thus, by Theorem 5.8, for μ-a.e. $x \in \mathbb{R}^n$ the limit

$$D_\mu \nu(x) = \lim_{r\to 0^+} \frac{\nu(B(x,r))}{\mu(B(x,r))} = \lim_{r\to 0^+} \frac{1}{\mu(B(x,r))} \int_{B(x,r)} u\,\mathrm{d}\mu$$

exists, and, for every Borel set $E \subset \mathbb{R}^n$,

$$\int_E u\,\mathrm{d}\mu = \nu(E) = \int_E D_\mu \nu\,\mathrm{d}\mu\,.$$

In particular, $u = D_\mu \nu$ μ-a.e. in $\mathbb{R}^n$, and (5.20) is proved. Now let $\mathbb{Q} = \{t_h\}_{h\in\mathbb{N}}$. For every $h \in \mathbb{N}$, there exists a μ-null set E_h such that

$$\lim_{r\to 0^+} \frac{1}{\mu(B(x,r))} \int_{B(x,r)} |u - t_h|^p\,\mathrm{d}\mu = |u(x) - t_h|^p\,, \qquad \forall x \in \mathbb{R}^n \setminus E_h\,.$$

If $E = \bigcup_{h\in\mathbb{N}} E_h$, then $\mu(E) = 0$ and for every $x \in \mathbb{R}^n \setminus E$ and $h \in \mathbb{N}$,

$$\int_{B(x,r)} |u(x) - u|^p \mathrm{d}\mu \le 2^{p-1}\Big(|u(x) - t_h|^p \mu(B(x,r)) + \int_{B(x,r)} |\, t_h - u|^p \mathrm{d}\mu\Big).$$

We divide by $\mu(B(x,r))$ and let $r \to 0^+$ to find

$$\lim_{r\to 0^+} \frac{1}{\mu(B(x,r))} \int_{B(x,r)} |u(x) - u|^p \mathrm{d}\mu \le 2^{p-1}|u(x) - t_h|^p,$$

for every $h \in \mathbb{N}$. We select a sequence $t_{h(k)} \to u(x)$ to conclude the proof. □

Exercise 5.18 If $u \in L^1_{\mathrm{loc}}(\mathbb{R}^n)$, then $u_\varepsilon \in C^\infty(\mathbb{R}^n)$, $\operatorname{spt} u_\varepsilon \subset \operatorname{spt} u + \varepsilon B$,

$$\nabla u_\varepsilon(x) = \int_{\mathbb{R}^n} \nabla \rho_\varepsilon(x - y) u(y) \mathrm{d}y\,, \quad x \in \mathbb{R}^n\,,$$

and $u_\varepsilon(E)$ is contained in the closed convex hull of $u(E)$ for every $E \subset \mathbb{R}^n$. Moreover, $u_\varepsilon \to u$ in $L^1_{\mathrm{loc}}(\mathbb{R}^n)$, $u_\varepsilon(x) \to u(x)$ at Lebesgue points x of u, and

$$\|u_\varepsilon\|_{L^1(B_R)} \le \|u\|_{L^1(B_{R+\varepsilon})}\,. \tag{5.21}$$

Exercise 5.19 Let $x \in \mathbb{R}^n$, $r_0 > 0$ and $\{C_r\}_{0<r<r_0}$ be a family of Borel sets such that, for some $\beta > \alpha > 0$, $B(x, \alpha r) \subset C_r \subset B(x, \beta r)$ for $0 < r < r_0$ (for example, $C_r = x + r A$, A open and bounded, $0 \in A$). If E is a Lebesgue measurable set in $\mathbb{R}^n$ and $t \in \{0, 1\}$, then $x \in E^{(t)}$ if and only if

$$\lim_{r\to 0^+} \frac{|E \cap C_r|}{|C_r|} = t\,.$$

This equivalence is easily seen to fail if $t \in (0, 1)$.

6

Two further applications of differentiation theory

6.1 Campanato's criterion

Campanato's criterion is a cornerstone in the regularity theory for variational problems, as it characterizes Hölder continuity in terms of the uniform decay of certain integral averages. We shall use this criterion in Section 26.2.

Theorem 6.1 (Campanato's criterion) *If $n \ge 1$, $p \in [1, \infty)$, $\gamma \in (0, 1]$, then there exists a constant $C(n, p, \gamma)$ with the following property. If $u \in L^p(B)$,*

$$(u)_{x,r} = \frac{1}{|B \cap B(x,r)|} \int_{B \cap B(x,r)} u\,, \qquad x \in B\,, r > 0\,,$$

and there exists a constant κ such that the uniform decay condition

$$\left(\frac{1}{r^n} \int_{B \cap B(x,r)} |u - (u)_{x,r}|^p \right)^{1/p} \le \kappa\, r^\gamma\,, \qquad \forall x \in B\,, \tag{6.1}$$

holds true, then there exists a function $\overline{u}\colon B \to \mathbb{R}$ with $\overline{u} = u$ a.e. on B and

$$|\overline{u}(x) - \overline{u}(y)| \le \kappa' |x - y|^\gamma\,, \qquad \forall x, y \in B\,. \tag{6.2}$$

where $\kappa' = C(n, p, \gamma)\, \kappa$.

Remark 6.2 It is easily seen that there exists a constant $c(n) > 0$ such that

$$c(n) r^n \le |B \cap B(x,r)| \le \omega_n\, r^n\,,$$

for every $x \in B$, $r > 0$. In particular, if $u \in L^p(B)$, then by Theorem 5.16

$$\lim_{r \to 0^+} \frac{1}{r^n} \int_{B \cap B(x,r)} |u - (u)_{x,r}|^p = 0\,,$$

for a.e. $x \in B$ (precisely, for every Lebesgue point x of $1_B\, u \in L^1(\mathbb{R}^n)$).

Remark 6.3 Conversely, if $u \in L^p(B)$, $\overline{u}\colon B \to \mathbb{R}$ satisfies (6.2) for a constant κ' and $\overline{u} = u$ a.e. on B, then by Jensen's inequality and $(\overline{u})_{x,r} = (u)_{x,r}$,

$$|\overline{u}(x) - (u)_{x,r}| \le \left(\fint_{B\cap B(x,r)} |\overline{u}(x) - \overline{u}(y)|^p \mathrm{d}y \right)^{1/p} \le \kappa' r^\gamma\,, \quad \forall x \in B\,,$$

which implies (6.1) with $\kappa = \kappa'$.

Proof of Theorem 6.1 Let $c(n)$ be the positive constant introduced in Remark 6.2. If we set $v_r(x) = (u)_{x,r}$, then for every $r < R$ and $x \in B$ we have

$$c(n)\, r^n |\, v_r(x) - v_R(x)|^p \le 2^{p-1}\left(\int_{B\cap B(x,r)} |u - (u)_{x,r}|^p + \int_{B\cap B(x,R)} |u - (u)_{x,R}|^p \right).$$

In particular, by (6.1), we obtain

$$|\, v_r(x) - v_R(x)| \le C(n,p)\kappa \left(\frac{R}{r}\right)^{n/p} R^\gamma\,, \qquad \forall r < R\,, x \in B\,. \tag{6.3}$$

If we set $r_k = 2^{-k} r$, and repeatedly apply (6.3), then we obtain

$$|\, v_{r_k}(x) - v_{r_h}(x)| \le \sum_{j=h}^{k-1} |\, v_{r_{j+1}}(x) - v_{r_j}(x)| \le C(n,p)\kappa \sum_{j=h}^{k-1} \left(\frac{1}{2^\gamma}\right)^j r^\gamma\,, \tag{6.4}$$

for $k > h \ge 0$ and $x \in B$. If we let $h = 0$ and $k \to \infty$ in (6.4), then by Theorem 5.16 (see Remark 6.2) we find that

$$|u(x) - v_r(x)| \le C(n,p,\gamma)\kappa\, r^\gamma\,, \tag{6.5}$$

whenever x is a Lebesgue point of u. Now let $x, y \in B$, set $r = |x - y|$, and consider the positive constant $\eta(n)$ such that $|B(x,r) \cap B(y,r)| = \eta(n) r^n$; as a consequence, there exists a constant $\eta'(n) < \eta(n)$ such that $\eta'(n) r^n \le |B(x,r) \cap B(y,r) \cap B|$. We thus obtain

$$\eta'(n) r^n |\, v_r(x) - v_r(y)|^p \le 2^{p-1}\left(\int_{B\cap B(x,r)} |u - (u)_{x,r}|^p + \int_{B\cap B(y,r)} |u - (u)_{y,r}|^p \right),$$

that, by (6.1), implies $|\, v_r(x) - v_r(y)| \le C(n,p)\kappa |x - y|^\gamma$. By the triangular inequality and by (6.5), if x and y are Lebesgue points of u in B, then

$$|u(x) - u(y)| \le C(n,p,\gamma)\,\kappa\,|x - y|^\gamma\,. \tag{6.6}$$

On the other hand, (6.4) implies that, for every $r > 0$, the sequence of continuous functions $\{v_{r_k}\}_{k\in\mathbb{N}}$ is a Cauchy sequence with respect to uniform convergence on B. By completeness, there exists a continuous function $\overline{u}\colon B \to \mathbb{R}$ which is the pointwise limit of $\{v_{r_k}\}_{k\in\mathbb{N}}$ on B. By Theorem 5.16, we have $\overline{u} = u$ at every Lebesgue point of u (in particular $\overline{u}$ is independent of r). By (6.6), $\overline{u}$ satisfies (6.2) whenever x and y are Lebesgue points of u. Finally, by continuity of $\overline{u}$, (6.2) holds for every $x, y \in B$. □

6.2 Lower dimensional densities of a Radon measure

Given a Radon measure μ on $\mathbb{R}^n$ and $s \in (0,n]$, we define the **upper s-dimensional density** $\theta_s^*(\mu) : \mathbb{R}^n \to [0,\infty]$ of μ as

$$\theta_s^*(\mu)(x) = \limsup_{r\to 0^+} \frac{\mu\big(\overline{B}(x,r)\big)}{\omega_s r^s}, \qquad x \in \mathbb{R}^n. \tag{6.7}$$

We note that, by Exercise 4.27, $\theta_s^*(\mu)$ is a Borel function. If $x \in \mathbb{R}^n$ is such that the limit in (6.7) exists, then we denote by $\theta_s(\mu)(x)$ this value, and call it the **s-dimensional density of μ at x**. If $\theta_s(\mu)(x)$ is defined at x, then closed balls may be replaced by open balls, that is,

$$\theta_s(\mu)(x) = \lim_{r\to 0^+} \frac{\mu(B(x,r))}{\omega_s r^s};$$

see Remark 5.7. Since $\omega_n r^n = |B(x,r)|$, looking at n-dimensional densities is equivalent to differentiating with respect to $\mathcal{L}^n$. Hence, the study of n-dimensional densities is fully addressed by the Lebesgue–Besicovitch differentiation theorem. The behavior of s-dimensional densities, when $s \in (0,n)$, is more complex. The following theorem and its corollary (which extend the identity $\theta_n(E) = 0$ a.e. on $\mathbb{R}^n \setminus E$ to arbitrary values of s) illustrate what can be concluded in full generality, and will be used in Chapters 11, 16, and 17.

Theorem 6.4 (Upper s-dimensional densities and comparison with $\mathcal{H}^s$) *If μ is a Radon measure on $\mathbb{R}^n$, M is a Borel set, and $s \in (0,n)$, then*

$$1 \le \theta_s^*(\mu) \quad \text{on } M \quad \Rightarrow \quad \mathcal{H}^s(M) \le \mu(M), \tag{6.8}$$

$$\theta_s^*(\mu) \le 1 \quad \text{on } M \quad \Rightarrow \quad \mu(M) \le 2^s \mathcal{H}^s(M). \tag{6.9}$$

Proof *Step one:* We prove (6.8). We may directly assume that $M \subset B_R$, for some $R > 0$. We first prove that $\theta_s^*(\mu) \ge 1$ on M implies $\mathcal{H}^s(M) < \infty$. Given $\delta > 0$, let us consider a family of closed balls

$$\mathcal{F} = \Big\{\overline{B}(x,r) : x \in M, 2r < \delta, \mu\big(\overline{B}(x,r)\big) \ge (1-\delta)\omega_s r^s\Big\}.$$

If $\{\mathcal{F}_i\}_{i=1}^{\xi(n)}$ are the subfamilies of $\mathcal{F}$ given by Theorem 5.1, then

$$\begin{aligned}\mathcal{H}^s_\delta(M) &\le \sum_{i=1}^{\xi(n)} \sum_{B\in\mathcal{F}_i} \omega_s \Big(\frac{\operatorname{diam}(B)}{2}\Big)^s \le \frac{1}{(1-\delta)} \sum_{i=1}^{\xi(n)} \sum_{B\in\mathcal{F}_i} \mu(B)\\ &\le \frac{\xi(n)}{(1-\delta)} \max_{1\le i\le \xi(n)} \mu\Big(\bigcup_{B\in\mathcal{F}_i} B\Big) \le \frac{\xi(n)\mu(B_{R+\delta})}{(1-\delta)}.\end{aligned}$$

We let $\delta \to 0^+$ to find $\mathcal{H}^s(M) < \infty$. Now let A be an open set with $M \subset A$, and define a covering $\mathcal{F}'$ of M as

$$\mathcal{F}' = \Big\{\overline{B}(x,r) : x \in M, 2r < \delta, \overline{B}(x,r) \subset A, \mu\big(\overline{B}(x,r)\big) \ge (1-\delta)\omega_s r^s\Big\}.$$

Since $\theta_s^*(\mu) \geq 1$ on M, $\mathcal{F}'$ satisfies assumption (5.9) of Corollary 5.5. Since $\mathcal{H}^s(M) < \infty$, $\mathcal{H}^s \llcorner M$ is a Radon measure and Corollary 5.5 gives a countable disjoint family $\mathcal{G} \subset \mathcal{F}'$ such that $\mathcal{H}^s(M \setminus \bigcup_{B \in \mathcal{G}} B) = 0$. Since $\mathcal{H}^s$ null sets are $\mathcal{H}^s_\delta$ null sets, we conclude that

$$\mathcal{H}^s_\delta(M) \leq \sum_{B \in \mathcal{G}} \omega_s \left(\frac{\operatorname{diam}(B)}{2} \right)^s \leq \frac{1}{(1-\delta)} \sum_{B \in \mathcal{G}} \mu(B) = \frac{1}{(1-\delta)} \mu\left(\bigcup_{B \in \mathcal{G}} B \right) \leq \frac{\mu(A)}{1-\delta}.$$

We let $\delta \to 0^+$, and recall (2.6), to prove that $\mathcal{H}^s(M) \leq \mu(M)$.

Step two: We prove (6.9). Given $\varepsilon > 0$, $\delta > 0$, we define a set M_δ as

$$M_\delta = \left\{ x \in M : \mu\left(\overline{B}(x,r)\right) \leq (1+\varepsilon)\omega_s r^s, \forall r \in (0,\delta) \right\}.$$

Let us consider a countable covering $\mathcal{F}$ of M_δ, with $\operatorname{diam}(F) \leq \delta$ and $F \cap M_\delta \neq \emptyset$ for every $F \in \mathcal{F}$ (cf. Exercise 1.3), and

$$\omega_s \sum_{F \in \mathcal{F}} \left(\frac{\operatorname{diam}(F)}{2} \right)^s \leq \mathcal{H}^s_\delta(M_\delta) + \delta.$$

For every $F \in \mathcal{F}$ there exists a closed ball B_F with center in M_δ such that $F \subset B_F$, and $\operatorname{diam}(B_F) = 2 \operatorname{diam}(F) \leq 2\delta$. Hence, $M_\delta \subset \bigcup_{F \in \mathcal{F}} B_F$, with

$$\begin{aligned} \mu(M_\delta) &\leq \sum_{F \in \mathcal{F}} \mu(B_F) \leq (1+\varepsilon)\omega_s \sum_{F \in \mathcal{F}} \left(\frac{\operatorname{diam}(B_F)}{2} \right)^s = (1+\varepsilon)\omega_s \sum_{F \in \mathcal{F}} \operatorname{diam}(F)^s \\ &\leq (1+\varepsilon) 2^s \left(\mathcal{H}^s_\delta(M_\delta) + \delta \right) \leq (1+\varepsilon) 2^s \left(\mathcal{H}^s_\delta(M) + \delta \right). \end{aligned}$$

Since $\theta_s^*(\mu) \leq 1$ on M, we have $\mu(M_\delta) \to \mu(M)$ as $\delta \to 0^+$. We first let $\delta \to 0$, and then $\varepsilon \to 0$, to prove that $\mu(M) \leq 2^s \mathcal{H}^s(M)$, as desired. □

Corollary 6.5 *If $s \in (0,n)$ and $M \subset \mathbb{R}^n$ is a Borel set with $\mathcal{H}^s(M \cap K) < \infty$ for every compact set K in $\mathbb{R}^n$, then for $\mathcal{H}^s$-a.e. $x \in \mathbb{R}^n \setminus M$,*

$$\lim_{r \to 0^+} \frac{\mathcal{H}^s(M \cap B(x,r))}{\omega_s r^s} = 0. \tag{6.10}$$

Proof We may directly assume that $\mathcal{H}^s(M) < \infty$. Given $\delta > 0$, let us now set $F_\delta = \{x \in \mathbb{R}^n \setminus M : \theta_s^*(\mu)(x) \geq \delta\}$, where $\mu = \mathcal{H}^s \llcorner M$. By Theorem 6.4,

$$\delta \mathcal{H}^s(F_\delta) \leq \mu(F_\delta) = \mathcal{H}^s(M \cap F_\delta) = 0.$$

Hence, $\mathcal{H}^s(F_\delta) = 0$. We let $\delta \to 0^+$ to prove (6.10). □

7

Lipschitz functions

The notion of Lipschitz function plays a special role in Geometric Measure Theory. Various metric properties which are characteristic of C^1-functions are also satisfied by Lipschitz functions. At the same time, the Lipschitz condition is stable under plain pointwise convergence and can be formulated in terms of set inclusions only, two features that make it particularly compatible with measure-theoretic arguments. In this chapter (and in the following) we address some fundamental properties of Lipschitz functions. In Section 7.1, we prove Kirszbraun's theorem, which guarantees the existence of Lipschitz-constant preserving extensions. In Sections 7.2 and 7.3 Lipschitz functions are shown to possess bounded distributional gradients and to be a.e. classically differentiable (Rademacher's theorem). Recall that, if $E \subset \mathbb{R}^n$ and $f\colon E \subset \mathbb{R}^n \to \mathbb{R}^m$, then is a **Lipschitz function on** E, provided

$$\mathrm{Lip}(f;E) = \sup\left\{\frac{|f(x)-f(y)|}{|x-y|} : x, y \in E\,, x \neq y\right\} < \infty\,.$$

We simply set $\mathrm{Lip}(f) = \mathrm{Lip}(f;\mathbb{R}^n)$. The geometric nature of the Lipschitz condition is suggested by the following remark.

Remark 7.1 If $\Gamma(f;\mathbb{R}^n)$ denotes the graph of $f\colon \mathbb{R}^n \to \mathbb{R}^m$ over $\mathbb{R}^n$, that is

$$\Gamma(f;\mathbb{R}^n) = \left\{(y, f(y)) \in \mathbb{R}^n \times \mathbb{R}^m : y \in \mathbb{R}^n\right\},$$

then f is a Lipschitz function on $\mathbb{R}^n$, provided that, for every $x \in \mathbb{R}^n$, the graph of f is contained in the "cone" of vertex $(x, f(x))$ and "opening" $\mathrm{Lip}(f)$, that is (see Figure 7.1),

$$\Gamma(f;\mathbb{R}^n) \subset \bigcap_{x\in\mathbb{R}^n} (x, f(x)) + \left\{(z, w) \in \mathbb{R}^n \times \mathbb{R}^m : |w| \le \mathrm{Lip}(f)|z|\right\}.$$

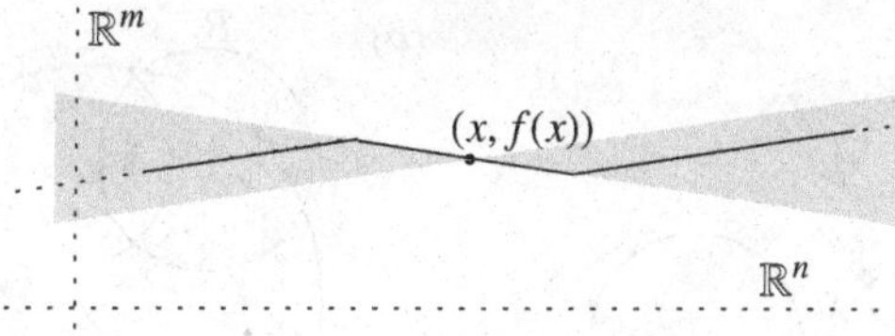

Figure 7.1 The Lipschitz condition as a family of set inclusions.

7.1 Kirszbraun's theorem

Theorem 7.2 (Kirszbraun's theorem) *If $E \subset \mathbb{R}^n$ and $f\colon E \to \mathbb{R}^m$ is a Lipschitz function, then there exists $g\colon \mathbb{R}^n \to \mathbb{R}^m$ such that $g = f$ on E and* $\mathrm{Lip}(g) = \mathrm{Lip}(f; E)$.

When $m = 1$, a (somewhat) explicit extension g of f is defined as

$$g(y) = \inf\left\{f(x) + \mathrm{Lip}(f)|y - x| : x \in E\right\}, \qquad y \in \mathbb{R}^n. \tag{7.1}$$

Having in mind Remark 7.1, (7.1) defines the maximal extension of f.

Lemma 7.3 (McShane's lemma) *If $E \subset \mathbb{R}^n$ and $f\colon E \to \mathbb{R}$ is a Lipschitz function on E, then the function $g\colon \mathbb{R}^n \to \mathbb{R}$ defined in (7.1) satisfies $g = f$ on E and* $\mathrm{Lip}(g) = \mathrm{Lip}(f; E)$.

Proof Clearly, $g \le f$ on E. Since $f(x) + \mathrm{Lip}(f)|y - x| \ge f(y)$ for every $x, y \in E$, minimizing over $x \in E$ we find $g \ge f$ on E. Now, if $x, y, z \in \mathbb{R}^n$, then

$$g(y) \le f(x) + \mathrm{Lip}(f; E)|y - x| \le \left(f(x) + \mathrm{Lip}(f; E)|z - x|\right) + \mathrm{Lip}(f; E)|z - y|.$$

Minimizing over $x \in E$ we find $g(y) \le g(z) + \mathrm{Lip}(f; E)|z - y|$, and then, by symmetry, $|g(y) - g(z)| \le \mathrm{Lip}(f; E)|y - z|$. □

When $m > 1$ we can extend each component $f^{(i)}$ of f by McShane's lemma, thus finding a Lipschitz function $g\colon \mathbb{R}^n \to \mathbb{R}^m$ with $g = f$ on E. This extension, however, will merely satisfy the non-optimal bound $\mathrm{Lip}(g) \le \sqrt{m}\,\mathrm{Lip}(f; E)$. We thus need a different strategy to prove Kirszbraun's theorem. We shall use the following geometric lemma; see Figure 7.2.

Lemma 7.4 *Given a finite collection of closed balls $\{\overline{B}(x_k, r_k)\}_{k=1}^N$ in $\mathbb{R}^n$, set*

$$C_t = \bigcap_{k=1}^N \overline{B}(x_k, t\,r_k), \qquad t \ge 0.$$

If $s = \inf\{t \ge 0\colon C_t \ne \emptyset\}$, then $s < \infty$ and C_s reduces to a single point x_0, which belongs to the convex hull of those x_k such that $|x_0 - x_k| = s\,r_k$.

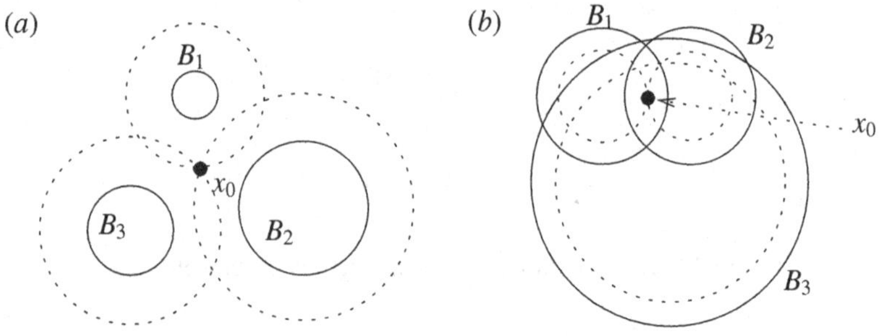

Figure 7.2 The situation in Lemma 7.4. In the picture, we have set $B_k = \overline{B}(x_k, r_k)$. We consider a finite family of closed balls and multiply the radius of each ball by a common factor $t \geq 0$. We let s be the minimal value of t such that the intersection of the deformed balls is non-empty, and note that this intersection reduces to a point x_0: (a) a situation where $s > 1$ and x_0 is a convex combination of x_1, x_2, and x_3; (b) a situation where $s < 1$ and s and x_0 are left unchanged if the ball $B(x_3, r_3)$ is removed from the family; as a consequence, in picture (b), x_0 is a convex combination of x_1 and x_2 only.

Proof *Step one:* Clearly, $s < \infty$ and $C_s \neq \emptyset$. To prove that C_s reduces to a point, let $y, z \in C_s$, so that $|y - x_k| \leq s\, r_k$ and $|z - x_k| \leq s\, r_k$ for every $k = 1, ..., N$, with $y \neq z$. Since balls are round, if $w = (y+z)/2$, then $|w - x_k| < s r_k$, that is

$$|w - x_k|^2 = \frac{|y - x_k|^2 + |z - x_k|^2}{2} - \frac{|y - z|^2}{4} \leq \left(s^2 - \frac{|y - z|^2}{4r^2}\right) r_k^2 ,$$

where we have set $r = \max\{r_k : 1 \leq k \leq N\}$. Therefore, if $y \neq z$, then C_t is non-empty for some $t < s$, a contradiction.

Step two: Up to permuting the x_k, we may assume that $|x_k - x_0| = s\, r_k$ if and only if $1 \leq k \leq M$ for some $M \leq N$. Replacing $\{\overline{B}(x_k, r_k)\}_{k=1}^N$ with $\{\overline{B}(x_k, r_k)\}_{k=1}^M$, we do not change s and x_0. We thus assume that $|x_k - x_0| = s\, r_k$ for $k = 1, \ldots, N$. If now $v \in S^{n-1}$ and $\varepsilon > 0$, then by construction of x_0 there exists x_k with

$$|x_k - x_0|^2 < |x_k - (x_0 + \varepsilon v)|^2 = |x_k - x_0|^2 + \varepsilon^2 - 2\varepsilon\, v \cdot (x_k - x_0) ,$$

i.e. $2v \cdot (x_k - x_0) \leq \varepsilon$. Hence, for every $v \in S^{n-1}$ there exists x_k such that $v \cdot (x_0 - x_k) \geq 0$. As a consequence, for every closed half-space H with $x_0 \in \partial H$ there exists $k \in \{1, \ldots, N\}$ such that $x_k \in H$, that is, $\{x_0\}$ and $\{x_k\}_{k=1}^N$ cannot be separated by a hyperplane. □

Proof of Theorem 7.2 *Step one:* We show that, if E is a proper subset of $\mathbb{R}^n$ and $y \in \mathbb{R}^n \setminus E$, then there exists $z \in \mathbb{R}^m$ such that, defining $g\colon E \cup \{y\} \to \mathbb{R}^m$ as

$$g(x) = \begin{cases} f(x), & \text{if } x \in E , \\ z, & \text{if } x = y , \end{cases}$$

then $\mathrm{Lip}(g; E \cup \{y\}) = \mathrm{Lip}(f; E)$. Setting $\mathrm{Lip}(f; E) = 1$, we thus seek $z \in \mathbb{R}^m$ with $|z - f(x)| \le |y - x|$ for every $x \in E$. In other words, we want the family

$$\left\{\overline{B}\big(f(x), |y - x|\big) : x \in E\right\}$$

of closed balls in $\mathbb{R}^m$ to have non-empty intersection. Since these balls are compact sets, we may further reduce to proving that, for every finite subset $\{x_k\}_{k=1}^N$ of E, the family

$$\left\{\overline{B}\big(f(x_k), |y - x_k|\big) : 1 \le k \le N\right\},$$

has non-empty intersection. Now, by Lemma 7.4, there exists $s \ge 0$ and $z \in \mathbb{R}^m$ such that, up to a permutation of the x_k and for some $M \in \mathbb{N}$, $1 \le M \le N$,

$$\{z\} = \bigcap_{k=1}^N \overline{B}\big(f(x_k), s\,|y - x_k|\big), \qquad |z - f(x_k)| = s|\,y - x_k|, \qquad z = \sum_{k=1}^M \lambda_k f(x_k),$$

where $\lambda_k > 0$ for $1 \le k \le M$ and $1 = \sum_{k=1}^M \lambda_k$. Thus, in order to conclude the proof, we only need to show that $s \le 1$. To this end,

$$\begin{aligned}
0 = 2\left|\sum_{k=1}^M \lambda_k \big(z - f(x_k)\big)\right|^2 &= 2\sum_{k,h=1}^M \lambda_k\lambda_h\big(z - f(x_k)\big)\cdot\big(z - f(x_h)\big)\\
&= \sum_{k,h=1}^M \lambda_k\lambda_h\Big(|z - f(x_k)|^2 + |z - f(x_h)|^2 - |f(x_k) - f(x_h)|^2\Big)\\
&\ge \sum_{k,h=1}^M \lambda_k\lambda_h\Big(s^2|\,y - x_k|^2 + s^2|\,y - x_h|^2 - |x_k - x_h|^2\Big)\\
&= \sum_{k,h=1}^M \lambda_k\lambda_h\Big(2s^2(y - x_k)\cdot(y - x_h) + (s^2 - 1)|x_k - x_h|^2\Big)\\
&= 2s^2\left|\sum_{k=1}^M \lambda_k(y - x_k)\right|^2 + (s^2 - 1)\sum_{k,h=1}^M \lambda_k\lambda_h|x_k - x_h|^2\,.
\end{aligned}$$

Either $M = 1$, and thus $s = 0$, or $M > 1$, and thus $s \le 1$.

Step two: Let $\mathcal{G}$ be the set of the pairs (g, F) where $E \subset F$, $g\colon F \to \mathbb{R}^m$, $g = f$ on E and $\mathrm{Lip}(g; F) \le \mathrm{Lip}(f; E)$. We set $(g_1, F_1) \preccurlyeq (g_2, F_2)$ if $F_1 \subset F_2$ and $g_2 = g_1$ on F_1. By the Hausdorff maximal principle, there exists a maximal element $(g, F) \in \mathcal{G}$ with respect to the ordering $\preccurlyeq$. If F were a proper subset of $\mathbb{R}^n$ then, by step one, (g, F) would not be maximal. Hence, $F = \mathbb{R}^n$. □

7.2 Weak gradients

The elementary Gauss–Green formula (apply Exercise 1.12 to $u\,\varphi$),

$$\int_{\mathbb{R}^n} \varphi \nabla u = -\int_{\mathbb{R}^n} u \nabla \varphi , \qquad \forall u \in C^1(\mathbb{R}^n), \varphi \in C^1_c(\mathbb{R}^n), \tag{7.2}$$

motivates the introduction of the **distributional gradient** Du of a function $u \in L^1_{\mathrm{loc}}(\mathbb{R}^n)$, as the linear functional $Du\colon C^\infty_c(\mathbb{R}^n) \to \mathbb{R}^n$,

$$\langle Du, \varphi \rangle = -\int_{\mathbb{R}^n} u \, \nabla \varphi , \qquad \varphi \in C^\infty_c(\mathbb{R}^n) .$$

Whenever Du is representable as integration of the test function φ against an L^1_{loc} vector field, that is, if there exists a vector field $T \in L^1_{\mathrm{loc}}(\mathbb{R}^n;\mathbb{R}^n)$ such that

$$\int_{\mathbb{R}^n} u \, \nabla \varphi = -\int_{\mathbb{R}^n} \varphi T , \qquad \forall \varphi \in C^\infty_c(\mathbb{R}^n), \tag{7.3}$$

we say that u has a **weak gradient** on $\mathbb{R}^n$. By Exercise 4.14, the weak gradient is uniquely determined as an element of $L^1_{\mathrm{loc}}(\mathbb{R}^n;\mathbb{R}^n)$. We denote it by ∇u as a classical gradient, since if $u \in C^1(\mathbb{R}^n)$, then the classical gradient ∇u of u is a weak gradient of u. Although it is very likely that our readers will be familiar with the basic properties of weak gradients, we shall gather in this chapter the few basic results that we are going to need in the sequel. The **local p-Sobolev space on $\mathbb{R}^n$**, $W^{1,p}_{\mathrm{loc}}(\mathbb{R}^n)$, consists of those functions $u \in L^p_{\mathrm{loc}}(\mathbb{R}^n)$ that admit a weak gradient $\nabla u \in L^p_{\mathrm{loc}}(\mathbb{R}^n;\mathbb{R}^n)$ $(1 \le p \le \infty)$. If $u \in W^{1,1}_{\mathrm{loc}}(\mathbb{R}^n)$ and $u_\varepsilon = u \star \rho_\varepsilon$, then, by definition (7.3) of weak gradient,

$$\nabla u_\varepsilon = (\nabla u) \star \rho_\varepsilon . \tag{7.4}$$

In particular, Exercise 5.18 gives $\nabla u_\varepsilon \to \nabla u$ in $L^1_{\mathrm{loc}}(\mathbb{R}^n;\mathbb{R}^n)$ for every $u \in W^{1,1}_{\mathrm{loc}}(\mathbb{R}^n)$. The following lemma is often useful.

Lemma 7.5 (Vanishing weak gradient) *If $u \in L^1_{\mathrm{loc}}(\mathbb{R}^n)$, A is open and connected, and*

$$\int_{\mathbb{R}^n} u \, \nabla \varphi = 0 , \qquad \forall \varphi \in C^\infty_c(A) ,$$

then there exists $c \in \mathbb{R}$ such that $u = c$ a.e. in A.

Proof We directly discuss the case $A = \mathbb{R}^n$. If $u_\varepsilon = u \star \rho_\varepsilon$, then

$$\nabla u_\varepsilon(x) = \int_{\mathbb{R}^n} u(y) \nabla \rho_\varepsilon(y - x) \mathrm{d}y = 0 , \qquad \forall x \in \mathbb{R}^n .$$

Since u_ε is smooth, there exists $c_\varepsilon \in \mathbb{R}$ such that $u_\varepsilon = c_\varepsilon$ on $\mathbb{R}^n$. As $u_\varepsilon \to u$ in $L^1_{\mathrm{loc}}(\mathbb{R}^n)$, there exists $c \in \mathbb{R}$ such that $c_\varepsilon \to c$. Thus, $u = c$ a.e. on $\mathbb{R}^n$. □

We say that $f\colon \mathbb{R} \to \mathbb{R}$ is a **piecewise affine function** if f is continuous and if there is a partition of $\mathbb{R}$ into finitely many intervals such that f is affine on each interval of the partition. Note that, if f is piecewise affine, then there exists a finite set F such that $f'(s)$ exists for every $s \in \mathbb{R} \setminus F$.

Lemma 7.6 (Chain rule for weak gradients) *If $f\colon \mathbb{R} \to \mathbb{R}$ is piecewise affine and $u \in W^{1,1}_{\rm loc}(\mathbb{R}^n)$, then $\nabla u = 0$ a.e. on $u^{-1}(F)$ and $(f \circ u) \in W^{1,1}_{\rm loc}(\mathbb{R}^n)$ with $\nabla(f \circ u) = (f' \circ u)\,\nabla u$.*

Proof *Step one:* We prove the lemma for $f \in C^1(\mathbb{R})$ with $C = \sup_{\mathbb{R}} |f'| < \infty$. Since $|f(s)| \le |f(0)| + C|s|$, we find $(f \circ u) \in L^1_{\rm loc}(\mathbb{R}^n)$ and $(f' \circ u)\,\nabla u \in L^1_{\rm loc}(\mathbb{R}^n; \mathbb{R}^n)$. Moreover, since $f \circ u_\varepsilon \in C^1(\mathbb{R}^n)$, we have

$$\int_{\mathbb{R}^n} (f \circ u_\varepsilon)\nabla\varphi = -\int_{\mathbb{R}^n} \varphi(f' \circ u_\varepsilon)\nabla u_\varepsilon\,, \qquad \forall \varphi \in C^\infty_{\rm c}(\mathbb{R}^n).$$

We let $\varepsilon \to 0^+$ and apply dominated convergence to conclude that

$$\int_{\mathbb{R}^n} (f \circ u)\nabla\varphi = -\int_{\mathbb{R}^n} \varphi(f' \circ u)\nabla u\,, \qquad \forall \varphi \in C^\infty_{\rm c}(\mathbb{R}^n).$$

Step two: We show that if $u \in W^{1,1}_{\rm loc}(\mathbb{R}^n)$ and $t \in \mathbb{R}$, then $\nabla u = 0$ a.e. on $\{u = t\} = \{x \in \mathbb{R}^n : u(x) = t\}$. We can assume $t = 0$. For every $\sigma > 0$, let us set

$$f_\sigma(s) = \max\left\{0, \sqrt{\sigma^2 + s^2} - \sigma\right\}, \qquad s \in \mathbb{R}.$$

Then $f_\sigma \in C^1(\mathbb{R})$ with $|f_\sigma{}'| \le 1$, $f_\sigma(s) \to s^+ = \max\{s, 0\}$ uniformly on $s \in \mathbb{R}$, and $f_\sigma{}'(s) \to 1_{(0,\infty)}(s)$ for every $s \in \mathbb{R}$. By step one

$$\int_{\mathbb{R}^n} (f_\sigma \circ u)\nabla\varphi = -\int_{\mathbb{R}^n} \varphi(f_\sigma{}' \circ u)\nabla u\,, \qquad \forall \varphi \in C^\infty_{\rm c}(\mathbb{R}^n).$$

By dominated convergence, as $\sigma \to 0$,

$$\int_{\mathbb{R}^n} u^+\nabla\varphi = -\int_{\mathbb{R}^n} \varphi(1_{(0,\infty)} \circ u)\nabla u\,, \qquad \forall \varphi \in C^\infty_{\rm c}(\mathbb{R}^n).$$

Thus $u^+ = \max\{u, 0\} \in W^{1,1}_{\rm loc}(\mathbb{R}^n)$, with $\nabla u^+ = 1_{\{u>0\}}\,\nabla u$. Similarly, $u^- = \max\{-u, 0\} \in W^{1,1}_{\rm loc}(\mathbb{R}^n)$, with $\nabla u^- = -1_{\{u<0\}}\,\nabla u$. Since $u = u^+ - u^-$, by linearity and a.e. uniqueness of weak gradients we find

$$\nabla u(x) = \left(1_{\{u>0\}}(x) + 1_{\{u<0\}}(x)\right)\nabla u(x)\,, \qquad \text{for a.e. } x \in \mathbb{R}^n.$$

Thus, $\nabla u = 0$ a.e. on $\{u = 0\}$.

Step three: Let f be piecewise affine and let F be the finite set of points where f is not differentiable. It is easily seen that $(f \circ u) \in L^1_{\rm loc}(\mathbb{R}^n)$ and that $(f' \circ u)\nabla u \in L^1_{\rm loc}(\mathbb{R}^n)$. By step two, $\nabla u = 0$ a.e. on $u^{-1}(F)$. Moreover, on suitably adapting the above argument, we see that $(f' \circ u)\nabla u$ is the weak gradient of $f \circ u$. □

7.3 Rademacher's theorem

We now apply the differentiation theory of Radon measures to prove that Lipschitz functions are a.e. differentiable and that they admit bounded weak gradients. Given $u\colon \mathbb{R}^n \to \mathbb{R}$ and $\tau \in S^{n-1}$ we define the **incremental ratios of** u **in the direction** τ, $\tau_h u\colon \mathbb{R}^n \to \mathbb{R}$ $(h \neq 0)$, by setting

$$\tau_h u(x) = \frac{u(x + h\,\tau) - u(x)}{h}\,, \qquad x \in \mathbb{R}^n\,.$$

By translation invariance of the Lebesgue measure, it is easily seen that

$$\int_{\mathbb{R}^n} u(x + h\,\tau)\,v(x)\,\mathrm{d}x = \int_{\mathbb{R}^n} u(x)\,v(x - h\tau)\,\mathrm{d}x\,,$$

if $u \in L^1_{\mathrm{loc}}(\mathbb{R}^n)$ and v is bounded and Borel measurable with compact support. Correspondingly, we have the following Gauss–Green type formula:

$$\int_{\mathbb{R}^n} v\,\tau_h u = -\int_{\mathbb{R}^n} u\,\tau_{-h} v\,. \tag{7.5}$$

We now prove that Lipschitz functions admit bounded weak gradients.

Proposition 7.7 (Weak gradient of a Lipschitz function) *If $f\colon \mathbb{R}^n \to \mathbb{R}^m$ is a Lipschitz function, then $f \in L^\infty_{\mathrm{loc}}(\mathbb{R}^n;\mathbb{R}^m)$ and f admits a weak gradient $\nabla f \in L^\infty(\mathbb{R}^n;\mathbb{R}^m \otimes \mathbb{R}^n)$.*

Proof It suffices to consider the case $m = 1$. Let $\tau \in S^{n-1}$ and let $h(k) \to 0^+$ as $k \to \infty$. The sequence $\{\tau_{h(k)} f\}_{k\in\mathbb{N}}$ is bounded in $L^\infty(\mathbb{R}^n)$, with $\|\tau_{h(k)} f\|_{L^\infty(\mathbb{R}^n)} \le \mathrm{Lip}(f)$ for every $k \in \mathbb{N}$. By Exercise 5.15, there exists $g_\tau \in L^\infty(\mathbb{R}^n)$ such that

$$\int_{\mathbb{R}^n} v\,g_\tau = \lim_{k\to\infty} \int_{\mathbb{R}^n} v\,\tau_{h(k)} f\,, \qquad \forall v \in L^1(\mathbb{R}^n)\,,$$

and $\|g_\tau\|_{L^\infty(\mathbb{R}^n)} \le \mathrm{Lip}(f)$. By (7.5), and since $\tau_{-h}\varphi \to \tau \cdot \nabla\varphi$ uniformly on $\mathbb{R}^n$,

$$\int_{\mathbb{R}^n} \varphi\,g_\tau = -\int_{\mathbb{R}^n} f\,(\tau \cdot \nabla\varphi)\,, \qquad \forall \varphi \in C^\infty_{\mathrm{c}}(\mathbb{R}^n)\,. \tag{7.6}$$

Defining $T \in L^\infty(\mathbb{R}^n;\mathbb{R}^n)$ by $T^{(i)} = g_{e_i}$ $(1 \le i \le n)$ we thus conclude that

$$\int_{\mathbb{R}^n} \varphi\,T = -\int_{\mathbb{R}^n} f\,\nabla\varphi\,, \qquad \forall \varphi \in C^\infty_{\mathrm{c}}(\mathbb{R}^n)\,. \qquad \square$$

Let us finally recall that $f\colon \mathbb{R}^n \to \mathbb{R}^m$ is **differentiable** at $x \in \mathbb{R}^n$ if there exists a linear map $\mathrm{d}f_x \in \mathbb{R}^m \otimes \mathbb{R}^n$ such that

$$\lim_{h\to 0} \frac{f(x + h\,\tau) - f(x)}{h} \to \mathrm{d}f_x\,[\tau]\,,$$

uniformly on $\tau \in S^{n-1}$. The map $\mathrm{d}f_x$ is called the **differential of** f **at** x.

Theorem 7.8 (Rademacher's theorem) *If $f\colon \mathbb{R}^n \to \mathbb{R}^m$ is a Lipschitz function and x is a Lebesgue point of the weak gradient ∇f, then f is differentiable at x (in particular, f is differentiable a.e. on $\mathbb{R}^n$), with*

$$\mathrm{d}f_x[\tau] = \nabla f(x)[\tau], \qquad \forall \tau \in \mathbb{R}^n .$$

Proof Let x be a Lebesgue point of ∇f, and define $g_h\colon \mathbb{R}^n \to \mathbb{R}^m$ ($h \neq 0$) as

$$g_h(\tau) = \frac{f(x + h\tau) - f(x)}{h}, \qquad \tau \in \mathbb{R}^n .$$

Then $\{g_h\}_{h\neq 0}$ is a family of Lipschitz functions on $\mathbb{R}^n$, with

$$g_h(0) = 0, \qquad \mathrm{Lip}(g_h) \leq \mathrm{Lip}(f), \qquad \nabla g_h(\tau) = \nabla f(x + h\tau), \quad \forall \tau \in \mathbb{R}^n .$$

As x is a Lebesgue point of ∇f, $\nabla g_h \to \nabla f(x)$ in $L^1(B; \mathbb{R}^m \otimes \mathbb{R}^n)$ for $h \to 0$, as

$$0 = \lim_{h\to 0} \frac{1}{h^n} \int_{B(x,h)} |\nabla f(z) - \nabla f(x)|\,\mathrm{d}z = \lim_{h\to 0} \int_B |\nabla g_h - \nabla f(x)| .$$

Now let $\{h(k)\}_{k\in\mathbb{N}}$ be such that $h(k) \to 0$ as $k \to \infty$. By the Ascoli–Arzelá theorem, there exist a Lipschitz function $g\colon \mathbb{R}^n \to \mathbb{R}^m$ and a subsequence $\overline{h}(k) \to 0$ as $k \to \infty$ such that $g_{\overline{h}(k)} \to g$ uniformly on compact sets of $\mathbb{R}^n$. We claim that

$$g(0) = 0, \qquad \nabla g(\tau) = \nabla f(x)[\tau] \quad \text{for a.e. } \tau \in B.$$

Indeed, since $\nabla g_h \to \nabla f(x)$ in $L^1(B; \mathbb{R}^m \otimes \mathbb{R}^n)$ as $h \to 0$, we find

$$\begin{aligned} -\int_{\mathbb{R}^n} \varphi \nabla g &= \int_{\mathbb{R}^n} g \nabla \varphi = \lim_{k\to\infty} \int_{\mathbb{R}^n} g_{\overline{h}(k)} \nabla \varphi = -\lim_{k\to\infty} \int_{\mathbb{R}^n} \varphi \nabla g_{\overline{h}(k)} \\ &= -\int_{\mathbb{R}^n} (\varphi(\tau)\, \nabla f(x)\,[\tau])\,\mathrm{d}\tau, \qquad \forall \varphi \in C^\infty_{\mathrm{c}}(B). \end{aligned}$$

If we finally set

$$g_0(\tau) = g(\tau) - \nabla f(x)\,[\tau], \qquad \tau \in \mathbb{R}^n,$$

then g_0 is a Lipschitz function with $g_0(0) = 0$ and $\int_{\mathbb{R}^n} g_0 \nabla \varphi = 0$ for every $\varphi \in C^\infty_{\mathrm{c}}(B)$. By Lemma 7.5, $g_0 = 0$ on B, that is, $g(\tau) = \nabla f(x)[\tau]$ for $\tau \in B$. By the arbitrariness of $\{h(k)\}_{k\in\mathbb{N}}$, we have thus proved that, as $h \to 0$, $g_h(\tau) \to \nabla f(x)[\tau]$ uniformly on $\tau \in B$. Thus, f is differentiable at x, with $\mathrm{d}f_x[\tau] = \nabla f(x)[\tau]$ for every $\tau \in \mathbb{R}^n$. □

8

Area formula

As already noticed in Chapter 7, a useful general principle about Lipschitz functions is that they possess the basic metric properties of C^1-functions. In this chapter we discuss the so-called area formula, better known in the C^1-framework as the change of variable formula. In its simplest incarnation, the area formula deals with an injective Lipschitz function $f\colon \mathbb{R}^n \to \mathbb{R}^m$, where $1 \le n \le m$. Given $E \subset \mathbb{R}^n$, by Proposition 3.5, $f(E)$ is (at most) n-dimensional in $\mathbb{R}^m$. The area formula provides a way to express $\mathcal{H}^n(f(E))$ in terms of integration over E of the **Jacobian** of f, that is defined as the bounded Borel function $Jf\colon \mathbb{R}^n \to [0,\infty]$,

$$Jf(x) = \begin{cases} \sqrt{\det(\nabla f(x)^* \nabla f(x))}\,, & \text{if } f \text{ is differentiable at } x\,; \\ +\infty\,, & \text{if } f \text{ is not differentiable at } x\,. \end{cases}$$

We notice that, with our convention, $\{x \in \mathbb{R}^n : Jf(x) < \infty\}$ coincides with the set of points $x \in \mathbb{R}^n$ at which f is differentiable. Thus, by Rademacher's theorem, $\{Jf < \infty\}$ has full Lebesgue measure in $\mathbb{R}^n$.

Theorem 8.1 (Area formula for injective maps) *If $f\colon \mathbb{R}^n \to \mathbb{R}^m$ $(1 \le n \le m)$ is an injective Lipschitz function and $E \subset \mathbb{R}^n$ is Lebesgue measurable, then*

$$\mathcal{H}^n\big(f(E)\big) = \int_E Jf(x)\,\mathrm{d}x\,, \tag{8.1}$$

and $\mathcal{H}^n \llcorner f(\mathbb{R}^n)$ is a Radon measure on $\mathbb{R}^m$.

Remark 8.2 If $n = 1$, then $\nabla f(x) = f'(x) \in \mathbb{R}^m$, $Jf(x) = |f'(x)|$ and Theorem 8.1 is a consequence of Theorem 3.8. If $f(E)$ is an n-dimensional C^1-image in $\mathbb{R}^m$, then Theorem 8.1 ensures that $\mathcal{H}^k(f(E))$ agrees with the classical notion of n-dimensional measure of $f(E)$; see (1.1).

Remark 8.3 If $g\colon \mathbb{R}^m \to [-\infty,\infty]$ is Borel measurable on $\mathbb{R}^m$ and either $g \ge 0$ or $g \in L^1(\mathbb{R}^m, \mathcal{H}^n \llcorner f(\mathbb{R}^n))$, then $g \circ f$ is Borel measurable on $\mathbb{R}^n$ and

$$\int_{f(\mathbb{R}^n)} g \, \mathrm{d}\mathcal{H}^n = \int_{\mathbb{R}^n} g(f(x))\, Jf(x)\, \mathrm{d}x\,. \tag{8.2}$$

Indeed, if $g \ge 0$, then $g = \sum_{h\in\mathbb{N}} c_h 1_{F_h}$, where $c_h \ge 0$ and $F_h \in \mathcal{B}(\mathbb{R}^m)$, $h \in \mathbb{N}$. Setting $E_h = f^{-1}(F_h)$, then $g \circ f = \sum_{h\in\mathbb{N}} c_h 1_{E_h}$ and, by (8.1),

$$\int_{\mathbb{R}^m} g \, \mathrm{d}\mathcal{H}^n = \sum_{h\in\mathbb{N}} c_h \mathcal{H}^n(F_h) = \sum_{h\in\mathbb{N}} c_h \int_{E_h} Jf = \int_{\mathbb{R}^n} (g \circ f)\, Jf\,.$$

If $g \in L^1(\mathbb{R}^m, \mathcal{H}^n \llcorner f(\mathbb{R}^n))$, then it suffices to notice that $g = g^+ - g^-$.

Before embarking on the proof of Theorem 8.1, which is presented in the following sections, we settle a simple measurability issue arising in (8.1).

Lemma 8.4 *If E is a Lebesgue measurable set in $\mathbb{R}^n$ and $f\colon \mathbb{R}^n \to \mathbb{R}^m$ ($1 \le n \le m$) is a Lipschitz function, then $f(E)$ is $\mathcal{H}^n$-measurable in $\mathbb{R}^m$.*

Proof We can assume that E is bounded, so that $|E| < \infty$. As E is Lebesgue measurable, there exist compact sets $\{K_h\}_{h\in\mathbb{N}}$ such that $K_h \subset E$ and $|E \setminus K_h| \to 0$. Since $f(K_h)$ is compact, the set $\bigcup_{h\in\mathbb{N}} f(K_h)$ is a Borel set. The $\mathcal{H}^n$-measurability of $f(E)$ thus follows by Proposition 3.5 and Theorem 3.10, as

$$\mathcal{H}^n\Big(f(E) \setminus \bigcup_{h\in\mathbb{N}} f(K_h)\Big) \le \mathcal{H}^n\Big(f\Big(E \setminus \bigcup_{h\in\mathbb{N}} K_h\Big)\Big) \le \mathrm{Lip}(f)^n \Big| E \setminus \bigcup_{h\in\mathbb{N}} K_h \Big| = 0\,. \qquad \square$$

8.1 Area formula for linear functions

In this section we prove the area formula for linear functions. With Notation 3 in force, we now introduce some results from Linear Algebra which will prove useful in our discussion. A linear function $T \in \mathbb{R}^m \otimes \mathbb{R}^n$ is an **orthogonal injection** if $(Tv)\cdot(Tw) = v \cdot w$ for every $v, w \in \mathbb{R}^n$. We denote by $\mathbf{O}(n,m)$ the set of orthogonal injections and set $\mathbf{O}(n) = \mathbf{O}(n,n)$. If $T \in \mathbf{O}(n,m)$, then $\ker T = \{0\}$; hence, $\mathbf{O}(n,m) = \emptyset$ if $n > m$. We also note that

$$\|T\| = \mathrm{Lip}(T) = 1\,, \qquad \forall T \in \mathbf{O}(n,m)\,.$$

If $T \in \mathbf{O}(n,m)$, then $T^*T = \mathrm{Id}_{\mathbb{R}^n}$. Therefore a linear function belonging to

$$\mathbf{O}^*(m,n) = \left\{T^* : T \in \mathbf{O}(n,m)\right\}$$

is called an **orthogonal projection**. If T^* is an orthogonal projection, then its kernel coincides with $T(\mathbb{R}^n)^\perp$ and $\|T^*\| = \mathrm{Lip}(T^*) = 1$. Moreover,

$$|T^*v - T^*w| = |v - w|\,, \qquad \forall v, w \in T(\mathbb{R}^n)\,.$$

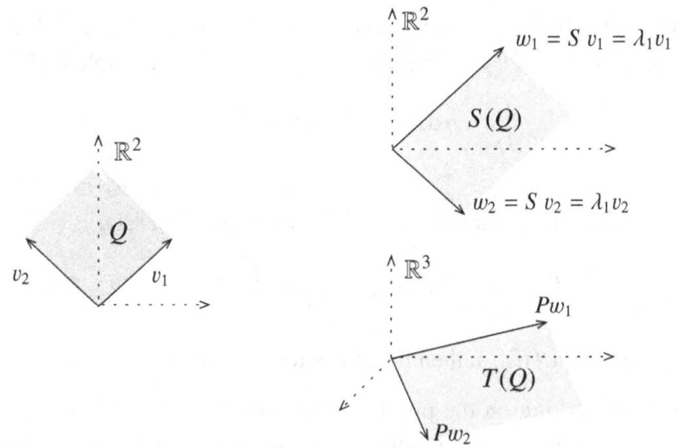

Figure 8.1 The polar decomposition theorem for a map $T \in \mathbb{R}^3 \otimes \mathbb{R}^2$. In this case $\lambda_1 = 3/2$ and $\lambda_2 = -1$. Note that $\mathcal{H}^2(T(Q)) = (3/2)|Q| = 3/2$ and that $JT = |\det S| = 3/2$.

The space $\mathbf{Sym}(n)$ of **symmetric linear functions** on $\mathbb{R}^n$ is defined by those $T \in \mathbb{R}^n \otimes \mathbb{R}^n$ such that $T^* = T$ (if, instead, $T^* = -T$, then T is called **antisymmetric**). We recall from Linear Algebra that, given $T \in \mathbf{Sym}(n)$, the **spectral theorem** ensures the existence of an orthonormal basis $\{v_i\}_{i=1}^n$ of $\mathbb{R}^n$ such that

$$T = \sum_{i=1}^{n} \lambda_i \, v_i \otimes v_i \,, \tag{8.3}$$

where $\lambda_i = v_i \cdot T v_i$. The action of orthogonal and symmetric linear functions is easily visualized. If $T \in \mathbf{O}(n,m)$, then T embeds in a linear and isometric way $\mathbb{R}^n$ into $\mathbb{R}^m$. If $T \in \mathbf{Sym}(n)$, then T acts on $\mathbb{R}^n$ by deforming the (mutually orthogonal) lines spanned by the v_i by the (possibly degenerate) factors λ_i. In turn, the **polar decomposition theorem** provides a clean way to visualize every linear function $T \in \mathbb{R}^m \otimes \mathbb{R}^n$ $(n \le m)$ as the composition of a symmetric linear function on $\mathbb{R}^n$, followed by an orthogonal injection of $\mathbb{R}^n$ into $\mathbb{R}^m$:

$$T = PS \,, \tag{8.4}$$

where $S \in \mathbf{Sym}(n)$ and $P \in \mathbf{O}(n,m)$; see Figure 8.1. Note that, if $n = m$ and $T \in \mathbf{GL}(n)$, then necessarily $S \in \mathbf{GL}(n)$. The polar decomposition theorem follows from the spectral theorem as follows. Since $T^*T \in \mathbf{Sym}(n)$, with $T^*Tv \cdot v \ge 0$ for every $v \in \mathbb{R}^n$, by the spectral theorem there exists $\{v_i\}_{i=1}^n$, an orthonormal basis of $\mathbb{R}^n$, such that $T^*T = \sum_{i=1}^n \lambda_i v_i \otimes v_i$ with $\lambda_i \ge 0$. If we set

$\sqrt{T^*T} = \sum_{i=1}^n \sqrt{\lambda_i}\, v_i \otimes v_i$, $I = \{i : \lambda_i > 0\}$ and define

$$w_i = \frac{T v_i}{\sqrt{\lambda_i}} \in \mathbb{R}^m, \qquad i \in I,$$

then, by construction, $\{w_i\}_{i\in I}$ is an orthonormal basis of $\mathbb{R}^m$. We complete $\{w_i\}_{i\in I}$ into an orthonormal basis $\{w_i\}_{i=1}^m$ of $\mathbb{R}^m$, and, finally, we define $P \in \mathbf{O}(n,m)$ by setting $Pv_i = w_i$, $1 \le i \le n$. It is straightforward to check that $T = P\sqrt{T^*T}$, thus proving the polar decomposition theorem. We now prove the area formula in the case of a linear map.

Theorem 8.5 *If $T \in \mathbb{R}^m \otimes \mathbb{R}^n$ $(1 \le n \le m)$, then for every $E \subset \mathbb{R}^n$*

$$\mathcal{H}^n(T(E)) = JT\,|E|. \tag{8.5}$$

Proof We shall prove the theorem by showing that

$$\mathcal{H}^n(T(E)) = \frac{\mathcal{H}^n(T(B))}{|B|}|E|, \qquad \forall E \subset \mathbb{R}^n, \tag{8.6}$$

$$JT = \frac{\mathcal{H}^n(T(B))}{|B|}. \tag{8.7}$$

For the sake of brevity, we set $\kappa = \mathcal{H}^n(T(B))/|B|$.

Step one: We prove (8.6). First, consider the case $\kappa = 0$. By definition of κ, and by linearity of T, for every $r > 0$ we have $\mathcal{H}^n(T(B_r)) = 0$. Hence $\mathcal{H}^n(T(\mathbb{R}^n)) = 0$, thus $\mathcal{H}^n(T(E)) = 0$ for every $E \subset \mathbb{R}^n$, and (8.6) holds true. We now let $\kappa > 0$, so that T is injective, and define an outer measure ν on $\mathbb{R}^n$,

$$\nu(E) = \mathcal{H}^n(T(E)), \qquad E \subset \mathbb{R}^n.$$

By Lemma 8.4 and Proposition 2.13, $\mathcal{H}^n \llcorner T(\mathbb{R}^n)$ is a Radon measure on $\mathbb{R}^m$. Since T maps compact sets into compect sets, and $\nu = (T^{-1})_\#(\mathcal{H}^n \llcorner T(\mathbb{R}^n))$, by Proposition 2.14, ν is a Radon measure on $\mathbb{R}^n$. By linearity and by definition of κ,

$$\begin{aligned}\nu(B(x,r)) &= \mathcal{H}^n(Tx + r\,T(B)) = \mathcal{H}^n(r\,T(B))\\ &= r^n\mathcal{H}^n(T(B)) = r^n\kappa\,|B| = \kappa|B(x,r)|, \qquad \forall x \in \mathbb{R}^n, r > 0.\end{aligned}$$

Thus $\nu \ll \mathcal{L}^n$, with $D_{\mathcal{L}^n}\nu = \kappa$ on $\mathbb{R}^n$. By the Lebesgue–Besicovitch differentiation theorem, $\nu = \kappa\,\mathcal{L}^n$ on $\mathcal{M}(\mathcal{L}^n)$. By Exercise 2.6, $\nu = \kappa\,\mathcal{L}^n$ on $\mathcal{P}(\mathbb{R}^n)$.

Step two: We prove (8.7). Let $T = PS$ as in (8.4). Since $P \in \mathbf{O}(n,m)$, we have $P^*P = \mathrm{Id}_{\mathbb{R}^n}$, $\mathrm{Lip}(P) = 1$, and $\mathrm{Lip}(P^*) = 1$. By Proposition 3.5, if $E \subset \mathbb{R}^n$,

$$|E| = |P^*P(E)| \le \mathrm{Lip}(P^*)^n\mathcal{H}^n(P(E)) \le \mathcal{H}^n(P(E)) \le \mathrm{Lip}(P)^n|E| = |E|,$$

that is $|E| = \mathcal{H}^n(P(E))$. In particular, if $Q \subset \mathbb{R}^n$ and we set $E = S(Q)$, then

$$\kappa = \frac{\mathcal{H}^n(T(Q))}{|Q|} = \frac{\mathcal{H}^n(P(E))}{|E|}\frac{|E|}{|Q|} = \frac{|S(Q)|}{|Q|}. \tag{8.8}$$

If $S = \sum_{i=1}^n \lambda_i v_i \otimes v_i$ is the spectral decomposition of S, then the cube

$$Q = \left\{x \in \mathbb{R}^n : |x \cdot v_i| \le \frac{1}{2}\right\},$$

with unit side lengths and faces orthogonal to the v_i, is mapped by S into

$$S(Q) = \left\{x \in \mathbb{R}^n : |x \cdot v_i| \le \frac{|\lambda_i|}{2}\right\},$$

a parallel cube with (possibly degenerate) side lengths given by the $|\lambda_i|$. Hence,

$$|S(Q)| = \prod_{i=1}^n |\lambda_i| = |\det S| = |\det S|\,|Q|,$$

and $\kappa = |\det S|$ by (8.8). We finally note that $T^* = S^*P^*$, $P^*P = \mathrm{Id}_{\mathbb{R}^n}$ and $S^*S = \sum_{i=1}^n \lambda_i^2 v_i \otimes v_i$, so that

$$JT = \sqrt{\det(T^*T)} = \sqrt{\det(S^*S)} = \left(\prod_{i=1}^n \lambda_i^2\right)^{1/2} = \prod_{i=1}^n |\lambda_i| = |\det S| = \kappa. \qquad \square$$

Remark 8.6 If $T \in \mathbf{GL}(n)$, then $(JT)J(T^{-1}) = J(TT^{-1}) = J(\mathrm{Id}_{\mathbb{R}^n}) = 1$. In particular $JT > 0$ and $J(T^{-1}) = (JT)^{-1}$. Let us also remark that

$$\|T^{-1}\|^{-n} \le JT \le \|T\|^n, \qquad \forall T \in \mathbf{GL}(n). \tag{8.9}$$

Indeed, by Proposition 3.5, we have $\mathcal{H}^n(T(B)) \le \|T\|^n|B|$, while, at the same time, $|B| = |T^{-1}(T(B))| \le \|T^{-1}\|^n\mathcal{H}^n(T(B))$. By a similar argument, if $n \le m$, $T_1 \in \mathbb{R}^m \otimes \mathbb{R}^n$ and $T_2 \in \mathbf{GL}(m)$, then

$$\|T_2^{-1}\|^{-n} JT_1 \le J(T_2T_1) \le \|T_2\|^n\, JT_1. \tag{8.10}$$

8.2 The role of the singular set $Jf = 0$

We now prove that the singular set $\{Jf = 0\}$ is mapped by f into an $\mathcal{H}^n$-negligible set. This is, of course, a necessary condition for (8.1) to hold.

Theorem 8.7 *If $f\colon \mathbb{R}^n \to \mathbb{R}^m$ ($1 \le n \le m$) is a Lipschitz function, then*

$$\mathcal{H}^n(f(E)) = 0,$$

where $E = \{x \in \mathbb{R}^n : Jf(x) = 0\}$.

Proof Given $k \in \mathbb{N}$, let us set $B_r^k = \{z \in \mathbb{R}^k : |z| < r\}$ for the ball of radius r and center at the origin in $\mathbb{R}^k$. If $F \subset \mathbb{R}^m$, then we set $I_\varepsilon(F) = \{x \in \mathbb{R}^m : \mathrm{dist}(x, F) < \varepsilon\}$ for the ε-neighborhood of F in $\mathbb{R}^m$.

Step one: Let D_s be a k-dimensional disk in $\mathbb{R}^m$ of radius $s > 0$, say

$$D_s = \left\{(z, y) \in \mathbb{R}^k \times \mathbb{R}^{m-k} : |z| < s\,, y = 0\right\}.$$

We claim that, if $1 \le k \le n-1$, then

$$\mathcal{H}^n_\infty(I_\delta(D_s)) \le C(n, s)\,\delta\,, \qquad \forall \delta \in (0,1)\,, \tag{8.11}$$

(where, of course, $C(n, s) \approx s^n$ as $s \to \infty$). Indeed, if we set

$$K = \left\{(z, y) \in \mathbb{R}^k \times \mathbb{R}^{m-k} : |z| < \delta s\,, |y| < \delta\right\} = B^k_{\delta s} \times B^{m-k}_\delta\,,$$

then there exists a finite covering $\mathcal{F}$ of $I_\delta(D_s)$ such that each $F \in \mathcal{F}$ is a translation of K, and the cardinality of $\mathcal{F}$ is bounded from above by $C\,\delta^{-k}$, for some positive constant C. Moreover, if $F \in \mathcal{F}$, then

$$\mathrm{diam}(F)^2 = \mathrm{diam}(K)^2 = \mathrm{diam}(B^k_{\delta s})^2 + \mathrm{diam}(B^{m-k}_\delta)^2 = 4\,\delta^2(1 + s^2)\,.$$

Since $n - k \ge 1$ and $\delta \in (0, 1)$, we conclude that

$$\mathcal{H}^n_\infty(I_\delta(D_s)) \le \omega_n \sum_{F \in \mathcal{F}} \left(\frac{\mathrm{diam}(F)}{2}\right)^n \le C(n)\,(1 + s^2)^{n/2}\,\delta^{n-k} \le C(n, s)\,\delta\,.$$

Step two: If $x \in E$, so that $Jf(x) = 0$, then $L_x = \nabla f(x)(\mathbb{R}^n)$ is a linear subspace of $\mathbb{R}^m$, with $k = \dim(L_x) \le n - 1 < m$. If $k \ge 1$, then $\nabla f(x)(B^n_r)$ is contained into a k-dimensional disk of radius $\mathrm{Lip}(f)\,r$ in $\mathbb{R}^m$ for every $r > 0$, that is

$$\nabla f(x)(B^n_r) \subset B^m_{\mathrm{Lip}(f)\,r} \cap L_x\,, \qquad \forall r > 0\,.$$

Hence, by (8.11), for every $\varepsilon \in (0, 1)$ and $r > 0$ we find

$$\begin{aligned}\mathcal{H}^n_\infty\Big(I_{\varepsilon r}(\nabla f(x)(B^n_r))\Big) &\le \mathcal{H}^n_\infty\Big(I_{\varepsilon r}(B^m_{\mathrm{Lip}(f)\,r} \cap L_x)\Big) = r^n \mathcal{H}^n_\infty\Big(I_\varepsilon(B^m_{\mathrm{Lip}(f)} \cap L_x)\Big)\\ &\le C(n, \mathrm{Lip}(f))\,r^n\,\varepsilon\,.\end{aligned}$$

If $k = 0$, then $\nabla f(x)(\mathbb{R}^n) = \{0\}$, and for every $\varepsilon \in (0, 1)$ and $r > 0$ we find

$$\mathcal{H}^n_\infty\Big(I_{\varepsilon r}(\nabla f(x)(B^n_r))\Big) = \mathcal{H}^n_\infty(B^m_{\varepsilon r}) \le \omega_n r^n \varepsilon^n \le \omega_n\,r^n\,\varepsilon\,.$$

Step three: If $x \in E$, and $\varepsilon \in (0, 1)$, then, as f is differentiable on E, there exists $r(\varepsilon, x) \in (0, 1)$ such that

$$|f(x + v) - f(x) - \nabla f(x)v| \le \varepsilon |\,v|\,,$$

whenever $|\,v| < r(\varepsilon, x)$. In particular, for every $r < r(\varepsilon, x)$ we have that

$$f(B^n(x, r)) \subset f(x) + I_{\varepsilon r}(\nabla f(x)(B^n_r))\,.$$

Since $Jf(x) = 0$, by step two we find that, if $r < r(\varepsilon, x)$, then

$$\mathcal{H}^n_\infty\Big(f(B^n(x,r)\Big) \le C(n, \mathrm{Lip}(f))\,\varepsilon r^n\,. \tag{8.12}$$

Given $R > 0$, the family of open balls

$$\mathcal{F} = \Big\{B^n(x,r) : x \in E \cap B^n_R\,, 0 < r < r(\varepsilon, x)\Big\}$$

has the bounded set $E \cap B^n_R$ as the set of its centers. Let $\{\mathcal{F}_i\}_{i=1}^{\xi(n)}$ be the subfamilies of $\mathcal{F}$ given by Besicovitch's covering theorem (see Remark 5.6). Since $E \cap B^n_R \subset \bigcup_{i=1}^{\xi(n)} \mathcal{F}_i$, with $\mathcal{F}_i$ countable and disjoint, by (8.12),

$$\begin{aligned}
\mathcal{H}^n_\infty\Big(f(E \cap B^n_R)\Big) &\le \sum_{i=1}^{\xi(n)} \sum_{B^n(x,r)\in\mathcal{F}_i} \mathcal{H}^n_\infty\Big(f(B^n(x,r))\Big) \le C\,\varepsilon \sum_{i=1}^{\xi(n)} \sum_{B^n(x,r)\in\mathcal{F}_i} r^n \\
&= \frac{C\,\varepsilon}{\omega_n} \sum_{i=1}^{\xi(n)} \sum_{B^n(x,r)\in\mathcal{F}_i} |B^n(x,r)| = \frac{C\,\varepsilon}{\omega_n} \sum_{i=1}^{\xi(n)} \Big| \bigcup_{B^n(x,r)\in\mathcal{F}_i} B^n(x,r) \Big| \\
&\le \frac{C\,\xi(n)\,\varepsilon}{\omega_n} \Big| I_1(E \cap B^n_R) \Big|\,,
\end{aligned}$$

where in the last inequality we have used the fact that $r(\varepsilon, x) \in (0,1)$, and where $C = C(n, \mathrm{Lip}(f))$. For $\varepsilon \to 0^+$, we find that $\mathcal{H}^n_\infty(f(E \cap B^n_R)) = 0$. By Proposition 3.4, $\mathcal{H}^n(f(E \cap B^n_R)) = 0$. We let $R \to \infty$ to conclude. □

8.3 Linearization of Lipschitz immersions

In this section we prove an important theorem concerning Lipschitz immersions, which will find its first application in the proof of (8.1), and will also play a crucial role in developing the theory of rectifiable sets. A basic technique from basic Calculus is to carry properties of linear functions to C^1 functions by exploiting the continuity of gradients to infer they are locally almost constant. Although this procedure makes no sense on Lipschitz functions, a beautiful idea due to Federer allows us to reformulate approximation via linear functions in this framework too. Roughly speaking, the idea is to fix a dense set $\{T_h\}_{h\in\mathbb{N}}$ in $\mathbb{R}^m \otimes \mathbb{R}^n$, to choose a parameter of approximation $\delta > 0$, and then to look at the Borel sets of those $x \in \mathbb{R}^n$ such that $\nabla f(x)$ is δ-close to a fixed T_h. In this way, we obtain a countable partition of the space such that ∇f is almost constant on each set in the partition. We now state this idea precisely, "linearizing" a Lipschitz function f on the set $\{Jf > 0\}$. Notice that, if $Jf(x) > 0$, then $\nabla f(x) = P_x S_x$, where $S_x \in \mathbf{Sym}(n) \cap \mathbf{GL}(n)$ and $P_x \in \mathbf{O}(n,m)$.

Theorem 8.8 (Lipschitz linearization) *If $f \colon \mathbb{R}^n \to \mathbb{R}^m$ $(1 \le n \le m)$ is a Lipschitz function, and*

$$F = \left\{ x \in \mathbb{R}^n : 0 < Jf(x) < \infty \right\},$$

then there exists a partition of F into Borel sets $\{F_h\}_{h \in \mathbb{N}}$ such that f is injective on each F_h. Moreover, for every $t > 1$, such a partition can be found with the property that, for every $h \in \mathbb{N}$, there exists an invertible linear function $S_h \in \mathbf{GL}(n)$ such that $f|_{F_h} \circ S_h^{-1}$ is almost an isometry of $\mathbb{R}^n$ into $\mathbb{R}^m$. Precisely, for every $x, y \in F_h$ and $v \in \mathbb{R}^n$,

$$t^{-1} | S_h x - S_h y | \le |f(x) - f(y)| \le t | S_h x - S_h y| \,, \tag{8.13}$$

$$t^{-1} | S_h v| \le |\nabla f(x) v| \le t | S_h v| \,, \tag{8.14}$$

$$t^{-n} J S_h \le Jf(x) \le t^n J S_h \,. \tag{8.15}$$

Proof It suffices to show that F can be *covered* by sets F_h having the properties stated above: indeed, once this has been done, we can replace each F_h with $F_h \setminus \bigcup_{k=1}^{h-1} F_k$ in order to define the desired *partition* of F. We also recall that, as is easily checked, if $T, S \in \mathbf{GL}(n)$ and $\|T - S\| \le \delta$, then

$$\|TS^{-1}\| \le 1 + \delta \|S^{-1}\| \,, \qquad \|ST^{-1}\| \le 1 + \delta \|T^{-1}\| \,. \tag{8.16}$$

We now choose $\varepsilon > 0$ so that $t^{-1} + \varepsilon < 1 < t - \varepsilon$, and a dense set (in the operator norm) $\mathcal{G}$ in $\mathbf{GL}(n)$. Correspondingly, for every $h \in \mathbb{N}$ and $S \in \mathcal{G}$, we define

$$F(S, h)$$

as the set of those $x \in F$ such that

$$(t^{-1} + \varepsilon) | S v| \le |\nabla f(x)\, v| \le (t - \varepsilon) | S v| \,, \qquad \forall v \in \mathbb{R}^n \,, \tag{8.17}$$

$$|f(x + v) - f(x) - \nabla f(x) v| \le \varepsilon | S v| \,, \qquad \forall v \in \mathbb{R}^n \,, |v| \le \frac{1}{h} \,. \tag{8.18}$$

Note that (8.17) implies automatically that

$$(t^{-1} + \varepsilon)^n JS \le Jf(x) \le (t - \varepsilon)^n JS \,, \qquad \forall x \in \textstyle\bigcup_{h \in \mathbb{N}} F(S, h) \,. \tag{8.19}$$

Indeed, as $S \in \mathbf{GL}(n)$, for every such x we have that

$$B_{t^{-1} + \varepsilon} \subset \nabla f(x) (S^{-1}(B)) \subset B_{t + \varepsilon} \,,$$

and thus, as required, $(t^{-1} + \varepsilon)^n \le J(\nabla f(x) S^{-1}) \le (t + \varepsilon)^n$. (Equivalently, one may argue as in the proof of (8.9).) Another relevant property of the sets $F(S, h)$ is that if $x, y \in F(S, h)$ and $|x - y| < h^{-1}$, then

$$|f(y) - f(x)| \le |\nabla f(x)(y - x)| + \varepsilon | S y - S x| \le t | S y - S x| \,, \tag{8.20}$$

$$|f(y) - f(x)| \ge |\nabla f(x)(y - x)| - \varepsilon | S y - S x| \ge t^{-1} | S y - S x| \,. \tag{8.21}$$

If now $\{x_j\}_{j\in\mathbb{N}}$ is a dense subset of F, and we relabel the sequence of sets

$$F(S,h) \cap B\left(x_j, \frac{1}{2h}\right), \qquad S \in \mathcal{G}, h, j \in \mathbb{N},$$

as $\{F_k\}_{k\in\mathbb{N}}$, then, by (8.19), (8.20), and (8.21), we see that (8.13) and (8.15) hold true on each F_k. We also notice that (8.14) holds trivially on each F_k by (8.17). We are left to prove that $F = \bigcup_{S\in\mathcal{G},h\in\mathbb{N}} F(S,h)$. Let $x \in F$, and consider the polar decomposition $\nabla f(x) = P_x S_x$. As $Jf(x) > 0$, we have $S_x \in \mathbf{GL}(n)$. In particular, by (8.16), we can find $S \in \mathcal{G}$ with

$$\|S_x S^{-1}\| \le t - \varepsilon, \quad \|S(S_x)^{-1}\| \le (t^{-1} + \varepsilon)^{-1}.$$

In this way we ensure that

$$|S_x v| \le (t - \varepsilon)|Sv|, \quad (t^{-1} + \varepsilon)|Sv| \le |S_x v|, \qquad \forall v \in \mathbb{R}^n,$$

that is (8.17), since $|\nabla f(x)v| = |P_x S_x v| = |S_x v|$. Concerning (8.18), the differentiability of f at x implies the existence of a modulus of continuity ω_x such that, whenever $|v| < h^{-1}$,

$$|f(x+v) - f(x) - \nabla f(x)v| \le \omega_x(h^{-1})|v| \le \omega_x(h^{-1})\|S^{-1}\|\,|Sv|.$$

We choose $h = h(x,S)$ so that $\omega_x(h^{-1})\|S^{-1}\| \le \varepsilon$, and prove (8.18). □

8.4 Proof of the area formula

We prove (8.1). Because $\mathcal{H}^n(f(E)) \le \mathrm{Lip}(f)^n|E|$ for every $E \subset \mathbb{R}^n$, both sides of (8.1) are zero whenever $\mathcal{H}^n(E) = 0$. Therefore, by Rademacher's theorem, we can reduce to proving (8.1) on a set E over which f is differentiable. Moreover, by Theorem 8.7, we can directly assume that

$$E \subset F = \left\{x \in \mathbb{R}^n : 0 < Jf(x) < \infty\right\}. \tag{8.22}$$

We now fix $t > 1$ and consider the partition $\{F_k\}_{k\in\mathbb{N}}$ of F given by Theorem 8.8. We see E as the union of the disjoint sets $F_k \cap E$, $k \in \mathbb{N}$, so that, by the global injectivity of f, we have that $f(E)$ is the disjoint union of the $\mathcal{H}^n$-measurable sets $f(F_k \cap E)$, $k \in \mathbb{N}$. Therefore, by Proposition 3.5 and the linear case of the area formula (8.5), we find that

$$\begin{aligned}\mathcal{H}^n(f(E)) &= \sum_{k\in\mathbb{N}} \mathcal{H}^n\big(f(F_k \cap E)\big) = \sum_{k\in\mathbb{N}} \mathcal{H}^n\Big(\big(f|_{F_k} \circ S_k^{-1}\big)\big(S_k(F_k \cap E)\big)\Big)\\ &\le \sum_{k\in\mathbb{N}} \mathrm{Lip}\big((f|_{F_k}) \circ S_k^{-1}\big)^n\, |S_k(F_k \cap E)|\\ &\le t^n \sum_{k\in\mathbb{N}} JS_k\, |F_k \cap E|\\ &\le t^{2n} \sum_{k\in\mathbb{N}} \int_{F_k\cap E} Jf(x)\,dx = t^{2n} \int_E Jf(x)\,dx\,, \end{aligned} \tag{8.23}$$

where we have also applied the fact that, thanks to the upper bound in (8.13), the Lipschitz norm of $f|_{F_k} \circ S_k^{-1}$ over $S_k(F_k)$ is controlled by t. In a similar way, the lower bound in (8.13) implies that the Lipschitz norm of $S_k \circ (f|_{F_k})^{-1}$ over $f(F_k)$ is controlled by t, so that, by an analogous argument,

$$\begin{aligned}\int_E Jf(x)\,dx &= \sum_{k\in\mathbb{N}} \int_{E\cap F_k} Jf(x)\,dx \le t^n \sum_{k\in\mathbb{N}} JS_k\, |E \cap F_k|\\ &= t^n \sum_{k\in\mathbb{N}} \Big|[S_k \circ (f|_{F_k})^{-1}](f(E \cap F_k))\Big|\\ &\le t^{2n} \sum_{k\in\mathbb{N}} \mathcal{H}^n(f(E \cap F_k)) = t^{2n}\mathcal{H}^n(f(E))\,. \end{aligned} \tag{8.24}$$

We thus prove (8.1) by letting $t \to 1^+$ in (8.23) and (8.24). By Lemma 8.4, $f(\mathbb{R}^n)$ is $\mathcal{H}^n$-measurable, while (8.1) implies $\mathcal{H}^n \llcorner f(\mathbb{R}^n)$ to be locally finite. By Proposition 2.13, $\mathcal{H}^n \llcorner f(\mathbb{R}^n)$ is a Radon measure.

8.5 Area formula with multiplicities

The area formula (8.1) was proved for injective Lipschitz maps. Its right-hand side $\int_E Jf$ is clearly well defined even when f is not injective, and in this case it can still be expressed as an integration over $f(E)$, provided we take into account the *multiplicity function* M of f over E, $M : \mathbb{R}^m \to \mathbb{N} \cup \{+\infty\}$,

$$M(y) = \mathcal{H}^0\big(E \cap \{f = y\}\big)\,, \qquad y \in \mathbb{R}^m\,,$$

where $\{f = y\} = \{x \in \mathbb{R}^n : f(x) = y\}$. Notice that $f(E) = \{M > 0\} = \{M \ge 1\}$.

Theorem 8.9 (Area formula with multiplicities) *If $f : \mathbb{R}^n \to \mathbb{R}^m$ ($1 \le n \le m$) is a Lipschitz map, and E is a Lebesgue measurable set of $\mathbb{R}^n$, then the*

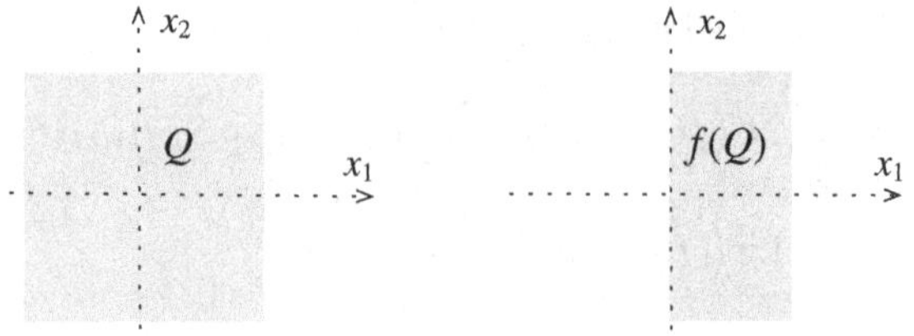

Figure 8.2 A local isometry f which folds $\mathbb{R}^2$ onto $\{x_1 > 0\}$: hence, $f(\mathbb{R}^2) = \{x_1 \geq 0\}$ and $\mathcal{H}^0(f^{-1}(x)) = 2$ for every $x \in \{x_2 > 0\}$.

multiplicity function M of f over E is $\mathcal{H}^n$-measurable on $\mathbb{R}^m$, and

$$\int_{\mathbb{R}^m} \mathcal{H}^0\big(E \cap \{f = y\}\big)\, \mathrm{d}\mathcal{H}^n(y) = \int_E Jf(x)\, \mathrm{d}x\,. \tag{8.25}$$

Remark 8.10 The role of the multiplicity function M in (8.25) is to compensate for possible "overlap effects" in the image of f. This is easily understood if we look at local isometries. A Lipschitz function $f\colon \mathbb{R}^n \to \mathbb{R}^m$ is a **local isometry** provided $\nabla f(x) \in \mathbf{O}(n, m)$ for a.e. $x \in \mathbb{R}^n$ (note that, necessarily, $1 \leq n \leq m$, and that isometries, defined in Remark 3.7 are local isometries). If f is a local isometry, then $Jf = 1$ a.e. on $\mathbb{R}^n$, and the right-hand side of (8.25) is equal to $|E|$. Thus, for an injective local isometry f,

$$\mathcal{H}^n(f(E)) = |E|\,.$$

However, local isometries are not necessarily injective. The function $f\colon \mathbb{R}^2 \to \mathbb{R}^2$,

$$f(x_1, x_2) = \begin{cases} (x_1, x_2)\,, & \text{if } x_1 > 0\,, \\ (-x_1, x_2)\,, & \text{if } x_1 \leq 0\,, \end{cases}$$

defines a local isometry of $\mathbb{R}^2$ into $\mathbb{R}^2$. If $Q = (-1, 1)^2$, then $|Q| = 4$, but $\mathcal{H}^2(f(Q)) = 2 \neq |Q| = \int_Q Jf$. In this case, the multiplicity function of f over Q is constantly equal to 2, and thus (8.25) holds true; see Figure 8.2.

Proof *Step one:* We prove that M is $\mathcal{H}^n$-measurable. Indeed, let $\mathcal{Q}_k$ be the standard diadic partition of $\mathbb{R}^n$ by half-open/half-closed cubes of side length 2^{-k}, and let $M_k\colon \mathbb{R}^m \to \mathbb{N} \cup \{+\infty\}$ be defined as

$$M_k(y) = \sum_{Q \in \mathcal{Q}_k} 1_{f(E \cap Q)}(y)\,, \qquad y \in \mathbb{R}^m\,.$$

By Lemma 8.4, $f(E \cap Q)$ is $\mathcal{H}^n$-measurable in $\mathbb{R}^m$ for every $Q \in \mathcal{Q}_k$, and thus M_k is $\mathcal{H}^n$-measurable on $\mathbb{R}^m$. Moreover, $M_k(y)$ equals the number of elements of $\mathcal{Q}_k$ containing a point $x \in E$ which is mapped into y by f. Thus, $M_k \leq M_{k+1}$, $M_k \to M$ pointwise on $\mathbb{R}^n$ as $k \to \infty$, and M is $\mathcal{H}^n$-measurable.

Step two: We show that, for every Lebesgue measurable set $E \subset \mathbb{R}^n$,

$$\int_{\mathbb{R}^m} \mathcal{H}^0\big(E \cap \{f = y\}\big)\, d\mathcal{H}^n(y) \le \mathrm{Lip}(f)^n |E|\,. \tag{8.26}$$

(As a consequence, in proving (8.25) we may freely modify E on subsets of measure zero.) To prove (8.26), notice that $\sum_{Q\in\mathcal{Q}_k} |E \cap Q| = |E|$ for every $k \in \mathbb{N}$, and, by monotone convergence and Proposition 3.5,

$$\int_{\mathbb{R}^m} M\, \mathrm{d}\mathcal{H}^n = \lim_{k\to\infty} \int_{\mathbb{R}^m} M_k\, \mathrm{d}\mathcal{H}^n = \lim_{k\to\infty} \sum_{Q\in\mathcal{Q}_k} \mathcal{H}^n\big(f(E\cap Q)\big)$$
$$\le \mathrm{Lip}(f)^n \liminf_{k\to\infty} \sum_{Q\in\mathcal{Q}_k} |E\cap Q| = \mathrm{Lip}(f)^n|E|\,.$$

Step three: We now prove (8.25). By step two and Theorem 8.7, we can directly assume that $E \subset F$, where $F = \{x \in \mathbb{R}^n : 0 < Jf(x) < \infty\}$. By Theorem 8.8, there exists a partition $\{F_k\}_{k\in\mathbb{N}}$ of F such that f is injective on each F_k. Then $\{E \cap F_k\}_{k\in\mathbb{N}}$ is a partition of E such that f is injective on each $E \cap F_k$, and by (8.1) we find that

$$\int_{\mathbb{R}^m} \mathcal{H}^0\big(E \cap \{f = y\}\big)\, \mathrm{d}\mathcal{H}^n(y) = \int_{\mathbb{R}^m} \sum_{k\in\mathbb{N}} \mathcal{H}^0\big(E \cap F_k \cap \{f = y\}\big)\, \mathrm{d}\mathcal{H}^n(y)$$
$$= \sum_{k\in\mathbb{N}} \mathcal{H}^n\big(f(E\cap F_k)\big) = \sum_{k\in\mathbb{N}} \int_{E\cap F_k} Jf(x)\, \mathrm{d}x$$
$$= \int_E Jf(x)\, \mathrm{d}x\,.$$

□

Corollary 8.11 *If $f\colon \mathbb{R}^n \to \mathbb{R}^m$ $(n \le m)$ is a Lipschitz function, $g\colon \mathbb{R}^n \to [-\infty,\infty]$ is a Borel function, and either $g \ge 0$ or $g \in L^1(\mathbb{R}^m, \mathcal{H}^n \llcorner f(E))$, then*

$$\int_{\mathbb{R}^m} \Big(\int_{\{f=y\}} g\, \mathrm{d}\mathcal{H}^0\Big) \mathrm{d}\mathcal{H}^n(y) = \int_{\mathbb{R}^n} g(x)\, Jf(x)\, \mathrm{d}x\,. \tag{8.27}$$

Proof If, for example, g is non-negative, there exists a sequence of Borel sets $\{E_h\}_{h\in\mathbb{N}}$ and $\{c_h\}_{h\in\mathbb{N}} \subset (0,\infty)$ such that $g = \sum_{h\in\mathbb{N}} c_h\, 1_{E_h}$. By (8.25),

$$\int_{\mathbb{R}^n} g(x)\, Jf(x)\, \mathrm{d}x = \sum_{h\in\mathbb{N}} c_h \int_{E_h} Jf(x)\, \mathrm{d}x$$
$$= \sum_{h\in\mathbb{N}} c_h \int_{\mathbb{R}^m} \mathcal{H}^0\big(E_h \cap \{f = y\}\big)\, \mathrm{d}\mathcal{H}^n(y)$$
$$= \int_{\mathbb{R}^m} \sum_{h\in\mathbb{N}} \Big(\int_{\{f=y\}} c_h\, 1_{E_h} d\mathcal{H}^0\Big) \mathrm{d}\mathcal{H}^n(y)$$
$$= \int_{\mathbb{R}^m} \Big(\int_{\{f=y\}} g d\mathcal{H}^0\Big) d\mathcal{H}^n(y)\,.$$

□

Exercise 8.12 If $f\colon \mathbb{R}^n \to \mathbb{R}^m$ $(1 \le n \le m)$ is a Lipschitz function, E is a Lebesgue measurable set in $\mathbb{R}^n$, and F is a Borel set in $\mathbb{R}^m$, then

$$f_{\#}(\mathcal{L}^n \llcorner E)(F) = \left|f^{-1}(F) \cap E \cap \{Jf = 0\}\right| + \int_{F \cap f(E \cap \{Jf>0\})} \left(\int_{\{f=y\}} \frac{d\mathcal{H}^0(x)}{Jf(x)} \right) d\mathcal{H}^n(y),$$

In particular, if f is injective and $Jf = 1$ a.e. on E, then we have

$$f_{\#}(\mathcal{L}^n \llcorner E) = \mathcal{H}^n \llcorner f(E) \qquad \text{on } \mathcal{P}(\mathbb{R}^n).$$

9

Gauss–Green theorem

The classical Gauss–Green theorem on open sets with C^1-boundary plays a fundamental role in the theory of sets of finite perimeter. We shall build its proof on a nice application of the area formula to codimension one graphs (Section 9.1), and then generalize it to the case of open sets whose boundaries fail to be of class C^1 due to the presence of an $\mathcal{H}^{n-1}$-negligible set (Section 9.3). **Throughout this chapter we use Notation 4.**

9.1 Area of a graph of codimension one

Given $u\colon \mathbb{R}^{n-1} \to \mathbb{R}$ and $G \subset \mathbb{R}^{n-1}$, we define the **graph of** u **over** G as

$$\Gamma(u; G) = \Big\{x \in \mathbb{R}^n : \mathbf{q}x = u(\mathbf{p}x)\,, \mathbf{p}x \in G\Big\}\,,$$

and set for brevity $\Gamma(u) = \Gamma(u; \mathbb{R}^{n-1})$. As a simple consequence of the area formula we find the following theorem.

Theorem 9.1 (Area of a graph of codimension one) *If $u\colon \mathbb{R}^{n-1} \to \mathbb{R}$ is a Lipschitz function, then for every Lebesgue measurable set G in $\mathbb{R}^{n-1}$,*

$$\mathcal{H}^{n-1}\Big(\Gamma(u; G)\Big) = \int_G \sqrt{1 + |\nabla' u(z)|^2}\, dz\,. \tag{9.1}$$

In fact, $\mathcal{H}^{n-1} \llcorner \Gamma(u)$ is a Radon measure on $\mathbb{R}^n$, for every $\varphi \in C^0_c(\mathbb{R}^n)$,

$$\int_{\Gamma(u)} \varphi\, d\mathcal{H}^{n-1} = \int_{\mathbb{R}^{n-1}} \varphi(z, u(z)) \sqrt{1 + |\nabla' u(z)|^2}\, dz\,. \tag{9.2}$$

Proof If $v \neq 0$, then $v = |v| w_1$, $|w_1| = 1$, and, introducing an orthonormal basis $\{w_i\}_{i=1}^n$ of $\mathbb{R}^n$, we find $\mathrm{Id} + v \otimes v = (1 + |v|^2) w_1 \otimes w_1 + \sum_{i=2}^n w_i \otimes w_i$; thus

$$\det(\mathrm{Id} + v \otimes v) = 1 + |v|^2\,, \qquad \forall v \in \mathbb{R}^n\,. \tag{9.3}$$

Now let $f\colon \mathbb{R}^{n-1} \to \mathbb{R}^n$ be the injective Lipschitz function defined as

$$f(z) = (z, u(z)), \qquad z \in \mathbb{R}^{n-1}.$$

Since $\Gamma(u; G) = f(G)$ for every $G \subset \mathbb{R}^n$, $\mathcal{H}^{n-1} \llcorner \Gamma(u)$ is a Radon measure by Theorem 8.1. We now show that $Jf = \sqrt{1 + |\nabla' u|^2}$ on $\mathbb{R}^{n-1}$, so that (9.1) and (9.2) will follow by (8.1) and (8.2) respectively. To this end, we compute

$$\nabla f = \sum_{i=1}^{n-1} \Big(e_i + (\partial_i u)\, e_n \Big) \otimes e_i ,$$

and recall that $(a \otimes b)(c \otimes d) = (b \cdot c)\,(a \otimes d)$ for $a, b, c, d \in \mathbb{R}^n$, to find

$$\begin{aligned}(\nabla f)^*(\nabla f) &= \Big(\sum_{i=1}^{n-1} e_i \otimes (e_i + (\partial_i u)\, e_n) \Big)\Big(\sum_{j=1}^{n-1} \Big(e_j + (\partial_j u)\, e_n\Big) \otimes e_j \Big) \\ &= \sum_{i,j=1}^{n-1} \Big(\delta_{i,j} + (\partial_i u)(\partial_j u)\Big)\, e_i \otimes e_j = \mathrm{Id} + (\nabla' u) \otimes (\nabla' u).\end{aligned}$$

By (9.3), we conclude that $Jf = \sqrt{1 + |\nabla' u|^2}$, as desired. □

9.2 Gauss–Green theorem on open sets with C^1-boundary

Let E be an open set in $\mathbb{R}^n$ and let $k \in \mathbb{N} \cup \{\infty\}$, $k \ge 1$. We say that E has **C^k-boundary** (or **smooth boundary**, if $k = \infty$) if for every $x \in \partial E$ there exist $r > 0$ and $\psi \in C^k(B(x, r))$ with $\nabla\psi(y) \neq 0$ for every $y \in B(x, r)$ and

$$B(x, r) \cap E = \Big\{ y \in B(x, r) : \psi(y) < 0 \Big\}, \tag{9.4}$$

$$B(x, r) \cap \partial E = \Big\{ y \in B(x, r) : \psi(y) = 0 \Big\}. \tag{9.5}$$

The **outer unit normal** ν_E to E is then defined locally as

$$\nu_E(y) = \frac{\nabla\psi(y)}{|\nabla\psi(y)|}, \qquad \forall y \in B(x, r) \cap \partial E .$$

This definition is independent of the choice of ψ and r, therefore ν_E can be considered as a vector field on the whole ∂E, with $\nu_E \in C^{k-1}(\partial E; S^{n-1})$.

Remark 9.2 If E is an open set with C^1-boundary, then $\mathcal{H}^{n-1} \llcorner \partial E$ is a Radon measure on $\mathbb{R}^n$. Indeed, by the implicit function theorem, if $x \in \partial E$ and $r > 0$ is the same as in (9.4) and (9.5), then there exist $s > 0$ and a function $u \in C^1(\mathbf{D}(\mathbf{p}x, s))$ such that $\mathbf{C}(x, s) \subset B(x, r)$ and, up to rotation,

$$\mathbf{C}(x, s) \cap E = \Big\{ y \in \mathbf{C}(x, s) : \mathbf{q}y > u(\mathbf{p}y) \Big\}, \tag{9.6}$$

$$\mathbf{C}(x, s) \cap \partial E = \Big\{ y \in \mathbf{C}(x, s) : \mathbf{q}y = u(\mathbf{p}y) \Big\}. \tag{9.7}$$

Hence, $\mathcal{H}^{n-1} \llcorner (\mathbf{C}(x,s) \cap \partial E) = \mathcal{H}^{n-1} \llcorner \Gamma(u; \mathbf{D}(\mathbf{p}x, s))$, where the right-hand side defines a Radon measure on $\mathbb{R}^n$ by Theorem 9.1. Starting from this remark it is easily seen that $\mathcal{H}^{n-1} \llcorner \partial E$ is a Radon measure on $\mathbb{R}^n$. Let us also notice that, having expressed $\mathbf{C}(x,s) \cap E$ as the epigraph of u over $\mathbf{D}(\mathbf{p}x, s)$, by the chain rule we infer the following formula for the outer unit normal ν_E of E:

$$\nu_E(y) = \frac{(\nabla' u(\mathbf{p}y), -1)}{\sqrt{1 + |\nabla' u(\mathbf{p}y)|^2}}, \qquad \forall y \in \mathbf{C}(x,s) \cap \partial E. \tag{9.8}$$

Theorem 9.3 (Gauss–Green theorem) *If E is an open set with C^1-boundary, then for every $\varphi \in C^1_c(\mathbb{R}^n)$,*

$$\int_E \nabla \varphi(x) \mathrm{d}x = \int_{\partial E} \varphi \, \nu_E \, \mathrm{d}\mathcal{H}^{n-1}. \tag{9.9}$$

Equivalently, the divergence theorem holds true:

$$\int_E \operatorname{div} T(x) \mathrm{d}x = \int_{\partial E} T \cdot \nu_E \, \mathrm{d}\mathcal{H}^{n-1}, \qquad \forall T \in C^1_c(\mathbb{R}^n; \mathbb{R}^n). \tag{9.10}$$

Exercise 9.4 (Perimeter and volume of the unit ball) $\mathcal{H}^{n-1}(S^{n-1}) = n\omega_n$.

Proof of Theorem 9.3 *Step one:* Given $x \in \partial E$, up to rotation, we may consider $r, s > 0$ and u as in Remark 9.2. We claim that

$$\int_E \nabla \varphi = \int_{\partial E} \varphi \nu_E \, \mathrm{d}\mathcal{H}^{n-1}, \qquad \forall \varphi \in C^1_c(\mathbf{C}(x,s)). \tag{9.11}$$

Indeed, given $\delta > 0$, we define a Lipschitz function $f_\delta \colon \mathbf{C}(x,s) \to \mathbb{R}$ by setting

$$f_\delta(y) = \begin{cases} 1, & \text{if } \mathbf{q}y > u(\mathbf{p}y) + \delta, \\ 0, & \text{if } \mathbf{q}y < u(\mathbf{p}y) - \delta, \\ (2\delta)^{-1}(\mathbf{q}y - u(\mathbf{p}y) + \delta), & \text{if } |u(\mathbf{p}y) - \mathbf{q}y| < \delta. \end{cases}$$

Since $f_\delta \to 1_{\mathbf{C}(x,s)\cap E}$ in $L^1(\mathbf{C}(x,s))$ as $\delta \to 0^+$, by Proposition 7.7

$$\int_E \nabla \varphi = \int_{E \cap \mathbf{C}(x,s)} \nabla \varphi = \lim_{\delta \to 0^+} \int_{\mathbf{C}(x,s)} f_\delta \nabla \varphi = -\lim_{\delta \to 0^+} \int_{\mathbf{C}(x,s)} \varphi \nabla f_\delta. \tag{9.12}$$

Let us now set (see Figure 9.1)

$$F_\delta = \left\{ y \in \mathbf{C}(x,s) : |\mathbf{q}y - u(\mathbf{p}y)| < \delta \right\},$$

and notice that $F_\delta = \{y \in \mathbf{C}(x,s) : \nabla f_\delta \neq 0\}$, with

$$\nabla f_\delta(y) = (2\delta)^{-1}\big(-\nabla' u(\mathbf{p}y), 1 \big), \qquad \forall y \in F_\delta.$$

By Fubini's theorem we have

$$\int_{\mathbf{C}(x,s)} \varphi \nabla f_\delta = \int_{F_\delta} \varphi \nabla f_\delta = \int_{\mathbf{D}(\mathbf{p}x,s)} \big(-\nabla' u(z), 1 \big) \left(\frac{1}{2\delta} \int_{u(z)-\delta}^{u(z)+\delta} \varphi(z,t) \, \mathrm{d}t \right) \mathrm{d}z.$$

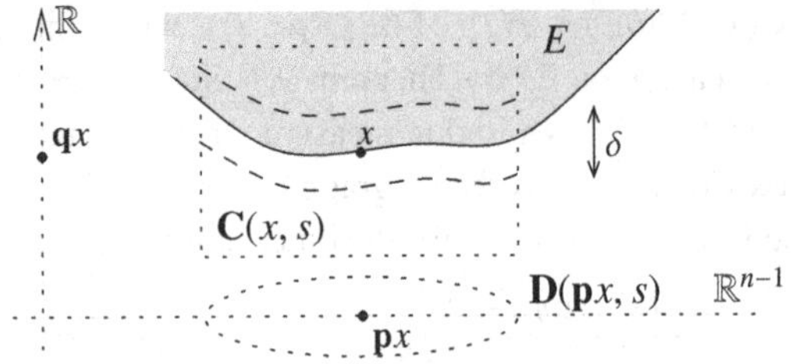

Figure 9.1 Proof of the Gauss–Green theorem. The region enclosed by the dashed curves is the set F_δ.

By continuity, for every $z \in \mathbf{D}(\mathbf{p}x, s)$,

$$\lim_{\delta\to 0^+} \frac{1}{2\delta}\int_{u(z)-\delta}^{u(z)+\delta} \varphi(z,t)\mathrm{d}t = \varphi(z,u(z))\,.$$

Finally, by dominated convergence, (9.8), and Theorem 9.1

$$\begin{aligned}
-\lim_{\delta\to 0^+}\int_{\mathbf{C}(x,s)} \varphi\nabla f_\delta &= -\int_{\mathbf{D}(\mathbf{p}x,s)} \varphi(z,u(z))\Big(-\nabla' u(z),1\Big)\,\mathrm{d}z\\
&= \int_{\mathbf{D}(\mathbf{p}x,s)} \varphi(z,u(z))\,\nu_E(z,u(z))\,\sqrt{1+|\nabla' u(z)|^2}\,\mathrm{d}z\\
&= \int_{\mathbf{C}(x,s)\cap\partial E} \varphi\,\nu_E\,\mathrm{d}\mathcal{H}^{n-1} = \int_{\partial E}\varphi\nu_E\,\mathrm{d}\mathcal{H}^{n-1}\,.
\end{aligned}$$

Step two: We conclude the proof by a standard argument based on partitions of unity. Let $\varphi \in C^1_c(\mathbb{R}^n)$ be given, and let A be an open set such that $\operatorname{spt}\varphi\cap\partial E\subset A$. By compactness and thanks to step one, there exist finitely many points $\{x_k\}_{k=1}^N \subset A\cap\partial E$ and finitely many open balls $\{B(x_k,s_k)\}_{k=1}^N$ with $B(x_k,s_k)\subset A$ such that, for every $\zeta\in C^1_c(B(x_k,s_k))$, $1\le k\le N$,

$$\operatorname{spt}\varphi\cap\partial E\subset\bigcup_{k=1}^N B(x_k,s_k)\,,\qquad \int_E \nabla(\zeta\varphi) = \int_{\partial E}\zeta\varphi\nu_E\,\mathrm{d}\mathcal{H}^{n-1}\,.$$

We now consider $\{\zeta_k\}_{k=1}^N$ with $\zeta_k\in C^1_c(B(x_k,s_k);[0,1])$, and $\sum_{k=1}^N\zeta_k = 1$ on A, and then construct $\zeta_0\in C^1_c(E;[0,1])$ such that $\sum_{k=0}^N\zeta_k = 1$ on $E\cup A$. Since $\zeta_0\varphi\in C^1_c(E)$ we have $0=\int_{\mathbb{R}^n}\nabla(\zeta_0\varphi) = \int_E\nabla(\zeta_0\varphi)$. Hence, by step one,

$$\int_E\nabla\varphi = \sum_{k=0}^N\int_E\nabla(\zeta_k\varphi) = \sum_{k=1}^N\int_{\partial E}\zeta_k\varphi\nu_E\,\mathrm{d}\mathcal{H}^{n-1} = \int_{\partial E}\varphi\nu_E\,\mathrm{d}\mathcal{H}^{n-1}\,. \qquad \square$$

Remark 9.5 (Open sets with Lipschitz or polyhedral boundary) An open set $E\subset\mathbb{R}^n$ has **Lipschitz boundary** if for every $x\in\partial E$ there exist $s>0$ and a Lipschitz function $u\colon \mathbb{R}^{n-1} \to \mathbb{R}$ such that (9.6) and (9.7) hold true; it has **polyhedral boundary** if every such u is finitely piecewise affine. If $G \subset \mathbb{R}^{n-1}$ is the set of points of differentiability of u, and $f(z) = (z, u(z))$, $z \in \mathbb{R}^{n-1}$, then we may define a unit vector field $\nu_E\colon f(G) \to \mathbb{R}^n$ by setting

$$\nu_E(z, u(z)) = \frac{(\nabla' u(z), -1)}{\sqrt{1 + |\nabla' u(z)|^2}}\,, \qquad z \in G\,, \tag{9.13}$$

where $\mathcal{H}^{n-1}((\mathbf{C}(x, s) \cap \partial E) \setminus f(G)) = 0$ by Rademacher's theorem and the area formula (8.1). Moreover, the same argument used in step one of the proof of Theorem 9.3 shows that (9.11) holds true with this definition of ν_E. By looking at a locally finite covering of ∂E, we thus conclude that if E is an open set with Lipschitz boundary, then $\mathcal{H}^{n-1} \llcorner \partial E$ is a Gauss–Green measure, and the Gauss–Green formula (9.9) holds true on E with a unit vector field ν_E which is defined $\mathcal{H}^{n-1}$-a.e. on ∂E, and which locally satisfies (9.13).

9.3 Gauss–Green theorem on open sets with almost C^1-boundary

We now prove a useful generalization of the Gauss–Green theorem. An open set $E \subset \mathbb{R}^n$ has **almost C^1-boundary** if there exists a closed set $M_0 \subset \partial E$, with

$$\mathcal{H}^{n-1}(M_0) = 0\,, \tag{9.14}$$

such that, for every $x \in M = \partial E \setminus M_0$, there exist $s > 0$ and $\psi \in C^1(B(x, s))$ with the property that

$$B(x, s) \cap E = \big\{y \in B(x, s) : \psi(y) < 0\big\}\,, \tag{9.15}$$

$$B(x, s) \cap \partial E = B(x, s) \cap M = \big\{y \in B(x, s) : \psi(y) = 0\big\}\,. \tag{9.16}$$

We call M the **regular part** of ∂E (note that M is a C^1-hypersurface; see page 96). The outer unit normal to E is defined as a continuous vector field $\nu_E \in C^0(M; S^{n-1})$, through the local representations

$$\nu_E(y) = \frac{\nabla\psi(y)}{|\nabla\psi(y)|}\,, \qquad y \in B(x, s) \cap M\,. \tag{9.17}$$

Theorem 9.6 *If E is an open set in $\mathbb{R}^n$ with almost C^1-boundary, and M is the regular part of ∂E, then for every $\varphi \in C^1_c(\mathbb{R}^n)$*

$$\int_E \nabla\varphi = \int_M \varphi\,\nu_E \, d\mathcal{H}^{n-1}\,. \tag{9.18}$$

Proof By (9.15), (9.16), the implicit function theorem, Theorem 9.1, and Proposition 2.13, we see that $\mathcal{H}^{n-1} \llcorner M$ is a Radon measure on $\mathbb{R}^n$. Next, given $x \in M$ and $s > 0$ such that (9.15) and (9.16) hold true, then, by repeating the proof of the Gauss–Green theorem,

$$\int_E \nabla\varphi = \int_M \varphi \nu_E \, d\mathcal{H}^{n-1}, \qquad \forall \varphi \in C_c^1(B(x,s)). \tag{9.19}$$

Now let $\varepsilon \in (0,1)$ and $\varphi \in C_c^1(\mathbb{R}^n)$. Since $\mathcal{H}^{n-1}(M_0) = 0$ there exists a countable cover $\{F_k\}_{k\in\mathbb{N}}$ of M_0 by sets F_k with $F_k \cap M_0 \neq \emptyset$ and

$$\operatorname{diam}(F_k) = \varepsilon_k < \varepsilon, \qquad \sum_{k\in\mathbb{N}} \varepsilon_k^{n-1} < \varepsilon. \tag{9.20}$$

Hence, for every $k \in \mathbb{N}$, we may choose $y_k \in F_k \cap M_0$ such that $F_k \subset B(y_k, \varepsilon_k)$. The compact set $\operatorname{spt}\varphi \cap \partial E$ is thus covered by the union of the family of open balls $\{B(y_k, \varepsilon_k)\}_{k\in\mathbb{N}}$ with the family of open balls $\mathcal{F} = \{B(x,s) : x \in M\}$ defined by (9.15) and (9.16). By compactness, there exist finite subfamilies $\{B(y_k, \varepsilon_k)\}_{k\in I}$ and $\{B(x_h, s_h)\}_{h\in J} \subset \mathcal{F}$ such that

$$\operatorname{spt}\varphi \cap \partial E \subset \bigcup_{k\in I} B(y_k, \varepsilon_k) \cup \bigcup_{h\in J} B(x_h, s_h).$$

Correspondingly, for every $k \in I$ and $h \in J$, we may find $\eta_k \in C_c^1(B(y_k, \varepsilon_k))$ with $0 \le \eta_k \le 1$, and $\zeta_h \in C_c^1(B(x_h, s_h))$ with $0 \le \zeta_h \le 1$, such that

$$|\nabla \eta_k| \le \frac{C}{\varepsilon_k}, \qquad \text{on } \mathbb{R}^n, \tag{9.21}$$

$$\sum_{k\in I} \eta_k + \sum_{h\in J} \zeta_h = 1, \qquad \text{on } \operatorname{spt}\varphi \cap \partial E. \tag{9.22}$$

As in step two of the proof of Theorem 9.3 we thus see that

$$\int_E \nabla\varphi = \sum_{k\in I} \int_E \nabla(\eta_k \varphi) + \sum_{h\in J} \int_E \nabla(\zeta_h \varphi).$$

On the one hand, by (9.21) and by (9.20), we find that

$$\Big| \sum_{k\in I} \int_E \nabla(\eta_k\varphi) \Big| \le \sum_{k\in I} \int_{B(y_k,\varepsilon_k)} |\nabla\varphi| + |\varphi||\nabla\eta_k| \le C(n,\varphi) \sum_{k\in\mathbb{N}} \varepsilon_k^{n-1} \le C(n,\varphi)\varepsilon;$$

on the other hand, by (9.19),

$$\sum_{h\in J} \int_E \nabla(\zeta_h\varphi) = \sum_{h\in J} \int_M \zeta_h \varphi \nu_E \, d\mathcal{H}^{n-1}.$$

In conclusion,

$$\Big| \int_E \nabla\varphi - \int_M \varphi\nu_E \, d\mathcal{H}^{n-1} \Big| \le C(n,\varphi)\varepsilon + \Big| \int_M \Big(1 - \sum_{h\in J} \zeta_h\Big) \varphi \nu_E \, d\mathcal{H}^{n-1} \Big|. \tag{9.23}$$

Let us now set $A_\varepsilon = \bigcup_{k\in I} B(y_k, \varepsilon_k)$. By (9.22), $\sum_{h\in J} \zeta_h = 1$ on $M \setminus A_\varepsilon$, hence,

$$\left| \int_M \Big(1 - \sum_{h\in J} \zeta_h\Big) \varphi \nu_E \, \mathrm{d}\mathcal{H}^{n-1} \right| \le \sup_{\mathbb{R}^n} |\varphi| \, \mathcal{H}^{n-1}(M \cap A_\varepsilon) . \tag{9.24}$$

Since A_ε is bounded, $\mathcal{H}^{n-1}(M \cap A_\varepsilon) < \infty$. Thus $\bigcap_{\varepsilon>0} A_\varepsilon = M_0$ implies

$$\lim_{\varepsilon\to 0^+} \mathcal{H}^{n-1}(M \cap A_\varepsilon) = \mathcal{H}^{n-1}(M \cap M_0) = 0 .$$

We let $\varepsilon \to 0^+$ in (9.23) and (9.24) to conclude the proof of (9.18). □

10

Rectifiable sets and blow-ups of Radon measures

We shall now introduce the notion of rectifiable set, which provides a generalization of the notion of surface of primary importance in the study of geometric variational problems. We start by fixing some terminology. Given $k \in \mathbb{N}$, $1 \le k \le n-1$, $h \ge 1$, we shall say that M is a **k-dimensional (embedded) surface of class C^h** in $\mathbb{R}^n$ (or a **C^h-hypersurface** when $k = n-1$) if for every $x \in M$ there exist an open neighborhood A of x, an open set $E \subset \mathbb{R}^k$ and a bijection $f\colon E \to A \cap M$ with $f \in C^h(E)$ and $Jf > 0$ on E. Each map f is called a **coordinate mapping** of M. Notice that, in this way, M is relatively open in $\mathbb{R}^n$, and can be covered by countably many images $f(E)$, with f and E as above. The notion of a countably $\mathcal{H}^k$-rectifiable set is just a straight generalization of this concept to the measure-theoretic setting. Given a $\mathcal{H}^k$-measurable set $M \subset \mathbb{R}^n$, we say that M is **countably $\mathcal{H}^k$-rectifiable** if there exist countably many Lipschitz maps $f_h\colon \mathbb{R}^k \to \mathbb{R}^n$ such that

$$\mathcal{H}^k\Big(M \setminus \bigcup_{h\in\mathbb{N}} f_h(\mathbb{R}^k)\Big) = 0\,; \tag{10.1}$$

we say that M is **locally $\mathcal{H}^k$-rectifiable** provided $\mathcal{H}^k(K \cap M) < \infty$ for every compact set $K \subset \mathbb{R}^n$; finally, if $\mathcal{H}^k(M) < \infty$, then M is simply called **$\mathcal{H}^k$-rectifiable**. The notion of locally $\mathcal{H}^k$-rectifiable set is the most important to us. Indeed, whenever M is a countably $\mathcal{H}^k$-rectifiable set M, then

$$\mathcal{H}^k \llcorner M$$

is a regular Borel measure. However, $\mathcal{H}^k \llcorner M$ is a Radon measure if and only if M is locally $\mathcal{H}^k$-rectifiable, as is easily inferred from Proposition 2.13. Therefore, it is under the assumption of local $\mathcal{H}^k$-rectifiability on M that we have a natural identification between M and a Radon measure μ. In turn, as seen in Example 4.24, this identification lies at the basis of the measure-theoretic formulation of the notion of tangent space. Indeed, if M is locally $\mathcal{H}^k$-rectifiable and $\mu = \mathcal{H}^k \llcorner M$, then for $\mathcal{H}^k$-a.e. $x \in M$ there exists a k-dimensional plane π_x in $\mathbb{R}^n$ such that the blow-ups $\mu_{x,r}$ of μ at x weak-star converge to $\mathcal{H}^k \llcorner \pi_x$ as $r \to 0^+$, that is

$$\mathcal{H}^k \llcorner \left(\frac{M - x}{r} \right) \stackrel{*}{\rightharpoonup} \mathcal{H}^k \llcorner \pi_x \qquad \text{as } r \to 0^+ ; \tag{10.2}$$

recall Figure 4.1. A crucial fact is that the converse also holds true: if μ is a Radon measure on $\mathbb{R}^n$ concentrated on a Borel set M and such that for every $x \in M$ there exists a k-dimensional plane π_x such that the **k-dimensional blow-ups of μ** have the property that

$$\mu_{x,r} = \frac{(\Phi_{x,r})_\# \mu}{r^k} \stackrel{*}{\rightharpoonup} \mathcal{H}^k \llcorner \pi_x \qquad \text{as } r \to 0^+ , \tag{10.3}$$

(where, as usual, $\Phi_{x,r}(y) = (y - x)/r$, $y \in \mathbb{R}^n$), then M is locally $\mathcal{H}^k$-rectifiable and $\mu = \mathcal{H}^k \llcorner M$. We now describe the organization of this chapter. In Section 10.1, we apply Rademacher's theorem and the Lipschitz linearization Theorem 8.8 in order to exploit the huge freedom we have in choosing the maps f_h realizing the covering property (10.1). The maps f_h so selected will possess various properties, translating into a measure-theoretic language the requirements made on the coordinate mappings of a C^1-surface. Starting from this result, in Section 10.2 we shall prove the convergence property (10.2). Finally, the converse statement (10.3) is proved in Section 10.3.

10.1 Decomposing rectifiable sets by regular Lipschitz images

By McShane's lemma and by the regularity properties of Radon measures, M is countably $\mathcal{H}^k$-rectifiable if and only if there exist a Borel set $M_0 \subset \mathbb{R}^n$, countably many Lipschitz maps $f_h \colon \mathbb{R}^k \to \mathbb{R}^n$ and Borel sets $F_h \subset \mathbb{R}^k$ such that

$$M = M_0 \cup \bigcup_{h \in \mathbb{N}} f_h(F_h) , \qquad \mathcal{H}^k(M_0) = 0 . \tag{10.4}$$

This kind of decomposition is of course non-unique, and several properties can be imposed on the functions f_h by decreasing the sets F_h while increasing the $\mathcal{H}^k$-null set M_0. Indeed, if f_h satisfies a good property on a subset E_h of F_h and $|F_h \setminus E_h| = 0$ (here $|\cdot|$ is Lebesgue measure on $\mathbb{R}^k$), then we can replace F_h with E_h in (10.4) up to augmenting M_0 by the set $f(F_h \setminus E_h)$ (which is $\mathcal{H}^k$-null by Proposition 3.5). In Theorem 10.1 we put this remark into use, and provide a "good decomposition" result for countably $\mathcal{H}^k$-rectifiable sets. Before coming to this, we make the following definition. Given a Lipschitz function $f\colon \mathbb{R}^k \to \mathbb{R}^n$, and a bounded Borel set $E \subset \mathbb{R}^k$, we say that the pair (f, E) defines a **regular Lipschitz image** $f(E)$ in $\mathbb{R}^n$ if

(i) f is injective and differentiable on E, with $Jf(x) > 0$ for every $x \in E$;
(ii) every $x \in E$ is a point of density 1 for E;
(iii) every $x \in E$ is a Lebesgue point of ∇f.

In particular, we immediately deduce from (ii) and (iii) that

$$\lim_{r\to 0^+} \frac{|E \cap B(x,r)|}{\omega_k r^k} = 1\,, \qquad \lim_{r\to 0^+} \frac{1}{r^k}\int_{B(x,r)} |Jf(z) - Jf(x)|\,dz = 0\,,$$

for every $x \in E$. Indeed, if x is a Lebesgue point of ∇f then x is a Lebesgue point of Jf, since $\nabla f \in L^\infty_{\rm loc}(\mathbb{R}^k;\mathbb{R}^n\otimes\mathbb{R}^k)$ and the map $T \in \mathbb{R}^n \otimes \mathbb{R}^k \mapsto JT$ is continuous. We now show that we can always decompose a countably $\mathcal{H}^k$-rectifiable set by means of (almost flat) regular Lipschitz images.

Theorem 10.1 (Decomposition of rectifiable sets) *If M is countably $\mathcal{H}^k$-rectifiable in $\mathbb{R}^n$ and $t > 1$, then there exist a Borel set $M_0 \subset \mathbb{R}^n$, countably many Lipschitz maps $f_h\colon \mathbb{R}^k \to \mathbb{R}^n$ and bounded Borel sets $E_h \subset \mathbb{R}^k$ such that*

$$M = M_0 \cup \bigcup_{h\in\mathbb{N}} f_h(E_h)\,, \qquad \mathcal{H}^k(M_0) = 0\,.$$

Each pair (f_h, E_h) defines a regular Lipschitz image, with $\mathrm{Lip}(f_h) \le t$ *and*

$$\begin{aligned} t^{-1}|x-y| \le |f_h(x) - f_h(y)| &\le t|x-y|\,,\\ t^{-1}|v| \le |\nabla f_h(x)v| &\le t|v|\,,\\ t^{-k} \le Jf_h(x) &\le t^k\,,\end{aligned}$$

for every $x, y \in E_h$ and $v \in \mathbb{R}^k$.

Proof By Theorem 2.10, Theorem 8.7, Theorem 8.8, Rademacher's theorem, and Theorem 5.16 we find that

$$M = M_0 \cup \bigcup_{h\in\mathbb{N}} g_h(G_h)\,, \qquad \mathcal{H}^k(M_0) = 0\,,$$

where each (g_h, G_h) defines a regular Lipschitz image. Moreover, there exists $\{S_h\}_{h\in\mathbb{N}} \subset \mathbf{GL}(k)$ such that, for every $x, y \in G_h$ and $v \in \mathbb{R}^k$,

$$\begin{aligned} t^{-1}|S_h x - S_h y| \le |g_h(x) - g_h(y)| &\le t|S_h x - S_h y|\,,\\ t^{-1}|S_h v| \le |\nabla g_h(x)v| &\le t|S_h v|\,.\end{aligned}$$

Let us now define $E_h \subset \mathbb{R}^k$ and $f_h\colon E_h \to \mathbb{R}^k$ setting

$$E_h = S_h(G_h)\,, \qquad f_h = g \circ S_h^{-1}\,.$$

Then f_h is injective on E_h, with

$$t^{-1}|x-y| \le |f_h(x) - f_h(y)| \le t|x-y|\,,$$

for every $x, y \in E_h$. By Kirszbaum's theorem, we may extend $f_h\colon \mathbb{R}^k \to \mathbb{R}^n$ with $\mathrm{Lip}(f_h) \le t$. Since g_h was differentiable on G_h, we have that f_h is differentiable on E_h, with $\nabla f_h(x) = \nabla g(S_h^{-1}x) \circ S_h^{-1}$, so that

$$t^{-1}|v| \le |\nabla f_h(x)v| \le t|v|\,,$$

for every $x \in E_h$ and $v \in \mathbb{R}^k$. In particular, $t^{-k} \le Jf_h \le t^k$ on E_h. Since the pair (f_h, E_h) defines the regular Lipschitz image $f_h(E_h) = g_h(G_h)$ we are done. □

10.2 Approximate tangent spaces to rectifiable sets

Theorem 10.1 allows us to prove the existence (in a measure-theoretic sense) of tangent spaces to rectifiable sets. Define $\Phi_{x,r}\colon \mathbb{R}^n \to \mathbb{R}^n$ as $\Phi_{x,r}(y) = (y-x)/r$, $y \in \mathbb{R}^n$, so that, if μ is a Radon measure on $\mathbb{R}^n$ and $E \subset \mathbb{R}^n$ is Borel set, then

$$\frac{(\Phi_{x,r})_\# \mu\,(E)}{r^k} = \frac{\mu(x + r\,E)}{r^k}\,.$$

Theorem 10.2 (Existence of approximate tangent spaces) *If $M \subset \mathbb{R}^n$ is a locally $\mathcal{H}^k$-rectifiable set, then for $\mathcal{H}^k$-a.e. $x \in M$ there exists a unique k-dimensional plane π_x such that, as $r \to 0^+$,*

$$\frac{(\Phi_{x,r})_\#(\mathcal{H}^k \llcorner M)}{r^k} = \mathcal{H}^k \llcorner \left(\frac{M-x}{r}\right) \overset{*}{\rightharpoonup} \mathcal{H}^k \llcorner \pi_x\,, \tag{10.5}$$

that is,

$$\lim_{r\to 0^+} \frac{1}{r^k} \int_M \varphi\left(\frac{y-x}{r}\right) d\mathcal{H}^k(y) = \int_{\pi_x} \varphi\, d\mathcal{H}^k\,, \qquad \forall \varphi \in C^0_c(\mathbb{R}^n)\,. \tag{10.6}$$

In particular, $\theta_k(\mathcal{H}^k \llcorner M) = 1$ $\mathcal{H}^k$-a.e. on M, as

$$\lim_{r\to 0^+} \frac{\mathcal{H}^k(M \cap B(x,r))}{\omega_k r^k} = 1\,, \qquad \textit{for } \mathcal{H}^k\textit{-a.e. } x \in M. \tag{10.7}$$

Remark 10.3 If a k-dimensional plane π_x satisfies (10.5), then we set $\pi_x = T_x M$ and name it the **approximate tangent space to M at x**. The set of points $x \in M$ such that (10.5) holds true depends only on the Radon measure $\mu = \mathcal{H}^k \llcorner M$. It is a locally $\mathcal{H}^k$-rectifiable set in $\mathbb{R}^n$, which is left unchanged if we modify M on and by $\mathcal{H}^k$-null sets.

Lemma 10.4 *If $M = f(E)$ is a k-dimensional regular Lipschitz image in $\mathbb{R}^n$ and $z \in E$, then*

$$T_x M = \nabla f(z)(\mathbb{R}^k), \qquad x = f(z). \tag{10.8}$$

Proof of Lemma 10.4 If $\varphi \in C^0_c(\mathbb{R}^n)$, then we have

$$\begin{aligned}\frac{1}{r^k}\int_M \varphi \circ \Phi_{x,r}\, d\mathcal{H}^k &= \frac{1}{r^k}\int_M \varphi\Big(\frac{y-x}{r}\Big)\, d\mathcal{H}^k(y)\\ &= \frac{1}{r^k}\int_E \varphi\Big(\frac{f(w)-f(z)}{r}\Big) Jf(w)\, dw = \int_{\mathbb{R}^k} u_r,\end{aligned}$$

where we have defined $u_r \colon \mathbb{R}^k \to \mathbb{R}$ as

$$u_r(w) = 1_E(z + r\,w)\,\varphi\Big(\frac{f(z + r\,w) - f(z)}{r}\Big) Jf(z + r\,w), \qquad w \in \mathbb{R}^k.$$

Since z is a Lebesgue point of 1_E and Jf, and f is differentiable at z, we find that $u_r \to u_0$ in $L^1_{\mathrm{loc}}(\mathbb{R}^k)$ as $r \to 0^+$, where $u_0 \colon \mathbb{R}^k \to \mathbb{R}$ is defined as

$$u_0(w) = \varphi\big(\nabla f(z)w\big)\, Jf(z), \qquad w \in \mathbb{R}^k.$$

Moreover, $\|u_r\|_{L^\infty(\mathbb{R}^k)} \le \sup_{\mathbb{R}^n} |\varphi|\, \mathrm{Lip}(f)^k$, and, as we are going to prove below, there exist $r_0 > 0$ and $L > 0$ such that $\operatorname{spt} u_r \subset B_L$ for every $r \in (0, r_0)$; as a consequence, by dominated convergence and by the area formula,

$$\lim_{r\to 0^+} \frac{1}{r^k}\int_M \varphi \circ \Phi_{x,r}\, d\mathcal{H}^k = \int_{\mathbb{R}^k} \varphi\big(\nabla f(z)w\big)\, Jf(z)\, dw = \int_{\nabla f(z)(\mathbb{R}^k)} \varphi\, d\mathcal{H}^k,$$

and (10.8) is proved. We are left to prove the existence of r_0 and L as above. Since $Jf(z) > 0$ and f is differentiable at z, there exist $s_0, \lambda > 0$ such that

$$|f(z') - f(z)| \ge \lambda |z' - z|, \qquad \forall z' \in B(z, s_0). \tag{10.9}$$

At the same time, if $R > 0$ is such that $\operatorname{spt}\varphi \subset B_R$, then

$$|f(z + rw) - f(z)| \le rR, \qquad \forall w \in \operatorname{spt} u_r. \tag{10.10}$$

Thus, if $w \in \operatorname{spt} u_r$ and $r < s_0/R$, then $z + rw \in B(z, s_0)$ and, combining (10.9) and (10.10), we find $\lambda\,|w| \le R$. We set $L = R/\lambda$ and $r_0 = s_0/R$ to conclude. □

Proof of Theorem 10.2 *Step one:* We decompose $M = M_0 \cup \bigcup_{h\in\mathbb{N}} f_h(E_h)$ as in Theorem 10.1. If we let $M_h = f_h(E_h)$, then by Lemma 10.4 we find that

$$\lim_{r\to 0^+} \frac{1}{r^k}\int_{M_h} \varphi \circ \Phi_{x,r}\, d\mathcal{H}^k = \int_{\pi_x} \varphi\, d\mathcal{H}^k, \qquad \forall \varphi \in C^0_c(\mathbb{R}^n), \tag{10.11}$$

for every $x \in M_h$, where we have set $\pi_x = \nabla f_h(x)(\mathbb{R}^k)$. By Corollary 6.5, which we may apply since $\mathcal{H}^k \llcorner M$ is a Radon measure on $\mathbb{R}^n$, we have

$$\theta_k\big(\mathcal{H}^k \llcorner (M \setminus M_h)\big) = 0,$$

$\mathcal{H}^k$-a.e. on M_h. In particular, for $\mathcal{H}^k$-a.e. $x \in M_h$, we find that

$$\lim_{r\to 0^+} \frac{1}{r^k}\int_{M\setminus M_h} \varphi\circ\Phi_{x,r}\,\mathrm{d}\mathcal{H}^k = 0\,, \qquad \forall \varphi \in C^0_{\mathrm{c}}(\mathbb{R}^n)\,. \tag{10.12}$$

We combine (10.11) and (10.12) to prove (10.5).

Step two: Let $x \in M$ satisfy (10.5). Since $\pi_x \cap \partial B$ is a $(k-1)$-dimensional sphere, $\mathcal{H}^k \llcorner \pi_x(\partial B) = 0$. Thus, by Proposition 4.26, we find (10.7), as

$$\omega_k = \mathcal{H}^k \llcorner \pi_x(B) = \lim_{r\to 0^+} \frac{\mathcal{H}^k \llcorner M(\Phi_{x,r}^{-1}(B))}{r^k} = \lim_{r\to 0^+}\frac{\mathcal{H}^k(M\cap B(x,r))}{r^k}\,. \qquad \square$$

We conclude this section proving a frequently useful proposition.

Proposition 10.5 (Locality of approximate tangent spaces) *If M_1 and M_2 are locally $\mathcal{H}^k$-rectifiable sets in $\mathbb{R}^n$, then for $\mathcal{H}^k$-a.e. $x \in M_1 \cap M_2$,*

$$T_x M_1 = T_x M_2\,.$$

Proof By Corollary 6.5, for $\mathcal{H}^k$-a.e. $x \in M_1 \cap M_2$, we have that

$$\lim_{r\to 0^+} \frac{\mathcal{H}^k\big((M_1 \Delta M_2)\cap B(x,r)\big)}{r^k} = 0\,, \tag{10.13}$$

that is, $\mathcal{H}^k(K \cap ((M_1)_{x,r}\Delta(M_2)_{x,r})) \to 0$ for every compact set $K \subset \mathbb{R}^n$, where we have set $(M_i)_{x,r} = \Phi_{x,r}(M_i)$. If $\varphi \in C^0_{\mathrm{c}}(\mathbb{R}^n)$, then we have

$$\begin{aligned}&\left|\frac{1}{r^k}\int_{M_1}(\varphi\circ\Phi_{x,r})\,\mathrm{d}\mathcal{H}^k - \frac{1}{r^k}\int_{M_2}(\varphi\circ\Phi_{x,r})\,\mathrm{d}\mathcal{H}^k\right| \\ &\qquad \le \int_{(M_1)_{x,r}\Delta(M_2)_{x,r}} |\varphi|\,\mathrm{d}\mathcal{H}^k \le \sup_{\mathbb{R}^k}|\varphi|\,\mathcal{H}^k\Big(\mathrm{spt}\,\varphi\cap\big((M_1)_{x,r}\Delta(M_2)_{x,r}\big)\Big).\end{aligned}$$

Thus, if $x \in M_1 \cap M_2$, $T_x M_1$ and $T_x M_2$ exist, and (10.13) holds true, then $T_x M_1 = T_x M_2$. $\square$

Exercise 10.6 (Tangent space to a graph) If $u\colon \mathbb{R}^{n-1} \to \mathbb{R}$ is a Lipschitz function, and we define $f\colon \mathbb{R}^{n-1} \to \mathbb{R}^n$ as $f(z) = (z, u(z))$, $z \in \mathbb{R}^{n-1}$, then $\Gamma = f(\mathbb{R}^{n-1})$ is locally $\mathcal{H}^{n-1}$-rectifiable and, for a.e. $z \in \mathbb{R}^{n-1}$,

$$T_{f(z)}\Gamma = \nu(z)^\perp\,, \qquad \nu(z) = (-\nabla' u(z), 1)\,. \tag{10.14}$$

Hint: Apply Lemma 10.4.

Exercise 10.7 (Tangent space to a surface of revolution) If $r\colon \mathbb{R} \to (0,\infty)$ is a Lipschitz function, and $M = \{(z,t) \in \mathbb{R}^n : |z| = r(t), t \in \mathbb{R}\}$, then M is a $\mathcal{H}^{n-1}$-rectifiable set in $\mathbb{R}^n$ and, for $\mathcal{H}^{n-1}$-a.e. $x = (z,t) \in M$,

$$T_x M = \nu(x)^\perp\,, \qquad \nu(x) = (-z, r'(t))\,.$$

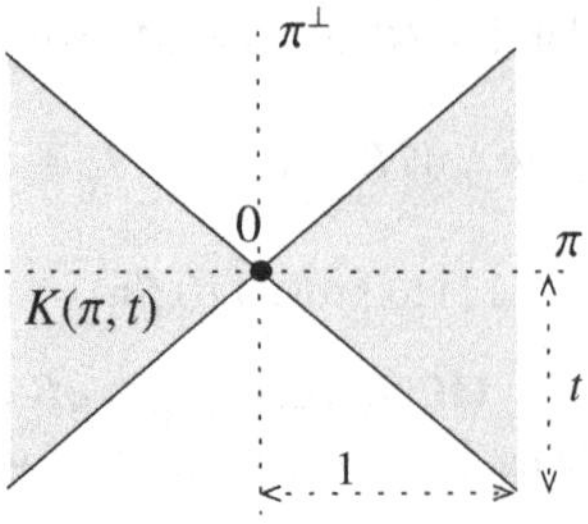

Figure 10.1 The cones $K(\pi, t)$ contain π for every $t > 0$. If $t \to 0^+$, $K(\pi, t)$ converges to π, while if $t \to \infty$, $K(\pi, t)$ converges to $\mathbb{R}^n$.

10.3 Blow-ups of Radon measures and rectifiability

We now prove a converse statement to Theorem 10.2 that is going to play an important role in studying the structure theory of sets of finite perimeter.

Theorem 10.8 (Rectifiability from convergence of the blow-ups) *If μ is a Radon measure on $\mathbb{R}^n$, M is a Borel set in $\mathbb{R}^n$, μ is concentrated on M, and, for every $x \in M$, there exists a k-dimensional plane π_x in $\mathbb{R}^n$ such that*

$$\frac{(\Phi_{x,r})_{\#}\mu}{r^k} \stackrel{*}{\rightharpoonup} \mathcal{H}^k \llcorner \pi_x ,$$

as $r \to 0^+$, then $\mu = \mathcal{H}^k \llcorner M$ and M is locally $\mathcal{H}^k$-rectifiable.

Aiming to prove Theorem 10.8, we now introduce a simple criterion for $\mathcal{H}^k$-rectifiability. Given a k-dimensional plane π in $\mathbb{R}^n$, we denote by $\mathbf{p}_\pi : \mathbb{R}^n \to \mathbb{R}^n$ and $\mathbf{p}_\pi^\perp : \mathbb{R}^n \to \mathbb{R}^n$ the orthogonal projections of $\mathbb{R}^n$ onto (respectively) π and $\pi^\perp$ (thus $\mathbf{p}_{\pi^\perp} = \mathbf{p}_\pi^\perp$), and define the cones $K(\pi, t)$, $t > 0$, as

$$K(\pi,t) = \left\{ y \in \mathbb{R}^n : |\mathbf{p}_\pi^\perp y| < t |\mathbf{p}_\pi y| \right\} = \left\{ y \in \mathbb{R}^n : |y| < \sqrt{1+t^2}\, |\mathbf{p}_\pi y| \right\}; \tag{10.15}$$

see Figure 10.1.

Proposition 10.9 (Rectifiability criterion) *If $M \subset \mathbb{R}^n$ is a compact set, π is a k-dimensional plane in $\mathbb{R}^n$, and there exist δ and t positive with*

$$M \cap B(x,\delta) \subset x + K(\pi,t), \qquad \forall x \in M, \tag{10.16}$$

then M is $\mathcal{H}^k$-rectifiable, since there exist finitely many Lipschitz maps $f_h : \mathbb{R}^k \to \mathbb{R}^n$, and compact sets $F_h \subset \mathbb{R}^k$ with

$$M = \bigcup_{h=1}^{N} f_h(F_h). \tag{10.17}$$

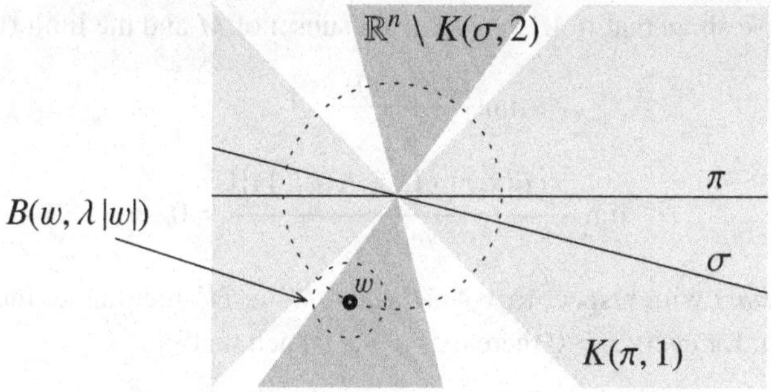

Figure 10.2 An illustration of (10.18).

Proof If $x_0 \in M$, and $x, y \in \overline{B}(x_0, \delta/2) \cap M$, then $y \in B(x, \delta) \cap M \subset x + K(\pi, t)$, that is, $|\mathbf{p}_\pi^\perp(y - x)| \le t\,|\mathbf{p}_\pi(y - x)|$. In particular, $\mathbf{p}_\pi$ is a bijection between the compact sets $\overline{B}(x_0, \delta/2) \cap M$ and $G_{x_0} = \mathbf{p}_\pi(M \cap \overline{B}(x_0, \delta/2))$. In other words, given $x_0 \in M$, there exist a compact set $G_{x_0} \subset \pi$ and an injective map $g_{x_0} \colon G_{x_0} \to \mathbb{R}^n$ such that $g_{x_0}(G_{x_0}) = \overline{B}(x_0, \delta/2) \cap M$ and

$$\mathbf{p}_\pi\, g_{x_0}(z) = z\,, \qquad |g_{x_0}(z) - g_{x_0}(w)| \le t\,|z - w|\,, \qquad \forall z, w \in G_{x_0}\,.$$

Since $\{B(x, \delta/2) : x \in M\}$ is an open covering of M, we find $\{x_h\}_{h=1}^N \subset M$ with

$$M = \bigcup_{h=1}^N \Big(M \cap \overline{B}(x_h, \delta/2)\Big) = \bigcup_{h=1}^N g_h(G_h)\,,$$

where $g_h \colon G_h \to \mathbb{R}^n$ are Lipschitz maps and $G_h \subset \pi$ are compact sets. If $P \in \mathbf{O}(k, n)$ is such that $P(\mathbb{R}^k) = \pi$ and we extend each g_h to π by McShane's lemma, then we conclude by setting $f_h = g_h \circ P$ and $F_h = P^{-1}(G_h)$. □

Proof of Theorem 10.8 If π and σ are k-dimensional planes in $\mathbb{R}^n$, we set

$$d(\pi, \sigma) = \|\,\mathbf{p}_\pi - \mathbf{p}_\sigma\| = \sup_{v \in S^{n-1}} |\,\mathbf{p}_\pi v - \mathbf{p}_\sigma v|\,.$$

The existence of $\lambda \in (0, 1)$ such that $d(\pi, \sigma) < \lambda$ implies (see Figure 10.2)

$$B\big(w, \lambda\,|w|\big) \cap K(\pi, 1) = \emptyset\,, \qquad \forall w \in \mathbb{R}^n \setminus K(\sigma, 2)\,, \tag{10.18}$$

is easily proved by arguing by contradiction. This said, we fix a finite family of k-dimensional planes $\{\sigma_h\}_{h=1}^N$ with the property that

$$\min_{1 \le h \le N} d(\sigma_h, \pi) < \lambda\,, \tag{10.19}$$

for every k-dimensional plane π in $\mathbb{R}^n$, and we divide the proof into three steps.

Step one: We show that if M' is a compact subset of M and the limit relations

$$\lim_{r\to 0^+} \frac{\mu(B(x,r))}{\omega_k\, r^k} = 1\,, \tag{10.20}$$

$$\lim_{r\to 0^+} \frac{\mu\Big(B(x,r) \setminus \big(x + K(\pi_x, 1)\big)\Big)}{\omega_k\, r^k} = 0 \tag{10.21}$$

hold *uniformly* with respect to $x \in M'$, then M' is $\mathcal{H}^k$-rectifiable. Indeed, by assumption, for every $\varepsilon > 0$ there exists $\delta > 0$ such that

$$\mu(B(x,r)) \ge (1-\varepsilon)\omega_k r^k\,, \tag{10.22}$$

$$\mu\Big(B(x,r) \setminus \big(x + K(\pi_x, 1)\big)\Big) \le \varepsilon\omega_k r^k\,, \tag{10.23}$$

for every $x \in M'$ and $r \in (0, 2\delta)$. We now claim that, if we set

$$M'_h = \big\{x \in M' : \mathrm{dist}(\sigma_h, \pi_x) < \lambda\big\}\,, \qquad 1 \le h \le N\,,$$

then by (10.22), (10.23), and by suitably choosing ε depending on k and λ,

$$B(x,\delta) \cap M'_h \subset x + K(\sigma_h, 2)\,, \qquad \forall x \in M'_h\,. \tag{10.24}$$

Indeed, if $x \in M'_h$, $y \in B(x,\delta) \cap M'_h$ but $y - x \in \mathbb{R}^n \setminus K(\sigma_h, 2)$, then by (10.18)

$$B\big(y, \lambda|\,y - x|\big) \subset \mathbb{R}^n \setminus \big(x + K(\pi_x, 1)\big)\,.$$

Since $\lambda \in (0,1)$, and thus $B(y, \lambda|\,y-x|) \subset B(x, 2|\,y-x|)$, we find that

$$B\big(y, \lambda|\,y - x|\big) \subset B\big(x, 2|\,y-x|\big) \setminus \big(x + K(\pi_x, 1)\big)\,.$$

Applying (10.23) (at x with $r = 2|\,y-x|$) and (10.22) (at y with $r = \lambda|\,y-x|$),

$$\varepsilon\omega_k|x-y|^k \ge (1-\varepsilon)\omega_k\lambda^k|x-y|^k\,,$$

a contradiction, as soon as ε is small enough with respect to k and λ. This proves (10.24). By Proposition 10.9, M'_h is thus $\mathcal{H}^k$-rectifiable for $h = 1, ..., N$. Since $M' \subset \bigcup_{h=1}^N M'_h$ by (10.19), M' is $\mathcal{H}^k$-rectifiable.

Step two: We prove that M is countably $\mathcal{H}^k$-rectifiable. First, we have

$$\lim_{r\to 0^+} \frac{\mu(B(x,r))}{\omega_k\, r^k} = 1\,, \tag{10.25}$$

$$\lim_{r\to 0^+} \frac{\mu\Big(B(x,r) \setminus \big(x + K(\pi_x, 1)\big)\Big)}{\omega_k\, r^k} = 0\,, \tag{10.26}$$

for every $x \in M$. Indeed, $(\mathcal{H}^k \llcorner \pi_x)(\partial B) = 0$, therefore, by Proposition 4.26,

$$\omega_k = \mathcal{H}^k(\pi_x \cap B) = \lim_{r\to 0^+} \frac{(\Phi_{x,r})_\# \mu\,(B)}{r^k} = \lim_{r\to 0^+} \frac{\mu(B(x,r))}{r^k}\,,$$

that is (10.25). We verify (10.26) analogously, as $\pi_x \subset K(\pi_x, 1)$. Second, given $R > 0$, since $\mu(M \cap B_R) < \infty$ we may apply Egoroff's theorem and Theorem 2.8 to prove the existence $M' \subset M \cap B_R$ compact, with

$$\mu\Big((M \cap B_R) \setminus M'\Big) < \frac{\mu\,(M \cap B_R)}{2},$$

and such that the limit relations (10.25) and (10.26) hold uniformly on M'. By step one, M' is $\mathcal{H}^k$-rectifiable. Iterating, we see that $M \cap B_R$ is countably $\mathcal{H}^k$-rectifiable. By the arbitrariness of R, M is countably $\mathcal{H}^k$-rectifiable.

Step three: By (10.25) and by Theorem 6.4, for every Borel set $E \subset M$,

$$\mathcal{H}^k(E) \le \mu(E) \le 2^k \mathcal{H}^k(E),$$

so that $\mathcal{H}^k(M \cap K) < \infty$ if $K \subset \mathbb{R}^n$ is compact. Thus, M is locally $\mathcal{H}^k$-rectifiable, $\mathcal{H}^k \llcorner M$ is a Radon measure, and $\mu \ll \mathcal{H}^k \llcorner M$. By Theorem 5.8,

$$\theta(x) = \lim_{r \to 0^+} \frac{\mu(B(x,r))}{\mathcal{H}^k(M \cap B(x,r))} \qquad \text{exists for } \mathcal{H}^k\text{-a.e. } x \in M,$$

and $\mu = \theta \mathcal{H}^k \llcorner M$ on $\mathcal{B}(\mathbb{R}^n)$. By (10.25) and (10.7), $\theta(x) = 1$ for $\mathcal{H}^k$-a.e. $x \in M$. Hence, $\mu = \mathcal{H}^k \llcorner M$ on $\mathcal{B}(\mathbb{R}^n)$, and, by Exercise 2.6, on $\mathcal{P}(\mathbb{R}^n)$. □

11

Tangential differentiability and the area formula

We now extend the area formula to rectifiable sets, proving that, if $M \subset \mathbb{R}^n$ is locally $\mathcal{H}^k$-rectifiable, $f\colon \mathbb{R}^n \to \mathbb{R}^m$ ($m \ge k$) is a Lipschitz function (injective on M), and $J^M f$ is the *tangential* Jacobian of f with respect to M, then

$$\mathcal{H}^k(f(M)) = \int_M J^M f \, \mathrm{d}\mathcal{H}^k .$$

We first prove this formula on k-dimensional C^1-surfaces (Section 11.1), and then address locally $\mathcal{H}^k$-rectifiable sets (Section 11.2). As an application, in Section 11.3 we extend the Gauss–Green theorem to C^2-hypersurfaces.

11.1 Area formula on surfaces

Let M be a k-dimensional C^1-surface in $\mathbb{R}^n$ and let $x \in M$. A function $f\colon \mathbb{R}^n \to \mathbb{R}^m$ is **tangentially differentiable with respect to** M **at** x, if there exists a linear function $\nabla^M f(x) \in \mathbb{R}^m \otimes T_x M$ such that, uniformly on $\{v \in T_x M\colon |v| = 1\}$,

$$\lim_{h \to 0} \frac{f(x + h\,v) - f(x)}{h} = \nabla^M f(x) v . \tag{11.1}$$

In other words, the restriction of f to $x + T_x M$ is differentiable at x. The **tangential Jacobian** of f with respect to M at x is then defined by

$$J^M f(x) = \sqrt{\det(\nabla^M f(x)^* \nabla^M f(x))} .$$

Remark 11.1 A function $f\colon \mathbb{R}^n \to \mathbb{R}^m$ may fail to be differentiable at every $x \in M$ while being tangentially differentiable with respect to M at every $x \in M$. For example, let $M = \{x_n = 0\} \subset \mathbb{R}^n$, $\varphi \in C^1(\mathbb{R}^{n-1}; \mathbb{R}^m)$, and set

$f(x) = \varphi(x') + |x_n| e$ for $e \in \mathbb{R}^m$ and $x = (x', x_n) \in \mathbb{R}^{n-1} \times \mathbb{R} = \mathbb{R}^n$. In this case,

$$\nabla^M f(x) = \sum_{i=1}^{n-1} \partial_i \varphi(x') \otimes e_i \,, \qquad J^M f(x) = J\varphi(x') \,, \qquad \forall x \in M \,.$$

Remark 11.2 If $f \in C^1(\mathbb{R}^n; \mathbb{R}^m)$, M is a k-dimensional C^1-surface in $\mathbb{R}^n$, and $x \in M$, then f is tangentially differentiable at x and $\nabla^M f(x)$ is the restriction of $\nabla f(x)$ at $T_x M$: thus, if $\{\tau_h\}_{h=1}^k$ is an orthonormal basis of $T_x M$ and $\{\nu_h\}_{h=1}^{n-k}$ is an orthonormal basis of $(T_x M)^\perp$, then

$$\nabla^M f(x) = \sum_{h=1}^k \Big(\nabla f(x) \tau_h\Big) \otimes \tau_h = \nabla f(x) - \sum_{h=1}^{n-k} \Big(\nabla f(x) \nu_h\Big) \otimes \nu_h \,.$$

Theorem 11.3 *If $M \subset \mathbb{R}^n$ is a k-dimensional C^1-surface and $f \in C^1(\mathbb{R}^n; \mathbb{R}^m)$ $(m \geq k)$ is injective, then*

$$\mathcal{H}^k(f(M)) = \int_M J^M f \, d\mathcal{H}^k \,.$$

Proof *Step one:* If V is a k-dimensional subspace of $\mathbb{R}^n$, $T_1 \in \mathbb{R}^n \otimes \mathbb{R}^k$ is such that $T_1(\mathbb{R}^k) = V$, and $T_2 \in \mathbb{R}^m \otimes V$ (so that $T_2 T_1 \in \mathbb{R}^m \otimes \mathbb{R}^k$), then

$$J(T_2 T_1) = J T_2 \, J T_1 \,. \tag{11.2}$$

Indeed, let us consider the polar decompositions $T_1 = P_1 S_1$ and $T_2 = P_2 S_2$, where $P_1 \in \mathbf{O}(k, n)$, $S_1 \in \mathbf{Sym}(k)$, $P_2 \in \mathbf{O}(V, m)$, and $S_2 \in \mathbf{Sym}(V)$. Then

$$(T_2 T_1)^* T_2 T_1 = T_1^* T_2^* T_2 T_1 = S_1 P_1^* S_2^2 P_1 S_1 = S_1 U S_1 \,,$$

where $U = P_1^* S_2^2 P_1 \in \mathbb{R}^k \otimes \mathbb{R}^k$ as $P_1(\mathbb{R}^k) = V$ by $T_1(\mathbb{R}^k) = V$. Thus,

$$J(T_2 T_1) = J S_1 \sqrt{\det(U)} = J T_1 \sqrt{\det(U)} \,.$$

We now prove $\det(U) = (J T_2)^2$. If $\{v_h\}_{h=1}^k$ is an orthonormal basis of V, then

$$S_2^2 = \sum_{h=1}^k \mu_h v_h \otimes v_h \,, \quad \mu_h \geq 0 \,.$$

Since $T_1(\mathbb{R}^k) = V$, if we set $w_h = P_1^{-1}(v_h)$, then $\{w_h\}_{h=1}^k$ is an orthonormal basis of $\mathbb{R}^k$, with $P_1 = \sum_{h=1}^k w_h \otimes v_h$. Thus $U = P_1^* S_2^2 P_1 = \sum_{h=1}^k \mu_h w_h \otimes w_h$, and $\det(U) = \det(S_2)^2 = (J T_2)^2$.

Step two: Since M is a k-dimensional C^1-surface, there exist $A_h \subset \mathbb{R}^k$ open and $g_h \in C^1(\mathbb{R}^k; \mathbb{R}^n)$ injective such that $M = \bigcup_{h \in \mathbb{N}} g_h(A_h)$. Since $T_x M = T_x(g_h(A_h))$

when $x \in M \cap g_h(A_h)$, we can directly assume $M = g(A)$ for $A \subset \mathbb{R}^k$ open and $g \in C^1(\mathbb{R}^k; \mathbb{R}^n)$ injective. Applying the area formula to $f \circ g \in C^1(\mathbb{R}^k; \mathbb{R}^m)$,

$$\mathcal{H}^k(f(M)) = \int_A J(f \circ g)(z)\, dz\,.$$

If $z \in A$, then $\nabla g(z)(\mathbb{R}^k) = T_{g(z)}M$, and, in particular,

$$\nabla(f \circ g)(z) = \nabla f(g(z))\nabla g(z) = \nabla^M f(g(z))\nabla g(z)\,.$$

By step one, $J(f \circ g) = ((J^M f) \circ g)Jg$ on A. Hence, by the area formula (8.1),

$$\mathcal{H}^k(f(M)) = \int_A \Big((J^M f) \circ g\Big)\, Jg = \int_{g(A)} J^M f \, d\mathcal{H}^k = \int_M J^M f \, d\mathcal{H}^k\,. \qquad \square$$

11.2 Area formula on rectifiable sets

Let M be a locally $\mathcal{H}^k$-rectifiable set in $\mathbb{R}^n$ and let $x \in M$ be such that the approximate tangent space T_xM exists. As in the C^1-case, we say that $f\colon \mathbb{R}^n \to \mathbb{R}^m$ is **tangentially differentiable with respect to** M **at** x if the restriction of f to $x + T_xM$ is differentiable at x. We now prove a Rademacher-type theorem concerning tangential differentiability on locally $\mathcal{H}^k$-rectifiable sets.

Theorem 11.4 *If M is a locally $\mathcal{H}^k$-rectifiable set, and $f\colon \mathbb{R}^n \to \mathbb{R}^m$ is a Lipschitz map, then $\nabla^M f(x)$ exists at $\mathcal{H}^k$-a.e. $x \in M$.*

As in proving Theorem 10.2, we first consider regular Lipschitz images.

Lemma 11.5 *If $M = g(E)$ is a k-dimensional regular Lipschitz image in $\mathbb{R}^n$, $f\colon \mathbb{R}^n \to \mathbb{R}^m$ is a Lipschitz function, and $f \circ g$ is differentiable at $z \in E$, then f is tangentially differentiable with respect to M at $x = g(z)$, with*

$$\nabla^M f(x) = \nabla(f \circ g)(z)\, \nabla g(z)^{-1} \qquad \text{on } T_xM = dg_z(\mathbb{R}^k). \tag{11.3}$$

Here, we have denoted by $\nabla g(z)^{-1}$ the inverse of $\nabla g(z)$ seen as an isomorphism between $\mathbb{R}^k$ and $T_xM = dg_z(\mathbb{R}^k)$.

Proof By Lemma 10.4, M admits the approximate tangent space $T_xM = dg_z(\mathbb{R}^k)$ at $x = g(z)$. If $w \in \mathbb{R}^k$ and $v = \nabla g(z)\, w$, then

$$\lim_{t \to 0} \frac{f(x + t\,v) - f(x)}{t} = \lim_{t \to 0} \frac{f(g(z) + t\,\nabla g(z)w) - f(g(z))}{t} = \lim_{t \to 0} \frac{f(g(z + tw)) - f(g(z))}{t}\,,$$

since f is a Lipschitz function and since $|g(z+tw) - g(z) - t\,\nabla g(x)\, w| = o(t)$. Since $f \circ g$ is differentiable at z, we thus find that f admits directional derivatives at

x along directions $v \in T_xM$, with

$$\frac{\partial f}{\partial v}(x) = \nabla(f \circ g)(z)\, w, \qquad w = \nabla g(z)^{-1} v. \tag{11.4}$$

Since $\nabla g(z)$ is a linear isomorphism between $\mathbb{R}^k$ and T_xM we find that

$$v \in T_xM \mapsto \frac{\partial f}{\partial v}(x)$$

is a linear map. Since f is a Lipschitz function it follows that

$$\lim_{t\to 0} \frac{f(x+tv) - f(x)}{t} = \frac{\partial f}{\partial v}(x) \qquad \text{uniformly on } v \in T_xM,\ |v| = 1,$$

which, by (11.4), is the thesis of the lemma. □

Proof of Theorem 11.4 By Proposition 10.5, if M_1 and M_2 are locally $\mathcal{H}^k$-rectifiable sets and $f\colon \mathbb{R}^n \to \mathbb{R}^m$ is tangentially differentiable at $\mathcal{H}^k$-a.e. $x \in M_1$ then $\nabla^{M_1} f(x) = \nabla^{M_2} f(x)$ at $\mathcal{H}^k$-a.e. $x \in M_1 \cap M_2$. Therefore the theorem follows by Theorem 10.1 and Lemma 11.5. □

Theorem 11.6 *If M is a locally $\mathcal{H}^k$-rectifiable set and $f\colon \mathbb{R}^n \to \mathbb{R}^m$ is a Lipschitz map with $1 \le k \le m$, then*

$$\int_{\mathbb{R}^m} \mathcal{H}^0\big(M \cap \{f = y\}\big)\, d\mathcal{H}^k(y) = \int_M J^M f\, d\mathcal{H}^k, \tag{11.5}$$

where $\{f = y\} = \{x \in \mathbb{R}^n : f(x) = y\}$. In particular, if f is injective on M, then

$$\mathcal{H}^k(f(M)) = \int_M J^M f\, d\mathcal{H}^k. \tag{11.6}$$

Proof By Theorem 10.1 we can directly assume that $M = g(E)$ is a regular Lipschitz image. In this case, by Lemma 10.4 and Lemma 11.5, we have that

$$T_xM = \nabla g(z)(\mathbb{R}^k), \qquad \nabla^M f(x) v = \nabla(f \circ g)(z) \nabla g(z)^{-1} v,$$

whenever $f \circ g$ is differentiable at $z \in E$ and $v \in T_xM$, so that, in particular,

$$\nabla(f \circ g)(z) = \nabla^M f(g(z)) \nabla g(z).$$

By step one in the proof of Theorem 11.3 and Rademacher's theorem,

$$J(f \circ g)(z) = J^M f(g(z)) Jg(z), \quad \text{for } \mathcal{H}^k\text{-a.e. } z \in E.$$

We apply the area formula to $f \circ g\colon \mathbb{R}^k \to \mathbb{R}^m$ and find that

$$\int_{\mathbb{R}^m} \mathcal{H}^0\big(E \cap \{f \circ g = y\}\big)\, d\mathcal{H}^k(y) = \int_E J(f \circ g)(z)\, dz.$$

Since $g(E \cap \{f \circ g = y\}) = M \cap \{f = y\}$, we conclude that

$$\int_{\mathbb{R}^m} \mathcal{H}^0\big(M \cap \{f = y\}\big)\, d\mathcal{H}^k(y) = \int_E J(f \circ g)(z)\, dz$$
$$= \int_E J^M f(g(z))\, Jg(z)\, dz = \int_M J^M f\, d\mathcal{H}^k . \qquad \square$$

Corollary 11.7 *If S is a locally $\mathcal{H}^{n-2}$-rectifiable set in $\mathbb{R}^{n-1}$, $u\colon \mathbb{R}^{n-1} \to \mathbb{R}$ is a Lipschitz function, $\Gamma = \{(z, u(z)) \in \mathbb{R}^n : z \in S\}$, $g\colon \mathbb{R}^n \to [-\infty, \infty]$ is a Borel function, and either $g \ge 0$ or $g \in L^1(\mathbb{R}^n, \mathcal{H}^{n-2} \llcorner \Gamma)$, then*

$$\int_\Gamma g\, d\mathcal{H}^{n-2} = \int_S \overline{g}\, \sqrt{1 + |\nabla^S u|^2}\, d\mathcal{H}^{n-2}, \tag{11.7}$$

where we have set $\overline{g}(z) = g(z, u(z))$, $z \in \mathbb{R}^{n-1}$.

Proof Consider the Lipschitz function $f\colon \mathbb{R}^{n-1} \to \mathbb{R}^n$ defined by $f(z) = (z, u(z))$, $z \in \mathbb{R}^{n-1}$, so that $\Gamma = f(S)$ is trivially a locally $\mathcal{H}^{n-2}$-rectifiable set in $\mathbb{R}^n$. By Theorem 11.6, we only have to prove that

$$J^S f = \sqrt{1 + |\nabla^S u|^2}, \qquad \mathcal{H}^{n-2}\text{-a.e. on } S.$$

Indeed, since $\nabla f = \mathrm{Id}_{\mathbb{R}^{n-1}} + e_n \otimes \nabla' u$ and $\nabla^S u = \nabla u - (\nabla u \cdot \nu_S)\nu_S$, where $\nu_S(z) \in (T_z S)^\perp$ for $\mathcal{H}^{n-2}$-a.e. $z \in S$, we have

$$\begin{aligned}
\nabla^S f &= \nabla f - (\nabla f\, \nu_S) \otimes \nu_S \\
&= \mathrm{Id}_{\mathbb{R}^{n-1}} + e_n \otimes \nabla' u - \nu_S \otimes \nu_S - (\nabla' u \cdot \nu_S)\, e_n \otimes \nu_S \\
&= \mathrm{Id}_{\nu_S^\perp} + e_n \otimes \nabla^S u,
\end{aligned}$$

so that $(\nabla^S f)^*(\nabla^S f) = \mathrm{Id}_{\nu_S^\perp} + (\nabla^S u) \otimes (\nabla^S u)$. We conclude by (9.3). $\square$

11.3 Gauss–Green theorem on surfaces

We finally introduce the natural extension of the Gauss–Green theorem to hypersurfaces in $\mathbb{R}^n$. The resulting formula will prove useful in understanding the geometric meaning of the first variation formula for perimeter in Chapter 17, and will play a crucial role in establishing an important necessary "boundary condition for minimality", known as Young's law, in Chapter 19. If $M \subset \mathbb{R}^n$ is a k-dimensional C^1-surface and $T \in C^1_c(\mathbb{R}^n; \mathbb{R}^n)$ we shall say that T is **tangential to** M if $T(x) \in T_x M$ for every $x \in M$, and that T is **normal to** M if, instead, $T(x) \in (T_x M)^\perp$ for every $x \in M$.

Theorem 11.8 (Gauss–Green theorem on surfaces) *If $M \subset \mathbb{R}^n$ is a C^2-hypersurface with boundary Γ, then there exist a normal vector field $\mathbf{H}_M \in C^0(M;\mathbb{R}^n)$ to M and a normal vector field $\nu_\Gamma^M \in C^1(\Gamma; S^{n-1})$ to Γ such that*

$$\int_M \nabla^M \varphi \, d\mathcal{H}^{n-1} = \int_M \varphi \, \mathbf{H}_M \, d\mathcal{H}^{n-1} + \int_\Gamma \varphi \, \nu_\Gamma^M \, d\mathcal{H}^{n-2}, \tag{11.8}$$

for every $\varphi \in C^1_c(\mathbb{R}^n)$. Moreover, if $T \in C^1_c(\mathbb{R}^n;\mathbb{R}^n)$ is normal to M, then

$$T \cdot \nu_\Gamma^M = 0 \qquad \text{on } \Gamma. \tag{11.9}$$

Remark 11.9 We say that M is a **C^h-hypersurface with boundary** Γ, if M is a C^h-hypersurface, Γ is the relative boundary of M, and, for every $x \in \Gamma$, there exist $r > 0$, an open set $E \subset \mathbb{R}^{n-1}$ with C^h-boundary, and a function $u \in C^h(\mathbb{R}^{n-1})$ such that, up to rotation and with Notation 4 in force,

$$\mathbf{C}(x,r) \cap M = \left\{(z, u(z)) : z \in \mathbf{D}(\mathbf{p}x, r) \cap E\right\}, \tag{11.10}$$

$$\mathbf{C}(x,r) \cap \Gamma = \left\{(z, u(z)) : z \in \mathbf{D}(\mathbf{p}x, r) \cap \partial E\right\}. \tag{11.11}$$

As it turns out, Γ is an $(n-2)$-dimensional C^h-surface (with empty relative boundary in $\mathbb{R}^n$). We also notice that at every relative interior point of M, that is, at every $x \in M$, condition (11.10) holds true with $E = \mathbb{R}^{n-1}$.

Remark 11.10 (Mean curvature vector) The vector field $\mathbf{H}_M$ is called the **mean curvature vector** to M. The definition of the **scalar mean curvature** $H_M \colon M \to \mathbb{R}$ of M depends on the mean curvature vector and the explicit choice of a unit normal vector field $\nu_M \colon M \to S^{n-1}$ to M through the formula

$$\mathbf{H}_M = H_M \, \nu_M .$$

If there exists a continuous unit normal vector field ν_M to M, then ν_M is an **orientation of** M, and M is **orientable**. In this case, H_M can be assumed continuous on M. Notice that Theorem 11.8 does not require M to be orientable.

Remark 11.11 By condition (11.9), ν_Γ^M is "tangential" to M, that is

$$\nu_\Gamma^M \cdot \nu_M = 0 \qquad \text{on } \Gamma. \tag{11.12}$$

This identity makes sense since for every $x \in \Gamma$ there always exist an open neighborhood A of x and a vector field $\nu_M \in C^0(A \cap M; S^{n-1})$ which is normal to M, and which may be extended by continuity up to $A \cap \Gamma$; see Figure 11.1.

Remark 11.12 (Divergence theorem on surfaces) Given a vector field $T \in C^1_c(\mathbb{R}^n;\mathbb{R}^n)$, we define the **tangential divergence** of T on M by the formula

$$\operatorname{div}^M T = \operatorname{div} T - (\nabla T \, \nu_M) \cdot \nu_M = \operatorname{trace}(\nabla^M T), \tag{11.13}$$

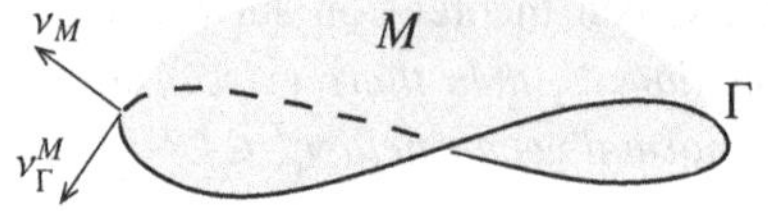

Figure 11.1 The normal to the boundary Γ of M induced through the tangential divergence theorem on M.

where $\nu_M\colon M \to S^{n-1}$ is any unit normal vector field to M. Discontinuously switching ν_M to $-\nu_M$ on part of M leaves $\mathrm{div}^M T$ unchanged. Hence, it is always $\mathrm{div}^M T \in C^0(M)$, even if M is not orientable. We note that (11.8) is equivalently reformulated as follows: for every $T \in C^1_c(\mathbb{R}^n;\mathbb{R}^n)$,

$$\int_M \mathrm{div}^M T \, d\mathcal{H}^{n-1} = \int_M T \cdot \mathbf{H}_M \, d\mathcal{H}^{n-1} + \int_\Gamma (T \cdot \nu_\Gamma^M) \, d\mathcal{H}^{n-2} . \qquad (11.14)$$

The "natural" proof of (11.14), in sketchy form, goes as follows. First, one remarks that, on tangential vector fields, (11.14) is a particular case of Stokes' theorem (in particular, the term $T \cdot \mathbf{H}_M$ automatically vanishes and the formula holds true even for C^1-surfaces). Secondly, if locally $\{\tau_h\}_{h=1}^{n-1} \subset S^{n-1}$ and $\{\kappa_h\}_{h=1}^{n-1} \subset \mathbb{R}$ are an orthonormal basis of principal directions and the corresponding principal curvatures, then for every normal vector field $T = \varphi\, \nu_M$ we have, locally,

$$\nabla T\, \tau_h = \sum_{h=1}^{n-1} \partial_h \varphi\, \nu_M + \sum_{h=1}^{n-1} \varphi\, \kappa_h\, \tau_h ,$$

so that $$\mathrm{div}^M T = \sum_{h=1}^{n-1} \tau_h \cdot (\nabla T\, \tau_h) = \varphi \sum_{h=1}^{n-1} \kappa_h = T \cdot \mathbf{H}_M ,$$

and (11.14) holds, trivially, as integration of a pointwise identity. Finally, the general case follows by linearity, since $T = T_0 + T_1$, where $T_0 = T - (T \cdot \nu_M)\, \nu_M$ is tangential to M and $T_1 = (T \cdot \nu_M)\, \nu_M$ is normal to M.

Proof of Theorem 11.8 By using partitions of unity, and up to rigid motions and homotheties, it suffices to prove (11.8) for $\varphi \in C^1_c(\mathbf{C})$, assuming that

$$\mathbf{C} \cap M = \Big\{(z, u(z)) : z \in \mathbf{D} \cap E\Big\}, \qquad (11.15)$$

$$\mathbf{C} \cap \Gamma = \Big\{(z, u(z)) : z \in \mathbf{D} \cap \partial E\Big\}, \qquad (11.16)$$

where $u \in C^2(\mathbb{R}^{n-1})$ and E is an open set with C^2-boundary in $\mathbb{R}^{n-1}$ (possibly, $E = \mathbb{R}^{n-1}$). An orientation of the C^2-surface $\mathbf{C} \cap M$ is then given by the vector field $\nu_M \in C^1(\mathbf{C} \cap M; S^{n-1})$, defined as

$$\overline{\nu_M} = \frac{(-\nabla' u, 1)}{\sqrt{1 + |\nabla' u|^2}}, \qquad \text{on } \mathbf{D} \cap E, \tag{11.17}$$

where, if $g\colon \mathbb{R}^n \to \mathbb{R}$, then we set $\overline{g}\colon \mathbb{R}^{n-1} \to \mathbb{R}$, $\overline{g}(z) = g(z, u(z))$ $(z \in \mathbb{R}^{n-1})$. Since $\mathbf{H}_M = H_M\, \nu_M$, we define $\mathbf{H}_M \in C^0(\mathbf{C} \cap M; \mathbb{R}^n)$ and $H_M \in C^0(\mathbf{C} \cap M)$ by taking into account (11.17), and by setting

$$\overline{H_M} = -\mathrm{div}'\left(\frac{\nabla' u}{\sqrt{1 + |\nabla' u|^2}}\right) \qquad \text{on } \mathbf{D} \cap E. \tag{11.18}$$

We now notice that $\overline{\varphi} \in C^1_c(\mathbf{D})$, with

$$\overline{\nabla\varphi \cdot \nu_M} = \frac{-\overline{\nabla'\varphi} \cdot \nabla' u + \overline{\partial_n \varphi}}{\sqrt{1 + |\nabla' u|^2}} \qquad \text{on } \mathbf{D} \cap E.$$

By Theorem 9.1, since $\nabla^M \varphi = \nabla\varphi - (\nabla\varphi \cdot \nu_M)\nu_M$, for $k = 1, ..., n-1$, we find

$$\begin{aligned}
e_n \cdot \int_M \nabla^M \varphi \, d\mathcal{H}^{n-1} &= \int_{\mathbf{D}\cap E} \left(\overline{\partial_n\varphi} + \frac{(\overline{\nabla'\varphi} \cdot \nabla' u - \overline{\partial_n \varphi})}{1 + |\nabla' u|^2}\right)\sqrt{1 + |\nabla' u|^2}\,,\\
e_k \cdot \int_M \nabla^M \varphi \, d\mathcal{H}^{n-1} &= \int_{\mathbf{D}\cap E} \left(\overline{\partial_k\varphi} - \frac{(\overline{\nabla'\varphi} \cdot \nabla' u - \overline{\partial_n \varphi})}{1 + |\nabla' u|^2}\,\partial_k u\right)\sqrt{1 + |\nabla' u|^2}\,,
\end{aligned} \tag{11.19}$$

Vertical component: Concerning (11.19), $\nabla'\overline{\varphi} = \overline{\nabla'\varphi} + \overline{\partial_n\varphi}\nabla' u$ gives

$$\left(\overline{\partial_n\varphi} + \frac{(\overline{\nabla'\varphi} \cdot \nabla' u - \overline{\partial_n \varphi})}{1 + |\nabla' u|^2}\right)\sqrt{1 + |\nabla' u|^2} = \frac{\nabla' u}{\sqrt{1 + |\nabla' u|^2}} \cdot \nabla'\overline{\varphi}\,,$$

and thus, by the divergence theorem (9.10) and since $\overline{\varphi} = 0$ on $\partial\mathbf{D}$,

$$\begin{aligned}
e_n \cdot \int_M & \nabla^M \varphi \, d\mathcal{H}^{n-1}\\
&= -\int_{\mathbf{D}\cap E} \overline{\varphi}\, \mathrm{div}'\left(\frac{\nabla' u}{\sqrt{1 + |\nabla' u|^2}}\right) + \int_{\mathbf{D}\cap\partial E} \overline{\varphi} \frac{\nabla' u \cdot \nu_E}{\sqrt{1 + |\nabla' u|^2}}\, d\mathcal{H}^{n-2}\\
&= e_n \cdot \int_M \varphi\, \mathbf{H}_M \, d\mathcal{H}^{n-1} + e_n \cdot \int_\Gamma \varphi\, \nu^M_\Gamma \, d\mathcal{H}^{n-2}\,,
\end{aligned}$$

provided we define $e_n \cdot \overline{\nu^M_\Gamma}$ on $\mathbf{C} \cap \Gamma$ by the formula

$$e_n \cdot \overline{\nu^M_\Gamma} = \frac{\nabla' u \cdot \nu_E}{\sqrt{1 + |\nabla' u|^2}\,\sqrt{1 + |\nabla^S u|^2}}\,, \tag{11.20}$$

where $S = \mathbf{D} \cap \partial E$ and Corollary 11.7 has been taken into account.

Horizontal components: By Theorem 9.1 and the divergence theorem,

$$
\begin{aligned}
e_k \cdot \int_M \varphi\, \mathbf{H}_M \, d\mathcal{H}^{n-1} &= \int_{\mathbf{D}\cap E} \overline{\varphi}\, \partial_k u \, \mathrm{div}'\Big(\frac{\nabla' u}{\sqrt{1+|\nabla' u|^2}}\Big) \\
&= -\int_{\mathbf{D}\cap E} \frac{\nabla' u}{\sqrt{1+|\nabla' u|^2}} \cdot \nabla'(\overline{\varphi}\, \partial_k u) + \int_{\mathbf{D}\cap \partial E} \overline{\varphi}\, \partial_k u \frac{\nabla' u \cdot \nu_E}{\sqrt{1+|\nabla' u|^2}} \\
&= -\int_{\mathbf{D}\cap E} \overline{\varphi}\, \partial_k\big(\sqrt{1+|\nabla' u|^2}\big) + \partial_k u \frac{\overline{\nabla' \varphi} \cdot \nabla' u}{\sqrt{1+|\nabla' u|^2}} + \partial_k u\, \overline{\partial_n \varphi} \frac{|\nabla' u|^2}{\sqrt{1+|\nabla' u|^2}} \\
&\quad + \int_{\mathbf{D}\cap \partial E} \overline{\varphi}\, \partial_k u \frac{\nabla' u \cdot \nu_E}{\sqrt{1+|\nabla' u|^2}} \,.
\end{aligned}
$$

From (11.19) and by the Gauss–Green theorem, we thus conclude that

$$
\begin{aligned}
& e_k \cdot \int_M (\nabla^M \varphi - \varphi \mathbf{H}_M)\, d\mathcal{H}^{n-1} \\
&= \int_{\mathbf{D}\cap E} (\overline{\partial_k \varphi} + \overline{\partial_n \varphi}\, \partial_k u) \sqrt{1+|\nabla' u|^2} + \overline{\varphi}\, \partial_k\Big(\sqrt{1+|\nabla' u|^2}\Big) \\
&\quad - \int_{\mathbf{D}\cap \partial E} \overline{\varphi}\, \partial_k u \frac{\nabla' u \cdot \nu_E}{\sqrt{1+|\nabla' u|^2}} \\
&= \int_{\mathbf{D}\cap E} \partial_k\Big(\overline{\varphi} \sqrt{1+|\nabla' u|^2}\Big) - \int_{\mathbf{D}\cap \partial E} \overline{\varphi}\, \partial_k u \frac{\nabla' u \cdot \nu_E}{\sqrt{1+|\nabla' u|^2}} \\
&= \int_{\mathbf{D}\cap \partial E} \overline{\varphi} \Big(\sqrt{1+|\nabla' u|^2} e_k - \frac{\partial_k u\, \nabla' u}{\sqrt{1+|\nabla' u|^2}}\Big) \cdot \nu_E \, d\mathcal{H}^{n-2} = e_k \cdot \int_\Gamma \varphi\, \nu_\Gamma^M \, d\mathcal{H}^{n-2} \,,
\end{aligned}
$$

provided we define $e_k \cdot \nu_\Gamma^M$ on $\mathbf{C} \cap \Gamma$ and for $k = 1, ..., n-1$ as

$$
e_k \cdot \overline{\nu_\Gamma^M} = \Big(\sqrt{1+|\nabla' u|^2} e_k - \frac{\partial_k u\, \nabla' u}{\sqrt{1+|\nabla' u|^2}}\Big) \cdot \frac{\nu_E}{\sqrt{1+|\nabla^S u|^2}} \,, \tag{11.21}
$$

and, once again, we take Corollary 11.7 into account.

Geometric properties of ν_Γ^M*:* The proof of the theorem is completed by checking through (11.17), (11.20), and (11.21), that ν_Γ^M is a unit vector which is orthogonal to ν_M and which is normal to Γ. □

Notes

Virtually all books in Geometric Measure Theory (GMT) are opened by a *vulgata editio* of [Fed69, Chapters Two and Three]. We make no exception to this rule, overlapping with various monographs, like [Sim83, Fal86, EG92, AFP00, GMS98a, Mat95], that readers may use as a source for further results and bibliographical information. We now make a few remarks on the contents of Part I.

We have considered Radon measures on $\mathbb{R}^n$, and not on more general ambient spaces, since nothing more is needed to develop the whole theory of the book, which includes existence, characterization, symmetrization, regularity, and analysis of singularities results for minimizers in a wide list of geometric variational problems. (In fact, Radon measures on a proper open set A of $\mathbb{R}^n$ are needed in some minor aspects of the compactness theory for perimeter minimizers; see Part III. This trivial generalization is developed through Exercises 2.18, 4.19, and 4.37.) However, Radon measures on more general ambient spaces play an important role in other parts of GMT, as well as in other areas of Mathematics. The reader may thus desire to clarify the natural assumptions under which the various results presented in this part hold true. To this end, reading [Fed69, Chapter Two] is a perfect starting point, as well as a good occasion to get acquainted with Federer's peculiar and fascinating writing style.

In Section 1.2 we have presented Vitali's example. In fact, in higher dimensions, one can show that the Lebesgue measure fails to be *finitely* additive on $\mathcal{P}(\mathbb{R}^n)$, by means of the famous Banach–Tarski paradox; see, for example, [Str79].

In Chapter 3 we have introduced the notion of Hausdorff dimension. Although sets with non-integer Hausdorff dimension and, more generally, fractal sets, play no role in this book, they are an important subject of study in GMT. We refer readers to Hutchinson [Hut81], Falconer [Fal86, Fal90, Fal97], and Mattila [Mat95] for more information. In particular, Section 3.2 is based on [Fal86].

I think this is a bit surprising, but, usually, Campanato's criterion (Section 6.1) is not presented as one of the major applications of the Besicovitch–Lebesgue differentiation theorem. Hölder regularity of solutions to second order elliptic equations in divergence form can be elegantly derived starting from Campanato's criterion; see, for example, the very accessible account by Giaquinta and Martinazzi [GM05, Chapter 5].

In discussing the area formula, we have deliberately avoided the use of the Cauchy–Binet formula, which appears in several accounts on the subject as a tool for computing Jacobians; see, for example, [EG92, Section 3.2, Theorem 4], [AFP00, Proposition 2.69], [KP08, Theorem 1.5.2]. Indeed, by exploiting the tensor representation of linear functions, and by choosing suitable orthonormal frames, it turns out that, at least for the computations needed in this book, it suffices to apply simple identities like $\det(\mathrm{Id} + v \otimes v) = 1 + |v|^2$, see Equation (9.3).

Our presentation of rectifiable sets, so to say, just scratches the tip of the iceberg. The beautiful theory of one-dimensional rectifiable sets, as originally developed by Besicovitch and collaborators, is found in [Fal86]. In higher dimension, the reader may consult in particular [AFP00, Chapter 2], [Mat95], and [DL08], in addition to more classical references such as [Fed69, Chapter 3] and [Sim83, Chapter 3].

PART TWO

Sets of finite perimeter

Synopsis

The starting point of the theory of sets of finite perimeter is a generalization of the Gauss–Green theorem based on the notion of vector-valued Radon measure. Precisely, we say that a Lebesgue measurable set $E \subset \mathbb{R}^n$ is a set of locally finite perimeter if there exists a $\mathbb{R}^n$-valued Radon measure μ_E on $\mathbb{R}^n$, called the Gauss–Green measure of E, such that the generalized Gauss–Green formula

$$\int_E \nabla\varphi = \int_{\mathbb{R}^n} \varphi \, d\mu_E \,, \qquad \forall \varphi \in C^1_c(\mathbb{R}^n)\,, \tag{1}$$

holds true. The total variation measure $|\mu_E|$ of μ_E induces the notions of relative perimeter $P(E; F)$ of E with respect to a set $F \subset \mathbb{R}^n$, and of (total) perimeter $P(E)$ of E, defined as

$$P(E; F) = |\mu_E|(F)\,, \qquad P(E) = |\mu_E|(\mathbb{R}^n)\,.$$

In particular, E is a set of finite perimeter if and only if $P(E) < \infty$. These definitions are motivated by the classical Gauss–Green theorem, Theorem 9.3. Indeed, if E is an open set with C^1-boundary with outer unit normal $\nu_E \in C^0(\partial E; S^{n-1})$, then Theorem 9.3 implies

$$\int_E \nabla\varphi = \int_{\partial E} \varphi \, \nu_E \, d\mathcal{H}^{n-1}\,, \qquad \forall \varphi \in C^1_c(\mathbb{R}^n)\,, \tag{2}$$

and thus E is a set of locally finite perimeter with

$$\mu_E = \nu_E \, \mathcal{H}^{n-1} \llcorner \partial E\,, \qquad |\mu_E| = \mathcal{H}^{n-1} \llcorner \partial E\,, \tag{3}$$

$$P(E; F) = \mathcal{H}^{n-1}(F \cap \partial E)\,, \qquad P(E) = \mathcal{H}^{n-1}(\partial E)\,, \tag{4}$$

for every $F \subset \mathbb{R}^n$; see Figure 1. One of the main themes of this part of the book is showing that these definitions lead to a geometrically meaningful generalization of the notion of open set with C^1-boundary, with natural and powerful applications to the study of geometric variational problems.

We start this programme in Chapter 12, where the link with the theory of Radon measures established by (1) is exploited to deduce some basic lower semicontinuity and compactness theorems for sequences of sets of locally finite perimeter; see Sections 12.1 and 12.4. In particular, these results make it

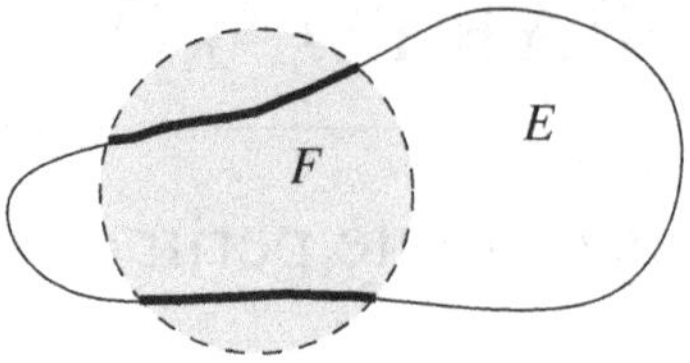

Figure 1 The perimeter $P(E; F)$ of E relative to F is the $(n-1)$-dimensional measure of the intersection of the (reduced) boundary of E with F.

possible to apply the Direct Method in order to prove the existence of minimizers in several geometric variational problems, see Section 12.5.

In Chapter 13 we discuss the possibility of approximating sets of finite perimeter by sequences of open sets with smooth boundary. The resulting approximation theorems appear often as useful technical devices, but also possess another merit. Indeed, generally speaking, they imply the coincidence of the minimum values of the different formulations of the same variational problems that are obtained by minimizing either among sets of finite perimeter or among open sets with C^1-boundary. Another relevant content of Chapter 13 is the *coarea formula*, which is a generalization of Fubini's theorem of ubiquitous importance in Geometric Measure Theory.

In Chapter 14 we study the Euclidean isoperimetric problem: given $m > 0$, minimize perimeter among sets of volume m, namely

$$\inf\Big\{P(E) : |E| = m\Big\}.$$

Exploiting the lower semicontinuity, compactness, and approximation theorems developed in the two previous chapters, together with the notion of Steiner symmetrization, we shall characterize Euclidean balls as the (unique) minimizers in the Euclidean isoperimetric problem. A remarkable feature of this result and, more generally, of the results from the first three chapters of this part, is that they are only based on the tools from basic Measure Theory and Functional Analysis set forth in Chapters 1–4, and that they are obtained without any knowledge on the geometric structure of arbitrary sets of finite perimeter.

We next turn to the following, fundamental question: does the validity of (1) imply a set of locally finite perimeter E to possess, in some suitable sense, a $(n-1)$-dimensional boundary and outer unit normal allowing us, for example, to generalize (2), (3), and (4)? The first important remark here is that the notion of topological boundary is of little use in answering this question. Indeed, if E is of locally finite perimeter and E' is equivalent to E (i.e., $|E\Delta E'| = 0$), then, as the left-hand side of (1) is left unchanged by replacing E with E', we have that E' is a set of locally finite perimeter too, with $\mu_E = \mu_{E'}$. Of course, the topological boundaries of E and E' may be completely different (for example, even if E is an open set with C^1-boundary, we may take $E' = E \cup \mathbb{Q}^n$, and have $\partial E' = \mathbb{R}^n$, $\mu_{E'} = \nu_E \mathcal{H}^{n-1} \llcorner \partial E$). For this reason, when dealing with sets of finite perimeter, it is always useful to keep in mind the possibility of making modifications on and/or by sets of measure zero to find a representative which "minimizes the size of the topological boundary". In other words, if E is of locally finite perimeter, then we always have $\operatorname{spt}\mu_E \subset \partial E$, and we can always find E' equivalent to E such that $\operatorname{spt}\mu_{E'} = \partial E'$; see Proposition 12.19. But even with these specifications in mind, we have to face the existence of sets of finite perimeter $E \subset \mathbb{R}^n$, $n \geq 2$, with $|E| < \infty$, $|\operatorname{spt}\mu_E| > 0$, and thus, in particular, $\mathcal{H}^{n-1}(\operatorname{spt}\mu_E) = \infty$; see Example 12.25. In conclusion, even after the suitable "minimization of size", the topological boundary of a set of finite perimeter may have Hausdorff dimension equal to that of its ambient space!

The key notion to consider in order to understand the geometric structure of sets of finite perimeter is that of *reduced boundary*, which may be explained as follows. If E is an open set with C^1-boundary, then the continuity of the outer unit normal ν_E allows us to characterize $\nu_E(x)$ in terms of the Gauss–Green measure $\mu_E = \nu_E \mathcal{H}^{n-1} \llcorner \partial E$ as

$$\nu_E(x) = \lim_{r \to 0^+} \fint_{B(x,r)\cap \partial E} \nu_E \, d\mathcal{H}^{n-1} = \lim_{r\to 0^+} \frac{\mu_E(B(x,r))}{|\mu_E|\,(B(x,r))}, \qquad \forall x \in \partial E\,.$$

If now E is a generic set of locally finite perimeter, then $|\mu_E|\,(B(x,r)) > 0$ for every $x \in \operatorname{spt}\mu_E$ and $r > 0$, and thus it makes sense to define the reduced boundary $\partial^* E$ of E as the set of those $x \in \operatorname{spt}\mu_E$ such that the limit

$$\lim_{r\to 0^+} \frac{\mu_E(B(x,r))}{|\mu_E|\,(B(x,r))} \quad \text{exists and belongs to } S^{n-1}. \tag{5}$$

In analogy with the regular case, the Borel vector field $\nu_E \colon \partial^* E \to S^{n-1}$ defined in (5) is called the *measure-theoretic outer unit normal to E*. The reduced boundary and the measure-theoretic outer unit normal depend on E only through its Gauss–Green measure, and are therefore left unchanged by modifications of E on and/or by a set of measure zero. It also turns out that $\partial^* E$ has the structure of an $(n-1)$-dimensional surface, that ν_E has a precise geometric meaning as the outer unit normal to E, and that (3) and (4) hold true on generic sets of finite perimeter by replacing topological boundaries and classical outer unit normals with reduced boundaries and measure-theoretic outer unit normals. Precisely, the following statements from *De Giorgi's structure theory*, presented in Chapter 15, hold true:

(i) The Gauss–Green measure μ_E is obtained by integrating ν_E against the restriction of $\mathcal{H}^{n-1}$ to $\partial^* E$, that is,

$$\mu_E = \nu_E \mathcal{H}^{n-1} \llcorner \partial^* E\,, \qquad |\mu_E| = \mathcal{H}^{n-1} \llcorner \partial^* E\,,$$
$$P(E;F) = \mathcal{H}^{n-1}(F \cap \partial^* E)\,, \qquad P(E) = \mathcal{H}^{n-1}(\partial^* E)\,,$$

for every $F \subset \mathbb{R}^n$, and the Gauss–Green formula (1) takes the form

$$\int_E \nabla\varphi = \int_{\partial^* E} \varphi\, \nu_E \, \mathrm{d}\mathcal{H}^{n-1}\,, \qquad \forall \varphi \in C^1_{\mathrm{c}}(\mathbb{R}^n)\,.$$

(ii) If $x \in \partial^* E$, then $\nu_E(x)$ is orthogonal to $\partial^* E$ at x, in the sense that

$$\mathcal{H}^{n-1} \llcorner \Big(\frac{\partial^* E - x}{r}\Big) \stackrel{*}{\rightharpoonup} \mathcal{H}^{n-1} \llcorner \nu_E(x)^\perp \qquad \text{as } r \to 0^+,$$

and it is an *outer* unit normal to E at x, in the sense that

$$\frac{E - x}{r} \stackrel{\mathrm{loc}}{\to} \big\{ y \in \mathbb{R}^n : y \cdot \nu_E(x) \le 0 \big\} \qquad \text{as } r \to 0^+\,.$$

(iii) The reduced boundary $\partial^* E$ is the union of (at most countably many) compact subsets of C^1-hypersurfaces is $\mathbb{R}^n$; more precisely, there exist at most countably many C^1-hypersurfaces M_h and compact sets $K_h \subset M_h$ with $T_x M_h = \nu_E(x)^\perp$ for every $x \in K_h$, such that

$$\partial^* E = N \cup \bigcup_{h \in \mathbb{N}} K_h\,, \qquad \mathcal{H}^{n-1}(N) = 0\,.$$

Statement (iii) implies of course that the reduced boundary of a set of locally finite perimeter is a locally $\mathcal{H}^{n-1}$-rectifiable set. In Chapter 16 we undertake the study of reduced boundaries and Gauss–Green measures in the light of the theory of rectifiable sets developed in Chapter 10. We prove *Federer's theorem*, stating the $\mathcal{H}^{n-1}$-equivalence between the reduced boundary of E, the set $E^{(1/2)}$ of its points of density one-half, and the essential boundary $\partial^{\mathrm{e}} E$, which is defined as the complement in $\mathbb{R}^n$ of $E^{(0)} \cup E^{(1)}$. This result proves a powerful tool, as sets of density points are much more easily manipulated than reduced boundaries. For example, it is starting from Federer's theorem that in Section 16.1 we prove some representation formulae for Gauss–Green measures of unions, intersections, and set differences of two sets of locally finite perimeter. These formulae allow us to easily "cut and paste" sets of finite perimeter, an operation which proves useful in building comparison sets for testing minimality conditions. As an application of these techniques, in Section 16.2 we prove upper and lower density estimates for reduced boundaries of local perimeter minimizers, which, combined with Federer's theorem, imply a first, mild, regularity property of local perimeter minimizers: the $\mathcal{H}^{n-1}$-equivalence between the reduced boundary and the support of the Gauss–Green measure, that is, as said, the topological boundary of "minimal size".

In Chapter 17 we apply the area formula of Chapter 8 to study the behavior of sets of finite perimeter under the action of one parameter families of diffeomorphisms. We compute the first and second variation formulae of perimeter, and introduce distributional formulations of classical first order necessary minimality conditions, like the vanishing mean curvature condition.

In Chapter 18 we present a refinement of the coarea formula from Chapter 13, which in turn allows us to discuss slicing of reduced boundaries. In particular, slicing by hyperplanes is discussed in some detail in Section 18.3.

We close Part II by briefly introducing two important examples of geometric variational problems which can be addressed in our framework. Precisely, in Chapter 19 we discuss the equilibrium problem for a liquid confined inside a given container, while in Chapter 20 we consider anisotropic surface energies and address the so-called Wulff problem, originating from the study of equlibrium shapes of crystals.

12

Sets of finite perimeter and the Direct Method

Let E be a Lebesgue measurable set in $\mathbb{R}^n$. We say that E is a **set of locally finite perimeter** in $\mathbb{R}^n$ if for every compact set $K \subset \mathbb{R}^n$ we have

$$\sup\left\{\int_E \operatorname{div} T(x)\,\mathrm{d}x : T \in C^1_c(\mathbb{R}^n;\mathbb{R}^n)\,,\ \operatorname{spt} T \subset K\,,\ \sup_{\mathbb{R}^n}|T| \le 1\right\} < \infty\,. \quad (12.1)$$

If this quantity is bounded independently of K, then we say that E is a **set of finite perimeter** in $\mathbb{R}^n$.

Proposition 12.1 *If E is a Lebesgue measurable set in $\mathbb{R}^n$, then E is a set of locally finite perimeter if and only if there exists a $\mathbb{R}^n$-valued Radon measure μ_E on $\mathbb{R}^n$ such that*

$$\int_E \operatorname{div} T = \int_{\mathbb{R}^n} T \cdot \mathrm{d}\mu_E\,, \qquad \forall T \in C^1_c(\mathbb{R}^n;\mathbb{R}^n)\,. \quad (12.2)$$

Moreover, E is a set of finite perimeter if and only if $|\mu_E|(\mathbb{R}^n) < \infty$.

Remark 12.2 Of course (12.2) is equivalent to

$$\int_E \nabla\varphi = \int_{\mathbb{R}^n} \varphi\,\mathrm{d}\mu_E\,, \qquad \forall \varphi \in C^1_c(\mathbb{R}^n)\,. \quad (12.3)$$

We call μ_E the **Gauss–Green measure** of E, and define the **relative perimeter** of E in $F \subset \mathbb{R}^n$, and the **perimeter** of E, as

$$P(E;F) = |\mu_E|(F)\,, \qquad P(E) = |\mu_E|(\mathbb{R}^n)\,.$$

By Exercise 4.14, μ_E is uniquely determined as a Radon measure on $\mathbb{R}^n$.

Remark 12.3 In connection with the terminology introduced in Section 7.2, we notice that a Lebesgue measurable set $E \subset \mathbb{R}^n$ is a set of locally finite perimeter if and only if the distributional gradient $D1_E$ of $1_E \in L^1_{\mathrm{loc}}(\mathbb{R}^n)$ can be represented as the integration with respect to the $\mathbb{R}^n$-valued Radon measure

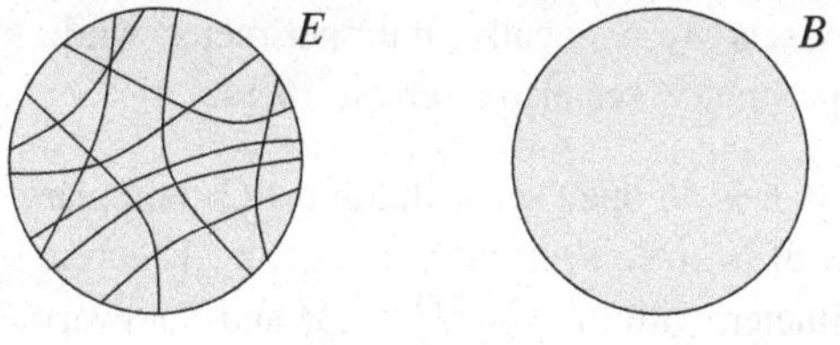

Figure 12.1 The set $E \subset \mathbb{R}^2$ is equivalent to the unit disk B. They both have distributional perimeter 2π, although we may arrange things so that $\mathcal{H}^1(\partial E)$ takes any value in $(2\pi, +\infty]$. In fact, $F = E \cup \mathbb{Q}^2$ is equivalent to B, has $\nu_B \mathcal{H}^1 \llcorner \partial B$ as its Gauss–Green measure, but is such that $|\partial F| = \infty$.

$-\mu_E$. Therefore we speak of *distributional perimeter* and we refer to (12.3) as the *distributional Gauss–Green theorem*.

Remark 12.4 If E is a set of (locally) finite perimeter in $\mathbb{R}^n$ and $|E\Delta F| = 0$, then F is a set of (locally) finite perimeter and $\mu_F = \mu_E$; the converse is also true, see Exercise 12.10. In particular, the perimeter $P(E)$ of E is invariant by modifications of E on and/or by a set of measure zero, although these modifications may wildly affect the size of its topological boundary of E; see Figure 12.1. Moreover, every set of Lebesgue measure zero is of finite perimeter, and has perimeter zero.

Proof of Proposition 12.1 Let E be a set of locally finite perimeter in $\mathbb{R}^n$, and consider the linear functional $L\colon C^1_c(\mathbb{R}^n;\mathbb{R}^n) \to \mathbb{R}$ defined by $\langle L, T\rangle = \int_E \operatorname{div} T(x)\mathrm{d}x$. For every compact set $K \subset \mathbb{R}^n$ there exists $C(K) \in \mathbb{R}$ such that $|\langle L, T\rangle| \le C(K)\sup_{\mathbb{R}^n}|T|$ whenever $\operatorname{spt} T \subset K$. Hence, L can be extended by density to a bounded continuous linear functional on $C^0_c(\mathbb{R}^n;\mathbb{R}^n)$, and the existence of μ_E follows by Riesz's theorem (Theorem 4.7). Clearly, if E is a set of finite perimeter then $|\mu_E|(\mathbb{R}^n) < \infty$. The converse implications are trivial. Indeed if $K \subset \mathbb{R}^n$ is compact, $T \in C^1_c(\mathbb{R}^n;\mathbb{R}^n)$ with $|T| \le 1$ on $\mathbb{R}^n$ and $\operatorname{spt} T \subset K$, then by (12.2) we have $\int_E \operatorname{div} T(x)\mathrm{d}x \le |\mu_E|(K)$. □

Example 12.5 By the Gauss–Green theorem, if $E \subset \mathbb{R}^n$ is an open (not necessarily bounded) set with C^1 boundary, then $\nu_E \mathcal{H}^{n-1} \llcorner \partial E$ is a $\mathbb{R}^n$-valued Radon measure on $\mathbb{R}^n$ such that (12.3) holds true. Hence E is a set of locally finite perimeter, with Gauss–Green measure $\mu_E = \nu_E \mathcal{H}^{n-1} \llcorner \partial E$, $P(E) = \mathcal{H}^{n-1}(\partial E)$, and $P(E; F) = \mathcal{H}^{n-1}(F \cap \partial E)$ for every $F \subset \mathbb{R}^n$.

Example 12.6 By Remark 9.5, an open set E with Lipschitz or polyhedral boundary is a of locally finite perimeter, with $P(E; F) = \mathcal{H}^{n-1}(F \cap \partial E)$ whenever $F \subset \mathbb{R}^n$. Moreover, if E is bounded, then E is of finite perimeter. In particular, convex sets are of locally finite perimeter, while *bounded* convex sets are of finite perimeter; see also Exercise 15.14.

Example 12.7 If E is an open set with almost C^1-boundary in $\mathbb{R}^n$, and if M is the regular part of ∂E (see Section 9.3), then, by Theorem 9.6, E is a set of locally finite perimeter, with $\mu_E = \nu_E\,\mathcal{H}^{n-1} \llcorner M$ and, for every $F \subset \mathbb{R}^n$,

$$P(E;F) = \mathcal{H}^{n-1}(F \cap M) = \mathcal{H}^{n-1}(F \cap \partial E)\,.$$

Exercise 12.8 (Scaling and translation) If $\lambda > 0$, $x \in \mathbb{R}^n$ and E is a set of finite perimeter in $\mathbb{R}^n$ then $x + \lambda\,E$ is a set of finite perimeter with

$$P(x + \lambda\,E) = \lambda^{n-1}\,P(E)\,.$$

More generally, if E is a set of locally finite perimeter in $\mathbb{R}^n$ then $x+\lambda E$ is a set of locally finite perimeter with $\mu_{x+\lambda E} = \Phi_\# \mu_E$, where $\Phi(y) = x + \lambda\,y$, $y \in \mathbb{R}^n$.

Exercise 12.9 (Complement) If E is a set of locally finite perimeter, then $\mathbb{R}^n \setminus E$ is a set of locally finite perimeter with

$$\mu_{\mathbb{R}^n \setminus E} = -\mu_E\,, \qquad P(E) = P(\mathbb{R}^n \setminus E)\,. \tag{12.4}$$

Hint: Apply Exercise 1.12.

Exercise 12.10 If E and F are sets of locally finite perimeter, then $\mu_E = \mu_F$ on $\mathcal{B}_b(\mathbb{R}^n)$ if and only if $|E \Delta F| = 0$. Characterize the case $\mu_E = -\mu_F$ on $\mathcal{B}_b(\mathbb{R}^n)$.

Exercise 12.11 If E is a set of locally finite perimeter in $\mathbb{R}^n$ and $Q \in \mathbf{O}(n)$, then $Q(E)$ is a set of locally finite perimeter in $\mathbb{R}^n$, $P(Q(E)) = P(E)$, and

$$\mu_{Q(E)}(F) = Q^*\,\mu_E(Q^*(F))\,, \qquad \forall F \in \mathcal{B}_b(\mathbb{R}^n)\,. \tag{12.5}$$

Hint: Given $\varphi \in C^1_c(\mathbb{R}^n)$, apply (12.2) to φ on $Q(E)$, change variables by (8.5), and consider that, if $\psi(x) = \varphi(Qx)$, then $\nabla\psi(x) = Q^*\nabla\varphi(Qx)$ for every $x \in \mathbb{R}^n$.

Exercise 12.12 Show that (12.2) holds true on a compactly supported Lipschitz vector field $T\colon \mathbb{R}^n \to \mathbb{R}^n$. *Hint:* If we set $T_\varepsilon = T \star \rho_\varepsilon$, then $T_\varepsilon \to T$ uniformly on $\mathbb{R}^n$, $\nabla T_\varepsilon \overset{*}{\rightharpoonup} \nabla T$ in $L^\infty(\mathbb{R}^n;\mathbb{R}^n)$, and the supports of the T_ε are uniformly bounded.

Proposition 12.13 (Sets of finite perimeter in $\mathbb{R}$) *A Lebesgue measurable set $E \subset \mathbb{R}$ is of locally finite perimeter if and only if it is equivalent to a countable union of (possibly unbounded) open intervals lying at mutually positive distance.*

Proof The "if" part being trivial, we focus on the "only if" assertion. We first assume that E is a of locally finite perimeter in $\mathbb{R}$, with $E \subset (a,\infty)$ for some $a \in \mathbb{R}$. In this way, it makes sense to define a Borel function $u\colon \mathbb{R} \to \mathbb{R}$ as

$$u(x) = \mu_E((-\infty, x))\,, \qquad x \in \mathbb{R}\,.$$

Indeed, since $|\mu_E|(-\infty, a) = 0$, the signed Radon measure μ_E is well defined (through the Jordan decomposition $\mu_E = \mu_E^+ - \mu_E^-$) on every Borel set in $\mathbb{R}$ which is bounded from above. By Fubini's theorem, if $\varphi \in C^1_c(\mathbb{R})$ then

$$\int_{\mathbb{R}} u\varphi' = \int_{\mathbb{R}} \varphi'(x)\mathrm{d}x \int_{-\infty}^{x} \mathrm{d}\mu_E = \int_{\mathbb{R}} \mathrm{d}\mu_E(y) \int_y^{\infty} \varphi' = -\int_{\mathbb{R}} \varphi \mathrm{d}\mu_E = -\int_E \varphi'\,.$$

In particular, $\int_{\mathbb{R}}(u + 1_E)\varphi' = 0$ for every $\varphi \in C^1_c(\mathbb{R})$. By Lemma 7.5, there exists $c \in \mathbb{R}$ such that $u + 1_E = c$ a.e. on $\mathbb{R}$. Since $1_E = 0$ on $(-\infty, a)$ and $|\mu_E|(-\infty, a) = 0$, we have $c = 0$, that is, $u(x) \in \{0, 1\}$ for a.e. $x \in \mathbb{R}$. Since u is left-continuous on $\mathbb{R}$, $u(x) \in \{0, 1\}$ for every $x \in \mathbb{R}$, that is, for some $F \subset \mathbb{R}$,

$$1_F(x) = u(x) = -\mu_E((-\infty, x))\,, \qquad \forall x \in \mathbb{R}\,.$$

In particular $|E\Delta F| = 0$, and, in turn,

$$\limsup_{h\to 0} |1_F(x+h) - 1_F(x)| \le |\mu_E|(\{x\})\,, \qquad \forall x \in \mathbb{R}\,.$$

Hence, $|\mu_E|(\{x\}) \ge 1$ at every discontinuity point x of 1_F. Since $|\mu_E|$ is locally finite in $\mathbb{R}$, the set of discontinuity points of 1_F is at most countable and has no accumulation points. Thus F is a countable union of (possibly unbounded) left-open/right-closed intervals, lying at mutually positive distance. The proof is concluded in the case when E is bounded from below. In the general case we apply the above argument to the sets $E_k = E \cap (-k, \infty)$ ($k \in \mathbb{N}$), which are of locally finite perimeter by Lemma 12.22 below. □

Exercise 12.14 If E is a set of locally finite perimeter in $\mathbb{R}$, then $|\mu_E| = \mathcal{H}^0 \llcorner E^{(1/2)}$, where $E^{(1/2)}$ is the set of points of density $1/2$ of E. In particular, $P(E;F) = \mathcal{H}^0(F \cap E^{(1/2)})$ for every Borel set $F \subset \mathbb{R}$.

12.1 Lower semicontinuity of perimeter

Given Lebesgue measurable sets $\{E_h\}_{h\in\mathbb{N}}$ and E in $\mathbb{R}^n$, we say that E_h **locally converges** to E, and write $E_h \overset{\mathrm{loc}}{\to} E$, if

$$\lim_{h\to\infty} \Big|K \cap (E\Delta E_h)\Big| = 0\,, \qquad \forall K \subset \mathbb{R}^n \text{ compact}\,.$$

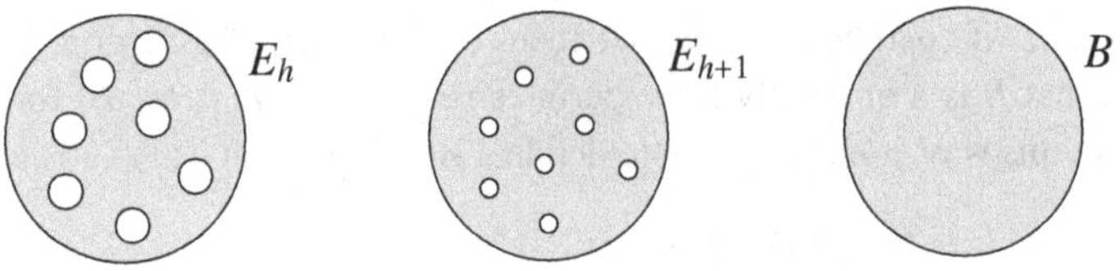

Figure 12.2 Inequality (12.8) may be strict. In this picture E_h is obtained by removing h disks of radius $h^{-\alpha}$, $\alpha \in (1/2, 1]$, from the unit disk $B \subset \mathbb{R}^2$. In this way $|B\Delta E_h| = h(\pi h^{-2\alpha}) \to 0$, $P(B) = 2\pi$ and $P(E_h) = 2\pi + h(2\pi h^{-\alpha}) \to \infty$ if $\alpha < 1$ or $P(E_h) \to 4\pi$ if $\alpha = 1$.

We simply say that E_h **converges** to E, and write $E_h \to E$, if

$$\lim_{h\to\infty} |E\Delta E_h| = 0\,.$$

The distributional perimeter is naturally lower semicontinuous with respect to local convergence. Indeed, if A is an open set on $\mathbb{R}^n$ and E is a set of locally finite perimeter in $\mathbb{R}^n$, then by Proposition 12.1 and (4.6) we find

$$P(E;A) = \sup\left\{ \int_E \operatorname{div} T(x)\,\mathrm{d}x \colon T \in C_c^\infty(A;\mathbb{R}^n)\,, \sup_{\mathbb{R}^n} |T| \le 1 \right\}. \tag{12.6}$$

Proposition 12.15 (Lower semicontinuity of perimeter) *If $\{E_h\}_{h\in\mathbb{N}}$ is a sequence of sets of locally finite perimeter in $\mathbb{R}^n$, with*

$$E_h \overset{\text{loc}}{\to} E\,, \qquad \limsup_{h\to\infty} P(E_h;K) < \infty\,, \tag{12.7}$$

for every compact set K in $\mathbb{R}^n$, then E is of locally finite perimeter in $\mathbb{R}^n$, $\mu_{E_h} \overset{*}{\rightharpoonup} \mu_E$ and, for every open set $A \subset \mathbb{R}^n$ we have

$$P(E;A) \le \liminf_{h\to\infty} P(E_h;A)\,. \tag{12.8}$$

Proof If A is open, $T \in C_c^\infty(A;\mathbb{R}^n)$ and $|T| \le 1$ on $\mathbb{R}^n$, then by (12.6),

$$\int_E \operatorname{div} T(x)\,\mathrm{d}x = \lim_{h\to\infty} \int_{E_h} \operatorname{div} T(x)\mathrm{d}x \le \liminf_{h\to\infty} P(E_h;A)\,. \tag{12.9}$$

By (12.6), (12.7), and by applying (12.9) with A bounded, we see that E is of locally finite perimeter in $\mathbb{R}^n$, and that (12.8) holds true (even if A is unbounded). By (12.3) and since $E_h \overset{\text{loc}}{\to} E$, we have that

$$\lim_{h\to\infty} \int_{\mathbb{R}^n} \varphi\,\mathrm{d}\mu_{E_h} = \int_{\mathbb{R}^n} \varphi\,\mathrm{d}\mu_E\,, \qquad \forall \varphi \in C_c^\infty(\mathbb{R}^n)\,. \tag{12.10}$$

By the density of $C_c^\infty(\mathbb{R}^n)$ into $C_c^0(\mathbb{R}^n)$ and by (12.7), we easily deduce (12.10) for every $\varphi \in C_c^0(\mathbb{R}^n)$, thus proving that $\mu_{E_h} \overset{*}{\rightharpoonup} \mu_E$. □

Exercise 12.16 (Locality of perimeter) If E and F are of locally finite perimeter in $\mathbb{R}^n$, A is open, and $|(E\Delta F)\cap A| = 0$, then $P(E;A) = P(F;A)$.

Exercise 12.17 If A is an open connected set in $\mathbb{R}^n$, E is of locally finite perimeter, and $P(E;A) = 0$, then either $|A\setminus E| = 0$ or $|E\cap A| = |A|$. *Hint:* Combine Lemma 7.5 and (12.6).

Exercise 12.18 If E is a Lebesgue measurable set in $\mathbb{R}^n$, $\{u_h\}_{h\in\mathbb{N}}\subset C^1_c(\mathbb{R}^n)$, $u_h\to 1_E$ in $L^1_{\rm loc}(\mathbb{R}^n)$, and, for every compact set K in $\mathbb{R}^n$,

$$\limsup_{h\to\infty}\int_K |\nabla u_h| < \infty\,,$$

then E is of locally finite perimeter in $\mathbb{R}^n$, with

$$P(E;A)\le \liminf_{h\to\infty}\int_A |\nabla u_h|\,,\qquad \text{for every } A\subset\mathbb{R}^n \text{ open.} \tag{12.11}$$

Hint: Use $\int_E \operatorname{div} T = \lim_{h\to\infty}\int_{\mathbb{R}^n} u_h \operatorname{div} T = -\lim_{h\to\infty}\int_{\mathbb{R}^n} T\cdot\nabla u_h$ for $T\in C^1_c(\mathbb{R}^n;\mathbb{R}^n)$.

12.2 Topological boundary and Gauss–Green measure

As seen in Remark 12.4, we may modify a set of locally finite perimeter E on and/or by a set of measure zero without changing its Gauss–Green measure, and, as a consequence, its perimeter. Such modifications may largely increase the topological boundary. In the following lemma it is shown how to modify E to "minimize" the size of the topological boundary.

Proposition 12.19 *If E is a set of locally finite perimeter in $\mathbb{R}^n$, then*

$$\operatorname{spt}\mu_E = \Big\{x\in\mathbb{R}^n : 0<|E\cap B(x,r)|<\omega_n r^n\ \forall r>0\Big\}\subset\partial E\,. \tag{12.12}$$

Moreover, there exists a Borel set F such that

$$|E\Delta F| = 0\,,\qquad \operatorname{spt}\mu_F = \partial F\,.$$

Proof *Step one:* If $x\in\mathbb{R}^n$ is such that $|E\cap B(x,r)| = 0$ for some $r>0$, then

$$0=\int_E \nabla\varphi = \int_{\mathbb{R}^n}\varphi\, d\mu_E\,,\qquad \forall\varphi\in C^\infty_c(B(x,r))\,.$$

Thus $|\mu_E|(B(x,r)) = 0$ and $x\notin\operatorname{spt}\mu_E$. Similarly, if $x\in\mathbb{R}^n$ and $|E\cap B(x,r)| = |B(x,r)|$, then $x\notin\operatorname{spt}\mu_E$, since

$$0=\int_{B(x,r)}\nabla\varphi = \int_E\nabla\varphi = \int_{\mathbb{R}^n}\varphi\, d\mu_E\,,\qquad\forall\varphi\in C^\infty_c(B(x,r))\,.$$

Finally, if $x \notin \operatorname{spt}\mu_E$, then $|\mu_E|(B(x,r)) = 0$ for some $r > 0$, and

$$0 = \int_{\mathbb{R}^n} \varphi \, d\mu_E = \int_E \nabla\varphi = \int_{\mathbb{R}^n} 1_E \nabla\varphi \,, \qquad \forall \varphi \in C_c^\infty(B(x,r)) \,.$$

By Lemma 7.5, there exists $c \in \mathbb{R}$ such that $1_E = c$ a.e. on $B(x,r)$. Necessarily, $c \in \{0,1\}$, and, correspondingly, $|E \cap B(x,r)| \in \{0, \omega_n r^n\}$. This proves (12.12).

Step two: Up to modifying E on a set of measure zero we may assume that E is a Borel set. We now construct a Borel set F with $|F \Delta E| = 0$ and

$$\partial F = \Big\{ x \in \mathbb{R}^n : 0 < |F \cap B(x,r)| < \omega_n r^n \text{ for every } r > 0 \Big\} .$$

To this end, let us define two disjoint open sets by setting

$$A_0 = \Big\{ x \in \mathbb{R}^n : \text{there exists } r > 0 \text{ s.t. } 0 = |E \cap B(x,r)| \Big\} ,$$
$$A_1 = \Big\{ x \in \mathbb{R}^n : \text{there exists } r > 0 \text{ s.t. } |E \cap B(x,r)| = \omega_n r^n \Big\} ,$$

and consider a sequence $\{x_h\}_{h\in\mathbb{N}} \subset A_0$ such that $A_0 \subset \bigcup_{h\in\mathbb{N}} B(x_h, r_h)$, $r_h > 0$, and $|E \cap B(x_h, r_h)| = 0$. Hence $|E \cap A_0| = 0$ and, by Exercise 12.9, we also have $|A_1 \setminus E| = 0$. Therefore, if we set

$$F = (A_1 \cup E) \setminus A_0 \,,$$

then F is a Borel set, with

$$|F \setminus E| \le |A_1 \setminus E| = 0 \,, \qquad |E \setminus F| \le |E \cap A_0| = 0 \,,$$

that is, $|E \Delta F| = 0$. By step one, $\mathbb{R}^n \setminus (A_0 \cup A_1) = \operatorname{spt}\mu_E = \operatorname{spt}\mu_F \subset \partial F$. At the same time, $\partial F \subset \mathbb{R}^n \setminus (A_0 \cup A_1)$, since, by construction,

$$A_1 \subset \mathring{F} \,, \qquad \overline{F} \subset \mathbb{R}^n \setminus A_0 \,.$$

□

12.3 Regularization and basic set operations

We study here the properties of convolutions of characteristic functions of sets of locally finite perimeter with regularizing kernels. In this way we obtain a simple but useful technical tool (see, for example, the proof of Lemma 12.22) and we gain some further insight on the notion of distributional perimeter. Let E be a Lebesgue measurable set, so that $1_E \in L^1_{\mathrm{loc}}(\mathbb{R}^n)$, and consider the ε-regularization $(1_E \star \rho_\varepsilon)$ of 1_E,

$$(1_E \star \rho_\varepsilon)(x) = \int_{\mathbb{R}^n} \rho_\varepsilon(x-y) 1_E(y) dy = \int_{E \cap B(x,\varepsilon)} \rho_\varepsilon(x-y) dy, \qquad x \in \mathbb{R}^n \,.$$

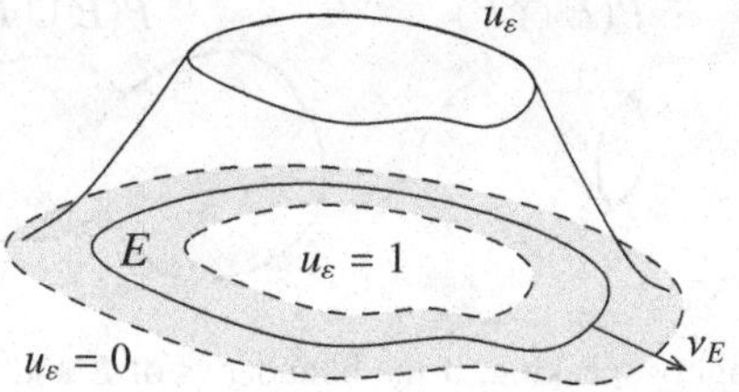

Figure 12.3 The ε-regularization of the characteristic function of an open set with smooth boundary. The ε-neighborhood of ∂E is painted in gray and corresponds to the set of those x such that $0 < u_\varepsilon(x) < 1$. Correspondingly $\nabla u_\varepsilon(x)$ is approximately $-(1/\varepsilon)\nu_E$ evaluated at the projection of x over ∂E.

Clearly, we have $0 \le (1_E \star \rho_\varepsilon) \le 1$, and, moreover,

$$(1_E \star \rho_\varepsilon)(x) = \begin{cases} 1, & \text{if } |B(x,\varepsilon) \setminus E| = 0, \\ 0, & \text{if } |B(x,\varepsilon) \cap E| = 0. \end{cases}$$

If E is an open set with smooth boundary, then we expect $\nabla(1_E \star \rho_\varepsilon)$ to satisfy

$$\nabla(1_E \star \rho_\varepsilon)(x) \approx -\varepsilon^{-1}\nu_E\big(\text{projection of } x \text{ on } \partial E\big), \qquad \text{if } \operatorname{dist}(x, \partial E) < \varepsilon,$$

and $\nabla(1_E \star \rho_\varepsilon)(x) = 0$ if $\operatorname{dist}(x, \partial E) > \varepsilon$. Hence, as $\varepsilon \to 0$, it should hold that

$$\int_{\mathbb{R}^n} |\nabla(1_E \star \rho_\varepsilon)(x)| dx \approx \frac{|\{y \in \mathbb{R}^n : \operatorname{dist}(y, \partial E) < \varepsilon\}|}{\varepsilon} \approx \frac{\varepsilon \mathcal{H}^{n-1}(\partial E)}{\varepsilon} = P(E). \tag{12.13}$$

We are now going to prove that, indeed, if E is a set of locally finite perimeter in $\mathbb{R}^n$, then (12.13) holds true (whether $P(E)$ is finite or not).

Proposition 12.20 *If E is a set of locally finite perimeter in $\mathbb{R}^n$, then*

$$(\mu_E)_\varepsilon = -\nabla(1_E \star \rho_\varepsilon)\mathcal{L}^n, \qquad \forall \varepsilon > 0, \tag{12.14}$$

$$-\nabla(1_E \star \rho_\varepsilon)\,\mathcal{L}^n \stackrel{*}{\rightharpoonup} \mu_E, \qquad |\nabla(1_E \star \rho_\varepsilon)|\,\mathcal{L}^n \stackrel{*}{\rightharpoonup} |\mu_E|, \tag{12.15}$$

as $\varepsilon \to 0^+$. If, conversely, E is a Lebesgue measurable set in $\mathbb{R}^n$ such that

$$\limsup_{\varepsilon \to 0^+} \int_K |\nabla(1_E \star \rho_\varepsilon)(x)|\, dx < \infty, \tag{12.16}$$

for every compact set K, then E is of locally finite perimeter.

Remark 12.21 By the Morse–Sard Lemma, for a.e. $t > 0$ the super-level set $\{1_E \star \rho_\varepsilon > t\}$ is an open set with smooth boundary. As suggested in Figure 12.3, as $\varepsilon \to 0^+$, these smooth sets converge to E; see Section 13.2.

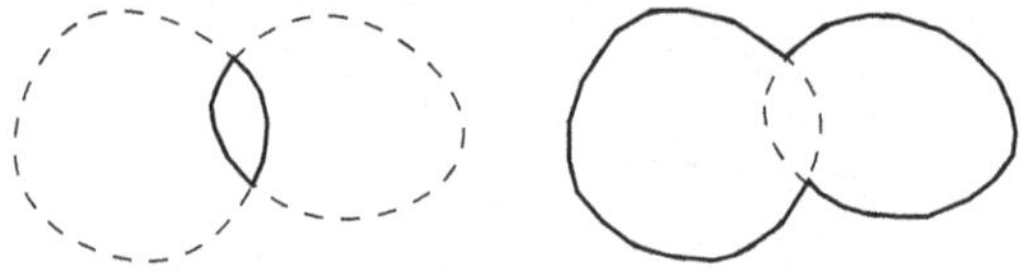

Figure 12.4 Roughly speaking, if the boundaries of E and F intersect on a set of null $(n-1)$-dimensional measure, then inequality (12.17) holds as an equality; see Theorem 16.3.

Proof of Proposition 12.20 By (12.3) we have that, for every $x \in \mathbb{R}^n$,

$$(\mu_E \star \rho_\varepsilon)(x) = \int_{\mathbb{R}^n} \rho_\varepsilon(x-y)\,\mathrm{d}\mu_E(y) = -\int_E \nabla\rho_\varepsilon(x-y)\,\mathrm{d}y = -\nabla(1_E \star \rho_\varepsilon)(x)\,.$$

By definition of $(\mu_E)_\varepsilon$, (12.14) follows. By Theorem 4.36 (applied to μ_E) and (12.14), we deduce (12.15). Conversely, if E is a Lebesgue measurable set in $\mathbb{R}^n$ such that (12.16) holds true, then by Corollary 4.34 there exist a $\mathbb{R}^n$-valued Radon measure μ on $\mathbb{R}^n$ and a sequence $\varepsilon_h \to 0^+$, such that $-\nabla(1_E \star \rho_{\varepsilon_h})\mathcal{L}^n \overset{*}{\rightharpoonup} \mu$. In particular, if $\varphi \in C^1_c(\mathbb{R}^n)$, then we have

$$\begin{aligned}
\int_{\mathbb{R}^n} \varphi\,\mathrm{d}\mu &= -\lim_{h\to\infty} \int_{\mathbb{R}^n} \varphi(x)\nabla(1_E \star \rho_{\varepsilon_h})(x)\,\mathrm{d}x \\
&= -\lim_{h\to\infty} \int_{\mathbb{R}^n} \varphi(x)\,\mathrm{d}x \int_{\mathbb{R}^n} 1_E(y)\nabla\rho_{\varepsilon_h}(x-y)\,\mathrm{d}y \\
&= -\lim_{h\to\infty} \int_E \mathrm{d}y \int_{\mathbb{R}^n} \varphi(x)\nabla\rho_{\varepsilon_h}(x-y)\,\mathrm{d}x \\
&= \lim_{h\to\infty} \int_E \mathrm{d}y \int_{\mathbb{R}^n} \rho_{\varepsilon_h}(x-y)\nabla\varphi(x)\,\mathrm{d}x = \lim_{h\to\infty} \int_E (\nabla\varphi)_{\varepsilon_h}(y)\,\mathrm{d}y = \int_E \nabla\varphi\,.
\end{aligned}$$

By Proposition 12.1, E is of locally finite perimeter in $\mathbb{R}^n$ and $\mu_E = \mu$. □

As an application of Proposition 12.20, we obtain the following useful result concerning unions and intersections of sets of finite perimeter; see Figure 12.4, and, for more general results, Theorem 16.3.

Lemma 12.22 *If E and F are sets of (locally) finite perimeter in $\mathbb{R}^n$, then $E \cup F$ and $E \cap F$ are sets of (locally) finite perimeter in $\mathbb{R}^n$, and, for $A \subset \mathbb{R}^n$ open,*

$$P(E \cup F; A) + P(E \cap F; A) \le P(E; A) + P(F; A)\,. \tag{12.17}$$

Proof If $u_\varepsilon = 1_E \star \rho_\varepsilon$, $v_\varepsilon = 1_F \star \rho_\varepsilon$, then $0 \le u_\varepsilon, v_\varepsilon \le 1$, $u_\varepsilon v_\varepsilon \to 1_{E\cap F}$ in $L^1_{\rm loc}(\mathbb{R}^n)$, and $w_\varepsilon = u_\varepsilon + v_\varepsilon - u_\varepsilon v_\varepsilon \to 1_{E\cup F}$ in $L^1_{\rm loc}(\mathbb{R}^n)$. Moreover,

$$\int_A |\nabla(u_\varepsilon v_\varepsilon)| \le \int_A v_\varepsilon|\nabla u_\varepsilon| + u_\varepsilon|\nabla v_\varepsilon|, \tag{12.18}$$

$$\int_A |\nabla w_\varepsilon| \le \int_A (1-v_\varepsilon)|\nabla u_\varepsilon| + (1-u_\varepsilon)|\nabla v_\varepsilon|,$$

whenever A is an open bounded set in $\mathbb{R}^n$. Adding up the two inequalities,

$$\int_A |\nabla(u_\varepsilon v_\varepsilon)| + \int_A |\nabla(w_\varepsilon)| \le \int_A |\nabla u_\varepsilon| + |\nabla v_\varepsilon|,$$

where the upper limit as $\varepsilon \to 0^+$ of the right-hand side is bounded above by $P(E;\overline{A}) + P(F;\overline{A}) < \infty$. By Exercise 12.18, $E\cap F$ and $E\cup F$ are of locally finite perimeter in $\mathbb{R}^n$, with

$$P(E\cup F;A) + P(E\cap F;A) \le P(E;\overline{A}) + P(F;\overline{A}), \tag{12.19}$$

for every bounded open set A. Now let A be any open set in $\mathbb{R}^n$, set $A_k = \{x \in A\cap B_k : \mathrm{dist}(x,\partial A) < k^{-1}\}$, $k\in\mathbb{N}$, and apply (12.19) to each A_k, to find

$$P(E\cup F;A_k) + P(E\cap F;A_k) \le P(E;A) + P(F;A).$$

Letting $k\to\infty$, the left-hand side converges to $P(E\cup F;A) + P(E\cap F;A)$. □

Exercise 12.23 If E and F are of (locally) finite perimeter then $E\setminus F$ is of (locally) finite perimeter with $P(E\setminus F;A) \le P(E;A) + P(F;A)$.

Exercise 12.24 If E is of finite perimeter in $\mathbb{R}^n$ and $E \subset F_1\cup F_2$ for $F_1, F_2 \subset \mathbb{R}^n$ such that $\mathrm{dist}(F_1,F_2) > 0$, then $E\cap F_1$ and $E\cap F_2$ are of finite perimeter, with $P(E) = P(E\cap F_1) + P(E\cap F_2)$. *Hint:* For $k = 1,2$, there exist open sets A_k with $A_1\cap A_2 = \emptyset$, and functions $\zeta_k \in C^1_c(A_k;[0,1])$ with $\zeta_k = 1$ on $\overline{F_k}$.

Example 12.25 (A "wild" set of finite perimeter) In $\mathbb{R}^n$, $n\ge 2$, given $\varepsilon > 0$ we may construct a set of finite perimeter $E\subset B$ such that

$$|E| \le \varepsilon \qquad |\mathrm{spt}\,\mu_E| \ge \omega_n - \varepsilon.$$

In particular, by Proposition 12.19, the topological boundary of any Lebesgue measurable set F equivalent to E will have positive Lebesgue measure, $|\partial F| > 0$, and thus, by Proposition 3.2 and Theorem 3.10, infinite $\mathcal{H}^{n-1}$-measure, $\mathcal{H}^{n-1}(\partial F) = \infty$. To this end, given $\{x_h\}_{h\in\mathbb{N}}$ dense in B and $\{r_h\}_{h\in\mathbb{N}} \subset (0,\varepsilon)$ such that $n\omega_n \sum_{h\in\mathbb{N}} r_h^{n-1} \le 1$, it suffices to consider the open set $E\subset B$ defined by

$$E = \bigcup_{h\in\mathbb{N}} B_h, \qquad \text{where } B_h = B(x_h, r_h).$$

Indeed, by Example 12.5, $P(B_h) = \mathcal{H}^{n-1}(\partial B_h) = n\omega_n\, r_h^{n-1}$, so that, by Lemma 12.22, for every $N \in \mathbb{N}$, $E_N = \sum_{h=1}^N B_h$ is of finite perimeter with

$$P(E_N) \le \sum_{h=1}^N P(B_h) \le n\omega_n \sum_{h\in\mathbb{N}} r_h^{n-1} \le 1\,.$$

Thus, as $|E| \le \omega_n < \infty$, we have $E_N \to E$ as $N \to \infty$. Hence, by Proposition 12.15, E is of finite perimeter with $P(E) \le 1$. We now notice that, being $\{x_h\}_{h\in\mathbb{N}}$ dense in B, we have $\overline{E} = \overline{B}$, and thus $|\,\partial E| = |\overline{E}| - |E| = \omega_n - |E|$. Since

$$|E| \le \omega_n \sum_{h\in\mathbb{N}} r_h^n \le \varepsilon\, n\omega_n \sum_{h\in\mathbb{N}} r_h^{n-1} \le \varepsilon\,,$$

we conclude that $|\,\partial E| \ge \omega_n - \varepsilon$. In fact, the stronger inequality $|\mathrm{spt}\,\mu_E| \ge \omega_n - \varepsilon$ holds true. Indeed, since $\{x_h\}_{h\in\mathbb{N}}$ is dense in B, we easily see that for every $x \in B$ and $r > 0$ it must be $|E \cap B(x,r)| > 0$; at the same time, exploiting the Lebesgue points theorem, see, in particular, (5.19), we see that for a.e. $x \in B \setminus E$ it must be that $|E \cap B(x,r)| < \omega_n r^n$ for every $r > 0$. Combining these two pieces of information with (12.12), we find that $|\mathrm{spt}\,\mu_E| = |B \setminus E|$, where, as seen above, $|B \setminus E| \ge \omega_n - \varepsilon$.

12.4 Compactness from perimeter bounds

Theorem 12.26 *If $R > 0$ and $\{E_h\}_{h\in\mathbb{N}}$ are sets of finite perimeter in $\mathbb{R}^n$, with*

$$\sup_{h\in\mathbb{N}} P(E_h) < \infty\,, \tag{12.20}$$

$$E_h \subset B_R\,, \qquad \forall h \in \mathbb{N}\,, \tag{12.21}$$

then there exist E of finite perimeter in $\mathbb{R}^n$ and $h(k) \to \infty$ as $k \to \infty$, with

$$E_{h(k)} \to E\,, \qquad \mu_{E_{h(k)}} \stackrel{*}{\rightharpoonup} \mu_E\,, \qquad E \subset B_R\,.$$

Proof *Step one:* We show that if $Q(x,r) = x + (0,r)^n$ and $u \in C^1(\mathbb{R}^n)$, then

$$\int_{Q(x,r)} \big|u - (u)_{Q(x,r)}\big| \le \sqrt{n}\, r \int_{Q(x,r)} |\nabla u|\,, \tag{12.22}$$

where $(u)_{Q(x,r)} = r^{-n}\int_{Q(x,r)} u$. By a change of variables and up to adding a constant to u, we reduce to considering the case $Q(x,r) = (0,1)^n = Q$ and $(u)_Q = 0$. Finally, since $\sum_{i=1}^n |x_i| \le \sqrt{n}\sqrt{\sum_{i=1}^n x_i^2}$, it suffices to show

$$\int_Q |u| \le \sum_{i=1}^n \int_Q |\partial_i u|\,. \tag{12.23}$$

In the case $n = 1$ there exists $x_0 \in Q$ such that $u(x_0) = (u)_Q = 0$, so that $|u(x)| = |u(x) - u(x_0)| \le \int_Q |u'|$ for every $x \in Q$, and (12.23) is proved.We now

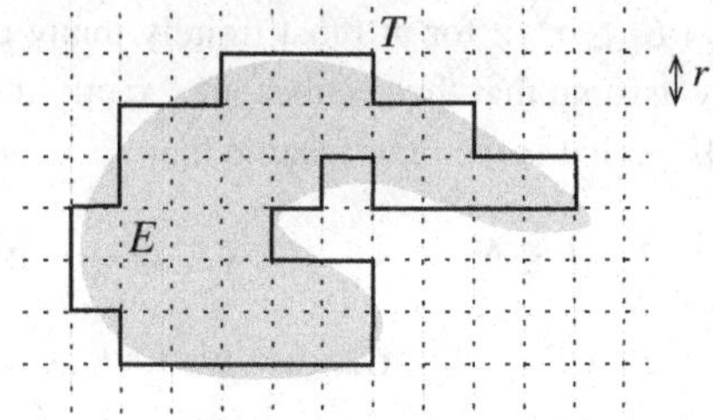

Figure 12.5 We obtain a set T as in (12.24) from a partition of $\mathbb{R}^n$ into cubes of side length r as the union of those cubes Q such that $|E \cap Q| \geq |Q|/2$.

let $n \geq 2$, set $x = (x_1, x') \in \mathbb{R} \times \mathbb{R}^{n-1}$, and define $v(x_1) = \int_{(0,1)^{n-1}} u(x_1, x')\mathrm{d}x'$. Since $\int_{(0,1)} v = \int_Q u = 0$, arguing by induction we find

$$\begin{aligned}\int_Q |u| &= \int_{(0,1)} \mathrm{d}x_1 \int_{(0,1)^{n-1}} |u(x)|\, \mathrm{d}x' \\ &\leq \int_{(0,1)} \mathrm{d}x_1 \int_{(0,1)^{n-1}} |u(x_1, x') - v(x_1)|\, \mathrm{d}x' + \int_{(0,1)} |v(x_1)|\, \mathrm{d}x_1 \\ &\leq \int_{(0,1)} \mathrm{d}x_1 \sum_{i=2}^{n} \int_{(0,1)^{n-1}} |\partial_i u|\, \mathrm{d}x' + \int_{(0,1)} |v'(x_1)|\, \mathrm{d}x_1 \\ &\leq \sum_{i=1}^{n} \int_Q |\partial_i u|.\end{aligned}$$

Step two: If E is a set of finite perimeter in $\mathbb{R}^n$ with $|E| < \infty$, then for every $r > 0$ there exists a finite union T of disjoint cubes of side length r with

$$|E \Delta T| \leq \sqrt{n} r P(E)\,; \tag{12.24}$$

see Figure 12.5. Indeed, let $\{Q_h\}_{h\in\mathbb{N}}$ be a disjoint family of open cubes of side length r such that $\bigcup_{h\in\mathbb{N}} \overline{Q_h} = \mathbb{R}^n$. If $\varepsilon > 0$ and $u = (1_E \star \rho_\varepsilon)$, then by step one

$$\int_{\mathbb{R}^n} |\nabla u| = \sum_{h\in\mathbb{N}} \int_{Q_h} |\nabla u| \geq \frac{1}{\sqrt{n} r} \sum_{h\in\mathbb{N}} \int_{Q_h} \left| u - (u)_{Q_h} \right|.$$

Letting $\varepsilon \to 0$, we find that

$$\begin{aligned}\sqrt{n} r P(E) &\geq \sum_{h\in\mathbb{N}} \int_{Q_h} \left|1_E - (1_E)_{Q_h}\right| = \sum_{h\in\mathbb{N}} \int_{Q_h} \left|1_E - \frac{|Q_h \cap E|}{r^n}\right| \\ &= \sum_{h\in\mathbb{N}} |E \cap Q_h| \left|1 - \frac{|Q_h \cap E|}{r^n}\right| + |Q_h \setminus E| \left|\frac{|Q_h \cap E|}{r^n}\right| \\ &= 2 \sum_{h\in\mathbb{N}} \frac{|E \cap Q_h||Q_h \setminus E|}{r^n}.\end{aligned}$$

Since $|E| < \infty$, $|Q_h \cap E| \ge r^n/2$ for at most finitely many cubes Q_h. Up to a permutation, we may assume that these cubes are exactly the first N elements of the sequence $\{Q_h\}_{h\in\mathbb{N}}$, that is we may assume that

$$|Q_h \cap E| \ge \frac{r^n}{2} \quad \text{if } 1 \le h \le N, \qquad |Q_h \setminus E| \ge \frac{r^n}{2} \quad \text{if } h \ge N+1\,.$$

As a consequence, if we let $T = \bigcup_{h=1}^N Q_h$, then we find, as required,

$$\sqrt{n}rP(E) \ge \sum_{h=1}^{N} |Q_h \setminus E| + \sum_{h=N+1}^{\infty} |Q_h \cap E| = |T \setminus E| + |E \setminus T| = |T \Delta E|\,.$$

Step three: The set $X = \{E \in \mathcal{M}(\mathcal{L}^n) : |E| < \infty\}$ is a complete metric space endowed with the distance $d(E,F) = |E\Delta F| = \|1_E - 1_F\|_{L^1(\mathbb{R}^n)}$ (here we identify E and F provided $|E\Delta F| = 0$). We now claim that each set $Y_{R,p} \subset X$ defined as

$$Y_{R,p} = \Big\{E \in \mathcal{M}(\mathcal{L}^n) : E \subset B_R\,, P(E) \le p\Big\}, \qquad R, p \in (0,\infty)\,,$$

is d-compact. By Proposition 12.15, $Y_{R,p}$ is closed. By a standard diagonal argument, we are thus left to prove that $Y_{R,p}$ is totally bounded: for every $\sigma > 0$ there exist $M \in \mathbb{N}$ and $\{T_h\}_{h=1}^M \subset X$ with

$$\min_{1\le h\le M} d(E, T_h) \le \sigma\,, \qquad \forall E \in Y_{R,p}\,.$$

Indeed, let $r > 0$ be such that $\sqrt{n}\,r\,p \le \sigma$, and let $\{Q_h\}_{h\in\mathbb{N}}$ be the family of cubes associated with r as in step two. The family $\{S_h\}_{h=1}^N$ of the cubes from $\{Q_h\}_{h\in\mathbb{N}}$ intersecting B_R is finite, thus the family $\{T_h\}_{h=1}^M$ of the finite unions of cubes from $\{S_h\}_{h=1}^N$ is finite too. By step two, for every $E \in Y_{R,p}$ there exists T_h such that $|E\Delta T_h| \le \sqrt{n}\,r\,p \le \sigma$, as required.

Step four: By assumption, $\{E_h\}_{h\in\mathbb{N}} \subset Y_{R,p}$ for some $R, p > 0$. By step three, there exist $E \subset B_R$ and a sequence $h(k) \to \infty$ as $k \to \infty$, such that $E_{h(k)} \to E$. By Proposition 12.15, E is a set of finite perimeter in $\mathbb{R}^n$ and $\mu_{E_h} \overset{*}{\rightharpoonup} \mu_E$. □

We cannot conclude the compactness of a sequence of sets from the perimeter bound (12.20) only. For example, if $\{x_h\}_{h\in\mathbb{N}} \subset \mathbb{R}^n$ is such that $|x_h| \to \infty$, then the sequence $E_h = B(x_h, 1)$ satisfies $P(E_h) = n\omega_n$ for every $h \in \mathbb{N}$, while $|E\Delta E_h| \to 2\omega_n$ as $h \to \infty$ for every Lebesgue measurable set E with $|E| = w_n$. Thus, $\{E_h\}_{h\in\mathbb{N}}$ does not admit any converging subsequence. It is clear, however, that $\{E_h\}_{h\in\mathbb{N}}$ locally converges to the empty set, so that compactness with respect to local convergence still holds. At the same time, it is often useful to consider sequences of sets that are only of locally finite perimeter, and that for this reason, are expected to converge at most locally (see, for example, Chapter 15 and, in particular, Theorem 15.5). In these situations the following corollary of Theorem 12.26 is particularly useful.

Corollary 12.27 *If $\{E_h\}_{h\in\mathbb{N}}$ are sets of locally finite perimeter in $\mathbb{R}^n$ with*

$$\sup_{h\in\mathbb{N}} P(E_h; B_R) < \infty\,, \qquad \forall R > 0\,, \tag{12.25}$$

then there exist E of locally finite perimeter and $h(k) \to \infty$ as $k \to \infty$, with

$$E_{h(k)} \overset{\text{loc}}{\to} E\,, \qquad \mu_{E_{h(k)}} \overset{*}{\rightharpoonup} \mu_E\,.$$

Proof *Step one:* If E is of locally finite perimeter and $R > 0$, then

$$P(E \cap B_R) \le P(E; B_R) + P(B_R)\,. \tag{12.26}$$

Indeed, given $R' < R$, let $v_\varepsilon \in C^\infty_c(B_{R'})$ be such that $0 \le v_\varepsilon \le 1$, $v_\varepsilon \to 1_{B_{R'}}$ in $L^1(\mathbb{R}^n)$, and $\int_{\mathbb{R}^n} |\nabla v_\varepsilon| \to P(B_{R'})$ as $\varepsilon \to 0^+$, and let $u_\varepsilon = 1_E \star \rho_\varepsilon$. By (12.18) and Exercise 12.18,

$$\begin{aligned}
P(E \cap B_{R'}) &\le \liminf_{\varepsilon\to 0^+} \int_{\mathbb{R}^n} |\nabla(u_\varepsilon v_\varepsilon)| \le \limsup_{\varepsilon\to 0^+} \int_{\mathbb{R}^n} |\nabla v_\varepsilon| + \int_{\mathbb{R}^n} v_\varepsilon |\nabla u_\varepsilon| \\
&\le P(B_{R'}) + \lim_{\varepsilon\to 0^+} \int_{B_{R'}} |\nabla u_\varepsilon| \le P(B_{R'}) + P(E; \overline{B_{R'}}) \\
&\le P(B_R) + P(E; B_R)\,.
\end{aligned}$$

Since $E \cap B_{R'} \overset{\text{loc}}{\to} E \cap B_R$ as $R' \to R$, by Proposition 12.15 we find (12.26).

Step two: By (12.25) and (12.26), and given $j \in \mathbb{N}$, we may apply Theorem 12.26 to $\{E_h \cap B_j\}_{h\in\mathbb{N}}$. By a standard diagonal argument, we find $h(k) \to \infty$ for $k \to \infty$ and sets of finite perimeter $\{F_j\}_{j\in\mathbb{N}}$ such that $E_{h(k)} \cap B_j \to F_j$ as $k \to \infty$. Up to null sets, $F_j \subset F_{j+1}$, so that $E_{h(k)}$ locally converges to $E = \bigcup_{j\in\mathbb{N}} F_j$. By Proposition 12.15, E is of locally finite perimeter and $\mu_{E_{h(k)}} \overset{*}{\rightharpoonup} \mu_E$. □

Remark 12.28 (Diameter bounds and compactness) If we replace the uniform boundedness assumption (12.21) in Theorem 12.26 with a uniform boundedness assumption on the *diameters* of the E_h, namely

$$\sup_{h\in\mathbb{N}} \operatorname{diam}(E_h) < \infty\,, \tag{12.27}$$

then we may still prove the existence of a set E of finite perimeter in $\mathbb{R}^n$, of $h(k) \to \infty$ as $k \to \infty$, and of $\{x_k\}_{k\in\mathbb{N}} \subset \mathbb{R}^n$, such that

$$x_k + E_{h(k)} \to E\,, \qquad \mu_{x_k+E_{h(k)}} \overset{*}{\rightharpoonup} \mu_E\,.$$

In checking condition (12.27), one wants to minimize the diameter of each E_h in its equivalence class. Since it is always true that $\operatorname{diam}(E) = \operatorname{diam}(\partial E)$, by Proposition 12.19 an optimal reformulation of (12.27) amounts in asking that

$$\sup_{h\in\mathbb{N}} \operatorname{diam}\big(\operatorname{spt}\mu_{E_h}\big) < \infty\,. \tag{12.28}$$

This remark is particularly effective in dimension $n = 2$, where perimeter and diameter are easily related. Indeed, it can be proved that if $E \subset \mathbb{R}^2$ is connected (or, more precisely, indecomposable; see Exercise 16.9), then $\operatorname{diam}\big(\operatorname{spt}\mu_E\big) \le P(E)$. In particular, (12.28) follows directly from (12.20) on sequences of indecomposable sets in the plane.

12.5 Existence of minimizers in geometric variational problems

There are several geometric variational problems for which existence of minimizers can be proved in the class of sets of finite perimeter. These existence results are achieved by means of the so-called *Direct Method*, consisting of first proving the compactness of an arbitrary minimizing sequence of competitors (by means of Theorem 12.26 or variants), and then showing the minimality of the limit set via lower semicontinuity (by Proposition 12.15 or variants). In this section, we apply the Direct Method for proving existence of minimizers in some variational problems of distinguished geometric character, like Plateau-type problems, relative isoperimetric problems, and prescribed mean curvature problems. Other geometric variational problems that will be discussed in the book include equilibrium problems for liquid drops confined in a given container (Chapter 19), problems involving anisotropic surface energies (Chapter 20), and isoperimetric problems with multiple chambers (Part IV).

Plateau-type problems: The classical Plateau problem, minimizing area among surfaces passing through a given curve, is one of the archetypical problems in Geometric Measure Theory. Although generalized formulations of this problem are more properly conceived with the aid of the theories of currents and varifolds, a simple formulation (with some implicit topological obstruction; see the notes to Part II) is also possible in our framework. Given a set $A \subset \mathbb{R}^n$, and a set E_0 of finite perimeter in $\mathbb{R}^n$, **the Plateau-type problem in A with boundary data E_0** amounts to minimizing $P(E)$ among those sets of finite perimeter E that coincide with E_0 outside A. Precisely, we consider

$$\gamma(A, E_0) = \inf\big\{P(E) : E \setminus A = E_0 \setminus A\big\}. \tag{12.29}$$

Roughly speaking, prescribing that $E \setminus A = E_0 \setminus A$ we impose $E_0 \cap \partial A$ as a "boundary condition" for the admissible sets E in (12.29). At the same time, the set A, being the region where E_0 can be modified to minimize perimeter, may act as an obstacle; see Figure 12.6. In general, we do not expect

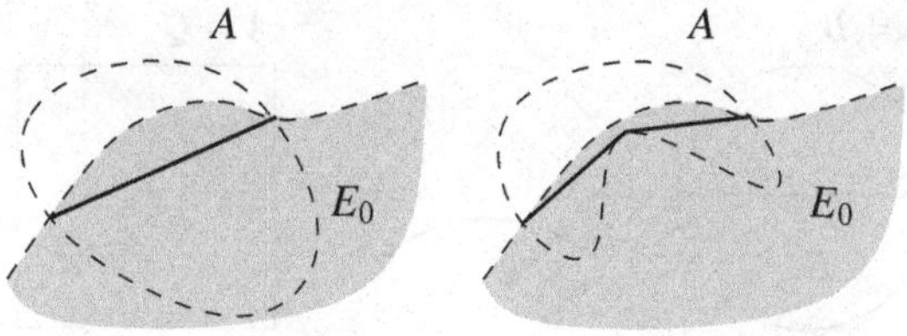

Figure 12.6 Minimizers in (12.29) may change if we modify A while keeping E_0 and $\partial A \cap \partial E_0$ fixed. On the left, the boundary in A of the minimizer in (12.29) is the segment spanned by $\partial A \cap \partial E_0$; on the right, a situation where A acts as an obstacle.

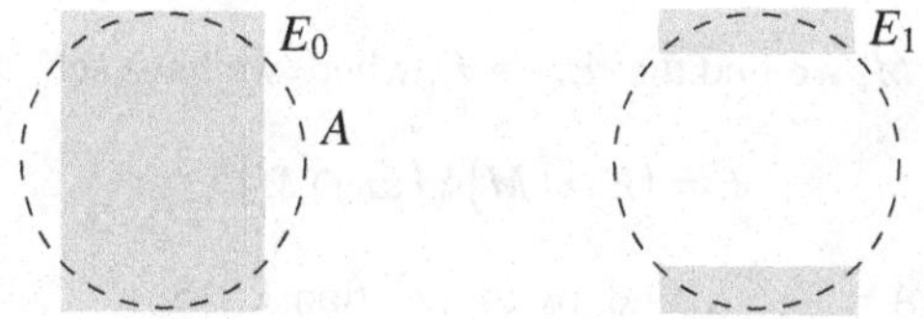

Figure 12.7 If $E_0 = \{x \in \mathbb{R}^2 : |x_2| < 1, |x_1| < 1/\sqrt{2}\}$, then both E_0 and $E_1 = E_0 \cap \{x \in \mathbb{R}^2 : 1 > |x_2| > 1/\sqrt{2}\}$ are minimizers in $\gamma(B; E_0)$.

uniqueness of minimizers for this problem; see Figure 12.7. Existence of minimizers is addressed as follows.

Proposition 12.29 (Existence of minimizers for the Plateau-type problem) *Let $A \subset \mathbb{R}^n$ be a bounded set and let E_0 be a set of finite perimeter in $\mathbb{R}^n$. Then there exists a set of finite perimeter E such that $E \setminus A = E_0 \setminus A$ and $P(E) \le P(F)$ for every F such that $F \setminus A = F \setminus E_0$. In particular, E is a minimizer in the variational problem (12.29).*

Proof Since E_0 itself is admissible in (12.29), we have $\gamma = \gamma(A, E_0) < \infty$. Let us now consider a minimizing sequence $\{E_h\}_{h \in \mathbb{N}}$ in (12.29),

$$E_h \setminus A = E_0 \setminus A, \qquad P(E_h) \le P(E_0), \qquad \lim_{h \to \infty} P(E_h) = \gamma.$$

If $M_h = E_h \Delta E_0 = (E_h \setminus E_0) \cup (E_0 \setminus E_h)$, then, by Lemma 12.22, M_h is a set of finite perimeter with

$$P(M_h) \le 2P(E_h) + 2P(E_0) \le 4P(E_0).$$

Since A is bounded and $M_h \subset A$, by Theorem 12.26 there exists a set of finite perimeter M such that, up to extracting a subsequence, we have $M_h \to M$. As

$$E_h = \big(E_0 \cup M_h\big) \setminus \big(E_0 \cap M_h\big),$$

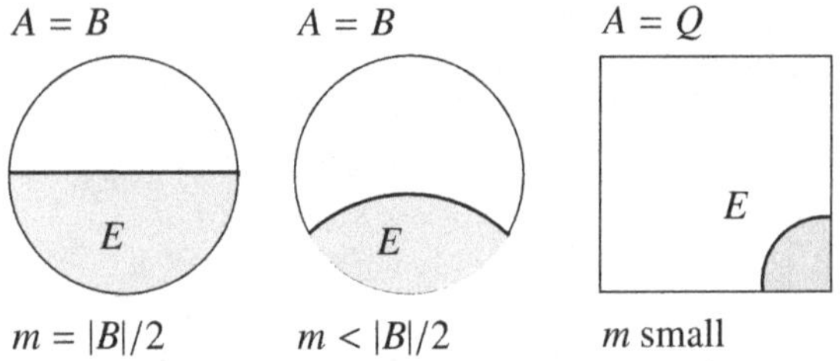

Figure 12.8 Some relative isoperimetric problems in the plane. As in Plateau-type problems, we cannot expect uniqueness of minimizers.

and since $M_h \to M$, we find that $E_h \to E$, where we have set

$$E = \big(E_0 \cup M\big) \setminus \big(E_0 \cap M\big).$$

In particular $E \setminus A = E_0 \setminus A$, and, by Proposition 12.15,

$$\gamma \le P(E) \le \liminf_{h\to\infty} P(E_h) = \gamma. \qquad \square$$

Relative isoperimetric problems: Given an open set $A \subset \mathbb{R}^n$, the **relative isoperimetric problem in** A amounts to the volume-constrained minimization of the relative perimeter in A, namely

$$\alpha(A, m) = \inf\Big\{P(E; A) : E \subset A, |E| = m\Big\}, \tag{12.30}$$

where $m \in (0, |A|)$ (and, possibly, we allow $|A| = \infty$); see Figure 12.8. A minimizer E in (12.30), normalized to obtain $\operatorname{spt}\mu_E = \partial E$ according to Proposition 12.19, is called **relative isoperimetric sets in** A. The case $A = \mathbb{R}^n$, of course, corresponds to the Euclidean isoperimetric problem, which is addressed in Chapter 14. Apart from their geometric interest, relative isoperimetric problems are also strictly related to the study of equilibrium shapes of a liquid confined in a given container, as discussed in Chapter 19. When A is bounded and has finite perimeter, the existence of minimizers is proved by the Direct Method along the following lines.

Proposition 12.30 (Existence of relative isoperimetric sets) *If A is an open bounded set of finite perimeter and $m \in (0, |A|]$, then there exists a set of finite perimeter $E \subset A$ such that $P(E; A) = \alpha(A, m)$ and $|E| = m$. In particular, E is a minimizer in the variational problem (12.30).*

Proof Let $E_t = A \cap \{x : x_1 < t\}$ $(t \in \mathbb{R})$. By a continuity argument, there exists $t \in \mathbb{R}$ such that $|E_t| = m$. By Lemma 12.22, $\alpha = \alpha(A, m) < \infty$. Now let $\{E_h\}_{h\in\mathbb{N}}$

be a minimizing sequence in (12.30), that is

$$E_h \subset A\,,\qquad |E_h| = m\,,\qquad \lim_{h\to\infty} P(E_h; A) = \alpha\,.$$

We now notice that

$$P(E_h) = P(E_h \cap A) \le P(E_h; A) + P(A)\,, \tag{12.31}$$

(in the case A is a ball, this was proved in (12.26); in the general case this follows, for example, by Theorem 16.3). By (12.31), $\sup_{h\in\mathbb{N}} P(E_h) < \infty$. Since A is bounded, by Theorem 12.26 there exists a set of finite perimeter $E \subset \mathbb{R}^n$ such that, up to extracting a subsequence, $E_h \to E$. In particular $E \subset A$ and $|E| = \lim_{h\to\infty} |E_h| = m$, so that, by Proposition 12.15,

$$\alpha \le P(E; A) \le \liminf_{h\to\infty} P(E_h; A) = \alpha\,. \qquad \square$$

Problems involving potential energies (prescribed mean-curvature problems): Interesting variational problems arise from the interaction between perimeter and potential energy terms. Given a Lebesgue measurable function $g\colon \mathbb{R}^n \to \mathbb{R} \cup \{+\infty\}$, we define the **potential energy** of E associated with g on the Lebesgue measurable set E as

$$\mathcal{G}(E) = \int_E g(x)\mathrm{d}x\,.$$

The minimization, under various side conditions, of the potential energy alone is, in general, trivial or easy to understand. Roughly speaking, there are two remarks one needs to keep in mind. First, competitors are confined inside the region $\{g < \infty\}$. Second, minimizers try to coincide with the sub-level set $\{g < t\}$ of g corresponding to the lowest value of t which is compatible with the side constraints. In other words, minimizers are shaped by the force field $-\nabla g$. For example, if we want to consider the action of gravity on subsets of $\mathbb{R}^3$ lying above a given horizontal plane, say $\{x_3 = 0\}$, then we set

$$g(x) = \begin{cases} x_3\,, & \text{if } x_3 > 0\,,\\ \infty\,, & \text{if } x_3 < 0\,. \end{cases}$$

The interaction between potential energy terms with perimeter leads to the formulation of quite rich and interesting variational problems. For some examples of physical interest, see, in particular, Chapter 19. A problem of geometric nature is the **prescribed mean curvature problem** associated with a Lebesgue measurable function $g\colon \mathbb{R}^n \to \mathbb{R}$ and an open set $A \subset \mathbb{R}^n$,

$$\inf\Big\{P(E) + \mathcal{G}(E) : E \subset A\Big\}\,. \tag{12.32}$$

The terminology used here arises from the fact that, if $g \in C^0(A)$, E is a minimizer in (12.32), and $A \cap \partial E$ a is a C^2-hypersurface, then the mean curvature H_E of E is equal to $-g$ in A; see Section 17.3 and Exercise 17.10. If g is positive, evidently, the problem is trivial, and the solution is the empty set. If, however, g takes negative values, then the problem will possess, in general, non-trivial minimizers. If $g \in L^1(A)$ and A is bounded, then the existence of minimizers is easily obtained by the Direct Method, as in our previous examples. One has only to take the following proposition into account.

Proposition 12.31 *If $g\colon \mathbb{R}^n \to \mathbb{R} \cup \{+\infty\}$ is a Lebesgue measurable function with $g^- \in L^1(F)$ for a Lebesgue measurable set $F \subset \mathbb{R}^n$, and $E_h \to E$, then*

$$\int_{E\cap F} g(x)\,\mathrm{d}x \le \liminf_{h\to\infty} \int_{E_h\cap F} g(x)\,\mathrm{d}x\,. \tag{12.33}$$

Proof By Fatou's lemma and since since $g^- \in L^1(F)$ we easily find that

$$\int_{E\cap F} g^+(x)\,\mathrm{d}x \le \liminf_{h\to\infty} \int_{E\cap F} g^+(x)\,\mathrm{d}x\,,$$

$$\int_{E\cap F} g^-(x)\,\mathrm{d}x = \lim_{h\to\infty} \int_{E_h\cap F} g^-(x)\,\mathrm{d}x\,.$$

We conclude by $g = g^+ - g^-$. □

Exercise 12.32 If $g\colon \mathbb{R}^n \to [0,\infty]$ is measurable and $E_h \stackrel{\text{loc}}{\to} E$, then

$$\mathcal{G}(E) \le \liminf_{h\to\infty} \mathcal{G}(E_h)\,. \tag{12.34}$$

Exercise 12.33 If A is a bounded, open set of finite perimeter, then the variational problem (12.32) admit minimizers. Show analogous existence results for the variational problems

$$\inf\Big\{P(E;A) + \mathcal{G}(E) : E \setminus A = E_0 \setminus A\Big\}, \tag{12.35}$$

$$\inf\Big\{P(E;A) + \mathcal{G}(E) : E \subset A\,, |E| = m\Big\}\,. \tag{12.36}$$

Exercise 12.34 If $g\colon \mathbb{R}^n \to [0,\infty)$ is a Lebesgue measurable function with $g(x) \to \infty$ as $|x| \to \infty$, and if A is a (possibly unbounded) open set with finite perimeter, then the variational problem

$$\inf\Big\{P(E;A) + \mathcal{G}(E) : E \subset A\,, |E| = m\Big\} \tag{12.37}$$

admits minimizers. *Hint:* If $E_h \stackrel{\text{loc}}{\to} E$ and for every $\varepsilon > 0$ there exists $R > 0$ such that $\sup_{h\in\mathbb{N}} |(\mathbb{R}^n \setminus B_R) \cap (E_h \Delta E)| < \varepsilon$, then $E_h \to E$.

12.6 Perimeter bounds on volume

In this section we shall discuss two useful inequalities providing a control on the volume of a set in terms of its perimeter, or of its relative perimeter inside a ball, stated respectively in Proposition 12.35 and Proposition 12.37. The corresponding *isoperimetric-type inequalities* are key tools in many of the arguments appearing in subsequent parts of the book. This terminology has its origin in the fact that obtaining these inequalities in sharp form (i.e., with optimal constants) is equivalent in solving one of the (relative) isoperimetric problems introduced in Section 12.5. For example, in Chapter 14 we shall address the Euclidean isoperimetric problem, corresponding to problem (12.30) for $A = \mathbb{R}^n$, and show that Euclidean balls are the only minimizers. This result is in turn equivalent to the *Euclidean isoperimetric inequality*: if $|E| < \infty$, then

$$P(E) \geq n\omega_n^{1/n}|E|^{(n-1)/n},$$

with equality if and only if $|E\Delta B(x,r)| = 0$ for some $x \in \mathbb{R}^n$ and $r > 0$. We may assert that the inequality is in sharp form due to the characterization of equality cases. In the following proposition we prove a non-sharp form of the Euclidean isoperimetric inequality, which has the advantage of admitting a much simpler proof, and which is sufficient in most of the occasions when one needs the isoperimetric inequality as a technical tool only.

Proposition 12.35 (A perimeter bound on volume) *If E is a bounded set of finite perimeter in $\mathbb{R}^n$ $(n \geq 2)$, then*

$$P(E) \geq |E|^{(n-1)/n}. \tag{12.38}$$

Remark 12.36 The assumption that E is bounded can be dropped; see Exercise 13.7. We also notice that if $E \subset B_R$ then we easily prove $n\,|E| \leq R\,P(E)$, by applying the divergence theorem (12.2) on E to $T(x) = x$.

Proof of Proposition 12.35 *Step one:* If $u \in C_c^\infty(\mathbb{R}^n)$, $u \geq 0$, then

$$\|\nabla u\|_{L^1(\mathbb{R}^n;\mathbb{R}^n)} \geq \|u\|_{L^{n/(n-1)}(\mathbb{R}^n)}. \tag{12.39}$$

Decomposing $\mathbb{R}^n$ as $\mathbb{R}^{n-1} \times \mathbb{R}$, with $x = (x', x_n)$ and $\nabla u = (\nabla' u, \partial_n u)$, we have

$$|u(x)| \leq \int_{\mathbb{R}} |\partial_n u(x', x_n)|\,\mathrm{d}x_n \leq \int_{\mathbb{R}} |\nabla u(x', x_n)|\,\mathrm{d}x_n, \quad \forall x \in \mathbb{R}^n. \tag{12.40}$$

Therefore, by Fubini's theorem,

$$\int_{\mathbb{R}^{n-1}} |u(x', x_n)|\,\mathrm{d}x' \leq \int_{\mathbb{R}^n} |\nabla u|, \quad \forall x_n \in \mathbb{R}. \tag{12.41}$$

If $n = 2$ then we may apply (12.40) twice, to find that

$$|u(x)|^2 \le \int_{\mathbb{R}} |\nabla u(x_1, t)|\, dt \int_{\mathbb{R}} |\nabla u(s, x_2)|\, ds\,, \quad \forall x \in \mathbb{R}^2\,.$$

Again by Fubini's theorem,

$$\int_{\mathbb{R}^2} |u|^2 \le \left(\int_{\mathbb{R}} dx_1 \int_{\mathbb{R}} |\nabla u(x_1, t)|\, dt\right)\left(\int_{\mathbb{R}} dx_2 \int_{\mathbb{R}} |\nabla u(s, x_2)|\, ds\right) = \left(\int_{\mathbb{R}^2} |\nabla u|\right)^2,$$

that is (12.39) for $n = 2$. We now let $n \ge 3$ and assume that (12.39) holds true in dimension $k = n - 1$. If we set $\lambda = 1/(n-1)$ then we have

$$\frac{n}{n-1} = \lambda + (1-\lambda)\frac{n-1}{n-2} = \lambda + (1-\lambda)\frac{k}{k-1}\,.$$

By Hölder's inequality

$$\begin{aligned}\int_{\mathbb{R}^n} u^{n/(n-1)} &= \int_{\mathbb{R}} dx_n \int_{\mathbb{R}^k} u^{\lambda}\, u^{(1-\lambda)[k/(k-1)]} dx' \\ &\le \int_{\mathbb{R}} \left(\int_{\mathbb{R}^k} u\, dx'\right)^{\lambda} \left(\int_{\mathbb{R}^k} u^{k/(k-1)} dx'\right)^{(1-\lambda)} dx_n\,. \end{aligned} \tag{12.42}$$

By (12.41) we have that

$$\left(\int_{\mathbb{R}^k} u(x', x_n)\, dx'\right)^{\lambda} \le \|\nabla u\|_{L^1(\mathbb{R}^n)}^{\lambda}\,. \tag{12.43}$$

On the other hand, since $u \ge 0$, we find that, for every $x_n \in \mathbb{R}$,

$$\left(x' \in \mathbb{R}^k \mapsto u(x', x_n)^{k/(k-1)}\right) \in C^\infty_c(\mathbb{R}^k)\,,$$

therefore, by the inductive hypothesis,

$$\begin{aligned}\left(\int_{\mathbb{R}^k} u(x', x_n)^{k/(k-1)} dx'\right)^{(1-\lambda)} &\le \left(\int_{\mathbb{R}^k} |\nabla' u(x', x_n)|\, dx'\right)^{k(1-\lambda)/(k-1)} \\ &= \int_{\mathbb{R}^k} |\nabla' u(x', x_n)|\, dx'\,, \end{aligned} \tag{12.44}$$

Since $|\nabla' u| \le |\nabla u|$, by (12.42), (12.43), and (12.44) we thus conclude

$$\int_{\mathbb{R}^n} u^{n/(n-1)} \le \|\nabla u\|_{L^1(\mathbb{R}^n)}^{\lambda} \int_{\mathbb{R}} dz \int_{\mathbb{R}^k} |\nabla' u(x, z)|\, dx \le \|\nabla u\|_{L^1(\mathbb{R}^n)}^{1+\lambda} = \|\nabla u\|_{L^1(\mathbb{R}^n)}^{n/(n-1)}\,.$$

Step two: If E is a bounded set of finite perimeter, and we apply step one to $u_\varepsilon = (1_E \star \rho_\varepsilon)$, then by Proposition 12.20 and Fatou's lemma we deduce

$$P(E)^{n/(n-1)} = \lim_{\varepsilon \to 0^+} \|\nabla u_\varepsilon\|_{L^1(\mathbb{R}^n;\mathbb{R}^n)}^{n/(n-1)} \ge \liminf_{\varepsilon \to 0^+} \int_{\mathbb{R}^n} |u_\varepsilon|^{n/(n-1)} \ge |E|\,. \qquad \square$$

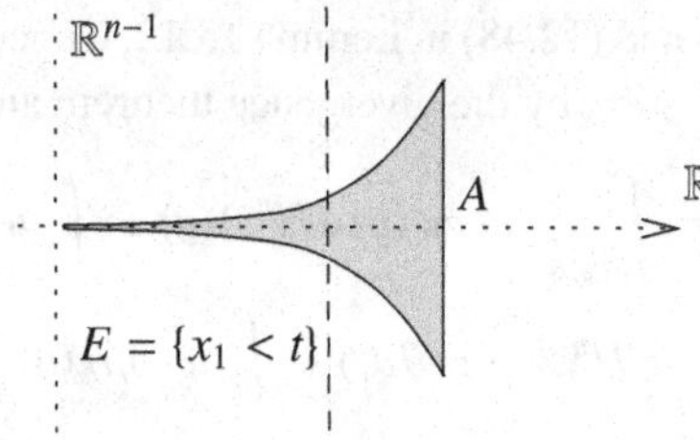

Figure 12.9 If $A = \{(x_1, x') \in \mathbb{R} \times \mathbb{R}^{n-1} : 0 < x_1 < r, |x'| < x_1^p\}$, $r > 0$, $p > 1$, and if $E_t = \{x : x_1 < t\}$, then $P(E_t; A) = \omega_{n-1} t^{p(n-1)}$ and $|E_t \cap A| = \omega_{n-1} \int_0^t s^{p(n-1)} ds = c(n,p) t^{p(n-1)+1}$ for every $t \in (0, r)$. In particular, for every $\alpha \in [1-1/n, 1)$, there exists $p > 1$ such that $P(E_t; A)|E_t \cap A|^{-\alpha} \to 0$ as $t \to 0^+$.

The second isoperimetric-type inequality we shall need is related to the relative isoperimetric problem (12.30) in the case when A is a ball. For this reason, it is usually called the *relative isoperimetric inequality on a ball.*

Proposition 12.37 (Local perimeter bound on volume) *If $n \ge 2$, $t \in (0, 1)$, $x \in \mathbb{R}^n$ and $r > 0$, then there exists a positive constant $c(n, t)$ such that*

$$P(E; B(x, r)) \ge c(n, t)|E \cap B(x, r)|^{(n-1)/n}, \tag{12.45}$$

for every set of locally finite perimeter E such that $|E \cap B(x, r)| \le t|B(x, r)|$.

Remark 12.38 By inequality (12.45) with $t = 1/2$, if $E \subset B(x, r)$, then

$$P(E; B(x, r)) \ge c(n) \min\Big\{|E \cap B(x, r)|, |B(x, r) \setminus E|\Big\}^{(n-1)/n}. \tag{12.46}$$

As consequence, referring to (12.30), by (12.46) we have

$$\alpha(B(x, r), m) \ge c(n) \min\Big\{m, |B(x, r)| - m\Big\}^{(n-1)/n}, \qquad \forall m \in \Big(0, |B(x, r)|\Big).$$

Remark 12.39 Some assumptions on A are required to have an inequality like $P(E; A) \ge c|E \cap A|^{(n-1)/n}$ whenever $|E \cap A| \le |A|/2$. For example, A must be connected, while ∂A should not present outward cusps; see Figure 12.9.

Proof of Proposition 12.37 Without loss of generality, we may assume that $x = 0$ and $r = 1$. We begin noticing that, for a.e. $r > 0$,

$$\mu_{E \cap B_r} = \mu_E \llcorner B_r + \mu_{B_r} \llcorner E\,, \tag{12.47}$$

$$P(E \cap B_r) = P(E; B_r) + \mathcal{H}^{n-1}(E \cap \partial B_r)\,, \tag{12.48}$$

$$r\,\mathcal{H}^{n-1}(E \cap \partial B_r) \le n|E \cap B_r| + r\,P(E; B_r)\,. \tag{12.49}$$

We shall prove (12.47) and (12.48) in Lemma 15.12. Concerning (12.49), since $\mathrm{div}\,(y) = n$ for every $y \in \mathbb{R}^n$, by the divergence theorem and (12.47),

$$
\begin{aligned}
n|E \cap B_r| &= \int_{E\cap\partial B_r} y \cdot \nu_{B_r}(y)\, \mathrm{d}\mathcal{H}^{n-1}(y) + \int_{B_r} y \cdot \mathrm{d}\mu_E(y)\\
&= r\,\mathcal{H}^{n-1}(E \cap \partial B_r) + \int_{B_r} y \cdot \mathrm{d}\mu_E(y)\\
&\geq r\,\mathcal{H}^{n-1}(E \cap \partial B_r) - r\,P(E; B_r)\,.
\end{aligned}
$$

We now prove (12.45). By contradiction, there exists $\{E_h\}_{h\in\mathbb{N}}$ with

$$
\lim_{h\to\infty} \frac{P(E_h; B)}{|E_h \cap B|^{(n-1)/n}} = 0\,, \qquad |E_h \cap B| \leq t\,|B|\,, \tag{12.50}
$$

and (12.49), (12.47), (12.48) hold for $E = E_h$ and a.e. $r > 0$. Clearly, by (12.50), $P(E_h; B) \to 0$. We claim that, correspondingly,

$$
\lim_{h\to\infty} |E_h \cap B| = 0\,. \tag{12.51}
$$

Indeed, given a sequence $\{r_k\}_{k\in\mathbb{N}}$ such that $r_k \to 1^-$ and (12.48) holds true for each r_k, then by Proposition 12.15, by $P(E_h; B_{r_k}) \leq P(E_h; B)$ and by $\mathcal{H}^{n-1}(E_h \cap \partial B_{r_k}) \leq n\omega_n r_k^{n-1}$, we find that

$$
P(E_h \cap B) \leq \liminf_{k\to\infty} P(E_h \cap B_{r_k}) \leq P(E_h; B) + n\omega_n\,.
$$

In particular, by Theorem 12.26, there exists $F \subset B$ of finite perimeter, such that, up to extracting a subsequence, $E_h \cap B \to F$. Since $P(E_h; B) \to 0$, by Proposition 12.15, $P(F; B) = 0$. By Exercise 12.17, and since $|F| \leq t|B| < |B|$, it must be that $|F| = 0$. Thus $|E_h \cap B| \to 0$, and the proof of (12.51) is completed. Adding $r\,P(E_h; B_r)$ to both sides of (12.49), and then dividing by $|E_h \cap B_r|^{(n-1)/n}$, by (12.48) we find that, for a.e. $r > 0$,

$$
n\,|E_h \cap B_r|^{1/n} + 2r \frac{P(E_h; B_r)}{|E_h \cap B_r|^{(n-1)/n}} \geq r \frac{P(E_h \cap B_r)}{|E_h \cap B_r|^{(n-1)/n}} \geq r\,, \tag{12.52}
$$

thanks also to (12.38). Since $P(E_h; B_r) \leq P(E_h; B)$ if $r < 1$, and $r \mapsto |E_h \cap B_r|$ defines a continuous function, if we let $r \to 1^-$ in (12.52), then we find

$$
n\,|E_h \cap B|^{1/n} + 2 \frac{P(E_h; B)}{|E_h \cap B|^{(n-1)/n}} \geq 1\,,
$$

which leads to a contradiction because of (12.50) and (12.51). □

13

The coarea formula and the approximation theorem

Let C^1 denote the family of the open sets of $\mathbb{R}^n$ with C^1-boundary. The classical notion of perimeter, intended as the $\mathcal{H}^{n-1}$-dimensional measure of the topological boundary, defines a functional $\sigma\colon C^1 \to [0,\infty]$,

$$\sigma(E) = \mathcal{H}^{n-1}(\partial E), \qquad E \in C^1 .$$

Distributional perimeter defines an extension of σ from C^1 to the family $\mathcal{M}(\mathcal{L}^n)$ of Lebesgue measurable sets of $\mathbb{R}^n$, that is $P\colon \mathcal{M}(\mathcal{L}^n) \to [0,\infty]$ is such that $P(E) = \sigma(E)$ whenever $E \in C^1$. Moreover, by Proposition 12.15, we know this extension is lower semicontinuous with respect to the local convergence of sets. In this chapter, we prove distributional perimeter to be the maximal lower semicontinuous extension of σ to $\mathcal{M}(\mathcal{L}^n)$, that is, that

$$P(E) = \inf\left\{ \liminf_{h\to\infty} P(E_h) : E_h \in C^1 , E_h \overset{\text{loc}}{\to} E\right\}, \qquad \forall E \in \mathcal{M}(\mathcal{L}^n) .$$

Thanks to Proposition 12.15, we just have to prove that for every $E \in \mathcal{M}(\mathcal{L}^n)$ with $P(E) < \infty$ there exists a sequence $\{E_h\}_{h\in\mathbb{N}} \subset C^1$ such that E_h locally converges to E with $P(E_h) \to P(E)$ as $h \to \infty$. This will indeed be the content of Theorem 13.8. As anticipated in Remark 12.21, we are going to construct the sequence E_h by selecting suitable super-level sets of the functions $1_E \star \rho_{\varepsilon_h}$, for $\varepsilon_h \to 0^+$. The coarea formula discussed in Section 13.1 will provide us with a tool to prove the convergence of the perimeters $P(E_h)$ to $P(E)$.

13.1 The coarea formula

If $u\colon \mathbb{R}^n \to \mathbb{R}$ is a smooth function, then, by the Morse–Sard lemma (see Lemma 13.15), the set $\{u = t\} = \{x \in \mathbb{R}^n : u(x) = t\}$ is a smooth hypersurface in $\mathbb{R}^n$ for a.e. $t \in \mathbb{R}$. It is often natural to look at the integral over $t \in \mathbb{R}$ of the

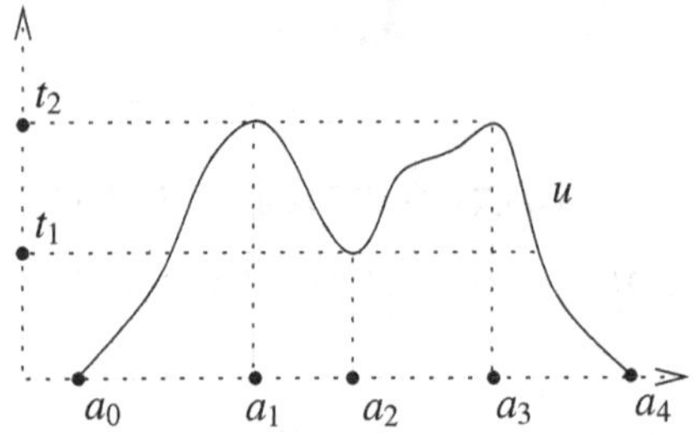

Figure 13.1 The coarea formula (13.1) for a non-negative Lipschitz function u with compact support on $\mathbb{R}$. By the fundamental theorem of calculus, $\int_{a_0}^{a_1}|u'| = \int_{a_3}^{a_4}|u'| = t_2$ and $\int_{a_1}^{a_2}|u'| = \int_{a_2}^{a_3}|u'| = t_2 - t_1$, so that $\int_{\mathbb{R}}|u'| = 4t_2 - 2t_1$. At the same time, $P(\{u > t\}) = 0$ if $t \in (-\infty, 0) \cup (t_2, \infty)$, $P(\{u > t\}) = 2$ if $t \in (0, t_1)$ and $P(\{u > t\}) = 4$ if $t \in (t_1, t_2)$, so that $\int_{\mathbb{R}} P(\{u > t\})\,\mathrm{d}t = 2t_1 + 4(t_2 - t_1) = 4t_2 - 2t_1$, as required.

$\mathcal{H}^{n-1}$-dimensional measure of the slices $E \cap \{u = t\}$ of a Borel set $E \subset \mathbb{R}^n$,

$$\int_{\mathbb{R}} \mathcal{H}^{n-1}\big(E \cap \{u = t\}\big)\,\mathrm{d}t\,,$$

which, by the *coarea formula*, coincides with the **total variation** of u over E,

$$\int_{\mathbb{R}} \mathcal{H}^{n-1}\big(E \cap \{u = t\}\big)\,\mathrm{d}t = \int_E |\nabla u|\,,$$

see Figure 13.1. This fact has a very clear justification on piecewise affine functions. If $e \in S^{n-1}$ and $u(x) = x \cdot e$, $x \in \mathbb{R}^n$, then by Fubini's theorem

$$\int_{\mathbb{R}} \mathcal{H}^{n-1}\big(E \cap \{u = t\}\big)\,\mathrm{d}t = |E|\,.$$

Therefore, given $\lambda \in \mathbb{R}$ and setting $u(x) = \lambda(x \cdot e)$, $x \in \mathbb{R}^n$, by a change of variable we easily see that

$$\int_{\mathbb{R}} \mathcal{H}^{n-1}\big(E \cap \{u = t\}\big)\,\mathrm{d}t = |\lambda||E| = \int_E |\nabla u|\,.$$

Finally, if u is a piecewise affine function on $\mathbb{R}^n$, then there exists a Borel partition $\{F_h\}_{h\in\mathbb{N}}$ of $\mathbb{R}^n$ with $u(x) = \lambda_h(x \cdot e_h)$ for $x \in F_h$, $\lambda_h \in \mathbb{R}$, $e_h \in S^{n-1}$, and

$$\begin{aligned}\int_{\mathbb{R}} \mathcal{H}^{n-1}\big(E \cap \{u = t\}\big)\,\mathrm{d}t &= \sum_{h\in\mathbb{N}} \int_{\mathbb{R}} \mathcal{H}^{n-1}\big(E \cap F_h \cap \{u = t\}\big)\,\mathrm{d}t \\ &= \sum_{h\in\mathbb{N}} |\lambda_h||E \cap F_h| = \int_E |\nabla u|\,,\end{aligned}$$

as desired. We now consider the case of a generic Lipschitz function.

Theorem 13.1 (Coarea formula) *If $u\colon \mathbb{R}^n \to \mathbb{R}$ is a Lipschitz function and $A \subset \mathbb{R}^n$ is open, then $t \in \mathbb{R} \mapsto P(\{u > t\}; A)$ is a Borel function on $\mathbb{R}$ with*

$$\int_A |\nabla u| = \int_{\mathbb{R}} P(\{u > t\}; A)\,dt \tag{13.1}$$

as elements of $[0, \infty]$.

Remark 13.2 By the Morse–Sard Lemma, if $u \in C^\infty(\mathbb{R}^n)$, then, for a.e. $t \in \mathbb{R}$, $\{u > t\}$ is an open set with smooth boundary. Hence $P(\{u > t\}; A) = \mathcal{H}^{n-1}(A \cap \{u = t\})$ for a.e. $t \in \mathbb{R}$ and for every $A \subset \mathbb{R}^n$ open. Combining (13.1) with Theorem 2.10, we conclude that, for every Borel set $E \subset \mathbb{R}^n$,

$$\int_E |\nabla u| = \int_{\mathbb{R}} \mathcal{H}^{n-1}\big(E \cap \{u = t\}\big)\,dt\,. \tag{13.2}$$

Example 13.3 (Super-level sets of Lipschitz functions are of finite perimeter) From (13.1), we immediately deduce that $u\colon \mathbb{R}^n \to \mathbb{R}$ is a locally Lipschitz function, then, for a.e. $t > 0$, the open set $\{u > t\}$ is of locally finite perimeter in $\mathbb{R}^n$. Moreover, if $\int_{\{u>s\}} |\nabla u| < \infty$ for some $s \in \mathbb{R}$, then for a.e. $t > s$ the open set $\{u > t\}$ is of finite perimeter in $\mathbb{R}^n$.

Example 13.4 As an application of Theorem 13.1 let us show that

$$|E \cap B(x, r)| = \int_0^r \mathcal{H}^{n-1}\big(E \cap \partial B(x, t)\big)\,dt\,, \tag{13.3}$$

for every Borel set E in $\mathbb{R}^n$. To this end let us apply (13.1) to the function $u(y) = |\,y - x|$, $y \in \mathbb{R}^n$. Since $|\nabla u| = 1$ on $\mathbb{R}^n$ by (13.1) we find that

$$|A \cap B(x, r)| = \int_0^\infty P\big(\{u > t\}; A \cap B(x, r)\big)\,dt\,,$$

for every open set A in $\mathbb{R}^n$. Since $\{u > t\} = \mathbb{R}^n \setminus B(x, t)$ by Example 12.5 we have that

$$|\mu_{\{u>t\}}| = \mathcal{H}^{n-1} \llcorner \partial B(x, t)\,,$$

for every $t > 0$. Thus $P(\{u > t\}; A) = \mathcal{H}^{n-1}(A \cap \partial B(x, t))$ and (13.3) is proved whenever E is an open set. Since the family $\mathcal{F}$ of the Borel sets E in $\mathbb{R}^n$ satisfying (13.3) is easily seen to be a σ-algebra on $\mathbb{R}^n$, and since $\mathcal{F}$ contains the open sets, we conclude that $\mathcal{F} = \mathcal{B}(\mathbb{R}^n)$, as claimed.

Exercise 13.5 Replace balls by cylinders in Example 13.4.

Remark 13.6 In the proof of Theorem 13.1 we are going to use the following *layer-cake formula*. If $u \in L^1(\mathbb{R}^n)$, $u \geq 0$, and $v \in L^\infty(\mathbb{R}^n)$, then

$$\int_{\mathbb{R}^n} u(x)\, v(x)\, \mathrm{d}x = \int_0^\infty \mathrm{d}t \int_{\{u>t\}} v(x)\, \mathrm{d}x\,. \tag{13.4}$$

Indeed, for every $x \in \mathbb{R}^n$,

$$u(x) = \int_{\mathbb{R}} 1_{(0,u(x))}(t)\mathrm{d}t = \int_{\mathbb{R}} 1_{(0,\infty)}(t)\, 1_{\{u>t\}}(x) = \int_0^\infty 1_{\{u>t\}}(x)\mathrm{d}t\,,$$

and thus, by Fubini's theorem,

$$\int_{\{u\geq 0\}} u(x)\, v(x)\, \mathrm{d}x = \int_{\{u\geq 0\}} v(x) \int_0^\infty 1_{\{u>t\}}(x)\, \mathrm{d}t = \int_0^\infty \mathrm{d}t \int_{\{u>t\}} v(x)\, \mathrm{d}x\,.$$

Proof of Theorem 13.1 *Step one:* If $T \in C^1_c(A;\mathbb{R}^n)$, then $\int_{\{u>t\}} \operatorname{div} T$ is a Borel measurable function of $t \in \mathbb{R}$. Indeed, it is the difference of two increasing functions of $t \in \mathbb{R}$, namely

$$\int_{\{u>t\}} \operatorname{div} T = \int_{\{u>t\}} (\operatorname{div} T)^+ - \int_{\{u>t\}} (\operatorname{div} T)^-\,.$$

If $\mathcal{F}$ is countable and dense in $C^\infty_c(A;\mathbb{R}^n)$, then, by (12.6),

$$P(\{u>t\};A) = \sup\left\{\int_{\{u>t\}} \operatorname{div} T : T \in \mathcal{F}\,, \sup_{\mathbb{R}^n} |T| \leq 1\right\}.$$

Since the supremum of countably many Borel functions is a Borel function, we have proved that $t \in \mathbb{R} \mapsto P(\{u>t\};A)$ is a Borel function.

Step two: We prove that if u is a non-negative Lipschitz function then

$$\int_A |\nabla u| \leq \int_0^\infty P(\{u>t\};A)\, \mathrm{d}t\,, \tag{13.5}$$

for every open set A in $\mathbb{R}^n$ (in particular, if the left-hand side is infinite then the right-hand side is infinite too). If $T \in C^\infty_c(A;\mathbb{R}^n)$, $|T| \leq 1$, then by (12.6)

$$\int_{\{u>t\}} \operatorname{div} T \leq P(\{u>t\};A)\,, \qquad t > 0\,.$$

By the distributional divergence theorem and by (13.4) (with $v = \operatorname{div} T$),

$$-\int_A \nabla u \cdot T = \int_{\mathbb{R}^n} u \operatorname{div} T = \int_0^\infty \mathrm{d}t \int_{\{u>t\}} \operatorname{div} T \leq \int_0^\infty P(\{u>t\};A)\, \mathrm{d}t\,. \tag{13.6}$$

Let K be a compact subset of A and define $S : \mathbb{R}^n \to \mathbb{R}^n$ as

$$S(x) = -1_{K\cap\{\nabla u \neq 0\}}(x) \frac{\nabla u(x)}{|\nabla u(x)|}\,, \qquad x \in \mathbb{R}^n\,,$$

so that S is a bounded Borel measurable vector field with $|S| \le 1$. For every $\varepsilon < \mathrm{dist}(K, \partial A)$ we have that $S_\varepsilon = (S \star \rho_\varepsilon) \in C^\infty_c(A; \mathbb{R}^n)$ with $|S_\varepsilon| \le 1$ and $S_\varepsilon(x) \to S(x)$ for a.e. $x \in \mathbb{R}^n$. We let $T = S_\varepsilon$ and $\varepsilon \to 0$ in (13.6) to find that

$$\int_K |\nabla u| \le \int_0^\infty P(\{u > t\}; A)\, dt\,,$$

(where the left-hand side is finite). Since K is arbitrary, we find (13.5).

Step three: We prove that if u is a non-negative Lipschitz function then

$$\int_A |\nabla u| \ge \int_0^\infty P(\{u > t\}; A)\, dt\,. \tag{13.7}$$

To this end we consider the increasing function $m\colon \mathbb{R} \to [0, \infty)$ defined as

$$m(t) = \int_{A \cap \{u \le t\}} |\nabla u|\,, \qquad t \in \mathbb{R}\,.$$

By Exercise 5.14 the classical derivative $m'(t)$ exists for a.e. $t \in \mathbb{R}$, and

$$\int_0^\infty m'(t)\, dt \le \lim_{t \to \infty} m(t) - \lim_{t \to -\infty} m(t) = \int_A |\nabla u|\,.$$

We are thus left to show that

$$m'(t) \ge P(\{u > t\}; A)\,, \qquad \text{for a.e. } t \ge 0\,. \tag{13.8}$$

Given $t \ge 0$ and $\varepsilon > 0$, define a piecewise affine function $\psi\colon [0, \infty) \to [0, 1]$ as

$$\psi(s) = \begin{cases} 1, & s \in [t + \varepsilon, \infty), \\ \varepsilon^{-1}(s - t), & s \in [t, t + \varepsilon), \\ 0, & s \in [0, t). \end{cases} \tag{13.9}$$

By Lemma 7.6, $\psi \circ u$ admits $(\psi' \circ u)\nabla u = -\varepsilon^{-1}\, 1_{(t,t+\varepsilon)}(u)\, \nabla u$ as its weak gradient on $\mathbb{R}^n$. If $T \in C^\infty_c(A; \mathbb{R}^n)$ with $|T| \le 1$, then

$$\begin{aligned} \int_A (\psi \circ u)\, \mathrm{div}\, T &= -\frac{1}{\varepsilon} \int_{A \cap \{t+\varepsilon > u > t\}} \nabla u \cdot T \\ &\le \frac{1}{\varepsilon} \int_{A \cap \{t+\varepsilon > u > t\}} |\nabla u| \le \frac{m(t + \varepsilon) - m(t)}{\varepsilon}\,. \end{aligned}$$

As $\varepsilon \to 0^+$ we find that, for a.e. $t > 0$,

$$\int_{A \cap \{u > t\}} \mathrm{div}\, T \le m'(t)\,,$$

which implies (13.8) by the arbitrariness of T.

Step four: Finally, let $u\colon \mathbb{R}^n \to \mathbb{R}$ be a Lipschitz function, and consider its positive and negative parts u^+ and u^-. By step two of Lemma 7.6,

$\nabla u^+ = 1_{\{u>0\}}\nabla u$ and $\nabla u^- = -1_{\{u<0\}}\nabla u$. By step two and step three, (13.1) holds true for u^+ and u^-. Hence,

$$\begin{aligned}\int_A |\nabla u| &= \int_A |\nabla u^+| + \int_A |\nabla u^-| \\ &= \int_0^\infty P(\{u^+ > t\}; A)\,dt + \int_0^\infty P(\{u^- > t\}; A)\,dt \\ &= \int_0^\infty P(\{u > t\}; A)\,dt + \int_0^\infty P(\{u < -t\}; A)\,dt \\ &= \int_0^\infty P(\{u > t\}; A)\,dt + \int_{-\infty}^0 P(\{u < t\}; A)\,dt. \end{aligned} \tag{13.10}$$

By (12.4) we have that $P(\{u < t\}; A) = P(\{u \geq t\}; A)$. Since, by Proposition 2.16, $|\{u = t\}| = 0$ for a.e. $t \in \mathbb{R}$, we thus have $P(\{u \geq t\}; A) = P(\{u > t\}; A)$ for a.e. $t \in \mathbb{R}$. Hence (13.1) follows from (13.10). □

Exercise 13.7 If E is a set of finite perimeter and $|E| < \infty$, then $P(E \cap B_{r_h}) \to P(E)$ on a suitable sequence $r_h \to \infty$ as $h \to \infty$. In particular, if we replace "E bounded" with "$|E| < \infty$", then (12.38) still holds true. *Hint:* Use (13.3) and (15.15) from Lemma 15.12. The solution is also contained in Remark 13.12.

13.2 Approximation by open sets with smooth boundary

We now prove the following fundamental approximation theorem.

Theorem 13.8 (Approximation by smooth sets) *A Lebesgue measurable set $E \subset \mathbb{R}^n$ is of locally finite perimeter if and only if there exists a sequence $\{E_h\}_{h\in\mathbb{N}}$ of open sets with smooth boundary in $\mathbb{R}^n$, and $\varepsilon_h \to 0^+$, such that*

$$E_h \overset{\text{loc}}{\to} E, \qquad \sup_{h\in\mathbb{N}} P(E_h; B_R) < \infty, \quad \forall R > 0,$$
$$|\mu_{E_h}| \overset{*}{\rightharpoonup} |\mu_E|, \qquad \partial E_h \subset I_{\varepsilon_h}(\partial E).$$

In particular, $P(E_h; F) \to P(E; F)$ whenever $P(E; \partial F) = 0$. Moreover,

(i) If $|E| < \infty$, then $E_h \to E$;
(ii) If $P(E) < \infty$, then $P(E_h) \to P(E)$.

Remark 13.9 In fact, if $|E| < \infty$ and $P(E) < \infty$ we can also assume that each set E_h is bounded; see Remark 13.12.

Remark 13.10 Having in mind Example 12.25, the inclusion $\partial E_h \subset I_{\varepsilon_h}(\partial E)$ may be of little use on generic sets of finite perimeter. However, it becomes

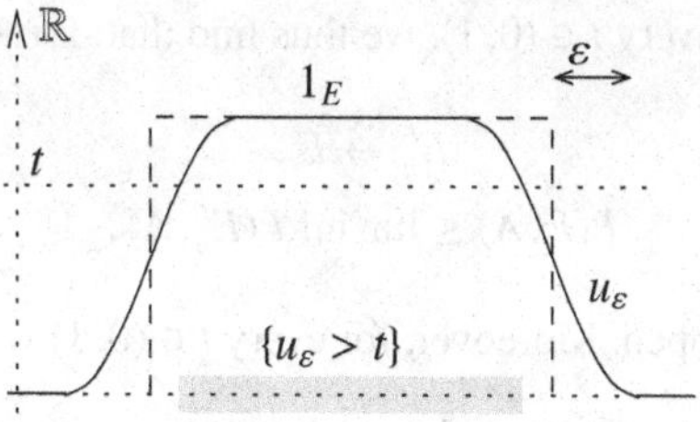

Figure 13.2 By the Morse–Sard lemma, for a.e. $t \in (0,1)$ the open set $\{u_\varepsilon > t\}$ has smooth boundary. By the coarea formula, it is close in perimeter to E.

useful when dealing with sets whose topological boundary is not too wild; for example, we shall use this information in the proof of Theorem 24.1.

Remark 13.11 In the proof of Theorem 13.8 we are going to use the following formula: if E is a bounded Borel set and $u, v \in L^1_{\mathrm{loc}}(\mathbb{R}^n)$, then

$$\int_E |u - v| = \int_{\mathbb{R}} \Big| E \cap \big(\{u > t\}\Delta\{v > t\}\big)\Big| \,\mathrm{d}t . \tag{13.11}$$

Indeed, by Fubini's theorem,

$$\begin{aligned}\int_{E\cap\{u>v\}} |u - v| &= \int_E 1_{\{u>v\}}(x)\mathrm{d}x \int_{\mathbb{R}} 1_{\{u>t\}}(x) 1_{\{v\le t\}}(x)\,\mathrm{d}t \\ &= \mathcal{L}^{n+1}\Big(\big\{(x,t) \in E \times \mathbb{R} \colon u(x) > t \ge v(x)\big\}\Big) \\ &= \int_{\mathbb{R}} \Big| E \cap \{u > t\} \setminus \{v > t\}\Big| \,\mathrm{d}t .\end{aligned}$$

Proof of Theorem 13.8 Given $\varepsilon > 0$, $\varepsilon_h \to 0^+$, $t \in (0,1)$, we set

$$u_\varepsilon = 1_E \star \rho_\varepsilon , \qquad u_h = u_{\varepsilon_h} , \qquad E_h^t = \{u_h > t\} .$$

In a nutshell, we are going to prove the theorem by suitably choosing $t \in (0,1)$, and then setting $E_h = E_h^t$; see Figures 13.2 and 12.3. Indeed, by the Morse–Sard lemma (Lemma 13.15 below), for a.e. $t \in (0,1)$,

$$\{E_h^t\}_{h\in\mathbb{N}} \text{ is a sequence of open sets with smooth boundary.} \tag{13.12}$$

Since $1_E \in L^1_{\mathrm{loc}}(\mathbb{R}^n)$ and $u_h \to 1_E$ in $L^1_{\mathrm{loc}}(\mathbb{R}^n)$, by (13.11) we also have

$$\{u_h > t\} \overset{\mathrm{loc}}{\to} \{1_E > t\} , \quad \text{for a.e. } t \in \mathbb{R} .$$

As $\{1_E > t\} = E$ for every $t \in (0,1)$, we thus find that, for a.e. $t \in (0,1)$,

$$E_h^t \overset{\text{loc}}{\to} E\,, \tag{13.13}$$

$$P(E;A) \le \liminf_{h\to\infty} P(E_h^t;A)\,, \tag{13.14}$$

whenever $A \subset \mathbb{R}^n$ is open. Moreover, for every $t \in (0,1)$ we have that

$$\partial E_h^t \subset I_{\varepsilon_h}(\partial E)\,. \tag{13.15}$$

Since $\partial E_h^t \subset \{u_h = t\}$, we only have to show the inclusion

$$\{0 < u_\varepsilon < 1\} \subset I_\varepsilon(\partial E)\,, \qquad \forall \varepsilon > 0\,. \tag{13.16}$$

To prove (13.16) we remark that if $x \in \mathbb{R}^n \setminus I_\varepsilon(\partial E)$ then $B(x,\varepsilon)$ is contained either in $\mathring{E}$ or in $\mathbb{R}^n \setminus \overline{E}$. Correspondingly we have either $|B(x,\varepsilon) \cap E| = |B(x,\varepsilon)|$ (and $u_\varepsilon(x) = 1$) or $|B(x,\varepsilon) \cap E| = 0$ (and $u_\varepsilon(x) = 0$), and (13.16) is proved. We now divide the proof into two steps.

Step one: We improve (13.14) by showing that, if A is an open set in $\mathbb{R}^n$, then, for a.e $t \in (0,1)$ (the set of exceptional values of t possibly depending on A),

$$P(E;A) = \liminf_{h\to\infty} P(E_h^t;A)\,. \tag{13.17}$$

Indeed, by (13.14) and Fatou's lemma,

$$P(E;A) \le \int_0^1 \liminf_{h\to\infty} P(E_h^t;A)\,dt \le \liminf_{h\to\infty} \int_0^1 P(E_h^t;A)\,dt\,.$$

Since $0 \le u_h \le 1$, we have $E_h^t = \emptyset$ if $t > 1$, and $E_h^t = \mathbb{R}^n$ if $t < 0$. Thus, by the coarea formula (13.1) and (12.15), we find that

$$P(E;A) = \lim_{h\to\infty} \int_A |\nabla u_h| = \lim_{h\to\infty} \int_{\mathbb{R}} P(E_h^t;A)\,dt\,.$$

In conclusion, it must be that

$$P(E;A) = \int_0^1 \liminf_{h\to\infty} P(E_h^t;A)\,dt\,,$$

which, combined with (13.14), immediately implies (13.17).

Step two: We conclude the proof of the theorem. Let $\{r_i\}_{i\in\mathbb{N}}$ be a sequence with $r_i \to \infty$. By step one we have that, for a.e. $t \in (0,1)$, (13.12), (13.13), and (13.15) hold true, with

$$P(E;B_{r_i}) = \liminf_{h\to\infty} P(E_h^t;B_{r_i})\,, \qquad \forall i \in \mathbb{N}\,.$$

By a diagonal argument we can find $t \in (0,1)$ and $h(k) \to \infty$ as $k \to \infty$ such that, if we set $E_k = E^t_{h(k)}$, then $\{E_k\}_{k\in\mathbb{N}}$ is a sequence of open sets with smooth boundary, locally converging to E, with $\partial E_k \subset I_{\delta_k}(\partial E)$, $\delta_k = \varepsilon_{h(k)}$, with

$$P(E;B_{r_i}) = \lim_{k\to\infty} P(E_k;B_{r_i}), \qquad \forall i \in \mathbb{N}. \tag{13.18}$$

Since E is a set of locally finite perimeter in $\mathbb{R}^n$, the left-hand side of (13.18) is bounded, and by a simple density argument we immediately see that $\mu_{E_k} \stackrel{*}{\rightharpoonup} \mu_E$ (cf. the proof of Proposition 12.15). Finally, by Exercise 4.31, (13.18) actually implies that $|\mu_{E_k}| \stackrel{*}{\rightharpoonup} |\mu_E|$, and we are done.

Step three: We are left to prove (i) and (ii). Concerning (i), if $|E| < \infty$, then $u_h \to 1_E$ in $L^1(\mathbb{R}^n)$, and in particular $E^t_h \to E$ for a.e. $t \in (0,1)$. Concerning (ii), if $P(E) < \infty$, then (13.17) directly implies that for a.e. $t \in (0,1)$,

$$P(E) = \liminf_{h\to\infty} P(E^t_h),$$

with finite left-hand side. We can thus conclude as in step two, avoiding the use of the diagonal argument, and achieving $P(E_k) \to P(E)$. □

Remark 13.12 (Approximation by bounded sets) Let us show that, if $|E| < \infty$ and $P(E) < \infty$, then there exists a sequence $\{E_h\}_{h\in\mathbb{N}}$ of *bounded* open sets with smooth boundary, such that $E_h \to E$ and $P(E_h) \to P(E)$. It will suffice to show the existence of a sequence $R_h \to \infty$ such that $E \cap B_{R_h} \to E$ and $P(E \cap B_{R_h}) \to P(E)$, and then to apply Theorem 13.8 to approximate each $E \cap B_{R_h}$. To this end, we first notice that $|E| < \infty$ implies $|E \setminus B_R| \to$ as $R \to \infty$, that is, $E \cap B_{R_h} \to E$ whenever $R_h \to \infty$. In order to achieve $P(E \cap B_{R_h}) \to P(E)$ we have to select R_h. To this end, we first notice that, since $P(E) < \infty$, we have

$$\lim_{R\to\infty} P(E;\mathbb{R}^n \setminus B_R) = 0.$$

At the same time, by Lemma 15.12 (whose proof does not use this remark!), for a.e. $R > 0$ we have that (see Figure 15.2),

$$P(E \cap B_R) = P(E;B_R) + \mathcal{H}^{n-1}(E \cap \partial B_R). \tag{13.19}$$

Finally, since $|E| < \infty$ and by the coarea formula (13.3), $\mathcal{H}^{n-1}(E \cap \partial B_R)$, as a function of $R > 0$, belongs to $L^1(0,\infty)$. Therefore, there exists $R_h \to \infty$ with

$$\lim_{h\to\infty} \mathcal{H}^{n-1}(E \cap \partial B_{R_h}) = 0,$$

and such that (13.19) holds true for $R = R_h$. We conclude that

$$|P(E \cap B_{R_h}) - P(E)| = P(E; \mathbb{R}^n \setminus B_{R_h}) + \mathcal{H}^{n-1}(E \cap \partial B_{R_h})$$

so that $P(E \cap B_{R_h}) \to P(E)$, as desired.

Remark 13.13 (Approximation by polyhedra) When E is a set of finite perimeter in $\mathbb{R}^n$ with $|E| < \infty$, we may also approximate E by a sequence of open bounded sets with polyhedral boundary; see Example 12.6. Indeed, let us first assume that E is bounded. In this case the sequence $\{u_h\}_{h\in\mathbb{N}}$ considered in the proof of Theorem 13.8 is contained in $C^1_c(\mathbb{R}^n)$. By affine interpolation, every $v \in C^1_c(\mathbb{R}^n)$ can be approximated by a sequence $\{v_h\}_{h\in\mathbb{N}}$ of piecewise affine functions with compact support, so that $v_h \to v$ in $L^1(\mathbb{R}^n)$ and $\int_{\mathbb{R}^n} |\nabla v_h| \to \int_{\mathbb{R}^n} |\nabla v|$. Therefore, we can repeat the proof of Theorem 13.8 with such a sequence $\{v_h\}_{h\in\mathbb{N}}$ in place of $\{u_h\}_{h\in\mathbb{N}}$. The approximating sequence for E will be selected among the sets $\{v_h > t\}$, which, for a.e. $t \in (0, 1)$, are bounded, open sets with polyhedral boundary. In this way, given a bounded set E of finite perimeter in $\mathbb{R}^n$, there exists a sequence $\{E_h\}_{h\in\mathbb{N}}$ of bounded open sets with polyhedral boundaries such that $E_h \to E$ and $P(E_h) \to P(E)$. By Remark 13.12, the same holds if, instead of assuming E bounded, we simply assume that $|E| < \infty$.

Exercise 13.14 If A is open and bounded, $P(A) < \infty$, and $m \in (0, |A|)$, then

$$\inf\Big\{P(E; A) : E \subset A\,, |E| = m\Big\} = \inf\Big\{P(F; A) : F \subset A\,, |F| = m\Big\},$$

where the minimization on the left-hand side is in the class of sets of finite perimeter, while the minimization on the right-hand side is in the class of open sets with smooth boundary. *Hint:* Prove the non-trivial inequality by Theorem 13.8.

13.3 The Morse–Sard lemma

For the sake of completeness we include a proof of the classical Morse–Sard lemma, in the particular case used in proving Theorem 13.8.

Lemma 13.15 (Morse–Sard lemma) *If $u \in C^\infty(\mathbb{R}^n)$ and $E = \{x \in \mathbb{R}^n : \nabla u(x) = 0\}$, then $|u(E)| = 0$. In particular, $\{u = t\} = \{x \in \mathbb{R}^n : u(x) = t\}$ is a smooth hypersurface in $\mathbb{R}^n$ for a.e. $t \in \mathbb{R}$.*

Proof Given $k \in \mathbb{N}$, let $\nabla^k u(x)$ denote the collection of kth order partial derivatives of u at x, and set

$$E_h = \bigcap_{k=1}^{h} \Big\{x \in \mathbb{R}^n : \nabla^k u(x) = 0\Big\}, \qquad h \in \mathbb{N}.$$

Step one: Let us prove that $|u(E_n)| = 0$, by showing that $|u(E_n \cap B_R)| = 0$ for every $R > 0$. By Taylor's formula, given $R > 0$, there exist positive constants C and r_0 such that,

$$|u(x) - u(y)| \le C|x - y|^{n+1}, \qquad \forall x \in E_n \cap B_R,\ y \in B(x, r_0). \tag{13.20}$$

Given $r \in (0, n^{-1/2} r_0)$, we cover $(-R, R)^n$ by disjoint cubes of side length r, and denote by $\{Q_k\}_{k=1}^N$ ($N = N(r)$) those cubes intersecting $E_n \cap B_R$. By construction, $N \le (2R/r)^n$ and, for every $k = 1, \dots, N$, there exists $x_k \in Q_k \cap E_n \cap B_R$. In particular, if $y \in Q_k$ then $|y - x_k| \le n^{1/2} r < r_0$, and, by (13.20),

$$|u(x_k) - u(y)| \le C(n)\, r^{n+1}, \qquad \forall y \in Q_k,$$

i.e., $u(Q_k) \subset \left[u(x_k) - C(n) r^{n+1}, u(x_k) + C(n) r^{n+1}\right]$. Hence,

$$|u(E_n \cap B_R)| \le \sum_{k=1}^{N} |u(Q_k)| \le N\, C(n)\, r^{n+1} \le C(n) R^n\, r.$$

We let $r \to 0^+$ to find that $|u(E_n \cap B_R)| = 0$, as required.

Step two: We now argue by induction over the dimension n. The case $n = 1$ follows by step one. Since $E_1 = E$ and $E_{h+1} \subset E_h$, we have

$$|u(E)| = |u(E_n)| + \sum_{k=1}^{n-1} |u(E_k \setminus E_{k+1})|. \tag{13.21}$$

By step one, we are left to show that,

$$|u(E_h \setminus E_{h+1})| = 0, \qquad 1 \le h \le n - 1. \tag{13.22}$$

We claim that, if $x \in E_h \setminus E_{h+1}$, then there exists A_x open with $x \in A_x$ and

$$\Big|u\big((E_h \setminus E_{h+1}) \cap A_x\big)\Big| = 0. \tag{13.23}$$

Notice that (13.22) follows from (13.23) by extracting a countable covering of $E_h \setminus E_{h+1}$ from $\{A_x\}_{x \in E_h \setminus E_{h+1}}$. We now prove (13.23). If $x \in E_h \setminus E_{h+1}$, then there exists a derivative of order h of u, denoted by g, such that $g(x) = 0$ and $\nabla g(x) \ne 0$. Hence, there exists A_x open, with $x \in A_x$, such that $G = \{x \in \mathbb{R}^n : g(x) = 0\} \cap A_x$ is a smooth hypersurface. Since $g = 0$ on E_h, we have

$$(E_h \setminus E_{h+1}) \cap A_x \subset G,$$

and thus (13.23) follows by showing that $|u(G)| = 0$. Indeed, up to restricting A_x, there exists a function $\psi \in C^\infty(\mathbb{R}^{n-1}; G)$ and an open set $F \subset \mathbb{R}^{n-1}$ such that $\psi(F) = G$. Thus $u \circ \psi \in C^\infty(\mathbb{R}^{n-1})$ with

$$\nabla(u \circ \psi) = (\nabla \psi)^* (\nabla u \circ \psi) = 0 \quad \text{on } F,$$

as $\psi(F) = G \subset E$. Therefore,

$$u(G) = (u \circ \psi)(F) \subset \left\{ y \in \mathbb{R}^{n-1} : \nabla(u \circ \psi)(y) = 0 \right\},$$

where this last set has measure zero by the inductive hypothesis. □

Exercise 13.16 If M is a k-dimensional smooth surface in $\mathbb{R}^n$, A is an open set with $M \subset A$, and $u \colon \mathbb{R}^n \to \mathbb{R}$ is such that $u \in C^\infty(A)$, then $\{u = t\} \cap M$ is a $(k-1)$-dimensional smooth surface for a.e. $t \in \mathbb{R}$.

14

The Euclidean isoperimetric problem

In this chapter we study the **Euclidean isoperimetric problem**,

$$\inf\left\{P(E) : |E| = m\right\}, \qquad m > 0, \tag{14.1}$$

characterizing its minimizers as the Euclidean balls of measure m. Taking into account that $P(B) = \omega_n$, this result will be equivalent to proving the following **Euclidean isoperimetric inequality**.

Theorem 14.1 *If E is a Lebesgue measurable set in $\mathbb{R}^n$ with $|E| < \infty$, then*

$$P(E) \ge n\omega_n^{1/n}|E|^{(n-1)/n}. \tag{14.2}$$

Equality holds if and only if $|E\Delta B(x,r)| = 0$ for some $x \in \mathbb{R}^n$, $r > 0$.

Remark 14.2 When $n = 1$, (14.2) reduces to $P(E) \ge 2$, with equality if and only if E is equivalent to an interval. Indeed, by Proposition 12.13, the assumptions $|E| < \infty$ and $P(E) < \infty$ imply that

$$E = \bigcup_{i=1}^{N}(a_i, b_i), \qquad b_i < a_{i+1},$$

for some $N \in \mathbb{N}$, $N \ge 1$, and up to a set of measure zero. In particular $P(E) = 2N \ge 2$, with equality if and only if $N = 1$, and thus E is a single interval.

In dimension $n \ge 2$ the problem changes completely. We present here the classical proof of (14.2) based on Steiner symmetrization, a geometric operation that we have already used in Section 3.3. In Section 14.1 we prove that perimeter is decreased under Steiner symmetrization (Steiner inequality), and identify some necessary conditions for equality. Finally, in Section 14.2, we prove Theorem 14.1.

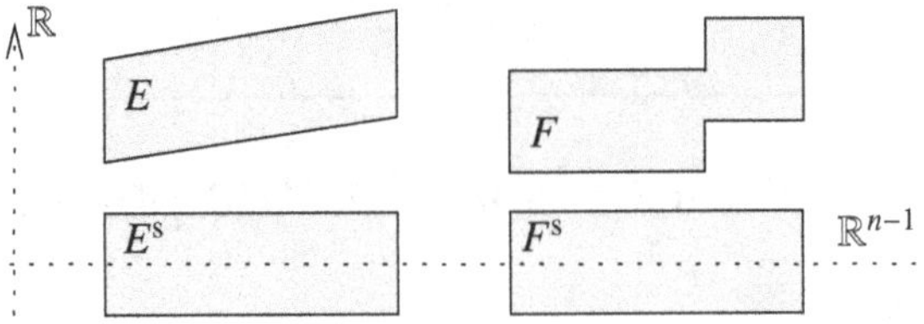

Figure 14.1 There could be a strict sign in the Steiner inequality even if the vertical slices E_z of E are all intervals.

Remark 14.3 The straightforward application of the Direct Method to the Euclidean isoperimetric problem does not even prove the existence of minimizers. Indeed, the compactness results from Section 12.4 only allow us to construct a minimizing sequence $\{E_h\}_{h\in\mathbb{N}}$ in (14.1) which is locally converging to some set E in $\mathbb{R}^n$. In particular, local convergence is not sufficient to imply that $|E| = m$, and one would have to exclude the possibility that $|E| < m$. These difficulties are solved here by exploiting the full symmetry of the problem, and a similar approach will be used in studying sessile liquid drops in Chapter 19. In dealing with the existence proof for minimizers in partitioning problems, we shall also learn to solve similar compactness issues without relying on symmetrization principles; see, in particular, Section 29.2.

14.1 Steiner inequality

We first recall the definition of Steiner symmetrization. Let us decompose $\mathbb{R}^n$, $n \geq 2$, as the product $\mathbb{R}^{n-1}\times\mathbb{R}$, with the projections $\mathbf{p}\colon \mathbb{R}^n \to \mathbb{R}^{n-1}$ and $\mathbf{q}\colon \mathbb{R}^n \to \mathbb{R}$, so that $x = (\mathbf{p}x, \mathbf{q}x)$ for $x \in \mathbb{R}^n$ (in particular, $\mathbf{q}x = x_n$). With every $z \in \mathbb{R}^{n-1}$ we associate the vertical slice $E_z \subset \mathbb{R}$ of E defined as

$$E_z = \Big\{t \in \mathbb{R}\colon (z,t) \in E\Big\},$$

and define the Steiner symmetrization E^s of E as

$$E^s = \left\{x \in \mathbb{R}^n : |\mathbf{q}x| \leq \frac{\mathcal{L}^1(E_{\mathbf{p}x})}{2}\right\},$$

By Fubini's theorem, $|E| = |E^s|$, and, as proved in Section 3.3, diameters are decreased under Steiner symmetrization. The same happens to perimeters.

Theorem 14.4 (Steiner inequality) *If E is a set of finite perimeter in $\mathbb{R}^n$, with $|E| < \infty$, then E^s is a set of finite perimeter in $\mathbb{R}^n$, with*

$$P(E^s) \leq P(E), \tag{14.3}$$

and, in fact, whenever A is an open set in $\mathbb{R}^{n-1}$,

$$P\big(E^s; A\times\mathbb{R}\big) \le P\big(E; A\times\mathbb{R}\big). \tag{14.4}$$

Moreover,

(i) *if equality holds in (14.3), then, for a.e.* $z \in \mathbb{R}^{n-1}$, *the vertical slice* E_z *is equivalent to an interval;*
(ii) *if E is equivalent to a convex set, then equality holds in (14.3) if and only if there exists* $c \in \mathbb{R}$ *such that E is equivalent to* $E^s + ce_n$.

We start with the following simple proposition, of independent interest.

Proposition 14.5 (Slicing perimeter by lines) *If E is a set of locally finite perimeter in* $\mathbb{R}^n$, *then, for a.e.* $z \in \mathbb{R}^{n-1}$, *the vertical slice* E_z *is a set of locally finite perimeter in* $\mathbb{R}$, *and, for* $I \subset \mathbb{R}$ *bounded and open, and* $H \subset \mathbb{R}^{n-1}$ *compact,*

$$\int_H P(E_z; I)\,\mathrm{d}z \le P\big(E; H\times\overline{I}\big). \tag{14.5}$$

If E is of finite perimeter, then, for a.e. $z \in \mathbb{R}^{n-1}$, E_z *is of finite perimeter, and*

$$\int_{\mathbb{R}^{n-1}} P(E_z)\,\mathrm{d}z \le P(E). \tag{14.6}$$

Proof Since $u_\varepsilon = 1_E \star \rho_\varepsilon \to 1_E$ in $L^1_{\mathrm{loc}}(\mathbb{R}^n)$, by Fubini's theorem there exists a sequence $\varepsilon_h \to 0^+$ such that, if we set $u_h = u_{\varepsilon_h}$, then, as $h \to \infty$,

$$u_h(z,\cdot) \to 1_{E_z} \quad \text{in } L^1_{\mathrm{loc}}(\mathbb{R}),$$

for a.e. $z \in \mathbb{R}^{n-1}$. If now $T \in C^1_c(\mathbb{R})$ with $K = \operatorname{spt} T$ and $|T| \le 1$, then we have

$$\begin{aligned}\left|\int_{E_z} T'\right| &= \lim_{h\to\infty}\left|\int_{\mathbb{R}} u_h(z,t)\,T'(t)\,\mathrm{d}t\right| = \lim_{h\to\infty}\left|\int_{\mathbb{R}} \partial_n u_h(z,t)\,T(t)\,\mathrm{d}t\right| \\ &\le \liminf_{h\to\infty}\int_K |\nabla u_h(z,t)|\,\mathrm{d}t,\end{aligned} \tag{14.7}$$

for a.e. $z \in \mathbb{R}^{n-1}$. Now let $H \subset \mathbb{R}^{n-1}$ be compact, and let $I \subset \mathbb{R}$ be an open bounded set. If we chose $K = \overline{I}$, then, having $|\nabla u_\varepsilon|\mathrm{d}\mathcal{L}^n \overset{*}{\rightharpoonup} |\mu_E|$, we find

$$\begin{aligned}\int_H \sup\left\{\left|\int_{E_z} T'\right| : T \in C^1_c(I), |T| \le 1\right\}\mathrm{d}z &\le \liminf_{h\to\infty}\int_{H\times K} |\nabla u_h| \\ &\le P\big(E; H\times K\big).\end{aligned}$$

In particular, for a.e. $z \in \mathbb{R}^{n-1}$ we find that

$$\sup\left\{\left|\int_{E_z} T'\right| : T \in C^1_c(I), |T| \le 1\right\} < \infty,$$

so that, by (12.1), E_z is of locally finite perimeter in $\mathbb{R}$. By (14.7), $P(E_z; I) \le \liminf_{h\to\infty} \int_I |\nabla u_h(z,t)|\,dt$, so that, by repeating the above argument,

$$\int_H P(E_z, I)\,dz \le P\big(E; H \times \overline{I}\big). \qquad \square$$

Proof of Theorem 14.4 We directly assume $A = \mathbb{R}^{n-1}$, as the general case will follow with minor modifications. If $u\colon \mathbb{R}^{n-1} \to \mathbb{R}$ and $G \subset \mathbb{R}^{n-1}$, we denote by $\Gamma(u, G) = \{(z,t) \in \mathbb{R}^n : z \in G\,, t = u(z)\}$ the graph of u over G.

Step one: Let us assume that E is a bounded set with polyhedral boundary, and that the outer unit normal to E (that is elementarily defined at $\mathcal{H}^{n-1}$-a.e. point of ∂E) is never orthogonal to e_n. By this assumption, and by the implicit function theorem, there exist a partition of the set $G = \{z \in \mathbb{R}^{n-1} : \mathcal{L}^1(E_z) > 0\}$ into finitely many $(n-1)$-dimensional polyhedral sets $\{G_h\}_{h=1}^M$ in $\mathbb{R}^{n-1}$,

$$G = \bigcup_{h=1}^M G_h\,,$$

and affine functions $v_h^k, u_h^k\colon G_h \to \mathbb{R}$, $1 \le h \le M$, $1 \le k \le N(h)$, with

$$\partial E = \bigcup_{h=1}^M \bigcup_{k=1}^{N(h)} \Gamma(u_h^k, G_h) \cup \Gamma(v_h^k, G_h)\,, \tag{14.8}$$

$$E = \bigcup_{h=1}^M \Big\{ (z,t) \in G_h \times \mathbb{R} : t \in \bigcup_{k=1}^{N(h)} \big(v_h^k(z), u_h^k(z)\big) \Big\}; \tag{14.9}$$

see Figure 14.2. By (14.9), if we set $m(z) = \mathcal{L}^1(E_z)$, $z \in \mathbb{R}^{n-1}$, then

$$m(z) = \sum_{k=1}^{N(h)} u_h^k(z) - v_h^k(z)\,, \qquad \forall z \in G_h\,, \tag{14.10}$$

so that m is affine on each G_h. Moreover, by (14.8), m is continuous, hence, piecewise affine, on $\mathbb{R}^{n-1}$. Since $E^s = \{(z,t) \in G \times \mathbb{R} \colon |t| < m(z)/2\}$, E^s is a bounded open set with polyhedral boundary. By Example 12.6, E and E^s are sets of finite perimeter, with $|\mu_E| = \mathcal{H}^{n-1} \llcorner \partial E$ and $|\mu_{E^s}| = \mathcal{H}^{n-1} \llcorner \partial E^s$. By the formula for the area of a graph, Theorem 9.1,

$$P(E^s) = \mathcal{H}^{n-1}(\partial E^s) = 2 \int_G \sqrt{1 + \left|\frac{\nabla m}{2}\right|^2} = \sum_{h=1}^M \int_{G_h} \sqrt{4 + |\nabla m|^2}\,,$$

$$P(E) = \sum_{h=1}^M \mathcal{H}^{n-1}\big(\partial E \cap (G_h \times \mathbb{R})\big) = \sum_{h=1}^M \int_{G_h} \sum_{k=1}^{N(h)} \sqrt{1 + |\nabla u_h^k|^2} + \sqrt{1 + |\nabla v_h^k|^2}\,.$$

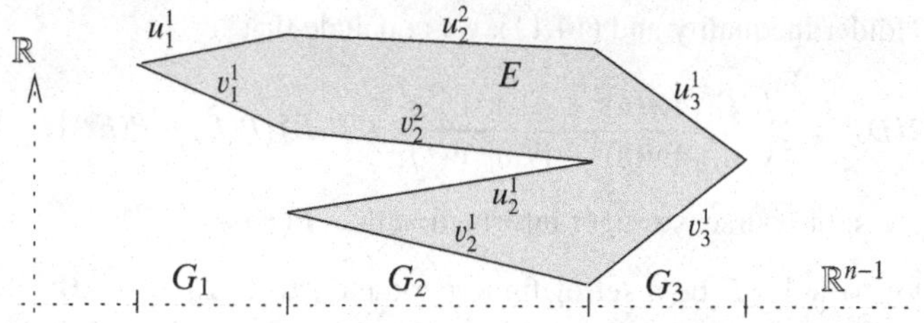

Figure 14.2 If E is a bounded open set with polyhedral boundary, with outer unit normal that is never orthogonal to e_n, then ∂E is parametrized as in (14.8). In particular, $u_h^k = v_h^k$ on $\partial G \cap \partial G_h$. Moreover, if $\mathcal{H}^{n-1}(\partial G_i \cap \partial G_h) > 0$ and $1 \le k \le N(h)$, then either $u_h^k = v_h^k$, or there exists $1 \le j \le N(i)$ such that $u_h^k = u_i^j$ and $v_h^k = v_i^j$ on $\partial G_i \cap \partial G_h$. These two properties, guaranteed by (14.8), imply the continuity of m on $\mathbb{R}^{n-1}$.

By (14.10), and by the convexity of $z \mapsto \sqrt{1+|z|^2}$, we find that, on G_h,

$$
\begin{aligned}
&\sum_{k=1}^{N(h)} \sqrt{1+|\nabla u_h^k|^2} + \sqrt{1+|\nabla v_h^k|^2} \\
&\ge 2\sum_{k=1}^{N(h)} \sqrt{1+\left|\frac{\nabla u_h^k - \nabla v_h^k}{2}\right|^2} = 2N(h)\left\{\frac{1}{N(h)}\sum_{k=1}^{N(h)} \sqrt{1+\left|\frac{\nabla u_h^k - \nabla v_h^k}{2}\right|^2}\right\} \\
&\ge 2N(h)\sqrt{1+\left|\frac{1}{N(h)}\sum_{k=1}^{N(h)}\frac{\nabla u_h^k - \nabla v_h^k}{2}\right|^2} = \sqrt{4N(h)^2+|\nabla m|^2}\,.
\end{aligned}
$$

Therefore, (14.3) is immediately deduced from

$$
P(E) \ge \sum_{h=1}^{M} \int_{G_h} \sqrt{4N(h)^2+|\nabla m(z)|^2}\,\mathrm{d}z\,, \tag{14.11}
$$

$$
P(E^s) = \sum_{h=1}^{M} \int_{G_h} \sqrt{4+|\nabla m(z)|^2}\,\mathrm{d}z\,, \tag{14.12}
$$

and $N(h) \ge 1$ for every h. We can actually deduce some more precise information. Let D be the set of those $z \in G$ such that E_z is not an interval, so that $N(h) \ge 2$ if and only if $G_h \cap D \ne \emptyset$. By (14.11) and (14.12),

$$
\begin{aligned}
P(E) - P(E^s) &\ge \sum_{h=1}^{M} \int_{G_h \cap D} \sqrt{4N(h)^2+|\nabla m|^2} - \sqrt{4+|\nabla m|^2} \\
&= \sum_{h=1}^{M} \int_{G_h \cap D} \frac{4(N(h)^2-1)}{\sqrt{4N(h)^2+|\nabla m|^2} + \sqrt{4+|\nabla m|^2}} \\
&\ge 2\sum_{h=1}^{M} \int_{G_h \cap D} \frac{1}{\sqrt{4N(h)^2+|\nabla m|^2}}\,.
\end{aligned}
$$

By the Hölder inequality and (14.11), we conclude that

$$2\mathcal{H}^{n-1}(D)^2 = 2\left(\int_D \frac{(4N(h)^2+|\nabla m|^2)^{1/4}}{(4N(h)^2+|\nabla m|^2)^{1/4}}\right)^2 \le P(E)\big(P(E)-P(E^s)\big), \quad (14.13)$$

which gives, of course, stronger information than $P(E) \ge P(E^s)$.

Step two: Now let E be a set of finite perimeter, with $|E| < \infty$. By Remark 13.13 there exists a sequence $\{E_h\}_{h\in\mathbb{N}}$ of bounded open sets with polyhedral boundary such that, as $h \to \infty$,

$$E_h \to E\,, \qquad P(E_h) \to P(E)\,. \quad (14.14)$$

Let us set $m_h(z) = \mathcal{L}^1((E_h)_z)$, $G_h = \{z \in \mathbb{R}^{n-1} : m_h(z) > 0\}$ (with a slight abuse of notation), and let D_h be the set of those $z \in \mathbb{R}^{n-1}$ such that $(E_h)_z \subset \mathbb{R}$ is not an interval. As ν_{E_h} takes only finitely many values, up to rotating each E_h by a rotation sufficiently close to the identity, we can assume that ν_{E_h} is never orthogonal to e_n while keeping (14.14). Applying step one to each E_h,

$$P(E_h^s) \le P(E_h)\,, \quad (14.15)$$

$$2\mathcal{H}^{n-1}(D_h)^2 \le P(E_h)\big(P(E_h) - P(E_h^s)\big)\,, \quad (14.16)$$

Moreover, by Fubini's theorem,

$$|E_h \Delta E| = \int_{\mathbb{R}^{n-1}} \mathcal{L}^1\big((E_h)_z \Delta E_z\big)\,dz \ge \int_{\mathbb{R}^{n-1}} |m_h(z) - m(z)|\,dz = |E_h^s \Delta E^s|\,. \quad (14.17)$$

In particular, $E_h^s \to E^s$, and by lower semicontinuity,

$$P(E^s) \le \liminf_{h\to\infty} P(E_h^s)\,.$$

As $P(E_h) \to P(E)$, by (14.15) we deduce (14.3). In fact, from (14.16),

$$2\limsup_{h\to\infty} \mathcal{H}^{n-1}(D_h) \le P(E)\big(P(E) - P(E^s)\big)\,.$$

In particular, if $P(E) = P(E^s)$, then $1_{D_h} \to 0$ in $L^1(\mathbb{R}^{n-1})$. Since (14.17) implies that $(E_h)_z \to E_z$ for a.e. $z \in \mathbb{R}^{n-1}$, as well as that $G_h \to G$, we may apply Propositions 12.15 and 14.5 to find that

$$\liminf_{h\to\infty} 1_{G_h\setminus D_h}(z) P((E_h)_z) \ge 1_G(z) P(E_z)\,,$$

for a.e. $z \in \mathbb{R}^{n-1}$. By Fatou's lemma,

$$\begin{aligned}\int_G P(E_z)\,dz &\le \liminf_{h\to\infty} \int_{G_h\setminus D_h} P\big((E_h)_z\big)\,dz = 2\liminf_{h\to\infty} \mathcal{H}^{n-1}(G_h \setminus D_h) \\ &= 2\mathcal{H}^{n-1}(G)\,.\end{aligned}$$

By the one-dimensional isoperimetric inequality, $P(E_z) \ge 2$ for a.e. $z \in \mathbb{R}^{n-1}$. Thus, $P(E_z) = 2$ for a.e. $z \in G$. By Proposition 12.13, for every such z, E_z is equivalent to a segment. In this way we have proved (i).

Step three: To conclude the proof of the theorem, let E be convex and such that $P(E) = P(E^s)$. Up to modifying E on a set of measure zero we may assume that E is open. We can find an open convex set $C \subset \mathbb{R}^{n-1}$ and pair of concave non-negative functions $\psi_1, \psi_2 : C \to [0, \infty)$ such that

$$E = \Big\{(z,t) \in C \times \mathbb{R} : \ -\psi_1(z) < t < \psi_2(z)\Big\}.$$

$$E^s = \Big\{(z,t) \in C \times \mathbb{R} : \ -\frac{\psi_1(z) + \psi_2(z)}{2} < t < \frac{\psi_1(z) + \psi_2(z)}{2}\Big\}.$$

Since ψ_1 and ψ_2 are locally Lipschitz in C, we easily prove that

$$P(E; G \times \mathbb{R}) = \int_G \sqrt{1 + |\nabla\psi_1|^2} + \sqrt{1 + |\nabla\psi_2|^2}, \tag{14.18}$$

$$P(E^s; G \times \mathbb{R}) = 2 \int_G \sqrt{1 + \left|\frac{\nabla\psi_1 + \nabla\psi_2}{2}\right|^2}, \tag{14.19}$$

whenever $G \subset\subset C$. If now $C' \subset\subset C$ is open, with $\mathcal{H}^{n-1}(\partial C') = 0$, then, by (14.18) and (14.19), we find $P(E; \partial C' \times \mathbb{R}) = P(E^s; \partial C' \times \mathbb{R}) = 0$. Hence, by applying (14.4) to the open sets C' and $C \setminus \overline{C'}$,

$$\begin{aligned} P(E) &= P\big(E; C' \times \mathbb{R}\big) + P\big(E; (C \setminus \overline{C'}) \times \mathbb{R}\big) \\ &\ge P\big(E^s; C' \times \mathbb{R}\big) + P\big(E^s; (C \setminus \overline{C'}) \times \mathbb{R}\big) = P(E^s). \end{aligned}$$

In particular, $P(E) = P(E^s)$ implies $P(E; C' \times \mathbb{R}) = P(E^s; C' \times \mathbb{R})$, that thanks to (14.18), (14.19), and the strict convexity of $z \in \mathbb{R}^{n-1} \mapsto \sqrt{1 + |z|^2}$ allows us to conclude that there exists $c \in \mathbb{R}$ such that $\psi_1 - \psi_2 = c$ on C'. By the arbitrariness of C' we conclude that c is independent of C' and that $E = E^s + c\, e_n$. □

The following lemma from [Fus04] allows us to fully exploit the necessary condition of equality (i) in the Steiner inequality.

Lemma 14.6 *If E is a set of locally finite perimeter in $\mathbb{R}^n$ such that, for a.e. $z \in \mathbb{R}^{n-1}$, the vertical slice $E_z \subset \mathbb{R}$ is equivalent to an interval, then the set of points of density one $E^{(1)}$ of E has the property that for every $z \in \mathbb{R}^{n-1}$, the vertical slice $(E^{(1)})_z$ is an interval.*

Proof Let us set for brevity $F = E^{(1)}$. We want to prove that if $x_k = (z_0, t_k) \in F$ $(k = 1, 2)$ for some $z_0 \in \mathbb{R}^{n-1}$ and for $t_1 < t_2$, then $x_0 = (z_0, t_0) \in F$ whenever $t_0 \in (t_1, t_2)$. Let us set $I_r(t) = (t - r, t + r)$ for $t \in \mathbb{R}$, $r > 0$, and let us fix

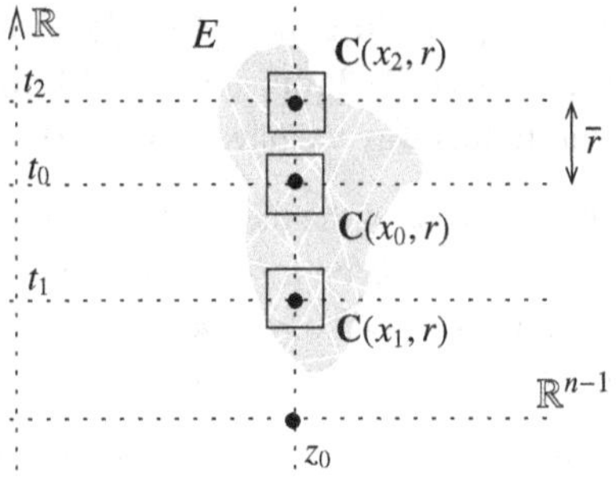

Figure 14.3 The situation in Lemma 14.6.

$\varepsilon \in (0, 1/2)$. Since $x_1, x_2 \in F$, $\theta_n(E)(x_1) = \theta_n(E)(x_2) = 1$. Thus there exists $\overline{r} > 0$ such that (see Exercise 5.19, Notation 4, and notice that $|\mathbf{C}_r| = 2\omega_{n-1}r^n$),

$$|E \cap \mathbf{C}(x_k, r)| \geq (1 - \varepsilon)\, 2\omega_{n-1}r^n\,, \tag{14.20}$$

$$I_r(t_k) \cap I_r(t_0) = \emptyset\,, \tag{14.21}$$

whenever $r \in (0, \overline{r})$ and $k = 1, 2$; see Figure 14.3. For $k = 1, 2$, let G_k be the set of those $z \in \mathbf{D}(z_0, r)$ such that E_z is an interval with $\mathcal{L}^1(E_z \cap I_r(t_k)) > 0$. By Fubini's theorem and by our assumption

$$|E \cap \mathbf{C}(x_k, r)| = \int_{\mathbf{D}(z_0,r)} \mathcal{L}^1\big(E_z \cap I_r(t_k)\big)\, dz = \int_{G_k} \mathcal{L}^1\big(E_z \cap I_r(t_k)\big)\, dz \leq 2r\,\mathcal{H}^{n-1}(G_k)\,.$$

By (14.20), $\mathcal{H}^{n-1}(G_k) \geq (1 - \varepsilon)\omega_{n-1}r^{n-1}$, and, since $G_1 \cup G_2 \subset \mathbf{D}(z_0, r)$,

$$\mathcal{H}^{n-1}(G_1 \cap G_2) \geq (1 - 2\varepsilon)\omega_{n-1}r^{n-1}\,. \tag{14.22}$$

Now let $z \in G_1 \cap G_2$: since E_z is equivalent to an interval and since $\mathcal{L}^1(E_z \cap I_r(t_k)) > 0$ for $k = 1, 2$, from (14.21) we immediately deduce that

$$\mathcal{L}^1(E_z \cap I_r(t_0)) = 2r\,.$$

Integrating this identity on $G_1 \cap G_2$ we thus find

$$|E \cap \mathbf{C}(x_0, r)| \geq \int_{G_1 \cap G_2} \mathcal{L}^1\big(E_z \cap I_r(t_0)\big)\, dz \geq 2r\,\mathcal{H}^{n-1}(G_1 \cap G_2) \geq (1-2\varepsilon)\, 2\omega_{n-1}r^n\,,$$

so that $\theta_n(E)(x_0) \geq 1 - 2\varepsilon$. By the arbitrariness of ε, $x_0 \in F$. □

14.2 Proof of the Euclidean isoperimetric inequality

Step one: Let $m, R > 0$, and consider the constrained isoperimetric problem

$$\inf\Big\{P(E) : E \subset B_R\,, |E| = m\Big\}\,. \tag{14.23}$$

If $m < \omega_n R^n$, the competition class is non-empty, and the existence of a minimizer E follows by the Direct Method as in Section 12.5. We now prove that E is equivalent to a ball. We may directly assume that $E = E^{(1)}$. Given $\nu \in S^{n-1}$, let E^s_ν be the Steiner symmetrization of E with respect to the hyperplane $\nu^\perp$. Since $|E^s_\nu| = |E|$, we have $P(E) \le P(E^s_\nu)$. By the Steiner inequality, $P(E) \ge P(E^s_\nu)$. Hence, $P(E) = P(E^s_\nu)$. By Theorem 14.4 (i) and Lemma 14.6 we conclude that for every $x \in \nu^\perp$, the one-dimensional slice

$$\Big\{s \in \mathbb{R} : x + s\nu \in E\Big\} \qquad \text{is a segment.}$$

If now $x, y \in E$, then, by applying the above argument to $\nu = (x-y)/|x-y|$, we see that the segment with endpoints at x and y is contained in E. Thus E is convex. By Theorem 14.4 (ii), for every $\nu \in S^{n-1}$ there exists $c_\nu \in \mathbb{R}$ such that

$$E = c_\nu \nu + E^s_\nu\,.$$

Let us now consider the set $F = -(c_{e_1}\, e_1 + \ldots + c_{e_n}\, e_n) + E$. Clearly, for every $\nu \in S^{n-1}$ there exists $d_\nu \in \mathbb{R}$ such that

$$F = d_\nu \nu + F^s_\nu\,.$$

By construction, $d_{e_k} = 0$ for every $k = 1, \ldots, n$, that is, F is invariant by reflection with respect to the coordinate hyperplanes $\{x_k = 0\}$, $k = 1, \ldots, n$. In particular, F is invariant under the map $x \mapsto -x$, so that $d_\nu = 0$ for every $\nu \in S^{n-1}$. Hence, F is convex and invariant by reflection with respect to any hyperplane through the origin, that is, F is a ball.

Step two: Since a ball of measure m has perimeter $n\omega_n^{1/n} m^{(n-1)/n}$, by step one

$$P(E) \ge n\omega_n^{1/n}|E|^{(n-1)/n}\,, \tag{14.24}$$

for every bounded set of finite perimeter E, with equality if and only if E is equivalent to a ball. By Remark 13.12, (14.24) holds in fact for every set of finite perimeter E with $|E| < \infty$. We conclude the proof by showing that, if E is a set of finite perimeter with $|E| < \infty$ and $E^{(1)}$ unbounded, then

$$P(E) > n\omega_n^{1/n}|E|^{(n-1)/n}\,.$$

Indeed, if this were not the case, then we would have $P(E) = P(E^s_\nu)$ for every $\nu \in S^{n-1}$. Arguing as in step one, $E^{(1)}$ would be a equivalent to a ball, against the assumption that $E^{(1)}$ is unbounded.

Exercise 14.7 (Cheeger sets) Given $p > 0$, $n \ge 2$, and A an open set in $\mathbb{R}^n$, the **p-Cheeger problem in** A is the variational problem

$$c(p, A) = \inf\left\{\frac{P(E)}{|E|^p} : E \subset A\right\}. \tag{14.25}$$

A minimizer E in (14.25) with (according to Proposition 12.19) $\operatorname{spt}\mu_E = \partial E$ is called a **p-Cheeger set of** A. If $p < (n-1)/n$, then, by scaling, $c(p, A) = 0$, and p-Cheeger sets cannot exist. If $p > (n-1)/n$ and A is bounded, then p-Cheeger sets exist, as is seen by applying the Direct Method and the isoperimetric inequality (12.38). If $p = (n-1)/n$, then, by Theorem 14.1, balls contained in A are the (only) p-Cheeger sets in A.

15

Reduced boundary and De Giorgi's structure theorem

The **reduced boundary** $\partial^* E$ of a set of locally finite perimeter E in $\mathbb{R}^n$ is the set of those $x \in \operatorname{spt}\mu_E$ such that the limit

$$\lim_{r \to 0^+} \frac{\mu_E(B(x,r))}{|\mu_E|(B(x,r))} \quad \text{exists and belongs to } S^{n-1}. \tag{15.1}$$

By Remark 5.7, we may define a Borel function $\nu_E \colon \partial^* E \to S^{n-1}$ by setting

$$\nu_E(x) = \lim_{r \to 0^+} \frac{\mu_E(B(x,r))}{|\mu_E|(B(x,r))}, \qquad x \in \partial^* E.$$

We call ν_E the **(measure-theoretic) outer unit normal to** E. By the Lebesgue–Besicovitch differentiation theorem, we have

$$\mu_E = \nu_E \, |\mu_E| \llcorner \partial^* E, \tag{15.2}$$

so that the distributional Gauss–Green theorem 12.3 takes the form

$$\int_E \nabla \varphi = \int_{\partial^* E} \varphi \, \nu_E \, \mathrm{d}|\mu_E|, \qquad \forall \varphi \in C^1_c(\mathbb{R}^n).$$

Remark 15.1 By Example 12.5 and (15.2), if E is an open set with C^1 boundary, then $\partial^* E = \partial E$ and the measure-theoretic outer unit normal coincides with classical notion of outer unit normal. This chapter is devoted to discussing in which way some basic geometric properties valid in the classical case continue to hold in our generalized measure-theoretic setting.

Remark 15.2 If E is a set of locally finite perimeter and $|E \Delta F| = 0$, then $\mu_E = \mu_F$ and therefore $\partial^* E = \partial^* F$. Indeed, the reduced boundary $\partial^* E$ is uniquely determined by the Gauss–Green measure μ_E of E.

Remark 15.3 By definition, $\partial^* E \subset \operatorname{spt}\mu_E$. By Proposition 12.19, $\operatorname{spt}\mu_E \subset \partial E$. Hence, the reduced boundary is always a subset of the topological boundary. In fact, by (15.2), the Gauss–Green measure μ_E is concentrated on $\partial^* E$, and hence on $\overline{\partial^* E}$. By definition of support, $\operatorname{spt}\mu_E \subset \overline{\partial^* E}$, and therefore

$$\operatorname{spt}\mu_E = \overline{\partial^* E}\,.$$

By Proposition 12.19, up to modifying E on a set of measure zero, we have that $\operatorname{spt}\mu_E = \partial E$. Therefore, **up to modification on sets of measure zero,**

$$\overline{\partial^* E} = \partial E\,. \tag{15.3}$$

Example 15.4 If $E \subset \mathbb{R}^2$ is a square with sides parallel to the coordinate axes, then the limit $\nu(x)$ in (15.1) exists for every $x \in \partial E$. However, $|\nu(x)| = 1$ if and only if x is not a vertex of E: indeed, if x is a vertex, then $|\nu(x)| = |(e_1 + e_2)/2| < 1$. Thus $\partial^* E$ is equal to ∂E minus the four vertexes of E.

We now introduce two fundamental results about reduced boundaries, describing, respectively, their local tangential properties (Theorem 15.5) and their structure of generalized hypersurfaces (Theorem 15.9). Local properties are studied by looking at the **blow-ups** $E_{x,r}$ of E:

$$E_{x,r} = \frac{E - x}{r} = \Phi_{x,r}(E)\,, \qquad x \in \mathbb{R}^n\,, r > 0\,,$$

where, as usual, $\Phi_{x,r}(y) = (y - x)/r$, $y \in \mathbb{R}^n$. By Lebesgue's points theorem (see, in particular, Example 5.17),

$$\begin{aligned} x \in E^{(1)} \qquad &\text{if and only if} \qquad E_{x,r} \stackrel{\text{loc}}{\to} \mathbb{R}^n \quad \text{as } r \to 0^+\,,\\ x \in E^{(0)} \qquad &\text{if and only if} \qquad E_{x,r} \stackrel{\text{loc}}{\to} \emptyset \quad \text{as } r \to 0^+\,. \end{aligned}$$

The behavior of blow-ups at boundary points is more interesting; see Figure 15.1. The following theorem, proved in Section 15.1, shows that the blow-ups at a reduce boundary point x locally converge to the negative half-space determined by the measure-theoretic outer unit normal $\nu_E(x)$. Moreover, convergence takes place at the level of Gauss–Green measures too.

Theorem 15.5 (Tangential properties of the reduced boundary) *If E is a set of locally finite perimeter in $\mathbb{R}^n$, and $x \in \partial^* E$, then*

$$E_{x,r} \stackrel{\text{loc}}{\to} H_x = \left\{ y \in \mathbb{R}^n : y \cdot \nu_E(x) \le 0 \right\} \quad \textit{as } r \to 0^+\,. \tag{15.4}$$

Similarly, if $\pi_x = \partial H_x = \nu_E(x)^\perp$, then, as $r \to 0^+$,

$$\mu_{E_{x,r}} \stackrel{*}{\rightharpoonup} \nu_E(x)\,\mathcal{H}^{n-1} \llcorner \pi_x\,, \qquad |\mu_{E_{x,r}}| \stackrel{*}{\rightharpoonup} \mathcal{H}^{n-1} \llcorner \pi_x\,. \tag{15.5}$$

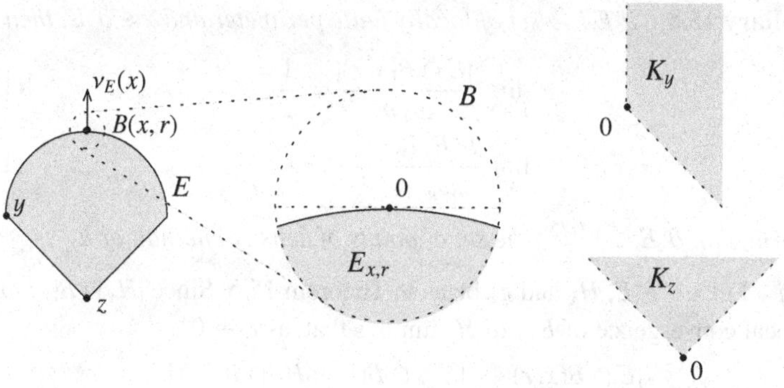

Figure 15.1 The blow-up $E_{x,r}$ inside B is obtained by translating $E \cap B(x,r)$ by $-x$, and then magnifying the resulting set by a factor $1/r$. If we perform this operation at a point $x \in \partial^* E$, then, as r decreases to zero, $E_{x,r}$ becomes increasingly closer to the negative half-space defined by $\nu_E(x)$, see (15.4), and the reduced boundary $\partial^* E_{x,r}$ flattens against $\nu_E(x)^\perp$, see (15.5). Blow-ups at points of the topological boundary which are not in the reduced boundary may still converge to some limit cone which may not be a half-space; for example, in the picture, $E_{y,r}$ and $E_{z,r}$ converge to the non-half-space cones K_y and K_z. There is no limit to the complexity of the behavior of blow-ups at points which are neither of density 1 or 0 and do not belong to $\partial^* E$; see Remark 15.7.

Remark 15.6 After proving in Theorem 15.9 that $\mu_E = \nu_E \mathcal{H}^{n-1} \llcorner \partial^* E$, we shall be able to state (15.5) in the more expressive form (see Example 4.24),

$$\nu_E \mathcal{H}^{n-1} \llcorner \Big(\frac{\partial^* E - x}{r} \Big) \overset{*}{\rightharpoonup} \nu_E(x) \mathcal{H}^{n-1} \llcorner \pi_x , \tag{15.6}$$

$$\mathcal{H}^{n-1} \llcorner \Big(\frac{\partial^* E - x}{r} \Big) \overset{*}{\rightharpoonup} \mathcal{H}^{n-1} \llcorner \pi_x , \tag{15.7}$$

Remark 15.7 If we exclude points x in $E^{(1)}$, $E^{(0)}$ and $\partial^* E$, the behavior of the blow-ups $E_{x,r}$ may be arbitrarily complex. For example, it is possible to construct an open set of finite perimeter E with a point $x \in \partial E$ having the following property: for every set of finite perimeter F there exists a sequence $r_h \to 0^+$ such that E_{x,r_h} locally converges to F as $h \to \infty$; see [Leo00].

Two straight consequences of Theorem 15.5 have great importance and deserve to be immediately stated. First, a set of locally finite perimeter has density one-half on its reduced boundary. Second, the relative perimeter of E inside balls $B(x,r)$ centered at $x \in \partial^* E$ is asymptotic to the measure of a $(n-1)$-dimensional ball of radius r as $r \to 0^+$.

Corollary 15.8 *If E is a set of locally finite perimeter and $x \in \partial^* E$, then*

$$\lim_{r\to 0^+} \frac{|E \cap B(x,r)|}{\omega_n\, r^n} = \frac{1}{2}\,, \tag{15.8}$$

$$\lim_{r\to 0^+} \frac{P(E;B(x,r))}{\omega_{n-1}\, r^{n-1}} = 1\,, \tag{15.9}$$

In particular, $\partial^* E \subset E^{(1/2)}$, the set of points of density one-half of E.

Proof Let $x \in \partial^* E$, H_x and π_x be as in Theorem 15.5. Since $|H_x \cap B| = \omega_n/2$, the local convergence of $E_{x,r}$ to H_x implies that, as $r \to 0^+$,

$$\frac{|E \cap B(x,r)|}{\omega_n\, r^n} = \frac{|E_{x,r} \cap B|}{\omega_n} \to \frac{|H_x \cap B|}{\omega_n} = \frac{1}{2}\,,$$

that is (15.8). Since $\pi_x \cap \partial B$ is an $(n-2)$-dimensional unit sphere, we have $\mathcal{H}^{n-1}(\pi_x \cap \partial B) = 0$. Thus, by (15.5) and Proposition 4.26(iii), as $r \to 0^+$,

$$|\mu_{E_{x,r}}|\,(B) \to \mathcal{H}^{n-1}(\pi_x \cap B) = \omega_{n-1}\,.$$

We find (15.9) as, by Lemma 15.11 below, $|\mu_{E_{x,r}}|\,(B) = r^{1-n}\, P(E;B(x,r))$. □

The second fundamental result about reduced boundaries is the following theorem, which we shall prove in Section 15.2. It asserts that reduced boundaries have the structure of generalized hypersurfaces, thus leading to a geometrically expressive reformulation of the distributional Gauss–Green theorem (12.3).

Theorem 15.9 (De Giorgi's structure theorem) *If E is a set of locally finite perimeter in $\mathbb{R}^n$, then the Gauss–Green measure μ_E of E satisfies*

$$\mu_E = \nu_E\, \mathcal{H}^{n-1} \llcorner \partial^* E\,, \qquad |\mu_E| = \mathcal{H}^{n-1} \llcorner \partial^* E\,, \tag{15.10}$$

and the generalized Gauss–Green formula holds true:

$$\int_E \nabla\varphi = \int_{\partial^* E} \varphi\, \nu_E\, d\mathcal{H}^{n-1}\,, \qquad \forall \varphi \in C^1_c(\mathbb{R}^n)\,. \tag{15.11}$$

Moreover, there exist countably many C^1-hypersurfaces M_h in $\mathbb{R}^n$, compact sets $K_h \subset M_h$, and a Borel set F with $\mathcal{H}^{n-1}(F) = 0$, such that

$$\partial^* E = F \cup \bigcup_{h \in \mathbb{N}} K_h\,,$$

and, for every $x \in K_h$, $\nu_E(x)^\perp = T_x M_h$, the tangent space to M_h at x.

Exercise 15.10 If E is of locally finite perimeter in $\mathbb{R}^n$ and $Q \in \mathbf{O}(n)$, then $Q(E)$ is a set of locally finite perimeter in $\mathbb{R}^n$, and

$$\nu_{Q(E)}(x) = Q\, \nu_E(Q^* x)\,, \qquad \forall x \in \partial^* Q(E) = Q(\partial^* E)\,.$$

Hint: Recall Exercise 12.11.

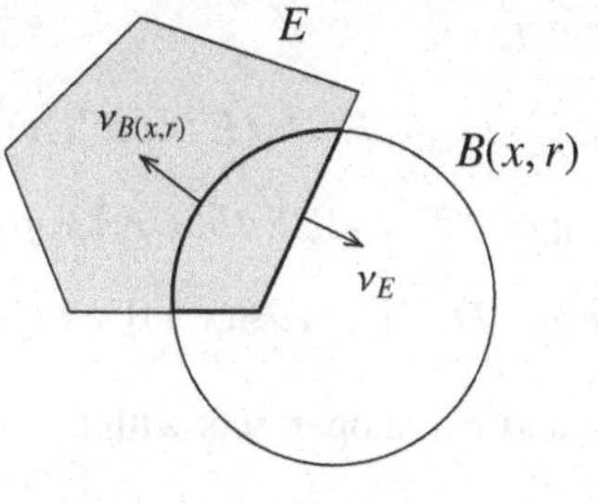

Figure 15.2 The boundary of $E \cap B(x,r)$ is the union of $E \cap \partial B(x,r)$, where we have $\nu_{E\cap B(x,r)} = \nu_{B(x,r)}$, and of $B(x,r) \cap \partial E$, where we have $\nu_{E\cap B(x,r)} = \nu_E$.

15.1 Tangential properties of the reduced boundary

This section is devoted to the proof of Theorem 15.5, which is introduced by three simple preparatory results of some independent interest.

Lemma 15.11 *If E is a set of locally finite perimeter in $\mathbb{R}^n$, $x \in \mathbb{R}^n$ and $r > 0$, then $E_{x,r}$ is a set of finite perimeter in $\mathbb{R}^n$ with*

$$\mu_{E_{x,r}} = \frac{(\Phi_{x,r})_\# \mu_E}{r^{n-1}} . \tag{15.12}$$

Proof If $\varphi \in C^1_c(\mathbb{R}^n)$ and $\varphi_{x,r} = \varphi \circ \Phi_{x,r}$, then $\nabla \varphi_{x,r} = r^{-1}(\nabla\varphi \circ \Phi_{x,r})$ and

$$\int_{E_{x,r}} \nabla\varphi = \frac{1}{r^n}\int_E (\nabla\varphi \circ \Phi_{x,r}) = \frac{1}{r^{n-1}}\int_{\mathbb{R}^n} \varphi_{x,r} \mathrm{d}\mu_E = \frac{1}{r^{n-1}}\int_{\mathbb{R}^n} \varphi \, \mathrm{d}(\Phi_{x,r})_\#\mu_E ,$$

by (2.10). Since $r^{1-n}(\Phi_{x,r})_\#\mu_E$ is a Radon measure, $E_{x,r}$ is a set of locally finite perimeter with $\mu_{E_{x,r}} = r^{1-n}(\Phi_{x,r})_\#\mu_E$. □

We have already seen, in Lemma 12.22, that the intersection of sets of (locally) finite perimeter defines a set of (locally) finite perimeter. In the proof of Theorem 15.5 we shall need, in fact, a more precise information, namely, we shall need a formula for the Gauss–Green measure of the intersection $E \cap B(x,r)$ of a set of locally finite perimeter E with a ball $B(x,r)$. Although the expected result, as depicted in Figure 15.2, is geometrically obvious, there are some measure-theoretic subtleties to be taken into account. For this reason, we shall deduce a formula for $\mu_{E\cap B(x,r)}$ for a.e. value of r only, excluding in particular those values of r such that $P(E; \partial B(x,r)) > 0$. (In Theorem 16.3 we shall see how to remove this kind of *caveat*.)

Lemma 15.12 (Intersection with a ball) *If E is a set of locally finite perimeter in $\mathbb{R}^n$ and $x \in \mathbb{R}^n$, then, for every $r > 0$, $E \cap B(x,r)$ is a set of finite perimeter* *in* $\mathbb{R}^n$. *Moreover, for a.e.* $r > 0$,

$$\mu_{E\cap B(x,r)} = \nu_{B(x,r)}\,\mathcal{H}^{n-1}\llcorner\big(E\cap\partial B(x,r)\big) + \mu_E \llcorner B(x,r)\,, \tag{15.13}$$

$$|\mu_{E\cap B(x,r)}| = \mathcal{H}^{n-1}\llcorner\big(E\cap\partial B(x,r)\big) + |\mu_E|\llcorner B(x,r)\,, \tag{15.14}$$

$$P(E\cap B(x,r)) = \mathcal{H}^{n-1}\big(E\cap\partial B(x,r)\big) + P(E;B(x,r))\,. \tag{15.15}$$

Proof *Step one:* If E and F are open sets with C^1-boundary, and

$$P(E;\partial F) = P(F;\partial E) = \mathcal{H}^{n-1}(\partial E\cap\partial F) = 0\,, \tag{15.16}$$

then $E\cap F$ (which is a set of locally finite perimeter by Lemma 12.22) satisfies

$$\mu_{E\cap F} = \nu_E\,\mathcal{H}^{n-1}\llcorner(F\cap\partial E) + \nu_F\,\mathcal{H}^{n-1}\llcorner(E\cap\partial F)\,. \tag{15.17}$$

Indeed, since $\mu_E = \nu_E\,\mathcal{H}^{n-1}\llcorner\partial E$ and $\mu_F = \nu_F\,\mathcal{H}^{n-1}\llcorner\partial F$ (see Example 12.5), and $\mu_{E\cap F}$ is concentrated on $\partial(E\cap F)$, by (12.17) we find

$$|\mu_{E\cap F}| \le |\mu_E| + |\mu_F| = \mathcal{H}^{n-1}\llcorner(\partial E\cup\partial F)\,,$$

which, combined with (15.16), implies $|\mu_{E\cap F}|(\partial E\cap\partial F) = 0$. Hence, $\mu_{E\cap F}$ is concentrated on $\partial(E\cap F)\setminus(\partial E\cap\partial F) = (E\cap\partial F)\cup(F\cap\partial E)$, that is,

$$\mu_{E\cap F} = \mu_{E\cap F}\llcorner\big(E\cap\partial F\big) + \mu_{E\cap F}\llcorner\big(F\cap\partial E\big)\,. \tag{15.18}$$

By the divergence theorem, for every $T\in C^1_c(\mathbb{R}^n;\mathbb{R}^n)$ we have that

$$\int_{E\cap F}\operatorname{div} T = \int_{E\cap\partial F} T\cdot d\mu_{E\cap F} + \int_{F\cap\partial E} T\cdot d\mu_{E\cap F}\,. \tag{15.19}$$

If we test (15.19) with $T\in C^1_c(E;\mathbb{R}^n)$, then the second term on the right-hand side of (15.19) vanishes, and

$$\int_{E\cap\partial F} T\cdot d\mu_{E\cap F} = \int_{E\cap F}\operatorname{div} T = \int_F \operatorname{div} T = \int_{\mathbb{R}^n} T\cdot d\mu_F = \int_E T\cdot d\mu_F\,.$$

By the arbitrariness of $T\in C^1_c(E;\mathbb{R}^n)$, we find $\mu_{E\cap F}\llcorner(E\cap\partial F) = \mu_F\llcorner E$. In the same way, testing with $T\in C^1_c(F;\mathbb{R}^n)$, we find $\mu_{E\cap F}\llcorner(F\cap\partial E) = \mu_E\llcorner F$. We combine these identities with (15.18) to achieve the proof of (15.17).

Step two: Without loss of generality, and in order to simplify the notation, we set $x = 0$. Up to changing E with $E\cap B_R$ – and then proving (15.13) and (15.14) for a.e. $r\in(0,R)$ – we can directly assume that $|E| < \infty$, so apply Theorem 13.8 and find a sequence $\{E_h\}_{h\in\mathbb{N}}$ of open sets with C^1-boundary such that $E_h\to E$ and $|\mu_{E_h}|\overset{*}{\rightharpoonup}|\mu_E|$. In particular, by (13.3),

$$|E\Delta E_h| = \int_0^\infty \mathcal{H}^{n-1}\big((E\Delta E_h)\cap\partial B_r\big)\,dr\,,$$

and thus, thanks also to Proposition 2.16, we find that, for a.e. $r > 0$,

$$\lim_{h\to\infty} \mathcal{H}^{n-1}\Big((E\Delta E_h)\cap \partial B_r\Big) = 0\,, \tag{15.20}$$

$$P(E;\partial B_r) = P(E_h;\partial B_r) = 0\,, \qquad \forall h \in \mathbb{N}\,. \tag{15.21}$$

Let r satisfy (15.20) and (15.21). By (15.21) and by step one, we have

$$\mu_{E_h\cap B_r} = \mu_{E_h} \llcorner B_r + \mu_{B_r} \llcorner E_h\,, \qquad \forall h \in \mathbb{N}\,. \tag{15.22}$$

We show that $\mu_{B_r} \llcorner E_h \overset{*}{\rightharpoonup} \mu_{B_r} \llcorner E$ *as* $h \to \infty$: this is easily checked from (15.20) and the fact that, by Example 12.5, $\mu_{B_r} = \nu_{B_r}\mathcal{H}^{n-1}\llcorner\partial B_r$.

We show that $\mu_{E_h\cap B_r} \overset{*}{\rightharpoonup} \mu_{E\cap B_r}$ *as* $h \to \infty$: indeed, by (15.22),

$$\limsup_{h\to\infty} P(E_h \cap B_r) \le \limsup_{h\to\infty} P(E_h; B_r) + P(B_r) \le P(E; \overline{B_r}) + P(B_r) < \infty\,,$$

so that $E_h \cap B_r \to E \cap B_r$ implies our claim.

We show that $\mu_{E_h}\llcorner B_r \overset{*}{\rightharpoonup} \mu_E\llcorner B_r$ *as* $h \to \infty$: if $s > r$ and $\psi_s \in C^0_c(B_s;[0,1])$ with $\psi_s = 1$ on B_r, then by $|\mu_{E_h}| \overset{*}{\rightharpoonup} |\mu_E|$ and by (4.29), for every $\varphi \in C^0_c(\mathbb{R}^n)$,

$$\left|\int_{B_r} \varphi\, \mathrm{d}\mu_E - \int_{\mathbb{R}^n} \varphi\psi_s\, \mathrm{d}\mu_E\right| \le \sup_{\mathbb{R}^n}|\varphi|\, |\mu_E|(\overline{B_s}\setminus B_r)\,,$$

$$\limsup_{h\to\infty}\left|\int_{B_r} \varphi\, \mathrm{d}\mu_{E_h} - \int_{\mathbb{R}^n} \varphi\psi_s\, \mathrm{d}\mu_{E_h}\right| \le \sup_{\mathbb{R}^n}|\varphi|\, |\mu_E|(\overline{B_s}\setminus B_r)\,.$$

Since $\int_{\mathbb{R}^n} \varphi\psi_s\, \mathrm{d}\mu_{E_h} \to \int_{\mathbb{R}^n} \varphi\psi_s\, \mathrm{d}\mu_E$ as $h \to \infty$, letting first $h \to \infty$ and then $s \to r^+$, and exploiting (15.21), we find as required that, for every $\varphi \in C^0_c(\mathbb{R}^n)$,

$$\limsup_{h\to\infty}\left|\int_{B_r} \varphi\, \mathrm{d}\mu_{E_h} - \int_{B_r} \varphi\, \mathrm{d}\mu_E\right| \le 2\sup_{\mathbb{R}^n}|\varphi|\, |\mu_E|(\partial B_r) = 0\,.$$

In conclusion, since $\mu_{B_r} = \nu_{B_r}\mathcal{H}^{n-1}\llcorner\partial B_r$, the above three claims allow us to let $h \to \infty$ in (15.22) to deduce (15.13). Again by (15.21), the vector-valued Radon measures on the right-hand side of (15.13) are mutually singular, so that (15.14) and (15.15) follow immediately from (15.13). □

The following exercises are not needed in the proof of Theorem 15.5, but they provide nice applications of the ideas presented in Lemma 15.12.

Exercise 15.13 (Intersection with half-spaces) If $H_t = \{x \in \mathbb{R}^n : x\cdot e < t\}$ for some $e \in S^{n-1}$, $t \in \mathbb{R}$, and E is a set of finite perimeter in $\mathbb{R}^n$, then $E \cap H_t$ is a set of finite perimeter in $\mathbb{R}^n$, and, for a.e. $t \in \mathbb{R}$,

$$\mu_{E\cap H_t} = \mu_E \llcorner H_t + e\,\mathcal{H}^{n-1}\llcorner\Big(E \cap \partial H_t\Big)\,.$$

In particular, for a.e. $t \in \mathbb{R}$ we have

$$\mathcal{H}^{n-1}\Big(E\cap\partial H_t\Big) \le P(E;H_t)\,, \qquad P(E\cap H_t) \le P(E)\,. \tag{15.23}$$

Hint: To characterize $\mu_{E\cap H_t}$ argue as in Lemma 15.12. To prove (15.23), apply the divergence theorem to the constant vector field $T(x) = e$, $x \in \mathbb{R}^n$ (here a truncation argument is required; see Proposition 19.22).

Exercise 15.14 (Convex sets are of locally finite perimeter) Show that, if E is a set of finite perimeter, and if K is a convex set, then $E \cap K$ is a set of finite perimeter, with $P(E \cap K) \le P(E)$. Refine the argument to show that every convex set is of locally finite perimeter, and that every bounded convex set is of finite perimeter. *Hint:* A set is convex if and only if its closure is the intersection of countably many closed half-spaces.

As a last preparatory lemma we show that a set of locally finite perimeter with constant measure-theoretic outer unit normal is equivalent to a half-space. We are going to meet similar characterization results for balls, cones and cylinders in Exercise 15.19, Proposition 28.8, and Lemma 28.13.

Proposition 15.15 (Characterization of half-spaces) *If F is a set of locally finite perimeter in $\mathbb{R}^n$ and $\nu \in S^{n-1}$ is such that $\nu_F(y) = \nu$ for $|\mu_F|$-a.e. $y \in \partial^* F$, then there exists $\alpha \in \mathbb{R}$ such that F is equivalent to the half-space*

$$\left\{z \in \mathbb{R}^n : z \cdot \nu < \alpha\right\}.$$

Proof Let $u_\varepsilon = 1_F \star \rho_\varepsilon \in C^\infty(\mathbb{R}^n)$, $\varepsilon > 0$. If $T \in C^1_c(\mathbb{R}^n, \mathbb{R}^n)$, then by Fubini's theorem $\int_{\mathbb{R}^n} u_\varepsilon \mathrm{div}\, T = \int_F \mathrm{div}\,(T \star \rho_\varepsilon)$. Since $\int_{\mathbb{R}^n} \mathrm{div}\,(u_\varepsilon T) = 0$, we find

$$-\int_{\mathbb{R}^n} \nabla u_\varepsilon \cdot T = \int_F \mathrm{div}\,(T \star \rho_\varepsilon) = \int_{\mathbb{R}^n} (T \star \rho_\varepsilon) \cdot \mathrm{d}\mu_F = \int_{\partial^* F} \nu \cdot (T \star \rho_\varepsilon)\, \mathrm{d}|\mu_F|. \tag{15.24}$$

If $T = \varphi \nu'$ for $\varphi \in C^1_c(\mathbb{R}^n)$ and $\nu' \in S^{n-1} \cap \nu^\perp$, then (15.24) gives

$$-\int_{\mathbb{R}^n} \frac{\partial u_\varepsilon}{\partial \nu'} \varphi = 0 \qquad \forall \varphi \in C^1_c(\mathbb{R}^n) \ \Rightarrow\ \frac{\partial u_\varepsilon}{\partial \nu'} = 0 \quad \text{on } \mathbb{R}^n. \tag{15.25}$$

If, instead, we test (15.24) with $T = \varphi \nu$, $\varphi \in C^1_c(\mathbb{R}^n)$, then we find

$$-\int_{\mathbb{R}^n} \frac{\partial u_\varepsilon}{\partial \nu} \varphi = \int_{\mathbb{R}^n} \varphi\, \mathrm{d}|\mu_E|,$$

where the right-hand side is non-negative for every $\varphi \ge 0$; by Exercise 4.5,

$$\frac{\partial u_\varepsilon}{\partial \nu} \le 0 \qquad \text{on } \mathbb{R}^n. \tag{15.26}$$

By (15.25) and (15.26), $u_\varepsilon(y)$ is a decreasing function of the variable $(y \cdot \nu)$, that is, there exists $f_\varepsilon \in C^\infty(\mathbb{R}; [0,1])$ with $f'_\varepsilon \le 0$ on $\mathbb{R}$ such that

$$(1_F \star \rho_\varepsilon)(y) = u_\varepsilon(y) = f_\varepsilon(y \cdot \nu), \qquad \forall y \in \mathbb{R}^n, \tag{15.27}$$

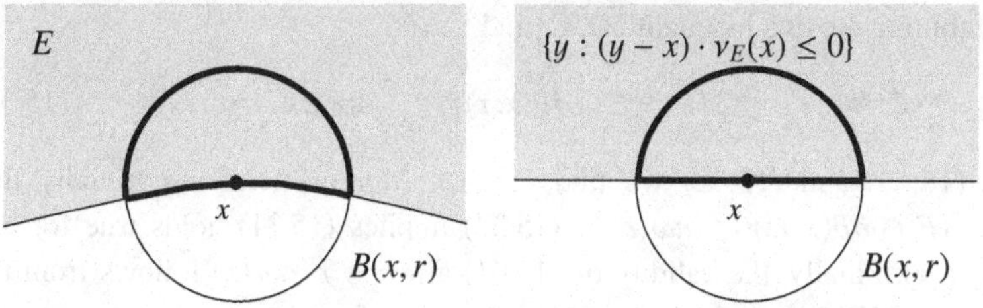

Figure 15.3 Since $x \in \partial^* E$ we expect $E \cap B(x,r)$ to approximate $\{y \in B(x,r) : (y-x)\cdot \nu_E(x) \le 0\}$ as $r \to 0^+$. Therefore $P(E \cap B(x,r))$ should be close to the perimeter of a half-ball of radius r that is smaller than the perimeter of a ball of radius r. In turn, this last quantity is approximately $2\mathcal{H}^{n-1}(E \cap \partial B(x,r))$, and we come to (15.30).

If we let $\varepsilon \to 0^+$ in (15.27), then we find $\lim_{\varepsilon\to 0} f_\varepsilon(t) \in \{0,1\}$ for a.e. $t \in \mathbb{R}$; hence, there exists $\alpha \in \mathbb{R}$ such that $f_\varepsilon(t) \to 1_{(-\infty,\alpha)}(t)$ for a.e. $t \in \mathbb{R}$, that is

$$(1_F \star \rho_\varepsilon)(y) = u_\varepsilon(y) \to 1_{(-\infty,\alpha)}(y\cdot\nu)\,, \qquad \text{for a.e. } y \in \mathbb{R}^n.$$

We conclude since $(1_F \star \rho_\varepsilon) \to 1_F$ in $L^1_{\mathrm{loc}}(\mathbb{R}^n)$. □

Proof of Theorem 15.5 Let $x \in \partial^* E$. By Lemma 15.12 we have

$$\mu_{E\cap B(x,r)} = \nu_{B(x,r)}\,\mathcal{H}^{n-1} \llcorner (E \cap \partial B(x,r)) + \mu_E \llcorner B(x,r)\,, \tag{15.28}$$

$$P(E \cap B(x,r)) = \mathcal{H}^{n-1}(E \cap \partial B(x,r)) + P(E; B(x,r))\,, \tag{15.29}$$

for a.e. $r > 0$. We now divide the proof into four steps.

Step one: We claim the existence of $r(x)$ and $C(n)$ positive such that

$$P(E \cap B(x,r)) \le 3\,\mathcal{H}^{n-1}\big(E \cap \partial B(x,r)\big), \qquad \text{for a.e. } r < r(x), \tag{15.30}$$

$$P(E; B(x,r)) \le C(n)\, r^{n-1}\,, \qquad \forall r < r(x); \tag{15.31}$$

see Figure 15.3. By (15.28), if $\varphi \in C^1_c(\mathbb{R}^n)$ is such that $\varphi = 1$ on $\overline{B}(x,r)$, then

$$\begin{aligned} 0 &= \int_{E\cap B(x,r)} \nabla\varphi = \int_{\mathbb{R}^n} \varphi\, d\mu_{E\cap B(x,r)} = \int_{E\cap\partial B(x,r)} \varphi\,\nu_{B(x,r)} d\mathcal{H}^{n-1} + \int_{B(x,r)} \varphi\, d\mu_E \\ &= \int_{E\cap\partial B(x,r)} \nu_{B(x,r)} d\mathcal{H}^{n-1} + \mu_E(B(x,r)), \end{aligned}$$

that is, $|\mu_E(B(x,r))| \le \mathcal{H}^{n-1}(E \cap \partial B(x,r))$ for a.e. $r > 0$. At the same time, since $x \in \partial^* E$, by (15.1) there exists $r(x) > 0$ such that

$$P(E; B(x,r)) \le 2|\mu_E(B(x,r))|\,, \qquad \forall r < r(x)\,.$$

Combining the two inequalities, we find

$$P(E; B(x,r)) \le 2\mathcal{H}^{n-1}(E \cap \partial B(x,r)), \qquad \text{for a.e. } r < r(x). \tag{15.32}$$

By (15.29) and (15.32) we find (15.30). Moreover, having trivially that $\mathcal{H}^{n-1}(E \cap \partial B(x,r)) \le n\omega_n r^{n-1}$, (15.32) implies (15.31) holds true for a.e. $r < r(x)$. Finally, the validity of (15.31) at *every* $r < r(x)$ follows from the fact that $P(E; B(x,r))$ is an increasing function of r.

Step two: We prove two rough lower bounds on the n-dimensional density ratios of E and $\mathbb{R}^n \setminus E$ at x; precisely, we show that

$$\frac{|E \cap B(x,r)|}{r^n} \ge \frac{1}{(3n)^n}, \qquad \forall r < r(x), \tag{15.33}$$

$$\frac{|(\mathbb{R}^n \setminus E) \cap B(x,r)|}{r^n} \ge \frac{1}{(3n)^n}, \qquad \forall r < r(x). \tag{15.34}$$

By (12.4), $\partial^* E = \partial^*(\mathbb{R}^n \setminus E)$. Thus it suffices to prove (15.33). To this end, we define an increasing function $m\colon (0,\infty) \to [0,\infty)$ by setting,

$$m(r) = |E \cap B(x,r)|, \qquad r > 0.$$

By (13.3), m is absolutely continuous with $m(0) = 0$ and

$$m'(r) = \mathcal{H}^{n-1}(E \cap \partial B(x,r)), \qquad \text{for a.e. } r > 0.$$

Moreover, $m > 0$: indeed, $x \in \partial^* E \subset \operatorname{spt}\mu_E$ implies $|\mu_E|(B(x,r)) > 0$, which, by (15.29), gives $P(E \cap B(x,r)) > 0$; we conclude as $m(r) = 0$ would imply $P(E \cap B(x,r)) = 0$. By (12.38) and (15.30) we thus find

$$\begin{aligned} m(r)^{(n-1)/n} &= |E \cap B(x,r)|^{(n-1)/n} \le P(E \cap B(x,r)) \\ &\le 3\mathcal{H}^{n-1}(E \cap \partial B(x,r)) = 3\, m'(r), \qquad \text{for a.e. } r < r(x). \end{aligned}$$

Dividing by $m(r)^{(n-1)/n} > 0$, we thus find

$$\frac{1}{3} \le m'(r)\, m(r)^{-1+1/n} = n(m^{1/n})'(r),$$

which gives $m(r)^{1/n} \ge r/3n$ for every $r < r(x)$, as desired.

Step three: We prove that $E_{x,r}$ locally converges to H_x as $r \to 0^+$. To this end, it suffices to show that, if $r_h \to 0^+$, then, up to extracting a subsequence, E_{x,r_h} locally converges to H_x as $h \to \infty$. By (15.12) and (15.31), for every $R > 0$,

$$P(E_{x,r}; B_R) = \frac{P(E; B(x, rR))}{r^{n-1}} \le C(n) R^{n-1}, \qquad \forall r < \frac{r(x)}{R}.$$

Given $r_h \to 0^+$ as $h \to \infty$, by applying Corollary 12.27 to $\{E_{x,r_h}\}_{h\in\mathbb{N}}$ we show the existence of a set of locally finite perimeter F such that

$$E_{x,r_h} \overset{\text{loc}}{\to} F\,, \qquad \mu_{E_{x,r_h}} \overset{*}{\rightharpoonup} \mu_F\,.$$

(This is true, of course, up to extracting a subsequence.) We now claim that

$$|\mu_{E_{x,r_h}}| \overset{*}{\rightharpoonup} |\mu_F| \tag{15.35}$$

$$\nu_F(y) = \nu_E(x)\,, \qquad \text{for } |\mu_F|\text{-a.e. } y \in \partial^* F. \tag{15.36}$$

Indeed, up to extracting a further subsequence, there exists a Radon measure λ such that $|\mu_{E_{x,r_h}}| \overset{*}{\rightharpoonup} \lambda$. By Proposition 4.30 (i), for a.e. $R > 0$ (precisely, for every $R > 0$ such that $\lambda(\partial B_R) = 0$), we have

$$\lim_{h\to\infty} \mu_{E_{x,r_h}}(B_R) = \mu_F(B_R)\,. \tag{15.37}$$

At the same time, since $x \in \partial^* E$, by (15.12) we find

$$\lim_{r\to 0^+} \frac{\mu_{E_{x,r}}(B_R)}{|\mu_{E_{x,r}}|(B_R)} = \lim_{r\to 0^+} \frac{\mu_E(B(x,rR))}{|\mu_E|(B(x,rR))} = \nu_E(x)\,,$$

so that, recalling of course that $P(E_{x,r};B_R) = |\mu_{E_{x,r}}|(B_R)$,

$$\lim_{r\to 0^+} \frac{P(E_{x,r};B_R)}{\nu_E(x)\cdot\mu_{E_{x,r}}(B_R)} = 1\,. \tag{15.38}$$

By Proposition 12.15, (15.38), and (15.37), we conclude that, for a.e. $R > 0$,

$$\begin{aligned} P(F;B_R) &\le \liminf_{h\to\infty} P(E_{x,r_h};B_R) = \lim_{h\to\infty} \nu_E(x)\cdot\mu_{E_{x,r_h}}(B_R) \\ &= \nu_E(x)\cdot\mu_F(B_R) \le |\mu_F(B_R)| \le |\mu_F|(B_R) = P(F;B_R)\,. \end{aligned}$$

In particular, we find that, for a.e. $R > 0$,

$$|\mu_F|(B_R) = \lim_{h\to\infty} |\mu_{E_{x,r_h}}|(B_R)\,, \tag{15.39}$$

$$|\mu_F|(B_R) = \nu_E(x)\cdot\mu_F(B_R)\,. \tag{15.40}$$

By (15.39) and Exercise 4.31, we deduce (15.35). By (15.40),

$$0 = \int_{B_R} \Big(1 - \nu_E(x)\cdot\nu_F(y)\Big)\, d|\mu_F|(y)\,, \qquad \text{for a.e. } R > 0\,,$$

which implies (15.36) as $1 - \nu_E(x)\cdot\nu_F(y) \ge 0$. By (15.36) and Proposition 15.15 there exists $\alpha \in \mathbb{R}$ such that

$$\Big|F\Delta\big\{y \in \mathbb{R}^n : \nu_E(x)\cdot y < \alpha\big\}\Big| = 0\,. \tag{15.41}$$

If $\alpha < 0$, then $F \subset H_x$ and $|F \cap B_\alpha| = 0$, so that

$$0 = \frac{|F\cap B_\alpha|}{|B_\alpha|} = \lim_{h\to\infty} \frac{|E_{x,r_h}\cap B_\alpha|}{|B_\alpha|} = \lim_{h\to\infty} \frac{|E\cap B(x,r_h\alpha)|}{|B(x,r_h\alpha)|}\,,$$

in contradiction with (15.33). Similarly, $\alpha > 0$ would contradict (15.34). Thus $\alpha = 0$ and $F = H_x$ by (15.41), as required.

Step four: We have thus proved that, as $r \to 0^+$, $E_{x,r}$ locally converges to H_x, $\mu_{E_{x,r}} \stackrel{*}{\rightharpoonup} \mu_{H_x}$, and $|\mu_{E_{x,r}}| \stackrel{*}{\rightharpoonup} |\mu_{H_x}|$. We conclude the proof of the theorem since, by the Gauss–Green theorem, $\mu_{H_x} = \nu_E(x)\mathcal{H}^{n-1} \llcorner \partial H_x$. □

Remark 15.16 (The "isoperimetric differential inequality argument") In step two of the proof of Theorem 15.5 we have used the Euclidean isoperimetric inequality to deduce a differential inequality of the form $C\, m'(r) \ge m(r)^{(n-1)/n}$ for the monotone function $m(r) = |E \cap B(x,r)|$, and then, as an immediate consequence, the density estimates (15.33) and (15.34).This is a very common and powerful kind of application of isoperimetric-type inequalities, usually related to the proof of density, boundedness, or decay estimates. For further examples, see Theorem 16.14, Theorem 19.24, Theorem 21.11, Lemma 29.10, Lemma 29.12, and Lemma 30.2.

15.2 Structure of Gauss–Green measures

We now prove Theorem 15.9, combining the local convergence properties proved in Theorem 15.5 with the following celebrated result.

Theorem (Whitney's extension theorem) *If $C \subset \mathbb{R}^n$ is closed, and $u\colon C \to \mathbb{R}$ and $T\colon C \to \mathbb{R}^n$ are continuous functions, then there exists $v \in C^1(\mathbb{R}^n)$ such that $u = v$ and $T = \nabla v$ on C if and only if, for every compact set $K \subset C$,*

$$\lim_{\delta \to 0^+} \sup\left\{\frac{|u(y) - u(x) - T(x)\cdot(y-x)|}{|x-y|} : 0 < |x-y| \le \delta,\ x, y \in K\right\} = 0\,. \tag{15.42}$$

Corollary *If $K \subset \mathbb{R}^n$ is a compact set, $T\colon K \to \mathbb{R}^n$ is continuous, and*

$$\lim_{\delta \to 0^+} \sup\left\{\frac{|T(x)\cdot(y-x)|}{|x-y|} : 0 < |x-y| \le \delta\,, x, y \in K\right\} = 0\,, \tag{15.43}$$

then there exists $v \in C^1(\mathbb{R}^n)$ such that $K \subset \{x \in \mathbb{R}^n : v(x) = 0\}$ and $T = \nabla v$ on K. In particular, if $T \ne 0$ on K, then K is contained in a C^1-hypersurface.

Proof of Theorem 15.9 If we introduce the following alternative notation to the one used in Theorem 15.5, setting, for every $x \in \partial^* E$,

$$H_x^- = \left\{y \in \mathbb{R}^n : (y-x)\cdot\nu_E(x) \le 0\right\}, \quad H_x^+ = \left\{y \in \mathbb{R}^n : (y-x)\cdot\nu_E(x) \ge 0\right\},$$

then we easily deduce from Theorem 15.5 and Corollary 15.8 that

$$\lim_{r\to 0^+}\frac{|B(x,r)\cap H_x^+\cap E|}{\omega_n r^n}=0\,, \tag{15.44}$$

$$\lim_{r\to 0^+}\frac{|B(x,r)\cap H_x^-\cap E|}{\omega_n r^n}=\frac{1}{2}\,. \tag{15.45}$$

Clearly, it suffices to prove the theorem with $E\cap B_R$ ($R>0$ arbitrary) in place of E. Since each $E\cap B_R$ is of finite perimeter by Lemma 15.12, we may directly assume that E is of finite perimeter. We now divide the proof into three steps.

Step one: We prove the existence of a $|\mu_E|$-negligible set F such that

$$\partial^* E = F\cup\bigcup_{h\in\mathbb{N}} K_h\,, \tag{15.46}$$

with K_h compact, $K_h\subset M_h$, where M_h is a C^1-hypersurface M_h and $\nu_E(x)^\perp = T_x M_h$ for every $x\in K_h$. Since $|\mu_E|(\partial^* E)=P(E)<\infty$, by applying Egoroff's theorem to the limit relations (15.44) and (15.45) on the set $\partial^* E$ and with respect to $|\mu_E|$, we prove the existence of $|\mu_E|$-measurable sets $F_i\subset\partial^* E$, and of increasing functions $\tau_i\colon[0,\infty)\to[0,\infty)$ with $\tau_i(0^+)=0$, such that

$$\frac{|B(x,r)\cap H_x^+\cap E|}{\omega_n r^n}\le\tau_i(r)\,,\qquad\left|\frac{|B(x,r)\cap H_x^-\cap E|}{\omega_n r^n}-\frac{1}{2}\right|\le\tau_i(r)\,,$$

for every $x\in F_i$ and $|\mu_E|(\partial^* E\setminus\bigcup_{i\in\mathbb{N}}F_i)=0$. By Lusin's theorem, for every $i\in\mathbb{N}$ there exist a sequence of compact sets $\{C_{i,j}\}_{j\in\mathbb{N}}$ such that ν_E is continuous on $C_{i,j}$ and $|\mu_E|(F_i\setminus\bigcup_{j\in\mathbb{N}}C_{i,j})=0$. In conclusion, there exists a sequence of compact sets $\{K_h\}_{h\in\mathbb{N}}$ with $K_h\subset\partial^* E$ such that

(i) if $F=\partial^* E\setminus\bigcup_{h\in\mathbb{N}}K_h$, then $|\mu_E|(F)=0$;

(ii) for every $h\in\mathbb{N}$, ν_E is continuous on K_h;

(iii) for every $h\in\mathbb{N}$, there exists an increasing function $\sigma_h\colon[0,\infty)\to[0,\infty)$ with $\sigma_h(0^+)=0$ such that, for every $x\in K_h$,

$$\frac{|B(x,r)\cap H_x^+\cap E|}{\omega_n r^n}\le\sigma_h(r)\,,\qquad\left|\frac{|B(x,r)\cap H_x^-\cap E|}{\omega_n r^n}-\frac{1}{2}\right|\le\sigma_h(r)\,.$$

Each pair (K_h,ν_E) satisfies (15.43): precisely, for every $\varepsilon\in(0,1)$ there exists $\delta>0$ (depending on ε, n and h) such that

$$\begin{cases}|x-y|\le\delta\,,\\ x,y\in K_h\,,\end{cases}\qquad\Rightarrow\qquad|\nu_E(x)\cdot(y-x)|\le\varepsilon|x-y|\,. \tag{15.47}$$

Indeed, if $x,y\in K_h$ and $\nu_E(x)\cdot(y-x)>\varepsilon|x-y|$, then we have

$$B\big(y,\varepsilon|x-y|\big)\subset B\big(x,2|x-y|\big)\cap H_x^+\,; \tag{15.48}$$

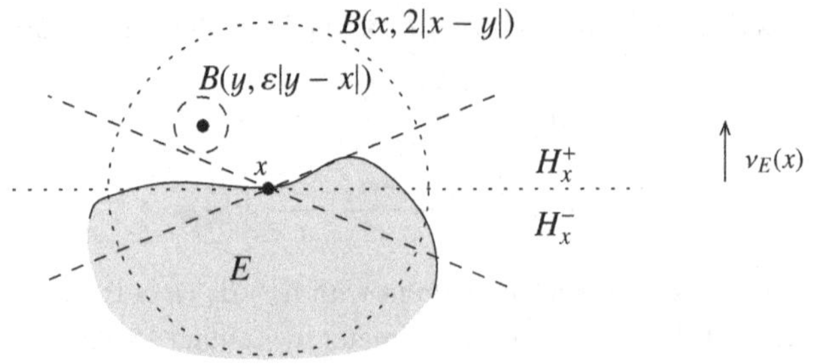

Figure 15.4 The situation described in (15.48). The dashed cone represents the set of those $z \in \mathbb{R}^n$ such that $|\nu_E(x) \cdot (z-x)| \le \varepsilon|z-x|$, which we expect to contain $B(x,\delta) \cap K_h$ provided δ is small enough.

see Figure 15.4. By property (iii) and (15.48), we thus find

$$\begin{aligned}\Big(\frac{\omega_n}{2} - \sigma_h\big(\varepsilon|x-y|\big)\Big)\varepsilon^n\,|x-y|^n &\le \Big|B\big(y,\varepsilon|x-y|\big) \cap H_y^- \cap E\Big| \\ &\le \Big|B\big(y,\varepsilon|x-y|\big) \cap E\Big| \le \Big|B\big(x,2|x-y|\big) \cap H_x^+ \cap E\Big| \\ &\le \sigma_h\big(2|x-y|\big)\,2^n\,|x-y|^n\,.\end{aligned}$$

If $|x-y| \le \delta$, then $(\omega_n - 2\,\sigma_h(\varepsilon\delta))\varepsilon^n \le 2^{n+1}\sigma_h(2\delta)$ which leads to a contradiction as soon as $\delta \to 0^+$ with ε and h fixed. Thus for every $\varepsilon \in (0,1)$ there exists $\delta > 0$ such that $|x-y| \le \delta$, $x, y \in K_h$, implies $\nu_E(x) \cdot (y-x) \le \varepsilon|x-y|$. The complementary estimate is derived similarly, and (15.47) is proved. Hence, by the corollary to Whitney's extension theorem, for each $h \in \mathbb{N}$ there exists a C^1-hypersurface M_h such that $K_h \subset M_h$ and $\nu_E(x)^\perp = T_x M_h$ for every $x \in K_h$.

Step two: We prove that for every Borel set $G \subset \mathbb{R}^n$ we have

$$\mathcal{H}^{n-1}(G \cap \partial^* E) \le C(n)\,|\mu_E|\,(G)\,, \tag{15.49}$$

that is, $\mathcal{H}^{n-1} \llcorner \partial^* E \le C(n)\,|\mu_E|$ on $\mathcal{B}(\mathbb{R}^n)$. Indeed, let A be an open set with $G \cap \partial^* E \subset A$, and consider the family $\mathcal{F}$ of those balls $B(x,r) \subset A$ such that

$$x \in G \cap \partial^* E\,, \qquad 0 < r < \delta\,, \qquad |\mu_E|\,(B(x,r)) \ge \frac{\omega_{n-1} r^{n-1}}{2}\,.$$

By (15.9), $\mathcal{F}$ is a covering of $G \cap \partial^* E$. By Besicovitch's covering theorem, there exist finitely many subfamilies $\{\mathcal{F}_h\}_{h=1}^{\xi(n)}$ of $\mathcal{F}$, with each $\mathcal{F}_h$ disjoint and at most countable, such that $\bigcup_{h=1}^{\xi(n)} \mathcal{F}_h$ is a covering of $G \cap \partial^* E$. Hence, by definition of

$\mathcal{H}^{n-1}_\delta$ and construction of $\mathcal{F}$

$$\mathcal{H}^{n-1}_\delta(G \cap \partial^* E) \le \sum_{h=1}^{\xi(n)} \sum_{B(x,r)\in\mathcal{F}_h} \omega_{n-1} r^{n-1} \le 2 \sum_{h=1}^{\xi(n)} \sum_{B(x,r)\in\mathcal{F}_h} |\mu_E|(B(x,r))$$
$$\le 2\xi(n)|\mu_E|(A).$$

We let first $\delta \to 0^+$, and then use the arbitrariness of A and the fact that $|\mu_E|$ is concentrated on $\partial^* E$, to infer (15.49) with $C(n) = 2\xi(n)$.

Step three: We prove that $\mathcal{H}^{n-1} \llcorner \partial^* E = |\mu_E|$ on $\mathcal{P}(\mathbb{R}^n)$. By (15.49), $\mathcal{H}^{n-1} \llcorner \partial^* E$ is locally finite. Hence, by Proposition 2.13, $\mathcal{H}^{n-1} \llcorner \partial^* E$ is a Radon measure, and, by Exercise 2.6, it will suffice to prove that $\mathcal{H}^{n-1} \llcorner \partial^* E = |\mu_E|$ on $\mathcal{B}(\mathbb{R}^n)$. Since $|\mu_E|(F) = 0$, by (15.46) and (15.49) we are thus left to prove that

$$\mathcal{H}^{n-1} \llcorner K_h = |\mu_E| \llcorner K_h \qquad \text{on } \mathcal{B}(\mathbb{R}^n), \tag{15.50}$$

for every $h \in \mathbb{N}$. By the Lebesgue–Besicovitch differentiation theorem, there exist Borel sets $Y_h \subset M_h$ with $|\mu_E|(Y_h) = 0$ such that

$$\mathcal{H}^{n-1} \llcorner M_h = \theta |\mu_E| + \mathcal{H}^{n-1} \llcorner Y_h, \tag{15.51}$$

where $\theta = D_{|\mu_E|} \mathcal{H}^{n-1} \llcorner M_h \colon \mathbb{R}^n \to \mathbb{R}$ satisfies

$$\theta(x) = \lim_{r\to 0^+} \frac{\mathcal{H}^{n-1}(M_h \cap B(x,r))}{|\mu_E|(B(x,r))}, \qquad \text{for } |\mu_E|\text{-a.e. } x \in \partial^* E. \tag{15.52}$$

By Theorem 9.1, if $x \in M_h$ then

$$\lim_{r\to 0^+} \frac{\mathcal{H}^{n-1}(M_h \cap B(x,r))}{\omega_{n-1} r^{n-1}} = 1.$$

At the same time, by (15.9), if $x \in \partial^* E$ then

$$\lim_{r\to 0^+} \frac{|\mu_E|(B(x,r))}{\omega_{n-1} r^{n-1}} = 1.$$

Since $K_h \subset M_h \cap \partial^* E$ we thus find from (15.52) that $\theta = 1$ on K_h. In particular, if G is a Borel set with $G \subset K_h$, then (15.51) implies (15.50):

$$\mathcal{H}^{n-1}(G) = \mathcal{H}^{n-1}(M_h \cap G) = |\mu_E|(G) + \mathcal{H}^{n-1}(G \cap Y_h) = |\mu_E|(G),$$

where we have used $\mathcal{H}^{n-1}(G \cap Y_h) \le \mathcal{H}^{n-1}(Y_h \cap \partial^* E) \le C(n)|\mu_E|(Y_h) = 0$. □

Exercise 15.17 If E is a set of locally finite perimeter and $x \in \partial^* E$ then

$$\nu_E(x) = \lim_{r\to 0^+} \frac{1}{\omega_{n-1} r^{n-1}} \int_{B(x,r)\cap\partial^* E} \nu_E \, d\mathcal{H}^{n-1}.$$

Hint: Combine the definition of $\partial^* E$ with (15.9) and with $\mu_E = \nu_E \mathcal{H}^{n-1} \llcorner \partial^* E$.

Exercise 15.18 (Stripes) If E is a set of locally finite perimeter in $\mathbb{R}^n$ and there exists $\nu \in S^{n-1}$ with $\nu_E(x) = \pm\nu$ for $\mathcal{H}^{n-1}$-a.e. $x \in \partial^* E$, then there exists a set F of locally finite perimeter in $\mathbb{R}$ such that E is equivalent to

$$\nu^\perp \times \Big\{t\nu : t \in F\Big\}.$$

Hint: Repeat the proof of Proposition 15.15.

Exercise 15.19 (Characterization of balls) A set of locally finite perimeter E in $\mathbb{R}^n$ is equivalent to $B(x, r)$, $x \in \mathbb{R}^n$, $r > 0$, if and only if

$$\nu_E(y) = \frac{y - x}{|y - x|}, \qquad \text{for } \mathcal{H}^{n-1}\text{-a.e. } y \in \partial^* E. \tag{15.53}$$

16

Federer's theorem and comparison sets

In this chapter we study the reduced boundary by taking advantage of the theory of rectifiable sets developed in Chapter 10. Let us recall that if $k \in \mathbb{N}$, $1 \le k \le n-1$, then a set $M \subset \mathbb{R}^n$ is called locally $\mathcal{H}^k$-rectifiable, provided that $\mathcal{H}^k \llcorner M$ is a Radon measure and M is covered, up to a $\mathcal{H}^k$-null set, by countably many images of Lipschitz maps from $\mathbb{R}^k$ to $\mathbb{R}^n$. In particular, by De Giorgi's structure theorem, the reduced boundary $\partial^* E$ of a set of locally finite perimeter is a locally $\mathcal{H}^{n-1}$-rectifiable set in $\mathbb{R}^n$. In Chapter 10, we have also proved an important rectifiability criterion for Radon measures. More precisely, in Theorem 10.8 we have proved that whenever μ is a Radon measure in $\mathbb{R}^n$, concentrated on a Borel set M, and such that for every $x \in M$ there exists a k-dimensional plane π_x in $\mathbb{R}^n$, with the property that, in the limit $r \to 0^+$,

$$\frac{(\Phi_{x,r})_\# \mu}{r^k} \stackrel{*}{\rightharpoonup} \mathcal{H}^k \llcorner \pi_x \,,$$

then M is locally $\mathcal{H}^k$-rectifiable, $\mu = \mathcal{H}^k \llcorner M$, and π_x is the approximate tangent space to M at x, denoted by $T_x M$. Combining these results with Theorem 15.5, we find an alternative proof of De Giorgi's identity $\mu_E = \nu_E \, \mathcal{H}^{n-1} \llcorner \partial^* E$ which avoids the use of Whitney's extension theorem.

Corollary 16.1 *If E is a set of (locally) finite perimeter, then $\partial^* E$ is (locally) $\mathcal{H}^{n-1}$-rectifiable and $\mu_E = \nu_E \mathcal{H}^{n-1} \llcorner \partial^* E$. Moreover, the approximate tangent space to $\partial^* E$ at $x \in \partial^* E$ agrees with the orthogonal space to the measure-theoretic outer unit normal to E at x, that is*

$$T_x(\partial^* E) = \nu_E(x)^\perp \,, \qquad \forall x \in \partial^* E \,. \tag{16.1}$$

Proof By Theorem 15.5, for every $x \in \partial^* E$ and as $r \to 0^+$, we have

$$\frac{(\Phi_{x,r})_\# |\mu_E|}{r^{n-1}} \stackrel{*}{\rightharpoonup} \mathcal{H}^{n-1} \llcorner \pi_x \,, \qquad \text{where } \pi_x = \nu_E(x)^\perp \,.$$

Since $|\mu_E|$ is concentrated on $\partial^* E$, by Theorem 10.8 we conclude that $\partial^* E$ is locally $\mathcal{H}^{n-1}$-rectifiable, with $\mu_E = \nu_E \, \mathcal{H}^{n-1} \llcorner \partial^* E$ and $T_x(\partial^* E) = \pi_x$. □

In Corollary 15.8, we have shown that the reduced boundary $\partial^* E$ is always contained in the set of points of density one-half $E^{(1/2)}$ of E. Let us now introduce the **essential boundary** $\partial^{\mathrm{e}} E$ of a Lebsegue measurable set $E \subset \mathbb{R}^n$,

$$\partial^{\mathrm{e}} E = \mathbb{R}^n \setminus \left(E^{(0)} \cup E^{(1)}\right).$$

We obviously have $E^{(1/2)} \subset \partial^{\mathrm{e}} E$. We now apply the density estimates of Theorem 6.4, to prove the $\mathcal{H}^{n-1}$-equivalence of the reduced boundary, the set of points of density one-half, and the essential boundary.

Theorem 16.2 (Federer's theorem) *If E is a set of locally finite perimeter in $\mathbb{R}^n$, then $\partial^* E \subset E^{(1/2)} \subset \partial^{\mathrm{e}} E$, with*

$$\mathcal{H}^{n-1}\left(\partial^{\mathrm{e}} E \setminus \partial^* E\right) = 0 . \tag{16.2}$$

Proof By the relative isoperimetric inequality (12.46), and since, trivially, $|E \cap B(x,r)| \le \omega_n^{1/n} \, r \, |E \cap B(x,r)|^{(n-1)/n}$, we find that

$$\frac{P(E; B(x,r))}{r^{n-1}} \ge c(n) \min\left\{\frac{|E \cap B(x,r)|}{r^n}, \frac{|B(x,r) \setminus E|}{r^n}\right\}.$$

Thus, $\theta^*_{n-1}(\mathcal{H}^{n-1} \llcorner \partial^* E)(x) = 0$ implies $x \in E^{(0)} \cup E^{(1)}$. In particular,

$$\partial^{\mathrm{e}} E \subset \left\{x \in \mathbb{R}^n : \theta^*_{n-1}\left(\mathcal{H}^{n-1} \llcorner \partial^* E\right)(x) > 0\right\},$$

so that

$$\partial^{\mathrm{e}} E \setminus \partial^* E \subset \left\{x \in \mathbb{R}^n \setminus \partial^* E : \theta^*_{n-1}\left(\mathcal{H}^{n-1} \llcorner \partial^* E\right)(x) > 0\right\}.$$

By Corollary 6.5, this last set is $\mathcal{H}^{n-1}$-negligible. □

16.1 Gauss–Green measures and set operations

We now combine Federer's theorem with Theorem 15.5 to characterize the Gauss–Green measures of $E \cap F$, $E \setminus F$ and $E \cup F$ (that, by Lemma 12.22, are sets of locally finite perimeter provided E and F are) in terms of μ_E and μ_F. Throughout the rest of this section we write $M_1 \approx M_2$ if M_1 and M_2 are Borel sets such that $\mathcal{H}^{n-1}(M_1 \Delta M_2) = 0$. By (16.2), $\partial^* E \approx E^{(1/2)} \approx \partial^{\mathrm{e}} E$ and

$$M \approx \left(M \cap E^{(1)}\right) \cup \left(M \cap E^{(0)}\right) \cup \left(M \cap E^{(1/2)}\right), \tag{16.3}$$

for every Borel set M and for every set of locally finite perimeter E.

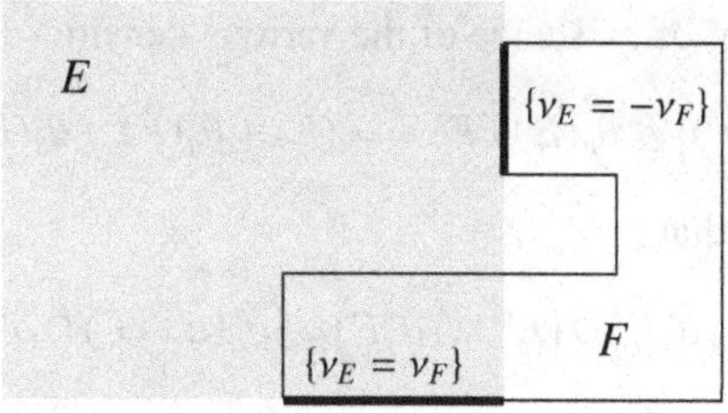

Figure 16.1 An illustration of Theorem 16.3; $\mathcal{H}^{n-1}(\{\nu_E = \nu_F\})$ contributes to the perimeter of the intersection and of the union of E and F, while $\mathcal{H}^{n-1}(\{\nu_E = -\nu_F\})$ contributes to the perimeter of $E \setminus F$ and $F \setminus E$.

Theorem 16.3 (Set operations on Gauss–Green measures) *If E and F are sets of locally finite perimeter, and we let*

$$\{\nu_E = \nu_F\} = \Big\{x \in \partial^* E \cap \partial^* F : \nu_E(x) = \nu_F(x)\Big\},$$
$$\{\nu_E = -\nu_F\} = \Big\{x \in \partial^* E \cap \partial^* F : \nu_E(x) = -\nu_F(x)\Big\},$$

then $E \cap F$, $E \setminus F$ and $E \cup F$ are sets of locally finite perimeter, with

$$\mu_{E\cap F} = \mu_E \llcorner F^{(1)} + \mu_F \llcorner E^{(1)} + \nu_E \mathcal{H}^{n-1} \llcorner \{\nu_E = \nu_F\}, \tag{16.4}$$
$$\mu_{E\setminus F} = \mu_E \llcorner F^{(0)} - \mu_F \llcorner E^{(1)} + \nu_E \mathcal{H}^{n-1} \llcorner \{\nu_E = -\nu_F\}, \tag{16.5}$$
$$\mu_{E\cup F} = \mu_E \llcorner F^{(0)} + \mu_F \llcorner E^{(0)} + \nu_E \mathcal{H}^{n-1} \llcorner \{\nu_E = \nu_F\}, \tag{16.6}$$

and

$$\partial^*(E \cap F) \approx \Big(F^{(1)} \cap \partial^* E\Big) \cup \Big(E^{(1)} \cap \partial^* F\Big) \cup \Big\{\nu_E = \nu_F\Big\}, \tag{16.7}$$
$$\partial^*(E \setminus F) \approx \Big(F^{(0)} \cap \partial^* E\Big) \cup \Big(E^{(1)} \cap \partial^* F\Big) \cup \Big\{\nu_E = -\nu_F\Big\}, \tag{16.8}$$
$$\partial^*(E \cup F) \approx \Big(F^{(0)} \cap \partial^* E\Big) \cup \Big(E^{(0)} \cap \partial^* F\Big) \cup \Big\{\nu_E = \nu_F\Big\}; \tag{16.9}$$

see Figure 16.3. Moreover, for every Borel set $G \subset \mathbb{R}^n$,

$$P(E\cap F; G) = P(E; F^{(1)}\cap G) + P(F; E^{(1)}\cap G) + \mathcal{H}^{n-1}\Big(\Big\{\nu_E = \nu_F\Big\}\cap G\Big), \tag{16.10}$$

$$P(E\setminus F; G) = P(E; F^{(0)}\cap G) + P(F; E^{(1)}\cap G) + \mathcal{H}^{n-1}\Big(\Big\{\nu_E = -\nu_F\Big\}\cap G\Big), \tag{16.11}$$

$$P(E\cup F; G) = P(E; F^{(0)}\cap G) + P(F; E^{(0)}\cap G) + \mathcal{H}^{n-1}\Big(\Big\{\nu_E = \nu_F\Big\}\cap G\Big). \tag{16.12}$$

Proof The proof is a rather lengthy (but rather simple) application of the structure theory. Setting for brevity $B_{x,r}$ in place of $B(x,r)$,

$$\begin{aligned}|(E\cup F)\cap B_{x,r}| &= |(E\setminus F)\cap B_{x,r}| + |(F\setminus E)\cap B_{x,r}| + |(E\cap F)\cap B_{x,r}| \\ &= |E\cap B_{x,r}| + |F\cap B_{x,r}| - |(E\cap F)\cap B_{x,r}|.\end{aligned}$$

Hence, at every point of existence of the various densities involved, we find

$$\max\left\{\theta_n(E),\theta_n(F)\right\}\le\theta_n(E\cup F)=\theta_n(E)+\theta_n(F)-\theta_n(E\cap F)\,.\qquad(16.13)$$

Step one: We claim that

$$\partial^*(E\cap F)\approx\left(F^{(1)}\cap\partial^*E\right)\cup\left(E^{(1)}\cap\partial^*F\right)\cup\left(\partial^*(E\cap F)\cap\partial^*E\cap\partial^*F\right),\qquad(16.14)$$

and that the three sets on the right-hand side of (16.14) have $\mathcal{H}^{n-1}$-negligible mutual intersections. Decomposing $M=(E\cap F)^{(1/2)}$ with respect to E by (16.3), and noticing that $(E\cap F)^{(1/2)}\cap E^{(0)}=\emptyset$ by $\theta_n(E\cap F)\le\theta_n(E)$, we find

$$(E\cap F)^{(1/2)}\approx\left((E\cap F)^{(1/2)}\cap E^{(1)}\right)\cup\left((E\cap F)^{(1/2)}\cap E^{(1/2)}\right).$$

Decomposing $M=(E\cap F)^{(1/2)}\cap E^{(1)}$ and $M=(E\cap F)^{(1/2)}\cap E^{(1/2)}$ with respect to F by (16.3), and using $(E\cap F)^{(1/2)}\cap F^{(0)}=\emptyset$, we find

$$\begin{aligned}(E\cap F)^{(1/2)}\ &\approx\left((E\cap F)^{(1/2)}\cap E^{(1)}\cap F^{(1)}\right)\cup\left((E\cap F)^{(1/2)}\cap E^{(1)}\cap F^{(1/2)}\right)\\&\quad\cup\left((E\cap F)^{(1/2)}\cap E^{(1/2)}\cap F^{(1)}\right)\cup\left((E\cap F)^{(1/2)}\cap E^{(1/2)}\cap F^{(1/2)}\right).\end{aligned}$$

By (16.13) we have that $(E\cap F)^{(1/2)}\cap E^{(1)}\cap F^{(1)}$ is empty. Again by (16.13),

$$(E^{(1)}\cap F^{(1/2)})\subset(E\cap F)^{(1/2)}\,,\qquad(F^{(1)}\cap E^{(1/2)})\subset(E\cap F)^{(1/2)}\,.\qquad(16.15)$$

We thus conclude that

$$(E\cap F)^{(1/2)}\approx\left(E^{(1)}\cap F^{(1/2)}\right)\cup\left(E^{(1/2)}\cap F^{(1)}\right)\cup\left((E\cap F)^{(1/2)}\cap E^{(1/2)}\cap F^{(1/2)}\right),$$

where the three sets on the right-hand side are mutually disjoint. The claim follows by Federer's theorem.

Step two: We prove (16.4). We set $\mu=\mu_{E\cap F}$ for the sake of brevity. Since $|\mu|=\mathcal{H}^{n-1}\llcorner\partial^*(E\cap F)$, (16.14) implies

$$\mu=\mu\llcorner\left(F^{(1)}\cap\partial^*E\right)+\mu\llcorner\left(E^{(1)}\cap\partial^*F\right)+\mu\llcorner\left(\partial^*(E\cap F)\cap\partial^*E\cap\partial^*F\right).\qquad(16.16)$$

Let us set $H[\nu]=\{y\in\mathbb{R}^n:y\cdot\nu<0\}$ ($\nu\in S^{n-1}$). If $x\in\partial^*(E\cap F)\cap F^{(1)}\cap\partial^*E$, then, by Theorem 15.5, as $r\to0^+$,

$$\begin{aligned}&(E\cap F)_{x,r}\overset{\text{loc}}{\to}H[\nu_{E\cap F}(x)]\,,\quad E_{x,r}\overset{\text{loc}}{\to}H[\nu_E(x)]\,,\quad F_{x,r}\overset{\text{loc}}{\to}\mathbb{R}^n\,,\\&(E\cap F)_{x,r}=E_{x,r}\cap F_{x,r}\overset{\text{loc}}{\to}H[\nu_E(x)]\cap\mathbb{R}^n=H[\nu_E(x)]\,,\end{aligned}$$

that is, $\nu_{E\cap F}(x)=\nu_E(x)$. Hence, by (16.15) and Federer's theorem,

$$\begin{aligned}\mu\llcorner\left(F^{(1)}\cap\partial^*E\right)&=\nu_{E\cap F}\,\mathcal{H}^{n-1}\llcorner\left(\partial^*(E\cap F)\cap F^{(1)}\cap\partial^*E\right)\\&=\nu_E\,\mathcal{H}^{n-1}\llcorner\left(F^{(1)}\cap\partial^*E\right).\end{aligned}\qquad(16.17)$$

By symmetry,

$$\mu \llcorner \left(E^{(1)} \cap \partial^* F\right) = \nu_F\, \mathcal{H}^{n-1} \llcorner \left(E^{(1)} \cap \partial^* F\right). \tag{16.18}$$

Finally, if $x \in \partial^*(E \cap F) \cap \partial^* E \cap \partial^* F$, then, again by Theorem 15.5, as $r \to 0^+$,

$$(E \cap F)_{x,r} \overset{\text{loc}}{\to} H[\nu_{E\cap F}(x)]\,, \quad E_{x,r} \overset{\text{loc}}{\to} H[\nu_E(x)]\,, \quad F_{x,r} \overset{\text{loc}}{\to} H[\nu_F(x)]\,,$$

$$(E \cap F)_{x,r} = E_{x,r} \cap F_{x,r} \overset{\text{loc}}{\to} H[\nu_E(x)] \cap H[\nu_F(x)]\,,$$

which implies $\nu_{E\cap F}(x) = \nu_E(x) = \nu_F(x)$. Therefore,

$$\partial^*(E \cap F) \cap \partial^* E \cap \partial^* F \subset \{\nu_E = \nu_F\}\,.$$

At the same time, if $x \in \{\nu_E = \nu_F\}$, then, by Theorem 15.5, as $r \to 0^+$,

$$(E \cap F)_{x,r} = E_{x,r} \cap F_{x,r} \to H[\nu_E(x)] \cap H[\nu_F(x)] = H[\nu_E(x)]\,,$$

which immediately implies $x \in (E \cap F)^{(1/2)}$. Thus, by Federer's theorem,

$$\partial^*(E \cap F) \cap \partial^* E \cap \partial^* F \approx \{\nu_E = \nu_F\}\,,$$

and, in particular,

$$\begin{aligned} \mu \llcorner \left(\partial^*(E \cap F) \cap \partial^* E \cap \partial^* F\right) &= \nu_{E\cap F}\, \mathcal{H}^{n-1} \llcorner \{\nu_E = \nu_F\} \\ &= \nu_E\, \mathcal{H}^{n-1} \llcorner \{\nu_E = \nu_F\}\,. \end{aligned} \tag{16.19}$$

By (16.16), (16.17), (16.18), and (16.19), we conclude the proof of (16.4).

Step three: We deduce (16.7) by (16.4) and the mutual singularity of the three Radon measures on the right-hand side of (16.4) (see Exercise 5.13). Since $(\mathbb{R}^n \setminus E)^{(0)} = E^{(1)}$, $(\mathbb{R}^n \setminus E)^{(1)} = E^{(0)}$, $\mu_E = -\mu_{\mathbb{R}^n \setminus E}$ (by (12.4)) and $E \setminus F = E \cap (\mathbb{R}^n \setminus F)$, by (16.4) we first deduce (16.5), and then (16.8). As $\mathbb{R}^n \setminus (E \cup F) = (\mathbb{R}^n \setminus E) \cap (\mathbb{R}^n \setminus F)$, we similarly prove (16.6) and (16.9). □

Exercise 16.4 Given $r > 0$ and $a < b$, the cylinder

$$E = \left\{(x', x_n) \in \mathbb{R}^n : |x'| < r\,, a < x_n < b\right\},$$

is a set of finite perimeter in $\mathbb{R}^n$, with

$$\begin{aligned} \mu_E &= \nu\, \mathcal{H}^{n-1} \llcorner \{(x', x_n) : |x'| = r\,, a < x_n < b\} \\ &\quad + e_n \mathcal{H}^{n-1} \llcorner \{(x', a) : |x'| < r\} - e_n \mathcal{H}^{n-1} \llcorner \{(x', b) : |x'| < r\}\,, \end{aligned}$$

where we have set $\nu(x', x_n) = (r^{-1}\, x', 0)$ for every $x \in \mathbb{R}^n$ with $|x'| = r$.

Exercise 16.5 (Gauss–Green measure of the symmetric difference) If E and F are of locally finite perimeter, then $E \Delta F$ is of locally finite perimeter and

$$\mu_{E\Delta F} = \mu_E \llcorner F^0 + \mu_F \llcorner E^0 - \mu_E \llcorner F^1 - \mu_F \llcorner E^1\,. \tag{16.20}$$

In particular, for every Borel set $G \subset \mathbb{R}^n$,

$$P(E\Delta F;G) = P(E;G \setminus \partial^* F) + P(F;G \setminus \partial^* E) \le P(E;G) + P(F;G).$$

Exercise 16.6 If E and F are sets of locally finite perimeter with $E \subset F$ then $\nu_E = \nu_F$ on $\partial^* F \cap \partial^* E$ and

$$\mu_E = \mu_E \llcorner F^{(1)} + \nu_F \llcorner \left(\partial^* F \cap \partial^* E\right).$$

In particular, $P(E) = P(E;F^{(1)}) + P(E;F^{(1/2)})$.

Exercise 16.7 (Perimeter and reflection through a hyperplane) Let H be an open half-space in $\mathbb{R}^n$, and let $Q_H \in \mathbf{O}(n)$ be the reflection of $\mathbb{R}^n$ with respect to ∂H. Given $E \subset H$, set $E_H = E \cup Q_H(E)$. Then

$$|E_H| = 2|E|, \qquad P(E_H) = 2\,P(E;H).$$

Hint: Show that if $x \in E^{(1/2)} \cap \partial H$ then x does not belong to $E_H^{(1/2)}$.

Exercise 16.8 (Relative isoperimetry in half-spaces) Show that the relative isoperimetric problem in the half-space admits half-balls as its only solutions. Precisely, for every $E \subset H$ with finite measure we have

$$P(E;H) \ge P(B(x,r);H),$$

whenever $x \in \partial H$ and $|B(x,r) \cap H| = |E|$; moreover, equality holds if and only if $|E\Delta B(x,r)| = 0$ for some $x \in \partial H$. In other words,

$$P(E;H) \ge n\left(\frac{\omega_n}{2}\right)^{1/n} |E|^{(n-1)/n}, \tag{16.21}$$

with equality if and only if E is equivalent to a half-ball centered on ∂H. *Hint:* Combine Exercise 16.7 and Theorem 14.1.

Exercise 16.9 (Indecomposability) A set of finite perimeter E is **indecomposable** if, whenever $E = E_1 \cup E_2$ with $|E_1 \cap E_2| = 0$ and $P(E) = P(E_1)+P(E_2)$, then either $|E_1| = 0$ or $|E_2| = 0$. Indecomposability plays the role of connectedness the theory of sets of finite perimeter. For example, show that if E and F are of finite perimeter, E is indecomposable and, up to Lebesgue and $\mathcal{H}^{n-1}$-null sets respectively, $F \subset E$ and $\partial^* F \subset \partial^* E$, then either $|F| = 0$ or $|E \setminus F| = 0$. *Hint:* Since $F \subset E$ we have $P(E) = P(F) + P(E \setminus F) - \mathcal{H}^{n-1}(E^{(1)} \cap \partial^* F)$.

Exercise 16.10 If A is an open, connected set with locally finite perimeter in $\mathbb{R}^n$, then A is indecomposable. *Hint:* Repeatedly apply Theorem 16.3 to deduce from $P(A) = P(E_1) + P(E_2)$, $A = E_1 \cup E_2$ and $|E_1 \cap E_2| = 0$, that $P(E_1;A) = P(E_2;A) = 0$. Conclude by Exercise 12.17.

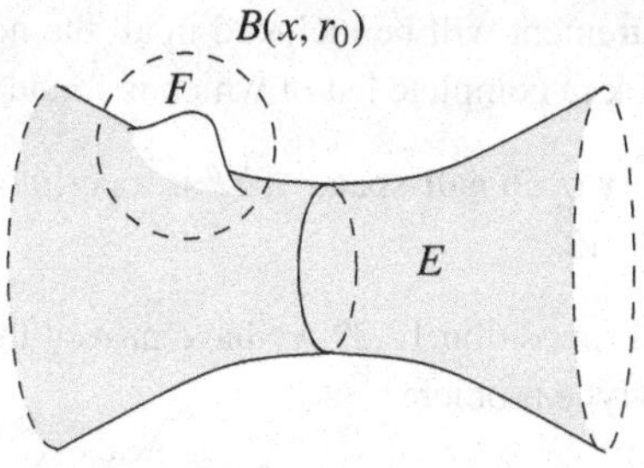

Figure 16.2 A "catenoid-like" local perimeter minimizer. Variations of E compactly supported in sufficiently small balls have larger perimeter.

16.2 Density estimates for perimeter minimizers

We shall have various occasions to use Theorem 16.3 in order to exploit minimality conditions against competitor sets defined through basic set operations. A first important application of this kind of argument is found in this section, where a comparison argument (depicted in Figure 16.3) is used to deduce uniform density estimates for perimeter minimizers. In turn, by Federer's theorem, these density estimates imply a mild regularity property, namely, the $\mathcal{H}^{n-1}$-equivalence of the reduced and topological boundaries of perimeter minimizers. In particular, this fact alone excludes the possibility for perimeter minimizers to present the wild singularities that generic sets of finite perimeter may show up (cf. Example 12.25). We define our terminology. Given an open bounded set A, and a set of locally finite perimeter E in $\mathbb{R}^n$, we say that E is a **perimeter minimizer in** A if $\operatorname{spt}\mu_E = \partial E$, and, whenever $E\Delta F \subset\subset A$,

$$P(E;A) \le P(F;A). \qquad (16.22)$$

This definition is extended to the case that A is not bounded (the case $A = \mathbb{R}^n$ being of particular importance; see Chapter 28), by replacing (16.22) with $P(E;A') \le P(F;A')$, for A' a bounded open set with $E\Delta F \subset\subset A' \subset A$. Finally, we say that E is a **local perimeter minimizer in** A **(at scale** r_0**)** if $\operatorname{spt}\mu_E = \partial E$ and (16.22) holds whenever $E\Delta F \subset\subset B(x,r_0)\cap A$ and $x \in A$; see Figure 16.2.

Remark 16.11 (Topological boundary of minimizers) If $P(E;A) \le P(F;A)$ whenever $E\Delta F \subset\subset A$, then, by Proposition 12.19, there exists E' equivalent to E and satisfying $\operatorname{spt}\mu_{E'} = \partial E'$. Thus, E' is a perimeter minimizer in A. The condition $\operatorname{spt}\mu_E = \partial E$ in the definition of perimeter minimizer is thus assumed, so to say, without loss of generality. It has the advantage of giving a precise geometric meaning to the notion of topological boundary of a perimeter minimizer. This requirement will be included in all the notions of minimality considered in this book (a complete list of which is found in the index).

Exercise 16.12 Every open half-space of $\mathbb{R}^n$ is a perimeter minimizer in $\mathbb{R}^n$. *Hint:* Recall Exercise 15.13.

Example 16.13 In Proposition 12.29 we have proved the existence of minimizers in the Plateau-type problem

$$\inf\Big\{P(E) : E \setminus A = E_0 \setminus A\Big\}, \tag{16.23}$$

for E_0 a given set of finite perimeter in $\mathbb{R}^n$, and for A a bounded set in $\mathbb{R}^n$. Let us show that any minimizer of (16.23), once modified according to Proposition 12.19 in order to have $\operatorname{spt}\mu_E = \partial E$, is also a perimeter minimizer in A. Indeed if $E \setminus A = E_0 \setminus A$, then, by Exercise 12.16, $P(E; \mathbb{R}^n \setminus \overline{A}) = P(E_0; \mathbb{R}^n \setminus \overline{A})$. Thus, for every E in the competition class of (16.23),

$$P(E) = P(E; A) + P(E; \partial A) + P(E_0; \mathbb{R}^n \setminus \overline{A}) .$$

If now E is a minimizer in (16.23) and $E \Delta F \subset\subset A$, then F is a competitor in (16.23) and, from $P(E) \le P(F)$, we deduce

$$P(E; A) + P(E; \partial A) \le P(F; A) + P(F; \partial A) .$$

Since $E \Delta F \subset\subset A$, by Exercise 12.16 we see that $\mu_E \llcorner A' = \mu_F \llcorner A'$ for some open set A' with $A \subset\subset A'$. Thus $P(E; \partial A) = P(F; \partial A)$, and we are done.

Theorem 16.14 (Density estimates for local perimeter minimizers) *For every $n \ge 2$, there exists a positive constant $c(n)$ with the following property. If A is an open set in $\mathbb{R}^n$ and E is a local perimeter minimizer in A at scale r_0, then for every ball $B(x, r) \subset\subset A$ with $x \in A \cap \partial E$ and $r < r_0$, we have*

$$\frac{1}{2^n} \le \frac{|E \cap B(x, r)|}{\omega_n r^n} \le 1 - \frac{1}{2^n} , \tag{16.24}$$

$$c(n) \le \frac{P(E; B(x, r))}{r^{n-1}} \le n\omega_n . \tag{16.25}$$

In particular,

$$\mathcal{H}^{n-1}\Big(A \cap (\partial E \setminus \partial^* E)\Big) = 0 . \tag{16.26}$$

Remark 16.15 We may take $c(n) = \omega_{n-1}$; see Section 17.4. The corresponding lower bound is evidently sharp, as it is saturated by half-spaces.

Proof By Proposition 12.19, since we are assuming $\operatorname{spt}\mu_E = \partial E$, we have

$$0 < |E \cap B(x, r)| < \omega_n r^n , \qquad \forall x \in \partial E , r > 0 . \tag{16.27}$$

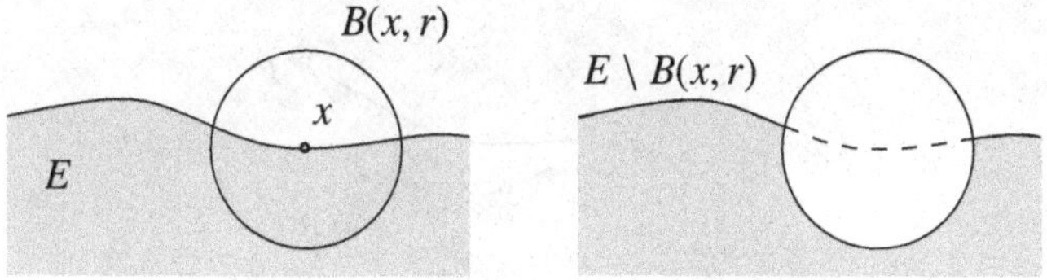

Figure 16.3 Proof of Theorem 16.14.

Let $x \in A \cap \partial E$, and set $d = \min\{\mathrm{dist}(x, \partial A), r_0\}$. By Proposition 2.16,

$$\mathcal{H}^{n-1}\big(\partial^* E \cap \partial B(x,r)\big) = 0\,, \qquad \text{for a.e. } r < d\,. \tag{16.28}$$

For every such r, let $F = E \setminus B(x, r)$. Since $E\Delta F \subset\subset B(x, s) \subset A$ for any $s \in (r, d)$, by local perimeter minimality, (16.28), and (16.11) we find

$$\begin{aligned} P(E; B(x,s)) &\le P(F; B(x,s)) = P\big(E \setminus B(x,r); B(x,s)\big) \\ &= \mathcal{H}^{n-1}\big(E^{(1)} \cap \partial B(x,r)\big) + P\big(E; B(x,s) \setminus \overline{B(x,r)}\big)\,. \end{aligned}$$

By letting $s \to r^+$ we thus find

$$P(E; B(x,r)) \le \mathcal{H}^{n-1}\big(E^{(1)} \cap \partial B(x,r)\big)\,. \tag{16.29}$$

We note that (16.29) implies the upper bound in (16.25) since $\mathcal{H}^{n-1}(E^{(1)} \cap \partial B(x_1 r)) \le P(B(x_1 r)) = n\omega_n r^{n-1}$. Moreover, by adding $\mathcal{H}^{n-1}(E^{(1)} \cap \partial B(x, r))$ to both sides of (16.29), and recalling that, by (16.10) and (16.28), $P(E \cap B(x, r)) = P(E; B(x, r)) + \mathcal{H}^{n-1}(E^{(1)} \cap \partial B(x, r))$, we find

$$P(E \cap B(x,r)) \le 2\,\mathcal{H}^{n-1}\big(E^{(1)} \cap \partial B(x,r)\big)\,,$$

so that, by the Euclidean isoperimetric inequality, applied to $E \cap B(x, r)$,

$$n\omega_n^{1/n} |B(x,r) \cap E|^{(n-1)/n} \le 2\,\mathcal{H}^{n-1}\big(E^{(1)} \cap \partial B(x,r)\big)\,. \tag{16.30}$$

By (13.3), the function $m\colon (0,\infty) \to [0,\infty)$, $m(s) = |E^{(1)} \cap B(x,s)| = |E \cap B(x,s)|$, $s > 0$, is absolutely continuous, with $m'(s) = \mathcal{H}^{n-1}(E^{(1)} \cap \partial B(x,s))$ for a.e. $s > 0$. Hence (16.30) implies

$$n\omega_n^{1/n} m(r)^{(n-1)/n} \le 2\,m'(r)\,, \qquad \text{for a.e. } r < d.$$

By (16.27), m is positive with $m(0^+) = 0$. Hence we may divide by $m^{(n-1)/n}$ and integrate the resulting inequality on $(0, r)$, to find $\omega_n^{1/n}\, r \le 2\,m(r)^{1/n}$ for *every* $r < d$; this proves the lower bound in (16.24). Since $\mathbb{R}^n \setminus E$ is a local perimeter minimizer in A (with constant r_0) and the upper bound in (16.27) implies $|(\mathbb{R}^n \setminus E) \cap B(x, s)| > 0$ for every $s > 0$, we can repeat the argument with $\mathbb{R}^n \setminus E$ in place

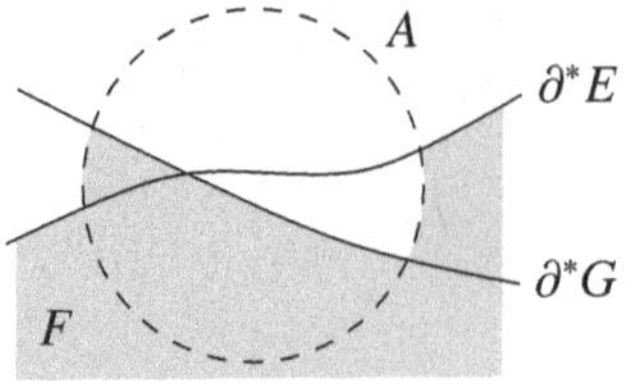

Figure 16.4 The sets E, G, and F of Theorem 16.16 in the case when the open set A is a ball. Roughly speaking, the boundary of F is made up by the boundary of E outside A, the boundary of G inside A and that part of the boundary of A that is contained in $E\Delta G$.

of E to deduce the upper bound in (16.24) as well. Finally, by (16.24) and the relative isoperimetric inequality (12.45), we find $P(E; B(x,r)) \ge c(n)\, r^{n-1}$ for every $r < d$, thus completing the proof of (16.25). Again by (16.24), $A \cap \partial E \subset A \cap \partial^e E$, and thus (16.26) follows from Federer's theorem, Theorem 16.2. □

We have just seen that a very effective way to test a perimeter minimality condition satisfied by a set E in some open set A' is by replacing E with a new set G inside a given region $A \subset\subset A'$. The following theorem, of later use in Part III, allows us to exploit the resulting minimality inequalities. Its technical proof does not introduce new ideas, and it may be safely skipped.

Theorem 16.16 (Comparison sets by replacement) *If E and G are sets of locally finite perimeter in $\mathbb{R}^n$ and A is an open set of finite perimeter such that*

$$\mathcal{H}^{n-1}\big(\partial^* A \cap \partial^* E\big) = \mathcal{H}^{n-1}\big(\partial^* A \cap \partial^* G\big) = 0\,, \tag{16.31}$$

then a set of locally finite perimeter F is defined by

$$F = \big(G \cap A\big) \cup \big(E \setminus A\big);$$

see Figure 16.4. Moreover, if $A \subset\subset A'$, A' open, then

$$P(F; A') = P(G; A) + P(E; A' \setminus \overline{A}) + \mathcal{H}^{n-1}\Big(\big(E^{(1)} \Delta G^{(1)}\big) \cap \partial^* A\Big)\,. \tag{16.32}$$

Corollary 16.17 *If A, E, and G are as in Theorem 16.16, and if E is a perimeter minimizer in some open set A' with $A \subset\subset A'$, then*

$$P(E; A) \le P(G; A) + \mathcal{H}^{n-1}\Big((E^{(1)} \Delta G^{(1)}) \cap \partial^* A\Big)\,. \tag{16.33}$$

Moreover, if $E\Delta G \subset\subset A$, then $P(E; A) \le P(G; A)$.

Proof of Corollary 16.17 Immediate from (16.31) and (16.32). □

Proof of Theorem 16.16 By Lemma 12.22, F is a set of locally finite perimeter. By Theorem 16.3, and in particular by (16.6),

$$\mu_F = \mu_{E\setminus A} \llcorner (G\cap A)^{(0)} + \mu_{G\cap A} \llcorner (E\setminus A)^{(0)} + \nu_{G\cap A}\mathcal{H}^{n-1} \llcorner \{\nu_{G\cap A} = \nu_{E\setminus A}\}. \tag{16.34}$$

We now study the various terms of this decomposition, to prove that

$$\mu_F = \mu_E \llcorner \big(\mathbb{R}^n \setminus \overline{A}\big) + \mu_G \llcorner A + \mu_A \llcorner \big(G^{(1)} \cap E^{(0)}\big) - \mu_A \llcorner \big(E^{(1)} \cap G^{(0)}\big). \tag{16.35}$$

Characterization of $\mu_{E\setminus A} \llcorner (G\cap A)^{(0)}$*:* By (16.31) and (16.5) we have that

$$\mu_{E\setminus A} = \mu_E \llcorner (\mathbb{R}^n \setminus \overline{A}) - \mu_A \llcorner E^{(1)}. \tag{16.36}$$

We now claim that

$$\big(\mu_E \llcorner (\mathbb{R}^n \setminus \overline{A})\big) \llcorner (G\cap A)^{(0)} = \mu_E \llcorner (\mathbb{R}^n \setminus \overline{A}), \tag{16.37}$$

$$\big(\mu_A \llcorner E^{(1)}\big) \llcorner (G\cap A)^{(0)} = \mu_A \llcorner \big(E^{(1)} \cap G^{(0)}\big). \tag{16.38}$$

By $\mathbb{R}^n \setminus \overline{A} = A^{(0)} \subset (G\cap A)^{(0)}$ we prove (16.37). To prove (16.38), note that $G^{(0)} \cap \partial^* A \subset (G\cap A)^{(0)} \cap \partial^* A$, and, by Theorem 16.2 and (16.31), for $\mathcal{H}^{n-1}$-a.e. $z \in (G\cap A)^{(0)} \cap \partial^* A$ we have $z \in G^{(1)} \cup G^{(0)}$; if $z \in G^{(1)}$, then from

$$\theta_n(G\cap A) + \theta_n(G\cup A) = \theta_n(G) + \theta_n(A),$$

we find $\theta_n(G\cup A) \ge 3/2$, a contradiction. Thus $(G\cap A)^{(0)} \cap \partial^* A$ is $\mathcal{H}^{n-1}$-equivalent to $G^{(0)} \cap \partial^* A$, and (16.38) is proved.

Characterization of $\mu_{G\cap A} \llcorner (E\setminus A)^{(0)}$*:* By (16.31) and (16.4) we have that

$$\mu_{G\cap A} = \mu_G \llcorner A + \mu_A \llcorner G^{(1)}. \tag{16.39}$$

We claim that

$$\big(\mu_G \llcorner A\big) \llcorner (E\setminus A)^{(0)} = \mu_G \llcorner A, \tag{16.40}$$

$$\big(\mu_A \llcorner G^{(1)}\big) \llcorner (E\setminus A)^{(0)} = \mu_A \llcorner (G^{(1)} \cap E^{(0)}). \tag{16.41}$$

By $A \subset (E\setminus A)^{(0)}$ we find (16.40), while (16.41) follows by proving the $\mathcal{H}^{n-1}$-equivalence of $E^{(0)} \cap \partial^* A$ and $(E\setminus A)^{(0)} \cap \partial^* A$. Indeed, on the one hand we have $E^{(0)} \cap \partial^* A \subset (E\setminus A)^{(0)} \cap \partial^* A$, while on the other hand, by Theorem 16.2 and (16.31), we have that $\mathcal{H}^{n-1}$-a.e. $z \in (E\setminus A)^{(0)} \cap \partial^* A$ satisfies either $z \in E^{(1)}$ or $z \in E^{(0)}$. The case $z \in E^{(1)}$ leads to a contradiction with the identity

$$\theta_n\big(E \cup (\mathbb{R}^n \setminus A)\big) + \theta_n(E\setminus A) = \theta_n(E) + \theta_n(\mathbb{R}^n \setminus A).$$

Hence $(E\setminus A)^{(0)} \cap \partial^* A$ is $\mathcal{H}^{n-1}$-equivalent to $E^{(0)} \cap \partial^* A$, and (16.41) is proved.

Conclusion of the proof of (16.35): By (16.34), (16.36), (16.37), (16.38), (16.39), (16.40), and (16.41), to prove (16.35) it suffices to show that

$$\mathcal{H}^{n-1}\big(\partial^* F \cap \partial^*(G\cap A) \cap \partial^*(E\setminus A)\big) = 0. \tag{16.42}$$

By (16.36) and (16.39) we know that, up to $\mathcal{H}^{n-1}$-negligible sets,

$$\partial^*(G \cap A) = \big(A \cap \partial^* G\big) \cup \big(G^{(1)} \cap \partial^* A\big),$$
$$\partial^*(E \setminus A) = \big(E^{(1)} \cap \partial^* A\big) \cup \big(\partial^* E \setminus \overline{A}\big).$$

By Theorem 16.2 and (16.31) we immediately deduce that

$$\mathcal{H}^{n-1}\big(\partial^* F \cap \partial^*(G \cap A) \cap \partial^*(E \setminus A)\big) = \mathcal{H}^{n-1}\big(F^{(1/2)} \cap E^{(1)} \cap G^{(1)} \cap \partial^* A\big);$$

thus (16.42) follows, as $F^{(1/2)} \cap E^{(1)} \cap G^{(1)} \cap \partial^* A = \emptyset$ by

$$\begin{aligned}\theta_n(F) &= \theta_n(G \cap A) + \theta_n(E \setminus A) = \theta_n(G) - \theta_n(G \setminus A) + \theta_n(E) - \theta_n(E \cap A)\\ &\geq \theta_n(G) + \theta_n(E) - 2\theta_n(A).\end{aligned}$$

We finally prove (16.32): Indeed, by (16.35) we find

$$\begin{aligned}P(F; A') = P\big(E; A' \setminus \overline{A}\big) + P(G; A)&\\ + \mathcal{H}^{n-1}\big(\partial^* A \cap \big((G^{(1)} \cap E^{(0)}) \cup (E^{(1)} \cap G^{(0)})\big)\big)&.\end{aligned} \tag{16.43}$$

By (16.31), since $\mathcal{H}^{n-1}(\mathbb{R}^n \setminus (E^{(0)} \cup E^{(1)} \cup \partial^* E)) = 0$,

$$\mathcal{H}^{n-1}\big(\partial^* A \cap \big(G^{(1)} \cap E^{(0)}\big)\big) = \mathcal{H}^{n-1}\big(\partial^* A \cap \big(G^{(1)} \setminus E^{(1)}\big)\big),$$

which, combined with (16.43), immediately gives (16.32). □

17

First and second variation of perimeter

One of the most fundamental ideas in the Calculus of Variations, giving the name to the discipline itself, is that of deriving necessary conditions for minimality from the basic rules of Calculus by looking at curves of competitors which "pass through" a given candidate minimizer. Indeed, this is the usual procedure used to derive the Euler–Lagrange equations (first order necessary minimality conditions) and stability inequalities (second order necessary minimality conditions). Let us examine this idea in the case of a perimeter minimizer E into some open set A. We construct a curve of competitors "passing through" E by fixing a compactly supported smooth vector field $T \in C_c^\infty(A;\mathbb{R}^n)$, and noticing that, for small values of a real parameter t, the maps

$$f_t : \mathbb{R}^n \to \mathbb{R}^n\,, \qquad f_t(x) = x + t\,T(x)\,, \quad x \in \mathbb{R}^n\,,$$

define a one-parameter family of diffeomorphisms of $\mathbb{R}^n$, with $f_0(x) = x$ and $f_t = f_0$ outside the support of T. Therefore we have $E \Delta f_t(E) \subset\subset A$ whenever t is small enough, and we may infer (up to differentiability issues) the necessary conditions to the perimeter minimality of E in A,

$$\frac{d}{dt}\Big|_{t=0} P(f_t(E);A) = 0\,, \tag{17.1}$$

$$\frac{d^2}{dt^2}\Big|_{t=0} P(f_t(E);A) \geq 0\,. \tag{17.2}$$

We thus want to formulate these conditions into explicit, exploitable forms, so as to derive valuable information about perimeter minimizers. This chapter is devoted to the analysis of this problem. In Sections 17.1–17.3 we set up the needed framework to justify the required differentiations. In Section 17.4 we study the basic properties of stationary sets for the perimeter, that are those sets satisfying (17.1). The results of Section 17.3 allow us to show that stationary sets have vanishing mean curvature (in a distributional sense). As a

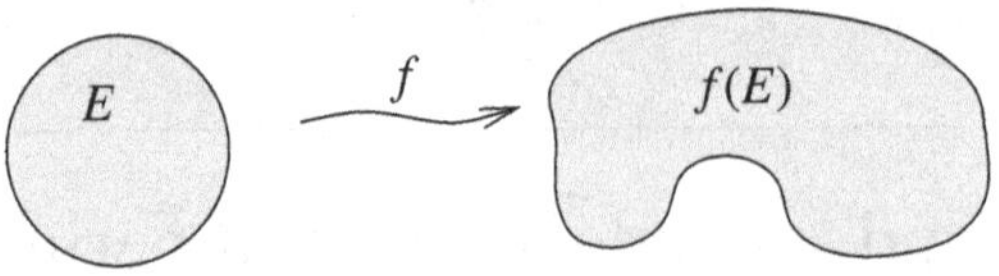

Figure 17.1 Deforming a set of finite perimeter E by a diffeomorphism f. The image of the reduced boundary of E is $\mathcal{H}^{n-1}$-equivalent to the reduced boundary of $f(E)$. Moreover, (17.5) ensures that the outer unit normal to $f(E)$ is obtained from the outer unit normal to E as in the smooth case (17.3).

classical application of the vanishing mean curvature condition, we prove the monotonicity of $(n-1)$-dimensional density ratios of reduced boundaries of stationary sets, which in turn implies a sharp lower density estimate on topological boundaries, as well as the $\mathcal{H}^{n-1}$-equivalence of reduced and topological boundaries. In Section 17.5 we consider volume-constrained minimizers of the perimeter (such as, for example, minimizers of the relative isoperimetric problem (12.30)), and show by a first variation argument that they satisfy (in the distributional sense) a constant mean curvature condition. Finally, in Section 17.6, we present the second order condition (17.2) in the case when E is an open set with C^2-boundary.

17.1 Sets of finite perimeter and diffeomorphisms

We say that $f\colon \mathbb{R}^n \to \mathbb{R}^n$ is a **diffeomorphism of** $\mathbb{R}^n$ if f is smooth, bijective, and has a smooth inverse $g = f^{-1}$. If E is an open set with C^1-boundary, then $f(E)$ is still an open set with C^1-boundary. In the notation of Chapter 9, from $f(\{\psi = 0\}) = \{\psi \circ g = 0\}$ and $\nabla(\psi \circ g) = (\nabla g)^*[(\nabla\psi) \circ g]$, we find

$$\nu_{f(E)}(y) = \frac{\nabla g(y)^* \nu_E(g(y))}{|\nabla g(y)^* \nu_E(g(y))|}, \qquad \forall y \in \partial f(E) = f(\partial E). \tag{17.3}$$

Similar conclusions hold for sets of locally finite perimeter.

Proposition 17.1 (Diffeomorphic images of sets of finite perimeter) *If E is a set of locally finite perimeter in $\mathbb{R}^n$ and f is a diffeomorphism of $\mathbb{R}^n$ with $g = f^{-1}$, then $f(E)$ is a set of locally finite perimeter in $\mathbb{R}^n$ with*

$$\mathcal{H}^{n-1}\big(f(\partial^* E)\Delta\partial^* f(E)\big) = 0, \tag{17.4}$$

$$\int_{\partial^* f(E)} \varphi\, \nu_{f(E)} d\mathcal{H}^{n-1} = \int_{\partial^* E} (\varphi \circ f)\, Jf\, (\nabla g \circ f)^* \nu_E\, d\mathcal{H}^{n-1}, \tag{17.5}$$

for every $\varphi \in C^0_c(\mathbb{R}^n)$. In particular, for every Borel set $F \subset \mathbb{R}^n$,

$$\mathcal{H}^{n-1}\big(F \cap \partial^* f(E)\big) = \int_{g(F)\cap\partial^* E} Jf\, \big|(\nabla g \circ f)^*\, \nu_E\big|\, d\mathcal{H}^{n-1}. \tag{17.6}$$

Remark 17.2 Of course (17.5) may take the more cumbersome form

$$\mu_{f(E)} = f_\#\Big(Jf\,(\nabla g \circ f)^*\mu_E\Big).$$

Proof of Proposition 17.1 We first remark that if $u_h \to u$ in $L^1_{\rm loc}(\mathbb{R}^n)$, then $u_h \circ g \to u \circ g$ in $L^1_{\rm loc}(\mathbb{R}^n)$. Indeed, if K is compact, then $g(K)$ is compact, and thus the area formula implies

$$\int_K |u_h \circ g - u \circ g| = \int_{g(K)} |u_h - u|\,Jf \le \mathrm{Lip}\big(f; g(K)\big)^n \int_{g(K)} |u_h - u|\,.$$

Now define a tensor field $Gf \in C^\infty(\mathbb{R}^n; \mathbb{R}^n \otimes \mathbb{R}^n)$ as

$$Gf = Jf\,(\nabla g \circ f)^*\,, \qquad g = f^{-1}\,.$$

If $u \in C^1(\mathbb{R}^n)$, $T \in C^1_{\rm c}(\mathbb{R}^n; \mathbb{R}^n)$, $v = u \circ g$, and $S = T \circ g$, then by (8.2)

$$\begin{aligned}\int_{\mathbb{R}^n} v \,\mathrm{div}\, S &= -\int_{\mathbb{R}^n} S \cdot \nabla v = -\int_{\mathbb{R}^n} S \cdot \Big((\nabla g)^*(\nabla u \circ g)\Big)\\ &= -\int_{\mathbb{R}^n} T(x) \cdot \Big(Gf(x)\nabla u(x)\Big)\,\mathrm{d}x\,. \end{aligned} \tag{17.7}$$

If $u = (1_E) \star \rho_\varepsilon$, then, by our initial remark, $v = u \circ g \to 1_{f(E)}$ in $L^1_{\rm loc}(\mathbb{R}^n)$ as $\varepsilon \to 0^+$. Hence, by (17.7) and Proposition 12.20,

$$\int_{f(E)} \mathrm{div}\, S = \int_{\partial^* E} T(x) \cdot \Big(Gf(x)\nu_E(x)\Big)\,\mathrm{d}\mathcal{H}^{n-1}(x)\,,$$

so that, by (12.6), $f(E)$ is of locally finite perimeter with

$$\int_{\partial^* f(E)} S \cdot \nu_{f(E)}\,\mathrm{d}\mathcal{H}^{n-1} = \int_{\partial^* E} T(x) \cdot \Big(Gf(x)\nu_E(x)\Big)\,\mathrm{d}\mathcal{H}^{n-1}(x)\,. \tag{17.8}$$

Take $S = (\varphi\, e) \star \rho_\varepsilon$ in (17.8) for $\varphi \in C^0_{\rm c}(\mathbb{R}^n)$ and $e \in S^{n-1}$; letting $\varepsilon \to 0^+$,

$$e \cdot \int_{\partial^* f(E)} \varphi\, \nu_{f(E)}\,\mathrm{d}\mathcal{H}^{n-1} = e \cdot \int_{\partial^* E} (\varphi \circ g)\, Gf \nu_E\,\mathrm{d}\mathcal{H}^{n-1}\,, \qquad \forall e \in S^{n-1}\,,$$

which is (17.5). By (17.5) and an approximation argument,

$$\int_{F \cap \partial^* f(E)} \nu_{f(E)}\mathrm{d}\mathcal{H}^{n-1} = \int_{g(F)\cap\partial^* E} Gf \nu_E\,\mathrm{d}\mathcal{H}^{n-1}\,, \tag{17.9}$$

for every Borel set $F \subset \mathbb{R}^n$. Taking total variations in (17.9), we prove (17.6). We now prove that $f(E^{(1)}) = f(E)^{(1)}$ and $f(E^{(0)}) = f(E)^{(0)}$, so as to deduce $\partial^{\rm e} f(E) = f(\partial^{\rm e} E)$, and thus, by Federer's theorem, (17.4). As $(\mathbb{R}^n \setminus E)^{(0)} = E^{(1)}$ and f is a bijection, it suffices to show $f(E^{(0)}) = f(E)^{(0)}$. If we set

$$L_x = \mathrm{Lip}(g; B(f(x), 1))\,, \qquad M_x = \mathrm{Lip}(f; B(x, L_x))\,, \qquad x \in \mathbb{R}^n\,,$$

then $g(B(f(x),r)) \subset B(x, L_x r)$ for every $r < 1$ and thus, by the area formula,

$$|f(E) \cap B(f(x),r)| = \int_{E\cap g(B(f(x),r))} Jf \le M_x^n\, |E \cap B(x, L_x r)|\,,$$

for every $x \in \mathbb{R}^n$, $r < 1$; in particular, $x \in E^{(0)}$ implies $f(x) \in f(E)^{(0)}$. □

Exercise 17.3 If $\{f_h\}_{h\in\mathbb{N}}$ is a sequence of diffeomorphisms such that $f_h \to \mathrm{Id}$ uniformly on (compact sets of) $\mathbb{R}^n$, then, for every Lebesgue measurable set $E \subset \mathbb{R}^n$, $f_h(E) \to E$ (resp. $f_h(E) \overset{\mathrm{loc}}{\to} E$) as $h \to \infty$. *Hint:* Use Theorem 4.3 to approximate 1_E and apply the area formula.

17.2 Taylor's expansion of the determinant close to the identity

In the following lemma we provide the second order Taylor's expansion of the determinant close to the identity. We warn the reader that the first order expansion – that can be derived by a consistent simplification of the argument below – is going to be sufficient for our aims until Section 17.6.

Lemma 17.4 *If* $Z \in \mathbb{R}^n \otimes \mathbb{R}^n$, $\mathrm{Id} = \mathrm{Id}_{\mathbb{R}^n}$, *and* $Z^2 = Z \circ Z$, *then*

$$(\mathrm{Id} + tZ)^{-1} = \mathrm{Id} - tZ + t^2 Z^2 + \mathrm{O}(t^3)\,, \tag{17.10}$$

$$\det(\mathrm{Id} + tZ) = 1 + t\,\mathrm{trace}(Z) + \frac{t^2}{2}\Big(\mathrm{trace}(Z)^2 - \mathrm{trace}(Z^2)\Big) + \mathrm{O}(t^3)\,. \tag{17.11}$$

Proof *Step one:* The linear function $C = \mathrm{Id} + tZ$ is invertible for suitably small values of t. Then $\mathrm{Id} = CC^{-1} = C^{-1} + tZC^{-1}$, that is $C^{-1} = \mathrm{Id} - tZC^{-1}$. Hence (17.10) follows by repeatedly applying this identity,

$$C^{-1} = \mathrm{Id} - tZC^{-1} = \mathrm{Id} - tZ + t^2Z^2C^{-1} = \mathrm{Id} - tZ + t^2Z^2 - t^3Z^3C^{-1}\,.$$

Step two: We now set $f_Z(t) = \det(\mathrm{Id} + tZ)$ and prove (17.11) in the case that Z is a symmetric tensor. Indeed, in this case we have that $Z = \sum_{i=1}^n \lambda_i v_i \otimes v_i$ for an orthonormal basis $\{v_i\}_{i=1}^n$ of $\mathbb{R}^n$ and for $\lambda_i \in \mathbb{R}$. Thus,

$$f_Z(t) = \prod_{i=1}^n (1 + t\lambda_i) = 1 + t\sum_{i=1}^n \lambda_i + t^2 \sum_{1\le i,j\le n} \lambda_i\lambda_j + \mathrm{O}(t^3)\,.$$

This is (17.11), since $\mathrm{trace}(Z) = \sum_{i=1}^n \lambda_i$, $Z^2 = \sum_{i=1}^n \lambda_i^2 v_i \otimes v_i$, and thus

$$\mathrm{trace}(Z)^2 - \mathrm{trace}(Z^2) = \Big(\sum_{i=1}^n \lambda_i\Big)^2 - \sum_{i=1}^n \lambda_i^2 = 2\sum_{1\le i,j\le n} \lambda_i\lambda_j\,.$$

Step three: We now decompose $Z = X + Y$, where $X = (Z + Z^*)/2$ is symmetric, and $Y = (Z - Z^*)/2$ is antisymmetric. Let us define $\Phi \colon \mathbb{R}^2 \to \mathbb{R}$ as

$$\Phi(s,t) = \det\left(\mathrm{Id} + t\,X + s\,Y\right), \qquad s,t \in \mathbb{R}.$$

We are now going to show that

$$\frac{\partial \Phi}{\partial t}(0,0) = \mathrm{trace}(Z), \qquad \frac{\partial^2 \Phi}{\partial t^2}(0,0) = \mathrm{trace}(Z)^2 - \mathrm{trace}(X^2), \quad (17.12)$$

$$\frac{\partial \Phi}{\partial s}(0,0) = \frac{\partial^2 \Phi}{\partial s \partial t}(0,0) = 0, \qquad \frac{\partial^2 \Phi}{\partial s^2}(0,0) = -\mathrm{trace}(Y^2). \quad (17.13)$$

Clearly, (17.12) follows by step two and $\mathrm{trace}(Z) = \mathrm{trace}(X)$. We now prove (17.13). To this end we note that $S_t = \mathrm{Id} + t\,X$ is a symmetric tensor, hence

$$\begin{aligned}\Phi(s,t) &= \det(S_t + s\,Y) = \det\left((S_t + s\,Y)^*\right) = \det\left(S_t + s\,(Y)^*\right)\\ &= \det(S_t - s\,Y) = \Phi(-s,t),\end{aligned}$$

and, in particular,

$$\frac{\partial \Phi}{\partial s}(0,t) = 0, \qquad \frac{\partial^2 \Phi}{\partial s \partial t}(0,0) = 0,$$

$$\Phi(s,0)^2 = \Phi(s,0)\Phi(-s,0) = \det\left((\mathrm{Id} + s\,Y)(\mathrm{Id} - s\,Y)\right) = \det(\mathrm{Id} - s^2 Y^2).$$

Since Y^2 is a symmetric linear function, by step two we have that

$$\Phi(s,0)^2 = 1 - s^2 \mathrm{trace}(Y^2) + \mathrm{O}(s^4).$$

Differentiating this formula twice at $s = 0$, and recalling that $\Phi(0,0) = 1$ and $(\partial\Phi/\partial s)(0,0) = 0$, we find

$$\frac{\partial^2 \Phi}{\partial s^2}(0,0) = -\mathrm{trace}(Y^2),$$

thus proving (17.13). Finally, $f_Z(t) = \Phi(t,t)$, so that, by (17.12) and (17.13),

$$f_Z(t) = 1 + t\,\mathrm{trace}(Z) + \frac{t^2}{2}\left(\mathrm{trace}(Z)^2 - \mathrm{trace}(X^2) - \mathrm{trace}(Y^2)\right) + \mathrm{O}(t^3),$$

that implies (17.11), since

$$\mathrm{trace}(Z^2) = \mathrm{trace}(X^2 + Y^2 + XY + YX) = \mathrm{trace}(X^2) + \mathrm{trace}(Y^2);$$

indeed, $XY + YX$ is an antisymmetric tensor,

$$(XY + YX)^* = (Y^* X^* + X^* Y^*) = -YX - XY = -(XY + YX). \qquad \square$$

17.3 First variation of perimeter and mean curvature

A **one parameter family of diffeomorphisms of** $\mathbb{R}^n$ is a smooth function

$$(x,t) \in \mathbb{R}^n \times (-\varepsilon, \varepsilon) \mapsto f(t,x) = f_t(x) \in \mathbb{R}^n \,, \qquad \varepsilon > 0 \,,$$

such that, for each fixed $|t| < \varepsilon$, $f_t : \mathbb{R}^n \to \mathbb{R}^n$ is a diffeomorphism of $\mathbb{R}^n$. Given an open set A in $\mathbb{R}^n$, we say that $\{f_t\}_{|t|<\varepsilon}$ is a **local variation in** A if it defines a one-parameter family of diffeomorphisms such that

$$f_0(x) = x \,, \qquad \forall x \in \mathbb{R}^n \,, \tag{17.14}$$

$$\Big\{x \in \mathbb{R}^n : f_t(x) \neq x\Big\} \subset\subset A \,, \qquad \forall |t| < \varepsilon \,. \tag{17.15}$$

It is easily seen that, if $\{f_t\}_{|t|<\varepsilon}$ is a local variation in A, then

$$f_t(E)\Delta E \subset\subset A \,, \qquad \forall E \subset \mathbb{R}^n \,, \tag{17.16}$$

and the following Taylor's expansions holds uniformly on $\mathbb{R}^n$,

$$f_t(x) = x + t\,T(x) + \mathrm{O}(t^2) \,, \qquad \nabla f_t(x) = \mathrm{Id} + t\,\nabla T(x) + \mathrm{O}(t^2) \,, \tag{17.17}$$

where $T \in C^\infty_c(A;\mathbb{R}^n)$ is the **initial velocity of** $\{f_t\}_{|t|<\varepsilon}$,

$$T(x) = \frac{\partial f_t}{\partial t}(x,0) \,, \qquad x \in \mathbb{R}^n \,. \tag{17.18}$$

Conversely, starting from $T \in C^\infty_c(A;\mathbb{R}^n)$, there are two general ways to construct a local variation $\{f_t\}_{|t|<\varepsilon}$ in A having T as its initial velocity. The first method, completely elementary, consists of setting

$$f_t(x) = x + t\,T(x) \,, \qquad x \in \mathbb{R}^n \,. \tag{17.19}$$

The second method relies on standard ODE theory, and consists of solving the Cauchy problems (parametrized with respect to the initial condition $x \in \mathbb{R}^n$)

$$\frac{\partial}{\partial t} f(t,x) = T(f(t,x)) \,, \qquad x \in \mathbb{R}^n \,, \tag{17.20}$$

$$f(0,x) = x \,, \qquad x \in \mathbb{R}^n \,, \tag{17.21}$$

for small values of t. In both cases, we say that $\{f_t\}_{|t|<\varepsilon}$ is a **local variation associated with** T. We now compute the **first variation of perimeter** (relative to A) with respect to local variations $\{f_t\}_{|t|<\varepsilon}$ in A, that is, we aim to compute

$$\frac{\mathrm{d}}{\mathrm{d}t}\Big|_{t=0} P(f_t(E);A) \,, \qquad \text{for } T \in C^\infty_c(A;\mathbb{R}^n) \text{ given.}$$

Theorem 17.5 (First variation of perimeter) *If A is an open set in $\mathbb{R}^n$, E is a set of locally finite perimeter, and $\{f_t\}_{|t|<\varepsilon}$ is a local variation in A, then*

$$P(f_t(E);A) = P(E;A) + t \int_{\partial^* E} \mathrm{div}_E T \, \mathrm{d}\mathcal{H}^{n-1} + \mathrm{O}(t^2) \,, \tag{17.22}$$

where T is the initial velocity of $\{f_t\}_{|t|<\varepsilon}$ and $\mathrm{div}_E T\colon \partial^* E \to \mathbb{R}$,

$$\mathrm{div}_E T(x) = \mathrm{div}\, T(x) - \nu_E(x)\cdot \nabla T(x)\nu_E(x)\,, \qquad x\in\partial^* E\,, \tag{17.23}$$

is a Borel function called the **boundary divergence of T on E.**

Remark 17.6 (Mean curvature vector and perimeter) The results from Section 11.3 provide an important geometric insight into the first variation formula of perimeter (17.22). Indeed, by Theorem 11.8, if E is an open set with C^2-boundary, then by applying Theorem 11.8 to $M = \partial E$ we find that

$$\int_{\partial E} \mathrm{div}^{\partial E}\, T\, \mathrm{d}\mathcal{H}^{n-1} = \int_{\partial E} T\cdot \mathbf{H}_{\partial E}\, \mathrm{d}\mathcal{H}^{n-1}\,, \qquad \forall T\in C^1_c(\mathbb{R}^n;\mathbb{R}^n)\,, \tag{17.24}$$

where $\mathrm{div}^{\partial E}$ denotes the tangential divergence of T with respect to ∂E, and where $\mathbf{H}_{\partial E}$ is the mean curvature vector to ∂E. By Remark 11.10, ν_E induces a continuous determination of the scalar mean curvature $H_{\partial E}$ of ∂E such that $\mathbf{H}_{\partial E} = H_{\partial E}\nu_E$. By Remark 11.12, $\mathrm{div}_E\, T = \mathrm{div}^{\partial E}\, T$ on ∂E; similarly, we shall directly set $\mathbf{H}_E = \mathbf{H}_{\partial E}$ and $H_E = H_{\partial E}$. With these conventions, the first variation of perimeter on open sets with C^2-boundary takes the form

$$\frac{\mathrm{d}}{\mathrm{d}t}\Big|_{t=0} P(f_t(E);A) = \int_{\partial E} (T\cdot\nu_E) H_E\, \mathrm{d}\mathcal{H}^{n-1}\,. \tag{17.25}$$

Remark 17.7 If E is of locally finite perimeter, then the **(distributional) mean curvature vector** of E in A open is the functional $\mathbf{H}_E\colon C^\infty_c(A;\mathbb{R}^n)\to\mathbb{R}$, defined by the formula

$$\langle \mathbf{H}_E, T\rangle = \int_{\partial^* E} \mathrm{div}_E T\, \mathrm{d}\mathcal{H}^{n-1}\,, \qquad T\in C^\infty_c(A;\mathbb{R}^n)\,. \tag{17.26}$$

Note that $\mathbf{H}_E = \mathbf{H}_{\mathbb{R}^n\setminus E}$. By Remark 17.6, if $A\cap\partial E$ is a C^2-hypersurface in $\mathbb{R}^n$, then $\mathbf{H}_E$ defines a signed Radon measure on A, with

$$\mathbf{H}_E = H_E\,\nu_E\,\mathcal{H}^{n-1} \llcorner \big(A\cap\partial E\big)\,.$$

We say that E has **(locally summable) distributional (scalar) mean curvature in** A, if there exists $H\in L^1_{\mathrm{loc}}(A\cap\partial^* E;\mathcal{H}^{n-1})$ such that

$$\int_{\partial^* E} \mathrm{div}_E T\, \mathrm{d}\mathcal{H}^{n-1} = \int_{\partial^* E} (T\cdot\nu_E)\, H\, \mathrm{d}\mathcal{H}^{n-1}\,, \qquad \forall T\in C^\infty_c(A;\mathbb{R}^n)\,. \tag{17.27}$$

In this case, H is uniquely defined $\mathcal{H}^{n-1}$-a.e. on $A\cap\partial^* E$, and we set $H_E = H$, so that $\mathbf{H}_E$ turns out to be a signed Radon measure on A, with

$$\mathbf{H}_E = H_E\,\nu_E\,\mathcal{H}^{n-1} \llcorner \big(A\cap\partial^* E\big)\,.$$

Note that $H_E = -H_{\mathbb{R}^n \setminus E}$. Sets of locally finite perimeter may not have a locally summable distributional mean curvature. For example, if E is an angle in $\mathbb{R}^2$ with $\partial^* E = \bigcup_{i=1,2}\{t\,v_i : t > 0\}$ ($v_i \in S^1$), then

$$\mathbf{H}_E = -(v_1 + v_2)\,\delta_0\,, \tag{17.28}$$

that is, $\mathbf{H}_E$ is a signed Radon measure singular to $\mathcal{H}^{n-1} \llcorner \partial^* E$. By looking at countable disjoint unions of suitably chosen disks we may even construct a set of finite perimeter E such that $\mathbf{H}_E$ is not a signed Radon measure in $\mathbb{R}^2$.

Proof of Theorem 17.5 By Proposition 17.1, in particular by (17.6),

$$P(f_t(E); A) = \int_{A \cap \partial^* E} Jf_t \left|(\nabla g_t \circ f_t)^* \, \nu_E\right| d\mathcal{H}^{n-1}\,, \qquad g_t = (f_t)^{-1}\,,$$

so that $P(f_t(E); A)$ is a smooth function of t in a neighborhood of $t = 0$. Since $\nabla f_t = \mathrm{Id} + t\nabla T + \mathrm{O}(t^2)$, by Lemma 17.4 we have

$$\nabla g_t \circ f_t = (\nabla f_t)^{-1} = \mathrm{Id} - t\,\nabla T + \mathrm{O}(t^2)\,, \tag{17.29}$$

$$Jf_t = 1 + t\,\mathrm{div}\,T + \mathrm{O}(t^2)\,, \tag{17.30}$$

uniformly on $\mathbb{R}^n$ as $t \to 0$. In particular,

$$\begin{aligned} |(\nabla g_t \circ f_t)^* \nu_E|^2 &= |\nu_E - t\,(\nabla T)^* \nu_E|^2 + \mathrm{O}(t^2) = 1 - 2\,t\nu_E \cdot \Big((\nabla T)^* \nu_E\Big) + \mathrm{O}(t^2) \\ &= 1 - 2\,t\,\nu_E \cdot (\nabla T\,\nu_E) + \mathrm{O}(t^2)\,, \end{aligned}$$

and thus we conclude, as required, that

$$Jf_t \left|(\nabla g_t \circ f_t)^* \nu_E\right| = 1 + t\Big(\mathrm{div}\,T - \nu_E \cdot (\nabla T \nu_E)\Big) + \mathrm{O}(t^2)\,. \qquad \square$$

We now compute the first variation of a potential energy $\mathcal{G}(E) = \int_E g(x)\mathrm{d}x$.

Proposition 17.8 (First variation of potential energy) *If E is a set of locally finite perimeter in $\mathbb{R}^n$, $|E| < \infty$, $g \in C^0(\mathbb{R}^n)$, A is open, and $\{f_t\}_{|t|<\varepsilon}$ is a local variation in A with initial velocity T, then*

$$\int_{f_t(E)} g = \int_E g + t \int_{\partial^* E} g\,(T \cdot \nu_E) \mathrm{d}\mathcal{H}^{n-1} + \mathrm{o}(t)\,.$$

In fact, in the case of Lebesgue measure we find that

$$|f_t(E)| = |E| + t \int_{\partial^* E} (T \cdot \nu_E) \mathrm{d}\mathcal{H}^{n-1} + \mathrm{O}(t^2)\,. \tag{17.31}$$

Proof If $g \in C^1(\mathbb{R}^n)$, then by the area formula and Lemma 17.4,

$$\begin{aligned}
\mathcal{G}(f_t(E)) - \mathcal{G}(E) &= \int_E g\big(x + t\,T(x) + \mathrm{O}(t^2)\big)\,Jf_t(x) - g(x)\,\mathrm{d}x \\
&= \int_E (g + t\,\nabla g \cdot T)(1 + t\,\mathrm{div}\,T) - g + \mathrm{O}(t^2) \\
&= t\int_E \mathrm{div}\,(g\,T) + \mathrm{O}(t^2) = t\int_{\partial^* E} g\,(T \cdot \nu_E)\,d\mathcal{H}^{n-1} + \mathrm{O}(t^2),
\end{aligned}$$

which gives the Taylor's expansion of Lebesgue measure (17.31). Now let $g \in C^0(\mathbb{R}^n)$, $\tilde{g} \in C^1(\mathbb{R}^n)$, and $\sigma = \sup_{\mathbb{R}^n} |g - \tilde{g}| > 0$. Then,

$$\left|\frac{\mathcal{G}(f_t(E)) - \mathcal{G}(E)}{t} - \frac{1}{t}\Big(\int_{f_t(E)} \tilde{g} - \int_E \tilde{g}\Big)\right| \le \frac{|f_t(E)\Delta E|}{|t|}\,\sigma\,.$$

By Lemma 17.9 below, there exist positive constants C and $\varepsilon_0 < \varepsilon$ such that $|f_t(E)\Delta E| \le C\,|t|$ whenever $|t| < \varepsilon_0$. Since $\tilde{g} \in C^1(\mathbb{R}^n)$,

$$\begin{aligned}
\int_{\partial^* E} \tilde{g}\,(T \cdot \nu_E)\,\mathrm{d}\mathcal{H}^{n-1} - C\,\sigma &\le \liminf_{t\to 0} \frac{\mathcal{G}(f_t(E)) - \mathcal{G}(E)}{t} \\
&\le \limsup_{t\to 0} \frac{\mathcal{G}(f_t(E)) - \mathcal{G}(E)}{t} \le \int_{\partial^* E} \tilde{g}\,(T \cdot \nu_E)\,d\mathcal{H}^{n-1} + C\,\sigma\,.
\end{aligned}$$

Since $\mathrm{spt}\,T$ is compact, by dominated convergence, $\int_{\partial^* E} \tilde{g}\,(T \cdot \nu_E)\,\mathrm{d}\mathcal{H}^{n-1} \to \int_{\partial^* E} g\,(T \cdot \nu_E)\,\mathrm{d}\mathcal{H}^{n-1}$ as $\sigma \to 0^+$. □

Lemma 17.9 *If E is a set of locally finite perimeter in $\mathbb{R}^n$, A is open, and $\{f_t\}_{|t|<\varepsilon}$ is a local variation in A, then there exist positive constants C and $\varepsilon_0 < \varepsilon$ such that, if K is a compact set with $\{x \ne f_t(x)\} \subset K \subset A$, then*

$$|f_t(E)\Delta E| \le C\,|t|\,P(E;K)\,. \tag{17.32}$$

Proof Set $g_t = f_t^{-1}$, and let $S \in C_c^\infty(A;\mathbb{R}^n)$ denote the initial velocity of the local variation $\{g_t\}_{|t|<\varepsilon}$. If $\Phi_{s,t}(x) = s\,x + (1 - s)\,g_t(x)$ for $x \in \mathbb{R}^n$, $s \in (0,1)$, $|t| < \varepsilon$, then positive constants C and $\varepsilon_0 < \varepsilon$ exist such that, if $s \in (0,1)$, then

$$J\Phi_{s,t}(x) \ge \frac{1}{2}\,, \qquad |x - g_t(x)| \le C\,|t|\,, \qquad \forall x \in \mathbb{R}^n\,, |t| < \varepsilon_0\,, \tag{17.33}$$

and $\{\Phi_{s,t}\}_{|t|<\varepsilon_0}$ is a local variation in A. By the fundamental theorem of Calculus, Fubini's theorem, and the area formula, we thus find

$$\begin{aligned}
\int_{\mathbb{R}^n} |u(x) - u(g_t(x))|\,\mathrm{d}x &\le C|t|\int_K \mathrm{d}x \int_0^1 |\nabla u(\Phi_{s,t}(x))|\,\mathrm{d}s \\
&= C|t|\int_0^1 \mathrm{d}s \int_K \frac{|\nabla u(y)|}{J\Phi_{s,t}(\Phi_{s,t}^{-1}(y))}\,\mathrm{d}y \le 2C|t|\int_K |\nabla u|\,.
\end{aligned}$$

We now set $u = u_\varepsilon = 1_E \star \rho_\varepsilon$, and let $\varepsilon \to 0^+$ to deduce (17.32) by dominated convergence and Propositions 12.20 and 4.26. □

Exercise 17.10 (Prescribed mean curvature problem) If E is a minimizer in the variational problem (12.32) with $g \in C^0(A)$, then E has distributional mean curvature in A equal to $-g$.

Remark 17.11 If $g \in L^1(A)$ and E is a minimizer in the variational problem (12.32), then $-g$ is called a **variational mean curvature of** E **in** A. If g is continuous, then, by Exercise 17.10, $-g$ is both the variational and the distributional mean curvature of E in A. By the Direct Method, if A is bounded and $g \in L^1(A)$, then there exists a set of finite perimeter which has $-g$ as a variational mean curvature. Conversely, in [BGT87] it is shown that if E is of finite perimeter, then there exists $g \in L^1(\mathbb{R}^n)$ such that $-g$ is a variational mean curvature of E; see also [GM94]. Finally, we refer to [BGM03] for further results on the relationship between variational and distributional mean curvatures.

Exercise 17.12 If $n \ge 2$, $p > (n-1)/n$, A is an open bounded set, and E is a p-Cheeger set of A (see Exercise 14.7), then

$$\int_{\partial^* E} \mathrm{div}_E\, T \, d\mathcal{H}^{n-1} = -\frac{p\,P(E)}{|E|} \int_{\partial^* E} (T \cdot \nu_E)\, d\mathcal{H}^{n-1}\,, \qquad \forall T \in C^\infty_c(A;\mathbb{R}^n)\,,$$

that is, H_E is constantly equal to $-p\,P(E)/|E|$ on $A \cap \partial^* E$.

Exercise 17.13 If M is a locally $\mathcal{H}^{n-1}$-rectifiable set in $\mathbb{R}^n$ and $\{f_t\}_{|t|<\varepsilon}$ is a local variation with initial velocity $T \in C^\infty_c(A;\mathbb{R}^n)$, then

$$\frac{d}{dt}\Big|_{t=0} \mathcal{H}^{n-1}(f_t(M)) = \int_M \mathrm{div}^M T \, d\mathcal{H}^{n-1}\,.$$

17.4 Stationary sets and monotonicity of density ratios

We say that a set of locally finite perimeter E is **stationary for perimeter** in an open set A if $\mathrm{spt}\,\mu_E = \partial E$ (see Remark 16.11), and

$$\frac{d}{dt}\Big|_{t=0} P(f_t(E);A) = 0\,, \tag{17.34}$$

whenever $\{f_t\}_{|t|<\varepsilon}$ is a local variation in A. Clearly, a perimeter minimizer in A is stationary for perimeter in A, although the converse is not always true; see Figure 17.2. By the results in the previous section, stationary sets have vanishing distributional mean curvature.

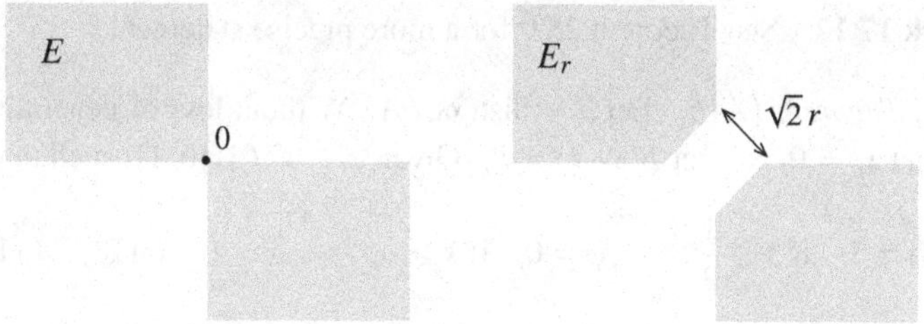

Figure 17.2 A stationary set for perimeter which is not a local perimeter minimizer in $\mathbb{R}^2$. Note that $0 \in E^{(1/2)} \setminus \partial^* E$, and that $\theta_1(\partial E)(0) = 2$, so that (17.42) can hold with strict sign outside $\partial^* E$.

Corollary 17.14 (Vanishing mean curvature) *A set of locally finite perimeter E is stationary for perimeter in the open set A if and only if*

$$\int_{\partial^* E} \mathrm{div}_E T \, d\mathcal{H}^{n-1} = 0 , \qquad \forall T \in C^1_c(A; \mathbb{R}^n) . \tag{17.35}$$

In particular, E has vanishing distributional mean curvature in A.

Proof Immediate from Theorem 17.5. □

Remark 17.15 Exploiting (17.35) and (17.28), it is easily seen that the cone $E = \{x \in \mathbb{R}^2 : x_1 x_2 > 0\}$ is stationary for perimeter in $\mathbb{R}^2$. Notice that E is not a local perimeter minimizer in $\mathbb{R}^2$; see Figure 17.2. Indeed, the sets $E_r = \{x \in E : |x_1 - x_2| > r\}$ $(r > 0)$ satisfy $E \Delta E_r \subset B_r$ and $P(E_r) < P(E)$, so that E admits variations in balls of arbitrarily small size with smaller perimeter. Note that none of the E_r can be obtained as a diffeomorphic image of E.

The vanishing mean curvature condition (17.35) is exploited by testing it with vector fields T which try to capture the action of a simple, fundamental geometric deformation of $\partial^* E$. We shall see many examples of this procedure, especially in Part III. We now use this technique to prove the monotonicity of $(n-1)$-dimensional density ratios of reduced boundaries of stationary sets, that we shall use in Corollary 17.18 to prove the identity $\mathcal{H}^{n-1}(A \cap (\partial E \setminus \partial^* E)) = 0$, and to show the existence of $(n-1)$-dimensional densities everywhere on the topological boundary.

Theorem 17.16 (Monotonicity of density ratios) *If E is stationary for perimeter in the open set A and $x_0 \in A$, then the density ratios*

$$\frac{P(E; B(x_0, r))}{\omega_{n-1} r^{n-1}}$$

are increasing on $r \in (0, \mathrm{dist}(x_0, \partial A))$.

Remark 17.17 See Theorem 28.9 for a more precise statement.

Proof of Theorem 17.16 Let $d = \mathrm{dist}(x_0, \partial A)$. Without loss of generality, assume that $x_0 = 0$, and set $B(x_0, r) = B_r$. Given $\varphi \in C^\infty(\mathbb{R}; [0,1])$ such that

$$\varphi(s) = 1 \quad \text{if } s \le \frac{1}{2}, \qquad \varphi = 0 \quad \text{if } s \ge 1, \qquad \varphi' \le 0 \quad \text{on } \mathbb{R}, \tag{17.36}$$

let us consider the function $\Phi \in C^\infty((0,d))$ defined by setting

$$\Phi(r) = \int_{\partial^* E} \varphi\Big(\frac{|x|}{r}\Big)\, d\mathcal{H}^{n-1}(x), \qquad r \in (0,d). \tag{17.37}$$

We claim that $r^{1-n}\,\Phi(r)$ is increasing on $r \in (0,d)$. Indeed, if we define

$$T_r \in C^1_c(A; \mathbb{R}^n), \qquad T_r(x) = \varphi\Big(\frac{|x|}{r}\Big) x, \qquad x \in \mathbb{R}^n, \tag{17.38}$$

then we can test (17.35) on T_r; taking into account that

$$\begin{aligned}
\nabla T_r &= \varphi\Big(\frac{|x|}{r}\Big)\mathrm{Id} + \frac{|x|}{r}\varphi'\Big(\frac{|x|}{r}\Big)\frac{x}{|x|} \otimes \frac{x}{|x|}, \forall x \in \mathbb{R}^n, \\
\operatorname{div} T_r &= n\,\varphi\Big(\frac{|x|}{r}\Big) + \frac{|x|}{r}\varphi'\Big(\frac{|x|}{r}\Big), && \forall x \in \mathbb{R}^n, \\
\nu_E \cdot \nabla T_r \nu_E &= \varphi\Big(\frac{|x|}{r}\Big) + \frac{|x|}{r}\varphi'\Big(\frac{|x|}{r}\Big)\frac{(x \cdot \nu_E(x))^2}{|x|^2}, && \forall x \in \partial^* E, \\
\operatorname{div}_E T_r &= (n-1)\varphi\Big(\frac{|x|}{r}\Big) + \frac{|x|}{r}\varphi'\Big(\frac{|x|}{r}\Big)\Big(1 - \frac{(x \cdot \nu_E(x))^2}{|x|^2}\Big), && \forall x \in \partial^* E,
\end{aligned}$$

we thus find the identity

$$\begin{aligned}
(n-1)\int_{\partial^* E} \varphi\Big(\frac{|x|}{r}\Big) d\mathcal{H}^{n-1} &+ \int_{\partial^* E} \frac{|x|}{r}\varphi'\Big(\frac{|x|}{r}\Big) d\mathcal{H}^{n-1} \\
&= \int_{\partial^* E} \frac{|x|}{r}\varphi'\Big(\frac{|x|}{r}\Big)\frac{(x \cdot \nu_E(x))^2}{|x|^2}\, d\mathcal{H}^{n-1}.
\end{aligned} \tag{17.39}$$

Since $\varphi' \le 0$ on $\mathbb{R}$, (17.39) implies $(n-1)\Phi(r) - \Phi'(r)/r \le 0$ for $r \in (0,d)$, that is, $(r^{1-n}\Phi(r))' \ge 0$ for $r \in (0,d)$. Thus $r^{1-n}\Phi(r)$ is increasing on $r \in (0,d)$ for every φ as in (17.36). If we now select a sequence $\{\varphi_h\}_{h\in\mathbb{N}} \subset C^\infty(\mathbb{R}; [0,1])$ with each φ_h satisfying (17.36), and such that φ_h monotonically converges to $1_{(-\infty,1)}$ as $h \to \infty$, then by monotone convergence we find

$$P(E; B_r) = \lim_{h\to\infty} \int_{\partial^* E} \varphi_h\Big(\frac{|x|}{r}\Big) d\mathcal{H}^{n-1}(x) = \lim_{h\to\infty} \Phi_h(r).$$

We thus deduce that $r^{1-n} P(E; B_r)$ is increasing on $r \in (0,d)$. □

Corollary 17.18 (Density estimates for stationary sets) *If E is stationary for perimeter in the open set A, then*

$$P(E; B(x,r)) \ge \omega_{n-1} r^{n-1}, \tag{17.40}$$

for every $x \in A \cap \partial E$ and $B(x,r) \subset A$. In particular,

$$\mathcal{H}^{n-1}\big(A \cap (\partial E \setminus \partial^* E)\big) = 0, \tag{17.41}$$

where $\theta_{n-1}(\partial E)$ and $\theta_{n-1}(\partial^* E)$ exist and coincide on $A \cap \partial E$, and

$$\theta_{n-1}(\partial E)(x) \ge 1, \qquad \forall x \in A \cap \partial E, \tag{17.42}$$

$$\theta_{n-1}(\partial E)(x) = 1, \qquad \forall x \in A \cap \partial^* E. \tag{17.43}$$

Proof By Theorem 17.16, we define an increasing function of r by setting

$$\gamma(x,r) = \frac{P(E; B(x,r))}{\omega_{n-1} r^{n-1}}, \qquad x \in A, r < \mathrm{dist}(x, \partial A).$$

Since, by (15.9), $\theta_{n-1}(\partial^* E)(x) = \gamma(x, 0^+) = 1$ if $x \in A \cap \partial^* E$, we thus find

$$\gamma(x,r) \ge 1, \qquad \forall x \in A \cap \partial^* E, B(x,r) \subset A. \tag{17.44}$$

If $x \in A \cap (\partial E \setminus \partial^* E)$, then by (15.3) there exists $\{x_h\}_{h\in\mathbb{N}} \subset A \cap \partial^* E$ such that $x_h \to x$. By (17.44), we thus find

$$P(E; B(x,r)) = \mathcal{H}^{n-1}\big(B(x,r) \cap \partial^* E\big) = \lim_{h\to\infty} \mathcal{H}^{n-1}\big(B(x_h,r) \cap \partial^* E\big) \ge \omega_{n-1} r^{n-1},$$

so that $\gamma(x,r) \ge 1$ for every $x \in A \cap \partial E$, $B(x,r) \subset A$, and (17.40) is proved. If we now set $\mu = \mathcal{H}^{n-1} \llcorner (A \cap \partial^* E)$, then, by (17.40), $\theta^*_{n-1}(\mu) \ge 1$ on $A \cap \partial E$. By Theorem 6.4 we thus find

$$\mathcal{H}^{n-1}\big(A \cap \partial^* E\big) = \mu\big(A \cap \partial E\big) \ge \mathcal{H}^{n-1}\big(A \cap \partial E\big),$$

which immediately implies (17.41). Next, we notice that $\theta_{n-1}(\partial^* E)(x)$ exists at every $x \in A \cap \partial E$, since in this case Theorem 17.16 implies

$$\inf_{r>0} \gamma(x,r) = \lim_{r\to 0^+} \gamma(x,r) = \theta_{n-1}(\partial^* E)(x).$$

By (17.41), $\theta_{n-1}(\partial E)(x) = \theta_{n-1}(\partial^* E)(x)$ at every $x \in A \cap \partial E$, with $\theta_{n-1}(\partial E) \ge 1$ thanks to (17.40). If $x \in A \cap \partial^* E$, then (15.9) gives $\theta_{n-1}(\partial E)(x) = 1$. □

Exercise 17.19 If $E \subset \mathbb{R}^n$ is of locally finite perimeter, A is open, and for every $K \subset\subset A$ there exist positive constants c_K and r_K such that $P(E; B(x,r)) \ge c_K r^{n-1}$ for every $x \in K \cap \partial E$ and $0 < r < \min\{r_K, \mathrm{dist}(x, \partial A)\}$, then

$$\mathcal{H}^{n-1}\big(A \cap (\partial E \setminus \partial^* E)\big) = 0.$$

Hint: Apply Theorem 6.4 with $\mu = \mathcal{H}^{n-1} \llcorner \partial^* E$ and $M = K \cap (\partial E \setminus \partial^* E)$.

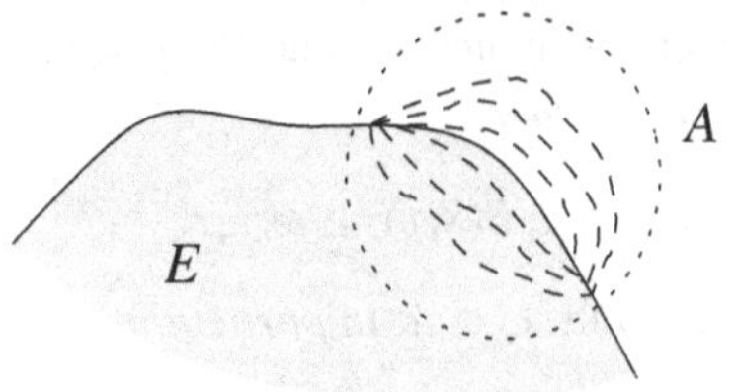

Figure 17.3 The situation in Lemma 17.21. Since $P(E;A) \neq 0$ there exists a vector field $T \in C_c^\infty(A;\mathbb{R}^n)$ that we can use to "move" $A \cap \partial^* E$. The local variations $f_t(x) = x + t\,T(x)$ associated with T allow us to increase or decrease volume by a certain maximal amount σ_0, that depends on E and A through T. The corresponding perimeter variations are proportional to the volume variations, through a constant C that, again, depends on E and A through T.

17.5 Volume-constrained perimeter minimizers

We now apply first variation arguments to study minimizers in relative isoperimetric problems. Given A open and E of finite perimeter in $\mathbb{R}^n$, we say that E is a **volume-constrained perimeter minimizer in** A, if $\operatorname{spt}\mu_E = \partial E$ and

$$P(E;A) \le P(F;A)\,, \tag{17.45}$$

whenever $|E \cap A| = |F \cap A|$ and $E \Delta F \subset\subset A$. By arguing as in Example 16.13, we see that minimizers in the relative isoperimetric problem (12.30) are volume-constrained perimeter minimizers. We now prove that volume-constrained perimeter minimizers have constant distributional mean curvature.

Theorem 17.20 (Constant mean curvature) *If E is a volume-constrained minimizer in the open set A, then there exists $\lambda \in \mathbb{R}$ such that*

$$\int_{\partial^* E} \operatorname{div}_E T \, d\mathcal{H}^{n-1} = \lambda \int_{\partial^* E} (T \cdot \nu_E)\, d\mathcal{H}^{n-1}\,, \qquad \forall T \in C_c^\infty(A;\mathbb{R}^n)\,. \tag{17.46}$$

In particular, E has distributional mean curvature in A constantly equal to λ.

The following lemma is the key technical tool used to obtain this result. Given a set of finite perimeter E and an open set A with $A \cap \partial^* E \neq \emptyset$, we change the volume of E by a prescribed (suitably small) amount, at the cost of a proportional perimeter variation; see Figure 17.3.

Lemma 17.21 (Volume-fixing variations) *If E is a set of finite perimeter and A is an open set such that $\mathcal{H}^{n-1}(A \cap \partial^* E) > 0$, then there exist $\sigma_0 = \sigma_0(E,A) > 0$ and $C = C(E,A) < \infty$ such that for every $\sigma \in (-\sigma_0, \sigma_0)$ we can find a set of finite perimeter F with $F \Delta E \subset\subset A$ and*

$$|F| = |E| + \sigma, \qquad |P(F;A) - P(E;A)| \le C|\sigma|\,.$$

Proof Since $\mathcal{H}^{n-1}(A \cap \partial^* E) > 0$ there exists $T \in C_c^\infty(A; \mathbb{R}^n)$ such that

$$\gamma = \int_{\partial^* E} (T \cdot \nu_E) \, d\mathcal{H}^{n-1} > 0 .$$

Let $\{f_t\}_{|t|<\varepsilon}$ be a local variation associated with T. Since $\gamma > 0$ and T has compact support, by (17.22) and (17.31) we may find $\varepsilon_0 > 0$ such that $f_t(E)\Delta E \subset\subset A$ and $|f_t(E)| = |E| + t\gamma + O(t^2)$ is increasing on $t \in (-\varepsilon_0, \varepsilon_0)$, with

$$\Big| |f_t(E)| - |E| \Big| \geq \frac{\gamma}{2} |t| , \qquad \forall |t| < \varepsilon_0 ,$$

$$\Big| P(f_t(E); A) - P(E; A) \Big| \leq 2 \left| \int_{\partial^* E} \operatorname{div}_E T \, d\mathcal{H}^{n-1} \right| |t| , \qquad \forall |t| < \varepsilon_0 .$$

If $\sigma_0 > 0$ is such that $(|E| - \sigma_0, |E| + \sigma_0) \subset \{|f_t(E)| : t \in (-\varepsilon_0, \varepsilon_0)\}$, and

$$C = \frac{4}{\gamma} \left| \int_{\partial^* E} \operatorname{div}_E T \, d\mathcal{H}^{n-1} \right| ,$$

then for every $|\sigma| < \sigma_0$ there exists $|t| < t_0$ such that $F = f_t(E)$ has all the required properties. □

Proof of Theorem 17.20 *Step one:* We prove that there exists $r_0 > 0$ such that if $T \in C_c^\infty(A; \mathbb{R}^n)$ with $\operatorname{spt} T \subset\subset B(x, r_0)$ for some $x \in A$, and

$$\int_{\partial^* E} (T \cdot \nu_E) d\mathcal{H}^{n-1} = 0 , \tag{17.47}$$

then

$$\int_{\partial^* E} \operatorname{div}_E T \, d\mathcal{H}^{n-1} = 0 . \tag{17.48}$$

In other words: if T produces a zero first order volume variation of E, then, by volume-constrained minimality, T produces a zero first order perimeter variation of E. Indeed, let us consider $r_0 > 0$ to be such that

$$\Big(A \cap \partial^* E\Big) \setminus B(z, r_0) \neq \emptyset , \qquad \forall z \in A . \tag{17.49}$$

Given $T \in C_c^1(A; \mathbb{R}^n)$ with $\operatorname{spt} T \subset B(x, r_0)$ for some $x \in A$ and with (17.47) in force, by Proposition 2.16 we find $r < r_0$ such that

$$\operatorname{spt} T \subset\subset B(x, r) , \qquad \mathcal{H}^{n-1}\Big(\partial^* E \cap \partial B(x, r)\Big) = 0 , \tag{17.50}$$

By (17.49) and, again, by Proposition 2.16, there exist $y \in A \cap \partial^* E$ and $s > 0$ with $B(y, s) \cap B(x, r) = \emptyset$ and

$$\mathcal{H}^{n-1}\Big(\partial^* E \cap \partial B(y, s)\Big) = 0 . \tag{17.51}$$

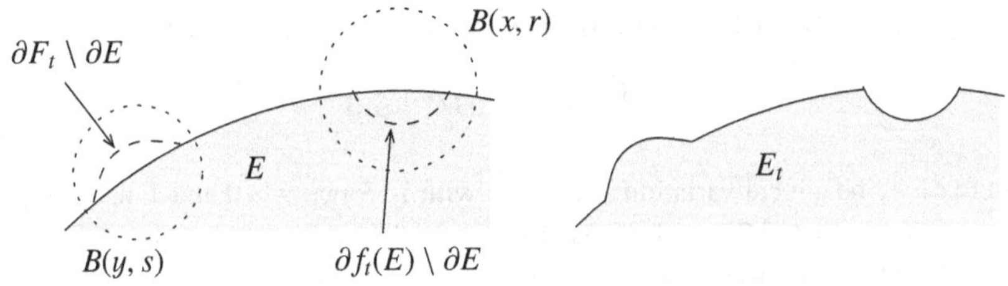

Figure 17.4 A comparison set E_t used in the proof of Theorem 17.20. The set F_t is a variation of E supported in $B(y,s)$ which compensates the volume deficit between $f_t(E)$ and E.

Now let σ_0 and C denote the constants associated by Lemma 17.21 with E in the open set $B(y,s)$, and let $\{f_t\}_{|t|<\varepsilon}$ be a local variation in $B(x,r)$ associated with T. By (17.22), (17.31), and (17.47), we find that

$$|f_t(E)| = |E| + \mathrm{O}(t^2)\,, \tag{17.52}$$

$$P(f_t(E);A) = P(E;A) + t\int_{\partial^* E} \mathrm{div}_E T\,\mathrm{d}\mathcal{H}^{n-1} + \mathrm{O}(t^2)\,. \tag{17.53}$$

If we set $\sigma(t) = |E| - |f_t(E)|$, then, up to decreasing ε, we find $|\sigma(t)| < \sigma_0$ for every $|t| < \varepsilon$. Hence, by Lemma 17.21, for every volume fraction $\sigma(t)$ corresponding to $|t| < \varepsilon$ we can construct F_t with $E\Delta F_t \subset\subset B(y,s)$ and

$$|F_t| - |E| = \sigma(t) = |E| - |f_t(E)|\,, \tag{17.54}$$

$$|P(F_t;B(y,s)) - P(E;B(y,s))| \le C|\sigma(t)| = \mathrm{O}(t^2)\,. \tag{17.55}$$

We finally test the volume-constrained minimality of E against the competitors

$$E_t = \big(f_t(E)\cap B(x,r)\big)\cup\big(F_t\cap B(y,s)\big)\cup\Big(E\setminus\big(B(x,r)\cup B(y,s)\big)\Big),$$

defined for $|t|<\varepsilon$; see Figure 17.4. These sets are indeed competitors for E, as $|E| = |E_t|$ by (17.54):

$$\begin{aligned}|E_t| - |E| &= |E_t\cap B(x,r)| - |E\cap B(x,r)| + |E_t\cap B(y,s)| - |E\cap B(y,s)|\\ &= |f_t(E)\cap B(x,r)| - |E\cap B(x,r)| + |F_t\cap B(y,s)| - |E\cap B(y,s)|\\ &= |f_t(E)| - |E| + |F_t| - |E| = 0\,.\end{aligned}$$

By the volume-constrained minimality of E, (16.32) (which is applied also taking into account (17.50) and (17.51)), (17.53), and (17.55),

$$\begin{aligned}0 &\le P(E_t;A) - P(E;A)\\ &\le P(f_t(E);B(x,r)) - P(E;B(x,r)) + P(F_t;B(y,s)) - P(E;B(y,s))\\ &= t\int_{\partial^* E}\mathrm{div}_E T\,\mathrm{d}\mathcal{H}^{n-1} + \mathrm{O}(t^2)\,,\end{aligned}$$

which, of course, gives (17.48).

Step two: Up to further decreasing the value of r_0, we may assume that

$$\big(A \cap \partial^* E\big) \setminus \big(B(x, r_0) \cup B(y, r_0)\big) \neq \emptyset\,, \qquad \forall x, y \in A\,. \tag{17.56}$$

Now let $T_1, T_2 \in C^\infty_c(A; \mathbb{R}^n)$ such that, for $h = 1, 2$,

$$\operatorname{spt} T_h \subset\subset B(x_h, r_0)\,, \qquad \int_{\partial^* E} (T_h \cdot \nu_E) d\mathcal{H}^{n-1} \neq 0\,.$$

By Proposition 2.16, we may find $r < r_0$ such that, for $h = 1, 2$,

$$\operatorname{spt} T_h \subset\subset B(x_h, r)\,, \qquad \mathcal{H}^{n-1}\Big(\partial^* E \cap \big(B(x_1, r) \cup B(x_2, r)\big)\Big) = 0\,. \tag{17.57}$$

Finally, define $T \in C^\infty_c(A; \mathbb{R}^n)$ by setting

$$T = T_1 - \frac{\int_{\partial^* E} (T_1 \cdot \nu_E) d\mathcal{H}^{n-1}}{\int_{\partial^* E} (T_2 \cdot \nu_E) d\mathcal{H}^{n-1}}\, T_2\,.$$

We have $\int_{\partial^* E} (T \cdot \nu_E) d\mathcal{H}^{n-1} = 0$. Hence, exploiting (17.56) and (17.57), and up to replacing $B(x, r)$ with $B(x_1, r) \cup B(x_2, r)$ everywhere, we may repeat the argument of step one to prove that $\int_{\partial^* E} \operatorname{div}_E T \, d\mathcal{H}^{n-1} = 0$, that is

$$\frac{\int_{\partial^* E} \operatorname{div}_E T_1 d\mathcal{H}^{n-1}}{\int_{\partial^* E} (T_1 \cdot \nu_E) d\mathcal{H}^{n-1}} = \frac{\int_{\partial^* E} \operatorname{div}_E T_2 d\mathcal{H}^{n-1}}{\int_{\partial^* E} (T_2 \cdot \nu_E) d\mathcal{H}^{n-1}}\,.$$

Therefore there exists $\lambda \in \mathbb{R}$ such that (17.46) holds true for every $T \in C^\infty_c(A; \mathbb{R}^n)$ such that $\operatorname{spt} T \subset\subset B(x, r_0)$ for some $x \in A$. Now let T be a generic vector field in $C^\infty_c(A; \mathbb{R}^n)$, and let $\{B(z_k, r_0)\}_{k=1}^N$ be a finite cover of $\operatorname{spt} T$ by open balls centered in A. Using a partition of unity $\{\zeta_k\}_{k=1}^N$ with $\zeta_k \in C^\infty_c(B(z_k, r_0))$ and $\sum_{k=1}^N \zeta_k = 1$ on an open neighborhood of $\operatorname{spt} T$, and exploiting the linearity of the boundary divergence operator, we thus find

$$\begin{aligned}\int_{\partial^* E} \operatorname{div}_E T \, d\mathcal{H}^{n-1} &= \sum_{k=1}^N \int_{\partial^* E} \operatorname{div}_E (\zeta_k T) \, d\mathcal{H}^{n-1} \\ &= \lambda \sum_{k=1}^N \int_{\partial^* E} \zeta_k (T \cdot \nu_E) \, d\mathcal{H}^{n-1} = \lambda \int_{\partial^* E} (T \cdot \nu_E) \, d\mathcal{H}^{n-1}\,. \qquad \square\end{aligned}$$

17.6 Second variation of perimeter

We conclude this chapter by computing the second variation formula for perimeter on open sets with C^2-boundary, a result we shall apply in Section 28.3 in the proof of Simons' theorem. We begin our discussion by recalling some facts concerning the signed distance function from an open set with C^2 boundary; see [Giu84, Appendix A], [AM98]. If E is an open set with C^2-boundary in A, then there exists an open set A' with $A\cap\partial E\subset A'\subset A$ such that the **signed distance function** $s_E\colon\mathbb{R}^n\to\mathbb{R}$ of E,

$$s_E(x)=\begin{cases}\operatorname{dist}(x,\partial E)\,, & x\in\mathbb{R}^n\setminus E\,,\\ -\operatorname{dist}(x,\partial E)\,, & x\in E\,,\end{cases}$$

satisfies $s_E\in C^2(A')$. We may thus define a vector field $N_E\in C^1(A';\mathbb{R}^n)$ and a tensor field $A_E\in C^0(A';\mathbf{Sym}(n))$ by setting

$$N_E=\nabla s_E\,,\qquad A_E=\nabla^2 s_E\,,\qquad \text{on } A'\,.$$

It turns out that N_E is an extension to A' of the outer unit normal ν_E to E, with the property that $|N_E|=1$ on A'. Moreover, for every $x\in A\cap\partial E$ there exist $r>0$, vector fields $\{\tau_h\}_{h=1}^{n-1}\subset C^1(B(x,r);S^{n-1})$, and functions $\{\kappa_h\}_{h=1}^{n-1}\subset C^0(B(x,r))$, such that $\{\tau_h(y)\}_{h=1}^{n-1}$ is an orthonormal basis of $T_y\partial E$ for every $y\in B(x,r)\cap\partial E$, $\{\tau_h(y)\}_{h=1}^{n-1}\cup\{N_E(y)\}$ is an orthonormal basis of $\mathbb{R}^n$ for every $y\in B(x,r)$, and

$$A_E(y)=\sum_{h=1}^{n-1}\kappa_h(y)\,\tau_h(y)\otimes\tau_h(y)\,,\qquad\forall y\in B(x,r)\,.$$

From this information we easily verify that on $B(x,r)$

$$A_E\,N_E=0\,,\qquad N_E\cdot(A_E\,e)=0,\qquad\forall e\in\mathbb{R}^n\,. \tag{17.58}$$

In geometric terms, if $y\in B(x,r)\cap\partial E$, then $A_E(y)$, seen as a symmetric tensor on $T_y\partial E\otimes T_y\partial E$, is the **second fundamental form of** ∂E at y, while $\{\tau_h(y)\}_{h=1}^{N}\subset S^{n-1}\cap T_y\partial E$ and $\{\kappa_h(y)\}_{h=1}^{n-1}$ are, respectively, the **principal directions** and the **principal curvatures** of ∂E at y. In particular, the (scalar) mean curvature H_E of the C^2-hypersurface $A\cap\partial E$ (defined by the orientation induced by ν_E) is locally representable as

$$H_E(y)=\sum_{h=1}^{n-1}\kappa_h(y)\,,\qquad\forall y\in B(x,r)\cap\partial E\,,$$

while the squared norm of the second fundamental form satisfies

$$|A_E(y)|^2=\sum_{h=1}^{n-1}\kappa_h(y)^2\,,\qquad\forall y\in B(x,r)\cap\partial E\,.$$

We finally remark that, from a geometric viewpoint, we may define A_E as a continuous symmetric tensor field on the whole $A\cap\partial E$. We now prove a representation formula for the **second variation of perimeter**.

Theorem 17.22 (Second variation of perimeter) *If E is an open set with C^2-boundary in the open set A, $\zeta \in C^\infty_c(A)$, and $\{f_t\}_{|t|<\varepsilon}$ is a local variation associated with the normal vector field $T = \zeta N_E \in C^1_c(A;\mathbb{R}^n)$, then*

$$\frac{d^2}{dt^2}\Big|_{t=0} P(f_t(E);A) = \int_{\partial E} |\nabla_E \zeta|^2 + \left(H_E^2 - |A_E|^2\right)\zeta^2 \, d\mathcal{H}^{n-1}, \qquad (17.59)$$

where $\nabla_E\zeta = \nabla\zeta - (\nu_E\cdot\nabla\zeta)\nu_E$ denotes the tangential gradient of ζ with respect to the boundary of E. In particular, if E is a perimeter minimizer in A, then

$$\int_{\partial E} |\nabla_E\zeta|^2 - |A_E|^2\zeta^2 \, d\mathcal{H}^{n-1} \ge 0, \qquad (17.60)$$

for every $\zeta \in C^\infty_c(A)$.

Proof By Proposition 17.1, $f_t(E)$ is a set of locally finite perimeter in $\mathbb{R}^n$ and $P(f_t(E);A)$ is a smooth function of t in a neighborhood of $t = 0$, with

$$P(f_t(E);A) = \int_{A\cap\partial E} Jf_t \left|(\nabla g_t \circ f_t)^*\nu_E\right| d\mathcal{H}^{n-1}, \qquad g_t = (f_t)^{-1}. \qquad (17.61)$$

We thus need to compute the second order terms in the Taylor's expansions of Jf_t and $|(\nabla g_t\circ f_t)^*\nu_E|$. To begin with,

$$\nabla T = \zeta A_E + N_E \otimes \nabla\zeta,$$
$$(\nabla T)^2 = \zeta^2 A_E^2 + \zeta(N_E\otimes\nabla\zeta)A_E + (N_E\cdot\nabla\zeta)N_E\otimes\nabla\zeta,$$

where, in computing $(\nabla T)^2$, we have taken into account that, by (17.58),

$$A_E(N_E\otimes\nabla\zeta)e = (e\cdot\nabla\zeta)A_E N_E = 0, \qquad \forall e\in\mathbb{R}^n. \qquad (17.62)$$

As a consequence, we find that

$$(\nabla T)^* = \zeta A_E + \nabla\zeta\otimes N_E, \qquad (17.63)$$
$$((\nabla T)^2)^* = \zeta^2 A_E^2 + \zeta A_E(\nabla\zeta\otimes N_E) + (N_E\cdot\nabla\zeta)\nabla\zeta\otimes N_E, \qquad (17.64)$$
$$\mathrm{trace}(\nabla T) = \zeta H_E + N_E\cdot\nabla\zeta, \qquad (17.65)$$
$$\mathrm{trace}((\nabla T)^2) = \zeta^2|A_E|^2 + (N_E\cdot\nabla\zeta)^2. \qquad (17.66)$$

We notice that, in proving (17.66), we have again used (17.62) to deduce that

$$\mathrm{trace}((N_E\otimes\nabla\zeta)A_E) = \mathrm{trace}(A_E(N_E\otimes\nabla\zeta)) = 0.$$

Finally, we remark that, by (17.58), $A_E^2 N_E = 0$, so that

$$(\nabla T)^* N_E = \nabla\zeta, \qquad ((\nabla T)^2)^* N_E = \zeta A_E\nabla\zeta + (N_E\cdot\nabla\zeta)\nabla\zeta.$$

We now compute the second order Taylor's expansion of the integrand in (17.61). By Lemma 17.4, (17.63), and (17.64), we have that

$$
\begin{aligned}
(\nabla g_t \circ f_t)^* N_E &= ((\nabla f_t)^{-1})^* N_E = ((\mathrm{Id} + t\,\nabla T)^{-1})^* N_E \\
&= N_E - t\,(\nabla T)^* N_E + t^2\,((\nabla T)^2)^* N_E + \mathrm{O}(t^3) \\
&= N_E - t\,\nabla\zeta + t^2\left(\zeta A_E \nabla\zeta + (N_E\cdot\nabla\zeta)\nabla\zeta\right) + \mathrm{O}(t^3). \qquad (17.67)
\end{aligned}
$$

Now, if $v_0, v_1, v_2 \in \mathbb{R}^n$, with $|\,v_0| = 1$, then

$$
|\,v_0 + t v_1 + t^2 v_2|^2 = 1 + 2t(v_0\cdot v_1) + t^2\left(|\,v_1|^2 + 2(v_0\cdot v_2)\right) + \mathrm{O}(t^3). \qquad (17.68)
$$

Combining (17.67) and (17.68) we thus find

$$
|\,(\nabla g_t \circ f_t)^* N_E\,|^2 = 1 - 2t(N_E\cdot\nabla\zeta) + t^2\left(|\nabla\zeta|^2 + 2(N_E\cdot\nabla\zeta)^2\right) + \mathrm{O}(t^3).
$$

Finally, from Taylor's expansions,

$$
\sqrt{1+\varepsilon} = 1 + \frac{\varepsilon}{2} - \frac{\varepsilon^2}{8} + \mathrm{O}(\varepsilon^3), \qquad \sqrt{1 + a\,t + b\,t^2} = 1 + t\,\frac{a}{2} + t^2\left(\frac{b}{2} - \frac{a^2}{8}\right) + \mathrm{O}(t^3),
$$

we conclude that

$$
\left|(\nabla g_t \circ f_t)^* N_E\right| = 1 - t(N_E\cdot\nabla\zeta) + \frac{t^2}{2}\left(|\nabla\zeta|^2 + (N_E\cdot\nabla\zeta)^2\right) + \mathrm{O}(t^3). \qquad (17.69)
$$

Concerning $J f_t$, by Lemma 17.4, (17.65), and (17.66), we find that

$$
\begin{aligned}
J f_t &= 1 + t\,\mathrm{trace}(\nabla T) + \frac{t^2}{2}\left(\mathrm{trace}(\nabla T)^2 - \mathrm{trace}((\nabla T)^2)\right) + \mathrm{O}(t^3) \qquad (17.70) \\
&= 1 + t\left(\zeta H_E + N_E\cdot\nabla\zeta\right) + \frac{t^2}{2}\left(\zeta^2(H_E^2 - |A_E|^2) + 2\zeta H_E(N_E\cdot\nabla\zeta)\right) + \mathrm{O}(t^3).
\end{aligned}
$$

By (17.69) and (17.70), since $|\nabla_E\zeta|^2 = |\nabla\zeta|^2 - (N_E\cdot\nabla\zeta)^2$,

$$
J f_t |\,(\nabla g_t \circ f_t)^* N_E| = 1 + t\,\zeta H_E + \frac{t^2}{2}\left(|\nabla^{\partial E}\zeta|^2 + \zeta^2(H_E^2 - |A_E|^2)\right) + \mathrm{O}(t^3),
$$

and (17.59) is proved. Finally, if E is a perimeter minimizer in A, then by Corollary 17.35 we have $H_M = 0$ on $A\cap\partial E$; moreover, since $f_t(E)\Delta E \subset\subset A$, we have that $\{f_t(E)\}_{|t|<\varepsilon}$ is an admissible family of competitors for the perimeter minimality of E. Hence the second variation of perimeter on E is non-negative along $T = \zeta\,N_E$, and (17.60) is proved. □

18

Slicing boundaries of sets of finite perimeter

In this chapter we consider the problem of computing the perimeter of a set as an integral of the Hausdorff measures of the lower dimensional sections (through a given slicing function) of its boundary. The basic tool here is the extension of the coarea formula to rectifiable sets (Section 18.2), which in turn is based on a revised version of the coarea formula (13.1) (Section 18.1). In Section 18.3 we study slicing by hyperplanes, which will prove to be a useful technical tool in the study of sessile liquid drops (Chapter 19), as well as in the regularity theory of Part III; see, in particular, the proof of Theorem 22.8. In fact, in this last context, we shall also apply slicing by cylinders (proof of Theorem 24.1), and spheres (proof of Theorem 28.4).

18.1 The coarea formula revised

In Section 13.1 we proved the coarea formula

$$\int_A |\nabla u| = \int_{\mathbb{R}} P(\{u > t\}; A)\, dt\,, \tag{18.1}$$

for every Lipschitz function $u\colon \mathbb{R}^n \to \mathbb{R}$, and open set $A \subset \mathbb{R}^n$. By Remark 13.2, if $u \in C^\infty(\mathbb{R}^n)$ and $\{u = t\} = \{x \in \mathbb{R}^n : u(x) = t\}$, then (18.1) implies

$$\int_E |\nabla u| = \int_{\mathbb{R}} \mathcal{H}^{n-1}\big(E \cap \{u = t\}\big)\, dt\,, \tag{18.2}$$

for every Borel set $E \subset \mathbb{R}^n$. We now show that (18.2) holds true even if u is merely a Lipschitz function.

Theorem 18.1 (Coarea formula revised) *If $u\colon \mathbb{R}^n \to \mathbb{R}$ is a Lipschitz function, then, for every Borel set $E \subset \mathbb{R}^n$,*

$$\int_E |\nabla u| = \int_{\mathbb{R}} \mathcal{H}^{n-1}\big(E \cap \{u = t\}\big)\,dt\,. \tag{18.3}$$

Remark 18.2 By Example 13.3, if $u\colon \mathbb{R}^n \to \mathbb{R}$ is a Lipschitz function, then for a.e. $t \in \mathbb{R}$ the super-level set $\{u > t\}$ is of locally finite perimeter in $\mathbb{R}^n$. Hence, by (18.1) and De Giorgi's structure theorem, for every $A \subset \mathbb{R}^n$ open,

$$\int_A |\nabla u| = \int_{\mathbb{R}} \mathcal{H}^{n-1}\big(A \cap \partial^*\{u > t\}\big)\,dt\,. \tag{18.4}$$

Since the right-hand side of (18.4) defines a Radon measure on $\mathbb{R}^n$, it is immediate to extend (18.4) by replacing the open set A with a generic Borel set E. Clearly, (18.3) is stronger than (18.4), since $\{u = t\}$ is the topological boundary of $\{u > t\}$, and there is no immediate reason to assert the $\mathcal{H}^{n-1}$-equivalence of $\partial\{u > t\}$ and $\partial^*\{u > t\}$. Indeed, the $\mathcal{H}^{n-1}$-equivalence of topological and reduced boundaries of a.e. super-level set of a Lipschitz function is an interesting corollary of Theorem 18.1.

Exercise 18.3 If E is of finite perimeter and $u\colon \mathbb{R}^n \to \mathbb{R}$ is a Lipschitz function, then for a.e. $t \in \mathbb{R}$,

$$P(E \cap \{u > t\}) = P(E; \{u > t\}) + \mathcal{H}^{n-1}\big(E \cap \{u = t\}\big)\,.$$

Hint: Use (16.10), Proposition 2.16, and the $\mathcal{H}^{n-1}$-equivalence of $E \cap \{u = t\}$ and $E^{(1)} \cap \{u = t\}$ for a.e. $t \in \mathbb{R}$.

We preface three lemmas to the proof of Theorem 18.1. We start by showing the Lebesgue measurability of the slice function $M_E\colon \mathbb{R} \to [0, \infty]$,

$$M_E(t) = \mathcal{H}^{n-1}\big(E \cap \{u = t\}\big)\,, \qquad t \in \mathbb{R}\,,$$

associated by u with any Lebesgue measurable set $E \subset \mathbb{R}^n$. Moreover, we prove that the right-hand side of (18.3) is left unchanged if E is modified on a set of Lebesgue measure zero (note that the left-hand side trivially has this property).

Lemma 18.4 (Measurability of the slice function and sets of measure zero) *If $u\colon \mathbb{R}^n \to \mathbb{R}$ is a Lipschitz function and E is a Lebesgue measurable set in $\mathbb{R}^n$, then the slice function M_E is Lebesgue measurable on $\mathbb{R}$, with*

$$\int_{\mathbb{R}} \mathcal{H}^{n-1}\big(E \cap \{u = t\}\big)\,dt \le C(n)\,\mathrm{Lip}(u)\,|E|\,. \tag{18.5}$$

Proof For every $\delta \in (0, \infty]$, we define $M_{E,\delta}\colon \mathbb{R} \to [0, \infty]$ as

$$M_{E,\delta}(t) = \mathcal{H}^{n-1}_\delta\big(E \cap \{u = t\}\big)\,, \qquad t \in \mathbb{R}\,.$$

Step one: We prove that, if $M_{E,\delta}$ is Lebesgue measurable on $\mathbb{R}$, then

$$\int_{\mathbb{R}} \mathcal{H}^{n-1}_\delta\big(E \cap \{u = t\}\big)\,dt \le C(n)\,\mathrm{Lip}(u)\,|E|\,. \tag{18.6}$$

Indeed, if $\mathcal{F}$ is a covering of E by balls of diameter at most δ and $t \in \mathbb{R}$, then

$$\mathcal{H}^{n-1}_\delta\big(E \cap \{u = t\}\big) \le \omega_{n-1} \sum_{F \in \mathcal{F}} \left(\frac{\mathrm{diam}(F)}{2}\right)^{n-1} 1_{u(F)}(t)\,.$$

As $u(F) \subset \mathbb{R}$, we have $\mathcal{L}^1(u(F)) \le \mathrm{diam}(u(F)) \le \mathrm{Lip}(u)\,\mathrm{diam}(F)$, and thus

$$\begin{aligned}\int_{\mathbb{R}} \mathcal{H}^{n-1}_\delta\big(E \cap \{u = t\}\big)\,dt &\le \omega_{n-1} \sum_{F \in \mathcal{F}} \left(\frac{\mathrm{diam}(F)}{2}\right)^{n-1} \mathcal{L}^1(u(F))\\ &\le C(n)\,\mathrm{Lip}(u) \sum_{F \in \mathcal{F}} \left(\frac{\mathrm{diam}(F)}{2}\right)^{n}.\end{aligned}$$

Minimizing in $\mathcal{F}$, the right-hand side becomes $C(n)\,\mathcal{H}^n_\delta(E)$, and $\mathcal{H}^n_\delta(E) = |E|$.

Step two: We remark on the following property of Hausdorff measures of step δ. Let $\{K_h\}_{h\in\mathbb{N}}$ be a sequence of compact sets, all contained in some fixed ball of $\mathbb{R}^n$, and let K be a compact set containing the limit points of sequences $\{x_{h(k)}\}_{k\in\mathbb{N}}$ such that $x_{h(k)} \in K_{h(k)}$ and $h(k) \to \infty$. Then

$$\limsup_{h\to\infty} \mathcal{H}^s_\delta(K_h) \le \mathcal{H}^s_\delta(K)\,,$$

for every $s \in [0, \infty)$ and $\delta \in (0, \infty]$. Indeed given $\varepsilon > 0$, then by compactness K admits a (finite) cover $\mathcal{F}$ by open sets with diameter less than δ, such that

$$\varepsilon + \mathcal{H}^s_\delta(K) \ge \omega_s \sum_{F \in \mathcal{F}} \left(\frac{\mathrm{diam}(F)}{2}\right)^{s}.$$

By the convergence assumption of the K_h to K, it is easily seen that $\mathcal{F}$ is a cover of K_h for h large enough. Correspondingly, $\varepsilon + \mathcal{H}^n_\delta(K) \ge \mathcal{H}^s_\delta(K_h)$.

Step three: We show that if E is open or compact, then M_E is Borel measurable on $\mathbb{R}$, and (18.5) holds true. Indeed, let $\delta \in (0, \infty]$. If E is compact, then by step two $M_{E,\delta}$ is upper semicontinuous, and thus Borel measurable on $\mathbb{R}$. If E is open, then there exists an increasing sequence of compact sets $\{E_h\}_{h\in\mathbb{N}}$, such that $E = \bigcup_{h\in\mathbb{N}} E_h$: in particular, as $M_{E,\delta} = \lim_{h\to\infty} M_{E_h,\delta}$ on $\mathbb{R}$, it turns out that $M_{E,\delta}$ is Borel measurable on $\mathbb{R}$. In both cases, we finally deduce (18.5) by Fatou's Lemma and by step one,

$$\int_{\mathbb{R}} M_E(t)\,dt \le \liminf_{\delta\to 0^+} \int_{\mathbb{R}} M_{E,\delta}(t)\,dt \le C(n)\,\mathrm{Lip}(u)\,|E|\,.$$

Step four: We conclude the proof of the lemma. Let E be Lebesgue measurable. For every $h \in \mathbb{N}$ there exist a compact set K_h and an open set A_h such that

$K_h \subset E_h \subset A_h$ and $|A_h \setminus E_h| \to 0$ as $h \to \infty$. By step three, we know that (18.5) holds on each A_h and K_h, thus leading to

$$0 \le \int_{\mathbb{R}} \mathcal{H}^{n-1}\big(A_h \cap \{u = t\}\big)\,\mathrm{d}t - \int_{\mathbb{R}} \mathcal{H}^{n-1}\big(K_h \cap \{u = t\}\big)\,\mathrm{d}t$$
$$\le C(n)\mathrm{Lip}(u)|A_h \setminus K_h| \to 0\,, \qquad \text{as } h \to \infty\,.$$

As $\mathcal{H}^{n-1}(A_h \cap \{u = t\}) \ge M_E(t) \ge \mathcal{H}^{n-1}(K_h \cap \{u = t\})$ we deduce that

$$M_E(t) = \lim_{h\to\infty} \mathcal{H}^{n-1}\big(K_h \cap \{u = t\}\big)\,,$$

for a.e. $t \in \mathbb{R}^m$. Therefore M_E is Lebesgue measurable and, by step one and Fatou's lemma, (18.5) holds true. □

We now address the role of the set $\{\nabla u = 0\}$ in (18.3), proving that, for a.e. $t \in \mathbb{R}$, we have $\mathcal{H}^{n-1}(\{\nabla u = 0\} \cap \{u = t\}) = 0$. The reason is that the integral

$$\int_{\mathbb{R}} \mathcal{H}^{n-1}\big(\{\nabla u = 0\} \cap \{u = t\}\big)\,\mathrm{d}t\,,$$

is performed on $t \in u(\{\nabla u = 0\})$, which is expected to have zero Lebesgue measure as, close to a point x where $\nabla u(x) = 0$, u is almost constant.

Lemma 18.5 (Slicing the set of critical points) *If $u\colon \mathbb{R}^n \to \mathbb{R}$ is a Lipschitz function and $E = \{x \in \mathbb{R}^n : \nabla u(x) = 0\}$, then for a.e. $t \in \mathbb{R}$*

$$\mathcal{H}^{n-1}\big(E \cap \{u = t\}\big) = 0\,.$$

Proof *Step one:* By step three in the proof of Lemma 18.4, by setting

$$\mu(F) = \int_{\mathbb{R}} \mathcal{H}^{n-1}_{\infty}\big(F \cap \{u = t\}\big)\,\mathrm{d}t\,,$$

we define a countably subadditive set function on Borel sets F of $\mathbb{R}^n$. Let us prove that there exists a constant $C = C(n)$ such that, for every $\varepsilon > 0$ and $x \in E$, there exists $r(\varepsilon, x) > 0$ with

$$\mu(\overline{B}(x,r)) \le C\varepsilon\, r^n\,, \qquad \forall r \in (0, r(\varepsilon, x))\,. \tag{18.7}$$

Indeed, let us define $u_{x,r}\colon B \to \mathbb{R}$ as $u_{x,r}(z) = r^{-1}(u(x + rz) - u(x))$, $z \in B$. We then rescale u in the range and in the domain, to obtain

$$\mu(\overline{B}(x,r)) = \int_{\mathbb{R}} \mathcal{H}^{n-1}_{\infty}\big(\overline{B}(x,r) \cap \{u = t\}\big)\,\mathrm{d}t$$
$$= r^{n-1} \int_{\mathbb{R}} \mathcal{H}^{n-1}_{\infty}\big(\big\{z \in \overline{B} : u(x + rz) = t\big\}\big)\,\mathrm{d}t$$
$$= r^n \int_{\mathbb{R}} \mathcal{H}^{n-1}_{\infty}\big(\overline{B} \cap \{u_{x,r} = s\}\big)\,\mathrm{d}s\,.$$

As $r \to 0^+$, by the differentiability of u at x we know that $u_{x,r}(z) \to \nabla u(x)z = 0$ uniformly as $z \in \overline{B}$. Thus for every $\varepsilon > 0$ there exists $r(\varepsilon, x) > 0$ such that

$$\overline{B} \cap \{u_{x,r} = s\} = \emptyset\,, \qquad \forall r \in (0, r(\varepsilon, x))\,, |s| > \varepsilon\,.$$

Hence, $\mu(\overline{B}(x, r)) \le 2\mathcal{H}^{n-1}_\infty(\overline{B})\,\varepsilon\, r^n$ for $r \in (0, r(\varepsilon, x))$. This is (18.7), with $C(n) = 2\mathcal{H}^{n-1}_\infty(\overline{B})$.

Step two: Let $K \subset E$ be compact, let $\varepsilon > 0$, and consider the covering $\mathcal{F}$ of K by non-degenerate, closed balls defined as

$$\mathcal{F} = \Big\{\overline{B}(x, r) : 0 < r < \min\{r(\varepsilon, x), 1\}, x \in K\Big\},$$

with $r(\varepsilon, x)$ as in step one. By Corollary 5.2, there exists a countable disjoint subfamily $\mathcal{F}'$ of $\mathcal{F}$ such that, keeping (18.7) in mind,

$$\begin{aligned}\mu(K) &\le \xi(n) \sum_{B\in\mathcal{F}_h} \mu(B) \le \xi(n)\, C\varepsilon \sum_{B\in\mathcal{F}_h} r(B)^n = \frac{\xi(n)\, C\,\varepsilon}{\omega_n}\Big|\bigcup\{B : B \in \mathcal{F}_h\}\Big| \\ &\le \frac{\xi(n)\, C\,\varepsilon}{\omega_n}|I_1(K)|\,,\end{aligned}$$

where $I_1(K)$ is the one-neighborhood of K. In particular, $\mu(K) = 0$, and, by Proposition 3.4, $\mathcal{H}^{n-1}(K \cap \{u = t\}) = 0$ for a.e. $t \in \mathbb{R}$. Now let K_h be a sequence of compact sets contained in E such that $|E \setminus \bigcup_{h\in\mathbb{N}} K_h| = 0$. If $E_0 = E \setminus \bigcup_{h\in\mathbb{N}} K_h$, then $\mathcal{H}^{n-1}((E \setminus E_0) \cap \{u = t\}) = 0$ for a.e. $t \in \mathbb{R}$, while, on the other hand, $\mathcal{H}^{n-1}(E_0 \cap \{u = t\}) = 0$ for a.e. $t \in \mathbb{R}$ thanks to (18.5). □

A difficulty in extending the chain rule $\nabla(g \circ f) = ((\nabla g) \circ f)\nabla f$ for vector-valued C^1-functions to the case of Lipschitz functions is that the image of f may lie in the set of non-differentiability of g, without f being constant. In the case $f = g^{-1}$, this problem cannot occur and the chain rule follows easily.

Lemma 18.6 *If $f, g\colon \mathbb{R}^n \to \mathbb{R}^n$ are Lipschitz functions and $E = \{x \in \mathbb{R}^n : g(f(x)) = x\}$ then, for a.e. $z \in E$,*

$$\nabla g(f(z))\nabla f(z) = \mathrm{Id}_{\mathbb{R}^n}\,.$$

Proof *Step one:* We show that, if $h\colon \mathbb{R}^n \to \mathbb{R}$ is a Lipschitz function and $t \in \mathbb{R}$, then $\nabla h(x) = 0$ at a.e. $x \in \{h = t\}$. Indeed, this follows from step two in the proof of Lemma 7.6, Proposition 7.7, and Rademacher's theorem.
Step two: Let G_f and G_g denote the sets of points of non-differentiability of f and g, and let F be the set of those $z \in E$ such that f is differentiable at z and g is differentiable at $f(z)$. In this way, if $z \in E \setminus F$, then either $z \in G_f$, or $f(z) \in G_g$ with $z = g(f(z))$, that is, $z \in g(G_g)$. By Rademacher's theorem and

Proposition 3.5, $|G_f| = |g(G_g)| = 0$. Hence, $|E \setminus F| = 0$. By the standard chain rule for differentiable functions, $g \circ f$ is differentiable at $z \in F$, with

$$\nabla(g \circ f)(z) = \nabla g(f(z))\nabla f(z).$$

We thus conclude by applying step one to the components of $z - g(f(z))$. □

We are now ready for the proof of Theorem 18.1. We decompose $\mathbb{R}^n$ as the product $\mathbb{R}^{n-1} \times \mathbb{R}$, setting $x = (z, t)$ for $x \in \mathbb{R}^n$; see Notation 4. We shall consider the projection map $\mathbf{p} : \mathbb{R}^n \to \mathbb{R}^{n-1}$ together with its adjoint $\mathbf{p}^* : \mathbb{R}^{n-1} \to \mathbb{R}^n$, which of course satisfies $\mathbf{p}^* z = (z, 0)$ for every $z \in \mathbb{R}^{n-1}$. Finally, given $E \subset \mathbb{R}^n$, $z \in \mathbb{R}^{n-1}$ and $t \in \mathbb{R}$, we shall consider the vertical and horizontal slices of E,

$$E_z = \{t \in \mathbb{R} : (z, t) \in E\}, \qquad E^t = \{z \in \mathbb{R}^{n-1} : (z, t) \in E\}.$$

Proof of Theorem 18.1 *Step one:* We reduce to proving (18.3) in the case when

$$\frac{\partial u}{\partial x_n}(x) \neq 0 \qquad \forall x \in E, \tag{18.8}$$

and the function $f : \mathbb{R}^n \to \mathbb{R}^n$ defined as

$$f(x) = (\mathbf{p}x, u(x)), \qquad x \in \mathbb{R}^n$$

is injective on E, with Lipschitz inverse $(f|_E)^{-1} : f(E) \to \mathbb{R}^n$. Indeed, let E be a generic Borel set in $\mathbb{R}^n$. By Lemma 18.4 and Lemma 18.5, in proving (18.3), we can directly assume that E is contained in the set G_0 of those points $x \in \mathbb{R}^n$ such that u is differentiable at x with $|\nabla u(x)| > 0$. Clearly, there exists a partition of G_0 into sets $\{G_k\}_{k=1}^n$ where

$$G_k \subset \left\{ x \in G_0 : \frac{\partial u}{\partial x_k}(x) \neq 0 \right\}, \qquad k = 1, \dots, n.$$

Thus we can assume without loss of generality that (18.8) holds true. Since

$$\nabla f = \sum_{h=1}^{n-1} e_h \otimes e_h + e_n \otimes \nabla u, \tag{18.9}$$

we find that $Jf > 0$ on E, as

$$\det(\nabla f) = \partial_n u, \qquad Jf = |\partial_n u|. \tag{18.10}$$

By Theorem 8.8, there exists a Borel partition $\{E_h\}_{h \in \mathbb{N}}$ of E such that f is injective on every E_h and $(f|_{E_h})^{-1} : f(E_h) \to \mathbb{R}^n$ is a Lipschitz function.

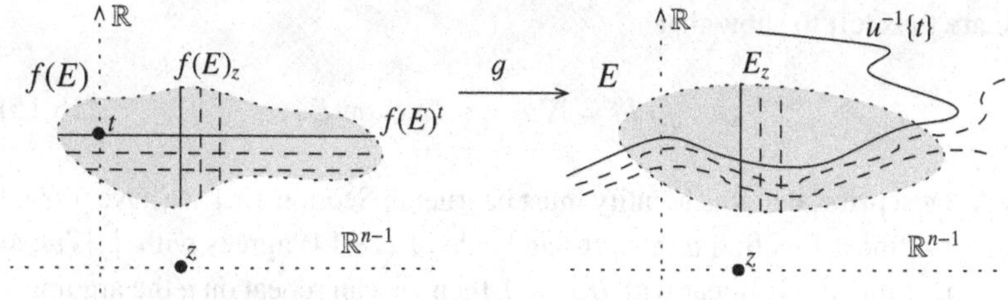

Figure 18.1 The function $g\colon \mathbb{R}^n \to \mathbb{R}^n$ maps the vertical slices $f(E)_z$ of $f(E)$ into the corresponding vertical slices E_z of E, and maps the horizontal slices $f(E)^t$ of $f(E)$ into the slices $E \cap \{u = t\}$ of E through u. By this change of variable we see that the coarea formula is indeed a consequence of the area formula (applied twice) and of Fubini's theorem.

Step two: Under the position of step one, there exists a Lipschitz function $g\colon \mathbb{R}^n \to \mathbb{R}^n$ such that $g = (f|_E)^{-1}$ on $f(E)$. In particular, by Lemma 18.6,

$$\nabla g(f(x)) = \nabla f(x)^{-1} \qquad \text{for a.e. } x \in E\,. \tag{18.11}$$

where, from (18.9), we see that

$$(\nabla f)^{-1} = \sum_{h=1}^{n-1} e_h \otimes e_h + \frac{1}{\partial_n u} e_n \otimes e_n - e_n \otimes \frac{\nabla' u}{\partial_n u}\,. \tag{18.12}$$

Given $t \in \mathbb{R}$, let us now define $g_t\colon \mathbb{R}^{n-1} \to \mathbb{R}^n$ as

$$g_t(z) = g(z,t)\,, \qquad z \in \mathbb{R}^{n-1}\,;$$

see Figure 18.1. For every $t \in \mathbb{R}$, g_t is a Lipschitz function, with

$$\begin{aligned} &E \cap \{u = t\} = g_t(f(E)^t)\,,\\ &\nabla' g_t(z) = \nabla g(z,t) \circ \mathbf{p}^*\,, \qquad \text{for a.e. } z \in \mathbb{R}^{n-1}. \end{aligned} \tag{18.13}$$

By the area formula applied to g_t on the horizontal slice $f(E)^t$ of $f(E)$,

$$\mathcal{H}^{n-1}\big(E \cap \{u = t\}\big) = \int_{f(E)^t} J g_t(z)\,\mathrm{d}z\,,$$

for every $t \in \mathbb{R}$. If we set for clarity $h(z,t) = J g_t(z)$, then by Fubini's theorem,

$$\int_{\mathbb{R}} \mathcal{H}^{n-1}\big(E \cap \{u = t\}\big)\,\mathrm{d}t = \int_{f(E)} h(x)\,\mathrm{d}x\,.$$

Finally, by the area formula applied to f on E, we conclude that

$$\int_{\mathbb{R}} \mathcal{H}^{n-1}\big(E \cap \{u = t\}\big)\,\mathrm{d}t = \int_E h(f(x)) J f(x) \mathrm{d}x\,. \tag{18.14}$$

We are thus left to show that

$$(h \circ f)\, Jf = |\nabla u|\,, \qquad \text{a.e. on } E\,. \tag{18.15}$$

We know a priori that this identity must be true: in Section 13.1 we have proved that for a linear function u the left-hand side of (18.14) agrees with $\int_E |\nabla u|$; at the same time, if u is linear with $\partial_n u \neq 0$, then we can repeat on u the argument leading to (18.14); combining these two remarks, the identity $(h \circ f)\, Jf = |\nabla u|$ is proved whenever we start with u linear and such that $\partial_n u \neq 0$. This suffices to deduce (18.15) in the general case, as, in fact, (18.15) only concerns the gradient of u at a fixed $x \in E$, that is to say, it only concerns the linear function defined by $\nabla u(x)$. We can also check (18.15) by a direct computation. Indeed, by (18.11) and (18.12) we find

$$\nabla g(f(x)) \circ \mathbf{p}^* = (\nabla f(x))^{-1} \circ \mathbf{p}^* = \sum_{h=1}^{n-1} e_h \otimes e_h - e_n \otimes \frac{\nabla' u(x)}{\partial_n u(x)}\,,$$

for a.e. $x \in E$. In particular,

$$\left(\nabla g(f(x)) \circ \mathbf{p}^*\right)^* \left(\nabla g(f(x)) \circ \mathbf{p}^*\right) = \mathrm{Id}_{\mathbb{R}^{n-1}} + \frac{\nabla' u(x) \otimes \nabla' u(x)}{\partial_n u(x)^2}\,,$$

so that, by (18.13), and by the formula $\det(\mathrm{Id} + v \otimes v) = 1 + |v|^2$,

$$h \circ f = \sqrt{1 + \frac{|\nabla' u|^2}{(\partial_n u)^2}} \qquad \text{on } E.$$

By (18.10), $Jf = |\partial_n u|$, and thus we conclude that $(h \circ f) Jf = |\nabla u|$ on E. □

Exercise 18.7 (An implicit function theorem) If $u \colon \mathbb{R}^n \to \mathbb{R}$ is a Lipschitz function, then $\{u = t\}$ is locally $\mathcal{H}^{n-1}$-rectifiable for a.e. $t \in \mathbb{R}$. *Hint:* Recall Example 13.3 and that $\mathcal{H}^{n-1}(\{u = t\} \Delta \partial^* \{u > t\}) = 0$ for a.e. $t \in \mathbb{R}$.

18.1.1 Further extensions of the coarea formula

The coarea formula for real-valued Lipschitz functions (18.3) is easily extended to the case of Lipschitz functions $f \colon \mathbb{R}^n \to \mathbb{R}^m$, $m \le n$, with, possibly, $m \ge 2$. Precisely, in this case if E is a Lebesgue measurable set in $\mathbb{R}^n$, then the function $y \in \mathbb{R}^m \mapsto \mathcal{H}^{n-m}(E \cap \{f = y\})$ is Lebesgue measurable on $\mathbb{R}^m$, and

$$\int_{\mathbb{R}^m} \mathcal{H}^{n-m}\Big(E \cap \{f = y\}\Big)\,dy = \int_E Cf(x)\,dx\,, \tag{18.16}$$

where the **coarea factor** $Cf: \mathbb{R}^n \to [0, \infty]$ of f is the Borel function

$$Cf(x) = \begin{cases} \sqrt{\det(\nabla f(x)\nabla f(x)^*)}\,, & \text{if } f \text{ is differentiable at } x\,; \\ +\infty\,, & \text{if } f \text{ is not differentiable at } x\,. \end{cases}$$

If $f = P \in \mathbf{O}^*(n,m)$ is an orthogonal projection (so that $CP = 1$), then (18.16) reduces to Fubini's theorem; if $f = T \in \mathbb{R}^m \otimes \mathbb{R}^n$ is a generic linear map, then we reduce to the case of an orthogonal projection thanks to the area formula for linear maps, Theorem 8.5, and by applying the polar decomposition (8.4) to T^*, to find $T = SP$ for $S \in \mathbf{Sym}(m)$ and $P \in \mathbf{O}^*(m,n)$ (note that $CT = JS$). In turn, in the case $m \geq 2$, the proof of (18.16) on a generic Lipschitz function f is obtained by the same argument we have presented for the case $m = 1$.

18.2 The coarea formula on $\mathcal{H}^{n-1}$-rectifiable sets

In view of applications to the theory of sets of finite perimeter, we now want to extend the coarea formula to the case where we slice a locally $\mathcal{H}^{n-1}$-rectifiable set in place of a Borel set. Let us recall that, by Theorem 11.4, a Lipschitz function $u: \mathbb{R}^n \to \mathbb{R}$ is tangentially differentiable at a $\mathcal{H}^{n-1}$-a.e. point of a locally $\mathcal{H}^{n-1}$-rectifiable set M. Its tangential gradient $\nabla^M u$ satisfies

$$\nabla^M u(x) = \nabla u(x) - \Big(\nabla u(x) \cdot \nu(x)\Big)\,\nu(x)\,,$$

whenever ∇u and $\nabla^M u$ exists at x, and $\nu(x) \in S^{n-1}$ is such that $\nu(x)^\perp = T_x M$.

Theorem 18.8 (Coarea formula on locally $(n-1)$-recitifiable sets) *If M is a locally $\mathcal{H}^{n-1}$-rectifiable set in $\mathbb{R}^n$ and $u: \mathbb{R}^n \to \mathbb{R}$ is a Lipschitz function, then*

$$\int_{\mathbb{R}} \mathcal{H}^{n-2}\Big(M \cap \{u = t\}\Big)\,dt = \int_M |\nabla^M f|\,d\mathcal{H}^{n-1}\,. \tag{18.17}$$

In particular, if $g: M \to [-\infty, \infty]$ is a Borel function and either $g \geq 0$ or $g \in L^1(\mathbb{R}^n, \mathcal{H}^{n-1} \llcorner M)$, then

$$\int_{\mathbb{R}} dt \int_{M\cap\{u=t\}} g\,d\mathcal{H}^{n-2} = \int_M g\,|\nabla^M f|\,d\mathcal{H}^{n-1}\,. \tag{18.18}$$

Proof The idea is to decompose the rectifiable set M into almost flat pieces by Theorem 10.1, and then to apply the coarea formula of Theorem 18.1 to each piece. More precisely, given $t > 1$, we apply Theorem 10.1 to find that

$M = M_0 \cup \bigcup_{h\in\mathbb{N}} M_h$ with $\mathcal{H}^{n-1}(M_0) = 0$ and $M_h = g_h(E_h)$, where $g_h : \mathbb{R}^{n-1} \to \mathbb{R}^n$ and $E_h \subset \mathbb{R}^{n-1}$ define $(n-1)$-dimensional regular Lipschitz images, with

$$t^{-1}|v| \le |\nabla g_h(x)v| \le t|v|, \qquad \forall x \in E_h, v \in \mathbb{R}^{n-1}. \tag{18.19}$$

If now $z \in E_h$ is such that $(u \circ g_h)$ is differentiable at z, and u is differentiable at $g_h(z)$ (in particular, this happens for a.e. $z \in E_h$), then

$$\nabla(u \circ g_h)(z) = \nabla u(g_h(z)) \circ \nabla g_h(z) = \nabla^{M_h} u(g_h(z)) \circ \nabla g_h(z),$$

and thus, by (18.19),

$$t^{-1}\,|\nabla^{M_h} u(g_h(z))| \le |\nabla(u \circ g_h)(z)| \le t\,|\nabla^{M_h} u(g_h(z))|. \tag{18.20}$$

By Exercise 18.7, the sets $E(h,t) = E_h \cap \{u \circ g_h = t\}$ are $\mathcal{H}^{n-2}$-rectifiable in $\mathbb{R}^{n-1}$ for a.e. $t \in \mathbb{R}$. Hence, by the area formula for rectifiable sets (11.5),

$$\mathcal{H}^{n-2}\big(M_h \cap \{u = t\}\big) = \int_{E(h,t)} J^{E(h,t)} g_h \,\mathrm{d}\mathcal{H}^{n-2}.$$

By (18.19) we have that

$$t^{-(n-2)}\,\mathcal{H}^{n-2}\big(E(h,t)\big) \le \mathcal{H}^{n-2}\big(M_h \cap \{u = t\}\big) \le t^{n-2}\mathcal{H}^{n-2}\big(E(h,t)\big),$$

while the coarea formula (18.3) applied to $u \circ g_h : \mathbb{R}^{n-1} \to \mathbb{R}$ implies

$$\int_{\mathbb{R}} \mathcal{H}^{n-2}\big(E(h,t)\big)\,\mathrm{d}t = \int_{E_h} |\nabla(u \circ g_h)|. \tag{18.21}$$

Thus,

$$t^{-(n-2)} \int_{E_h} |\nabla(u \circ g_h)| \le \int_{\mathbb{R}} \mathcal{H}^{n-2}\big(M_h \cap \{u = t\}\big)\,\mathrm{d}t \le t^{n-2} \int_{E_h} |\nabla(u \circ g_h)|. \tag{18.22}$$

By the area formula applied to $g_h : \mathbb{R}^{n-1} \to \mathbb{R}^n$ on the set E_h, we have

$$\int_{M_h} |\nabla^M u|\,\mathrm{d}\mathcal{H}^{n-1} = \int_{M_h} |\nabla^{M_h} u|\,\mathrm{d}\mathcal{H}^{n-1} = \int_{E_h} |\nabla^{M_h} u(g_h)|\, J g_h,$$

so that, again by (18.19),

$$t^{-(n-1)} \int_{E_h} |\nabla^{M_h} u(g_h)| \le \int_{M_h} |\nabla^M u|\,\mathrm{d}\mathcal{H}^{n-1} \le t^{n-1} \int_{E_h} |\nabla^{M_h} u(g_h)|. \tag{18.23}$$

We conclude by (18.20), (18.22), and (18.23), and letting $t \to 1^+$. □

Remark 18.9 (Coarea formula on locally $\mathcal{H}^k$-rectifiable sets) If M is a locally $\mathcal{H}^k$-rectifiable set in $\mathbb{R}^n$, $f : \mathbb{R}^n \to \mathbb{R}^m$ is a Lipschitz function, and $k \le m$, then the following extension of (18.17) holds true,

$$\int_{\mathbb{R}^m} \mathcal{H}^{k-m}\big(M \cap \{f = y\}\big)\,\mathrm{d}\mathcal{H}^m(y) = \int_M C^M f\,\mathrm{d}\mathcal{H}^k, \tag{18.24}$$

where the **coarea factor** $C^M f$ **of** f **relative to** M is the Borel function

$$C^M f(x) = \det\sqrt{\nabla^M f(x)\nabla^M f(x)^*}, \quad \mathcal{H}^k\text{-a.e. } x \in M.$$

To prove (18.24) one argues as above, the only difference being that (18.16) is applied in (18.21) in place of (18.3).

Exercise 18.10 If M is a $\mathcal{H}^k$-rectifiable set in $\mathbb{R}^n$, then $M \times \mathbb{R}^h$ is a $\mathcal{H}^{k+h}$-rectifiable set in $\mathbb{R}^n \times \mathbb{R}^h$ and $(\mathcal{H}^k \llcorner M) \times \mathcal{L}^h = \mathcal{H}^{k+h} \llcorner (M \times \mathbb{R}^h)$.

18.3 Slicing perimeters by hyperplanes

We now apply the coarea formula for rectifiable sets to the study of the slicing by hyperplanes of a set of finite perimeter. Decomposing $\mathbb{R}^n$ as $\mathbb{R}^{n-1} \times \mathbb{R}$, with projections $\mathbf{p} : \mathbb{R}^n \to \mathbb{R}^{n-1}$ and $\mathbf{q} : \mathbb{R}^n \to \mathbb{R}$, as in Notation 4, we denote by E_t the horizontal slice of $E \subset \mathbb{R}^n$,

$$E_t = \left\{ z \in \mathbb{R}^{n-1} : (z,t) \in E \right\}, \qquad t \in \mathbb{R}.$$

If E is a set of locally finite perimeter, then applying Theorem 18.8 with $M = \partial^* E$ and $u(x) = \mathbf{q}x$, $x \in \mathbb{R}^n$, and by also taking into account that

$$|\nabla^M u(x)| = \sqrt{1 - (\nu_E(x) \cdot e_n)^2} = |\mathbf{p}\nu_E(x)|, \qquad \text{for } \mathcal{H}^{n-1}\text{-a.e. } x \in \partial^* E,$$

we find that for every Borel function $g \colon \mathbb{R}^n \to [-\infty, \infty]$ with either $g \ge 0$ or $g \in L^1(\mathbb{R}^n, \mathcal{H}^{n-1} \llcorner \partial^* E)$, it holds that

$$\int_{\partial^* E} g \, |\mathbf{p}\nu_E| \, d\mathcal{H}^{n-1} = \int_{\mathbb{R}} dt \int_{(\partial^* E)_t} g \, d\mathcal{H}^{n-2}. \tag{18.25}$$

Starting from this slicing formula, we now prove the following theorem.

Theorem 18.11 (Slicing boundaries by hyperplanes) *If E is a set of locally finite perimeter in $\mathbb{R}^n$ then, for a.e. $t \in \mathbb{R}$, the horizontal section E_t of E is a set of locally finite perimeter in $\mathbb{R}^{n-1}$, with*

$$\mathcal{H}^{n-2}\left(\partial^* E_t \Delta (\partial^* E)_t\right) = 0, \tag{18.26}$$

$$\mathbf{p}\nu_E(z,t) \neq 0, \qquad \text{for } \mathcal{H}^{n-2}\text{-a.e. } z \in (\partial^* E)_t, \tag{18.27}$$

and

$$\mu_{E_t} = \frac{\mathbf{p}\nu_E(\cdot,t)}{|\mathbf{p}\nu_E(\cdot,t)|} \mathcal{H}^{n-2} \llcorner (\partial^* E)_t. \tag{18.28}$$

Moreover, if E has finite Lebesgue measure and

$$\mathcal{H}^{n-1}\left(\left\{ x \in \partial^* E : \nu_E(x) = \pm e_n \right\}\right) = 0, \tag{18.29}$$

then $v_E(t) = \mathcal{H}^{n-1}(E_t)$ ($t \in \mathbb{R}$) is such that $v_E \in W^{1,1}_{\mathrm{loc}}(\mathbb{R})$, with

$$v_E'(t) = -\int_{(\partial^* E)_t} \frac{\mathbf{q}\nu_E(z,t)}{|\mathbf{p}\nu_E(z,t)|} \, d\mathcal{H}^{n-2}(z), \qquad \text{for a.e. } t \in \mathbb{R}. \tag{18.30}$$

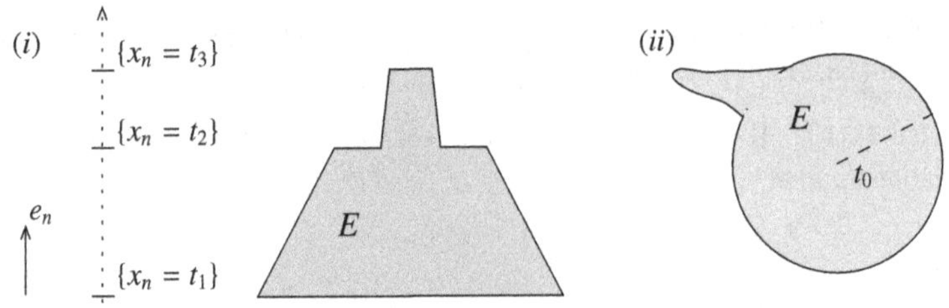

Figure 18.2 (i) The function $v(t) = \mathcal{H}^{n-1}(E_t)$ jumps at t_0, t_1 and t_2 and it is absolutely continuous on (t_0, t_1) and on (t_1, t_2). Note that the size of $|v'|$ depends on the projection of ν_E along e_n. (ii) With reference to the slicing by spheres described in Remark 18.14, this is an example where the slice function $v(t) = \mathcal{H}^{n-1}(E \cap \partial B_t)$ jumps at $t = t_0$.

Remark 18.12 Assumption (18.29) amounts to asking that $\partial^* E$ does not contain "horizontal faces". If we drop this assumption then we easily construct examples in which v_E has jump discontinuities; see Figure 18.2.

Remark 18.13 Theorem 18.11 is easily extended to include slicing by k-dimensional planes, with $1 \le k \le n-2$ and $n \ge 3$. To this end, it suffices to employ (18.24) in place of (18.17). We shall not need this generalization.

Proof of Theorem 18.11 *Step one:* The same argument used in proving Proposition 14.5 shows that, for a.e. $t \in \mathbb{R}$, E_t is a set of locally finite perimeter in $\mathbb{R}^{n-1}$. Let $\Sigma = \{x \in \partial^* E : \mathbf{p}\nu_E(x) = 0\}$. By (18.25) (with $g = 1_\Sigma$),

$$\int_{\mathbb{R}} \mathcal{H}^{n-2}(\Sigma_t)\,dt = \int_{\mathbb{R}} dt \int_{(\partial^* E)_t} 1_\Sigma \, d\mathcal{H}^{n-2} = \int_{\partial^* E} 1_\Sigma \, |\mathbf{p}\nu_E| \, d\mathcal{H}^{n-1} = 0\,.$$

In particular, for a.e. $t \in \mathbb{R}$ we have

$$\mathcal{H}^{n-2}(\Sigma_t) = 0\,, \tag{18.31}$$

that is, (18.27) holds true. We now prove (18.28). Taking (18.31) into account, and setting $M = \partial^* E \setminus \Sigma$, it is enough to prove that, for a.e. $t \in \mathbb{R}$, the Gauss–Green measure μ_{E_t} of E_t satisfies

$$\mu_{E_t} = \frac{\mathbf{p}\nu_E(\cdot, t)}{|\mathbf{p}\nu_E(\cdot, t)|} \mathcal{H}^{n-2} \llcorner M_t\,, \tag{18.32}$$

or, equivalently, that for a.e. $t \in \mathbb{R}$ and for every $T \in C^1_c(\mathbb{R}^{n-1}; \mathbb{R}^{n-1})$, we have

$$\int_{E_t} \mathrm{div}'\, T(z)\,dz = \int_{M_t} T(z) \cdot \frac{\mathbf{p}\nu_E(z, t)}{|\mathbf{p}\nu_E(z, t)|} d\mathcal{H}^{n-2}(z)\,, \tag{18.33}$$

where, clearly, div' denotes the divergence operator on vector fields in $\mathbb{R}^{n-1}$. Indeed, given $\varphi \in C^1_c(\mathbb{R})$, if we define $S \in C^1_c(\mathbb{R}^n;\mathbb{R}^n)$ as

$$S(x) = (\varphi(t)\,T(z), 0)\,, \qquad x = (z,t) \in \mathbb{R}^n\,,$$

then we have $e_n \cdot S(x) = 0$ and

$$\mathrm{div}\,S(x) = \varphi(t)\,\mathrm{div}'T(z)\,, \qquad \forall x = (z,t) \in \mathbb{R}^n\,.$$

In particular, by Fubini's theorem and the divergence theorem

$$\begin{aligned}\int_{\mathbb{R}} \varphi(t)\mathrm{d}t \int_{E_t} \mathrm{div}'T(z)\mathrm{d}z &= \int_E \mathrm{div}\,S = \int_{\partial^* E} S \cdot \nu_E \,\mathrm{d}\mathcal{H}^{n-1} \\ &= \int_{\partial^* E} S \cdot \mathbf{p}\nu_E \,\mathrm{d}\mathcal{H}^{n-1} = \int_M S \cdot \mathbf{p}\nu_E \,\mathrm{d}\mathcal{H}^{n-1} \\ &= \int_{\mathbb{R}} \varphi(t)\mathrm{d}t \int_{M_t} T(z) \cdot \frac{\mathbf{p}\nu_E(z,t)}{|\mathbf{p}\nu_E(z,t)|}\,\mathrm{d}\mathcal{H}^{n-2}(z)\,,\end{aligned}$$

where we have applied (18.25) again. By the arbitrariness of φ, we find (18.33), and (18.28) is proved. Finally (18.26) is proved by passing to total variations in (18.28), which gives $\mathcal{H}^{n-2} \llcorner (\partial^* E)_t = |\mu_{E_t}| = \mathcal{H}^{n-1} \llcorner \partial^* E_t$.

Step two: If $|E| < \infty$, then by Fubini's theorem $v_E \in L^1(\mathbb{R})$, with $|E| = \|v_E\|_{L^1(\mathbb{R})}$. We now prove (18.30). Given $\varphi \in C^1_c(\mathbb{R})$ we define a bounded vector field $S \in C^1(\mathbb{R}^n;\mathbb{R}^n)$, setting $S(x) = \varphi(t)\,e_n$, $x = (z,t) \in \mathbb{R}^n$, and notice that

$$\mathrm{div}\,S(x) = \varphi'(t)\,, \qquad \forall x = (z,t) \in \mathbb{R}^n\,.$$

By Fubini's theorem, the divergence theorem, and $\mathcal{H}^{n-1}(\Sigma) = 0$,

$$\begin{aligned}\int_{\mathbb{R}} \varphi'(t)\,v_E(t)\,\mathrm{d}t &= \int_E \mathrm{div}\,S = \int_{\partial^* E} S \cdot \nu_E \,\mathrm{d}\mathcal{H}^{n-1} \\ &= \int_M \varphi(\mathbf{q}x)\big(e_n \cdot \nu_E(x)\big)\,\mathrm{d}\mathcal{H}^{n-1}(x) = \int_{\partial^* E} g\,|\mathbf{p}\nu_E|\,\mathrm{d}\mathcal{H}^{n-1}\,,\end{aligned}$$

where $g\colon \mathbb{R}^n \to \mathbb{R}$ is the (possibly unbounded) Borel function defined as

$$g(x) = 1_M(x)\,\varphi(\mathbf{q}x)\,\frac{\mathbf{q}\nu_E(x)}{|\mathbf{p}\nu_E(x)|}\,, \qquad x \in \mathbb{R}^n\,.$$

Since $|E| < \infty$ and φ' is bounded and compactly supported, it turns out that $\int_{\mathbb{R}} \varphi'\,v_E$ is finite. In particular, both $\int_M g^+\,|\mathbf{p}\nu_E|\,\mathrm{d}\mathcal{H}^{n-1}$ and $\int_M g^-\,|\mathbf{p}\nu_E|\,\mathrm{d}\mathcal{H}^{n-1}$ are finite. If we thus apply (18.25) to g^+ and g^- we conclude that

$$\int_{\mathbb{R}} \varphi'(t)\,v_E(t)\,\mathrm{d}t = \int_{\mathbb{R}} \varphi(t)\mathrm{d}t \int_{M_t} \frac{\mathbf{q}\nu_E(z,t)}{|\mathbf{p}\nu_E(z,t)|}\,\mathrm{d}\mathcal{H}^{n-2}(z)\,.$$

By the arbitrariness of φ, $v_E \in W^{1,1}_{\mathrm{loc}}(\mathbb{R})$, with v_E' as in (18.30). □

Remark 18.14 We may also prove the absolute continuity of other slice functions, including slicing by spheres or cylinders (which arise, respectively, by setting $u(x) = |x|$ or $u(x) = |\mathbf{p}x|$ in the following considerations). Indeed, let $u\colon \mathbb{R}^n \to \mathbb{R}$ be a Lipschitz function with $|\nabla u| = 1$ a.e. on $\mathbb{R}^n$ and with range given by the (possibly unbounded) interval (a, b), and let $v_E : (a, b) \to [0, \infty]$,

$$v_E(t) = \mathcal{H}^{n-1}\Big(E \cap \{u = t\}\Big), \qquad t \in (a, b).$$

By the coarea formula (18.3), $v_E \in L^1(\mathbb{R})$ if $|E| < \infty$. Now, if the singular set

$$\Sigma = \Big\{x \in \partial^* E : \nu_E(x) = \pm\nabla u(x)\Big\},$$

is $\mathcal{H}^{n-1}$-negligible and $u \in C^2(\mathbb{R})$, then $v_E \in W^{1,1}_{\rm loc}(\mathbb{R})$ and, for a.e. $t \in (a, b)$,

$$v'(t) = -\int_{\{u=t\}\cap\partial^* E} \frac{(\nu_E \cdot \nabla u)}{\sqrt{1 - (\nu_E \cdot \nabla u)^2}}\, d\mathcal{H}^{n-2} + \int_{\{u=t\}\cap E} \Delta u \, d\mathcal{H}^{n-1}. \qquad (18.34)$$

To see this we replace (18.25) in step two of the proof of Theorem 18.11 by

$$\int_{\partial^* E} g\, \sqrt{1 - (\nu_E \cdot \nabla u)^2}\, d\mathcal{H}^{n-1} = \int_a^b dt \int_{\{u=t\}\cap\partial^* E} g\, d\mathcal{H}^{n-2}, \qquad (18.35)$$

(which, in turn, follows immediately from (18.17)), and we notice that

$$\int_a^b \varphi'(t)\, v_E(t)\, dt = \int_E \mathrm{div}\,(w\,\nabla u) - w\,\Delta u,$$

where $w(x) = \varphi(\mathbf{q}x)$, $x \in \mathbb{R}^n$. The first term on the right-hand side is now characterized by the divergence theorem and (18.35) as before, while, concerning the second term, it suffices to apply (18.3) to find

$$\int_E w\Delta u = \int_a^b \varphi(t)\, dt \int_{\{u=t\}\cap E} \Delta u \, d\mathcal{H}^{n-1}.$$

Exercise 18.15 If E is a set of locally finite perimeter, $g : \mathbb{R}^n \to [0, \infty]$ is a Borel function and $\Sigma = \{x \in \partial^* E : \nu_E(x) = \pm e_n\}$, then

$$\int_{\partial^* E} g\, d\mathcal{H}^{n-1} = \int_\Sigma g\, d\mathcal{H}^{n-1} + \int_{\mathbb{R}} dt \int_{\partial^* E_t} \frac{g(z, t)}{|\mathbf{p}\nu_E(z, t)|}\, d\mathcal{H}^{n-2}(z). \qquad (18.36)$$

19

Equilibrium shapes of liquids and sessile drops

In this chapter we study the equilibrium shapes of a liquid confined in a given container, with particular emphasis on the model problem of the sessile liquid drop. This is a classical subject, with a huge interdisciplinary literature. Therefore we shall provide here only a rough introduction to this theory, referring the reader to the beautiful book [Fin86] for a more complete account. At the same time, the study of these problems will allow us to provide significant and instructive applications of the various methods and ideas developed so far.
Since the work of Gauss, the problem is studied through the introduction of a free energy functional. Precisely, if a liquid occupies a region E inside a given container A (mathematically, E will be a set of finite perimeter and A an open set with sufficiently smooth boundary), then its free energy is given by

$$\sigma\big(P(E;A) - \beta\, P(E;\partial A)\big) + \int_E g(x)\,\mathrm{d}x\,;$$

see Figure 19.1. Here, $\sigma > 0$ denotes the surface tension at the interface between the liquid and the other medium (be it another liquid or gas) filling A. The coefficient β is called the **relative adhesion coefficient** between the fluid and the bounding solid walls of the recipient, and for reasons to be soon clarified (see Theorem 19.8), it satisfies $|\beta| \le 1$. The term $-\sigma\beta\, P(E;\partial A)$ is called the **wetting energy**. Finally, the third term denotes a potential energy acting on the liquid, which is typically assumed to be the gravitational energy $g(x) = g\rho\, x_n$, where ρ is the (constant) density of the (incompressible) liquid, and g is the acceleration of gravity. In the following, in order to simplify the notation, we shall always set $\sigma = 1$.

The free energy functional is usually minimized under a prescribed volume constraint $|E| = m$. The corresponding variational problems will be the object of our discussion. We will start in Section 19.1, applying the Direct Method to prove an existence result in bounded containers, and deriving the

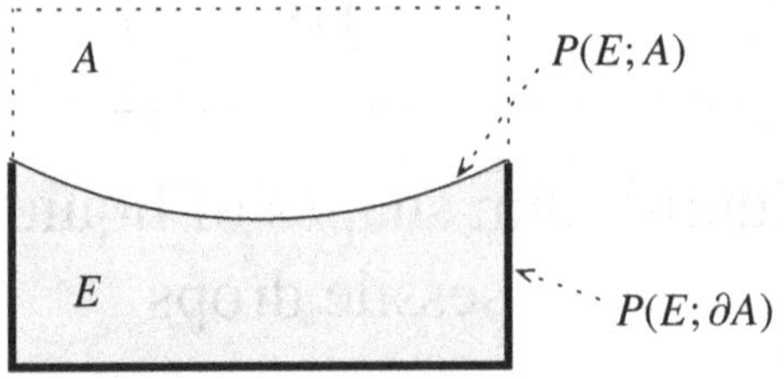

Figure 19.1 The equilibrium shape of a liquid inside a container A. The total surface energy is not directly proportional to the perimeter E, as the contribution of the interface between the liquid and the container is weighted by the constant β.

corresponding stationarity conditions, including the boundary stationarity condition known as Young's law. We shall then focus on the case that A is a half-space, thus setting the stage for the study of sessile liquid drops. In Section 19.2 we present the Schwartz inequality, a powerful geometric tool that we will apply in Section 19.3 to solve a constrained relative isoperimetric problem in the half-space, and in Section 19.5 to prove a useful symmetrization principle. In Section 19.4 we combine these results to characterize equilibrium configurations (in half-spaces) in the absence of gravity. This result will provide the starting point for the analysis of equilibrium configurations in the presence of gravity, the so-called sessile drops, in Section 19.6. In particular, we shall prove the existence, boundedness, and rotational symmetry of sessile drops. These results, combined with the regularity theory of Part III, will provide the reader with a full justification of the symmetry and regularity assumptions usually adopted in the classical literature on this problem.

19.1 Existence of minimizers and Young's law

We present here an existence result for equilibrium shapes, together with interior and boundary stationarity conditions. Given $\beta \in \mathbb{R}$, an open set $A \subset \mathbb{R}^n$, and a set of finite perimeter $E \subset A$, we shall set

$$\mathcal{F}_\beta(E; A) = P(E; A) - \beta\, P(E; \partial A) \tag{19.1}$$

for the total surface energy, and denote by

$$\mathcal{G}(E) = \int_E g(x)\, dx \tag{19.2}$$

the potential energy associated with a given Borel function $g\colon \mathbb{R}^n \to \mathbb{R}$.

19.1.1 Lower semicontinuity and existence of minimizers

We start by discussing the lower semicontinuity of the total surface energy $\mathcal{F}_\beta$ with respect to L^1-convergence of sets. A necessary condition for lower semicontinuity is that $|\beta| \le 1$; see Remark 19.2. When $|\beta| \le 1$, lower semicontinuity holds true quite trivially in the case $\beta \le 0$ (Proposition 19.1), and with some additional assumption on A if $\beta > 0$ (Proposition 19.3).

Proposition 19.1 *If $\beta \in [-1, 0]$, A is an open set of finite perimeter in $\mathbb{R}^n$, $\{E_h\}_{h\in\mathbb{N}}$ and E are set of finite perimeter contained in A, and $E_h \to E$, then*

$$\mathcal{F}_\beta(E; A) \le \liminf_{h\to\infty} \mathcal{F}_\beta(E_h; A). \tag{19.3}$$

Proof By Exercise 16.6, for every $E \subset A$ we have

$$P(E) = P(E; A) + P(E; \partial A), \qquad P(E; \partial A) \le P(A), \tag{19.4}$$

$$P(E) \le \mathcal{F}_\beta(E; A) + (1 + |\beta|)P(A). \tag{19.5}$$

Without loss of generality, let us assume the right-hand side of (19.3) to be finite. By (19.5), we thus find that $\sup_{h\in\mathbb{N}} P(E_h)$ is finite. For this reason, the convergence of the E_h to E implies that $\mu_{E_h} \overset{*}{\rightharpoonup} \mu_E$, and, by Proposition 4.30,

$$\liminf_{h\to\infty} P(E_h) \ge P(E), \qquad \liminf_{h\to\infty} P(E_h; A) \ge P(E; A). \tag{19.6}$$

We are thus left to exploit the identity

$$\begin{aligned}\mathcal{F}_\beta(E; A) &= (1+\beta)\, P(E; A) - \beta\big(P(E; A) + P(E; \partial A)\big)\\ &= (1+\beta)\, P(E; A) - \beta\, P(E),\end{aligned}$$

and the non-negativity of $1+\beta$ and $-\beta$ to deduce (19.3) from (19.6). □

Remark 19.2 If $|\beta| > 1$, then the lower semicontinuity inequality (19.3) may fail. Indeed, let us set $A = B$, $E_h = B_{r_h}$, $F_h = B \setminus B_{r_h}$ for $r_h \to 1^-$, so that $E_h \to B$ and $F_h \to \emptyset$. If $\beta < -1$, then

$$\mathcal{F}_\beta(E_h; B) = n\omega_n\, r_h^{n-1} \to n\omega_n < -n\omega_n\beta = \mathcal{F}_\beta(B; B),$$

and lower semicontinuity fails. If $\beta > 1$, then

$$\mathcal{F}_\beta(F_h; B) = n\omega_n(r_h^{n-1} - \beta) \to n\omega_n(1-\beta) < 0 = \mathcal{F}_\beta(\emptyset; B),$$

and, again, lower semicontinuity is disproved.

Proposition 19.3 *Let A be a bounded open set with finite perimeter with the property that, for every sufficiently small $\delta > 0$, a compactly supported Lipschitz vector field $T_\delta : \mathbb{R}^n \to \mathbb{R}^n$ exists such that $|T_\delta| \le 1$ on $\mathbb{R}^n$ and*

$$T_\delta \cdot \nu_A = 1 \quad \text{on } \partial^* A, \qquad T_\delta = 0 \quad \text{on } A \setminus \overline{A_\delta}, \tag{19.7}$$

where $A_\delta = \{x \in A : \mathrm{dist}(x, \partial A) < \delta\}$. *Then the same conclusions of Proposition 19.1 holds true with* $\beta \in (0,1]$ *too.*

Remark 19.4 The validity of (19.7) is simple to check on specific examples. Moreover, if A is a bounded open set with C^2-boundary, then T_δ may be constructed starting from the gradient of the signed distance function from A.

Proof of Proposition 19.3 If F is a set of finite perimeter contained in A, then by the divergence theorem (as in Exercise 12.12) and Exercise 16.6,

$$\int_F \mathrm{div}\, T_\delta = \int_{A\cap\partial^* F} T_\delta \cdot \nu_F \, d\mathcal{H}^{n-1} + \int_{\partial A\cap\partial^* F} T_\delta \cdot \nu_A \, d\mathcal{H}^{n-1}.$$

Exploiting (19.7), we thus find

$$P(F;\partial A) \le P(F;A_\delta) + C(\delta)\,|F|, \qquad \forall F \subset A, \tag{19.8}$$

(where $C(\delta) = \sup_{\mathbb{R}^n} |\nabla T_\delta| \to \infty$ as $\delta \to 0^+$). If now $\{E_h\}_{h\in\mathbb{N}}$ are sets of finite perimeter contained in A and $E_h \to E$, then, as in the proof of Proposition 19.1, we find $\mu_{E_h} \stackrel{*}{\rightharpoonup} \mu_E$. Applying (19.8) to $F = E_h \Delta E$, while taking also into account Exercise 16.5 and Exercise 16.6, we find

$$\begin{aligned}|P(E_h;\partial A) - P(E;\partial A)| &\le \mathcal{H}^{n-1}\Big(\partial A \cap (\partial^* E_h \Delta \partial^* E)\Big) = P(E_h\Delta E;\partial A)\\ &\le P(E_h\Delta E;A_\delta) + C(\delta)|E_h\Delta E|,\\ &\le P(E_h;A_\delta) + P(E;A_\delta) + C(\delta)|E_h\Delta E|.\end{aligned}$$

In particular, if $0 \le \beta \le 1$,

$$\begin{aligned}\mathcal{F}_\beta(E_h;A) - \mathcal{F}_\beta(E;A) &\ge P(E_h;A) - P(E;A) - |P(E_h;\partial A) - P(E;\partial A)|\\ &\ge P(E_h;A\setminus\overline{A_\delta}) - P(E;A) - P(E;A_\delta) - C(\delta)|E_h\Delta E|.\end{aligned}$$

By Proposition 12.15, since $A \setminus \overline{A_\delta}$ is open, letting $h \to \infty$, we find

$$\liminf_{h\to\infty} \mathcal{F}_\beta(E_h;A) \ge \mathcal{F}_\beta(E;A) + P(E;A\setminus\overline{A_\delta}) - P(E;A) - P(E;A_\delta),$$

where the right-hand side converges to $\mathcal{F}_\beta(E;A)$ as $\delta \to 0^+$. □

Theorem 19.5 (Existence of minimizers in bounded containers) *If* $|\beta| \le 1$, $g \in L^1_{\mathrm{loc}}(\mathbb{R}^n)$, *A is an open bounded set of finite perimeter in* $\mathbb{R}^n$ *(satisfying (19.7) in the case* $\beta > 0$*), and* $m \in (0, |A|)$*, then there exists a minimizer in*

$$\gamma = \inf\Big\{\mathcal{F}_\beta(E;A) + \mathcal{G}(E) : E \subset A, |E| = m\Big\}. \tag{19.9}$$

Remark 19.6 In the case $\beta = 0$, $g = 0$, the variational problem (19.9) reduces to the relative isoperimetric problem (12.30). By (19.4), if $\beta = -1$ and $g = 0$, then (19.9) is the **constrained isoperimetric problem in** A,

$$\inf\Big\{P(E) : E \subset A, |E| = m\Big\}. \tag{19.10}$$

The minimizers E in (19.10) satisfying $\operatorname{spt}\mu_E = \partial E$ (according to Proposition 12.19), are called **constrained isoperimetric sets** in A. These sets may behave quite differently from relative isoperimetric sets, cf. Figure 12.8 with Figure 19.2 below. We notice that every p-Cheeger set of A (see Exercises 14.7–17.12) is necessarily a constrained isoperimetric set in A.

Proof of Theorem 19.5 As shown in Proposition 12.30, the competition class in non-empty, so that $\gamma < \infty$. In fact, we have $\gamma \in \mathbb{R}$, since

$$\mathcal{F}_\beta(E;A) + \mathcal{G}(E) \ge -|\beta|\, P(A) - \int_A g(x)\mathrm{d}x\,.$$

Let $\{E_h\}_{h\in\mathbb{N}}$ be a minimizing sequence in (19.9). By (19.5) we deduce that $\sup_{h\in\mathbb{N}} P(E_h)$ is finite. Since A is bounded, by Theorem 12.26, there exists a set $E \subset A$ such that, up to subsequences, $E_h \to E$, so that, evidently, $|E| = m$. We conclude that E is a minimizer thanks to Proposition 19.1, Proposition 19.3, and Proposition 12.31. □

19.1.2 Stationarity conditions

Adopting the methods from Chapter 17, we are now going to prove that the mean curvature of the interior interface of an equilibrium configuration equals the potential energy plus a constant additive factor. The wetting energy plays no role in this result, and indeed no restriction on β is required (of course, when $|\beta| > 1$, non-existence phenomena are expected to occur).

Theorem 19.7 (Interior stationarity condition) *If $\beta \in \mathbb{R}$, A is open, $g \in C^0(A)$, $E \subset A$ has finite perimeter and measure, and*

$$\mathcal{F}_\beta(E;A) + \mathcal{G}(E) \le \mathcal{F}_\beta(F;A) + \mathcal{G}(F)\,,$$

for every $F \subset A$ with $|E| = |F|$, then there exists a constant $\lambda \in \mathbb{R}$ such that

$$\int_{A\cap\partial^* E} \operatorname{div}_E T\, \mathrm{d}\mathcal{H}^{n-1} = \int_{A\cap\partial^* E} (-g+\lambda)\,(T\cdot\nu_E)\,\mathrm{d}\mathcal{H}^{n-1}\,,$$

for every $T \in C^1_c(A;\mathbb{R}^n)$. In particular, there exists $\lambda \in \mathbb{R}$ such that E has distributional mean curvature equal to $-g+\lambda$ in A.

Proof If $T \in C^\infty_c(A;\mathbb{R}^n)$ and $\{f_t\}_{|t|<\varepsilon}$ is a local variation associated with T, then $\{x \in \mathbb{R}^n : x \ne f_t(x)\} \subset\subset A$ gives, for ε small enough, $f_t(E) \subset A$ for every $|t| < \varepsilon$. By Proposition 17.1, $\partial^*(f_t(E))$ is $\mathcal{H}^{n-1}$-equivalent to $f_t(\partial^* E)$, while $f_t(\partial^* E) \cap \partial A = \partial^* E \cap \partial A$. We thus find $P(f_t(E);\partial A) = P(E;\partial A)$ for every $|t| < \varepsilon$, that is, the wetting energy is constant along $\{f_t(E)\}_{|t|<\varepsilon}$. Taking Proposition 17.8 into account, and arguing as in Theorem 17.20, we conclude the proof of the theorem. □

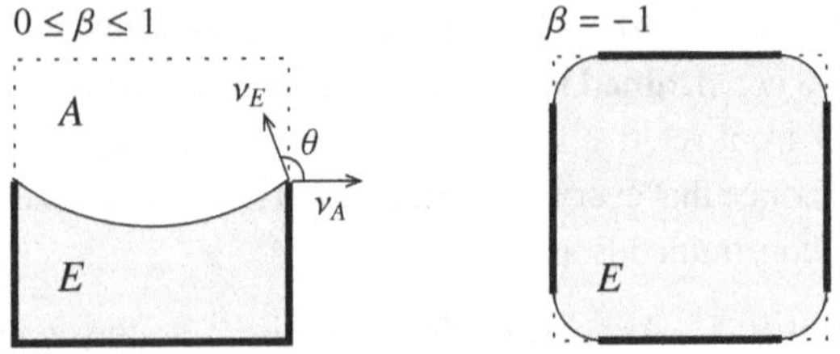

Figure 19.2 The constant mean curvature condition and Young's law reduce the minimization of the total surface energy to the class of sets E such that $A \cap \partial E$ is made up of countably many circular arcs meeting ∂A at a fixed angle θ such that $\cos\theta = -\beta$. In the case of constrained isoperimetric sets, depicted on the right, and corresponding to the choice $\beta = -1$, $A \cap \partial E$ meets ∂A tangentially, and ∂E tries to overlap with ∂A as much as possible.

We now discuss a stationarity condition for equilibrium shapes at boundary points, known as *Young's law*; see Figure 19.2. For its derivation, we will need to assume the regularity of the interior interface up to the boundary. The proof will use local variations associated with vector fields which act tangentially on ∂A, as well as Theorem 11.8.

Theorem 19.8 (Young's law) *If $\beta \in \mathbb{R}$, $g \in L^1(\mathbb{R}^n)$, A is an open set with C^1-boundary in $\mathbb{R}^n$, $E \subset A$ is an open set with finite perimeter and measure, $A \cap \partial E$ is a C^2-hypersurface with boundary, and*

$$\mathcal{F}_\beta(E;A) + \mathcal{G}(E) \le \mathcal{F}_\beta(F;A) + \mathcal{G}(F), \tag{19.11}$$

for every $F \subset A$ with $|F| = |E|$, then

$$\nu_E \cdot \nu_A = -\beta, \qquad on \qquad \mathrm{bdry}\,(A \cap \partial E). \tag{19.12}$$

In particular, necessarily $|\beta| \le 1$.

Remark 19.9 Young's law is insensitive to the presence of the potential energy. The contact angle of a liquid drop at equilibrium on a horizontal plane is determined by the capillarity effects only, which turn out to be much stronger than gravity effects. Gravity, of course, influences the equilibrium shape away from the contact plane, for example, by flattening the liquid drop at its top.

Remark 19.10 By Remark 19.6, Young's law implies that a relative isoperimetric set E in A, with $A \cap \partial E$ regular enough, intersects the boundary of A orthogonally; see Figure 12.8. Similarly, a Cheeger set E in A, with $A \cap \partial E$ regular enough, has to intersect the boundary of A tangentially; see Figure 19.2. In the planar case $n = 2$ one always has enough regularity to apply Young's law to these examples, as follows from the regularity theory of Part III.

Proof of Theorem 19.8 *Step one:* We show that if $T \in C^\infty_c(\mathbb{R}^n;\mathbb{R}^n)$ is tangent to ∂A and preserves volume at first order, that is,

$$T(x)\cdot\nu_A(x) = 0\,, \qquad \forall x \in \partial A\,, \tag{19.13}$$

$$\int_{\partial^* E} T\cdot\nu_E \,\mathrm{d}\mathcal{H}^{n-1} = 0\,, \tag{19.14}$$

(see Proposition 17.8), then there exists a one-parameter family of diffeomorphisms $h\colon (-\varepsilon,\varepsilon)\times\mathbb{R}^n \to \mathbb{R}^n$ having T as initial velocity, such that

$$h_t(E) \subset A\,, \qquad |h_t(E)| = |E|\,, \qquad \forall |t| < \varepsilon\,.$$

To show this we start by constructing a local variation $\{f_t\}_{|t|<\varepsilon}$ having T as its initial velocity, by solving Cauchy's problems,

$$\frac{\partial}{\partial t} f(t,x) = T(f(t,x))\,, \qquad x \in \mathbb{R}^n\,, \tag{19.15}$$

$$f(0,x) = x\,, \qquad x \in \mathbb{R}^n\,. \tag{19.16}$$

Since, locally, ∂A is the level set of a scalar function, we deduce from (19.13) that $f_t(\partial A) \subset \partial A$ for every $|t| < \varepsilon$. Exploiting the uniqueness in the Cauchy problem, we see that $f_t(A) \subset A$, and, in particular, that $f_t(E) \subset A$, for every $|t| < \varepsilon$. We now want to modify $\{f_t\}_{|t|<\varepsilon}$ into a volume-preserving local variation, without losing the confinement property in A. To this end we employ a trick which, according to Bolza, has its roots with Weierstrass himself. Precisely, we consider a vector field which, at first order, increases the measure of E, specifically, we consider $S \in C^\infty_c(A;\mathbb{R}^n)$ such that

$$\int_{A\cap\partial E} S\cdot\nu_E \,\mathrm{d}\mathcal{H}^{n-1} > 0\,. \tag{19.17}$$

(The existence of S follows by $A\cap\partial E \neq \emptyset$.) Up to decreasing ε, we may define a two-parameter family of diffeomorphisms $g\colon (-\varepsilon,\varepsilon)^2\times\mathbb{R}^n \to \mathbb{R}^n$, setting

$$g(t,s,x) = f(t,x) + s\,S(x) = g_{t,s}(x)\,, \qquad |t|,|s| < \varepsilon\,, x \in \mathbb{R}^n\,.$$

Notice that, since $\operatorname{spt} S \subset\subset A$ and up to decreasing ε, we may assume that

$$g_{t,s}(E) \subset A\,, \qquad \forall |t|,|s| < \varepsilon\,. \tag{19.18}$$

Correspondingly, let us consider the map $\phi\colon (-\varepsilon,\varepsilon)^2 \to (0,\infty)$,

$$\phi(t,s) = |g_{t,s}(E)| = \int_E Jg_{t,s}(x)\,\mathrm{d}x\,, \qquad |t|,|s| < \varepsilon\,.$$

Clearly, $\phi(0,0) = |E|$. Since we have

$$\nabla g_{t,s}(x) = \mathrm{Id} + t\,T(x) + s\,S(x) + \mathrm{o}\Big(\sqrt{t^2+s^2}\Big)\,, \tag{19.19}$$

uniformly on $\mathbb{R}^n$, by Lemma 17.4, (19.14), and (19.17) we find

$$\frac{\partial\phi}{\partial t}(0,0) = \int_E \operatorname{div} T = \int_{\partial^* E} T\cdot\nu_E = 0\,, \tag{19.20}$$

$$\frac{\partial\phi}{\partial s}(0,0) = \int_E \operatorname{div} S = \int_{\partial^* E} S\cdot\nu_E > 0\,. \tag{19.21}$$

By the implicit function theorem, up to further decreasing the value of ε, there exists a smooth function $\gamma\colon(-\varepsilon,\varepsilon)\to\mathbb{R}$ such that, for every $|t|<\varepsilon$,

$$\phi(t,\gamma(t)) = |E|\,, \qquad \gamma(0) = 0\,. \tag{19.22}$$

In particular,

$$0 = \frac{\partial\phi}{\partial t}(0,0) + \frac{\partial\phi}{\partial s}(0,0)\,\gamma'(0)\,,$$

which, by (19.20) and (19.21), gives $\gamma'(0) = 0$. If we set $h(t,x) = g(t,\gamma(t),x)$, then by (19.22) and (19.18), $|h_t(E)| = |E|$ and $h_t(E)\subset A$ for every $|t|<\varepsilon$. Moreover, by (19.19) and $\gamma'(0)=0$, $\{h_t\}_{|t|<\varepsilon}$ has initial velocity T.

Step two: Given T and h_t as in step one, we now apply the minimality inequality (19.11) to deduce that

$$\frac{\mathrm{d}}{\mathrm{d}t}\Big|_{t=0}\Big(\mathcal{F}_\beta(h_t(E);A) + \int_{h_t(E)} g\Big) = 0\,. \tag{19.23}$$

By Proposition 17.8 and Theorem 17.5 we find

$$\begin{aligned}
\frac{\mathrm{d}}{\mathrm{d}t}\Big|_{t=0}\int_{h_t(E)} g &= \int_{\partial^* E} g\,(T\cdot\nu_E)\mathrm{d}\mathcal{H}^{n-1}\,,\\
\frac{\mathrm{d}}{\mathrm{d}t}\Big|_{t=0} P(h_t(E);A) &= \int_{\partial^* E}\operatorname{div}_E T\,\mathrm{d}\mathcal{H}^{n-1}\\
&= \int_{A\cap\partial E}\operatorname{div}_E T\,\mathrm{d}\mathcal{H}^{n-1} + \int_{\partial A\cap\partial E}\operatorname{div}_E T\,\mathrm{d}\mathcal{H}^{n-1}\,,
\end{aligned}$$

where in the last identity we have applied Exercise 16.6 to deduce that

$$\mu_E = \nu_E\,\mathcal{H}^{n-1}\llcorner(A\cap\partial E) + \nu_A\,\mathcal{H}^{n-1}\llcorner(\partial^* E\cap\partial A)\,. \tag{19.24}$$

Let us now consider the C^2-hypersurface with boundary $M = A\cap\partial E$, and the C^1-hypersurface with boundary $N = \partial E\cap\partial A$. Since $\mathrm{bdry}\,M = \mathrm{bdry}\,N$, we set $\Gamma = \mathrm{bdry}\,M = \mathrm{bdry}\,N$ for the common boundary of M and N. If we use ν_E and ν_A to define the orientation of M and N respectively, and denote by ν_Γ^M and ν_Γ^N

the induced orientations on Γ, then by Theorem 11.8,

$$\begin{aligned}\int_{A\cap\partial E}\operatorname{div}_E T\,\mathrm{d}\mathcal{H}^{n-1} &= \int_M \operatorname{div}^M T\,\mathrm{d}\mathcal{H}^{n-1}\\ &= \int_M H_M\,(T\cdot\nu_E)\,\mathrm{d}\mathcal{H}^{n-1} + \int_\Gamma T\cdot\nu_\Gamma^M\,\mathrm{d}\mathcal{H}^{n-2}\\ &= \int_{A\cap\partial E} H_E\,(T\cdot\nu_E)\,\mathrm{d}\mathcal{H}^{n-1} + \int_\Gamma T\cdot\nu_\Gamma^M\,\mathrm{d}\mathcal{H}^{n-2}\,.\end{aligned}$$

At the same time, taking into account that T is tangential to ∂A by (19.13),

$$\int_{\partial A\cap\partial E}\operatorname{div}_E T\,\mathrm{d}\mathcal{H}^{n-1} = \int_N \operatorname{div}^N T\,\mathrm{d}\mathcal{H}^{n-1} = \int_\Gamma T\cdot\nu_\Gamma^N\,\mathrm{d}\mathcal{H}^{n-2}\,,$$

which, of course, is interesting to us because

$$\frac{\mathrm{d}}{\mathrm{d}t}\Big|_{t=0} P(h_t(E);\partial A) = \int_{\partial E\cap\partial A}\operatorname{div}_E T\,\mathrm{d}\mathcal{H}^{n-1}\,.$$

Therefore, from (19.23), we deduce that

$$0 = \int_{A\cap\partial E}(H_E+g)\,(T\cdot\nu_E)\,\mathrm{d}\mathcal{H}^{n-1} + \int_\Gamma T\cdot(\nu_\Gamma^M - \beta\,\nu_\Gamma^N)\,\mathrm{d}\mathcal{H}^{n-2}\,.$$

By (19.14), since $H_E + g$ is constant on $A\cap\partial E$ (Theorem 19.7), we thus find

$$\int_\Gamma T\cdot(\nu_\Gamma^M - \beta\,\nu_\Gamma^N)\,\mathrm{d}\mathcal{H}^{n-2} = 0\,, \tag{19.25}$$

whenever $T\in C_c^\infty(\mathbb{R}^n;\mathbb{R}^n)$ satisfies (19.13) and (19.14). We now remark that for every $T_0\in C_c^\infty(\mathbb{R}^n;\mathbb{R}^n)$ satisfying $T_0\cdot\nu_A = 0$ on ∂A, there exist $s>0$ and $S_0\in C_c^\infty(A;\mathbb{R}^n)$ such that $T = S_0 + s\,T_0\in C_c^\infty(\mathbb{R}^n;\mathbb{R}^n)$ satisfies (19.13) and (19.14). By (19.25), we thus conclude that

$$\int_\Gamma T_0\cdot(\nu_\Gamma^M - \beta\,\nu_\Gamma^N)\,\mathrm{d}\mathcal{H}^{n-2} = 0\,, \tag{19.26}$$

whenever $T_0\in C_c^\infty(\mathbb{R}^n;\mathbb{R}^n)$ and $T_0\cdot\nu_A = 0$ on ∂A. In particular, for every such vector field, we have $T_0\cdot\nu_\Gamma^M = T_0\cdot((\nu_\Gamma^M\cdot\nu_\Gamma^N)\,\nu_\Gamma^N)$, so that (19.26), combined with Exercise 4.14, will finally imply $\beta = \nu_\Gamma^M\cdot\nu_\Gamma^N = -\nu_E\cdot\nu_A$ on Γ. □

19.2 The Schwartz inequality

We now begin laying the ground for the study of sessile liquid drops in Section 19.6, introducing the notion of Schwartz symmetrization. Given a Lebesgue measurable set $E\subset\mathbb{R}^n$ with $|E|<\infty$, we denote by

$$E_t = \{z\in\mathbb{R}^{n-1} : (z,t)\in E\}\,,\qquad t\in\mathbb{R}\,,$$

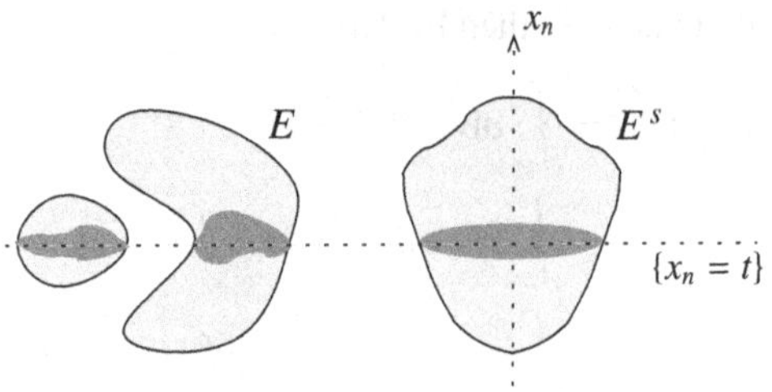

Figure 19.3 Schwartz symmetrization.

the horizontal slices of E, and consider the function $v_E \in L^1(\mathbb{R})$ defined as

$$v_E(t) = \mathcal{H}^{n-1}(E_t), \qquad t \in \mathbb{R}.$$

With Notation 4 in force, we define the Lebesgue measurable set

$$E^* = \left\{x \in \mathbb{R}^n : |\mathbf{p}x| < \left(\frac{v_E(\mathbf{q}x)}{\omega_{n-1}}\right)^{1/(n-1)}\right\}, \tag{19.27}$$

known as the **Schwartz symmetrization E^* of E**; see Figure 19.3. The horizontal slices E_t^* are $(n-1)$-dimensional open balls in $\mathbb{R}^{n-1}$, with

$$v_{E^*}(t) = \mathcal{H}^{n-1}(E_t^*) = \mathcal{H}^{n-1}(E_t) = v_E(t), \qquad \forall t \in \mathbb{R}.$$

By Fubini's theorem, $|E| = |E^*|$. Concerning perimeters, we have the following natural variant of the Steiner inequality (Theorem 14.4).

Theorem 19.11 (Schwartz inequality) *If E is a set of finite perimeter in $\mathbb{R}^n$ with $|E| < \infty$, then E^* is a set of finite perimeter in $\mathbb{R}^n$ and*

$$P(E) \geq P(E^*). \tag{19.28}$$

If equality holds in (19.28), then, for a.e. $t \in \mathbb{R}$, E_t is $\mathcal{H}^{n-1}$-equivalent to an $(n-1)$-dimensional ball, and $\mathbf{q}\nu_E$ is $\mathcal{H}^{n-2}$-a.e. constant on $\partial^* E_t$.

Remark 19.12 These necessary conditions for equality in (19.28) are not sufficient. This is seen by looking at sets E whose horizontal boundary

$$\Sigma_E = \left\{x \in \partial^* E : \nu_E(x) = \pm e_n\right\} \tag{19.29}$$

satisfies $\mathcal{H}^{n-1}(\Sigma_E) > 0$; see Figure 19.4.

Remark 19.13 By construction, E^* is invariant by rotations around the e_n-axis, meaning that $QE^* = E^*$ whenever $Q \in \mathbf{O}(n)$ and $Qe_n = e_n$. By Exercise 15.10, for every $Q \in \mathbf{O}(n)$ such that $Qe_n = e_n$,

$$\nu_{E^*}(Qx) = Q\,\nu_{E^*}(x), \qquad \forall x \in \partial^* E^*.$$

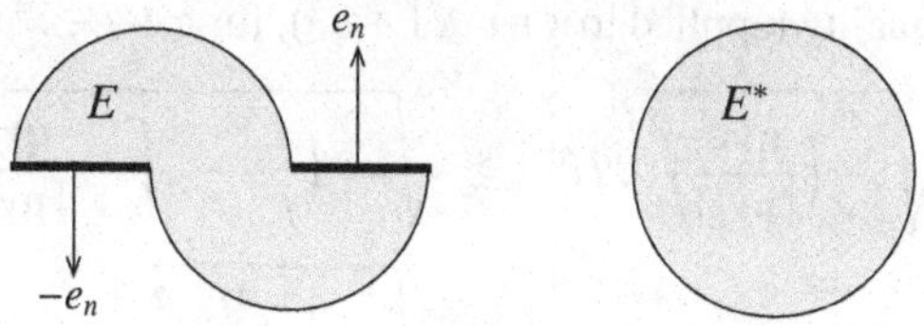

Figure 19.4 If the horizontal boundary of E is not $\mathcal{H}^{n-1}$-negligible, then $P(E) > P(E^*)$ may hold even if the horizontal slices E_t are an $(n-1)$-dimensional ball and the vertical projection of ν_E is constant along $\partial^* E_t$ for a.e. t.

Thus $\mathbf{q}\nu_{E^*}$ is constant on the $(n-2)$-dimensional sphere ∂E_t^* for every $t \in \mathbb{R}$.

Exercise 19.14 (Contractivity of Schwartz rearrangement) If E and F are Lebesgue measurable sets, then $|E^* \Delta F^*| \le |E \Delta F|$. *Hint:* Recall (14.17).

Proof of Theorem 19.11 For every Lebesgue measurable set $E \subset \mathbb{R}^n$, let us consider the singular set Σ_E, defined as in (19.29), and the function $p_E : \mathbb{R} \to [0, \infty]$,

$$p_E(t) = P(E_t), \qquad t \in \mathbb{R}.$$

By Theorem 18.11, if E is a set of locally finite perimeter, then for a.e. $t \in \mathbb{R}$, E_t is a set of locally finite perimeter, and $p_E(t) = \mathcal{H}^{n-2}(\partial^* E_t)$.

Step one: We claim that, if $\mathcal{H}^{n-1}(\Sigma_E) = 0$, then

$$P(E) \ge \int_{\mathbb{R}} \sqrt{v_E'(t)^2 + p_E(t)^2}\, dt, \tag{19.30}$$

with equality if and only if for a.e. $t \in \mathbb{R}$ there exists $c(t) \in \mathbb{R}$ such that

$$\mathbf{q}\nu_E(z,t) = c(t), \qquad \text{for } \mathcal{H}^{n-2}\text{-a.e. } z \in \partial^* E_t. \tag{19.31}$$

Indeed, applying (18.36) to $g(x) = 1_{\partial^* E}(x)$, $x = (z,t) \in \mathbb{R}^n$, and by taking into account that $\mathcal{H}^{n-1}(\Sigma_E) = 0$, we find that

$$P(E) = \int_{\mathbb{R}} dt \int_{\partial^* E_t} \frac{d\mathcal{H}^{n-2}(z)}{|\mathbf{p}\nu_E(z,t)|},$$

which, by $1 = |\nu_E| = \sqrt{(\mathbf{q}\nu_E)^2 + |\mathbf{p}\nu_E|^2}$, can be written as

$$P(E) = \int_{\mathbb{R}} dt \int_{\partial^* E_t} \sqrt{1 + \left(\frac{\mathbf{q}\nu_E(z,t)}{|\mathbf{p}\nu_E(z,t)|}\right)^2}\, d\mathcal{H}^{n-2}(z).$$

By Jensen's inequality (applied to $s \mapsto \sqrt{1+s^2}$), for a.e. $t \in \mathbb{R}$ we find

$$\frac{1}{p_E(t)}\int_{\partial^* E_t}\sqrt{1+\Big(\frac{\mathbf{q}\nu_E}{|\mathbf{p}\nu_E|}\Big)^2}\,d\mathcal{H}^{n-2} \ge \sqrt{1+\Big(\frac{1}{p_E(t)}\int_{\partial^* E_t}\frac{\mathbf{q}\nu_E}{|\mathbf{p}\nu_E|}\,d\mathcal{H}^{n-2}\Big)^2}$$
$$= \sqrt{1+\Big(\frac{v_E'(t)}{p_E(t)}\Big)^2},$$

where the last identity follows from (18.30). Hence, (19.30) is proved. Moreover, by strict convexity, equality holds in (19.30) if and only if, for a.e. $t \in I$,

$$\frac{\mathbf{q}\nu_E}{|\mathbf{p}\nu_E|} = \frac{\mathbf{q}\nu_E}{\sqrt{1-(\mathbf{q}\nu_E)^2}}$$

is $\mathcal{H}^{n-2}$-a.e. constant on $\partial^* E_t$. This proves (19.31).

Step two: We show that if E is such that $v_E \in W^{1,1}(\mathbb{R})$, then E^* is a set of finite perimeter in $\mathbb{R}^n$. We first notice that, if $v \in C^1_c(\mathbb{R})$ with $v \ge 0$ on $\mathbb{R}$, then

$$E(v) = \Big\{x \in \mathbb{R}^n : |z| < \Big(\frac{v(t)}{\omega_{n-1}}\Big)^{1/(n-1)}\Big\}$$

is an open set with almost C^1-boundary (see Section 9.3) and with

$$\nu_{E(v)}(x) = \frac{(-z, r'(t))}{\sqrt{r(t)^2 + r'(t)^2}}, \qquad \forall x = (z,t) \in \partial E(v) \setminus L,$$

where we have set $r(t) = (v(t)/\omega_{n-1})^{n-1}$, $t \in \mathbb{R}$, and where L denotes the e_n-axis. In particular, $\Sigma_{E(v)} \subset L$, and thus $\mathcal{H}^{n-1}(\Sigma_{E(v)}) = 0$. By step one, since $\mathbf{q}\nu_{E(v)}$ is constant on $\partial E(v) \setminus L$, we conclude that

$$P(E(v)) = \int_{\mathbb{R}} \sqrt{v'(t)^2 + p_{E(v)}(t)^2}\,dt, \tag{19.32}$$

where, in turn,

$$p_{E(v)}(t) = (n-1)\omega_{n-1}^{1/(n-1)} v(t)^{(n-2)/(n-1)}, \qquad \forall t \in \mathbb{R}. \tag{19.33}$$

If now E is a set of finite perimeter with $v_E \in W^{1,1}(\mathbb{R})$, then there exists a sequence $\{v_h\}_{h\in\mathbb{N}} \subset C^1_c(\mathbb{R})$ such that $v_h \to v_E$ in $L^1(\mathbb{R})$, and $\sup_{h\in\mathbb{N}} \int_{\mathbb{R}} |v_h'| < \infty$. Since E_t^* and $E(v_h)_t$ are concentric balls, by Fubini's theorem we have

$$|E^* \Delta E(v_h)| = \int_{\mathbb{R}} |v_E - v_h|, \tag{19.34}$$

and thus $E(v_h) \to E^*$ as $h \to \infty$. At the same time, since $\sup_{\mathbb{R}} |v_h| \le \int_{\mathbb{R}} |v_h'|$, we deduce from (19.32) and (19.33) that

$$\sup_{h\in\mathbb{N}} P(E(v_h)) < \infty. \tag{19.35}$$

Combining (19.34) and (19.35), by Proposition 12.15 we conclude that E^* is a set of finite perimeter, as required.

Step three: We show that, if E is a bounded open set with polyhedral boundary and if $\mathcal{H}^{n-1}(\Sigma_E) = 0$, then E^* is a set of finite perimeter in $\mathbb{R}^n$, with

$$P(E^*) = \int_{\mathbb{R}} \sqrt{v_E'(t)^2 + p_{E^*}(t)^2}\,dt\,. \tag{19.36}$$

Moreover, (19.28) holds true and, in fact,

$$2\,P(E)\big(P(E) - P(E^*)\big) \geq \Big(\int_{\mathbb{R}} p_E^2 - p_{E^*}^2\Big)^2\,. \tag{19.37}$$

We start by proving that, under the above assumptions, $\mathrm{Lip}(v_E) < \infty$. First, since ν_E takes only finitely many values and $\mathcal{H}^{n-1}(\Sigma_E) = 0$, there exists $\varepsilon > 0$ such that $|\,\mathbf{q}\nu_E| < 1 - \varepsilon$ on $\partial^* E$. In particular,

$$\sup_{\partial^* E} \frac{|\,\mathbf{q}\nu_E|}{|\,\mathbf{p}\nu_E|} < \infty\,. \tag{19.38}$$

Second, since $E \subset B_R$ for some $R > 0$ and $\partial^* E$ is contained in the union of finitely many hyperplanes in $\mathbb{R}^n$, it turns out that $\partial^* E_t$ is contained in the union of finitely many hyperplanes in $\mathbb{R}^{n-1}$, intersected with an $(n-1)$-dimensional ball of radius R. Hence,

$$\sup_{t\in\mathbb{R}} p_E(t) < \infty\,. \tag{19.39}$$

Since $\mathcal{H}^{n-1}(\Sigma_E) = 0$, by Theorem 18.11 we have $v_E \in W^{1,1}_{\mathrm{loc}}(\mathbb{R})$ with (18.30) in force. Combining (18.30) with (19.38) and (19.39) we thus find $\mathrm{Lip}(v_E) < \infty$. We now prove (19.36). Since $\mathrm{Lip}(v_E) < \infty$,

$$\partial E^* = \big\{(z,t) \in \mathbb{R}^n : |z| = r(t)\big\}\,, \qquad \text{with} \qquad r(t) = \Big(\frac{v_E(t)}{\omega_{n-1}}\Big)^{n-1}\,, \qquad t \in \mathbb{R}\,,$$

by Exercise 10.7, we see that ∂E^* is a $\mathcal{H}^{n-1}$-rectifiable set, with $T_x(\partial E^*) = (-z, r'(t))^\perp$ for $\mathcal{H}^{n-1}$-a.e. $x = (z,t) \in \partial E^*$. Since, by step two, E^* is a set of finite perimeter, by Proposition 10.5 we deduce that

$$\nu_{E^*}(x) = \frac{(-z, r'(t))}{\sqrt{r(t)^2 + r'(t)^2}}\,, \qquad \text{for } \mathcal{H}^{n-1}\text{-a.e. } x = (z,t) \in \partial^* E^*.$$

In particular, $\mathcal{H}^{n-1}(\Sigma_{E^*}) = 0$. Since $\mathbf{q}\nu_{E^*}$ is constant on $\partial^* E$ (by the above formula or by Remark 19.13), we deduce (19.36) from step one. Now, by the Euclidean isoperimetric inequality in $\mathbb{R}^{n-1}$, we have

$$p_E(t) \geq p_{E^*}(t)\,, \qquad \text{for a.e. } t \in \mathbb{R}\,, \tag{19.40}$$

so that (19.28) immediately follows (under the current assumptions on E) by combining (19.30) and (19.36). Finally, from (19.30), (19.36), and

$$\sqrt{a^2+b^2}-\sqrt{a^2+c^2}\geq\frac{b^2-c^2}{2\sqrt{a^2+b^2}}\,,\qquad a,b,c\in\mathbb{R}\,,$$

and by the Hölder inequality we deduce (19.37), as

$$\left(\int_{\mathbb{R}}\sqrt{p_E^2-p_{E^*}^2}\right)^2\leq P(E)\int_{\mathbb{R}}\frac{p_E^2-p_{E^*}^2}{\sqrt{(v_E')^2+p_E^2}}\leq 2\,P(E)\big(P(E)-P(E^*)\big).$$

Step four: We finally prove the theorem. By Remark 13.13, there exists a sequence $\{E_h\}_{h\in\mathbb{N}}$ of bounded open sets with polyhedral boundary such that $E_h\to E$ and $|\mu_{E_h}|\overset{*}{\rightharpoonup}|\mu_E|$. Since ν_{E_h} takes only finitely many values on ∂E_h, for every $h\in\mathbb{N}$ we can find $Q_h\in\mathbf{O}(n)$ such that $\|Q_h-\mathrm{Id}\|<1/h$ and $\nu_{Q_h(E_h)}\neq\pm e_n$ on $\partial Q(E_h)$. By Exercise 15.10 and by Exercise 17.3, up to replace E_h with $Q_h(E_h)$, we may thus assume that Σ_{E_h} is empty for every $h\in\mathbb{N}$. In particular, by step three, $P(E_h)\geq P(E_h^*)$. Since $E_h^*\to E^*$ by Exercise 19.14, by lower semicontinuity we find that

$$P(E)=\lim_{h\to\infty}P(E_h)\geq\liminf_{h\to\infty}P(E_h^*)\geq P(E^*)\,,$$

that is, E^* is a set of finite perimeter and (19.28) holds true. If now equality holds in (19.28), then we also have $P(E_h^*)\to P(E^*)$ and thus, by (19.37),

$$0=\lim_{h\to\infty}2\,P(E_h)\big(P(E_h)-P(E_h^*)\big)\geq\limsup_{h\to\infty}\left(\int_{\mathbb{R}}\sqrt{p_{E_h}^2-p_{E_h^*}^2}\right)^2.$$

In particular, $|p_{E_h}-p_{E_h^*}|\to 0$ a.e. on $\mathbb{R}$ as $h\to\infty$. Since $v_{E_h}\to v_E$ in $L^1(\mathbb{R})$ and

$$p_{E_h^*}=\tau(n)(v_{E_h})^{(n-2)/(n-1)}\,,\qquad p_{E^*}=\tau(n)(v_E)^{(n-2)/(n-1)}\,,$$

(where $\tau(n)=(n-1)\omega_{n-1}^{1/(n-1)}$), we find $p_{E_h}\to p_{E^*}$ a.e. on $\mathbb{R}$. At the same time, for a.e. $t\in\mathbb{R}$, $(E_h)_t\to E_t$ in $\mathbb{R}^{n-1}$. By lower semicontinuity, for a.e. $t\in\mathbb{R}$,

$$p_E(t)=P(E_t)\leq\liminf_{h\to\infty}P((E_h)_t)=\lim_{h\to\infty}p_{E_h}(t)=p_{E^*}(t)\,,$$

that is, E_t is optimal in the Euclidean isoperimetric inequality in $\mathbb{R}^{n-1}$. By Theorem 14.1, it is $\mathcal{H}^{n-1}$-equivalent to an $(n-1)$-dimensional ball. □

19.3 A constrained relative isoperimetric problem

We have already seen (in Exercise 16.8) that half-balls are the only minimizers in the relative isoperimetric problem in a half-space. We now consider a variant of the relative isoperimetric problem in the half-space, in which an additional "trace" constraint is added. Precisely, given $\sigma > 0$, and setting $H = \{x_n > 0\}$, we shall consider the variational problem

$$\inf\Big\{P(E;H) : E \subset H\,,\ |E| = 1\,,\ P(E;\partial H) = \sigma\Big\}\,, \tag{19.41}$$

and characterize the intersection of (suitably positioned) balls with H as the only minimizers in (19.41).

Theorem 19.15 *If $\sigma > 0$, then E is a minimizer in (19.41) if and only if, up to horizontal translations, it is equivalent to the set*

$$F_\sigma = B(s\,e_n, r) \cap H\,, \tag{19.42}$$

where $s \in \mathbb{R}$ and $r > 0$ are uniquely determined by the constraints

$$|F_\sigma| = 1\,, \qquad P(F_\sigma;\partial H) = \sigma\,. \tag{19.43}$$

We preface to the proof of this theorem three simple propositions.

Proposition 19.16 *If E and F are two set of finite perimeter with $E \subset H = \{x_n > 0\}$ and $F \subset \{x_n < 0\}$, then*

$$\begin{aligned}\mu_{E\cup F} &= \mu_E \llcorner H + \mu_F \llcorner \{x_n < 0\}\\ &\quad + e_n\,\mathcal{H}^{n-1} \llcorner \Big((F^{(1/2)} \setminus E^{(1/2)}) \cap \partial H\Big)\\ &\quad - e_n\,\mathcal{H}^{n-1} \llcorner \Big((E^{(1/2)} \setminus F^{(1/2)}) \cap \partial H\Big).\end{aligned}$$

In particular, $P(E \cup F;\partial H) = \mathcal{H}^{n-1}((E^{(1/2)} \Delta F^{(1/2)}) \cap \partial H)$.

Proof Apply Theorem 16.3 and Federer's theorem, Theorem 16.2. □

Proposition 19.17 *If $E \subset H$ is of locally finite perimeter, then $P(E;\partial H) = P(E^*;\partial H)$.*

Proof Consider the **precise Schwartz symmetrization** $F^\star$ of $F \subset \mathbb{R}^n$,

$$F^\star = \left\{x \in \mathbb{R}^n : |z| < \left(\frac{u_F(t)}{\omega_{n-1}}\right)^{1/(n-1)}\right\},$$

where $u_F : \mathbb{R} \to [0,\infty]$ is given by

$$u_F(t) = \mathcal{H}^{n-1}\Big((F^{(1)})_t\Big) = \mathcal{H}^{n-1}\Big((F^{(1)}) \cap \{x_n = t\}\Big)\,, \qquad t \in \mathbb{R}\,.$$

If $|F\Delta G| = 0$, then $F^{(1)} = G^{(1)}$, $u_F = u_G$ on $\mathbb{R}$, and $F^\star = G^\star$ (while $v_F = v_G$ only a.e. on $\mathbb{R}$, and $|F^*\Delta G^*| = 0$). Now let $Q \in \mathbf{O}(n)$ denote the reflection of $\mathbb{R}^n$ with respect to ∂H, and set $E_H = E \cup Q(E)$. Since $E \subset H$, we find

$$E^{(1/2)} \cap \partial H = (E_H)^{(1)} \cap \partial H\,. \tag{19.44}$$

Hence, by Federer's theorem (Theorem 16.2),

$$
\begin{aligned}
P(E;\partial H) = \mathcal{H}^{n-1}((E_H)^{(1)}\cap\partial H) &= u_{E_H}(0)\\
(\text{as } u_{E_H} = u_{(E_H)^\star} \text{ on } \mathbb{R}) \qquad &= u_{(E_H)^\star}(0)\\
(\text{as } |(E_H)^\star\Delta(E_H)^*| = 0) \qquad &= u_{(E_H)^*}(0)\\
(\text{as } |(E_H)^*\Delta(E^*)_H| = 0) \qquad &= u_{(E^*)_H}(0) = P(E^*;\partial H)\,,
\end{aligned}
$$

where the last identity follows from (19.44) (with E^* in place of E). □

Proposition 19.18 *If $E \subset H$ is a set of finite perimeter, equivalent to its Schwartz symmetrization E^*, then there exists $r_0 > 0$ such that*

$$
E^{(1/2)} \cap \partial H = \mathbf{D}_{r_0}\,. \tag{19.45}
$$

Proof *Step one:* Given $r > 0$ and $e \in S^{n-1}$, we claim that the function $\varphi(\lambda) = |B(\lambda\, e, r) \setminus B|$, $\lambda > 0$, is increasing on $(0,\infty)$. The claim is obvious if $n = 1$. We now argue by induction, assuming the claim holds in dimension $n-1$. Assuming without loss of generality that $e = (v, 0)$ for some $v \in \mathbb{R}^{n-1}$, by Fubini's theorem we find

$$
\begin{aligned}
|B(\lambda\, e, r)\setminus B| &= \Big|\Big(B(\lambda\, e, r)\setminus B\Big)\cap\Big\{|x_n| > \min\{r,1\}\Big\}\Big|\\
&\quad + \int_{-\min\{r,1\}}^{\min\{r,1\}} \mathcal{H}^{n-1}\Big(\mathbf{D}\Big(\lambda\, v,\ \sqrt{r^2-t^2}\Big)\setminus\mathbf{D}\Big(0,\ \sqrt{1-t^2}\Big)\Big)\,dt\,.
\end{aligned}
$$

The first term vanishes if $r \le 1$, and it is constant in λ if $r > 1$; the second term is the integral in dt of functions $\Phi(t,\lambda)$ which are increasing in λ by the inductive hypothesis. The claim is proved.

Step two: By assumption, $Q(E)$ is equivalent to E for every $Q \in \mathbf{O}(n)$ such that $Qe_n = e_n$. Hence, $Q(E^{(1/2)}) = E^{(1/2)}$ for every such Q, and

$$
x \in E^{(1/2)} \cap \partial H \qquad \text{if and only if} \qquad (|\mathbf{p}x|v, x_n) \in E^{(1/2)} \cap \partial H\,, \quad \forall v \in S^{n-1}\,.
$$

We are thus left to show that if $x \in E^{(1/2)} \cap \partial H$, then $\lambda\, x \in E^{(1/2)} \cap \partial H$ for every $\lambda \in (0,1)$. Indeed, if $x = (z, 0) \in \partial H$, then by Fubini's theorem and $E \subset H$

$$
\begin{aligned}
\left|\frac{1}{2} - \frac{|E\cap B(\lambda\, x, r)|}{\omega_n\, r^n}\right| &= \frac{|(H\cap B(\lambda x, r))\setminus E|}{\omega_n\, r^n}\\
&= \frac{1}{\omega_n\, r^n}\int_0^r \mathcal{H}^{n-1}\Big(\mathbf{D}(\lambda z,\ \sqrt{r^2-t^2})\setminus\mathbf{D}_{r_E(t)}\Big)\,dt\,,
\end{aligned}
$$

where $r_E(t) = (v_E(t)/\omega_{n-1})^{1/(n-1)}$. By step one (applied in $\mathbb{R}^{n-1}$) and $\lambda < 1$,

$$
\left|\frac{1}{2} - \frac{|E\cap B(\lambda\, x, r)|}{\omega_n\, r^n}\right| \le \left|\frac{1}{2} - \frac{|E\cap B(x, r)|}{\omega_n\, r^n}\right|\,.
$$

Thus, $x \in E^{(1/2)} \cap \partial H$ implies $\lambda x \in E^{(1/2)} \cap \partial H$. □

Proof of Theorem 19.15 *Step one:* We first show that the set F_σ defined by (19.42) and (19.43) is a minimizer in (19.41). Let $E \subset H$, with $|E| = 1$, $P(E) < \infty$, and $P(E; \partial H) = \sigma$. By Proposition 19.17, we have that $P(E^*; \partial H) = P(E; \partial H) = \sigma$. Moreover, by Proposition 19.18, there exists $r_0 > 0$ such that

$$(E^*)^{(1/2)} \cap \partial H = \mathbf{D}_{r_0}\,. \tag{19.46}$$

Necessarily, $\mathcal{H}^{n-1}(\mathbf{D}_{r_0}) = \sigma$. Hence, by construction, the ball $B(s\,e_n, r)$ used in the definition of F_σ satisfies

$$B(s\,e_n, r) \cap \partial H = \mathbf{D}_{r_0}\,. \tag{19.47}$$

In particular, if we set

$$G = \big(E^* \cap H\big) \cup \big(B(s\,e_n, r) \cap \{x_n < 0\}\big),$$

then by (19.46), (19.47), and Proposition 19.16 we find that

$$P(G) = P(E^*; H) + P(B(s\,e_n, r); \{x_n < 0\})\,.$$

By the Euclidean isoperimetric inequality, since $|G| = |B(s\,e_n, r)|$, we also have $P(G) \ge P(B(s\,e_n, r))$, that is, $P(E^*; H) \ge P(B(s\,e_n, r); H) = P(F_\sigma; H)$. By the Schwartz inequality, we finally conclude that $P(E; H) \ge P(F_\sigma; H)$.

Step two: If $s < t$ and E is a set of finite perimeter with $E = E^{(1)}$ and

$$\mathcal{H}^{n-1}(E_t) < \infty\,, \quad \mathcal{H}^{n-1}(E_s) < \infty\,, \quad \mathcal{H}^{n-1}(E_t \cap E_s) = 0\,, \tag{19.48}$$

$$\mathcal{H}^{n-1}((\partial^* E)_t) = \mathcal{H}^{n-1}((\partial^* E)_s) = 0\,, \tag{19.49}$$

then

$$\mathcal{H}^{n-1}(E_t) + \mathcal{H}^{n-1}(E_s) \le P(E; \{s < x_n < t\})\,. \tag{19.50}$$

Indeed, by (19.48) there exists $\{\varphi_h\}_{h\in\mathbb{N}} \subset C^1_c(\mathbb{R}^{n-1}; [-1, 1])$ such that

$$\varphi_h \to 1_{E_t} - 1_{E_s}\,, \qquad \text{in } L^1(\mathbb{R}^{n-1}), \tag{19.51}$$

as $h \to \infty$. Correspondingly, define $\{T_h\}_{h\in\mathbb{N}} \subset C^1_c(\mathbb{R}^n; \mathbb{R}^n)$ by

$$T_h(x) = \varphi_h(x')\,\psi(x_n)\,e_n\,, \qquad x \in \mathbb{R}^n\,, h \in \mathbb{N}\,,$$

where $\psi \in C^1_c(\mathbb{R}; [0, 1])$ and $\psi = 1$ on a neighborhood of (s, t). In this way,

$$\operatorname{div} T_h = 0\,, \qquad \text{on } \{s < x_n < t\}\,, \tag{19.52}$$

for every $h \in \mathbb{N}$. The set $F = E \cap \{s < x_n < t\}$ is of finite perimeter, with

$$\mu_F = \mu_E \llcorner \big\{s < x_n < t\big\} + e_n\,\mathcal{H}^{n-1} \llcorner E_t - e_n\,\mathcal{H}^{n-1} \llcorner E_s\,.$$

by (16.4), (19.49), and since $E = E^{(1)}$. If we apply the divergence theorem to each T_h on F, then by (19.52) we find

$$\int_{\{s<x_n<t\}\cap\partial^* E} T_h \cdot \nu_E \, d\mathcal{H}^{n-1} = \int_{E_t} \varphi_h - \int_{E_s} \varphi_h \, .$$

We let $h \to \infty$ while taking into account (19.51) and the fact that, by construction, $|T_h| \le 1$ on $\mathbb{R}^n$, to deduce (19.50).

Step three: We finally prove that if E is a minimizer in (19.41), then E is equivalent to $z + F_\sigma$ for some $z \in \partial H$. Without loss of generality, we assume that $E = E^{(1)}$. Repeating the argument in step one, and recalling Theorem 14.1 and Theorem 19.11, we easily deduce that E^* is equivalent to F_σ and that, for a.e. $t > 0$, $z_t + E_t$ is equivalent to E_t^* for some $z_t \in \partial H$. In particular, there exist two sequences $\{t_h\}_{h\in\mathbb{N}}$ with $t_h \to 0^+$ and $\{z_h\}_{h\in\mathbb{N}} \subset \partial H$, such that

$$\mathcal{H}^{n-1}\Big((\partial^* E)_{t_h}\Big) = 0 \, , \qquad \mathcal{H}^{n-1}\Big((E + z_h)_{t_h} \Delta (F_\sigma)_{t_h}\Big) = 0 \, , \qquad \forall h \in \mathbb{N} \, .$$

Were $\{z_h\}_{h\in\mathbb{N}}$ unbounded, and up to extracting a subsequence, $\{E_{t_h}\}_{h\in\mathbb{N}}$ would be a disjoint family of balls with $\mathcal{H}^{n-1}(E_{t_h}) = \mathcal{H}^{n-1}((F_\sigma)_{t_h}) = \sigma$ for every $h \in \mathbb{N}$, thus contradicting $P(E; H) < \infty$ by a repeated application of (19.50). Therefore we may also assume without loss of generality that $z_h \to z$ as $h \to \infty$, where $z \in \partial H$. Finally, let us consider the sets of finite perimeter

$$\begin{aligned} G_h &= \Big((E + z_h) \cap \{x_n > t_h\}\Big) \cup \Big(B(s\, e_n, r) \cap \{x_n < t_h\}\Big) \, , \\ G_0 &= \Big(E + z\Big) \cup \Big(B(s\, e_n, r) \cap \{x_n < 0\}\Big) \, . \end{aligned} \tag{19.53}$$

By Proposition 19.16 and by our choice of t_h we easily find that

$$P(G_h) = P(E; \{x_n > t_h\}) + P(B(t\, e_n, r); \{x_n < t_h\}) \, ,$$

so that

$$P(B(s\, e_n, r)) = \lim_{h\to\infty} P(G_h) \, .$$

At the same time, since $z_h \to z$ and $t_h \to 0^+$, we also have $G_h \to G_0$ with $|G_h| = |B(s\, e_n, r)|$ for every $h \in \mathbb{N}$. Hence $|G_0| = |B(s\, e_n, r)|$ and, lower semicontinuity and the Euclidean isoperimetric inequality,

$$P(B(s\, e_n, r)) \le P(G_0) \le \liminf_{h\to\infty} P(G_h) = P(B(s\, e_n, r)) \, .$$

By Theorem 14.1 and (19.53), G_0 is equivalent to $B(s\, e_n, r)$. In particular, $E + z$ is equivalent to F_σ, as required. □

19.4 Liquid drops in the absence of gravity

We now characterize equilibrium shapes of liquid drops confined in a half-space in the absence of gravity. We thus consider the variational problems

$$\psi(\beta) = \inf\Big\{\mathcal{F}_\beta(E;H) : E \subset H\,, P(E) < \infty\,, |E| = 1\Big\}, \tag{19.54}$$

where $H = \{x_n > 0\}$ and $\mathcal{F}_\beta$ was defined in (19.1). The volume constraint is fixed to be $|E| = 1$ for the sake of definiteness, but it is easily modified by scaling. We first remark that the problem is trivial if $\beta \ge 1$, and that it reduces to the Euclidean isoperimetric problem if $\beta \le -1$. The remaining cases $\beta \in (-1, 1)$ are then addressed in Theorem 19.21.

Remark 19.19 (Non-existence of minimizers when $\beta \ge 1$) Given a sequence $R_h \to \infty$, we set $\varepsilon_h = (\omega_{n-1} R_h^{n-1})^{-1}$, and define a sequence $\{E_h\}_{h\in\mathbb{N}}$ of sets of finite perimeter, all contained in H, by setting

$$E_h = \Big\{x \in \mathbb{R}^n : 0 < x_n < \varepsilon_h\,, |\mathbf{p}x| < R_h\Big\}, \qquad h \in \mathbb{N}\,.$$

These cylinders satisfy $|E_h| = 1$ by construction, moreover

$$\begin{aligned}\mathcal{F}_\beta(E_h;H) &= (1-\beta)\,\omega_{n-1}R_h^{n-1} + (n-1)\omega_{n-1}R_h^{n-2}\varepsilon_h\\ &= (1-\beta)\,\omega_{n-1}R_h^{n-1} + \frac{(n-1)}{R_h}\,.\end{aligned}$$

In particular, if $\beta > 1$, then $\mathcal{F}_\beta(E_h;H) \to -\infty$ as $h \to \infty$, so that $\psi(\beta) = -\infty$, and no minimizer in (19.54) exists. In the case $\beta = 1$, we simply have $\mathcal{F}_\beta(E_h;H) \to 0$ as $h \to \infty$, showing that $\psi(1) \le 0$. However, by Proposition 19.22 below, we have $\mathcal{F}_1(E;H) > 0$ for every competitor E, so that $\psi(1) = 0$ and minimizers cannot exist in this case either.

Remark 19.20 (Balls are the only minimizers when $\beta \le -1$) We claim that if $r > 0$ is such that $\omega_n r^n = 1$, then

$$\mathcal{F}_\beta(E;H) \ge \mathcal{F}_\beta(B(x,r);H)\,, \tag{19.55}$$

for every $\beta \le -1$, $E \subset H$ with $|E| = 1$, and $x \in \mathbb{R}^n$ such that $x_n \ge r$; moreover, equality holds if and only if E is equivalent to a ball. Indeed, by (19.4) (with $A = H$), since $\beta \le -1$, and by the Euclidean isoperimetric inequality,

$$\begin{aligned}\mathcal{F}_\beta(E;H) &\ge P(E;H) + P(E;\partial H) = P(E)\\ &\ge P(B(x,r)) = \mathcal{F}_\beta(B(x,r);H)\,,\end{aligned}$$

where in the last identity we have used $x_n \ge r$ to infer $P(B(x,r);\partial H) = 0$. This proves (19.55). Moreover, if equality holds in (19.55), then we have $P(E) = P(B(x,r))$, and thus E is equivalent to a ball thanks to Theorem 14.1.

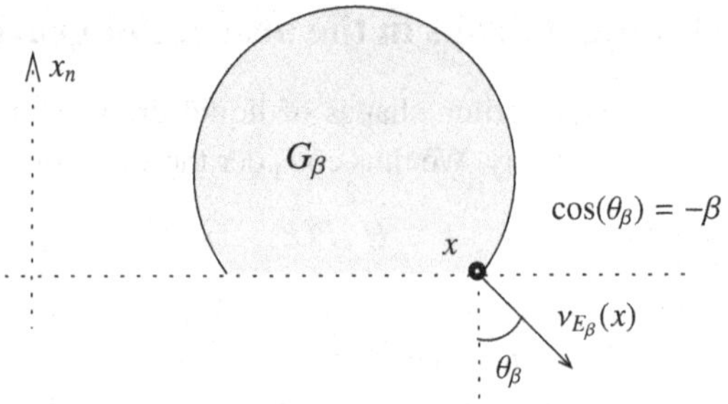

Figure 19.5 Given $\beta \in (-1, 1)$, a minimizer (unique up to horizontal translations) G_β in (19.54) is obtained by suitably intersecting a ball with center on the e_n-axis with the half-space H. For $\beta \to (-1)^+$, minimizers converge to balls contained in H and tangent to ∂H, in accordance with Remark 19.20. For $\beta \to 1^-$, we have $r_\beta \to +\infty$, and minimizers flatten out against ∂H, mimicking the minimizing sequence constructed in Remark 19.19.

Theorem 19.21 (Liquid drops in the absence of gravity) *For every $\beta \in (-1, 1)$, there exists a unique $\sigma(\beta) > 0$ with the following property: a set of finite perimeter $E \subset H$ with $|E| = 1$ is a minimizer in the variational problem (19.54) if and only if, up to horizontal translation, E is equivalent to the set*

$$G_\beta = F_{\sigma(\beta)}\,,$$

where F_σ, $\sigma > 0$, is defined in (19.42) and (19.43). Moreover,

$$\nu_{G_\beta} \cdot e_n = \beta\,, \qquad \text{on } \mathrm{bdry}\,(H \cap \partial G_\beta). \tag{19.56}$$

We preface the following proposition to the proof of Theorem 19.21.

Proposition 19.22 *If E is a set of finite perimeter in $\mathbb{R}^n$, $|E| < \infty$, then*

$$\mathcal{H}^{n-1}\Big(E^{(1)} \cap \{x_n = t\}\Big) < P(E; \{x_n > t\})\,, \tag{19.57}$$

for every $t \in \mathbb{R}$ such that $|E \cap \{x_n > t\}| > 0$. In particular, if $E \subset H$ and $0 < |E| < \infty$, then $P(E; \partial H) < P(E; H)$ and

$$\mathcal{F}_\beta(E; H) \geq \frac{1-\beta}{2}\,P(E)\,, \qquad \forall |\beta| \leq 1\,. \tag{19.58}$$

Proof *Step one:* By Theorem 16.3, if we prove $P(F; \partial J) < P(F; J)$ for $F = E \cap \{x_n > t\}$ and $J = \{x_n > t\}$, then (19.57) follows. In other words, it suffices to prove that $P(E; \partial H) < P(E; H)$ whenever $E \subset H$ and $0 < |E| < \infty$. To this end, we consider $\varphi_R \in C^\infty(\mathbb{R}; [0, 1])$ with $\varphi_R = 1$ on $(-\infty, R]$, $\varphi_R = 0$ on $[R + 1, \infty)$ and $|\varphi_R'| \leq 2$, and define $T_R \in C_c^1(\mathbb{R}^n; \mathbb{R}^n)$ $(R > 0)$ as

$$T_R(x) = \varphi_R(|x|)\,e_n\,, \qquad x \in \mathbb{R}^n\,.$$

In this way, for every $x \in \mathbb{R}^n$,

$$\nabla T_R(x) = \frac{\varphi_R'(|x|)}{|x|}\, e_n \otimes x\,, \qquad \operatorname{div} T_R(x) = \frac{\varphi_R'(|x|)\, x_n}{|x|}\,.$$

By Exercise 16.6, $\nu_E = \nu_H = -e_n$ $\mathcal{H}^{n-1}$-a.e. on $\partial^* E \cap \partial H$. Hence,

$$\begin{aligned}\int_E \frac{\varphi_R'(|x|)\, x_n}{|x|}\, \mathrm{d}x &= \int_{\partial^* E} T_R \cdot \nu_E \, \mathrm{d}\mathcal{H}^{n-1} \\ &= \int_{H\cap\partial^* E} \varphi_R\, (\nu_E \cdot e_n)\, \mathrm{d}\mathcal{H}^{n-1} - \int_{\partial^* E\cap\partial H} \varphi_R \, \mathrm{d}\mathcal{H}^{n-1}\,, \qquad (19.59)\end{aligned}$$

by the divergence theorem. Since $|E| < \infty$, we certainly have, as $R \to \infty$,

$$\omega(R) = \left| \int_E \frac{\varphi_R'(|x|)\, x_n}{|x|}\, \mathrm{d}x \right| \le 2 \left| E \cap \big(B_{R+1} \setminus B_R\big) \right| \to 0\,.$$

Therefore we may regroup terms in (19.59) to find

$$\begin{aligned}-\omega(R) + \int_{\partial^* E\cap\partial H} \varphi_R \, \mathrm{d}\mathcal{H}^{n-1} &= \int_{H\cap\partial^* E} \varphi_R\, (\nu_E \cdot e_n)\, \mathrm{d}\mathcal{H}^{n-1} \\ &\le \int_{H\cap\partial^* E} |\nu_E \cdot e_n| \, \mathrm{d}\mathcal{H}^{n-1}\,,\end{aligned}$$

which implies, by the properties of φ_R,

$$-\omega(R) + P(E; \partial H \cap B_R) \le P(E; H) - \int_{H\cap\partial^* E} \big(1 - |\nu_E \cdot e_n|\big)\, \mathrm{d}\mathcal{H}^{n-1}\,.$$

We now let $R \to \infty$ (recall that $P(E; \partial H \cap B_R) = |\mu_E|(\partial H \cap B_R) \to |\mu_E|(\partial H)$ by Exercise 1.8) to find that

$$P(E; \partial H) \le P(E; H) - \int_{H\cap\partial^* E} \big(1 - |\nu_E \cdot e_n|\big)\, \mathrm{d}\mathcal{H}^{n-1}\,.$$

We have thus proved that either $P(E; \partial H) < P(E; H)$, or

$$\nu_E = \pm\, e_n\,, \qquad \mathcal{H}^{n-1}\text{-a.e. on } H \cap \partial^* E\,.$$

In the latter case, by Exercise 15.18, there exists a set of finite perimeter $I \subset (0, \infty)$ such that $E \cap H = I \times \mathbb{R}^{n-1}$. Since $|E| > 0$, it must be that $|I| > 0$, and then $|E| = \infty$, against our assumptions. Thus, $P(E; \partial H) < P(E; H)$.

Step two: By Exercise 16.6, $P(E) = P(E; H) + P(E; \partial H)$, so that

$$\mathcal{F}_\beta(E; H) = \frac{1+\beta}{2}\big(P(E; H) - P(E; \partial H)\big) + \frac{1-\beta}{2}\, P(E)\,.$$

Since $P(E; H) > P(E; \partial H)$ we immediately deduce (19.58). □

Proof of Theorem 19.21 Thanks to Theorem 19.15,

$$\psi(\beta) = \inf\left\{P(F_\sigma; H) - \beta\sigma : \sigma > 0\right\}. \tag{19.60}$$

By Proposition 19.22, $P(F_\sigma; H) > P(F_\sigma; \partial H) = \sigma$, so that

$$P(F_\sigma; H) - \beta\sigma > (1-\beta)\sigma, \qquad \forall \sigma > 0.$$

In particular, there exists $\sigma_0 \ge 0$ such that

$$\psi(\beta) = \mathcal{F}_\beta(F_{\sigma_0}; H). \tag{19.61}$$

In fact, there is a unique σ_0 with this property, since, by Young's law (Theorem 19.8), (19.61) implies $\nu_{F_{\sigma_0}} \cdot e_n = \beta$ on bdry $(H \cap \partial F_{\sigma_0})$, a condition that uniquely determines σ_0 in terms of β in the family $\{F_\sigma\}_{\sigma \ge 0}$. Moreover, since $\beta > -1$, it must be that $\sigma_0 > 0$, as claimed. □

19.5 A symmetrization principle

In this section we apply the Schwartz inequality, together with some elementary arguments based on symmetrization by reflection, in order to prove an a priori symmetry result for equilibrium shapes of the liquid drop free energy in strips. Given $T \in (0, \infty]$, we shall work in the strip

$$S_T = \left\{0 < x_n < T\right\}.$$

Of course, in the case $T = \infty$, $S_T = H = \{x_n > 0\}$. We define $\mathcal{F}_\beta(E; S_T)$ and $\mathcal{G}(E)$ as in (19.1) and (19.2).

Theorem 19.23 (Symmetrization principle for liquid drops in strips) *If $\beta \in \mathbb{R}$, $g \in L^1(\mathbb{R}^n)$, $E \subset S_T$ is a set of finite perimeter with $0 < |E| < \infty$ and*

$$\mathcal{F}_\beta(E; S_T) + \mathcal{G}(E) \le \mathcal{F}_\beta(F; S_T) + \mathcal{G}(E), \tag{19.62}$$

for every $F \subset S_T$ with $|E| = |F|$, then there exists $z \in \mathbb{R}^{n-1}$ such that E is equivalent to $z + E^*$.

Proof By Exercise 16.6 we have

$$\mu_E = \nu_E \mathcal{H}^{n-1} \llcorner (S_T \cap \partial^* E) \tag{19.63}$$
$$-e_n \mathcal{H}^{n-1} \llcorner \left(E^{(1/2)} \cap \left\{x_n = 0\right\}\right) + e_n \mathcal{H}^{n-1} \llcorner \left(E^{(1/2)} \cap \left\{x_n = T\right\}\right),$$

so that, in particular,

$$P(E) = P(E; S_T) + P(E; \partial S_T). \tag{19.64}$$

When $T = \infty$ and $S_T = H$, we have of course $\{x_n = T\} = \emptyset$.

Step one: We show that for a.e. $t \in \mathbb{R}$ the horizontal slice E_t of E is $\mathcal{H}^{n-1}$-equivalent to $z_t + E_t^*$ for some $z_t \in \mathbb{R}^{n-1}$. Indeed, by Fubini's theorem,

$$\int_E g(x_n)\,\mathrm{d}x = \int_0^T g(t)\,\mathcal{H}^{n-1}(E_t)\,\mathrm{d}t = \int_0^T g(t)\,\mathcal{H}^{n-1}(E_t^*)\,\mathrm{d}t = \int_{E^*} g(x_n)\,\mathrm{d}x\,.$$

Since $E^* \subset S_T$, from (19.62) we find $\mathcal{F}_\beta(E;S_T) \le \mathcal{F}_\beta(E^*;S_T)$. By Proposition 19.17, $P(E;\partial S_T) = P(E^*;\partial S_T)$, so that

$$P(E;S_T) \le P(E^*;S_T)\,.$$

Adding $P(E;\partial S_T) = P(E^*;\partial S_T)$ to both sides, and taking (19.64) into account, we thus find that $P(E) \le P(E^*)$. By the Schwartz inequality, $P(E) = P(E^*)$, and thus Theorem 19.11 implies our claim.

Step two: We show that, up to a horizontal translation along the e_1-axis, E may be assumed symmetric under reflection with respect to the hyperplane $\{x_1 = 0\}$. Indeed, let $Q_1 \in \mathbf{O}(n)$ denote the reflection of $\mathbb{R}^n$ with respect to $\{x_1 = 0\}$, and translate E along the e_1-axis to have

$$\left|E \cap \{x_1 > 0\}\right| = \left|E \cap \{x_1 < 0\}\right| = \frac{|E|}{2}\,.$$

Correspondingly, consider the sets of finite perimeter E^+ and E^- defined as

$$E^+ = \big(E \cap \{x_1 > 0\}\big) \cup Q_1\big(E \cap \{x_1 > 0\}\big),$$
$$E^- = \big(E \cap \{x_1 < 0\}\big) \cup Q_1\big(E \cap \{x_1 < 0\}\big);$$

see Figure 19.6. Since $E^+, E^- \subset S_T$, with $|E^+| = |E^-| = |E|$, we have

$$\min\left\{\mathcal{F}(E^+), \mathcal{F}(E^-)\right\} \ge \mathcal{F}(E)\,, \tag{19.65}$$

where we have set $\mathcal{F}(E) = \mathcal{F}_\beta(E;S_T) + \mathcal{G}(E)$ for the sake of brevity. We easily see that $(E^+)^{(1/2)} \cap \{x_1 = 0\}$ is empty, so that, by Federer's theorem and by directly applying the divergence theorem we compute

$$\mu_{E^+} = \nu_E\,\mathcal{H}^{n-1} \llcorner \big(\{x_1 > 0\} \cap \partial^* E\big) + Q_1 \nu_E\,\mathcal{H}^{n-1} \llcorner \big(\{x_1 < 0\} \cap Q_1 \partial^* E\big).$$

In particular,

$$\frac{\mathcal{F}_\beta(E^+;S_T)}{2} = P\big(E; S_T \cap \{x_1 > 0\}\big) - \beta\, P\big(E; \partial S_T \cap \{x_1 > 0\}\big).$$

Taking into account the analogous identity for E^-, we find that

$$\mathcal{F}(E) = \frac{\mathcal{F}(E^+) + \mathcal{F}(E^-)}{2} + P\big(E; S_T \cap \{x_1 = 0\}\big), \tag{19.66}$$

(notice that $P(E;\partial S_T \cap \{x_1 = 0\}) = 0$, as $\mathcal{H}^{n-1}(\partial S_T \cap \{x_1 = 0\}) = 0$). Combining (19.65) and (19.66) we thus find that $\mathcal{F}(E^+) = \mathcal{F}(E^-) = \mathcal{F}(E)$. In particular,

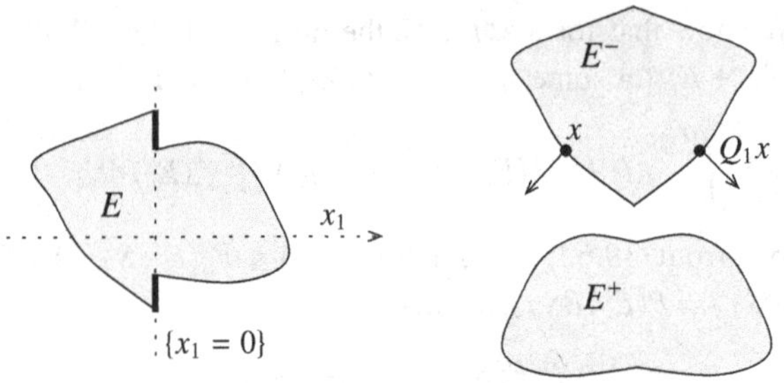

Figure 19.6 A set E which is cut into two halves of equal measure by the hyperplane $\{x_1 = 0\}$. Clearly, $P(E)$ is the sum of $P(E;\{x_1 > 0\}) = P(E^+)/2$, $P(E;\{x_1 < 0\}) = P(E^-)/2$, plus a possible contribution coming from $\partial^* E \cap \{x_1 = 0\}$, which we have depicted by bold segments, and which is "invisible" for E^+ and E^-. We also notice that $P(E;\partial H)$ is equal to the sum of $P(E;\partial H \cap \{x_1 > 0\})$ and $P(E;\partial H \cap \{x_1 < 0\})$, since $\partial H \cap \{x_1 = 0\}$ is an $(n-2)$-dimensional plane in $\mathbb{R}^n$. Finally, $x \in \partial^* E^+$ if and only if $Q_1 x \in \partial^* E^+$, with $\nu_{E^+}(Q_1 x) = Q_1 \nu_E(x)$; see Exercise 15.10.

E^+ satisfies the same minimality property of E. Applying step one to E^+, we see that for a.e. $t \in \mathbb{R}$ the horizontal slice $(E^+)_t$ is $\mathcal{H}^{n-1}$-equivalent to an $(n-1)$-dimensional ball, whose center lies necessarily on $\{x_1 = 0\}$ (indeed, $Q_1 E^+ = E^+$). Since, by construction and by step one,

$$(E^+)_t \cap \{x_1 > 0\} = E_t \cap \{x_1 > 0\} = (z_t + E_t^*) \cap \{x_1 > 0\},$$

with $z_t \in \mathbb{R}^{n-1}$, we conclude that, necessarily, $z_t \in \mathbb{R}^{n-1} \cap \{x_1 = 0\}$ for a.e. $t \in \mathbb{R}$. In particular, for a.e. $t \in \mathbb{R}$, E_t is $\mathcal{H}^{n-1}$-equivalent to an $(n-1)$-dimensional ball centered on $\{x_1 = 0\}$, that is, E is symmetric under reflection with respect to $\{x_1 = 0\}$, as claimed.

Step three: Up to performing suitable translations along the first $(n-1)$-coordinate axes, we can safely assume that

$$\left|E \cap \{x_k > 0\}\right| = \left|E \cap \{x_k < 0\}\right| = \frac{|E|}{2}, \qquad \forall k = 1, \dots, n-1.$$

By step two we conclude that for a.e. $t \in \mathbb{R}$, E_t is $\mathcal{H}^{n-1}$-equivalent to an $(n-1)$-dimensional ball with center on

$$\mathbb{R}^{n-1} \cap \bigcap_{k=1}^{n-1} \{x_k = 0\} = \{0\},$$

that is, for a.e. $t \in \mathbb{R}$, E_t is $\mathcal{H}^{n-1}$-equivalent to E_t^*, as required. □

19.6 Sessile liquid drops

We finally discuss the sessile liquid drop problem,

$$\inf\left\{\mathcal{F}_\beta(E;H)+g\int_E x_n\,\mathrm{d}x : E\subset H\,,P(E)<\infty\,,|E|=m\right\},\qquad(19.67)$$

that is the equilibrium problem for a liquid drop sitting on a horizontal (hyper)plane under the action of gravity.

Theorem 19.24 (Sessile liquid drops) *If $\beta\in(-1,1)$, $g>0$ and $m>0$, then there exist minimizers in the variational problem (19.67). Every such minimizer is equivalent to a bounded set, which, up to translation, it is equivalent to its Schwartz symmetrization.*

Remark 19.25 (Regularity, Young's law, and uniqueness) In Part III we shall complete this result by showing that if E is a minimizer in (19.67), then $H\cap\partial E$ is an analytic hypersurface. This regularity property makes it possible to fully exploit the axial symmetry of the liquid drop, as it allows us to write down the Euler–Lagrange equation of the problem as a classical ODE satisfied by the one-dimensional profile of the drop. The analysis of this ODE shows that $H\cap\partial E$ is analytic up to the boundary (so that Young's law is satisfied at $\mathrm{bdry}\,(H\cap\partial E)$), and that, given m and g, the minimizing profile is unique (thus providing a global uniqueness result up to horizontal translations).

Remark 19.26 When $\beta\ge 1$ we see that (19.67) admits no minimizer by looking at the same minimizing sequence constructed in Remark 19.19.

Proposition 19.27 *If $\beta\in[-1,1)$ and $\{E_h\}_{h\in\mathbb{N}}$ is a sequence of sets of finite perimeter with $E_h\to E$, then*

$$\mathcal{F}_\beta(E;H)\le\liminf_{h\to\infty}\mathcal{F}_\beta(E_h;H)\,.\qquad(19.68)$$

Proof We may directly assume the right-hand side of (19.68) to be finite. In this way, by (19.58) and since $\beta<1$, we find that $\sup_{h\in\mathbb{N}}P(E_h)<\infty$ and $\mu_{E_h}\rightharpoonup\mu_E$. Hence, if $\beta\in[-1,0]$, then (19.68) follows from the identity

$$\mathcal{F}_\beta(E;H)=(1+\beta)P(E;H)-\beta\,P(E)\,.$$

If, instead, $\beta\in[0,1)$, then, by arguing as in the proof of Proposition 19.3, we can reduce to proving that

$$P(F;\partial H)\le P\Big(F;\big\{0<x_n<\delta\big\}\Big)+C(\delta)|F|\,,\qquad\forall F\subset H\,.\qquad(19.69)$$

To this end, let us fix a Lipschitz function $\varphi\colon[0,\infty)\to[0,1]$ with $\varphi=1$ on $[0,1]$, $\varphi=0$ on $[2,\infty)$ and $|\varphi'|\le 1$ everywhere. Correspondingly, for every

$\delta > 0$ and $R > 0$, we may define a compactly supported Lipschitz vector field $T_{\delta,R}\colon \mathbb{R}^n \to \mathbb{R}^n$ by setting

$$T_{\delta,R}(x) = -\varphi\left(\frac{|x_n|}{\delta}\right)\varphi\left(\frac{|\mathbf{p}x|}{R}\right) e_n\,, \qquad x \in \mathbb{R}^n\,.$$

This vector field satisfies $T_{\delta,R} \cdot \nu_H \ge 0$ on ∂H, and $T_{\delta,R} \cdot \nu_H = 1$ on $\mathbf{D}_R \cap \partial H$. Since $|T_{\delta,R}| \le 1$ on $\mathbb{R}^n$ and, in fact, $|\nabla T_{\delta,R}| \le C/\delta$ (provided $\delta < R$), from

$$\int_F \operatorname{div} T_{\delta,R} = \int_{H\cap\partial F} T_{\delta,R} \cdot \nu_F \, d\mathcal{H}^{n-1} + \int_{\partial H\cap\partial F} T_{\delta,R} \cdot \nu_H \, d\mathcal{H}^{n-1}\,,$$

we conclude that

$$P\big(F; \mathbf{D}_R \cap \partial H\big) \le P\big(F; \{0 < x_n < \delta\}\big) + \frac{C}{\delta}\,|F|\,.$$

We just let $R \to \infty$ in this inequality to prove (19.69). □

Proof of Theorem 19.24 *Step one:* The symmetrization principle of Theorem 19.23 allows us to restrict the competition class to those sets $E \subset H$ with $P(E) < \infty$ and $|E| = m$ which are equivalent to their Schwartz symmetrizations. In particular, denoting by γ the infimum in (19.67), we may consider a sequence $\{E_h\}_{h\in\mathbb{N}}$ of sets of finite perimeter contained in H, with $|E_h| = m$ and

$$E_h = \left\{x \in \mathbb{R}^n : x_n > 0\,, |\mathbf{p}x| < \left(\frac{v_h(x_n)}{\omega_{n-1}}\right)^{1/(n-1)}\right\}, \tag{19.70}$$

where $\{v_h\}_{h\in\mathbb{N}} \subset L^1([0,\infty))$, $v_h \ge 0$, and

$$\gamma = \lim_{h\to\infty} \mathcal{F}_\beta(E_h; H) + g\int_{E_h} x_n \, dx\,. \tag{19.71}$$

Since $\gamma < \infty$, $\beta \in (-1,1)$ and, by (19.58),

$$\mathcal{F}_\beta(E_h; H) + g\int_{E_h} x_n \, dx \ge \mathcal{F}_\beta(E_h; H) \ge \frac{1-\beta}{2}\, P(E_h)\,, \tag{19.72}$$

$\{P(E_h)\}_{h\in\mathbb{N}}$ is bounded. By Corollary 12.27, up to extracting a subsequence,

$$E_h \overset{\text{loc}}{\to} E\,, \qquad \mu_{E_h} \overset{*}{\rightharpoonup} \mu_E\,,$$

where $E \subset H$ is a set of finite perimeter. By (19.57),

$$P(E_h) \ge P(E_h; \{x_n > t\}) \ge \mathcal{H}^{n-1}\Big((E_h)^{(1)} \cap \big\{x_n = t\big\}\Big) = v_{(E_h)^{(1)}}(t)\,, \qquad \forall t > 0\,,$$

so that, by (19.70), (19.72), and $E_h \overset{\text{loc}}{\to} E$,

$$E\,, E_h \subset \mathbf{D}_R \times (0,\infty)\,, \tag{19.73}$$

where

$$R = \left(\frac{2\gamma}{\omega_{n-1}(1-\beta)}\right)^{1/(n-1)}.$$

At the same time, for every $M > 0$ we have

$$|E_h \Delta E| \le \left|\left(E_h \Delta E\right) \cap \left(\mathbf{D}_R \times (0, M)\right)\right| + \frac{g}{M}\left(\int_{E_h} x_n \, dx + \int_E x_n \, dx\right),$$

so that, by local convergence, (19.71), and Proposition 12.31,

$$\limsup_{h\to\infty} |E_h \Delta E| \le \frac{2\gamma}{M}, \qquad \forall M > 0.$$

In particular, $E_h \to E$ and $|E| = m$, so that E is a competitor in (19.67). By Proposition 19.27 and Proposition 12.31, E is a minimizer.

Step two: As an intermediate step in proving the boundedness of E, we notice the following property of the equilibrium shapes in the absence of gravity G_β constructed in Theorem 19.21. Precisely, for every $\beta \in (-1, 1)$ there exists a positive constant τ, depending on n and β only, such that

$$P(F) \ge (1+\tau)\mathcal{F}_\beta(r\,G_\beta) \tag{19.74}$$

for every $F \subset H$ with $|F| = |r\,G_\beta| = r^n$. Indeed, if B_s is a ball with $P(B_s; \partial H) = 0$ and $|B_s| = |F|$, then by the Euclidean isoperimetric inequality we have

$$P(F) \ge P(B_s) = \mathcal{F}_\beta(B_s) = (1+\tau)\mathcal{F}_\beta(r\,G_\beta),$$

where, by definition,

$$\tau = \frac{\mathcal{F}_\beta(B_s)}{\mathcal{F}_\beta(r\,G_\beta)} - 1.$$

Since $\beta > -1$, we find that $\tau > 0$ by the uniqueness part of Theorem 19.21.

Step three: We show that E is equivalent to a bounded set. Taking (19.73) into account, we reduce to proving the boundedness of the support $[0, a]$ of the decreasing function $m_E\colon (0,\infty) \to [0, m]$,

$$m_E(t) = |E \cap \{x_n > t\}|, \qquad t > 0,$$

which, by Fubini's theorem, is absolutely continuous with

$$m_E'(t) = -v_E(t), \qquad \text{for a.e. } t > 0. \tag{19.75}$$

In fact, we shall prove that

$$a \le t_0 + \varepsilon\left(\frac{2\,m}{\omega_n}\right)^{1/n}, \qquad t_0 = m^{1/n}\,h_\beta, \tag{19.76}$$

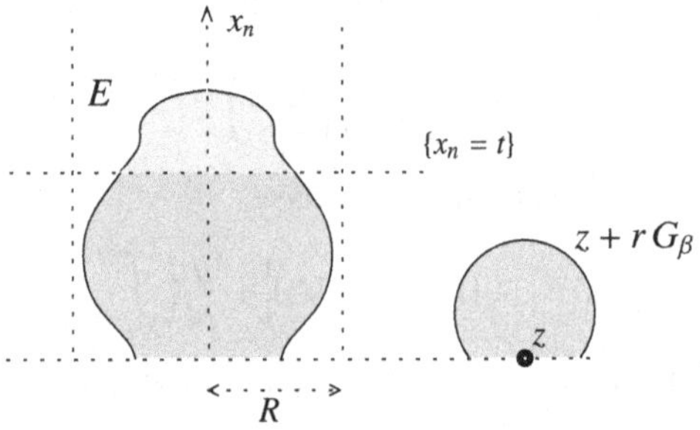

Figure 19.7 The set E (equivalent to E^*), and the competitor $F(t)$ used to prove the boundedness of E (depicted in darker gray).

where $h_\beta = \sup\{x_n : x \in G_\beta\}$. To prove this let us consider the comparison sets

$$F(t) = \left(E \cap \left\{x_n < t\right\}\right) \cup \left(z + r G_\beta\right), \qquad t > t_0 ,$$

where $z \in \mathbb{R}^{n-1}$, $|z| > R$, and $r = r(t)$ is such that $m_E(t) = |r G_\beta|$, so as to guarantee $|E| = |F(t)|$; see Figure 19.7. Since we also have $F(t) \subset H$, by minimality,

$$\begin{aligned} &P(E; H) - \beta P(E; \partial H) + g \int_E x_n \qquad (19.77)\\ &\le P\left(E \cap \left\{x_n < t\right\}; H\right) - \beta P(E; \partial H) + \mathcal{F}_\beta(r G_\beta; H) + g \int_{F(t)} x_n \, dx . \end{aligned}$$

By (16.10), for a.e. $s > 0$,

$$P(E; H) - P(E \cap \{x_n < s\}; H) = P(E; \{x_n > s\}) - |m_E'(s)| ,$$

while, by step two and since $|r G_\beta| = |E \cap \{x_n > t\}|$, we have

$$\mathcal{F}_\beta(r G_\beta; H) \le (1 - \varepsilon) P(E \cap \{x_n > t\}) ,$$

where $\varepsilon > 0$ depends on n and β only (with $\varepsilon \to 0^+$ if $\beta \to (-1)^+$). Taking also into account that, by $t \ge t_0$,

$$\begin{aligned} \int_E x_n \, dx - \int_{F(t)} x_n \, dx &= \int_{E \cap \{x_n > t\}} x_n \, dx - \int_{r G_\beta} x_n \, dx \\ &\ge \left(t - m^{1/n} h_\beta\right) m_E(t) \ge 0 , \end{aligned}$$

we finally deduce from (19.77) that

$$P(E; \{x_n > t\}) \le (1 - \varepsilon) P(E \cap \{x_n > t\}) + |m_E'(t)| .$$

Adding $|m'_E(t)|$ to both sides of the inequality, and recalling that, by (16.10),

$$P(E; \{x_n > s\}) + |m'_E(s)| = P(E \cap \{x_n > s\}), \qquad \text{for a.e. } s > 0,$$

we conclude that, for a.e. $t > t_0$, and by the Euclidean isoperimetric inequality,

$$2\,|m'_E(t)| \geq \varepsilon\, P(E \cap \{x_n > t\}) \geq n\omega_n^{1/n}\, \varepsilon\, m_E(t)^{(n-1)/n}.$$

Since $m_E > 0$ on $[0, a)$, we thus find $(-2\, m_E^{1/n})' \geq \omega_n^{1/n} \varepsilon$ a.e. on $[t_0, a)$. Integrating this inequality on $t \in [m^{1/n}\, h_\beta, t)$, we find

$$m_E(t)^{1/n} \leq m_E(t_0)^{1/n} - \varepsilon\omega_n^{1/n} \frac{(t - t_0)}{2},$$

which implies (19.76). □

20

Anisotropic surface energies

The surface tension energy of a liquid can be modeled, as seen in the previous chapter, by the perimeter of the region occupied by the liquid itself. Something similar happens in the study of solid crystals with sufficiently small grains, although in this case the crystalline structure of the material will lead us to observe a surface tension energy of anisotropic character. In mathematical terms, an anisotropic surface tension energy may be introduced by considering a bounded, positive Borel function Φ on S^{n-1} (the anisotropic surface tension) and, correspondingly, by associating with every bounded open set $E \subset \mathbb{R}^n$ with C^1-boundary its **Φ-surface energy**,

$$\int_{\partial E} \Phi(\nu_E(x))\, d\mathcal{H}^{n-1}(x) . \tag{20.1}$$

In the isotropic case (Φ is constant), Φ-surface energy and perimeter coincide. However, the surface tension Φ corresponding to a given crystal agrees with the maximum of finitely many affine functions, and thus shows a genuinely anisotropic nature. In this chapter we provide a brief introduction to anisotropic surface energies. In Section 20.1 we discuss the basic properties of anisotropic surface energies, and prove the existence of minimizers in some anisotropic variational problems. In Section 20.2 we address the *Wulff problem*, the natural anisotropic counterpart of the Euclidean isoperimetric problem. Noticeably, the Wulff problem will provide us with an example of a physically meaningful geometric variational problem which minimizers possess singularities.

20.1 Basic properties of anisotropic surface energies

We shall say that $\Phi \colon \mathbb{R}^n \to [0, \infty]$ is **one-homogeneous** if

$$\Phi(x) = |x|\, \Phi\Big(\frac{x}{|x|}\Big), \qquad \forall x \in \mathbb{R}^n \setminus \{0\} .$$

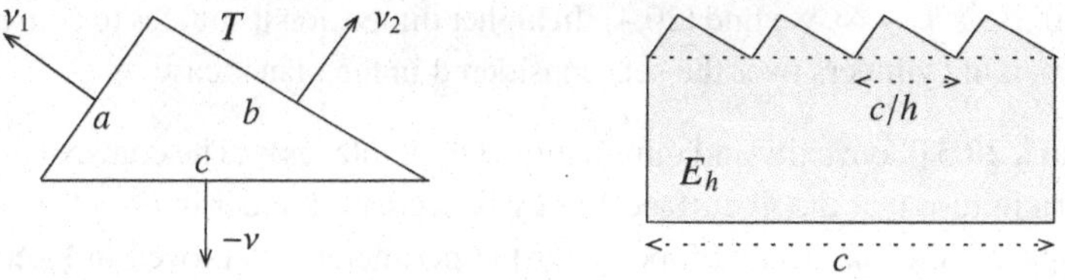

Figure 20.1 A sequence of sets used in showing that convexity of Φ is a necessary condition to lower semicontinuity of $\mathbf{\Phi}$.

For a Borel, one-homogeneous function Φ on $\mathbb{R}^n$, F a Borel set F, and E of locally finite perimeter, we define the **Φ-surface energy of E relative to F** as

$$\mathbf{\Phi}(E; F) = \int_{F\cap\partial^* E} \Phi(\nu_E(x))\, d\mathcal{H}^{n-1}(x) \in [0, \infty] . \tag{20.2}$$

The **Φ-surface energy of** E is $\mathbf{\Phi}(E) = \mathbf{\Phi}(E; \mathbb{R}^n)$. The convexity of Φ implies the lower semicontinuity of $\mathbf{\Phi}$ (we shall prove this theorem in Section 20.3).

Theorem 20.1 *If $\Phi: \mathbb{R}^n \to [0, \infty]$ is one-homogeneous, convex, and lower semicontinuous, then*

$$\mathbf{\Phi}(E; A) \le \liminf_{h\to\infty} \mathbf{\Phi}(E_h; A) , \tag{20.3}$$

whenever $A \subset \mathbb{R}^n$ is open, and $\{E_h\}_{h\in\mathbb{N}}$, E are sets of locally finite perimeter with $E_h \overset{\text{loc}}{\to} E$ and $\mu_{E_h} \overset{*}{\rightharpoonup} \mu_E$.

Remark 20.2 (Convexity is necessary to lower semicontinuity) In fact, convexity is also a necessary condition for the lower semicontinuity inequality (20.3). By homogeneity, convexity is equivalent to subadditivity,

$$\Phi(x_1 + x_2) \le \Phi(x_1) + \Phi(x_2) , \qquad \forall x_1, x_2 \in \mathbb{R}^n , \tag{20.4}$$

which, in turn, may be deduced by testing (20.3) on suitable sequences. For example, in the planar case $n = 2$, let $x_1 = a\,\nu_1$ and $x_2 = b\,\nu_2$, for $a, b > 0$ and $\nu_1, \nu_2 \in S^{n-1}$, and let $c > 0$ and $\nu \in \Sigma^{n-1}$ be such that $x_1 + x_2 = c\,\nu$, so that a, b, and c are the side lengths of a triangle T oriented by ν_1, ν_2, and $-\nu$; see Figure 20.1. We now let E be a rectangle with a side of length c oriented by ν, and define the sets E_h by attaching to that side h-many copies of $h^{-1}\,T$. In this way,

$$\begin{aligned}\mathbf{\Phi}(E) - \mathbf{\Phi}(E_h) &= c\,\Phi(\nu) - \sum_{k=1}^{h} \frac{a}{h}\,\Phi(\nu_1) + \frac{b}{h}\,\Phi(\nu_2)\\ &= c\,\Phi(\nu) - a\,\Phi(\nu_1) - b\,\Phi(\nu_2) .\end{aligned}$$

By (20.3), as $h \to \infty$, we find (20.4). In higher dimension it suffices to construct finite height cylinders over the sets considered in the planar case.

Remark 20.3 (Convexity and minimality of hyperplanes) The convexity of Φ implies in turn that the $\mathbf{\Phi}$-surface energy is decreased under intersection with half-spaces (the analogous property for the perimeter was proved in Exercise 15.13 and Proposition 19.22). Precisely, if E is a set of finite perimeter, H is an open half-space, and A is an open set in $\mathbb{R}^n$, then

$$\mathbf{\Phi}(E;A) \ge \mathbf{\Phi}(E \cap H;A)\,, \tag{20.5}$$

whenever $E \setminus H \subset\subset A$. Let us directly assume that $A = \mathbb{R}^n$. By Theorem 16.3,

$$\mu_{E\setminus H} = \nu_E\,\mathcal{H}^{n-1} \llcorner \big((\partial^* E)\setminus \overline{H}\big) - \nu_H\,\mathcal{H}^{n-1} \llcorner \big(E^{(1)} \cap \partial H\big) + \nu_E\,\mathcal{H}^{n-1} \llcorner \big\{\nu_E = -\nu_H\big\},$$

$$\mu_{E\cap H} = \nu_E\,\mathcal{H}^{n-1} \llcorner \big(H \cap \partial^* E\big) + \nu_H\,\mathcal{H}^{n-1} \llcorner \big(E^{(1)} \cap \partial H\big) + \nu_E\,\mathcal{H}^{n-1} \llcorner \big\{\nu_E = \nu_H\big\}.$$

By Exercise 15.17, $\int_{\partial^*(E\setminus H)} \nu_{E\setminus H}\, d\mathcal{H}^{n-1} = 0$. Hence, ν_H being constant,

$$\nu_H\,\mathcal{H}^{n-1}\big((E^{(1)} \cap \partial H) \cup \{\nu_E = -\nu_H\}\big) = \int_{\partial^* E\setminus \overline{H}} \nu_E\, d\mathcal{H}^{n-1}\,,$$

and Jensen's inequality implies

$$\begin{aligned}\int_{(E^{(1)}\cap\partial H)\cup\{\nu_E=-\nu_H\}} \Phi(\nu_H)\, d\mathcal{H}^{n-1} &= \Phi\Big(\int_{\partial^* E\setminus\overline{H}} \nu_E\, d\mathcal{H}^{n-1}\Big)\\ &\le \int_{\partial^* E\setminus\overline{H}} \Phi(\nu_E)\, d\mathcal{H}^{n-1}\,.\end{aligned}$$

Adding $\int_{H\cap\partial^* E} \Phi(\nu_E)\, d\mathcal{H}^{n-1}$ to both sides of this inequality, and taking into account the formula for $\mu_{E\cap H}$, we thus find, as required,

$$\mathbf{\Phi}(E \cap H) + \int_{\{\nu_E=-\nu_H\}} \Phi(\nu_E) + \Phi(\nu_H)\, d\mathcal{H}^{n-1} \le \mathbf{\Phi}(E)\,.$$

Remark 20.4 (Strict convexity, non-uniqueness, and regularity) We now make a further remark on (20.5). If Φ is *strictly* convex and $|E \setminus H| > 0$, so that $\mathcal{H}^{n-1}(\partial^* E \setminus \overline{H}) > 0$, then we have a strict sign in the application of Jensen's inequality, and, as a consequence, (20.5) holds with *strict* sign; in particular, half-spaces uniquely minimize $\mathbf{\Phi}$ with respect to their compact variations. This basic geometric property is, in a certain sense, a necessary requirement to expect a certain degree of regularity in variational problems involving $\mathbf{\Phi}$ (see also the notes to Part III). For example, in the planar case $n = 2$, let us consider the Plateau-type variational problem

$$\inf\Big\{\mathbf{\Phi}(E) : E \setminus (-1,1)^2 = \{x_1 + x_2 < 0\} \setminus (-1,1)^2\Big\},$$

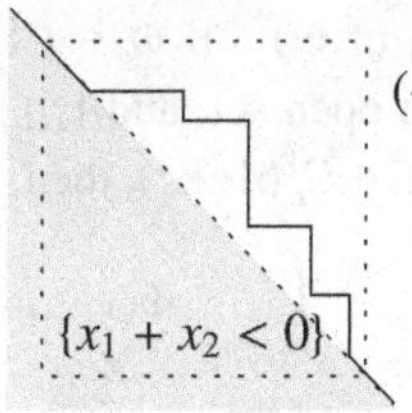

Figure 20.2 Lack of strict convexity of Φ is compatible with rough non-uniqueness phenomena. Boundaries of minimizers may not be of class C^1.

for $\Phi(x) = |x_1| + |x_2|$ ($x \in \mathbb{R}^2$). In this case, we easily construct admissible sets E with $\nu_E(x) \in \{e_1, e_2\}$ for $\mathcal{H}^1$-a.e. $x \in (-1,1)^2 \cap \partial^* E$ such that $\mathbf{\Phi}(E) = \mathbf{\Phi}(H)$, where $H = \{x_1 + x_2 < 0\}$ is the half-plane minimizer, see Figure 20.2.

Remark 20.5 (Convexity and existence of minimizers) A one-homogeneous function $\Phi\colon \mathbb{R}^n \to [0,\infty]$ is **coercive** if there exists $c > 0$ such that $\Phi(x) \geq c\,|x|$ for every $x \in \mathbb{R}^n$. If $\Phi\colon \mathbb{R}^n \to [0,\infty]$ is lower semicontinuous, one-homogeneous, coercive and convex, A is a bounded open set with finite perimeter, E_0 is of locally finite perimeter, and $m > 0$, then the variational problems

$$\inf\Big\{\mathbf{\Phi}(E) : E \setminus A = E_0 \setminus A\Big\},$$
$$\inf\Big\{\mathbf{\Phi}(E;A) : E \subset A\,, |E| = m\Big\},$$

admit minimizers. Indeed, by coercivity of Φ, for every sequence $\{E_h\}_{h\in\mathbb{N}}$,

$$\sup_{h\in\mathbb{N}} \mathbf{\Phi}(E_h;A) < \infty \qquad \Rightarrow \qquad \sup_{h\in\mathbb{N}} P(E_h;A) < \infty\,.$$

Hence, by Theorem 20.1, we may deduce the existence of minimizers in the above problems by applying the Direct Method as in the proofs of Proposition 12.29 and Proposition 12.30.

We close this section with a continuity theorem for the Φ-surface energy, whose proof is postponed to Section 20.3. We notice that, in the same spirit Chapter 13, and combining Theorem 13.8, Theorem 20.1, and Theorem 20.6, one can show that $\mathbf{\Phi}$ is the maximal lower semicontinuous extension of the functional defined in (20.1) on open bounded sets with C^1-boundary.

Theorem 20.6 (Continuity of $\mathbf{\Phi}$) *If $\Phi\colon S^{n-1} \to [0,\infty)$ is continuous, then*

$$\mathbf{\Phi}(E) = \lim_{h\to\infty} \mathbf{\Phi}(E_h)\,, \tag{20.6}$$

for $\{E_h\}_{h\in\mathbb{N}}$, E sets of finite perimeter, $E_h \overset{\text{loc}}{\to} E$, $|\mu_{E_h}| \overset{*}{\rightharpoonup} |\mu_E|$ and $P(E_h) \to P(E)$.

Exercise 20.7 (First variation of **Φ**) If $\Phi \in C^1(S^{n-1})$, E is a set of locally finite perimeter in $\mathbb{R}^n$, A is an open set, and $\{f_t\}_{|t|<\varepsilon}$ is a local variation in A, associated to the vector field $T \in C^1_c(A;\mathbb{R}^n)$, then

$$\frac{\mathrm{d}}{\mathrm{d}t}\Big|_{t=0} \Phi(f_t(E);A) = \int_{A\cap\partial^* E} \Phi(\nu_E)\mathrm{div}_{E,\Phi}\, T \, \mathrm{d}\mathcal{H}^{n-1},$$

where the boundary Φ-divergence of T with respect to E is defined on $\partial^* E$ as

$$\begin{aligned}\mathrm{div}_{E,\Phi}\, T &= \mathrm{div}\, T - \nabla\Phi(\nu_E)\cdot\frac{(\nabla T)^*\nu_E}{\Phi(\nu_E)}\\ &= \mathrm{trace}\Big(\Big(\mathrm{Id} - \nabla\Phi(\nu_E)\otimes\frac{\nu_E}{\Phi(\nu_E)}\Big)\nabla T\Big).\end{aligned}$$

Hint: Apply Proposition 17.1, (17.29), and (17.30).

20.2 The Wulff problem

The anisotropic version of the Euclidean isoperimetric problem is known as the **Wulff problem**,

$$\inf\Big\{\Phi(E) : E \subset \mathbb{R}^n, |E| = m\Big\}, \qquad m > 0, \tag{20.7}$$

that we consider for $\Phi\colon \mathbb{R}^n \to [0,\infty)$ one-homogeneous, convex, and coercive. The Wulff problem is named after the German crystallographer Georg Wulff, who first guessed, in [Wul01], how to describe minimizers in (20.7) in terms of the surface tension Φ. Precisely, up to translations and scaling, the unique minimizer in (20.7) is given by the bounded, open convex set

$$W_\Phi = \bigcap_{y\in S^{n-1}} \Big\{x\in\mathbb{R}^n : x\cdot y < \Phi(y)\Big\}, \tag{20.8}$$

known as the **Wulff shape of Φ**. In this section we will prove the minimality of the Wulff shape in the Wulff problem; for more information on the subtler issue of uniqueness, see the notes to Part II.

Theorem 20.8 (Wulff theorem) *If $\Phi\colon\mathbb{R}^n\to[0,\infty)$ is one-homogenous, convex and coercive, then for every $m > 0$ there exists $r > 0$ such that $m = r^n|W_\Phi|$, and for every $x_0 \in \mathbb{R}^n$ the open, bounded convex set $x_0 + r\,W_\Phi$ is a minimizer in the Wulff problem (20.7).*

Example 20.9 If $\Phi(x) = c\,|x|$ for some $c > 0$ and every $x \in \mathbb{R}^n$, then the Wulff problem reduces to the Euclidean isoperimetric problem. If, for example, $\Phi(x) = \max\{|x_1|,|x_2|\}$, $x\in\mathbb{R}^2$, so that $\{\Phi < 1\}$ is a square with center at the origin and side length 2, then the Wulff shape W_Φ is obtained from $\{\Phi < 1\}$ by a

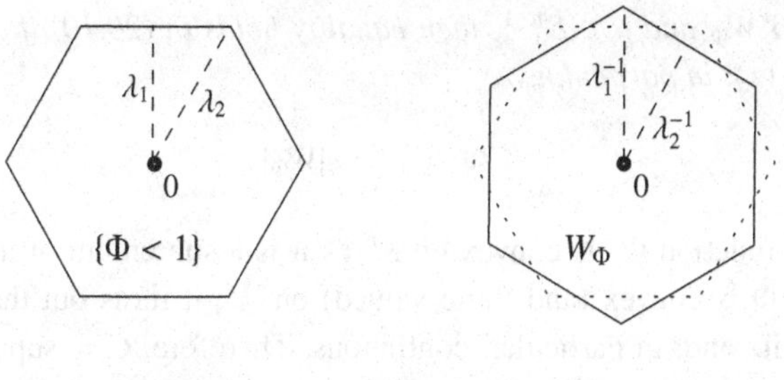

Figure 20.3 The construction of W_Φ starting from $\{\Phi < 1\}$.

rotation of 90 degrees. If $\Phi(x) = \sqrt{(x_1/a)^2 + (x_2/b)^2}$, $x \in \mathbb{R}^2$, so that $\{\Phi < 1\}$ is an ellipse with semi-axes of length a and b respectively, then W_Φ is the ellipse with semi-axes of length $1/a$ and $1/b$. The construction of W_Φ starting from $\{\Phi < 1\}$ is illustrated in Figure 20.3. Finally, let us notice that every open, bounded convex set $K \subset \mathbb{R}^n$ which contains the origin is the Wulff shape of a one-homogeneous, convex, coercive function $\Phi\colon \mathbb{R}^n \to [0, \infty)$. Indeed one easily checks that $K = W_\Phi$ for

$$\Phi(x) = \sup\left\{x \cdot y : y \in K\right\}, \qquad x \in \mathbb{R}^n .$$

The following proposition gathers some basic properties of Wulff shapes. In the terminology of Convex Analysis, the function Φ^* introduced in (20.9) is called the **convex conjugate** of Φ and the Cauchy–Schwartz type inequality (20.10) is the **Fenchel inequality**. In the classical book [Roc70] these kinds of assertion are proved in greater generality (and, concerning statement (iii), without any reference to reduced boundaries).

Proposition 20.10 (Properties of Wulff shapes) *If $\Phi\colon \mathbb{R}^n \to [0, \infty)$ is one-homogeneous, convex, and coercive, and if we define $\Phi^*\colon \mathbb{R}^n \to [0, \infty)$ as*

$$\Phi^*(x) = \sup\left\{x \cdot y : \Phi(y) < 1\right\}, \qquad x \in \mathbb{R}^n , \tag{20.9}$$

then the following properties hold true.

(i) The function Φ^* is one-homogeneous, convex, and coercive on $\mathbb{R}^n$, and there exist positive constants c and C such that

$$c\,|x| \le \Phi(x) \le C\,|x|, \qquad \forall x \in \mathbb{R}^n ,$$
$$\frac{|x|}{C} \le \Phi^*(x) \le \frac{|x|}{c}, \qquad \forall x \in \mathbb{R}^n ,$$
$$x \cdot y \le \Phi^*(x)\,\Phi(y), \qquad \forall x, y \in \mathbb{R}^n . \tag{20.10}$$

(ii) $W_\Phi = \{\Phi^* < 1\}$; in particular, W_Φ is an open, bounded convex set, with

$$B_c \subset W_\Phi \subset B_C .$$

(iii) If $x \in \partial^* W_\Phi$ and $y \in S^{n-1}$, then equality holds in (20.10) if and only if $y = \nu_{W_\Phi}(x)$; in particular,

$$\mathbf{\Phi}(W_\Phi) = n|W_\Phi|\,. \tag{20.11}$$

Proof The function Φ^* is convex on $\mathbb{R}^n$ as it is a supremum of affine functions. Since Φ is convex (and finite valued) on $\mathbb{R}^n$, it turns out that Φ is locally Lipschitz and, in particular, continuous. Therefore, $C = \sup_{S^{n-1}} \Phi$ is finite, and the various assertions in statement (i) are easily proved. The identity $W_\Phi = \{\Phi^* < 1\}$ is also easily checked and, combined with (i), it completes the proof of (ii). In particular, W_Φ is a set of finite perimeter in $\mathbb{R}^n$ (Example 12.6 or Exercise 15.14). Given $y \in S^{n-1}$, let us now set $H[y] = \{z \in \mathbb{R}^n : z \cdot y \le 0\}$. If $x \in \partial^* W_\Phi$ and $y \in S^{n-1}$ then equality holds in (20.10) if and only if

$$W_\Phi - x \subset H[y]\,. \tag{20.12}$$

We now argue as follows. Since $(W_\Phi)_{x,r} \overset{\text{loc}}{\to} H[\nu_{W_\Phi}(x)]$ as $r \to 0^+$, and (20.12) implies $(W_\Phi)_{x,r} \subset H[y]$ for every $r > 0$, if equality holds in (20.10), then

$$H[\nu_{W_\Phi}(x)] \subset H[y]\,,$$

which necessarily gives $y = \nu_{W_\Phi}(x)$. If, conversely, $y = \nu_{W_\Phi}(x)$ but equality does not hold in (20.10), then by (20.12) and since $W_\Phi - x$ is open, there exists an open ball $B(z,s) \subset\subset (W_\Phi - x) \setminus H[y]$. In particular, by convexity, the convex hull $K(z,s)$ of $\{0\} \cup B(z,s)$ satisfies $K(z,s) \subset (W_\Phi - x) \setminus H[y]$. We thus find

$$\frac{|W_\Phi \cap B(x,r)|}{r^n} \ge \Big|(W_\Phi)_{x,r} \cap H[y] \cap B\Big| + \Big|K\Big(\frac{z}{r}, \frac{s}{r}\Big) \cap B\Big|\,.$$

Since $x \in \partial^* W_\Phi$ and $y = \nu_{W_\Phi}(x)$, by Corollary 15.8 the left-hand side of this inequality, as well as the first term on the right-hand side, tends to $1/2$ as $r \to 0^+$. At the same time, the second term on the right-hand side is easily seen to be positively bounded from below as soon as r is large enough, thus leading to the desired contradiction. Finally, we prove (20.11) by applying the divergence theorem to the identity map on W_Φ. Indeed, by the characterization of equality cases in (20.10), we see that

$$\begin{aligned} n|W_\Phi| &= \int_{W_\Phi} \mathrm{div}\,(x)\,\mathrm{d}x = \int_{\partial^* W_\Phi} x \cdot \nu_{W_\Phi}(x)\,\mathrm{d}\mathcal{H}^{n-1} \\ &= \int_{\partial^* W_\Phi} \Phi^*(x)\Phi(\nu_{W_\Phi}(x))\,\mathrm{d}\mathcal{H}^{n-1} = \mathbf{\Phi}(W_\Phi)\,, \end{aligned} \tag{20.13}$$

where the inclusion $\partial^* W_\Phi \subset \partial W_\Phi = \{\Phi^* = 1\}$ has also been used. □

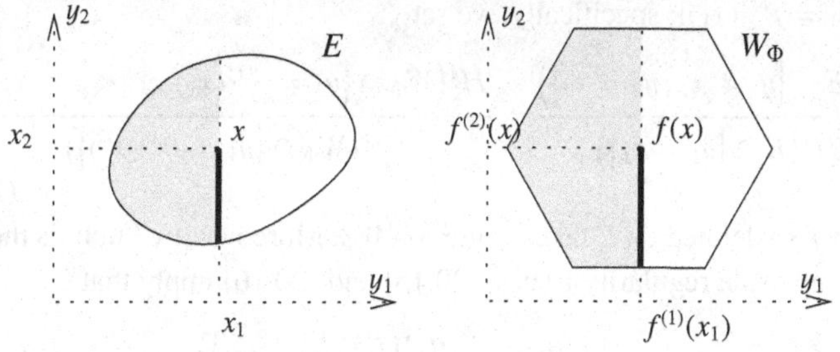

Figure 20.4 The construction of the Knothe map f between E and W_Φ. The relative area of the grey zone inside E is the same as the relative area of the grey zone inside W_Φ. The relative length of the bold line inside $E \cap \{y_1 = x_1\}$ is the same as the relative length of the bold line inside $W_\Phi \cap \{y_1 = f^{(1)}(x_1)\}$.

We now introduce the proof of Theorem 20.8. By (20.11), if E has finite and positive measure, and if $r > 0$ is such that $r^n|W_\Phi| = |E|$, then

$$\mathbf{\Phi}(x_0 + r\,W_\Phi) = \mathbf{\Phi}(W_\Phi)\,r^{n-1} = n\,|W_\Phi|^{1/n}|E|^{(n-1)/n}\,.$$

Theorem 20.8 is thus equivalent to the **Wulff inequality**: for every set E of finite perimeter and volume,

$$\mathbf{\Phi}(E) \ge n\,|W_\Phi|^{1/n}|E|^{(n-1)/n}\,. \tag{20.14}$$

We shall prove (20.14) by a particularly effective argument due to Gromov [MS86] (we refer to the notes for further approaches). For the sake of clarity, we first introduce informally the argument in the planar case $n = 2$. First, we construct the **Knothe map** [Kno57] between E and W_Φ,

$$f : E \to W_\Phi\,,$$

which is defined as follows. Setting $f = (f^{(1)}, f^{(2)})$ and given $x \in E$, we define $f^{(1)}(x) \in \mathbb{R}$ so that the relative area of $E \cap \{y_1 < x_1\}$ inside E equals the relative area of $W_\Phi \cap \{x_1 < f^{(1)}(x)\}$ inside W_Φ; that is, we set

$$\frac{\Big|\{y \in E : y_1 < x_1\}\Big|}{|E|} = \frac{\Big|\{y \in W_\Phi : y_1 < f^{(1)}(x)\}\Big|}{|W_\Phi|}\,; \tag{20.15}$$

see Figure 20.4. In particular, $f^{(1)}(x) = f^{(1)}(x_1)$, and the intersection of E and the line $\{y_1 = t\}$ is mapped by f into the line $\{y_1 = f^{(1)}(t)\}$. Next, we define $f^{(2)}(x)$ so that the relative length of $E \cap \{y_1 = x_1\,, y_2 < x_2\}$ inside $E \cap \{y_1 = x_1\}$ is equal to the relative length of $W_\Phi \cap \{y_1 = f^{(1)}(x_1), y_2 < f^{(2)}(x)\}$ inside

$W_\Phi \cap \{y_1 = f^{(1)}(x_1)\}$; specifically, we set

$$\frac{\mathcal{H}^1\Big(E \cap \big\{y_1 = x_1\,, y_2 < x_2\big\}\Big)}{\mathcal{H}^1\Big(E \cap \big\{y_1 = x_1\big\}\Big)} = \frac{\mathcal{H}^1\Big(W_\Phi \cap \big\{y_1 = f^{(1)}(x_1)\,, y_2 < f^{(2)}(x)\big\}\Big)}{\mathcal{H}^1\Big(W_\Phi \cap \big\{y_1 = f^{(1)}(x_1)\big\}\Big)}\,. \tag{20.16}$$

The map f so defined on E takes values on W_Φ. Moreover, by Fubini's theorem and leaving aside regularity issues, (20.15) and (20.16) imply that

$$\partial_1 f^{(1)}(x_1) = \frac{|W_\Phi|}{|E|}\,\frac{\mathcal{H}^1\Big(E \cap \big\{y_1 = x_1\big\}\Big)}{\mathcal{H}^1\Big(W_\Phi \cap \big\{y_1 = f^{(1)}(x_1)\big\}\Big)}\,,$$

$$\partial_2 f^{(2)}(x) = \frac{\mathcal{H}^1\Big(W_\Phi \cap \big\{y_1 = f^{(1)}(x_1)\big\}\Big)}{\mathcal{H}^1\Big(E \cap \big\{y_1 = x_1\big\}\Big)}\,.$$

In particular, $\partial_2 f^{(1)} = 0$, and thus ∇f is a triangular tensor, with

$$\det \nabla f = \partial_1 f^{(1)} \partial_2 f^{(2)} = \frac{|W_\Phi|}{|E|}\,, \qquad \operatorname{div} f = \partial_1 f^{(1)} + \partial_2 f^{(2)}\,.$$

Since $\partial_k f^{(k)} \ge 0$, $k = 1, 2$, the arithmetic-geometric mean inequality thus gives

$$\left(\frac{|W_\Phi|}{|E|}\right)^{1/2} = (\det \nabla f)^{1/2} \le \frac{\operatorname{div} f}{2}\,.$$

Integrating this inequality over E, applying the divergence theorem and taking also into account that f takes values in $W_\Phi = \{\Phi^* < 1\}$, we thus find

$$\begin{aligned} 2|W_\Phi|^{1/2}|E|^{1/2} &\le \int_E \operatorname{div} f = \int_{\partial^* E} f \cdot \nu_E \, d\mathcal{H}^1 \\ &\le \int_{\partial^* E} \Phi^*(f)\Phi(\nu_E)\, d\mathcal{H}^1 \le \mathbf{\Phi}(E)\,, \end{aligned}$$

that is the Wulff inequality in the planar case $n = 2$. Generalizing this argument in higher dimension is not difficult, and it is also relatively easy to write down a formal uniqueness argument by tracking back equality cases. If we are also interested in rigor, however, then some delicate technical issues arise. It turns out that we may avoid such difficulties if instead of working with (the characteristic functions of) E and W_Φ, we apply Gromov's argument to a pair of compactly supported smooth densities u and v, and then pass to the limit $\varepsilon \to 0^+$ in the resulting inequalities for $u = 1_E \star \rho_\varepsilon$ and $v = 1_{W_\Phi} \star \rho_\varepsilon$.

Proof of Theorem 20.8 Given $x \in \mathbb{R}^n$ and $k = 1, ..., n$, we shall consider the half-spaces $H(x_k)$ and the hyperplanes $I(x_k)$ defined as follows,

$$H(x_k) = \big\{y \in \mathbb{R}^n : y_k < x_k\big\}\,, \qquad I(x_k) = \partial H(x_k) = \big\{y \in \mathbb{R}^n : y_k = x_k\big\}\,.$$

In step one, we construct Knothe maps between compactly supported densities of class C^1 and the unit cube of $\mathbb{R}^n$, which are used in step two to prove an inequality between any two given compactly supported densities of class C^1. We finally conclude by an easy approximation argument.

Step one: Given $\varphi \in C^1_c(\mathbb{R}^n)$, $\varphi \geq 0$, we construct a C^1-bijection g between $\{\varphi > 0\}$ and $(0,1)^n$, with $g \in C^1(\mathbb{R}^n;[0,1]^n)$, ∇g upper triangular and with non-negative diagonal entries, and such that

$$\det \nabla g = \frac{\varphi}{\int_{\mathbb{R}^n} \varphi}. \tag{20.17}$$

Indeed, for every $x \in \mathbb{R}^n$ it suffices to set

$$g^{(1)}(x) = \frac{\int_{H(x_1)} \varphi}{\int_{\mathbb{R}^n} \varphi} \in (0,1),$$

$$g^{(k)}(x) = \frac{\int_{H(x_k)\cap I(x_1)\cap\cdots\cap I(x_{k-1})} \varphi \, d\mathcal{H}^{n-k+1}}{\int_{I(x_1)\cap\cdots\cap I(x_{k-1})} \varphi \, d\mathcal{H}^{n-k+1}} \in (0,1),$$

where $k = 2, \dots, n$ and $g^{(k)}$ is the kth component of g. By construction, $g^{(k)}$ is increasing in the variable x_k. By Fubini's theorem, $g \in C^1(\mathbb{R}^n;[0,1]^n)$ with

$$\partial_k g^{(k)}(x) = \frac{\int_{I(x_1)\cap\cdots\cap I(x_k)} \varphi \, d\mathcal{H}^{n-k}}{\int_{I(x_1)\cap\cdots\cap I(x_{k-1})} \varphi \, d\mathcal{H}^{n-k+1}}, \qquad k = 1, \dots, n, \tag{20.18}$$

$$\partial_h g^{(k)}(x) = 0, \qquad k = 1, \dots, n-1, k < h \leq n, \tag{20.19}$$

where we have also used the fact that, if $k = 1, ..., n-1$, then $g^{(k)}$ does not depend on the variables $x_{k+1}, ..., x_n$. In particular, ∇g is upper triangular, with

$$\partial_k g^{(k)}(x) > 0, \qquad \forall x \in \operatorname{spt}\varphi, \tag{20.20}$$

$$\det \nabla g(x) = \prod_{k=1}^n \partial_k g^{(k)}(x) = \frac{\varphi(x)}{\int_{\mathbb{R}^n} \varphi}, \qquad \forall x \in \operatorname{spt}\varphi. \tag{20.21}$$

By (20.20) and (20.19), g defines a bijection between $\{\varphi > 0\}$ and $(0,1)^n$.

Step two: Given two functions $u, v \in C^1_c(\mathbb{R}^n)$ with $u, v \geq 0$ and $\int_{\mathbb{R}^n} u^{n/(n-1)} = \int_{\mathbb{R}^n} v^{n/(n-1)} = 1$, we shall now prove that

$$n \int_{\mathbb{R}^n} v \leq \max\left\{\Phi^*(x) : x \in \operatorname{spt} v\right\} \int_{\mathbb{R}^n} \Phi(-\nabla u). \tag{20.22}$$

Indeed, let g_u and g_v be the maps associated with $\varphi = u^{n'}$ and $\varphi = v^{n'}$ by the construction of step one. If we let $f = g_v^{-1} \circ g_u \colon \operatorname{spt} u \to \operatorname{spt} v$, then f turns to be a C^1 bijection between $\operatorname{spt} u$ and $\operatorname{spt} v$, whose gradient is upper triangular, with positive diagonal components (i.e. $\partial_k f^{(k)} > 0$ on $\operatorname{spt} u$), and such that

$$v(f(x))^{n'} \det \nabla f(x) = u(x)^{n'} . \tag{20.23}$$

Hence $\det \nabla f > 0$ on $\operatorname{spt} u$, so that $Jf = \det \nabla f$ on $\operatorname{spt} u$. By the area formula and (20.23), we thus have

$$\int_{\mathbb{R}^n} v = \int_{\operatorname{spt} u} v(f)\, Jf = \int_{\operatorname{spt} u} u\, (\det \nabla f)^{1/n} . \tag{20.24}$$

Since ∇f is upper triangular and $\partial_k f^{(k)} > 0$ on $\operatorname{spt} u$, we may apply the arithmetic–geometric mean inequality to find

$$(\det \nabla f)^{1/n} = \Big(\prod_{k=1}^{n} \partial_k f^{(k)} \Big)^{1/n} \le \frac{1}{n} \sum_{k=1}^{n} \partial_k f^{(k)} = \frac{\operatorname{div} f}{n} .$$

Coming back to (20.24), and applying the divergence theorem to the compactly supported vector field $u\, f$,

$$\begin{aligned} n \int_{\mathbb{R}^n} v &\le \int_{\operatorname{spt} u} u \operatorname{div} f = \int_{\mathbb{R}^n} u \operatorname{div} f = \int_{\mathbb{R}^n} -(\nabla u) \cdot f \\ &\le \max \Big\{ \Phi^*(x) : x \in \operatorname{spt} v \Big\} \int_{\mathbb{R}^n} \Phi(-\nabla u) , \end{aligned}$$

where in the last inequality we have applied the Cauchy–Schwartz inequality (20.10) and we have also taken into account that f takes its values in $\operatorname{spt} v$.

Step three: Given $\varepsilon > 0$ and a bounded set of finite perimeter $E \subset \mathbb{R}^n$, we now apply (20.22) to the functions

$$u_\varepsilon = \frac{1_E \star \rho_\varepsilon}{\| 1_E \star \rho_\varepsilon \|_{L^{n'}(\mathbb{R}^n)}} , \qquad v_\varepsilon = \frac{1_{W_\Phi} \star \rho_\varepsilon}{\| 1_{W_\Phi} \star \rho_\varepsilon \|_{L^{n'}(\mathbb{R}^n)}} .$$

Since $\operatorname{spt} v_\varepsilon \subset W_\Phi + \varepsilon B$ (where B denotes the Euclidean unit ball), and since $W_\Phi = \{\Phi^* < 1\}$, we easily see that

$$\lim_{\varepsilon \to 0^+} \max \Big\{ \Phi^*(x) : x \in \operatorname{spt} v_\varepsilon \Big\} = 1 .$$

At the same time, if $p \ge 1$ and $\varepsilon \to 0^+$, then $1_{W_\Phi} \star \rho_\varepsilon \to 1_{W_\Phi}$ in $L^p(\mathbb{R}^n)$, so that

$$\lim_{\varepsilon \to 0^+} \int_{\mathbb{R}^n} v_\varepsilon = \frac{|W_\Phi|}{|W_\Phi|^{1/n'}} = |W_\Phi|^{1/n} .$$

Finally, by Proposition 12.20, we have that $|\nabla u_\varepsilon| \mathcal{L}^n \stackrel{*}{\rightharpoonup} |E|^{-1/n'} |\mu_E|$. Thus, by Reshetnyak's continuity theorem (see, in particular, Exercise 20.13),

$$\lim_{\varepsilon \to 0^+} \int_{\mathbb{R}^n} \Phi(-\nabla u_\varepsilon) = \frac{1}{|E|^{1/n'}} \int_{\partial^* E} \Phi(\nu_E)\, d\mathcal{H}^{n-1} .$$

Hence, when E is a bounded set of finite perimeter, the Wulff inequality follows by passing to the limit as $\varepsilon \to 0^+$ in (20.22) applied to $u = u_\varepsilon$ and $v = v_\varepsilon$.

Step four: Let E be a set of finite perimeter and measure. By step three and by Theorem 16.3, for a.e. $R > 0$ we have

$$\begin{aligned} n|W_\Phi|^{1/n}|E \cap B_R|^{(n-1)/n} &\le \mathbf{\Phi}(E \cap B_R) \\ &= \int_{B_R \cap \partial^* E} \Phi(\nu_E)\, d\mathcal{H}^{n-1} + \int_{E \cap \partial B_R} \Phi(\nu_{B_R})\, d\mathcal{H}^{n-1} \\ &\le \mathbf{\Phi}(E) + C\,\mathcal{H}^{n-1}(E \cap \partial B_R). \end{aligned}$$

Considering a sequence $R_h \to \infty$ such that $\mathcal{H}^{n-1}(E \cap \partial B_{R_h}) \to 0$ and the above inequality holds true, we thus prove the Wulff inequality on E. □

20.3 Reshetnyak's theorems

In this section we prove two useful theorems about functionals defined on Radon measures, known as *Reshetnyak's theorems*, and apply them to prove Theorem 20.1 and Theorem 20.6 from Section 20.1. We consider a one-homogeneous Borel function $\Phi\colon \mathbb{R}^n \to [0,\infty]$, and for every $\mathbb{R}^n$-valued Radon measure ν on $\mathbb{R}^m$ and every Borel set $F \subset \mathbb{R}^m$, we define the **Φ-anisotropic total variation of ν on F** as

$$\mathbf{\Phi}(\nu; F) = \int_F \Phi\left(D_{|\nu|}\nu(x)\right) d|\nu|\,(x) \in [0,\infty].$$

Theorem 20.11 (Reshetnyak's lower semicontinuity theorem) *If $\Phi\colon \mathbb{R}^n \to [0,\infty]$ is one-homogeneous, lower semicontinuous and convex, then*

$$\mathbf{\Phi}(\nu; A) \le \liminf_{h\to\infty} \mathbf{\Phi}(\nu_h; A),$$

whenever $A \subset \mathbb{R}^m$ is open and $\{\nu_h\}_{h\in\mathbb{N}}$ and ν are $\mathbb{R}^n$-valued Radon measures with $\nu_h \stackrel{*}{\rightharpoonup} \nu$.

Theorem 20.12 (Reshetnyak's continuity theorem) *If $\Phi\colon S^{n-1} \to [0,\infty)$ is continuous and $\{\nu_h\}_{h\in\mathbb{N}}$ and ν are $\mathbb{R}^n$-valued Radon measures, then*

$$\mathbf{\Phi}(\nu) = \lim_{h\to\infty} \mathbf{\Phi}(\nu_h), \tag{20.25}$$

whenever $\nu_h \stackrel{*}{\rightharpoonup} \nu$, $|\nu_h|(\mathbb{R}^m) \to |\nu|(\mathbb{R}^m)$ and $|\nu|\,(\mathbb{R}^m) < \infty$.

Proof of Theorem 20.1 and Theorem 20.6 If E is of locally finite perimeter in $\mathbb{R}^n$ then the Φ-surface energy of E is equal to the Φ-anisotropic total variation of μ_E, that is

$$\mathbf{\Phi}(\mu_E; F) = \mathbf{\Phi}(E; F). \tag{20.26}$$

Indeed $|\mu_E| = \mathcal{H}^{n-1} \llcorner \partial^* E$ by Theorem 15.9 and $D_{|\mu_E|}\mu_E(x) = \nu_E(x)$ by definition of reduced boundary. Theorem 20.1 and Theorem 20.6 then follow from Theorem 20.11 and Theorem 20.12 by means of (20.26). □

Proof of Theorem 20.11 By the Hahn–Banach separation theorem [Roc70],

$$\Phi(x) = \sup_{i \in \mathbb{N}} x_i \cdot x\,, \quad \forall x \in \mathbb{R}^n\,, \tag{20.27}$$

for a sequence $\{x_i\}_{i\in\mathbb{N}} \subset \mathbb{R}^n$. Moreover, we may directly assume that

$$\liminf_{h\to\infty} \mathbf{\Phi}(\nu_h; A) = \lim_{h\to\infty} \mathbf{\Phi}(\nu_h; A) < \infty\,. \tag{20.28}$$

By (20.28), the sequence $\{\mu_h\}_{h\in\mathbb{N}}$ of Radon measures on $\mathbb{R}^m$ defined as

$$\mu_h = \Phi(D_{|\nu_h|}\nu_h)\,|\nu_h| \llcorner A\,,$$

satisfies $\sup_{h\in\mathbb{N}} \mu_h(\mathbb{R}^n) < \infty$. By Theorem 4.33 and Remark 4.35, $\mu_{h(k)} \overset{*}{\rightharpoonup} \mu$ and $|\nu_{h(k)}| \overset{*}{\rightharpoonup} \lambda$ as $h(k) \to \infty$, where μ and λ are Radon measures on $\mathbb{R}^m$. Differentiating μ with respect to $|\nu|$ (Theorem 5.8),

$$\mu = D_{|\nu|}\mu + \mu^s\,, \qquad \mu^s \perp |\nu|\,, \tag{20.29}$$

$$D_{|\nu|}\mu(z) = \lim_{r\to 0^+} \frac{\mu\,(B\,(z,r))}{|\nu|\,(B\,(z,r))}\,, \qquad \text{for } |\nu|\text{-a.e. } z \in \mathbb{R}^m\,. \tag{20.30}$$

By (20.28), Proposition 4.29 and (20.29),

$$\liminf_{h\to\infty} \mathbf{\Phi}(\nu_h; A) = \lim_{k\to\infty} \mu_{h(k)}(A) \geq \mu(A) \geq \int_A D_{|\nu|}\mu \,\mathrm{d}|\nu|\,.$$

Thus, we are only left to prove that

$$D_{|\nu|}\mu(x) \geq \Phi(D_{|\nu|}\nu(z))\,, \qquad \text{for } |\nu|\text{-a.e. } z \in A\,. \tag{20.31}$$

To this end, let us fix $z \in A$ such that (20.30) and

$$\lim_{r\to 0^+} \frac{\nu\,(B\,(z,r))}{|\nu|\,(B\,(z,r))} = D_{|\nu|}\nu(z) \tag{20.32}$$

hold true, and choose $r_j \to 0^+$ so that $B(z, r_j) \subset A$ for every $j \in \mathbb{N}$, and

$$\lim_{k\to\infty} \mu_{h(k)}(B(z,r_j)) = \mu(B(z,r_j))\,, \qquad \lim_{k\to\infty} \nu_{h(k)}(B(z,r_j)) = \nu(B(z,r_j))\,, \tag{20.33}$$

(this choice is possible by Proposition 2.16, Proposition 4.26, and Proposition 4.30). By (20.30), (20.33), (20.27), and (20.32), we thus find that

$$
\begin{aligned}
D_{|\nu|}\mu(z) &= \lim_{j\to\infty} \frac{\mu(B(z,r_j))}{|\nu|\,(B(z,r_j))} = \lim_{j\to\infty}\lim_{k\to\infty} \frac{\int_{B(z,r_j)} \Phi(D_{|\nu_{h(k)}|}\nu_{h(k)}) \mathrm{d}|\nu_{h(k)}|}{|\nu|\,(B(z,r_j))} \\
&\geq \lim_{j\to\infty}\lim_{k\to\infty} \frac{\int_{B(z,r_j)} (z_i \cdot D_{|\nu_{h(k)}|}\nu_{h(k)})\, \mathrm{d}|\nu_{h(k)}|}{|\nu|\,(B(z,r_j))} = \lim_{j\to\infty}\lim_{k\to\infty} \frac{z_i \cdot \nu_{h(k)}(B(z,r_j))}{|\nu|\,(B(z,r_j))} \\
&= \lim_{j\to\infty} \frac{z_i \cdot \nu(B(z,r_j))}{|\nu|\,(B(z,r_j))} = z_i \cdot D_{|\nu|}\nu(z)\,,
\end{aligned}
$$

for every $i \in \mathbb{N}$. By (20.27), we easily deduce (20.31). □

Proof of Theorem 20.12 By the continuity of Φ on S^{n-1} we easily prove the existence, for every $\delta > 0$, of $C(\delta) \geq 0$ such that

$$
|\Phi(z) - \Phi(w)| \leq \delta + C(\delta)\,|z - w|^2\,, \qquad \forall z, w \in S^{n-1}\,. \tag{20.34}
$$

Let us now consider $\{\psi_k\}_{k\in\mathbb{N}} \subset C^0_c(\mathbb{R}^m)$ with $0 \leq \psi_k \leq 1$, $\psi_k = 1$ on B_k, $\psi_k = 0$ on $\mathbb{R}^m \setminus B_{k+1}$. Since $|\nu_h| \stackrel{*}{\rightharpoonup} |\nu|$ and thus $|\nu|\,(\mathbb{R}^m) < \infty$, we have

$$
\lim_{k\to\infty} \sup_{h\in\mathbb{N}} \int_{\mathbb{R}^m} (1 - \psi_k)\, \mathrm{d}(|\nu_h| + |\nu|) = 0\,.
$$

As a consequence, since Φ is bounded on S^{n-1}, (20.25) will follow from

$$
\int_{\mathbb{R}^m} \Phi(D_{|\nu|}\nu)\, \psi_k\, \mathrm{d}|\nu| = \lim_{h\to\infty} \int_{\mathbb{R}^m} \Phi(D_{|\nu_h|}\nu_h)\, \psi_k\, \mathrm{d}|\nu_h|\,, \qquad \forall k \in \mathbb{N}\,. \tag{20.35}
$$

We now prove (20.35). Since $D_{|\nu|}\nu \in L^1(\mathbb{R}^m, |\nu|; \mathbb{R}^n)$, by Theorem 4.3 there exists $\{\varphi_i\}_{i\in\mathbb{N}}$ with $\varphi_i \to D_{|\nu|}\nu$ in $L^1(\mathbb{R}^m, |\nu|; \mathbb{R}^n)$. By dominated convergence,

$$
\lim_{i\to\infty} \int_{\mathbb{R}^m} \Big|\Phi(D_{|\nu|}\nu) - \Phi(\varphi_i)\Big|\, \psi_k\, \mathrm{d}|\nu| = 0\,. \tag{20.36}
$$

Moreover, assuming also without loss of generality that $|\varphi_i| \leq 1$ and by (20.34),

$$
\begin{aligned}
&\int_{\mathbb{R}^m} \Big|\Phi(D_{|\nu_h|}\nu_h) - \Phi(\varphi_i)\Big|\, \psi_k\, \mathrm{d}|\nu_h| \\
&\leq \delta \int_{\mathbb{R}^m} \psi_k\, \mathrm{d}|\nu_h| + C(\delta) \int_{\mathbb{R}^m} \Big(1 + |\varphi_i|^2 - 2(D_{|\nu_h|}\nu_h \cdot \varphi_i)\Big)\, \psi_k\, \mathrm{d}|\nu_h| \\
&\leq \delta\, |\nu_h|\,(\mathbb{R}^m) + 2\,C(\delta) \left(\int_{\mathbb{R}^m} \psi_k\, \mathrm{d}|\nu_h| - \int_{\mathbb{R}^m} \psi_k\, \varphi_i \cdot \mathrm{d}\nu_h \right).
\end{aligned}
$$

Since $\psi_k \varphi_i \in C^0_c(\mathbb{R}^m;\mathbb{R}^n)$, by $\nu_h \overset{*}{\rightharpoonup} \nu$ and $|\nu_h| \overset{*}{\rightharpoonup} |\nu|$ we have, for every $k, i \in \mathbb{N}$,

$$\limsup_{h\to\infty} \int_{\mathbb{R}^m} \Big|\Phi(D_{|\nu_h|}\nu_h) - \Phi(\varphi_i)\Big|\, \psi_k \,\mathrm{d}|\nu_h|$$
$$\le \delta\,|\nu|\,(\mathbb{R}^m) + 2\,C(\delta)\bigg(\int_{\mathbb{R}^m} \psi_k \,\mathrm{d}|\nu| - \int_{\mathbb{R}^m} \psi_k\,\varphi_i \cdot \mathrm{d}\nu\bigg)\,.$$

Letting $i \to \infty$ we thus find

$$\limsup_{i\to\infty}\limsup_{h\to\infty} \int_{\mathbb{R}^m} \Big|\Phi(D_{|\nu_h|}\nu_h) - \Phi(\varphi_i)\Big|\, \psi_k \,\mathrm{d}|\nu_h| \le C\,\delta\,|\nu|\,(\mathbb{R}^m)\,, \qquad (20.37)$$

for every $k \in \mathbb{N}$. An analogous argument shows that

$$\limsup_{i\to\infty} \int_{\mathbb{R}^m} \Big|\Phi(D_{|\nu|}\nu) - \Phi(\varphi_i)\Big|\, \psi_k \,\mathrm{d}|\nu| \le C\,\delta\,|\nu|\,(\mathbb{R}^m)\,, \qquad (20.38)$$

for every $k \in \mathbb{N}$. If we finally consider that $\psi_k\,\Phi\circ\varphi_i \in C^0_c(\mathbb{R}^m)$, then we find

$$\lim_{h\to\infty} \int_{\mathbb{R}^m} \Phi(\varphi_i)\,\psi_k \,\mathrm{d}|\nu_h| = \int_{\mathbb{R}^m} \Phi(\varphi_i)\,\psi_k \,\mathrm{d}|\nu|\,, \qquad \forall i,k \in \mathbb{N}\,,$$

which, combined with (20.37) and (20.38), gives (20.35). □

Exercise 20.13 If E is a set of finite perimeter in $\mathbb{R}^n$ and $\{u_h\}_{h\in\mathbb{N}} \subset C^1_c(\mathbb{R}^n)$ satisfy $u_h \to 1_E$ and $|\nabla u_h|\, d\mathcal{L}^n \overset{*}{\rightharpoonup} |\mu_E|$, then

$$\Phi(E) = \lim_{h\to\infty} \int_{\mathbb{R}^n} \Phi_0(-\nabla u_h)\,\mathrm{d}\mathcal{L}^n\,.$$

Notes

We gather here some comments on the material presented in Part II, also pointing out some related references. Our aim here is not to provide a complete description of the huge literature about sets of finite perimeter, but rather to make some remarks which should clarify the background material of this part of the book, and stimulate further reading.

Chapters 12–15 are an elaboration of De Giorgi's founding papers [DG54, DG55, DG58]. We have preferred to discuss the whole theory in terms of "the Gauss–Green measure μ_E of E" rather than of "the distributional derivative $D1_E$ of the characteristic function of E", because this terminology and notation looks more inspiring to us, and seems also to capture more closely the style of De Giorgi's original papers. Let us also mention that, in a series of five papers which appeared in Italian in 1952, Caccioppoli sketched several ideas anticipating later developments in Geometric Measure Theory. Among those, a notion of set with oriented boundary, which De Giorgi proved equivalent to that of set of finite perimeter (by Proposition 12.15, Theorem 13.8, and Remark 13.13; see also the beginning of Chapter 13).

The ingenious formulation of Plateau's problem on sets of finite perimeter (12.29) appears in [DG60, Section 1]. It has of course the implicit drawback of imposing topological restrictions on the admissible boundary data. As already noticed in the text, more natural settings for the formulation of Plateau's problem are provided by the theories of currents and varifolds.

In Section 14.1 we have proved some necessary condition for equality in the Steiner inequality, Theorem 14.4. The characterization of equality cases in symmetrization inequalities may be subtle; for the Steiner inequality, see [CCF05].

Following [DG55], we have proved De Giorgi's structure theorem, Theorem 15.9, by using Whitney's extension theorem. For a proof of this last result, see [EG92, Section 6.5]. A self-contained proof of Corollary 16.1, which is just a slightly weaker version of Theorem 15.9, is proved instead in Chapter 16 as a consequence of the theory of rectifiable sets from Part I.

The Gauss–Green measures formulae of unions, intersections, and set differences of sets of finite perimeter presented in Section 16.1, although well-known to experts, and implicitly stated, for example, in [AFP00, Example 3.97], seems to be absent in the standard references on sets of finite perimeter. They provide an extremely useful tool to transform badly drawn pictures into rigorous proofs, and to avoid the (sometimes geometrically obscure) procedure of first proving identities and inequalities on smooth functions, and later translating them onto sets via limiting procedures.

The notion of indecomposable set in Exercise 16.9 was introduced in [DM95, Definition 2.11] starting from Federer's notion of indecomposable current [Fed69, 4.2.25], and provides a natural generalization of the notion of open connected set in this framework. For example, connected open sets with finite perimeter are indecomposable, and every set of finite perimeter E is decomposed into countably many maximal indecomposable components $\{E_h\}_{h\in\mathbb{N}}$, with $P(E) = \sum_{h\in\mathbb{N}} P(E_h)$; see [ACMM01].

In Section 17.3 we have introduced the notion of distributional scalar mean curvature for a set of locally finite perimeter. The same kind of procedure works in higher codimension on generic rectifiable sets, and naturally leads, for example, to the notion of stationary varifold; see [Sim83, Chapters 3 and 4]. Lemma 17.21 is a toy-version of the volume-fixing variations theorem for minimizing clusters discussed in Section 29.6 and originating from [Alm76, Section VI].

The second variation formula for perimeter has several interesting applications. For example, a very elegant proof of the Euclidean isoperimetric inequality on bounded open sets with C^2-boundary was found by Wente [Wen91] (see also [BdC84]) by exploiting the non-negativity of the second derivative of $P(r(t)\,f_t(E))$ at $t = 0$, where $f_t(x) = x + t\,\nu_E(x)$ for $x \in \partial E$, and $r(t)$ is such that $|r(t)\,f_t(E)| = |E|$ for every $t > 0$. The second variation formula holds in fact on sets with "almost" C^2-boundary, in the sense that one can allow the presence of singular set Σ inside ∂E such that $\mathcal{H}^{n-3}(\Sigma) = 0$; see Sternberg and Zumbrun [SZ98]. This information can be used, for example, to prove the connectedness of the topological boundary of relative isoperimetric sets inside strictly convex open sets; see [SZ99]. As we shall see in Part III, the assumption $\mathcal{H}^{n-3}(\Sigma) = 0$ is usually verified by minimizers of reasonable variational problems involving perimeter.

Chapter 19 is mainly based on some papers by Giusti, Gonzalez, Massari, and Tamanini [Gon76, Gon77, GT77, GMT80, Giu80, Giu81]. Incredibly little is known on the geometric properties of minimizers in the liquid drop problem (19.9), or on its variant where $P(E; A)$ is replaced by an anisotropic surface energy term $\int_{A\cap\partial^* E} \Phi(\nu_E)\,d\mathcal{H}^{n-1}$. For example, if $A = \mathbb{R}^n$ and g has convex sub-level sets, then it would be reasonable to expect minimizers to be convex; but, in fact, even connectedness (or, say, indecomposability) is open. McCann [McC98], has shown that, in the planar case, minimizers have finitely many connected components lying at mutually positive distance, which are convex and minimize energy among convex sets with their same area. Convexity results have also been proved in the large volume regime, see [BCCN06, CC06], and (without convexity assumptions on the sub-level sets of the potential) in the small volume regime, see [FM11].

In Exercises 14.7–17.12 we have introduced p-Cheeger sets. If the ambient open set A is convex, then uniqueness and convexity of 1-Cheeger sets has been proved in [CCN07, AC09]. Moreover, uniqueness of 1-Cheeger sets holds up to arbitrarily small perturbations of A, as shown in [CCN10]. The uniqueness and convexity of constrained isoperimetric sets of volume m, see (14.23), has been proved in [ACC05] in the case that A is an open convex set with $C^{1,1}$-boundary, and $m \in [m_0, |A|)$, where m_0 denotes the volume of the 1-Cheeger set of A.

In Exercise 20.7, we have derived a first variation formula for regular anisotropic surface energy. The non-smooth case is considerably more delicate; see [BNP01a, BNP01b, BNR03].

The original proof of the optimality of Wulff shapes in the Wulff problem by Dinghas [Din44] was based on a celebrated argument involving "differentiation" of the Brunn–Minkowski inequality; see, for example, [Gar02, Section 5]. The uniqueness issue is more subtle, and has been settled, in the generality considered here, by Taylor [Tay78] and Fonseca and Müller [FM91]. Gromov's original argument [MS86], which is reproduced in Section 20.2 with a certain fidelity, does not lead to uniqueness. However, with the proper tools from Geometric Measure Theory at our disposal, Gromov's argument can be refined to achieve this goal too; see Brothers and Morgan [BM94] and [FMP10].

In proving Theorem 20.11 and Theorem 20.12, we follow the simple argument recently proposed by Spector [Spe11], which, in turn, provides a nice application of the *blow-up method* for lower semicontinuity introduced by Fonseca and Müller [FM92] in a different context.

PART THREE

Regularity theory and analysis of singularities

Synopsis

In this part we shall discuss the regularity of boundaries of those sets of finite perimeter which arise as minimizers in some of the variational problems considered so far. The following theorem exemplifies the kind of result we shall obtain. We recall from Section 16.2 that E is a local perimeter minimizer (at scale r_0) in some open set A, if $\operatorname{spt}\mu_E = \partial E$ (recall Remark 16.11) and

$$P(E;A) \le P(F;A), \tag{1}$$

whenever $E \Delta F \subset\subset B(x, r_0) \cap A$ and $x \in A$.

Theorem *If $n \ge 2$, A is an open set in $\mathbb{R}^n$, and E is a local perimeter minimizer in A, then $A \cap \partial^* E$ is an analytic hypersurface with vanishing mean curvature which is relatively open in $A \cap \partial E$, while the* **singular set** *of E in A,*

$$\Sigma(E;A) = A \cap (\partial E \setminus \partial^* E),$$

satisfies the following properties:

(i) if $2 \le n \le 7$, then $\Sigma(E;A)$ is empty;
(ii) if $n = 8$, then $\Sigma(E;A)$ has no accumulation points in A;
(iii) if $n \ge 9$, then $\mathcal{H}^s(\Sigma(E;A)) = 0$ for every $s > n - 8$.

These assertions are sharp: there exists a perimeter minimizer E in $\mathbb{R}^8$ such that $\mathcal{H}^0(\Sigma(E;\mathbb{R}^8)) = 1$; moreover, if $n \ge 9$, then there exists a perimeter minimizer E in $\mathbb{R}^n$ such that $\mathcal{H}^{n-8}(\Sigma(E;\mathbb{R}^n)) = \infty$.

The proof of this deep theorem, which will take all of Part III, is essentially divided into two parts. The first one concerns the regularity of the reduced boundary in A and, precisely, it consists of proving that the locally $\mathcal{H}^{n-1}$-rectifiable set $A \cap \partial^* E$ is, in fact, a $C^{1,\gamma}$-hypersurface for every $\gamma \in (0, 1)$. (As we shall see, its analiticity will then follow rather straightforwardly from standard elliptic regularity theory.) The second part of the argument is devoted to the analysis of the structure of the singular set $\Sigma(E;A)$. By the density estimates of Theorem 16.14, we already know that $\mathcal{H}^{n-1}(\Sigma(E;A)) = 0$. In order to improve this estimate, we shall move from the fact that, roughly speaking, the blow-ups $E_{x,r}$ of E at points $x \in \Sigma(E;A)$ will have to converge to *cones* which are local perimeter minimizers in $\mathbb{R}^n$, and which have their vertex at a singular point. Starting from this result, and discussing the possible existence of such singular minimizing cones, we shall prove the claimed estimates.

In fact, we shall not confine our attention to local perimeter minimizers, but we shall work instead in the broader class of (Λ, r_0)-*perimeter minimizers*. This is a generalization of the notion of local perimeter minimizer, which allows for the presence on the right-hand side of the minimality inequality (1) of a higher order term of the form $\Lambda\,|E\Delta F|$. The interest of this kind of minimality condition, originally introduced in a more general context and form by Almgren [Alm76], lies in the fact that, contrary to local perimeter minimality, it is satisfied by minimizers in geometric variational problems with volume-constraints and potential-type energies. At the same time, the smaller the scale at which the competitor F differs from E, the closer (Λ, r_0)-perimeter minimality is to plain local perimeter minimality, and thus the regularity theory and the analysis of singularities may be tackled in both cases with essentially the same effort.

In Chapter 21 we thus introduce (Λ, r_0)-perimeter minimality, we discuss its applicability in studying minimizers which arise from the variational problems presented in Part II, and prove the compactness theorem for sequences of (Λ, r_0)-perimeter minimizers. In Chapter 22 we introduce the fundamental notion of excess $\mathbf{e}(E, x, r)$, which is used to measure the integral oscillation of the measure-theoretic outer unit normal to E over $B(x, r) \cap \partial^* E$. We discuss the basic properties of the excess and prove that its smallness at a given point x and scale r implies the uniform proximity of $B(x, r) \cap \partial E$ to a hyperplane. Starting from this result, in Chapter 23, we show that the $\mathcal{H}^{n-1}$-rectifiable set $B(x, r) \cap \partial E$ can always be covered by the graph of a Lipschitz function u over an $(n-1)$-dimensional ball $\mathbf{D}_r$ of radius r, up to an error which is controlled by the size of $\mathbf{e}(E, x, r)$. Moreover, again in terms of the size of $\mathbf{e}(E, x, r)$, the function u is in fact close to minimizing the area integrand $\int_{\mathbf{D}_r} \sqrt{1 + |\nabla' u|^2}$, and $\int_{\mathbf{D}_r} |\nabla' u|^2$ is close to zero, so that, by Taylor's formula

$$\int_{\mathbf{D}_r} \sqrt{1 + |\nabla' u|^2} = \mathcal{H}^{n-1}(\mathbf{D}_r) + \frac{1}{2} \int_{\mathbf{D}_r} |\nabla' u|^2 + \dots,$$

u is in fact close to minimizing the Dirichlet integral $\int_{\mathbf{D}_r} |\nabla' u|^2$; that is, u is almost a *harmonic function*. Through the use of the *reverse Poincaré inequality* (Chapter 24), and exploiting some basic properties of harmonic functions, in Chapter 25 we use this information to prove some explicit decay estimates for the integral averages of ∇u which, in turn, are equivalent in proving the uniform decay of the excess $\mathbf{e}(E, x, r)$ in r. In Chapter 26 we exploit the decay of the excess to prove the $C^{1,\gamma}$-regularity of $A \cap \partial^* E$. As a by-product we obtain a characterization of the singular set $\Sigma(E; A)$ in terms of the excess, as well as a powerful C^1-convergence theorem for sequences of (Λ, r_0)-perimeter minimizers. The exposition of the regularity theory is concluded in Chapter 27, where the connection with elliptic equations in divergence form is used to improve the $C^{1,\gamma}$-regularity result on minimizers of specific variational problems. Finally, Chapter 28 is devoted to the study of singular sets and singular minimizing cones. We refer to the beginning of that chapter for a detailed overview of its contents.

NOTATION WARNING: Throughout this part we shall continuously adopt Notation 4. Moreover, we shall denote by $\mathbf{C}(x, r, \nu)$ the cylinder

$$\mathbf{C}(x, r, \nu) = x + \left\{ y \in \mathbb{R}^n : |y \cdot \nu| < r, |y - (y \cdot \nu)\nu| < r \right\},$$

where $x \in \mathbb{R}^n$, $r > 0$ and $\nu \in S^{n-1}$.

21

(Λ, r_0)-perimeter minimizers

Given an open set A and a set of locally finite perimeter E in $\mathbb{R}^n$, $n \geq 2$, we say that E is a **(Λ, r_0)-perimeter minimizer in** A provided $\operatorname{spt}\mu_E = \partial E$ (see Remark 16.11) and there exist two constants Λ and r_0 with

$$0 \leq \Lambda < \infty\,, \qquad r_0 > 0\,, \tag{21.1}$$

such that

$$P(E\,; B(x,r)) \leq P(F\,; B(x,r)) + \Lambda\,|E\Delta F|\,, \tag{21.2}$$

whenever $E\Delta F \subset\subset B(x,r) \cap A$ and $r < r_0$. In this chapter we gather the basic facts about (Λ, r_0)-perimeter minimizers. Examples are collected in Section 21.1, a comparison with local perimeter minimality is presented in Section 21.2, and the $C^{1,\gamma}$-regularity theorem is stated in Section 21.3. Section 21.4 deals with density estimates (analogous to the ones obtained in Section 16.2 for local perimeter minimizers), while in Section 21.5 we present a compactness theorem for sequences of (Λ, r_0)-perimeter minimizers.

Exercise 21.1 If E is a (Λ, r_0)-perimeter minimizer in A, then $\mathbb{R}^n \setminus E$ is a (Λ, r_0)-perimeter minimizer in A too.

21.1 Examples of (Λ, r_0)-perimeter minimizers

Local perimeter minimizers (at a given scale r_0) (see Section 16.2), are (Λ, r_0)-perimeter minimizers with $\Lambda = 0$. In particular, by Example 16.13, so are minimizers in the Plateau-type problem (12.29). When Λ is positive, the term $\Lambda\,|E\Delta F|$ in (21.2) appears as a higher order perturbation of local perimeter minimality, and allows us to cover further important examples.

Example 21.2 Given a *bounded* Borel function $g\colon \mathbb{R}^n \to \mathbb{R}$, if E is a minimizer in the prescribed mean curvature problem (12.32),

$$\inf\left\{P(E;A) + \int_E g(x)\mathrm{d}x\colon E \subset A\right\},$$

then E is a (Λ, r_0)-perimeter minimizer in A, with

$$\Lambda = \|g\|_{L^\infty(\mathbb{R}^n)}, \qquad r_0 \text{ arbitrary}.$$

Indeed, if F is such that $E\Delta F \subset\subset B(x,r) \cap A$, then

$$P(E;A) \le P(F;A) + \int_F g - \int_E g \le P(F;A) + \|g\|_{L^\infty(\mathbb{R}^n)}|E\Delta F|\,.$$

Since $P(E;A) - P(F;A) = P(E;B(x,r)) - P(F;B(x,r))$, we are done.

Example 21.3 In Section 17.5 we have seen that relative isoperimetric sets in an open set A are volume-constrained perimeter minimizers in A. Let us now show that if E is a volume-constrained perimeter minimizer in A, then it is a (Λ, r_0)-perimeter minimizer in A, for suitable values of Λ and r_0 depending on E and A only, and that, in fact, (21.2) holds in the stronger form

$$P(E\,;B(x,r)) \le P(F\,;B(x,r)) + \Lambda\Big||E| - |F|\Big|, \tag{21.3}$$

for every $E\Delta F \subset\subset B(x,r)\cap A$ and $r < r_0$. Indeed, let us consider $x_1, x_2 \in A\cap\partial E$ and $t_0 > 0$ such that, if we set $B_1 = B(x_1, t_0)$ and $B_2 = B(x_2, t_0)$, then we have $B_1 \cap B_2 = \emptyset$ and $B_1 \cup B_2 \subset\subset A$. By applying Lemma 17.21 to E with respect to the open sets B_1 and B_2, we find two positive constants σ_0 and C_0, implicitly depending on E and A, with the following property: given $|\sigma| < \sigma_0$, there exist two sets of finite perimeter F_1 and F_2 with

$$E\Delta F_k \subset\subset B_k\,, \quad |F_k| = |E| + \sigma\,, \quad \Big|P(E;B_k) - P(F_k;B_k)\Big| \le C_0\,|\sigma|\,. \tag{21.4}$$

We finally choose

$$\Lambda = C_0\,, \qquad r_0 = \min\left\{\frac{t_0}{2}, \frac{\sigma_0^{1/n}}{\omega_n}, t_1\right\},$$

where $t_1 = (|x_1 - x_2| - 2t_0)/2$ has the property that, if a ball of radius t_1 intersects B_1 (respectively, B_2), then it is disjoint from B_2 (resp., from B_1); see Figure 21.1. Let F be such that $E\Delta F \subset\subset B(x,r) \cap A$, where $r < r_0$. Since

$$\Big||E| - |F|\Big| \le |E\Delta F| \le \omega_n\,r^n < \omega_n\,r_0^n \le \sigma_0\,,$$

we can surely compensate for the (possible) volume deficit $\big||E|-|F|\big|$ between E and F by modifying F inside either B_1 or B_2. Precisely, thanks to the definition

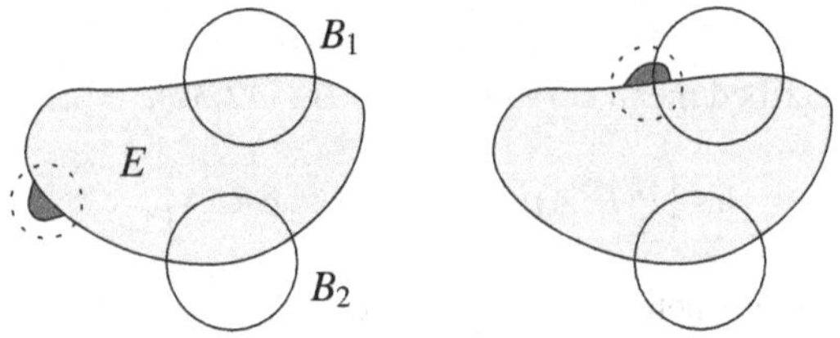

Figure 21.1 The volume change produced by any variation F of E that is compactly supported in a ball of radius r_0 can always be fixed by a further variation of the kind described in Lemma 17.21, supported either in B_1 or B_2. On the left, a variation F that can be compensated indifferently in B_1 or B_2; on the right, a variation F that forces us to deform E inside B_2.

of t_1, we may assume that $B(x, r)$ does not intersect B_1, set $\sigma = |E| - |F|$, and consider F_1 as in (21.4), so that, in particular,

$$E \Delta F_1 \subset\subset B_1\,, \qquad \left(E \Delta F\right) \subset\subset B(x, r) \subset\subset \mathbb{R}^n \setminus \overline{B_1}\,. \tag{21.5}$$

Since $\sigma = |F_1| - |E|$ by (21.4), if we define

$$G = \left(F \cap B(x, r)\right) \cup \left(F_1 \cap B_1\right) \cup \left(E \setminus \left(B(x, r) \cup B_1\right)\right),$$

then $|G| = |E|$ and $E \Delta G \subset\subset A$. By the volume-constrained minimality of E, $P(E; A) \le P(G; A)$, so that, by Exercise 12.16, (21.4), and (21.5),

$$\begin{aligned} P(E; A) &\le P(G; A \setminus \overline{B_1}) + P(G; B_1) + P(G; \partial B_1) \\ &= P(F; A \setminus \overline{B_1}) + P(F_1; B_1) + P(F; \partial B_1) \\ &\le P(F; A \setminus B_1) + P(E; B_1) + C_0\, |\sigma| = P(F; A) + C_0 \Big| |E| - |F| \Big|. \end{aligned}$$

Exercise 21.4 If E is a minimizer in the liquid drop problem (19.9) defined by A open and g bounded, then E is (Λ, r_0)-perimeter minimizer in A, with $\Lambda = C_0 + \|g\|_{L^\infty(A)}$, and with C_0 and r_0 depending on E and A.

Exercise 21.5 If $p > (n-1)/n$ and A is an open bounded set in $\mathbb{R}^n$, then every p-Cheeger set E of A (see Exercises 14.7–17.12) is a (Λ, r_0)-perimeter minimizer in A, with $\Lambda = c(p)P(E)/|E|$ and $r_0 = (|E|/2\omega_n)^{1/n}$ for an explicitly computable constant $c(p)$. *Hint:* Exploit $P(E) \le (|E|/|F|)^p P(F)$ on $E \Delta F \subset\subset B(x_0, r_0) \subset A$, discarding the trivial cases $|E| \le |F|$ and $P(E) \le P(F)$.

21.2 (Λ, r_0) and local perimeter minimality

We now compare (Λ, r_0)-perimeter minimality with local minimality. What is important to keep in mind here is that the term $\Lambda\, |E \Delta F|$ behave like a higher order perturbation in r. Indeed, $|E \Delta F| \le \omega_n r^n$, while $P(E; B(x, r))$ and

$P(F; B(x,r))$ behave like r^{n-1} for small r. In this way, E is "almost" a local perimeter minimizer in A, with an increasing precision at increasingly smaller scales. Another formulation of this idea is obtained by looking at blow-ups.

Remark 21.6 (Scaling of (Λ, r_0)-perimeter minimality) If E is a (Λ, r_0)-perimeter minimizer in A, then, for every $x \in \mathbb{R}^n$ and $r > 0$, $E_{x,r} = (E - x)/r$ is a (Λ', r_0')-perimeter minimizer in $A_{x,r}$, where

$$\Lambda' = \Lambda\, r\,, \qquad r_0' = \frac{r_0}{r}\,.$$

If $r < 1$, that is, we are zooming in at x, then $\Lambda' < \Lambda$ and $r_0' > r_0$. In other words, the blow-ups $E_{x,r}$ are closer than E to satisfying a standard perimeter minimality condition, and they achieve this at increasingly larger scales. Note that **the product Λr_0 is invariant under blow-up**, that is $\Lambda\, r_0 = \Lambda'\, r_0'$.

Remark 21.7 (About the size of the scale r_0) We may expect inequality (21.2) to be trivial if the scale r is too large with respect to Λ. Indeed, (21.2) implies that, whenever $E\Delta F \subset\subset B(x,r) \cap A$ and $r < r_0$,

$$\Big(1 - \frac{\Lambda\, r}{n}\Big) P(E; B(x,r)) \le \Big(1 + \frac{\Lambda\, r}{n}\Big) P(F; B(x,r))\,, \tag{21.6}$$

which is non-trivial if and only if $r < n/\Lambda$ (we refer to (21.6) as the **weak (Λ, r_0)-minimality condition**). Indeed, by applying the Euclidean isoperimetric inequality to $E\Delta F$, and by Exercise 16.5, we find that

$$\begin{aligned} |E\Delta F| &\le |E\Delta F|^{1/n}\, |E\Delta F|^{(n-1)/n} \le (\omega_n\, r^n)^{1/n} \frac{P(E\Delta F)}{n\omega_n^{1/n}} \\ &= \frac{r}{n}\, P(E\Delta F) = \frac{r}{n}\, P(E\Delta F; B(x,r)) \\ &\le \frac{r}{n} \Big(P(E; B(x,r)) + P(F; B(x,r))\Big), \end{aligned}$$

so that (21.6) follows from (21.2). Having this remark in mind, it is convenient to assume, **as we shall always do in the rest of the book**, that

$$\Lambda\, r_0 \le 1\,. \tag{21.7}$$

This caveat allows us to prove estimates with constants independent of Λ, although holding at scales $r < r_0$, with the size of r_0 bounded above by Λ^{-1}. The density estimates of Theorem 21.11 provide a first example of this idea.

21.3 The $C^{1,\gamma}$-reguarity theorem

We state here the fundamental regularity result for (Λ, r_0)-perimeter minimizers (for more precise statements; see Theorems 26.3 and 26.5).

Theorem 21.8 (The $C^{1,\gamma}$-regularity theorem) *If $n \geq 2$, $A \subset \mathbb{R}^n$ is an open set, and E is a (Λ, r_0)-perimeter minimizer in A with $\Lambda r_0 \leq 1$, then $A \cap \partial^* E$ is a $C^{1,\gamma}$-hypersurface for every $\gamma \in (0, 1/2)$, it is relatively open in $A \cap \partial E$, and it is $\mathcal{H}^{n-1}$-equivalent to $A \cap \partial E$.*

Remark 21.9 (The case of local perimeter minimizers) Proving Theorem 21.8 in the case of local perimeter minimizers is slightly simpler, and leads to a stronger statement where $\gamma \in (0, 1)$. A first obvious reason for these simplifications to happen is that, in this case, $\Lambda = 0$ and the perturbation $|E \Delta F|$ disappears. More interestingly, if E is a local perimeter minimizer in A at scale r_0, then the vanishing mean curvature condition

$$\int_{\partial^* E} \operatorname{div}_E T \, d\mathcal{H}^{n-1} = 0 \tag{21.8}$$

holds true for every $T \in C^1_c(B(x, r); \mathbb{R}^n)$ with $B(x, r) \subset A$ and $r < r_0$. In the course of the proof we shall highlight the various occasions in which the restriction to local perimeter minimizers allows for substantial simplifications through the use of (21.8).

Remark 21.10 (Higher regularity) The $C^{1,\gamma}$-regularity result of Theorem 21.8 can often be improved when dealing with explicit examples of (Λ, r_0)-perimeter minimizers, like the ones introduced in Section 21.1. This is possible by exploiting the stationarity conditions associated with the variational problem under consideration; see Chapter 27.

21.4 Density estimates for (Λ, r_0)-perimeter minimizers

The density estimates for local perimeter minimizers of Theorem 16.14 hold true for (Λ, r_0)-perimeter minimizers too, with identical proof.

Theorem 21.11 (Density estimates) *Given $n \geq 2$, there exists a positive constant $c(n)$ with the following property. If E is a (Λ, r_0)-perimeter minimizer in the open set $A \subset \mathbb{R}^n$ and $\Lambda r_0 \leq 1$, then*

$$\frac{1}{4^n} \leq \frac{|E \cap B(x, r)|}{\omega_n r^n} \leq 1 - \frac{1}{4^n}, \tag{21.9}$$

$$c(n) \leq \frac{P(E; B(x, r))}{r^{n-1}} \leq 3n\omega_n, \tag{21.10}$$

whenever $x \in A \cap \partial E$, $B(x,r) \subset A$, *and* $r < r_0$. *In particular,*

$$\mathcal{H}^{n-1}\Big(A \cap (\partial E \setminus \partial^* E)\Big) = 0\,. \tag{21.11}$$

Remark 21.12 Under the assumptions of Theorem 21.11, and since $\mathbf{C}_r \subset B_{\sqrt{2}\,r} \subset \mathbf{C}_{\sqrt{2}\,r}$, the following density estimates on cylinders hold true:

$$c(n)\, r^{n-1} \le P\Big(E; \mathbf{C}(x,r,\nu)\Big) \le 3\, 2^{(n-1)/2}\, r^{n-1}\,. \tag{21.12}$$

Proof of Theorem 21.11 The argument is a trivial adaptation of the analogous proof for local perimeter minimizers. We briefly recall it for the sake of clarity. We fix $x \in A \cap \partial E$, set $d = \min\{r_0, \mathrm{dist}(x, \partial A)\}$, and notice that the function $m\colon (0,d) \to \mathbb{R}$ defined by $m(r) = |B(x,r) \cap E|$, $0 < r < d$ satisfies

$$\begin{aligned} 0 < m(r) < \omega_n\, r^n\,, &\qquad \forall r \in (0,d)\,, \\ m'(r) = \mathcal{H}^{n-1}(E \cap \partial B(x,r))\,, &\qquad \text{for a.e. } r \in (0,d). \end{aligned} \tag{21.13}$$

Given $r \in (0,d)$ with $\mathcal{H}^{n-1}(\partial^* E \cap \partial B(x,r)) = 0$ and $s \in (r,d)$, one plugs the comparison set $F = E \setminus B(x,r)$ into the weak (Λ, r_0)-perimeter minimality condition (21.6) for E, to find that

$$\begin{aligned} &\Big(1 - \frac{\Lambda\, s}{n}\Big) P(E; B(x,s)) \\ &\le \Big(1 + \frac{\Lambda\, s}{n}\Big)\Big\{\mathcal{H}^{n-1}\Big(E^{(1)} \cap \partial B(x,r)\Big) + P\Big(E; B(x,s) \setminus \overline{B(x,r)}\Big)\Big\}. \end{aligned} \tag{21.14}$$

By first letting $s \to r^+$ in (21.14) and since $\mathcal{H}^{n-1}(E^{(1)} \cap \partial B(x_1 r)) \le P(B(x,r)) = n\omega_n r^{n-1}$, by (21.7) we deduce the upper bound in (21.10). Next, by adding $(1 - (\Lambda r/n))\mathcal{H}^{n-1}(E^{(1)} \cap \partial B(x,r))$ to both sides of (21.14), one finds

$$\Big(1 - \frac{\Lambda r}{n}\Big) P(E \cap B(x,r)) \le 2\,\mathcal{H}^{n-1}\Big(E^{(1)} \cap \partial B(x,r)\Big).$$

By the Euclidean isoperimetric inequality and thanks to (21.7), we have

$$\frac{n\omega_n^{1/n} m(r)^{(n-1)/n}}{2} \le 2\, m'(r)\,, \qquad \text{for a.e. } r \in (0,d).$$

By (21.13), we may integrate this differential inequality to prove the lower bound in (21.9). The upper bound in (21.9) is deduced by symmetry, since $\mathbb{R}^n \setminus E$ is a (Λ, r_0)-perimeter minimizer in A. The lower bound in (21.10) then follows again by (21.9) and the relative isoperimetric inequality in

Proposition 12.37. Finally, since (21.9) implies that $A \cap \partial E \subset A \cap \partial^e E$, (21.11) is consequence of Federer's theorem, Theorem 16.2. □

21.5 Compactness for sequences of (Λ, r_0)-perimeter minimizers

A versatile tool in the study of the regularity theory is provided by the compactness theorem for sequences of (Λ, r_0)-perimeter minimizers which is discussed in this section. A pre-compactness result follows quite directly from the upper density estimate in (21.12). The closure theorem is slightly more delicate, and requires the use of the minimality condition (21.2) on suitable comparison sets. For the assumption $\Lambda r_0 \le 1$, see Remark 21.7.

Proposition 21.13 (Pre-compactness for sequences of (Λ, r_0)-perimeter minimizers) *If $\{E_h\}_{h\in\mathbb{N}}$ is a sequence of (Λ, r_0)-perimeter minimizers in the open set $A \subset \mathbb{R}^n$ with $\Lambda r_0 \le 1$, then for every open set $A_0 \subset\subset A$ with $P(A_0) < \infty$ there exist $h(k) \to \infty$ as $k \to \infty$ and a set of finite perimeter $E \subset A_0$, such that*

$$A_0 \cap E_{h(k)} \to E\,, \qquad \mu_{A_0\cap E_{h(k)}} \stackrel{*}{\rightharpoonup} \mu_E\,.$$

Proof Given $x \in A_0$, let us consider a ball $B(x,r) \subset\subset A$ with $r < r_0$. By (16.10) and the upper density estimate (21.10),

$$P\big(E_h \cap B(x,r)\big) \le P(E_h; B(x,r)) + P(B(x,r)) \le 3n\omega_n\, r^{n-1} + n\omega_n r^{n-1}\,.$$

Hence, for $x \in A_0$, $B(x,r) \subset\subset A$ and $r < r_0$, we have

$$\sup_{h\in\mathbb{N}} P\big(E_h \cap B(x,r)\big) \le 4n\omega_n r_0^{n-1}\,. \tag{21.15}$$

We now cover A_0 by the union of finitely many balls $B_j = B(x_j, s_j) \subset\subset A$ with $x_j \in A_0$ and $s_j < r_0$ for $1 \le j \le N$, $N \in \mathbb{N}$. By (21.15), we may iteratively apply Theorem 12.26 to construct a subsequence $h(k) \to \infty$ as $k \to \infty$, and a family of sets of finite perimeter $\{F_j\}_{j=1}^N$, such that, for every $j = 1, ..., N$,

$$B_j \cap E_{h(k)} \to F_j\,, \qquad \mu_{B_j\cap E_{h(k)}} \stackrel{*}{\rightharpoonup} \mu_{F_j}\,,$$

as $k \to \infty$, and

$$F_i \cap B_i \cap B_j = F_j \cap B_i \cap B_j\,, \tag{21.16}$$

whenever $1 \le i < j \le N$. We conclude the proof by setting

$$E = A_0 \cap \bigcup_{j=1}^N F_j\,.$$

Indeed, setting for brevity $E_{h(k)} = E_k$, by (21.16), $A_0 \cap E_k \to E$. At the same time, by (21.15) and (16.10) (applied to $A_0 \cap E_k$ and to $E_k \cap B_j$),

$$P(A_0 \cap E_k) \le P(E_k; A_0) + P(A_0) \le \sum_{j=1}^{N} P(E_k; B_j) + P(A_0)$$
$$\le 4n\omega_n r_0^{n-1} N + P(A_0),$$

that is,

$$\sup_{k\in\mathbb{N}} |\mu_{A_0 \cap E_k}|(\mathbb{R}^n) < \infty. \tag{21.17}$$

Since $A_0 \cap E_k \to E$, we immediately deduce $\mu_{A_0 \cap E_{h(k)}} \stackrel{*}{\rightharpoonup} \mu_E$. □

Theorem 21.14 (Closure for sequences of (Λ, r_0)-perimeter minimizers) *If $\{E_h\}_{h\in\mathbb{N}}$ is a sequence of (Λ, r_0)-perimeter minimizers in the open set $A \subset \mathbb{R}^n$, with $\Lambda r_0 \le 1$, and if $A_0 \subset\subset A$ is an open set with $P(A_0) < \infty$ such that $A_0 \cap E_h \to E$ for a set of finite perimeter E, then E is a (Λ, r_0)-perimeter minimizer in A_0. Moreover,*

$$\mu_{A_0 \cap E_h} \stackrel{*}{\rightharpoonup} \mu_E, \tag{21.18}$$
$$|\mu_{E_h}| \stackrel{*}{\rightharpoonup} |\mu_E|, \qquad \text{in } A_0, \tag{21.19}$$

and, in particular,

(i) if $x_h \in A_0 \cap \partial E_h$, $x_h \to x$, and $x \in A_0$, then $x \in A_0 \cap \partial E$;
(ii) if $x \in A_0 \cap \partial E$, then there exists $\{x_h\}_{h\in\mathbb{N}} \subset A_0 \cap \partial E_h$ such that $x_h \to x$.

Remark 21.15 Notice that in (21.19) we have used the notion of weak-star convergence in an open set, which was introduced in Exercise 4.32. In particular, (21.19) means that

$$\int_{\partial^* E} \varphi \, d\mathcal{H}^{n-1} = \lim_{h\to\infty} \int_{\partial^* E_h} \varphi \, d\mathcal{H}^{n-1}, \qquad \forall \varphi \in C_c^0(A_0).$$

As a consequence, mimicking the proof of Proposition 4.26, we see that, if A' is an open set with $A' \subset A_0$ and K is a compact set with $K \subset A_0$, then

$$P(E; A') \le \liminf_{h\to\infty} P(E_h; A'), \tag{21.20}$$
$$P(E; K) \ge \limsup_{h\to\infty} P(E_h; K), \tag{21.21}$$

and, if G is a Borel set with $G \subset\subset A_0$ and $\mathcal{H}^{n-1}(\partial^* E \cap \partial G) = 0$, then

$$P(E; G) = \lim_{h\to\infty} P(E_h; G). \tag{21.22}$$

Remark 21.16 Statement (i) will be noticeably strengthened in the case that $x \in A_0 \cap \partial^* E$; see Theorem 26.6.

Proof of Theorem 21.14 *Step one:* We aim to prove that, given F such that $E\Delta F \subset\subset B(x,r) \cap A_0$ and $r < r_0$, then we have

$$P(E; B(x,r)) \le P(F; B(x,r)) + \Lambda\, |E\Delta F|\,. \tag{21.23}$$

To this end, we shall first modify the comparison set F outside $B(x,r)$ to construct a competitor for the (Λ, r_0)-perimeter minimality of E_h, and then pass to the limit in the corresponding minimality inequalities, exploiting the convergence of $E_h \cap A_0$ to E. We implement this idea as follows. Let us remark that, if $y \in A_0$, then for a.e. $r \in (0, d(y))$, $d(y) = \min\{r_0, \mathrm{dist}(y, \partial A_0)\}$, we have

$$\mathcal{H}^{n-1}\Big(\partial B(y,r) \cap \partial^* F\Big) = \mathcal{H}^{n-1}\Big(\partial B(y,r) \cap \partial^* E_h\Big) = 0\,, \quad \forall h \in \mathbb{N}\,, \tag{21.24}$$

$$\liminf_{h\to\infty} \mathcal{H}^{n-1}\Big(\partial B(y,r) \cap \Big(E^{(1)} \Delta E_h^{(1)}\Big)\Big) = 0\,. \tag{21.25}$$

Here, (21.24) is proved by Proposition 2.16, while by $B(y, d(y)) \subset A_0$, (5.19), and the coarea formula (13.3) we have

$$\begin{aligned} 0 &= \lim_{h\to\infty} \Big|B\Big(y, d(y)\Big) \cap \Big(E_h \Delta E\Big)\Big| = \lim_{h\to\infty} \Big|B\Big(y, d(y)\Big) \cap \Big(E_h^{(1)} \Delta E^{(1)}\Big)\Big| \\ &= \lim_{h\to\infty} \int_0^{d(y)} \mathcal{H}^{n-1}\Big(\partial B(y,r) \cap \Big(E^{(1)} \Delta E_h^{(1)}\Big)\Big)\, dr\,, \end{aligned}$$

which, in turn, implies (21.25) by Fatou's lemma. Since $E\Delta F$ is compactly contained in $B(x,r) \cap A_0$, we can find a finite family of balls $\{B_j\}_{j=1}^N$, $N \in \mathbb{N}$, with $B_j = B(y_j, r_j)$ for $y_j \in A_0$ and $r_j \in (0, d(y_j))$ such that

$$E\Delta F \subset\subset G \subset\subset B(x,r) \cap A_0\,, \qquad \text{where} \qquad G = \bigcup_{j=1}^N B_j\,,$$

and such that (21.24) and (21.25) hold true at $y = y_j$ and $r = r_j$. Let us thus consider the sets of finite perimeter defined by

$$F_h = \Big(F \cap G\Big) \cup \Big(E_h \setminus G\Big)\,, \qquad h \in \mathbb{N}\,;$$

see Figure 21.2 for the case $N = 1$. Since $\partial G \subset \bigcup_{j=1}^N \partial B_j$, by (21.24),

$$\mathcal{H}^{n-1}\Big(\partial G \cap \partial^* F\Big) = \mathcal{H}^{n-1}\Big(\partial G \cap \partial^* E_h\Big) = 0\,, \qquad \forall h \in \mathbb{N}\,. \tag{21.26}$$

Moreover, $E\Delta F \subset\subset G$ implies $E^{(1)} \cap \partial G = F^{(1)} \cap \partial G$, so that (21.25) gives

$$\liminf_{h\to\infty} \mathcal{H}^{n-1}\Big(\partial G \cap \Big(F^{(1)} \Delta E_h^{(1)}\Big)\Big) = 0\,. \tag{21.27}$$

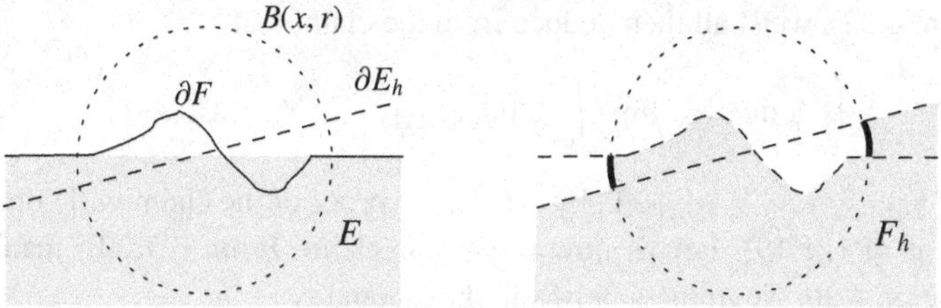

Figure 21.2 The comparison sets F_h used in the proof of Theorem 21.14. The set F is a compact variation of E into $B(x,r)$, but, in general, it is not a compact variation of E_h in $B(x,r)$. In computing the perimeter of F_h we have to take into account the extra perimeter contribution due to the surface $\partial B(x,r) \cap (E^{(1)} \Delta E_h^{(1)})$, represented by bold circular arcs in the picture.

Finally, we notice that $E_h \Delta F_h \subset G \subset\subset B(x,r) \cap A$, where $x \in A$ and $r < r_0$. By the (Λ, r_0)-minimality of the E_h in A we thus find

$$P(E_h; B(x,r)) \le P(F_h; B(x,r)) + \Lambda\,|E_h \Delta F_h|\,.$$

By (21.26) we may apply (16.32) to find that

$$\begin{aligned} P(E_h; B(x,r)) \le P(F; B(x,r)) &+ \mathcal{H}^{n-1}\Big(\partial G \cap \Big(E_h^{(1)} \Delta F^{(1)}\Big)\Big) \\ &+ \Lambda\Big|\Big(E_h \Delta F\Big) \cap G\Big|\,. \end{aligned} \tag{21.28}$$

Since $E_h \cap A_0 \to E$ and $G \subset A_0$, by (21.27) we find that

$$P(E; B(x,r)) \le \liminf_{h\to\infty} P(E_h; B(x,r)) \le P(F; B(x,r)) + \Lambda\,|(E \Delta F) \cap G|\,.$$

As $|(E\Delta F) \cap G| = |E \Delta F|$, we have proved (21.23), as required.

Step three: Arguing as in step two of the proof of Proposition 21.13, we find

$$\mu_{A_0 \cap E_h} \stackrel{*}{\rightharpoonup} \mu_E\,,$$

which is (21.18). We now claim that if a Radon measure μ on $\mathbb{R}^n$ satisfies

$$|\mu_{A_0 \cap E_h}| \stackrel{*}{\rightharpoonup} \mu\,, \tag{21.29}$$

then $\mu \llcorner A_0 = |\mu_E| \llcorner A_0$. Since every subsequence of $|\mu_{A_0 \cap E_h}|$ is compact in the weak-star convergence of Radon measures (this, thanks to (21.17) and

Theorem 4.33), we shall then deduce from the claim that

$$\int_{\mathbb{R}^n} \varphi \, \mathrm{d}|\mu_E| = \lim_{h\to\infty} \int_{\mathbb{R}^n} \varphi \, \mathrm{d}|\mu_{A_0\cap E_h}|, \qquad \forall \varphi \in C^0_c(A_0).$$

Having $|\mu_{A_0\cap E_h}| \llcorner A_0 = |\mu_{E_h}| \llcorner A_0$ by (16.4), the proof of the claim will complete the proof of (21.19). Let us now prove our claim. From (21.29), thanks to Proposition 4.30, we immediately find the inequality

$$\mu \geq |\mu_E| \quad \text{on } \mathcal{B}(\mathbb{R}^n). \tag{21.30}$$

Now let $B(x, s_0) \subset\subset A_0$ with $s_0 < r_0$. Arguing as in step two we define

$$F_h = \big(E \cap B(x,s)\big) \cup \big(E_h \setminus B(x,s)\big),$$

for $s \in (0, s_0)$ such that

$$\mathcal{H}^{n-1}\big(\partial^* E \cap \partial B(x,s)\big) = \mathcal{H}^{n-1}\big(\partial^* E_h \cap \partial B(x,s)\big) = 0, \qquad \forall h \in \mathbb{N},$$
$$\liminf_{h\to\infty} \mathcal{H}^{n-1}\big(\partial B(x,s) \cap (E_h^{(1)} \Delta E^{(1)})\big) = 0.$$

Once again, a.e. $s \in (0, s_0)$ will work. Since $E_h \Delta F_h \subset\subset B(x, s_0) \subset\subset A$, by repeating the argument used in step two,

$$P(E_h; B(x,s)) \leq P(E; B(x,s)) + \mathcal{H}^{n-1}\Big(\partial B(x,s) \cap \big(E_h^{(1)} \Delta E\big)\Big) + \Lambda |E_h \Delta F_h|.$$

Letting $h \to \infty$, by (4.30), and since $P(E_h; B(x,s)) = |\mu_{E_h\cap A_0}|(B(x,s))$ and $|E_h \Delta F_h| = |(E_h \Delta E) \cap B(x,s)|$, we thus find

$$\mu(B(x,s)) \leq |\mu_E|(B(x,s)).$$

By (21.30), if $B(x, s_0) \subset\subset A_0$, $s_0 < r_0$, then

$$|\mu_E|(B(x,s)) = \mu(B(x,s)), \qquad \text{for a.e. } s < s_0. \tag{21.31}$$

By Theorem 5.8, for μ-a.e. $x \in \operatorname{spt}\mu$ the ratio of $|\mu_E|(B(x,s))$ over $\mu(B(x,s))$ converges to $D_\mu|\mu_E|(x) \in \mathbb{R}$ as $s \to 0^+$. Hence (21.31) suffices to deduce

$$D_\mu|\mu_E|(x) = 1, \qquad \text{for } \mu\text{-a.e. } x \in A_0 \cap \operatorname{spt}\mu.$$

Since $|\mu_E| \ll \mu$, again by Theorem 5.8, we conclude that $\mu = |\mu_E|$ on $\mathcal{B}(A_0)$.

Step four: Property (ii) follows from (21.19) by a trivial adaptation of the proof of (4.31) from Proposition 4.26. Similarly, property (i) follows from (21.19) arguing as in Remark 4.28 and making use of the uniform lower density bounds for (Λ, r_0)-minimizers (21.10). Repeating this last argument for the sake of clarity, let us consider a sequence $\{x_h\}_{h\in\mathbb{N}} \subset A_0 \cap \partial E_h$ such that $x_h \to x$ for some $x \in A_0$. If $s > 0$ is such that $B(x, 2s) \subset\subset A_0$ and $2s < r_0$, then there exists

$h(s)$ such that, for every $h \geq h(s)$, we have $B(x_h, s) \subset B(x, 2s)$. Thus, the lower density estimate in (21.10) gives

$$P\big(E_h; \overline{B}(x, 2s)\big) \geq P\big(E_h; \overline{B}(x_h, s)\big) \geq c(n)\, s^{n-1}, \quad \forall h \geq h(s),$$

By (21.21), we thus conclude that, if $B(x, 2s) \subset\subset A_0$ and $2s < r_0$, then

$$P\big(E; \overline{B}(x, 2s)\big) \geq c(n)\, s^{n-1} > 0.$$

In particular, $x \in \operatorname{spt}\mu_E = \partial E$, and property (i) is proved. □

Remark 21.17 (Sequences of (Λ_h, r_h)-minimizers) Proposition 21.13 and Theorem 21.14 apply to sequences $\{E_h\}_{h\in\mathbb{N}}$ where each E_h is a (Λ_h, r_h)-perimeter minimizer in some open set A, provided

$$\limsup_{h\to\infty} \Lambda_h < \infty, \qquad \liminf_{h\to\infty} r_h > 0. \tag{21.32}$$

22

Excess and the height bound

We now introduce the notion of excess, a key concept in the regularity theory for (Λ, r_0)-perimeter minimizers. Given a set of locally finite perimeter E in $\mathbb{R}^n$, the **cylindrical excess** of E at the point $x \in \partial E$, at the scale $r > 0$, and with respect to the direction $\nu \in S^{n-1}$, is defined as

$$\begin{aligned}\mathbf{e}(E, x, r, \nu) &= \frac{1}{r^{n-1}} \int_{\mathbf{C}(x,r,\nu)\cap\partial^* E} \frac{|\nu_E(y) - \nu|^2}{2}\, d\mathcal{H}^{n-1}(y) \qquad (22.1)\\ &= \frac{1}{r^{n-1}} \int_{\mathbf{C}(x,r,\nu)\cap\partial^* E} \Big(1 - (\nu_E \cdot \nu)\Big)\, d\mathcal{H}^{n-1}\,;\end{aligned}$$

see Figure 22.1. The **spherical excess** of E at the point $x \in \partial E$ and at scale $r > 0$ is similarly defined as

$$\mathbf{e}(E, x, r) = \min_{\nu \in S^{n-1}} \frac{1}{r^{n-1}} \int_{B(x,r)\cap\partial^* E} \frac{|\nu_E(y) - \nu|^2}{2}\, d\mathcal{H}^{n-1}(y)\,. \qquad (22.2)$$

Hence, when considering the spherical excess at a given scale, we essentially minimize the cylindrical excess at that scale with respect to the direction. The fundamental result related to the notion of excess is that, if E is a (Λ, r_0)-perimeter minimizer, then the smallness of $\mathbf{e}(E, x, r, \nu)$ at some $x \in \partial E$ actually forces $\mathbf{C}(x, s, \nu) \cap \partial E$ (for some $s < r$) to agree with the graph (with respect to the direction ν) of a $C^{1,\gamma}$-function (see Theorem 26.1 for the case of local perimeter minimizers, and Theorem 26.3 for the general case). This theorem is proved through a long series of intermediate results, in which increasingly stronger conclusions are deduced from a small excess assumption. We begin this long journey in the next two chapters, where we shall prove, in particular, the so-called height bound, Theorem 22.8: if E is a (Λ, r_0)-perimeter minimizer in $\mathbf{C}(x, 4r, \nu)$ with $x \in \partial E$ and $\mathbf{e}(x, 4r, \nu)$ suitably small, then the uniform distance of $\mathbf{C}(x, r, \nu) \cap \partial E$ from the hyperplane passing through x and orthogonal to ν is bounded from above by $\mathbf{e}(x, 4r, \nu)^{1/2(n-1)}$.

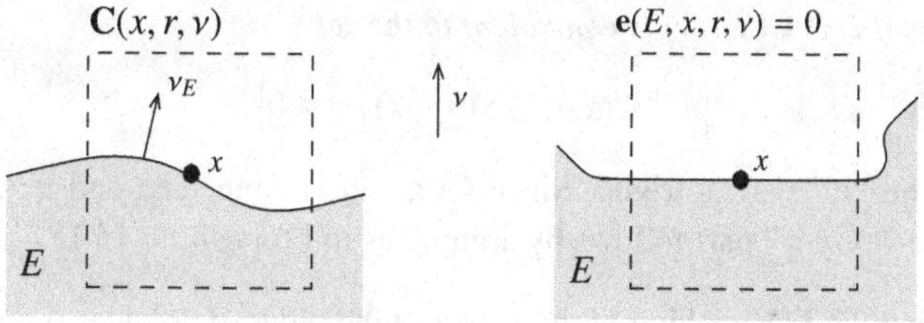

Figure 22.1 The cylindrical excess $\mathbf{e}(E,x,r,\nu)$ is the average L^2 oscillation from the given direction ν of the outer unit normal to E over the cylinder $\mathbf{C}(x,r,\nu)$. If $x \in \partial E$, then $\mathbf{e}(E,x,r,\nu) = 0$ if and only if E coincides in $\mathbf{C}(x,r,\nu)$ with the half-space passing through x, with outer unit normal equal to ν. We shall need to quantify the geometric consequences of the *smallness* of the cylindrical excess on (Λ, r_0)-perimeter minimizers.

22.1 Basic properties of the excess

In this section we gather some basic properties of the excess which hold true on generic sets of locally finite perimeter (Propositions 22.1–22.4), or which are somehow related to minimality (Proposition 22.5 and 22.6).

Proposition 22.1 (Scaling of the excess) *If E is a set of locally finite perimeter in $\mathbb{R}^n$, $x \in \partial E$, $r > 0$, $\nu \in S^{n-1}$, then*

$$\mathbf{e}(E,x,r,\nu) = \mathbf{e}(E_{x,r},0,1,\nu)\,, \qquad \mathbf{e}(E,x,r) = \mathbf{e}(E_{x,r},0,1)\,, \tag{22.3}$$

where, as usual, $E_{x,r} = (E-x)/r$.

Proof Since $|\nu - \nu_E|^2 = 2(1 - (\nu \cdot \nu_E))$, we have

$$\mathbf{e}(E,x,r,\nu) = \frac{|\mu_E|\,(\mathbf{C}(x,r,\nu)) - \nu \cdot \mu_E(\mathbf{C}(x,r,\nu))}{r^{n-1}}\,. \tag{22.4}$$

Hence, the first formula in (22.3) follows by Lemma 15.11. Similarly,

$$\begin{aligned}\mathbf{e}(E,x,r) &= \min_{\nu \in S^{n-1}} \frac{|\mu_E|\,(B(x,r)) - \nu \cdot \mu_E(B(x,r))}{r^{n-1}} \\ &= \frac{|\mu_E|\,(B(x,r))}{r^{n-1}}\Big(1 - \frac{|\mu_E(B(x,r))|}{|\mu_E|\,(B(x,r))}\Big),\end{aligned} \tag{22.5}$$

and (22.3) follows by applying Lemma 15.11 one more time. □

Proposition 22.2 (Zero excess implies being a half-space) *If E is a set of locally finite perimeter in $\mathbb{R}^n$, with* $\operatorname{spt}\mu_E = \partial E$, *$x \in \partial E$, $r > 0$, $\nu \in S^{n-1}$, then*

$$\mathbf{e}(E,x,r,\nu) = 0\,,$$

if and only if $E \cap \mathbf{C}(x, r, \nu)$ is equivalent to the set

$$\Big\{y \in \mathbf{C}(x, r, \nu) : (y - x) \cdot \nu \le 0\Big\}.$$

Proof The "if" part is trivial. Since $\mathbf{C}(x, r, \nu)$ is connected and $x \in \partial E = \operatorname{spt}\mu_E$, the "only if" part follows by arguing as in Proposition 15.15. □

Proposition 22.3 (Vanishing of the excess at the reduced boundary) *If E is a set of locally finite perimeter in $\mathbb{R}^n$ and $x \in \partial^* E$, then*

$$\lim_{r \to 0^+} \mathbf{e}(E, x, r) = 0. \tag{22.6}$$

Hence, given $\varepsilon > 0$, there exist $r > 0$ and $\nu \in S^{n-1}$ with $\mathbf{e}(E, x, r, \nu) \le \varepsilon$.

Proof By the definition of reduced boundary (15.1) and by (15.9) we have

$$\lim_{r \to 0^+} \frac{|\mu_E(B(x, r))|}{|\mu_E|(B(x, r))} = 1, \qquad \lim_{r \to 0^+} \frac{|\mu_E|(B(x, r))}{\omega_{n-1} r^{n-1}} = 1.$$

This limit relations, combined with (22.5), immediately imply (22.6). Finally, since $\mathbf{C}(x, r, \nu) \subset B(x, \sqrt{2}r)$ for every $\nu \in S^{n-1}$, we see that if $r > 0$ is such that $\mathbf{e}(E, x, r) \le \varepsilon$ for some $\varepsilon > 0$, then there exists $\nu \in S^{n-1}$ with

$$\mathbf{e}\Big(E, x, \frac{r}{\sqrt{2}}, \nu\Big) \le 2^{(n-1)/2} \varepsilon.$$

□

Proposition 22.4 (Excess at different scales) *If E is a set of locally finite perimeter in $\mathbb{R}^n$, $x \in \partial E$, $r > s > 0$, $\nu \in S^{n-1}$, then*

$$\mathbf{e}(E, x, s, \nu) \le \Big(\frac{r}{s}\Big)^{n-1} \mathbf{e}(E, x, r, \nu). \tag{22.7}$$

Proof Trivial. □

Proposition 22.5 (Excess and changes of direction) *For every $n \ge 2$, there exists a constant $C(n)$ with the following property. If E is a (Λ, r_0)-perimeter minimizer in the open set $A \subset \mathbb{R}^n$ with $\Lambda r_0 \le 1$, then*

$$\mathbf{e}(E, x, r, \nu) \le C(n) \Big(\mathbf{e}(E, x, \sqrt{2}\, r, \nu_0) + |\nu - \nu_0|^2\Big), \tag{22.8}$$

whenever $x \in A \cap \partial E$, $B(x, 2\,r) \subset\subset A$, $\nu, \nu_0 \in S^{n-1}$.

Proof Since $|\nu - \nu_E|^2/2 \le |\nu_0 - \nu_E|^2 + |\nu - \nu_0|^2$ and $\mathbf{C}(x, r, \nu) \subset \mathbf{C}(x, \sqrt{2}r, \nu_0)$,

$$\mathbf{e}(E, x, r, \nu) \le \frac{2}{r^{n-1}} \int_{\mathbf{C}(x, \sqrt{2}\, r, \nu_0) \cap \partial^* E} \frac{|\nu_E - \nu_0|^2}{2}\, d\mathcal{H}^{n-1} + \frac{P(E; \mathbf{C}(x, r, \nu))}{r^{n-1}} |\nu - \nu_0|^2.$$

We conclude by the upper density estimate in (21.12). □

Proposition 22.6 (Lower semicontinuity of the excess) *If A, A_0 are open sets in $\mathbb{R}^n$ with $A_0 \subset\subset A$, $P(A_0) < \infty$, and if $\{E_h\}_{h\in\mathbb{N}}$ is a sequence of (Λ, r_0)-perimeter minimizers in A with $\Lambda r_0 < 1$, and such that $A_0 \cap E_h \to E$, then, for every cylinder $\mathbf{C}(x, r, \nu) \subset\subset A_0$ we have*

$$\mathbf{e}(E, x, r, \nu) \le \liminf_{h\to\infty} \mathbf{e}(E_h, x, r, \nu) . \tag{22.9}$$

In fact, if $\mathbf{C}(x, r, \nu)$ is such that

$$\mathcal{H}^{n-1}\big(\partial^* E \cap \partial\mathbf{C}(x, r, \nu)\big) = 0 , \tag{22.10}$$

then we have exactly

$$\mathbf{e}(E, x, r, \nu) = \lim_{h\to\infty} \mathbf{e}(E_h, x, r, \nu) . \tag{22.11}$$

Proof *Step one:* By Theorem 21.14, $|\mu_{E_h}| \overset{*}{\rightharpoonup} |\mu_E|$ in A_0 and $\mu_{A_0\cap E_h} \overset{*}{\rightharpoonup} \mu_E$. If (22.10) holds true, then we have

$$|\mu_E|\,(\mathbf{C}(x, r, \nu)) = \lim_{h\to\infty} |\mu_{E_h}|\,(\mathbf{C}(x, r, \nu)) ,$$

$$\mu_E(\mathbf{C}(x, r, \nu)) = \lim_{h\to\infty} \mu_{A_0\cap E_h}(\mathbf{C}(x, r, \nu)) ,$$

thanks to (21.22) in the first case, and thanks to Proposition 4.30(i) in the second case. Since $\mathbf{C}(x, r, \nu) \subset\subset A_0$, we have $\mu_{A_0\cap E_h}(\mathbf{C}(x, r, \nu)) = \mu_{E_h}(\mathbf{C}(x, r, \nu))$, and thus (22.11) follows from (22.4).

Step two: Let us first remark that, as we may directly check from (22.1), the function $r \mapsto \mathbf{e}(E, x, r, \nu)$ is continuous from the left on $(0, \infty)$,

$$\mathbf{e}(E, x, r, \nu) = \lim_{s\to r^-} \mathbf{e}(E, x, s, \nu) . \tag{22.12}$$

By Proposition 2.16, we may have $\mathcal{H}^{n-1}(\partial^* E \cap \partial\mathbf{C}(x, r_k, \nu)) = 0$ on a sequence $r_k \to r^-$. Since $\mathbf{C}(x, r_k, \nu) \subset\subset A_0$ for every $k \in \mathbb{N}$, by (22.11) we find

$$\mathbf{e}(E, x, r_k, \nu) = \lim_{h\to\infty} \mathbf{e}(E_h, x, r_k, \nu) \le \left(\frac{r}{r_k}\right)^{n-1} \liminf_{h\to\infty} \mathbf{e}(E_h, x, r, \nu) .$$

Finally, we let $k \to \infty$, and obtain (22.9). □

Exercise 22.7 If E is a perimeter minimizer in A and $x \in \partial E$, then

$$\mathbf{e}(E, x, r, \nu) + \mathbf{e}(E, x, r, -\nu) \ge 2\,\omega_{n-1} , \tag{22.13}$$

for every $x \in \partial E$, $\nu \in S^{n-1}$, $\mathbf{C}(x, r, \nu) \subset\subset A$; thus $\mathbf{e}(E, x, r, \nu)$ and $\mathbf{e}(E, x, r, -\nu)$ cannot be simultaneously small, as the cylindrical excess detects orientation.

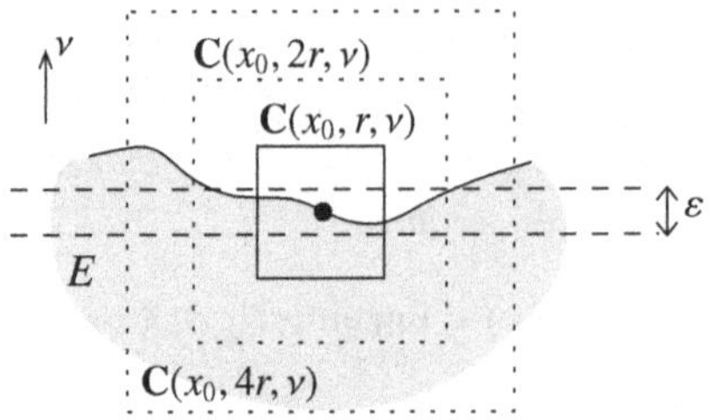

Figure 22.2 The height bound, Theorem 22.8, states the inclusion of $\mathbf{C}(x_0, r_0, \nu) \cap \partial E$ into an ε-neighborhood of the hyperplane $x_0 + \nu^\perp$, for $\varepsilon = \mathbf{e}(E, x_0, 4\, r_0, \nu)^{1/2(n-1)}$.

22.2 The height bound

We now begin to analyze some first consequences of a small cylindrical excess assumption. We aim to prove a fundamental estimate relating the height of a perimeter minimizer to its cylindrical excess. Precisely, we show that if E is a (Λ, r_0)-perimeter minimizer in the cylinder $\mathbf{C}(x_0, 4r, \nu)$, with $x_0 \in \partial E$ and $\mathbf{e}(E, x_0, 4r, \nu)$ small enough, then $\mathbf{e}(E, x_0, 4r, \nu)^{1/2(n-1)}$ controls the uniform distance of $\mathbf{C}(x_0, r, \nu) \cap \partial E$ from the $(n-1)$-dimensional space $x_0 + \nu^\perp$; see Figure 22.2. We directly state this theorem in the case $\nu = e_n$, where we set, for the sake of brevity,

$$\mathbf{e}(E, x_0, r, e_n) = \mathbf{e}_n(x_0, r)\,.$$

Theorem 22.8 (The height bound) *Given $n \ge 2$, there exist positive constants $\varepsilon_0(n)$ and $C_0(n)$ with the following property. If E is a (Λ, r_0)-perimeter minimizer in $\mathbf{C}(x_0, 4r_0)$ with*

$$\Lambda r_0 \le 1\,, \qquad x_0 \in \partial E\,, \qquad \mathbf{e}_n(x_0, 4r_0) \le \varepsilon_0(n)\,, \tag{22.14}$$

then (with $\mathbf{q}$ denoting the projection of $\mathbb{R}^{n-1} \times \mathbb{R}$ onto $\mathbb{R}$; see Notation 4)

$$\sup\left\{\frac{|\mathbf{q}y - \mathbf{q}x_0|}{r_0} : y \in \mathbf{C}(x_0, r_0) \cap \partial E\right\} \le C_0(n)\, \mathbf{e}_n(x_0, 4r_0)^{1/2(n-1)}\,. \tag{22.15}$$

Remark 22.9 As is unavoidable in regularity estimates, information over a ball or a cylinder of a given radius provides an estimate on a concentric ball or cylinder with radius decreased by a constant factor. It is therefore an arbitrary choice to deduce, from a small excess assumption in $\mathbf{C}(x_0, 4r_0)$, the height bound (22.15) in $\mathbf{C}(x_0, r_0)$. This choice should make some set inclusions and elementary inequalities used in later proofs easier to check.

The proof of the height bound is divided into several lemmas. The starting point is Lemma 22.10, where by a compactness argument based on Theorem 21.13 we deduce the geometric properties summarized in Figure 22.3 from

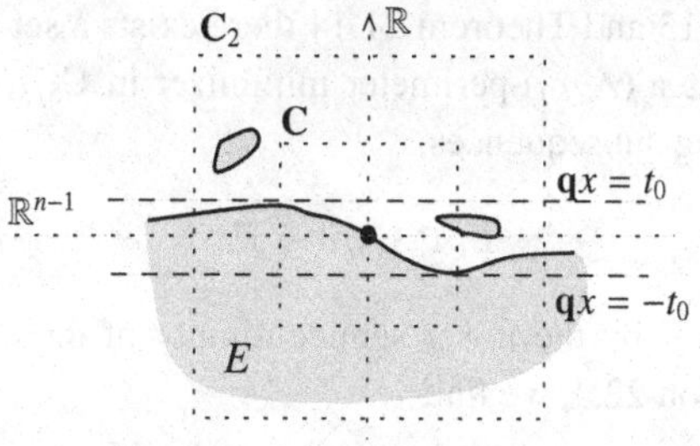

Figure 22.3 In Lemma 22.10 we show that if $t_0 \in (0,1)$, $0 \in \partial E$, and the cylindrical excess $\mathbf{e}(E, 0, 2, e_n)$ is sufficiently small (depending on t_0), then E fills $\mathbf{C} \cap \{\mathbf{q}x < -t_0\}$ and leaves $\mathbf{C} \cap \{\mathbf{q}x > t_0\}$ empty . Moreover, the topological boundary of E inside $\mathbf{C}$ lies in the stripe $\{|\,\mathbf{q}x| < t_0\}$.

the smallness of $\mathbf{e}(E, x, 2r, \nu)$. (This is the only step in the proof of Theorem 22.8 where minimality is used.) Next, in Lemma 22.11, these geometric properties are combined with the divergence theorem to introduce the notion of *excess measure*.

Lemma 22.10 (Small-excess position) *For every $n \geq 2$ and $t_0 \in (0,1)$, there exists a positive constant $\omega(n, t_0)$ with the following property. If E is a (Λ, r_0)-perimeter minimizer in $\mathbf{C}_2$ with $\Lambda\, r_0 \leq 1$, $0 \in \partial E$, and*

$$\mathbf{e}_n(2) \leq \omega(n, t_0)$$

(where we have set, $\mathbf{e}_n(s) = \mathbf{e}(E, 0, s, e_n)$, $s > 0$), then

$$|\,\mathbf{q}x| < t_0\,, \qquad \forall x \in \mathbf{C} \cap \partial E\,, \tag{22.16}$$

$$\Big|\Big\{x \in \mathbf{C} \cap E : \mathbf{q}x > t_0\Big\}\Big| = 0\,, \tag{22.17}$$

$$\Big|\Big\{x \in \mathbf{C} \setminus E : \mathbf{q}x < -t_0\Big\}\Big| = 0\,. \tag{22.18}$$

Proof Arguing by contradiction, let us consider $t_0 \in (0,1)$ and a sequence $\{E_h\}_{h\in\mathbb{N}}$ of (Λ, r_0)-perimeter minimizers in $\mathbf{C}_2$ such that

$$\Lambda\, r_0 \leq 1\,, \qquad \lim_{h\to\infty} \mathbf{e}(E_h, 0, 2, e_n) = 0\,, \qquad 0 \in \partial E_h \quad \forall h \in \mathbb{N}\,,$$

and at least *one* of the following conditions hold true for infinitely many $h \in \mathbb{N}$:

$$\text{either} \quad \Big\{x \in \mathbf{C} \cap \partial E_h : t_0 \leq |\,\mathbf{q}x| \leq 1\Big\} \neq \emptyset\,, \tag{22.19}$$

$$\text{or} \quad \Big|\Big\{x \in \mathbf{C} \cap E_h : \mathbf{q}x > t_0\Big\}\Big| > 0\,, \tag{22.20}$$

$$\text{or} \quad \Big|\Big\{x \in \mathbf{C} \setminus E_h : \mathbf{q}x < -t_0\Big\}\Big| > 0\,. \tag{22.21}$$

By Proposition 21.13 and Theorem 21.14 there exists a set of finite perimeter $F \subset \mathbf{C}_{5/3}$, which is a (Λ, r_0)-perimeter minimizer in $\mathbf{C}_{5/3}$, such that $0 \in \partial F$ and, up to extracting subsequences,

$$E_h \cap \mathbf{C}_{5/3} \to F\,. \tag{22.22}$$

Since $\mathbf{C}_{4/3} \subset\subset \mathbf{C}_{5/3}$, by the lower semicontinuity of the excess, Proposition 22.6, and Proposition 22.4, we find

$$\mathbf{e}\big(F, 0, 4/3, e_n\big) \le \liminf_{h\to\infty} \mathbf{e}\big(E_h, 0, 4/3, e_n\big) \le \Big(\frac{3}{2}\Big)^{n-1} \lim_{h\to\infty} \mathbf{e}\big(E_h, 0, 2, e_n\big) = 0\,.$$

Having $0 \in \partial F$ and $\mathbf{e}(F, 0, 4/3, e_n) = 0$, Proposition 22.2 implies that

$$F \cap \mathbf{C}_{4/3} \text{ is equivalent to } \mathbf{C}_{4/3} \cap \{\mathbf{q}x < 0\}. \tag{22.23}$$

If (22.19) were valid for infinitely many values of $h \in \mathbb{N}$, then, up to extracting a further subsequence, we may construct $\{x_h\}_{h\in\mathbb{N}}$ with $x_h \in \mathbf{C} \cap \partial E_h$, $t_0 \le |\mathbf{q}x_h| \le 1$ and $x_h \to x_0$, for some $x_0 \in \overline{\mathbf{C}} \cap \partial F$. In particular, it would be that

$$\mathbf{C}_{4/3} \cap \partial F \cap \Big\{|\mathbf{q}x| \ge t_0\Big\} \ne \emptyset\,,$$

in contradiction with (22.23). Therefore, there exists $h_0 \in \mathbb{N}$ such that

$$\Big\{x \in \mathbf{C} \cap \partial E_h : t_0 \le |\mathbf{q}x| \le 1\Big\} = \emptyset\,, \qquad \forall h \ge h_0\,.$$

Since (16.4) guarantees that

$$|\mu_{\mathbf{C}\cap E_h}| = |\mu_{\mathbf{C}}| \llcorner E_h^{(1)} + |\mu_{E_h}| \llcorner \Big(\mathbf{C} \cup \Big\{\nu_{E_h} = \nu_{\mathbf{C}}\Big\}\Big),$$

we thus find that, for every $h \ge h_0$,

$$|\mu_{\mathbf{C}\cap E_h}|\Big(\Big\{x \in \mathbf{C} \colon t_0 < |\mathbf{q}x| < 1\Big\}\Big) = 0\,.$$

By Proposition 7.5, $1_{\mathbf{C}\cap E_h}$ is equivalent to a constant on $\{x \in \mathbf{C} \colon t_0 < \mathbf{q}x < 1\}$; for the same reason, $1_{\mathbf{C}\cap E_h}$ is also equivalent to a (possibly different) constant on $\{x \in \mathbf{C} \colon -t_0 > \mathbf{q}x > -1\}$. By (22.22), necessarily $1_{\mathbf{C}\cap E_h} = 0$ a.e. on $\{x \in \mathbf{C} \colon t_0 < \mathbf{q}x < 1\}$, and $1_{\mathbf{C}\cap E_h} = 1$ a.e. on $\{x \in \mathbf{C} \colon -t_0 > \mathbf{q}x > -1\}$. In particular, this contradicts both (22.20) and (22.21). □

We now combine the divergence theorem with the geometric information gathered in the previous lemma. As in Section 18.3, we let

$$E_t = \Big\{z \in \mathbb{R}^{n-1} : (z, t) \in E\Big\}$$

denote the horizontal slice of E at height $t \in \mathbb{R}$.

Lemma 22.11 (Excess measure) *If E is a set of locally finite perimeter in $\mathbb{R}^n$, with $0 \in \partial E$, and such that, for some $t_0 \in (0,1)$,*

$$|\mathbf{q}x| < t_0\,, \qquad \forall x \in \mathbf{C} \cap \partial E\,, \tag{22.24}$$

$$\Big|\big\{x \in \mathbf{C} \cap E : \mathbf{q}x > t_0\big\}\Big| = 0\,, \tag{22.25}$$

$$\Big|\big\{x \in \mathbf{C} \setminus E : \mathbf{q}x < -t_0\big\}\Big| = 0\,. \tag{22.26}$$

then, setting for brevity $M = \mathbf{C} \cap \partial^* E$, we have

$$\mathcal{H}^{n-1}(G) \le \mathcal{H}^{n-1}\big(M \cap \mathbf{p}^{-1}(G)\big), \tag{22.27}$$

$$\mathcal{H}^{n-1}(G) = \int_{M \cap \mathbf{p}^{-1}(G)} (\nu_E \cdot e_n)\, \mathrm{d}\mathcal{H}^{n-1}\,, \tag{22.28}$$

$$\int_{\mathbf{D}} \varphi = \int_M \varphi(\mathbf{p}x)(\nu_E(x) \cdot e_n)\, \mathrm{d}\mathcal{H}^{n-1}(x)\,, \tag{22.29}$$

$$\int_{E_t \cap \mathbf{D}} \varphi = \int_{M \cap \{\mathbf{q}x > t\}} \varphi(\mathbf{p}x)(\nu_E(x) \cdot e_n)\, \mathrm{d}\mathcal{H}^{n-1}(x)\,, \tag{22.30}$$

for every Borel set $G \subset \mathbf{D}$, $\varphi \in C^0_c(\mathbf{D})$ and $t \in (-1,1)$. The set function

$$\begin{aligned}\zeta(G) &= P(E; \mathbf{C} \cap \mathbf{p}^{-1}(G)) - \mathcal{H}^{n-1}(G) \\ &= \mathcal{H}^{n-1}\big(M \cap \mathbf{p}^{-1}(G)\big) - \mathcal{H}^{n-1}(G)\,, \qquad G \subset \mathbb{R}^{n-1}\,,\end{aligned} \tag{22.31}$$

defines a Radon measure on $\mathbb{R}^{n-1}$, concentrated on **D**. *The Radon measure ζ is called the* **excess measure** *of E over* **D** *since $\zeta(\mathbf{D}) = \mathbf{e}_n(1)$.*

Remark 22.12 The lower bound (22.27) ensures that $\mathbf{C} \cap \partial^* E$ "leaves no holes" over **D**. If $\mathbf{e}_n(1)$ is small, the trivial upper bound $\zeta(G) \le \zeta(\mathbf{D}) = \mathbf{e}_n(1)$, implies that $\mathbf{C} \cap \partial^* E$ is "almost flat" over **D**, which amounts to saying that

$$\mathcal{H}^{n-1}(G) \le \mathcal{H}^{n-1}\big(\mathbf{C} \cap \partial^* E \cap \mathbf{p}^{-1}(G)\big) \le \mathcal{H}^{n-1}(G) + \mathbf{e}_n(1)\,,$$

for every Borel set $G \subset \mathbf{D}$; see Figure 22.4. In a similar way, starting from (22.30) we see that, for every $t \in (-1,1)$,

$$\mathcal{H}^{n-1}\big(E_t \cap \mathbf{D}\big) \le \mathcal{H}^{n-1}\big(M \cap \{\mathbf{q}x > t\}\big) \le \mathcal{H}^{n-1}\big(E_t \cap \mathbf{D}\big) + \mathbf{e}_n(1)\,.$$

Proof of Lemma 22.11 We first remark that, by a standard approximation argument, (22.30) implies (22.28), which, in turn, implies (22.27). We now prove (22.29) and (22.30). In doing this, again by a density argument, we can directly assume that $\varphi \in C^1_c(\mathbf{D})$. By Proposition 2.16, we have

$$\mathcal{H}^{n-1}\big(\partial^* E \cap (\partial \mathbf{D}_r \times \mathbb{R})\big) = 0\,, \tag{22.32}$$

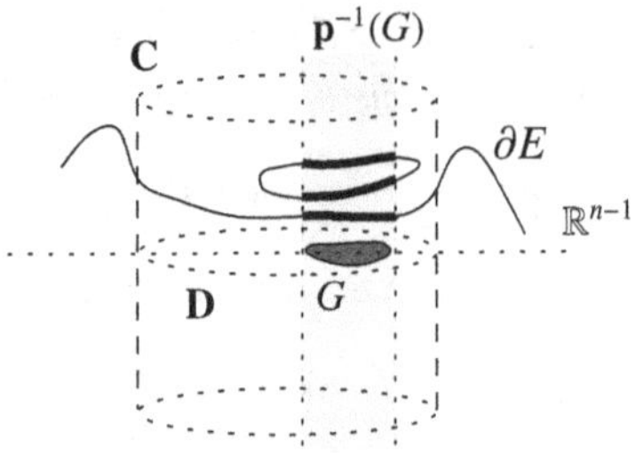

Figure 22.4 The geometric meaning of (22.27). The part of ∂E lying over $G \subset \mathbf{D}$ (that is represented with a bold line) covers G. The equality sign in (22.28) takes into account the possibility of cancelations in the integral on the right-hand side due to the orientation of ν_E.

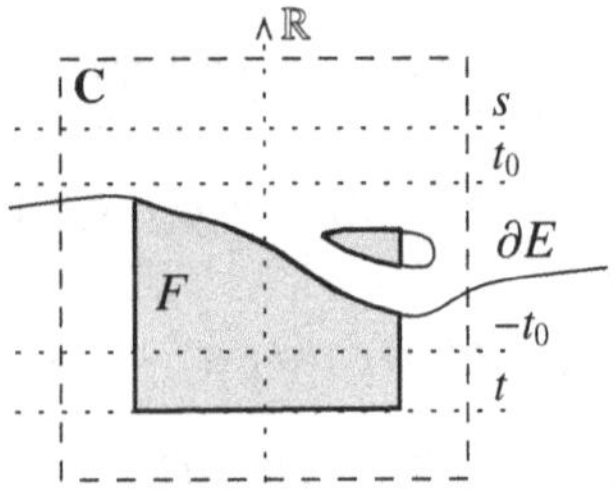

Figure 22.5 The set F in the proof of Lemma 22.11.

for a.e. $r \in (0, 1)$. By (22.25), (22.26), and thanks to Fubini's theorem,

$$\mathcal{H}^{n-1}\big(E \cap (\mathbf{D} \times \{s\})\big) = 0, \qquad \text{for a.e. } s \in (t_0, 1), \tag{22.33}$$

$$\mathcal{H}^{n-1}\big(E \cap (\mathbf{D} \times \{t\})\big) = \mathcal{H}^{n-1}(\mathbf{D}), \qquad \text{for a.e. } t \in (-1, -t_0). \tag{22.34}$$

We let $r \in (0, 1)$ and $s \in (t_0, 1)$ satisfy respectively (22.32) and (22.33). Given $t \in (-1, s)$, we define a set of finite perimeter F as

$$F = E \cap \big(\mathbf{D}_r \times (t, s)\big);$$

see Figure 22.5. By (16.4) and (22.32) we have a (geometrically obvious) formula for the Gauss–Green measure of F, namely

$$\mu_F = \mu_E \llcorner \big(\mathbf{D}_r \times (t, s)\big) + \mu_{\mathbf{D}_r \times (t,s)} \llcorner E\,.$$

If we set $\nu(x) = \mathbf{p}x/|\mathbf{p}x|$ for every $x \in \mathbb{R}^n$ such that $\mathbf{p}x \neq 0$ (so that $\nu(x)$ is the outer normal to the cylinder $\mathbf{D}_r \times \mathbb{R}$ at $x \in \partial\mathbf{D}_r \times \mathbb{R}$), then by Exercise 16.4,

$$\mu_{\mathbf{D}_r \times (t,s)} = e_n \mathcal{H}^{n-1} \llcorner \big(\mathbf{D}_r \times \{s\}\big) + \nu\, \mathcal{H}^{n-1} \llcorner \big(\partial\mathbf{D}_r \times (t, s)\big) - e_n \mathcal{H}^{n-1} \llcorner \big(\mathbf{D}_r \times \{t\}\big).$$

By (22.33) and since $\nu(x)\cdot e_n = 0$ for every $x \in \mathbb{R}^n$ with $\mathbf{p}x \neq 0$, we find that

$$e_n \cdot \mu_F = (e_n \cdot \nu_E)\mathcal{H}^{n-1} \llcorner \Big(\partial^* E \cap \big(\mathbf{D}_r \times (t,s)\big)\Big) - \mathcal{H}^{n-1} \llcorner \Big(E \cap (\mathbf{D}_r \times \{t\})\Big). \tag{22.35}$$

Hence, given $\varphi \in C^1_c(\mathbf{D})$ we may define a vector field $T \in C^1(\mathbb{R}^n;\mathbb{R}^n)$ by setting $T(x) = \varphi(\mathbf{p}x)e_n$, $x \in \mathbb{R}^n$. Since $\operatorname{div} T = 0$, the divergence theorem applied on F combined with (22.35) implies

$$\int_{E\cap(\mathbf{D}_r\times\{t\})} \varphi(\mathbf{p}x)\, d\mathcal{H}^{n-1}(x) = \int_{\partial^* E\cap(\mathbf{D}_r\times(t,s))} \varphi(\mathbf{p}x)\,(e_n\cdot\nu_E(x))\, d\mathcal{H}^{n-1}(x)\,.$$

We first let $r \to 1^-$ and then $s \to 1^-$ to prove (22.30), that is

$$\int_{E_t\cap\mathbf{D}} \varphi = \int_{E\cap(\mathbf{D}\times\{t\})} \varphi(\mathbf{p}x)\, d\mathcal{H}^{n-1}(x) = \int_{\partial^* E\cap(\mathbf{D}\times(t,1))} \varphi(\mathbf{p}x)\,(e_n\cdot\nu_E(x))\, d\mathcal{H}^{n-1}(x)$$
$$= \int_{M\cap\{\mathbf{q}x>t\}} \varphi(\mathbf{p}x)\,(e_n\cdot\nu_E(x))\, d\mathcal{H}^{n-1}(x)\,.$$

Finally, by letting $t \to (-1)^+$, and by (22.34), we prove (22.29). □

Proof of Theorem 22.8 *Step one:* By Remark 21.6 and Proposition 22.1, up to replacing E with $E_{x_0,2r_0} = (E - x_0)/2r_0$, we can reduce to the following situation: given a $(\Lambda', 1/2)$-perimeter minimizer E in $\mathbf{C}_2$, with

$$\frac{\Lambda'}{2} \le 1\,, \qquad 0 \in \partial E\,, \qquad \mathbf{e}_n(2) \le \varepsilon_0(n)\,,$$

we want to prove that

$$|\mathbf{q}x| \le C_0(n)\,\mathbf{e}_n(2)^{1/2(n-1)}\,, \qquad \forall x \in \mathbf{C}_{1/2} \cap \partial E\,. \tag{22.36}$$

where we are setting $\mathbf{e}_n(s) = \mathbf{e}(E, 0, s, e_n)$, $s > 0$. If we assume that

$$\varepsilon_0(n) \le \omega(n, 1/4)\,, \tag{22.37}$$

with $\omega(n, 1/4)$ as in Lemma 22.10, and set $M = \mathbf{C} \cap \partial E$, then we deduce by Lemma 22.10 that

$$|\mathbf{q}x| \le \frac{1}{4}\,, \qquad \forall x \in M\,, \tag{22.38}$$

and, by Lemma 22.11 and Remark 22.12, that

$$0 \le \mathcal{H}^{n-1}(M) - \mathcal{H}^{n-1}(\mathbf{D}) \le 2^{n-1}\,\mathbf{e}_n(2)\,, \tag{22.39}$$

$$0 \le \mathcal{H}^{n-1}\Big(M \cap \big\{\mathbf{q}x > t\big\}\Big) - \mathcal{H}^{n-1}\Big(E_t \cap \mathbf{D}\Big) \le 2^{n-1}\,\mathbf{e}_n(2) \tag{22.40}$$

for every $t \in (-1, 1)$, where we have also used the fact that $\mathbf{e}_n(1) \le 2^{n-1}\mathbf{e}_n(2)$ by Proposition 22.4. Starting from these estimates, and applying only the lower

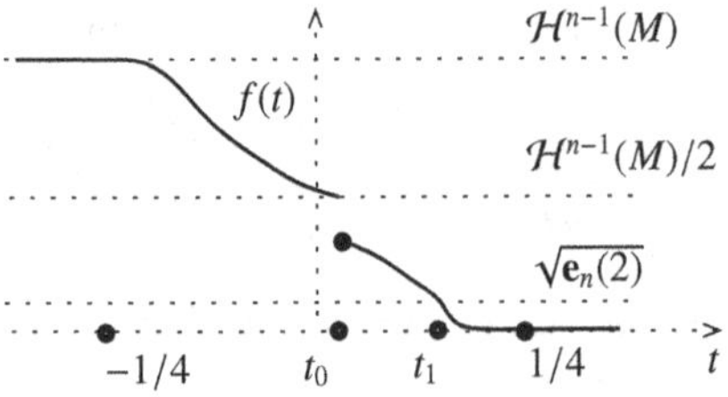

Figure 22.6 The function $f(t) = \mathcal{H}^{n-1}(M \cap \{\mathbf{q}x > t\}) = P(E; \mathbf{C} \cap \{\mathbf{q}x > t\})$, and the heights t_0 and t_1.

density estimate in (21.10) and the relative isoperimetric inequality (12.45), we are now going to prove (22.36).

Step two: By (22.37) and thanks to Lemma 22.10, the right continuous, decreasing function $f\colon (-1,1) \to [0, \mathcal{H}^{n-1}(M)]$, defined as

$$f(t) = \mathcal{H}^{n-1}\Big(M \cap \big\{\mathbf{q}x > t\big\}\Big), \qquad |t| < 1,$$

satisfies

$$f(t) = \mathcal{H}^{n-1}(M), \quad \forall t \in \Big(-1, -\frac{1}{4}\Big), \qquad f(t) = 0, \quad \forall t \in \Big(\frac{1}{4}, 1\Big).$$

Clearly, there certainly exists t_0, with $|t_0| < 1/4$, such that

$$\begin{aligned} f(t) &\le \frac{\mathcal{H}^{n-1}(M)}{2}, && \text{if } t \ge t_0, \\ f(t) &\ge \frac{\mathcal{H}^{n-1}(M)}{2}, && \text{if } t < t_0; \end{aligned} \tag{22.41}$$

see Figure 22.6. In the following two steps of the proof we are going to show

$$\mathbf{q}x - t_0 \le C(n)\, \mathbf{e}_n(2)^{1/2(n-1)}, \qquad \forall x \in \mathbf{C}_{1/2} \cap \partial E. \tag{22.42}$$

By applying the same argument with $\mathbb{R}^n \setminus E$ in place of E (this is possible by the choice of t_0 and Exercise 21.1) we shall then deduce that

$$t_0 - \mathbf{q}x \le C(n)\, \mathbf{e}_n(2)^{1/2(n-1)}, \qquad \forall x \in \mathbf{C}_{1/2} \cap \partial E. \tag{22.43}$$

Finally, having $0 \in \partial E$, we shall deduce (22.36) from (22.42), (22.43), and by the triangular inequality. We thus turn to the proof of (22.42).

Step three: Let us consider a height $t_1 \in (t_0, 1/4)$ such that M has small $\mathcal{H}^{n-1}$-dimensional measure above t_1, that is, we ask that

$$f(t) \le \sqrt{\mathbf{e}_n(2)}, \qquad \text{if } t \ge t_1. \tag{22.44}$$

We now prove that

$$\mathbf{q}y - t_1 \le C(n)\,\mathbf{e}_n(2)^{1/2(n-1)}\,, \qquad \forall y \in \mathbf{C}_{1/2} \cap \partial E\,. \tag{22.45}$$

Indeed, if $y \in \mathbf{C}_{1/2} \cap \partial E$ and $\mathbf{q}y > t_1$, then by (22.38)

$$B(y, \mathbf{q}y - t_1) \subset\subset \mathbf{C}_2\,, \qquad \text{with} \quad \mathbf{q}y - t_1 < \frac{1}{2}\,.$$

Since E is a $(\Lambda', 1/2)$-perimeter minimizer in $\mathbf{C}_2$, with $\Lambda'/2 \le 1$, we may apply the lower density estimate (21.10) on $B(y, \mathbf{q}y - t_1)$, to find that

$$c(n)(\mathbf{q}y - t_1)^{n-1} \le P\big(E; B\big(y, \mathbf{q}y - t_1\big)\big) \le f(t_1)\,,$$

where the inclusion $B(y, \mathbf{q}y - t_1) \subset \mathbf{C} \cap \{\mathbf{q}x > t_1\}$ was used. By (22.44), we immediately deduce (22.45).

Step four: We now show that

$$t_1 - t_0 \le C(n)\,\mathbf{e}_n(2)^{1/2(n-1)}\,. \tag{22.46}$$

This inequality, combined with (22.45), implies (22.42), which in turn, as explained in step two, allows us to conclude the proof of the theorem. To this end, let us recall from Theorem 18.11 that, for a.e. $t \in \mathbb{R}$,

$$\mathcal{H}^{n-2}\big(\partial^* E_t \Delta (\partial^* E)_t\big) = 0\,, \tag{22.47}$$

and that, for every Borel function $g \colon \mathbb{R}^n \to [0, \infty]$,

$$\int_{\partial^* E} g\sqrt{1 - (\nu_E \cdot e_n)^2}\, d\mathcal{H}^{n-1} = \int_{\mathbb{R}} dt \int_{\partial^* E_t} g\, d\mathcal{H}^{n-2}\,. \tag{22.48}$$

Here, $E_t = \{z \in \mathbb{R}^{n-1} : (z, t) \in E\}$ is the horizontal slice of E at height t. By (22.47), we find that, for a.e. $t \in \mathbb{R}$,

$$\mathcal{H}^{n-2}\big(\mathbf{D} \cap \partial^* E_t\big) = \mathcal{H}^{n-2}\big(\mathbf{D} \cap (\partial^* E)_t\big) = \mathcal{H}^{n-2}\big((\mathbf{C} \cap \partial^* E)_t\big) = \mathcal{H}^{n-2}(M_t)\,.$$

Hence, by (22.48) (with $g = 1_{\mathbf{C}}$), by Hölder inequality and (22.39),

$$\begin{aligned}
&\int_{-1}^{1} \mathcal{H}^{n-2}\big(\mathbf{D} \cap \partial^* E_t\big)\, dt \\
&= \int_M \sqrt{1 - (\nu_E \cdot e_n)^2}\, d\mathcal{H}^{n-1} \le \sqrt{2} \int_M \sqrt{1 - (\nu_E \cdot e_n)}\, d\mathcal{H}^{n-1} \\
&\le \sqrt{2\mathcal{H}^{n-1}(M)} \sqrt{\int_M 1 - (\nu_E \cdot e_n)\, d\mathcal{H}^{n-1}} \le C(n)\,\sqrt{\mathbf{e}_n(2)}\,,
\end{aligned}$$

where, again by (22.39), we have $\mathcal{H}^{n-1}(M) \le C(n)$. Thus,

$$\int_{-1}^{1} \mathcal{H}^{n-2}\big(\mathbf{D} \cap \partial^* E_t\big)\, dt \le C(n)\,\sqrt{\mathbf{e}_n(2)}\,. \tag{22.49}$$

By (22.40), (22.41), and (22.39), if $t \in [t_0, t_1)$, then

$$\mathcal{H}^{n-1}(E_t \cap \mathbf{D}) \le \mathcal{H}^{n-1}\big(M \cap \{\mathbf{q}x > t\}\big) \le \frac{\mathcal{H}^{n-1}(M)}{2} \le \frac{\mathcal{H}^{n-1}(\mathbf{D}) + 2^{n-1}\,\mathbf{e}_n(2)}{2} \le \frac{3}{4}\mathcal{H}^{n-1}(\mathbf{D}),$$

provided ε_0 is small enough. By applying the relative isoperimetric inequality (12.45) in the ball $\mathbf{D}$ to the set of finite perimeter $E_t \cap \mathbf{D}$, we find that

$$\mathcal{H}^{n-2}\big(\mathbf{D} \cap \partial^* E_t\big) = P\big(E_t \cap \mathbf{D}; \mathbf{D}\big) \ge c(n)\mathcal{H}^{n-1}(E_t \cap \mathbf{D})^{(n-2)/(n-1)}.$$

Combining this inequality with (22.49) we deduce that

$$C(n)\,\sqrt{\mathbf{e}_n(2)} \ge \int_{t_0}^{1} \mathcal{H}^{n-1}(E_t \cap \mathbf{D})^{(n-2)/(n-1)}\,dt. \tag{22.50}$$

Again by (22.40), and taking the definition (22.44) of t_1 into account, we find that, if $t \in [t_0, t_1)$ and $\varepsilon_0(n)$ is small enough,

$$\mathcal{H}^{n-1}(E_t \cap \mathbf{D}) \ge \mathcal{H}^{n-1}\big(M \cap \{\mathbf{q}x > t\}\big) - 2^{n-1}\,\mathbf{e}_n(2) \ge \sqrt{\mathbf{e}_n(2)} - 2^{n-1}\,\mathbf{e}_n(2) \ge c(n)\,\sqrt{\mathbf{e}_n(2)}.$$

Hence, (22.50) implies

$$C(n)\,\sqrt{\mathbf{e}_n(2)} \ge c(n)(t_1 - t_0)\Big(\sqrt{\mathbf{e}_n(2)}\Big)^{(n-2)/(n-1)},$$

that is (22.46). We thus achieve the proof of the height bound. □

23

The Lipschitz approximation theorem

The goal of this chapter is to prove that, if E is a (Λ, r_0)-perimeter minimizer in $\mathbf{C}(x_0, 9r)$, with $9r < r_0$ and $\mathbf{e}_n(x_0, r)$ small enough, then $\mathbf{C}(x_0, r) \cap \partial E$ is almost entirely covered by the graph of a Lipschitz function u, which turns out to posses suitable almost-minimality properties (related to the (Λ, r_0)-perimeter minimality of E). This is the content of the Lipschitz approximation theorem, Theorem 23.7, which is stated and proved in Section 23.3. Before coming to this, in Section 23.1 we discuss conditions under which the topological boundary of a set of finite perimeter E (normalized so that $\operatorname{spt}\mu_E = \partial E$) locally agrees with the graph of a Lipschitz function u, while Section 23.2 briefly introduces the minimality properties inherited by such a function u when E is a local perimeter minimizer (this second problem will be further discussed with more details in Chapter 27).

23.1 The Lipschitz graph criterion

In Theorem 21.8, we aim to prove that the reduced boundary of a (Λ, r_0)-perimeter minimizer is locally representable by the graph of a $C^{1,\gamma}$-function. In this section we present a simple criterion for a set of locally finite perimeter to be locally representable as the graph of a Lipschitz function: it suffices that the reduced boundary of the set is locally covered by a Lipschitz graph.

Theorem 23.1 (Lipschitz graph criterion) *If E is a set of locally finite perimeter in $\mathbb{R}^n$ with $\operatorname{spt}\mu_E = \partial E$ and $0 \in \partial E$, and if $u \colon \mathbb{R}^{n-1} \to \mathbb{R}$ is a Lipschitz function with* $\operatorname{Lip}(u) \le 1$, *such that*

$$\mathbf{C} \cap \partial^* E \subset \left\{(z, u(z)) : z \in \mathbb{R}^{n-1}\right\}, \tag{23.1}$$

then

$$\mathbf{C} \cap \partial E = \left\{(z, u(z)) : z \in \mathbf{D}\right\}, \tag{23.2}$$

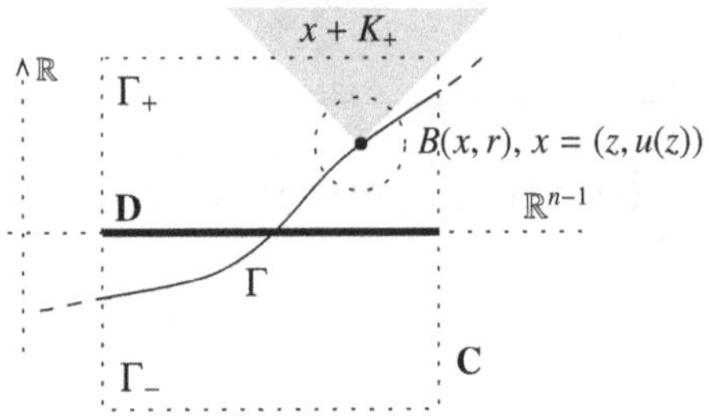

Figure 23.1 The situation in Theorem 23.1. The grey region is the cone $x + K_+$, considered in the proof of (23.2).

and either

$$\mathbf{C} \cap E = \Big\{(z, t) \in \mathbf{C} : z \in \mathbf{D}, -1 < t < u(z)\Big\} \tag{23.3}$$

or

$$\mathbf{C} \cap E = \Big\{(z, t) : z \in \mathbf{D}, u(z) < t < 1\Big\}. \tag{23.4}$$

Moreover, for every Borel set $G \subset \mathbf{D}$, we have

$$P\Big(E; \mathbf{C} \cap \mathbf{p}^{-1}(G)\Big) = \int_G \sqrt{1 + |\nabla' u(z)|^2}\, dz, \tag{23.5}$$

and, depending on which of (23.3) and (23.4) holds true, we have either

$$\nu_E(z, u(z)) = \frac{(\nabla' u(z), -1)}{\sqrt{1 + |\nabla' u(z)|^2}}, \tag{23.6}$$

or

$$\nu_E(z, u(z)) = \frac{(-\nabla' u(z), 1)}{\sqrt{1 + |\nabla' u(z)|^2}}, \qquad \textit{for a.e. } z \in \mathbf{D}. \tag{23.7}$$

Remark 23.2 It is not really necessary to assume that $0 \in \partial E$ and that $\mathrm{Lip}(u) \le 1$, but, to our taste, this choice leads to a more elegant statement, as it ensures that $|u(z)| < 1$ for every $z \in \mathbf{D}$. Note that, by McShane's lemma, condition (23.1) holds true if there exists $L < 1$ such that

$$|\,\mathbf{q}x - \mathbf{q}y| \le L\,|x - y|\,, \qquad \forall x, y \in \mathbf{C} \cap \partial^* E\,.$$

In this case, we easily construct a Lipschitz function $u\colon \mathbb{R}^{n-1} \to \mathbb{R}$, satisfying (23.1), and such that $\mathrm{Lip}(u) \le L/\sqrt{1 - L^2}$.

Proof of Theorem 23.1 Let us consider the open connected sets

$$\Gamma_+ = \Big\{(z, t) : z \in \mathbf{D}, u(z) < t < 1\Big\}, \qquad \Gamma_- = \Big\{(z, t) : z \in \mathbf{D}, -1 < t < u(z)\Big\},$$

and the closed set $\Gamma = \{(z, u(z)) : z \in \mathbf{D}\}$; see Figure 23.1. Since Γ is closed and $\overline{\partial^* E} = \mathrm{spt}\,\mu_E = \partial E$, by (23.1) we have

$$\mathbf{C} \cap \partial E \subset \Gamma\,. \tag{23.8}$$

Having $0 \in \partial E$, we deduce $u(0) = 0$. Hence, by $\mathrm{Lip}(u) \le 1$, we find $|u(z)| < 1$ on $\mathbf{D}$, and thus $\Gamma \subset \mathbf{C}$.

Step one: We prove that either (23.3) or (23.4) holds true. By (23.1), and since μ_E is concentrated on $\partial^* E$, we have $|\mu_E|(\Gamma_+) = |\mu_E|(\Gamma_-) = 0$. Thus, by Lemma 7.5, 1_E is (equivalent to a) constant both on Γ_+ and on Γ_-. If these constant values agree, then 1_E would be constant on $\mathbf{C}$. In particular, we would have $|\mu_E|(\mathbf{C}) = 0$, against the fact that $\mathbf{C} \cap \partial E$ is non-empty (as it contains 0). Thus, either $1_E = 1$ a.e. on Γ_+ and $1_E = 0$ a.e. on Γ_-, or $1_E = 0$ a.e. on Γ_+ and $1_E = 1$ a.e. on Γ_-, as required.

Step two: We prove (23.2). Having (23.8), we are left to prove

$$\Gamma \subset \mathbf{C} \cap \partial E \,. \tag{23.9}$$

To this end, we fix $x \in \Gamma$, and consider the cones

$$K_+ = \Big\{(z,t) : t > \mathrm{Lip}(u)|z|\Big\}\,, \qquad K_- = \Big\{(z,t) : t < -\mathrm{Lip}(u)|z|\Big\}\,.$$

By definition of Lip(u), we find that

$$\mathbf{C} \cap \Big(x + K_+\Big) \subset \Gamma_+\,, \qquad \mathbf{C} \cap \Big(x + K_-\Big) \subset \Gamma_-\,.$$

If (23.4) holds true, and $r > 0$ is such that $B(x,r) \subset \mathbf{C}$, then we find

$$\frac{|E \cap B(x,r)|}{\omega_n r^n} = \frac{|\Gamma_+ \cap B(x,r)|}{\omega_n r^n} \ge \frac{|(x+K_+) \cap B(x,r)|}{\omega_n r^n} = \frac{|K_+ \cap B|}{\omega_n} > 0\,,$$
$$\frac{|B(x,r) \setminus E|}{\omega_n r^n} = \frac{|\Gamma_- \cap B(x,r)|}{\omega_n r^n} \ge \frac{|(x+K_-) \cap B(x,r)|}{\omega_n r^n} = \frac{|K_- \cap B|}{\omega_n} > 0\,,$$

that is $0 < |E \cap B(x,r)| < \omega_n r^n$. By symmetry, we draw the same conclusion in the case that (23.3) holds true. By Proposition 12.19, we conclude that $\Gamma \subset \operatorname{spt}\mu_E = \partial E$, and (23.9) immediately follows.

Step three: By Theorem 9.1 and (23.2) we immediately deduce (23.5). By Exercise 10.6 we easily deduce the validity of (23.6) in the case when (23.3) holds true, or (23.7) in the complementary case. □

23.2 The area functional and the minimal surfaces equation

The discussion of the Lipschitz graph criterion in the previous section provides a good occasion to introduce some fundamental connections between the regularity theories for parametric and non-parametric variational problems. These connections are crucial both in the study of (Λ, r_0)-perimeter minimizers, and in particular in the proof of their $C^{1,\gamma}$-regularity (Theorem 21.8), as well as in developing the higher regularity theory for minimizers in geometric variational problems, discussed in Chapter 27. The starting point for understanding these fruitful ideas is provided by the following simple proposition.

Proposition 23.3 (Local minimizers of the area functional) *Under the assumptions of the Lipschitz graph criterion, Theorem 23.1, if E is further assumed to be a perimeter minimizer in* **C**, *then u is a local minimizer of the area functional in* **D**.

The **area functional** (over a ball B of $\mathbb{R}^n$) is defined as

$$\mathcal{A}(u;B) = \int_B \sqrt{1+|\nabla u(x)|^2}\,\mathrm{d}x\,, \tag{23.10}$$

for every Lipschitz function $u\colon \mathbb{R}^n \to \mathbb{R}$. We say u is a **local minimizer of the area functional in** B if for every compact set $K \subset B$ there exists $\varepsilon > 0$ with

$$\mathcal{A}(u;B) \le \mathcal{A}(u+\varphi;B)\,, \tag{23.11}$$

whenever $\varphi \in C^\infty_c(B)$, $\operatorname{spt}\varphi \subset K$ and $\sup|\varphi| \le \varepsilon$.

Proof of Proposition 23.3 Let us assume, without loss of generality, that (23.4) holds true. Let us fix $r \in (0,1)$, and prove that

$$\mathcal{A}(u;\mathbf{D}) \le \mathcal{A}(u+\varphi;\mathbf{D})\,, \tag{23.12}$$

whenever $\varphi \in C^1_c(\mathbf{D}_r)$ and $\sup|\varphi| < 1-r$. Indeed, in this case, by $u(0)=0$ and $\mathrm{Lip}(u) \le 1$, we find that $\sup_{\mathbf{D}_r}|u+\varphi| < 1$. In particular, if we set

$$F = \Big\{(z,t) : z \in \mathbf{D}\,, -1 < t < u(z)+\varphi(z)\Big\}\,, \tag{23.13}$$

then we have $E\Delta F \subset\subset \mathbf{C}$. By (23.5) and perimeter minimality of E,

$$\mathcal{A}(u;\mathbf{D}) = P(E;\mathbf{C}) \le P(F;\mathbf{C})\,. \tag{23.14}$$

A further application of Theorem 9.1 implies in turn that

$$P(F;\mathbf{C}) = \int_{\mathbf{D}} \sqrt{1+|\nabla' u + \nabla'\varphi|^2} = \mathcal{A}(u+\varphi;\mathbf{D})\,,$$

so that (23.12) follows from (23.14). □

We have thus established a link between the regularity theory for local perimeter minimizers and the regularity theory for Lipschitz minimizers of the area functional. As we shall see in the next section, an important part of the proof of Theorem 21.8 (which deals with the broader notion of (Λ, r_0)-minimality) is built on this connection. For these reasons it is now convenient to briefly recall the starting point of the regularity theory for Lipschitz minimizers of the area functional, which is the derivation of the minimal surfaces equation. We are in fact going to consider a generic integral functional

$$\mathcal{F}(u;B) = \int_B f(\nabla u(x))\,\mathrm{d}x\,,$$

associated with some convex function $f\colon \mathbb{R}^n \to \mathbb{R}$. We shall say that $u\colon \mathbb{R}^n \to \mathbb{R}$ is a **Lipschitz local minimizer of $\mathcal{F}$ in B** if u is a Lipschitz function and if for every compact set $K \subset B$ there exists $\varepsilon > 0$ such that

$$\mathcal{F}(u; B) \le \mathcal{F}(u + \varphi; B), \tag{23.15}$$

whenever $\varphi \in C_c^\infty(B)$, $\operatorname{spt}\varphi \subset K$ and $\sup|\varphi| \le \varepsilon$.

Theorem 23.4 (Euler–Lagrange equations) *If $f \in C^1(\mathbb{R}^n)$ is convex, then $u\colon \mathbb{R}^n \to \mathbb{R}$ is a Lipschitz local minimizer of $\mathcal{F}$ in B if and only if*

$$\int_B \nabla f(\nabla u) \cdot \nabla\varphi = 0, \qquad \forall \varphi \in C_c^\infty(B). \tag{23.16}$$

If, moreover, $f \in C^2(\mathbb{R}^n)$ and $u \in C^2(B)$, then (23.16) is equivalent to

$$-\operatorname{div}\Big(\nabla f\big(\nabla u(x)\big)\Big) = 0, \qquad \forall x \in B. \tag{23.17}$$

Equations (23.16) and (23.17) are called, respectively, the weak and strong form of the Euler–Lagrange equation of $\mathcal{F}$.

Example 23.5 (Dirichlet integral and harmonic functions) The **Dirichlet integral** $\mathcal{D}\colon W^{1,2}(B) \to [0, \infty)$, defined as

$$\mathcal{D}(u; B) = \frac{1}{2}\int_B |\nabla u|^2, \qquad u \in W^{1,2}(B),$$

corresponds to the choice $f(\xi) = |\xi|^2/2$, $\xi \in \mathbb{R}^n$, in Theorem 23.4. In fact, in this case, rather than at Lipschitz local minimizers, one looks at local minimizers in $W^{1,2}(B)$. The proof of Theorem 23.4, in this case, is particularly simple. Indeed, if u is a local minimizer of $\mathcal{D}$, then for every $\varphi \in C_c^\infty(B)$ there exists $\varepsilon > 0$ such that

$$0 \le \mathcal{D}(u + t\varphi; B) - \mathcal{D}(u; B) = t\int_B \nabla u \cdot \nabla\varphi + \frac{t^2}{2}\int_B |\nabla\varphi|^2, \qquad \forall |t| < \varepsilon.$$

In particular the weak form of the Euler–Lagrange equation (23.16),

$$\int_B \nabla u \cdot \nabla\varphi = 0, \quad \forall \varphi \in C_c^\infty(B), \tag{23.18}$$

holds true. The strong form of the Euler–Lagrange equation reduces to the **Laplace equation**,

$$-\Delta u(x) = 0, \qquad \forall x \in B. \tag{23.19}$$

A solution to (23.19) is called a **harmonic function**. Harmonic functions are going to play a crucial role in the proof of the regularity theorem for perimeter minimizers; see, in particular, Section 25.2. A weak solution of (23.18) is also called a harmonic function, since it is necessarily smooth, and thus solves (23.19); see, for example, [Eva98, Section 6.3.1].

Example 23.6 (Area functional and the minimal surfaces equation) In the case that $\mathcal{F}$ is the area functional $\mathcal{A}$, which corresponds to the choice $f(\xi) = \sqrt{1+|\xi|^2}$, the weak form of the Euler–Lagrange equation is

$$\int_B \frac{\nabla u}{\sqrt{1+|\nabla u|^2}} \cdot \nabla\varphi = 0\,, \qquad \forall \varphi \in C^\infty_{\rm c}(B)\,. \tag{23.20}$$

The strong form of (23.20) is called the **minimal surfaces equation**,

$$-\operatorname{div}\left(\frac{\nabla u(x)}{\sqrt{1+|\nabla u(x)|^2}}\right) = 0\,, \qquad \forall x \in B\,. \tag{23.21}$$

This partial differential equation expresses the vanishing of the mean curvature of the hypersurface defined by the graph of u over B. It is the expression in local coordinates of the vanishing mean curvature condition (21.8).

Proof of Theorem 23.4 It is the following argument which gives the name to a discipline itself, the Calculus of Variations. With every *variation* $\varphi \in C^\infty_{\rm c}(B)$ we associate a convex function $\Phi\colon \mathbb{R} \to \mathbb{R}$ by setting $\Phi(t) = \mathcal{F}(u + t\varphi)$, $t \in \mathbb{R}$. By dominated convergence we easily see that $\Phi \in C^1(\mathbb{R})$, with

$$\Phi'(0) = \int_B \nabla f(\nabla u(x)) \cdot \nabla\varphi(x)\,{\rm d}x\,.$$

If u is a Lipschitz local minimizer of $\mathcal{F}$ in B, then there exists $\varepsilon > 0$ such that $\Phi(0) \le \Phi(t)$ whenever $|t| < \varepsilon$. In particular, $\Phi'(0) = 0$, and (23.16) is proved. If, conversely, (23.16) holds true, then it amounts to say that $\Phi'(0) = 0$. The convexity of Φ then implies that $t = 0$ is in fact the absolute minimum of Φ on $\mathbb{R}$. By the arbitrariness of φ and a density argument, it follows that u is a Lipschitz local minimizer of $\mathcal{F}$ in B. Finally, when $f \in C^2(\mathbb{R}^n)$ and $u \in C^2(B)$, the equivalence between the strong and weak forms of the Euler–Lagrange equation immediately follows by the divergence theorem and the fundamental lemma of the Calculus of Variations (Exercise 4.14). □

23.3 The Lipschitz approximation theorem

The goal of this section is to prove that if E is a (Λ, r_0)-perimeter minimizer in a cylinder $\mathbf{C}(x_0, r_0)$, then at every sufficiently small scale r we can cover a large portion of $\mathbf{C}(x_0, r) \cap \partial E$ with the graph of a Lipschitz function u, which turns out to be "almost" harmonic. The accuracy of the corresponding estimates will depend on the size of the excess of E in $\mathbf{C}(x_0, r)$, and on the absolute size of the scale r, through the quantity $\Lambda\, r$. We directly state the theorem in the case $\nu = e_n$, and set for brevity

$$\mathbf{e}_n(x, s) = \mathbf{e}(E, x, s, e_n)\,.$$

Theorem 23.7 (Lipschitz approximation) *There exist positive constants $C_1(n)$, $\varepsilon_1(n)$, and $\delta_0(n)$ with the following property. If E is a (Λ, r_0)-perimeter minimizer in $\mathbf{C}(x_0, 9\,r)$ with*

$$\Lambda r_0 \le 1\,, \qquad 9r < r_0\,, \qquad x_0 \in \partial E\,, \qquad \mathbf{e}_n(x_0, 9\,r) \le \varepsilon_1(n)\,,$$

and if we set

$$M = \mathbf{C}(x_0, r) \cap \partial E\,, \qquad M_0 = \Big\{ y \in M : \sup_{0<s<8\,r} \mathbf{e}_n(y, s) \le \delta_0(n) \Big\}\,,$$

then there exists a Lipschitz function $u\colon \mathbb{R}^{n-1} \to \mathbb{R}$ with

$$\sup_{\mathbb{R}^{n-1}} \frac{|u|}{r} \le C_1(n)\, \mathbf{e}_n(x_0, 9r)^{1/2(n-1)}\,, \qquad \mathrm{Lip}(u) \le 1\,, \tag{23.22}$$

such that a suitable translation Γ of the graph of u over $\mathbf{D}_r$ contains M_0,

$$M_0 \subset M \cap \Gamma\,, \qquad \Gamma = x_0 + \Big\{ (z, u(z)) : z \in \mathbf{D}_r \Big\}\,, \tag{23.23}$$

and covers a large portion of M in terms of $\mathbf{e}_n(x_0, 9r)$, that is

$$\frac{\mathcal{H}^{n-1}(M \Delta \Gamma)}{r^{n-1}} \le C_1(n)\, \mathbf{e}_n(x_0, 9r)\,. \tag{23.24}$$

Moreover, u is "almost harmonic" in $\mathbf{D}_r$, in the sense that

$$\frac{1}{r^{n-1}} \int_{\mathbf{D}_r} |\nabla' u|^2 \le C_1(n)\, \mathbf{e}_n(x_0, 9r)\,, \tag{23.25}$$

$$\frac{1}{r^{n-1}} \Big| \int_{\mathbf{D}_r} \nabla' u \cdot \nabla' \varphi \Big| \le C_1(n) \sup_{\mathbf{D}_r} |\nabla' \varphi| \Big(\mathbf{e}_n(x_0, 9r) + \Lambda\, r \Big)\,, \tag{23.26}$$

for every $\varphi \in C^1_c(\mathbf{D}_r)$.

Remark 23.8 (Smallness of the excess at every scale) This theorem allows us to focus on the ultimate goal of the following sections, namely, showing that a small excess assumption at *a given* point, scale, and direction implies the uniform smallness of the excess at *every* point inside a smaller cylinder, with respect to *every* sufficiently small scale, and with respect to the *same* direction. Indeed, having such a result at hand, we would be able to say that the "good" set M_0 in Theorem 23.7 actually coincides with the whole boundary of E in the smaller cylinder, thus bridging between the small excess assumption and the Lipschitz graph criterion, Theorem 23.1. The minimality of E plays a crucial

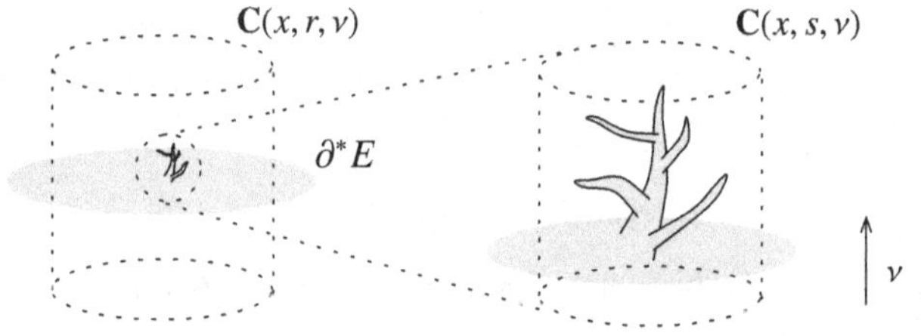

Figure 23.2 Since the excess is an integral quantity, given $r \gg s > 0$ we easily construct a set of finite perimeter E such that $\mathbf{e}(E, x, r, \nu) \approx (s/r)^{n-1} \approx 0$ while $\mathbf{e}(E, x, s, \nu)$ is as large as we wish. These behaviors are excluded in the presence of a minimality condition; indeed the smallness of $\mathbf{e}(E, x, 9\,r, \nu)$ implies the uniform smallness of $\mathbf{e}(E, y, r, \nu)$ for every $y \in \mathbf{C}(x, r, \nu) \cap \partial E$; see the proof of Theorem 26.3.

role in this kind of transferring of the small excess assumption, which cannot be expected to hold true for arbitrary sets; see Figure 23.2.

Remark 23.9 (Small gradient regime and deviation from harmonicity) At first glance, one may wonder why the "almost harmonicity" condition (23.26) should pop up in this theorem. The reason is the following. Since, roughly speaking, E is a local minimizer of the perimeter up to a rescaled error of size Λr on variations at scale r, it turns out that u, up to a rescaled error of size $\mathcal{H}^{n-1}(M \Delta \Gamma) + \Lambda r$, is a local minimizer of the area functional. However, in the small gradient regime (23.25), the area functional and the Dirichlet integral are close, that is

$$\int_{\mathbf{D}_r} \sqrt{1 + |\nabla' u|^2} \approx \mathcal{H}^{n-1}(\mathbf{D}_r) + \frac{1}{2} \int_{\mathbf{D}_r} |\nabla' u|^2 .$$

For this reason, the deviation of u from being harmonic turns out to be controlled in terms of $\mathcal{H}^{n-1}(M \Delta \Gamma) + \Lambda r$.

We divide the proof of Theorem 23.7 into two parts.

Proof of Theorem 23.7, Part I *Step one:* Up to replacing E with $E_{x_0,r}$ (and, correspondingly, u with $u_r(z) = r^{-1}\, u(r\,z)$, $z \in \mathbb{R}^{n-1}$), we can reduce to proving the following statement: if E is a (Λ', r_0')-perimeter minimizer in $\mathbf{C}_9$ with

$$\Lambda' = \Lambda r\,, \qquad r_0' = \frac{r_0}{r} > 9\,, \qquad \Lambda' r_0' \le 1\,, \qquad 0 \in \partial E\,,$$

and if we set $\mathbf{e}_n(y, s) = \mathbf{e}_n(E, y, s)$,

$$M = \mathbf{C} \cap \partial E\,, \qquad M_0 = \Big\{ y \in M : \sup_{0<s<8} \mathbf{e}_n(y, s) \le \delta_0(n) \Big\}\,,$$

then, provided $\mathbf{e}_n(0,9) \le \varepsilon_1(n)$, there exists a Lipschitz function $u\colon \mathbb{R}^{n-1} \to \mathbb{R}$ with $\mathrm{Lip}(u) \le 1$ such that

$$\sup_{\mathbb{R}^{n-1}} |u| \le C_1(n)\, \mathbf{e}_n(0,9)^{1/2(n-1)}, \tag{23.27}$$

$$M_0 \subset M \cap \Gamma, \qquad \Gamma = \Big\{(z, u(z)) : z \in \mathbf{D}\Big\}, \tag{23.28}$$

$$\mathcal{H}^{n-1}(M \Delta \Gamma) \le C_1(n)\, \mathbf{e}_n(0,9). \tag{23.29}$$

$$\int_{\mathbf{D}} |\nabla' u|^2 \le C_1(n)\, \mathbf{e}_n(0,9), \tag{23.30}$$

$$\left| \int_{\mathbf{D}} \nabla' u \cdot \nabla' \varphi \right| \le C_1(n) \sup_{\mathbf{D}} |\nabla' \varphi| \Big(\mathbf{e}_n(0,9) + \Lambda\, r\Big), \tag{23.31}$$

for every $\varphi \in C^1_c(\mathbf{D})$. Let $\varepsilon_0(n)$ and $C_0(n)$ denote the constants determined in Theorem 22.8, and assume that

$$\varepsilon_1(n) \le \varepsilon_0(n).$$

By the height bound we then infer that

$$\sup\Big\{|\,\mathbf{q}x| : x \in \mathbf{C}_2 \cap \partial E\Big\} \le C_0(n)\, \mathbf{e}_n(0,9)^{1/2(n-1)}, \tag{23.32}$$

while Lemma 22.10, Lemma 22.11, and Proposition 22.1 ensure that

$$0 \le \mathcal{H}^{n-1}\Big(M \cap \mathbf{p}^{-1}(G)\Big) - \mathcal{H}^{n-1}(G) \le \mathbf{e}_n(0,1) \le 9^{n-1}\, \mathbf{e}_n(0,9), \tag{23.33}$$

for every Borel set $G \subset \mathbf{D}$. Moreover, since, by construction, $\varepsilon_0(n) \le \omega(n, 1/4)$, Lemma 22.10 implies that

$$\left\{x \in \mathbf{C}_2 : \mathbf{q}x < -\frac{1}{4}\right\} \subset \mathbf{C}_2 \cap E \subset \left\{x \in \mathbf{C}_2 : \mathbf{q}x < \frac{1}{4}\right\}. \tag{23.34}$$

Step two: We show that M_0 is contained in the graph of a Lipschitz function u satisfying (23.27), (23.28), and (23.29). To this end, let us fix $y \in M_0$, $x \in M$, and consider the blow-up of E at scale $\|y - x\|$ centered at y, that is,

$$F = \frac{E - y}{\|y - x\|} = E_{y,\|y-x\|}.$$

Here $\|z\| = \max\{|\,\mathbf{p}z|, |\,\mathbf{q}z|\}$ ($z \in \mathbb{R}^n$), so that $\mathbf{C}(y, s) = \{z \in \mathbb{R}^n : \|z - y\| < s\}$, and $\|y - x\| < 2$. By Remark 21.6, we have that F is a (Λ'', r_0'')-perimeter minimizer in $\mathbf{C}_{9/\|x-y\|}$, with

$$\Lambda'' = \Lambda\, r\, \|x - y\|, \qquad r_0'' = \frac{r_0}{r\,\|x - y\|}.$$

Since $9/\|x - y\| > 4$, $\Lambda'' r_0'' \le 1$, $0 \in \partial F$, and, by definition of M_0,

$$\mathbf{e}_n(F, 0, 4) = \mathbf{e}_n\Big(y, 4\|y - x\|\Big) \le \delta_0(n),$$

then, provided we assume $\delta_0(n) \le \varepsilon_0(n)$, by the height bound we have

$$\sup\Big\{|\,\mathbf{q}w|: w \in \mathbf{C} \cap \partial F\Big\} \le C_0(n)\,\delta_0(n)^{1/2(n-1)}.$$

Testing this condition on $w = (x-y)/\|x-y\|$ we thus find that

$$|\,\mathbf{q}y - \mathbf{q}x| \le C_0(n)\,\delta_0^{1/2(n-1)}\|\,y - x\|. \tag{23.35}$$

If we now set

$$L(n) = C_0(n)\,\delta_0(n)^{1/2(n-1)},$$

and we choose $\delta_0(n)$ so that $L(n) < 1$, then we have $\|\,y - x\| = |\,\mathbf{p}(y-x)|$ in (23.35), which thus becomes

$$|\,\mathbf{q}y - \mathbf{q}x| \le L(n)\,|\,\mathbf{p}x - \mathbf{p}y|, \qquad \forall y \in M_0\,, x \in M\,. \tag{23.36}$$

By (23.36), $\mathbf{p}$ is invertible on M_0. Thus we can define a function $u\colon \mathbf{p}(M_0) \to \mathbb{R}$ such that $u(\mathbf{p}x) = \mathbf{q}x$ for every $x \in M_0$ and

$$|u(\mathbf{p}y) - u(\mathbf{p}x)| \le L(n)\,|\,\mathbf{p}y - \mathbf{p}x|, \qquad \forall x, y \in M_0\,. \tag{23.37}$$

Since $M_0 \subset M$, by (23.32) we also have

$$|u(\mathbf{p}x)| \le C_0(n)\,\mathbf{e}_n(0,9)^{1/2(n-1)}, \qquad \forall x \in M_0\,. \tag{23.38}$$

We extend u from $\mathbf{p}(M_0)$ to $\mathbb{R}^{n-1}$ by McShane's lemma. Hence, $u\colon \mathbb{R}^{n-1} \to \mathbb{R}$,

$$\mathrm{Lip}(u) \le L(n) < 1\,, \qquad M_0 \subset \Gamma = \Big\{(z, u(z)) : z \in \mathbf{D}\Big\}. \tag{23.39}$$

Moreover, thanks to (23.38) and up to truncating u, we may also assume that

$$\sup_{\mathbb{R}^{n-1}} |u| \le C_0(n)\,\mathbf{e}_n(0,9)^{1/2(n-1)}. \tag{23.40}$$

We have thus proved the validity of (23.27) and (23.28). We now prove (23.29). By definition of M_0, for every $y \in M \setminus M_0$ there exists $s \in (0,4)$ with

$$\delta_0(n)\,s^{n-1} < \int_{\mathbf{C}(y,s)\cap\partial E} \frac{|\nu_E - e_n|^2}{2}\,d\mathcal{H}^{n-1}. \tag{23.41}$$

By Corollary 5.2 to Besicovitch's covering theorem, we find disjoint balls $\{B(y_h, \sqrt{2}s_h)\}_{h\in\mathbb{N}}$ contained in $\mathbf{C}_9$, with y_h, s_h satisfying (23.41), and

$$\begin{aligned}
\mathcal{H}^{n-1}(M \setminus M_0) &\le \xi(n) \sum_{h\in\mathbb{N}} \mathcal{H}^{n-1}\Big((M \setminus M_0) \cap B(y_h, \sqrt{2}s_h)\Big) \\
&\le \xi(n) \sum_{h\in\mathbb{N}} \mathcal{H}^{n-1}\Big(M \cap B(y_h, \sqrt{2}s_h)\Big) \\
\text{by (21.10)} \qquad &\le C(n) \sum_{h\in\mathbb{N}} s_h^{n-1}.
\end{aligned}$$

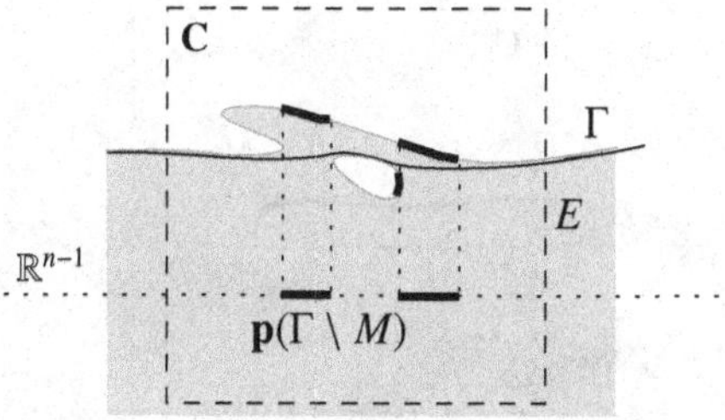

Figure 23.3 The situation in the proof of Theorem 23.7. The boundary of E is represented by a grey line, the graph Γ by a black line. Note that Γ may jump between two distinct connected components of $\mathbf{C} \cap \partial E$. The bold lines are used to represent both $\mathbf{p}(\Gamma \setminus M)$ and $\mathbf{p}^{-1}(\mathbf{p}(\Gamma \setminus M)) \subset M \setminus \Gamma$.

As the cylinders $\{\mathbf{C}(y_h, s_h)\}_{h\in\mathbb{N}}$ are mutually disjoint and contained in $\mathbf{C}_9$,

$$\mathcal{H}^{n-1}(M \setminus M_0) \le C(n) \sum_{h\in\mathbb{N}} \int_{\mathbf{C}(y_h,s_h)\cap\partial E} \frac{|\nu_E - e_n|^2}{2}\, d\mathcal{H}^{n-1} \le C(n)\,\mathbf{e}_n(0,9)\,. \tag{23.42}$$

Therefore, by $M \setminus \Gamma \subset M \setminus M_0$ and by (23.42) we find

$$\mathcal{H}^{n-1}(M \setminus \Gamma) \le C(n)\,\mathbf{e}_n(0,9)\,, \tag{23.43}$$

which is the "first half" of (23.29). To bound $\mathcal{H}^{n-1}(\Gamma \setminus M)$ we first remark that by Theorem 9.1, (23.39), and (23.33) we have that

$$\begin{aligned}\mathcal{H}^{n-1}(\Gamma \setminus M) &\le \sqrt{1 + \mathrm{Lip}(u)^2}\,\mathcal{H}^{n-1}\big(\mathbf{p}(\Gamma \setminus M)\big)\\ &\le \sqrt{2}\,\mathcal{H}^{n-1}\Big(M \cap \mathbf{p}^{-1}\big(\mathbf{p}(\Gamma \setminus M)\big)\Big).\end{aligned}$$

Since $M \cap \mathbf{p}^{-1}(\mathbf{p}(\Gamma \setminus M)) \subset M \setminus \Gamma$ (see Figure 23.3), by (23.43) we conclude that

$$\mathcal{H}^{n-1}(\Gamma \setminus M) \le \sqrt{2}\,\mathcal{H}^{n-1}(M \setminus \Gamma) \le C(n)\,\mathbf{e}_n(0,9)\,. \tag{23.44}$$

Combining (23.43) and (23.44) we have thus proved (23.29).

Step three: We prove (23.30). We first notice that, by (10.14), (16.1), and Proposition 10.5, for $\mathcal{H}^{n-1}$-a.e. $x \in M \cap \Gamma$ there exists $\lambda(x) \in \{-1, 1\}$ such that

$$\nu_E(x) = \lambda(x)\,\frac{(-\nabla' u(\mathbf{p}x), 1)}{\sqrt{1 + |\nabla' u(\mathbf{p}x)|^2}}\,; \tag{23.45}$$

see Figure 23.4. Taking into account that

$$\frac{|\nu_E - e_n|^2}{2} = 1 - (\nu_E \cdot e_n) \ge \frac{1 - (\nu_E \cdot e_n)^2}{2} = \frac{|\mathbf{p}\nu_E|^2}{2}\,,$$

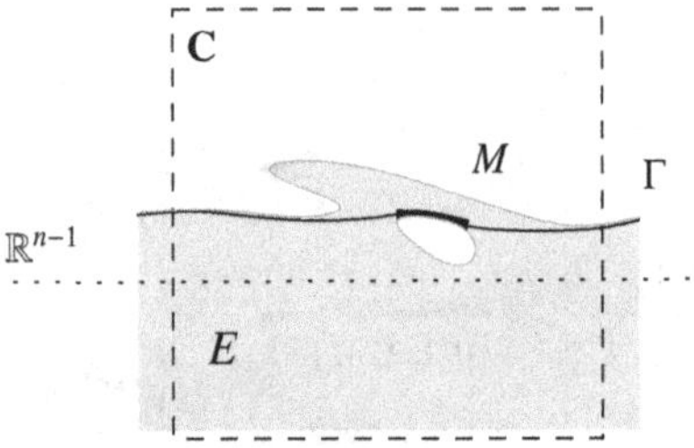

Figure 23.4 Where Γ and $M = \mathbf{C} \cap \partial E$ coincide the outer unit normal of E may be oriented either as $(-\nabla' u, 1)$ or as $(\nabla' u, -1)$. The bold line represents a part of $M \cap \Gamma$ where the second possibility occur. In (23.50) it is shown that this "bad" part of $M \cap \Gamma$ is controlled by the excess.

by (23.45) and Theorem 9.1, we find that

$$\begin{aligned}\mathbf{e}_n(0,1) &\ge \frac{1}{2}\int_{M\cap\Gamma} |\mathbf{p}\nu_E|^2 \, d\mathcal{H}^{n-1} \\ &= \frac{1}{2}\int_{M\cap\Gamma} \frac{|\nabla' u(\mathbf{p}x)|^2}{1+|\nabla' u(\mathbf{p}x)|^2} d\mathcal{H}^{n-1}(x) = \frac{1}{2}\int_{\mathbf{p}(M\cap\Gamma)} \frac{|\nabla' u(z)|^2}{\sqrt{1+|\nabla' u(z)|^2}} \, dz \,.\end{aligned}$$

Having $\mathrm{Lip}(u) \le 1$, we conclude that

$$\int_{\mathbf{p}(M\cap\Gamma)} |\nabla' u|^2 \le 2\sqrt{2}\, \mathbf{e}_n(0,1) \,. \qquad (23.46)$$

For the same reason we find

$$\int_{\mathbf{p}(M\Delta\Gamma)} |\nabla' u|^2 \le \mathcal{H}^{n-1}\big(\mathbf{p}(M\Delta\Gamma)\big) \le \mathcal{H}^{n-1}(M\Delta\Gamma) \,. \qquad (23.47)$$

By $\mathbf{e}_n(0,1) \le 9^{n-1}\mathbf{e}_n(0,9)$, (23.46), (23.47), and (23.29), we find (23.30).

Step four: We reduce the proof of the "almost-harmonicity" inequality (23.31) to the proof of the following "almost-vanishing mean curvature" condition:

$$\left| \int_{\mathbf{D}} \frac{\nabla' u \cdot \nabla' \varphi}{\sqrt{1+|\nabla' u|^2}} \right| \le C(n) \sup_{\mathbf{D}} |\nabla' \varphi| \left(\mathcal{H}^{n-1}(M\Delta\Gamma) + \Lambda\, r \right) , \qquad (23.48)$$

for every $\varphi \in C^\infty_c(\mathbf{D})$. To this end, let us notice that, since $\mathrm{Lip}(u) \le 1$ and $\sqrt{1+s} \le 1 + (s/2)$ for every $s > 0$, we have

$$\begin{aligned}\left| \frac{\nabla' u \cdot \nabla' \varphi}{\sqrt{1+|\nabla' u|^2}} - (\nabla' u \cdot \nabla' \varphi) \right| &= |\nabla' u \cdot \nabla' \varphi| \frac{\sqrt{1+|\nabla' u|^2} - 1}{\sqrt{1+|\nabla' u|^2}} \\ &\le |\nabla' \varphi| \frac{|\nabla' u|^2}{2\sqrt{1+|\nabla' u|^2}} \le |\nabla' \varphi| \frac{|\nabla' u|^2}{2} \,.\end{aligned}$$

In particular,

$$\int_{\mathbf{D}}\left|\frac{\nabla' u\cdot\nabla'\varphi}{\sqrt{1+|\nabla' u|^2}}-(\nabla' u\cdot\nabla'\varphi)\right|\le\sup_{\mathbf{D}}|\nabla'\varphi|\int_{\mathbf{D}}\frac{|\nabla' u|^2}{2}\,,\tag{23.49}$$

and (23.31) follows from (23.30) and (23.48), as claimed.

Step five: As a first step towards the proof of (23.48), we introduce the set

$$\Gamma_1=M\cap\Gamma\cap\big\{\lambda=1\big\}=\Big\{x\in M\cap\Gamma:\nu_E(x)=\frac{(-\nabla' u(\mathbf{p}x),1)}{\sqrt{1+|\nabla' u(\mathbf{p}x)|^2}}\Big\},$$

which is the "good" part of $M\cap\Gamma$ (see Figure 23.4), and show that a large portion of M is directly covered by Γ_1 alone, see (23.51) below. Indeed, by (23.45), if $x\in(M\cap\Gamma)\setminus\Gamma_1$, then $\nu_E(x)\cdot e_n\le 0$. Hence,

$$\mathbf{e}_n(1)\ge\int_{(M\cap\Gamma)\setminus\Gamma_1}1-(\nu_E\cdot e_n)\,d\mathcal{H}^{n-1}\ge\mathcal{H}^{n-1}\big((M\cap\Gamma)\setminus\Gamma_1\big),$$

so that, in particular,

$$\begin{aligned}\mathcal{H}^{n-1}(M\Delta\Gamma_1)&\le\mathcal{H}^{n-1}(M\setminus\Gamma_1)+\mathcal{H}^{n-1}(\Gamma\setminus M)\\&\le\mathcal{H}^{n-1}\big((M\cap\Gamma)\setminus\Gamma_1\big)+\mathcal{H}^{n-1}(M\setminus\Gamma)+\mathcal{H}^{n-1}(\Gamma\setminus M)\\&\le\mathbf{e}_n(1)+\mathcal{H}^{n-1}(M\Delta\Gamma)\,.\end{aligned}\tag{23.50}$$

Since $\mathbf{e}_n(0,1)\le 9^{n-1}\mathbf{e}_n(0,9)$, by (23.29) and (23.50), we conclude that

$$\mathcal{H}^{n-1}(M\Delta\Gamma_1)\le C(n)\,\mathbf{e}_n(0,9)\,.\tag{23.51}$$

We close here Part I of the proof of Theorem 23.7, which we now conclude in two different ways, depending on whether $\Lambda=0$ or not. Of course, this distinction is not strictly necessary, but it allows us to exemplify how to use the vanishing mean curvature condition when possible; see Remark 21.9. □

Conclusion of the proof of Theorem 23.7 in the case $\Lambda=0$ Since now E is a local perimeter minimizer in $\mathbf{C}_9$ at scale $r_0/r>9$, by Corollary 17.14 we know that, in particular,

$$\int_{\partial^* E}\operatorname{div}_E T\,d\mathcal{H}^{n-1}=0\,,\qquad\forall T\in C^1_c(\mathbf{C};\mathbb{R}^n)\,.\tag{23.52}$$

Given $\varphi\in C^\infty_c(\mathbf{D})$, we define a vector field $T\in C^\infty_c(\mathbf{C};\mathbb{R}^n)$ by setting

$$T(x)=\alpha(\mathbf{q}x)\,\varphi(\mathbf{p}x)\,e_n\,,\qquad x\in\mathbb{R}^n\,,\tag{23.53}$$

$$\alpha\in C^\infty_c((-1,1);[0,1])\,,\qquad\alpha(s)=1\quad\forall|s|<\frac{1}{4}\,.\tag{23.54}$$

In order to apply (23.52) we start by computing

$$\nabla T(x)=\alpha(\mathbf{q}x)\,e_n\otimes\nabla'\varphi(\mathbf{p}x)+\varphi(\mathbf{p}x)\alpha'(\mathbf{q}x)\,e_n\otimes e_n\,,\qquad\forall x\in\mathbb{R}^n\,,$$

so that, by (23.54), whenever $|\mathbf{q}x| < 1/4$,

$$\nabla T(x) = e_n \otimes \nabla'\varphi(\mathbf{p}x)\,, \qquad \operatorname{div} T(x) = e_n \cdot \nabla'\varphi(\mathbf{p}x) = 0\,.$$

Having in mind that, thanks to (23.34), $|\mathbf{q}x| < 1/4$ for every $x \in M$, and writing for simplicity $\nabla'\varphi$ in place of $(\nabla'\varphi) \circ \mathbf{p}$, we find

$$\operatorname{div}_E T = -(\nabla T \nu_E) \cdot \nu_E = -(\nu_E \cdot e_n)(\nu_E \cdot \nabla'\varphi)\,, \qquad \text{on } M\,.$$

If we thus apply (23.52) with T as in (23.53) we find

$$\int_M (\nu_E \cdot \nabla'\varphi)(\nu_E \cdot e_n)\, d\mathcal{H}^{n-1} = 0\,. \tag{23.55}$$

On the one hand we have

$$\left| \int_{M\setminus\Gamma_1} (\nu_E \cdot \nabla'\varphi)(\nu_E \cdot e_n)\, d\mathcal{H}^{n-1} \right| \le \sup_{\mathbf{D}} |\nabla'\varphi|\, \mathcal{H}^{n-1}(M \Delta \Gamma_1)\,; \tag{23.56}$$

on the other hand, by definition of Γ_1 and Theorem 9.1,

$$\begin{aligned} \int_{M\cap\Gamma_1} (\nu_E \cdot \nabla'\varphi)(\nu_E \cdot e_n)\, d\mathcal{H}^{n-1} &= \int_{M\cap\Gamma_1} \frac{\nabla' u(\mathbf{p}x) \cdot \nabla'\varphi(\mathbf{p}x)}{1 + |\nabla' u(\mathbf{p}x)|^2}\, d\mathcal{H}^{n-1} \\ &= \int_{\mathbf{p}(M\cap\Gamma_1)} \frac{\nabla' u \cdot \nabla'\varphi}{\sqrt{1 + |\nabla' u|^2}}\,, \end{aligned} \tag{23.57}$$

where, by Proposition 3.5,

$$\begin{aligned} \left| \int_{\mathbf{p}(M\Delta\Gamma_1)} \frac{\nabla' u \cdot \nabla'\varphi}{\sqrt{1 + |\nabla' u|^2}} \right| &\le \sup_{\mathbf{D}} |\nabla'\varphi|\, \mathcal{H}^{n-1}\big(\mathbf{p}(M\Delta\Gamma_1)\big) \\ &\le \sup_{\mathbf{D}} |\nabla'\varphi|\, \mathcal{H}^{n-1}(M\Delta\Gamma_1)\,. \end{aligned} \tag{23.58}$$

Since $\mathbf{D} = \mathbf{p}(M)$, by (23.55), (23.56), (23.57), and (23.58), we find

$$\left| \int_{\mathbf{D}} \frac{\nabla' u \cdot \nabla'\varphi}{\sqrt{1 + |\nabla' u|^2}} \right| \le 2 \sup_{\mathbf{D}} |\nabla'\varphi|\, \mathcal{H}^{n-1}(M\Delta\Gamma_1)\,,$$

which gives (23.48) (with $\Lambda = 0$) thanks to (23.51). □

The above argument concludes the proof of the Lipschitz approximation theorem in the case of local perimeter minimizers. In the general case, we cannot rely on the stationarity condition, but we need to consider "explicit" comparison sets in order to exploit the (Λ, r_0)-perimeter minimality condition. These comparison sets are of course obtained by the considering the local variations associated to the test vector field T defined in (23.53), as we now detail.

Proof of Theorem 23.7, Part II We are thus left to prove (23.48) on a given $\varphi \in C^1_c(\mathbf{D})$. To this end, we may safely multiply φ by a constant, to obtain

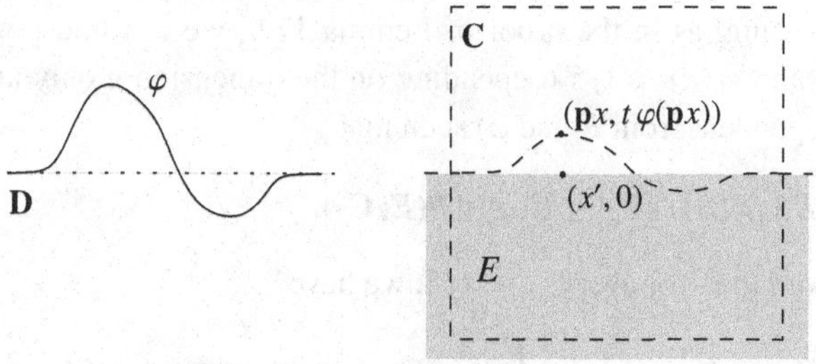

Figure 23.5 The set F_t associated with φ in the case that E is the half-space $\{x \in \mathbb{R}^n : \mathbf{q}x < 0\}$. In this picture, the boundary of F_t is depicted by a dashed curve, and the factor t is roughly equal to $1/2$.

$\sup_{\mathbb{R}^{n-1}} |\nabla'\varphi| = 1$. In particular, by the fundamental theorem of Calculus, and since $\varphi = 0$ on $\partial\mathbf{D}$, we shall also obtain $\sup_{\mathbb{R}^{n-1}} |\varphi| \le 1$. We now let α be as in (23.54), require that $|\alpha'| < 5$ on $\mathbb{R}$, to obtain

$$s \in \mathbb{R} \mapsto s + t\,\alpha(s) \text{ is invertible on } \mathbb{R}\,, \qquad \forall |t| < \frac{1}{5}\,, \tag{23.59}$$

and define a one-parameter family of diffeomorphisms $\{f_t\}_{|t|<1/5}$ by setting

$$f_t(x) = x + t\,\alpha(\mathbf{q}x)\,\varphi(\mathbf{p}x)\,e_n\,, \qquad x \in \mathbb{R}^n\,.$$

The fact that for every $|t| < 1/5$, f_t is a bijection on $\mathbb{R}^n$ is easily verified starting from (23.59) and $\sup_{\mathbb{R}^{n-1}} |\varphi| \le 1$. Since $\{x \in \mathbb{R}^n : f_t(x) \ne x\} \subset\subset \mathbf{C}$ and $f_t(\mathbf{C}) \subset\subset \mathbf{C}_2$ for $|t| < 1/5$, we conclude that

$$f_t(E)\Delta E = f_t(E \cap \mathbf{C})\Delta(E \cap \mathbf{C}) \subset \Big(f_t(E \cap \mathbf{C}) \setminus (E \cap \mathbf{C})\Big) \cup \mathbf{C} \subset\subset \mathbf{C}_2\,.$$

If we define a family of sets of finite perimeter $\{F_t\}_{|t|<1/5}$ as

$$F_t = \Big(f_t(E) \cap \mathbf{C}_2\Big) \cup \Big(E \setminus \mathbf{C}_2\Big)\,, \qquad |t| < \frac{1}{5}\,,$$

then each F_t is a competitor for the $(\Lambda r, r_0/r)$-perimeter minimality of E, and

$$P(E; \mathbf{C}_2) \le P(f_t(E); \mathbf{C}_2) + \Lambda\,r\,\Big|\Big(E\Delta f_t(E)\Big) \cap \mathbf{C}_2\Big|\,, \qquad \forall\,|t| < \frac{1}{5}\,; \tag{23.60}$$

see Figure 23.5. We now notice that f_t has the form $f_t(x) = x + t\,T(x)$, where $|T| \le 1$ on $\mathbb{R}^n$ and

$$|\nabla T(x)| = \sqrt{\alpha(\mathbf{q}x)^2|\nabla'\varphi(\mathbf{p}x)|^2 + \alpha'(\mathbf{q}x)^2\varphi(\mathbf{p}x)^2} \le 6\,, \qquad \forall x \in \mathbb{R}^n\,;$$

therefore, arguing as in the proof of Lemma 17.9, we can find positive constants $C(n)$ and $\varepsilon_0(n) < 1/5$ (depending on the dimension n only and, in particular, independent from E and φ) such that

$$\Big|\big(E\Delta f_t(E)\big)\cap \mathbf{C}_2\Big| \le C(n)|\,t|\,P(E;\mathbf{C}_2), \qquad \forall|\,t| < \varepsilon_0\,. \tag{23.61}$$

We now claim that, for every $|\,t| < 1/5$, we have

$$P(f_t(E);\mathbf{C}_2) \le P(E;\mathbf{C}_2) - t\int_M (\nu_E\cdot\nabla'\varphi)(\nu_E\cdot e_n)\,\mathrm{d}\mathcal{H}^{n-1} + C(n)\,t^2\,; \tag{23.62}$$

note that, by combining (23.60), (23.61), and (23.62) with the fact that $P(E;\mathbf{C}_2) \le C(n)$ by (21.12), we find

$$|t|\left|\int_M (\nu_E\cdot\nabla'\varphi)(\nu_E\cdot e_n)\,\mathrm{d}\mathcal{H}^{n-1}\right| \le C(n)\Big\{t^2 + \Lambda\,r\,|t|\Big\}, \qquad \forall|t| < \varepsilon_0\,,$$

which gives (23.48) by $\sup_{\mathbf{D}}|\nabla'\varphi| = 1$, (23.56), (23.57), and (23.58), and provided we take

$$0 < |\,t| < \min\Big\{\varepsilon_0(n)\,,\mathcal{H}^{n-1}(M\Delta\Gamma_1) + \Lambda r\Big\}.$$

We prove (23.62): by the area formula (11.6), and since $f_t(x) = x$ if $x \notin \mathbf{C}$,

$$P(f_t(E);\mathbf{C}_2) - P(E;\mathbf{C}_2) = \int_M \big(J^M f_t - 1\big)\,\mathrm{d}\mathcal{H}^{n-1}\,, \qquad |\,t| < \frac{1}{5}\,;$$

since $|\mathbf{q}x| < 1/4$ for every $x\in M$, by (23.54) we have $\nabla f_t = \mathrm{Id} + t\,e_n\otimes(\nabla'\varphi\circ\mathbf{p})$ on an open neighborhood of M; by Lemma 23.10 below, setting $\nabla'\varphi$ in place of $(\nabla'\varphi)\circ\mathbf{p}$, we have

$$J^M f_t = J^M(\mathrm{Id} + t\,e_n\otimes\nabla'\varphi) \le 1 - t\,(\nu_E\cdot e_n)(\nu_E\cdot\nabla'\varphi) + C(n)t^2\,|\nabla'\varphi|^2\,, \tag{23.63}$$

and thus (23.62) is proved. □

Lemma 23.10 *There exists a constant $C(n)$ with the following property. If $e\in S^{n-1}$, $v\in\mathbb{R}^n$, and M is a $(n-1)$-dimensional subspace of $\mathbb{R}^n$, then for every $t\in\mathbb{R}$ with $|\,t| < |\,v|^{-1}$ we have*

$$J^M(\mathrm{Id} + t\,e\otimes v) \le 1 + t\,(\mathbf{p}_M v\cdot e) + C(n)|\,t\,v|^2\,.$$

Here, $\mathbf{p}_M\colon\mathbb{R}^n\to M$ denotes the orthogonal projection of $\mathbb{R}^n$ onto M.

Proof Let us denote by Id_M the restriction of the identity tensor to M, and let us notice that the restriction of $e\otimes v$ to M reduces to $e\otimes\mathbf{p}_M v$. In this way,

$$\begin{aligned} J^M(\mathrm{Id} + t\,e\otimes v)^2 &= \det\Big((\mathrm{Id}_M + t\,e\otimes\mathbf{p}_M v)^*(\mathrm{Id}_M + t\,e\otimes\mathbf{p}_M v)\Big)\\ &= \det(\mathrm{Id}_M + t\,S)\,, \end{aligned}$$

where S is the symmetric tensor in $M \otimes M$ defined as

$$S = \mathbf{p}_M v \otimes e + e \otimes \mathbf{p}_M v + t\, \mathbf{p}_M v \otimes \mathbf{p}_M v\,.$$

We apply the spectral theorem to S, in order to find an orthonormal basis $\{\tau_i\}_{i=1}^{n-1}$ of M and $\{\lambda_i\}_{i=1}^{n-1} \subset \mathbb{R}$ such that $S = \sum_{i=1}^{n-1} \lambda_i \tau_i \otimes \tau_i$. In this way we find

$$\begin{aligned} J^M(\mathrm{Id} + t\, e \otimes v)^2 &= \det(\mathrm{Id}_M + t\, S) \\ &= \prod_{i=1}^{n-1}(1 + t\,\lambda_i) = 1 + t\,\mathrm{trace}(S) + \sum_{k=2}^{n-1} t^k \mu_k(S)\,, \end{aligned}$$

where $\mu_k(S)$ denotes the sum of all the possible products with k factors that one may obtain from $\{\lambda_i\}_{i=1}^{n-1}$ (in particular, $\mu_{n-1}(S) = \det(S)$). We now notice that $\mathrm{trace}(S) = 2(\mathbf{p}_M v \cdot e) + t|\,\mathbf{p}_M v|^2$, so that

$$J^M(\mathrm{Id} + t\, e \otimes v)^2 = 1 + 2t(\mathbf{p}_M v \cdot e) + t^2|\,\mathbf{p}_M v|^2 + \sum_{k=2}^{n-1} t^k \mu_k(S)\,.$$

At the same time, since $|\,t| < |\,v|^{-1}$,

$$\max_{1 \le i \le n-1} |\lambda_i| \le |S\,| \le 2|\,\mathbf{p}_M v \cdot e| + |\,t|\,|\,\mathbf{p}_M v|^2 \le 3\,|\,\mathbf{p}_M v|\,.$$

In particular, $|\mu_k(S)| \le c(n,k)|\,\mathbf{p}_M v|^k$, so that, for $k \ge 2$, $|\,t^k \mu_k(S)| \le |\,t\,v|^2$, and

$$J^M(\mathrm{Id} + t\, e \otimes v)^2 \le 1 + 2t(\mathbf{p}_M v \cdot e) + C(n)|\,t\,v|^2\,.$$

We conclude as $\sqrt{1+s} \le 1 + (s/2)$ whenever $s \ge -1$. □

24

The reverse Poincaré inequality

As explained in Remark 23.8, the Lipschitz approximation theorem reduces the regularity problem by showing that, by perimeter minimality, the smallness of the excess at a given scale implies the smallness of the excess at every smaller scale. To prove this, in Chapter 25, we shall need a reverse height bound, in which the excess is controlled through a sort of L^2-height. Precisely, we introduce the **cylindrical flatness** of a set of locally finite perimeter $E \subset \mathbb{R}^n$ at $x \in \mathbb{R}^n$ with respect to $\nu \in S^{n-1}$ at scale $r > 0$, as

$$\mathbf{f}(E, x, r, \nu) = \inf_{c \in \mathbb{R}} \frac{1}{r^{n-1}} \int_{\mathbf{C}(x,r,\nu)\cap\partial^* E} \frac{|(y - x) \cdot \nu - c|^2}{r^2} \, d\mathcal{H}^{n-1}(y) .$$

The flatness $\mathbf{f}(E, x, r, \nu)$ measures the L^2-average distance of $\partial^* E$ from the family of hyperplanes $\{y\colon (y - x) \cdot \nu = c\}$ $(c \in \mathbb{R})$ in the cylinder $\mathbf{C}(x, r, \nu)$. We now provide the required bound (24.2) on the excess in terms of the flatness. In the statement, $\omega(n, t_0)$ denotes the constant introduced in Lemma 22.10.

Theorem 24.1 (Reverse Poincaré inequality) *There exists a positive constant $C(n)$ with the following property. If E is a (Λ, r_0)-perimeter minimizer in $\mathbf{C}(x_0, 4r, \nu)$ with*

$$\Lambda r_0 \le 1 , \qquad x_0 \in \partial E , \qquad 4r < r_0 ,$$

and with

$$\mathbf{e}(E, x_0, 4r, \nu) \le \omega\Big(n, \frac{1}{8}\Big), \tag{24.1}$$

then

$$\mathbf{e}(E, x_0, r, \nu) \le C(n)\Big(\mathbf{f}(E, x_0, 2r, \nu) + \Lambda\, r\Big) . \tag{24.2}$$

Remark 24.2 For technical reasons, the proof of this result is a bit lengthy. However, since it contains no ideas which are going to be reused in other parts of the book, we strongly suggest skipping it on a first reading, directly moving to Chapter 25. This should allow the reader to focus on the remaining part of the proof of the $C^{1,\gamma}$-regularity theorem, and to clearly understand the use and the utility of (24.2) before entering into its proof.

We divide the proof of Theorem 24.1 into three main steps. First, in Lemma 24.6 we show how to modify an open set with smooth boundary inside a cylinder to construct suitable comparison set having ruled surfaces as their boundaries, which allow us to prove the basic excess-flatness estimate. Next, in Lemma 24.9 we combine this construction with an approximation argument, to prove a weak form of the reverse Poincaré inequality (24.2). Finally, in Section 24.4, we shall prove (24.2) from its weak form through a covering argument. We preface to this proof a short, hopefully clarifying, digression.

A digression on elliptic equations, stationarity, and non-oriented excess: Inequality (24.2) is usually called a reverse Poincaré (or a Caccioppoli type) inequality because of its analogy with the well-known Caccioppoli inequality for weak solutions to elliptic equations. We now make a short digression aimed to explain this analogy. Let us recall that a bounded symmetric tensor field $A \in L^\infty(B; \mathbf{Sym}(n))$ is **elliptic** on B with **ellipticity constant** $\lambda > 0$ if

$$(A(x)e) \cdot e \geq \lambda |e|^2 , \qquad \forall e \in \mathbb{R}^n ,$$

for a.e. $x \in B$. In this case, we shall always set

$$\Lambda = \|A\|_{L^\infty(B;\mathbf{Sym}(n))} .$$

Correspondingly, a function $u \in W^{1,2}_{\mathrm{loc}}(B)$ is a **weak solution of the elliptic equation defined by** A if

$$\int_B \Big(A(x)\nabla u(x)\Big) \cdot \nabla\varphi(x)\,\mathrm{d}x = 0 , \qquad \forall \varphi \in C^\infty_{\mathrm{c}}(B) , \tag{24.3}$$

or, which is equivalent by a standard density argument, if this last identity holds true for every $\varphi \in W^{1,2}(B)$ with compact support in B. If $A \in C^1(B; \mathbf{Sym}(n))$ and $u \in C^2(B)$, then, by the divergence theorem, (24.3) is equivalent to the classical elliptic equation in divergence form

$$-\operatorname{div}\Big(A(x)\,\nabla u(x)\Big) = 0 , \qquad \forall x \in B .$$

Of course, if $A(x) = \mathrm{Id}$ a.e. on B, then (24.3) reduces to the weak form (23.18) of the Laplace equation $-\Delta u = 0$, introduced in Example 23.5. One of the basic tools in the regularity theory for weak solutions of (24.3) is the following Caccioppoli inequality.

Proposition 24.3 (Caccioppoli inequality) *If $u \in W^{1,2}_{\rm loc}(B)$ is a weak solution of the elliptic equation (24.3), then*

$$\int_{B(x,r/2)} |\nabla u|^2 \le 16\left(\frac{\Lambda}{\lambda}\right)^2 \int_{B(x,r)} \frac{|u-(u)_{x,r}|^2}{r^2}\,, \tag{24.4}$$

whenever $B(x,r) \subset\subset B$ and $(u)_{x,r} = (\omega_n r^n)^{-1} \int_{B(x,r)} u$.

Proof Let $B(x,r) \subset\subset B$. If $c \in \mathbb{R}$ and $u \in W^{1,2}_{\rm loc}(B)$ is a weak solution of the elliptic equation (24.3), then $u + c \in W^{1,2}_{\rm loc}(B)$ and is a weak solution of the elliptic equation (24.3). Therefore we may assume without loss of generality that $(u)_{x,r} = 0$. Given $\varepsilon > 0$, let us now consider $\zeta \in C^\infty_c(B(x,r))$ such that

$$0 \le \zeta \le 1\,, \qquad \zeta = 1 \ \text{ on } \ B\left(x, \frac{r}{2}\right), \qquad |\nabla\zeta| \le \frac{2+\varepsilon}{r}\,.$$

One easily checks that $\varphi = \zeta^2 u \in W^{1,2}(B)$, with $\operatorname{spt}\varphi \subset\subset B(x,r) \subset\subset B$ and

$$\nabla\varphi = 2\zeta\, u\, \nabla\zeta + \zeta^2 \nabla u\,.$$

Therefore we can test (24.3) by φ. Taking the ellipticity and the boundedness conditions on A into account we thus find

$$\begin{aligned}\lambda \int_B \zeta^2|\nabla u|^2 &\le \int_B \zeta^2\big(A\,\nabla u\big)\cdot\nabla u = -\int_B 2\,u\,\zeta\big(A\,\nabla u\big)\cdot\nabla\zeta \\ &\le 2\Lambda \int_B \zeta|\nabla u|\, u|\nabla\zeta| \le 2\Lambda\Big(\int_B \zeta^2|\nabla u|^2\Big)^{1/2}\Big(\int_B u^2|\nabla\zeta|^2\Big)^{1/2},\end{aligned}$$

that is

$$\Big(\int_B \zeta^2|\nabla u|^2\Big)^{1/2} \le 2\,\frac{\Lambda}{\lambda}\Big(\int_B u^2|\nabla\zeta|^2\Big)^{1/2}.$$

By the defining properties of ζ we conclude that

$$\Big(\int_{B(x,r/2)} |\nabla u|^2\Big)^{1/2} \le 2(2+\varepsilon)\,\frac{\Lambda}{\lambda}\Big(\int_{B(x,r)} \frac{u^2}{r^2}\Big)^{1/2}. \qquad \square$$

For E of locally finite perimeter in $\mathbb{R}^n$, we define the **orientation-free cylindrical excess** of E at $x \in \mathbb{R}^n$ with respect to $\nu \in S^{n-1}$ at scale $r > 0$ as

$$\widehat{\mathbf{e}}(E,x,r,\nu) = \frac{1}{r^{n-1}} \int_{\mathbf{C}(x,r,\nu)\cap\partial^* E} (1 - (\nu_E\cdot\nu)^2)\,d\mathcal{H}^{n-1}\,. \tag{24.5}$$

This notion of excess is called "orientation-free", since it satisfies

$$\widehat{\mathbf{e}}(E,x,r,\nu) = \widehat{\mathbf{e}}(E,x,r,-\nu)\,.$$

In particular, $\widehat{\mathbf{e}}(E,x,r,\nu) = 0$ whenever $\mathbf{C}(x,r,\nu)\cap\partial^* E$ is the union of finitely many $(n-1)$-dimensional balls orthogonal to ν; see Figure 24.1. If we now

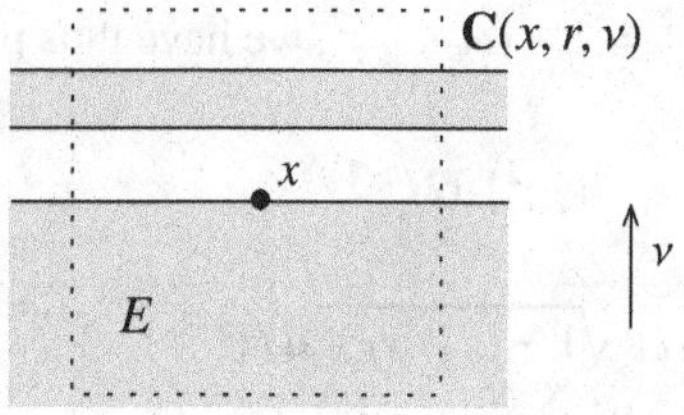

Figure 24.1 It may happen that $\widehat{\mathbf{e}}(E, x, r, \nu) = 0$ while $\mathbf{e}(E, x, r, \nu) > 0$ for some x, r, and ν, even if E a local perimeter minimizer.

make the perimeter stationarity condition play the role of the property of solving the weak elliptic equation (24.3) in the proof of the Caccioppoli inequality, then we find the following proposition, strongly resembling the reverse Poincaré inequality of Theorem 24.1.

Proposition 24.4 (Caccioppoli inequality for stationary sets) *If E is stationary for the perimeter in $\mathbf{C}(x_0, 2r, \nu)$, then*

$$\widehat{\mathbf{e}}(E, x_0, r, \nu) \le 16\,\mathbf{f}(E, x_0, 2\,r, \nu)\,. \tag{24.6}$$

Proof By taking into account the scaling properties of the flatness and of the orientation-free excess, and up to translation, we may directly assume that $x = 0$, $r = 1$, and $\nu = e_n$. We have thus reduced to proving that, if E is stationary for the perimeter in the cylinder $\mathbf{C}_2$, then

$$\int_{\mathbf{C}\cap\partial^* E} \left(1 - (e_n \cdot \nu_E)^2\right) d\mathcal{H}^{n-1} \le 16 \int_{\mathbf{C}_2\cap\partial^* E} |x_n - c|^2 d\mathcal{H}^{n-1}(x)\,, \tag{24.7}$$

for every $c \in \mathbb{R}$. Indeed, given $c \in \mathbb{R}$ and $\varepsilon > 0$, let $\zeta \in C^\infty_c(\mathbf{C}_2)$ be such that

$$0 \le \zeta \le 1\,, \qquad \zeta = 1 \quad \text{on} \quad \mathbf{C}\,, \qquad |\nabla\zeta| \le 2 + \varepsilon\,.$$

and correspondingly define

$$\varphi(x) = x_n - c\,, \quad S(x) = \varphi(x) e_n\,, \quad T(x) = \zeta(x)^2 S(x)\,, \quad x \in \mathbb{R}^n\,,$$

so that $T \in C^\infty_c(\mathbf{C}_2; \mathbb{R}^n)$. Since $\nabla S = e_n \otimes e_n$, we easily compute

$$\begin{aligned}
\nabla T &= 2\,\zeta\,\varphi\, e_n \otimes \nabla\zeta + \zeta^2 e_n \otimes e_n\,,\\
\operatorname{div} T &= 2\,\zeta\,\varphi (e_n \cdot \nabla\zeta) + \zeta^2\,,\\
\nu_E \cdot (\nabla T \nu_E) &= 2\,\zeta\,\varphi\,(\nu_E \cdot \nabla\zeta)(e_n \cdot \nu_E) + \zeta^2 (e_n \cdot \nu_E)^2\,.
\end{aligned}$$

By stationarity, $\int_{\partial^* E} \operatorname{div}_E T\, d\mathcal{H}^{n-1} = 0$, so that

$$0 = \int_{\partial^* E} \left(2\,\zeta\,\varphi\,\nabla\zeta \cdot \left(e_n - (e_n \cdot \nu_E)\nu_E\right) + \zeta^2 \left(1 - (e_n \cdot \nu_E)^2\right)\right) d\mathcal{H}^{n-1}\,.$$

Since $|e_n - (e_n \cdot \nu_E)\nu_E|^2 = 1 - (e_n \cdot \nu_E)^2$, we have thus proved

$$\begin{aligned}
&\int_{\partial^* E} \zeta^2 \left(1 - (e_n \cdot \nu_E)^2\right) d\mathcal{H}^{n-1} \\
&\le 2 \int_{\partial^* E} \zeta |\varphi| |\nabla \zeta| \sqrt{1 - (e_n \cdot \nu_E)^2}\, d\mathcal{H}^{n-1} \\
&\le 2 \Big(\int_{\partial^* E} |\varphi|^2 |\nabla \zeta|^2 \, d\mathcal{H}^{n-1} \Big)^{1/2} \Big(\int_{\partial^* E} \zeta^2 \left(1 - (e_n \cdot \nu_E)^2\right) d\mathcal{H}^{n-1} \Big)^{1/2},
\end{aligned}$$

which, arguing as in the proof of Proposition 24.3, implies (24.7). □

Remark 24.5 We may wonder if, in the case $\Lambda = 0$, (24.2) can be derived from (24.6). However, even if the orientation-free cylindrical excess is controlled by the "oriented" cylindrical excess, namely

$$\widehat{\mathbf{e}}(E, x, r, \nu) \le 2\, \mathbf{e}(E, x, r, \nu),$$

the reverse inequality may fail; see, e.g., Figure 24.1. For local perimeter minimizers with small (oriented) excess on $\mathbf{C}(x, r, \nu)$ it holds that

$$\mathbf{e}(E, x, r, \nu) \le C(n)\, \widehat{\mathbf{e}}(E, x, r, \nu).$$

This fact does not seem to admit a direct proof; see [Spa09, Proposition 13.7].

24.1 Construction of comparison sets, part one

The main step in the proof of the reverse Poincaré inequality (24.2) consists in the construction of suitable comparison sets. In Section 28.1 we shall prove the monotonicity of the ratio $r^{1-n} P(E; B(x, r))$ for a (Λ, r_0)-perimeter minimizer E by comparing the $(n-1)$-dimensional rectifiable set $B(x, r) \cap \partial^* E$ with the cone of vertex at x spanned by the $(n-2)$-dimensional "trace" $\partial E \cap \partial B(x, r)$ of ∂E on $\partial B(x, r)$; see Figure 28.1. We follow here a similar idea, constructing a comparison set having as its boundary an $(n-1)$-dimensional ball ("floating" at a variable height c) and an $(n-1)$-dimensional ruled surface passing the intersection of ∂E with the boundary of a cylinder; see Figure 24.2 and Figure 24.3. Throughout this section we set $\mathbf{K}_s = \mathbf{D}_s \times (-1, 1)$, that is

$$\mathbf{K}_s = \left\{ x \in \mathbb{R}^n : |\mathbf{p}x| < s, |\mathbf{q}x| < 1 \right\}. \tag{24.8}$$

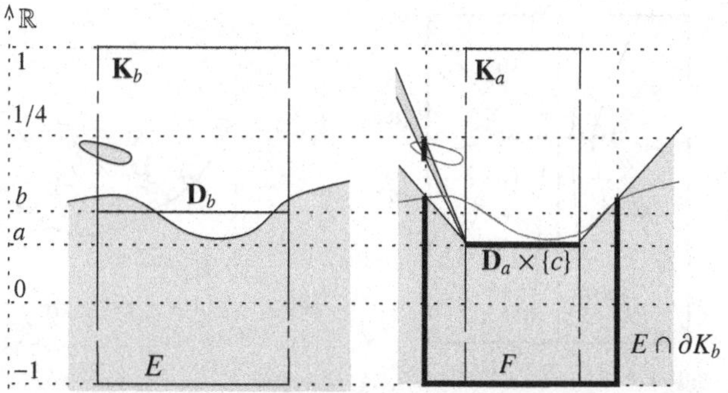

Figure 24.2 The construction of Lemma 24.6. There is no unit measure on the vertical axis. The set F is obtained by an affine interpolation between the $(n-1)$-dimensional disk $\mathbf{D}_a \times \{c\}$ and the trace $E \cap \partial \mathbf{K}_b$ of E inside $\partial \mathbf{K}_b$. Both objects are represented by bold lines.

Lemma 24.6 (Cone-like comparison sets, I) *If $0 < a < b$, $|c| < 1/4$ and E is an open set with smooth boundary, such that*

$$\partial \mathbf{K}_b \cap \partial E \text{ is an } (n-2)\text{-dimensional smooth surface}, \tag{24.9}$$

$$|\mathbf{q}x| < \frac{1}{4}, \qquad \forall x \in \mathbf{K}_b \cap \partial E, \tag{24.10}$$

$$\left\{x \in \mathbf{K}_b : \mathbf{q}x < -\frac{1}{4}\right\} \subset \mathbf{K}_b \cap E \subset \left\{x \in \mathbf{K}_b : \mathbf{q}x < \frac{1}{4}\right\}, \tag{24.11}$$

then there exists an open set F of locally finite perimeter in $\mathbb{R}^n$, satisfying the "boundary conditions"

$$F \cap \partial \mathbf{K}_b = E \cap \partial \mathbf{K}_b, \tag{24.12}$$

$$\mathbf{K}_a \cap F = \left\{x \in \mathbf{K}_a : \mathbf{q}x < c\right\}, \tag{24.13}$$

$$\left\{x \in \mathbf{K}_b : \mathbf{q}x < -\frac{1}{4}\right\} \subset \mathbf{K}_b \cap F \subset \left\{x \in \mathbf{K}_b : \mathbf{q}x < \frac{1}{4}\right\}, \tag{24.14}$$

and the perimeter estimate

$$\begin{aligned} P(F;\mathbf{K}_b) &\le \mathcal{H}^{n-1}(\mathbf{D}_a) \\ &\quad + \frac{\mathcal{H}^{n-1}(\mathbf{D}_b) - \mathcal{H}^{n-1}(\mathbf{D}_a)}{\mathcal{H}^{n-2}(\partial \mathbf{D}_b)} \int_{\partial \mathbf{K}_b \cap \partial E} \sqrt{1 + \left(\frac{\mathbf{q}x - c}{b-a}\right)^2}\, d\mathcal{H}^{n-2}(x). \end{aligned} \tag{24.15}$$

Remark 24.7 If, for some $|c| < 1/4$, $\partial \mathbf{K}_b \cap \partial E = \partial \mathbf{D}_b \times \{c\}$, then our construction gives $\mathbf{K}_b \cap \partial F = \mathbf{D}_b \times \{c\}$ and (24.15) takes the form $P(F; K_b) = \mathcal{H}^{n-1}(\mathbf{D}_b)$.

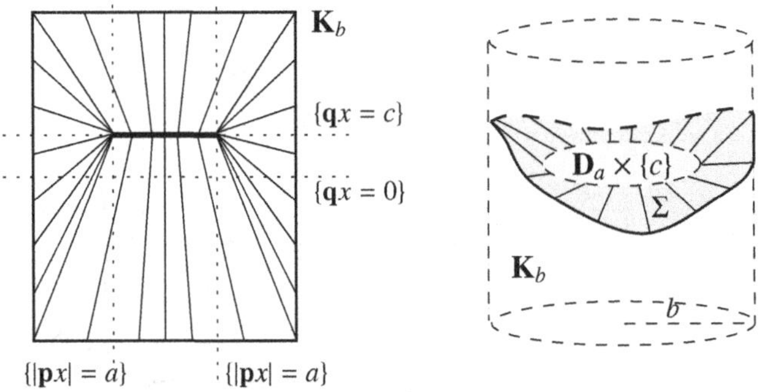

Figure 24.3 On the left: a schematic representation of the various segments $\Phi((0,1) \times \{x\})$ which are obtained as x varies in $\partial \mathbf{K}_b$. The bold segment at height c represents $\mathbf{D}_a \times \{c\}$. Note that $\pi = \Phi(0, \cdot)$ takes $\{x \in \partial \mathbf{K}_b : \mathbf{q}x = \pm 1\}$ onto $\overline{\mathbf{D}_a} \times \{c\}$ (with multiplicity two), and that π takes $\{x \in \partial \mathbf{K}_b : |\mathbf{p}x| = b\}$ onto $\partial \mathbf{D}_a \times \{c\}$. On the right: a three-dimensional picture of the part of the ruled surface Σ contained in $\mathbf{K}_b$: note that Σ matches the $(n-1)$-dimensional ball $\mathbf{D}_a \times \{c\}$, and passes through the $(n-2)$-dimensional smooth surface $\partial \mathbf{K}_b \cap \partial E$ (here represented by a bold line).

Proof of Lemma 24.6 We define $\theta \in (0,1)$ by $a = \theta b$, and consider the two smooth functions $\pi \colon \mathbb{R}^n \to \mathbb{R}^n$ and $\Phi \colon (0,\infty) \times \mathbb{R}^n \to \mathbb{R}^n$ defined as

$$\pi(x) = (\theta\, \mathbf{p}x, c)\,, \qquad x \in \mathbb{R}^n\,,$$
$$\Phi(t,x) = \pi(x) + t(x - \pi(x))\,, \qquad (t,x) \in (0,\infty) \times \mathbb{R}^n\,;$$

see Figure 24.3. The set F is then defined as

$$F = \Big\{\Phi(t,x) : t > 0\,, x \in E \cap \partial \mathbf{K}_b\Big\}\,;$$

see Figure 24.2. It is easily seen that F is open, with

$$\partial F = \Big(\overline{\mathbf{D}_a} \times \{c\}\Big) \cup \Sigma\,,$$

where Σ is the ruled-surface

$$\Sigma = \Big\{\Phi(t,x) : t > 0\,, x \in \partial E \cap \partial \mathbf{K}_b\Big\}\,,$$

and that (24.12), (24.13), and (24.14) hold true. We now divide the proof into two steps. We warn the reader that the most of the following considerations should be slightly modified, and actually, they are trivialized, in the case $n = 2$.

Step one: We may prove that F is a set of locally finite perimeter by Theorem 9.6 and Example 12.7 (we regard $M_0 = \partial \mathbf{D}_a \times \{c\}$ as the singular part of ∂F). To do this, we are just left to check that for every $x_0 \in \Sigma$, there exists $r_0 > 0$ and a function $\varphi \in C^1(B(x_0, r_0))$ such that

$$F \cap B(x_0, r_0) = \Big\{x \in B(x_0, r_0) : \varphi(x) < 0\Big\}\,, \tag{24.16}$$

$$\partial F \cap B(x_0, r_0) = \Big\{x \in B(x_0, r_0) : \varphi(x) = 0\Big\}\,. \tag{24.17}$$

Indeed, since E is an open set with smooth boundary, there exists $y_0 \in \partial E \cap \partial \mathbf{K}_b$, $t_0 > 0$, $\delta_0 > 0$, and $\psi \in C^1(B(y_0, \delta_0))$, such that $x_0 = \Phi(t_0, y_0)$ and

$$E \cap B(y_0, \delta_0) = \Big\{ y \in B(y_0, \delta_0) : \psi(y) < 0 \Big\},$$
$$\partial E \cap B(y_0, \delta_0) = \Big\{ y \in B(y_0, \delta_0) : \psi(y) = 0 \Big\}.$$

Up to decreasing the value of δ_0 we may further ask that Φ is a diffeomorphism between the open neighborhood $U(t_0, y_0)$ of (t_0, y_0) in $(0, \infty) \times \partial \mathbf{K}_b$:

$$U(t_0, y_0) = \Big\{ (t, y) \in (0, \infty) \times \partial \mathbf{K}_b : |t - t_0| < \delta_0 \,, y \in \partial \mathbf{K}_b \cap B(y_0, \delta_0) \Big\},$$

and the open neighborhood $V(x_0)$ of x_0 in $\mathbb{R}^n$ defined by

$$V(x_0) = \Phi(U(t_0, y_0)).$$

Hence, there exist smooth functions $f\colon V(x_0) \to \mathbb{R}$ and $g\colon V(x_0) \to \partial \mathbf{K}_b$, with

$$\Phi(f(x), g(x)) = x \,, \qquad \forall x \in V(x_0).$$

If we now set $\varphi = \psi \circ g$, then we immediately have

$$F \cap V(x_0) = \Big\{ x \in V(x_0) : \varphi(x) < 0 \Big\},$$
$$\partial F \cap V(x_0) = \Big\{ x \in V(x_0) : \varphi(x) = 0 \Big\}.$$

Taking r_0 to satisfy $B(x_0, r_0) \subset V(x_0)$ we have finally proved (24.16) and (24.17). Hence, as explained, it follows that F is an open set with almost C^1-boundary (hence, a set of locally finite perimeter in $\mathbb{R}^n$). In particular, there exists a continuous vector field $\nu\colon \Sigma \to S^{n-1}$, with

$$\nu(x)^\perp = T_x \Sigma \,, \qquad \forall x \in \Sigma \,, \tag{24.18}$$

such that

$$\mu_F = e_n \mathcal{H}^{n-1} \llcorner \Big(\mathbf{D}_a \times \{c\} \Big) + \nu \mathcal{H}^{n-1} \llcorner \Sigma \,. \tag{24.19}$$

Step two: We now estimate $P(F; \mathbf{K}_b)$. By (24.19),

$$P(F; \mathbf{K}_b) = \mathcal{H}^{n-1}(\mathbf{D}_a) + \mathcal{H}^{n-1}\Big(\Sigma \cap (\mathbf{K}_b \setminus \mathbf{K}_a)\Big).$$

Let us define a smooth hypersurface M in $\mathbb{R} \times \mathbb{R}^n$, by setting

$$M = (0,1) \times \Big(\partial \mathbf{K}_b \cap \partial E \Big) \subset \mathbb{R} \times \mathbb{R}^n \,.$$

Since $\Sigma \cap (\mathbf{K}_b \setminus \mathbf{K}_a) = \Phi(M)$, by Theorem 11.3 and Exercise 18.10,

$$\begin{aligned} \mathcal{H}^{n-1}(\Sigma \cap (\mathbf{K}_b \setminus \mathbf{K}_a)) &= \int_M J^M \Phi(x,t) \, d\mathcal{H}^{n-1}(x,t) \qquad (24.20) \\ &= \int_0^1 dt \int_{\partial \mathbf{K}_b \cap \partial E} J^M \Phi(x,t) \, d\mathcal{H}^{n-2}(x) \,. \end{aligned}$$

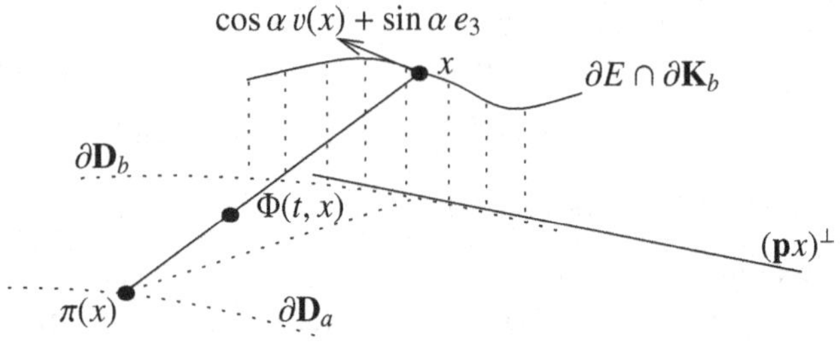

Figure 24.4 The situation in the proof of (24.21), in the case $n = 3$.

We now claim that

$$J^M\Phi(x,t) \le \sqrt{\Big((1-\theta)\mathbf{p}x\Big)^2 + (\mathbf{q}x-c)^2}\,\Big(\theta + t(1-\theta)\Big)^{n-2}. \qquad (24.21)$$

To show this, let $\{e_h\}_{h=0}^n$ be the canonical basis of $\mathbb{R}\times\mathbb{R}^n$. As

$$\Phi(t,x) = \Big((\theta + t(1-\theta))\,\mathbf{p}x, c + t\,(\mathbf{q}x - c)\Big), \qquad (t,x)\in(0,\infty)\times\mathbb{R}^n,$$

then we have, for every $(t,x)\in(0,\infty)\times\mathbb{R}^n$,

$$\nabla\Phi(t,x) = \Big((1-\theta)\mathbf{p}x, \mathbf{q}x - c\Big)\otimes e_0 + \Big(\theta + t(1-\theta)\Big)\sum_{h=1}^{n-1} e_h\otimes e_h + t\,e_n\otimes e_n\,.$$

Let us now fix $x\in\partial\mathbf{K}_b\cap\partial E$. Since $T_x(\partial\mathbf{K}_b\cap\partial E)\subset(\mathbf{p}x)^\perp$ and

$$\dim\Big(T_x(\partial\mathbf{K}_b\cap\partial E)\Big) = n-2 = \dim\Big(e_n^\perp\cap(\mathbf{p}x)^\perp\Big),$$

then we have $\dim(T_x(\partial\mathbf{K}_b\cap\partial E)\cap e_n^\perp)\ge n-3$. Hence, there exists an orthonormal basis $\{\tau_h(x)\}_{h=1}^{n-2}$ of $T_x(\partial\mathbf{K}_b\cap\partial E)$ such that

$$\tau_h(x)\cdot e_n = 0, \qquad \forall h = 1,\dots,n-3,$$
$$\tau_{n-2}(x) = \cos\alpha(x)\,\upsilon(x) + \sin\alpha(x)\,e_n\,,$$

for some $\alpha(x)\in[0,2\pi)$ and $\upsilon(x)\in S^{n-1}\cap e_n^\perp$; see Figure 24.4. In this way $\{e_0\}\cup\{\tau_h(x)\}_{h=1}^{n-2}$ is an orthonormal basis for $T_{(t,x)}M$, and moreover (setting for the sake of brevity $\nabla\Phi = \nabla\Phi(t,x)$, $\tau_h = \tau_h(x)$, $\alpha = \alpha(x)$, $\upsilon = \upsilon(x)$)

$$\nabla\Phi\,[\tau_h] = (\theta + t(1-\theta))\tau_h\,, \qquad 1\le h\le n-3,$$
$$\nabla\Phi\,[\tau_{n-2}] = (\theta + t(1-\theta))\,\cos\alpha\,\upsilon + t\,\sin\alpha\,e_n\,,$$
$$\nabla\Phi\,[e_0] = \Big((1-\theta)\mathbf{p}x, \mathbf{q}x - c\Big).$$

We thus compute the tangential gradient at (t,x) of Φ with respect to M as

$$\begin{aligned}\nabla^M\Phi &= \big(\nabla\Phi\,[e_0]\big)\otimes e_0 + \sum_{h=1}^{n-2}\big(\nabla\Phi[\tau_h]\big)\otimes\tau_h\\ &= \big((1-\theta)\mathbf{p}x,\mathbf{q}x-c\big)\otimes e_0 + \big(\theta+t(1-\theta)\big)\sum_{h=1}^{n-3}\tau_h\otimes\tau_h\\ &\quad+\big((\theta+t(1-\theta))\cos\alpha\,v+t\,\sin\alpha\,e_n\big)\otimes\tau_{n-2}\,,\end{aligned}$$

so that, in particular,

$$\begin{aligned}(\nabla^M\Phi)^*\,\nabla^M\Phi &= \big((1-\theta)\mathbf{p}x,\mathbf{q}x-c\big)^2\,e_0\otimes e_0 + \big(\theta+t(1-\theta)\big)^2\sum_{h=1}^{n-3}\tau_h\otimes\tau_h\\ &\quad+\big((\theta+t(1-\theta))\cos\alpha\,v+t\,\sin\alpha\,e_n\big)^2\,\tau_{n-2}\otimes\tau_{n-2}\,,\end{aligned}$$

and thus,

$$\begin{aligned}(J^M\Phi)^2 &= \big((1-\theta)\mathbf{p}x,\mathbf{q}x-c\big)^2\big(\theta+t(1-\theta)\big)^{2(n-3)}\\ &\quad\times\Big(\big(\theta+t(1-\theta)\big)\cos\alpha\,v+t\sin\alpha\,e_n\Big)^2\,. \end{aligned}\qquad(24.22)$$

If we take into account that $\theta+t(1-\theta)\ge t$, then we infer

$$\big((\theta+t(1-\theta))\,\cos\alpha\,v+t\,\sin\alpha\,e_n\big)^2\le\big(\theta+t(1-\theta)\big)^2,$$

and thus deduce (24.21) from (24.22). From (24.21) we finally compute

$$\begin{aligned}&\int_0^1 \mathrm{d}t\int_{\partial\mathbf{K}_b\cap\partial E} J^M\Phi(x,t)\mathrm{d}\mathcal{H}^{n-2}(x)\\ &\le\int_0^1(\theta+t(1-\theta))^{n-2}\,\mathrm{d}t\int_{\partial\mathbf{K}_b\cap\partial E}\sqrt{\big((1-\theta)b\big)^2+(\mathbf{q}x-c)^2}\,\mathrm{d}\mathcal{H}^{n-2}(x)\\ &=\frac{b\,(1-\theta^{n-1})}{n-1}\int_{\partial\mathbf{K}_b\cap\partial E}\sqrt{1+\Big(\frac{\mathbf{q}x-c}{(1-\theta)b}\Big)^2}\,\mathrm{d}\mathcal{H}^{n-2}(x)\,,\end{aligned}\qquad(24.23)$$

which in turn implies (24.15) if combined with (24.20), the definition of θ, and the fact that $\mathcal{H}^{n-2}(\partial\mathbf{D}_b)=(n-1)\omega_{n-1}b^{n-2}$. □

24.2 Construction of comparison sets, part two

In the following lemma we refine the construction of Lemma 24.6, through a selection argument for the radii a and b based on the coarea formula. In this way, loosely speaking, we derive an estimate for the perimeter of the comparison set in terms of the excess and the flatness of the original set. We recall that $\mathbf{K}_s = \mathbf{D}_s \times (-1, 1)$; see (24.8).

Lemma 24.8 (Cone-like comparison sets, II) *If $s > 0$ and E is an open set with smooth boundary, such that*

$$|\mathbf{q}x| < \frac{1}{4}, \quad \forall x \in \mathbf{K}_s \cap \partial E, \tag{24.24}$$

$$\left\{x \in \mathbf{K}_s : \mathbf{q}x < -\frac{1}{4}\right\} \subset \mathbf{K}_s \cap E \subset \left\{x \in \mathbf{K}_s : \mathbf{q}x < \frac{1}{4}\right\}, \tag{24.25}$$

then for every $\lambda \in (0, 1/4)$ and $|c| < 1/4$, there exist $r \in (2/3, 3/4)$ and an open set F, of locally finite perimeter in $\mathbb{R}^n$, satisfying the "boundary conditions"

$$F \cap \partial\mathbf{K}_{rs} = E \cap \partial\mathbf{K}_{rs}, \tag{24.26}$$

$$\mathbf{K}_{s/2} \cap \partial F = \mathbf{D}_{s/2} \times \{c\}, \tag{24.27}$$

and the "excess-flatness estimate"

$$P(F; \mathbf{K}_{rs}) - \mathcal{H}^{n-1}(\mathbf{D}_{rs}) \le C(n)\left\{\lambda\left(P(E; \mathbf{K}_s) - \mathcal{H}^{n-1}(\mathbf{D}_s)\right) + \frac{1}{\lambda}\int_{\mathbf{K}_s \cap \partial E} \frac{|\mathbf{q}x - c|^2}{s^2}\, d\mathcal{H}^{n-1}(x)\right\}. \tag{24.28}$$

In fact, given s, E, λ and c as above, there exists $I \subset (2/3, 3/4)$ with

$$|I| \ge \frac{1}{24},$$

such that, for every $r \in I$, there exists an open set F, of locally finite perimeter in $\mathbb{R}^n$, satisfying (24.26), (24.27), and (24.28).

Proof *Step one:* By (24.26) and (24.27), applying the divergence theorem as in the proof of Lemma 22.11, we see that the set function

$$\zeta(G) = P\left(E; \mathbf{p}^{-1}(G) \cap \mathbf{K}_s\right) - \mathcal{H}^{n-1}(G), \qquad G \subset \mathbf{D}_s,$$

defines a Radon measure on $\mathbb{R}^{n-1}$, concentrated on $\mathbf{D}_s$. Hence, ζ is monotone.

Step two: Let $\lambda \in (0, 1/4)$ and $|c| < 1/4$ be fixed. We introduce the Lipschitz function $u : \mathbb{R}^n \to \mathbb{R}$, $u(x) = |\mathbf{p}x|$ ($x \in \mathbb{R}^n$). Since $u \in C^\infty(\mathbb{R}^n \setminus \{x : \mathbf{p}x = 0\})$ and ∂E is a smooth hypersurface, by the Morse–Sard lemma (as formulated in Exercise 13.16), we find that, for a.e. $r \in (2/3, 3/4)$,

$$\partial E \cap \partial\mathbf{K}_{rs} \text{ is a } (n-2)\text{-dimensional smooth surface.} \tag{24.29}$$

Given $r \in (2/3, 3/4)$ such that (24.29) holds true, we now apply Lemma 24.6 to E, with the choices

$$b = rs\,, \qquad a = (1-\lambda)rs\,.$$

Correspondingly, we find an open set F_r of locally finite perimeter in $\mathbb{R}^n$ with

$$F_r \cap \partial\mathbf{K}_{rs} = E \cap \partial\mathbf{K}_{rs}\,, \tag{24.30}$$

$$\mathbf{K}_{(1-\lambda)rs} \cap F_r = \{x \in \mathbf{K}_{(1-\lambda)rs} : \mathbf{q}x < c\}\,, \tag{24.31}$$

$$\Big\{x \in \mathbf{K}_{rs} : \mathbf{q}x < -\frac{1}{4}\Big\} \subset \mathbf{K}_{rs} \cap F_r \subset \Big\{x \in \mathbf{K}_{rs} : \mathbf{q}x < \frac{1}{4}\Big\}, \tag{24.32}$$

(which imply (24.26) and (24.27)), and such that

$$\begin{aligned} &P(F_r; \mathbf{K}_{rs}) - \mathcal{H}^{n-1}(\mathbf{D}_{(1-\lambda)rs}) \\ &\le \frac{rs(1-(1-\lambda)^{n-1})}{n-1}\int_{\partial\mathbf{K}_{rs}\cap\partial E} 1 + \Big(\frac{\mathbf{q}x - c}{\lambda rs}\Big)^2 d\mathcal{H}^{n-2}(x)\,. \end{aligned}$$

(Here we have taken into account the form of the right-hand side of (24.15) appearing in (24.23), which holds with $\theta = 1-\lambda$, as well as the elementary inequality $\sqrt{1+t^2} \le 1+t^2$.) In particular, since $1-(1-\lambda)^{n-1} \le (n-1)\lambda$ and $\max\{r, r^{-1}\} = 3/2$,

$$\begin{aligned} &P(F_r; \mathbf{K}_{rs}) - \mathcal{H}^{n-1}(\mathbf{D}_{rs}) \\ &\le \frac{rs(1-(1-\lambda)^{n-1})}{n-1}\Big\{\int_{\partial\mathbf{K}_{rs}\cap\partial E} 1 + \Big(\frac{\mathbf{q}x - c}{\lambda rs}\Big)^2 d\mathcal{H}^{n-2}(x) - \mathcal{H}^{n-2}(\partial\mathbf{D}_{rs})\Big\} \\ &\le \frac{3}{2}\Big\{\lambda s\Big(\mathcal{H}^{n-2}\big(\partial\mathbf{K}_{rs} \cap \partial E\big) - \mathcal{H}^{n-2}(\partial\mathbf{D}_{rs})\Big) + \frac{1}{\lambda s}\int_{\partial\mathbf{K}_{rs}\cap\partial E}(\mathbf{q}x - c)^2\, d\mathcal{H}^{n-2}(x)\Big\}. \end{aligned} \tag{24.33}$$

Slicing ∂E by (18.35) through the level sets of $u(x) = |\mathbf{p}x|$, $x \in \mathbb{R}^n$,

$$\begin{aligned} \int_{2/3}^{3/4} \mathcal{H}^{n-2}\big(\partial E \cap \partial\mathbf{K}_{rs}\big)\, dr &= \frac{1}{s}\int_{\partial E\cap(\mathbf{K}_{3s/4}\setminus\mathbf{K}_{2s/3})} \sqrt{1-(\nu_E\cdot\nabla u)^2}\, d\mathcal{H}^{n-1} \\ &\le \frac{1}{s}P\big(E; \mathbf{K}_{3s/4} \setminus \mathbf{K}_{2s/3}\big), \\ \int_{2/3}^{3/4} \mathcal{H}^{n-2}(\partial\mathbf{D}_{rs})\, dr &= \frac{1}{s}\mathcal{H}^{n-1}\big(\mathbf{D}_{3s/4} \setminus \mathbf{D}_{2s/3}\big). \end{aligned}$$

By step one, $\zeta(\mathbf{D}_{3s/4} \setminus \mathbf{D}_{2s/3}) \le \zeta(\mathbf{D}_s)$, so that

$$\int_{2/3}^{3/4}\Big(\mathcal{H}^{n-2}\big(\partial E \cap \partial\mathbf{K}_{rs}\big) - \mathcal{H}^{n-2}(\partial\mathbf{D}_{rs})\Big)\, dr \le \frac{1}{s}\Big(P(E; \mathbf{K}_s) - \mathcal{H}^{n-1}(\mathbf{D}_s)\Big). \tag{24.34}$$

By a similar argument we readily see that

$$\int_{2/3}^{3/4} \mathrm{d}r \int_{\partial E\cap\partial\mathbf{K}_{rs}} (\mathbf{q}x-c)^2\,\mathrm{d}\mathcal{H}^{n-2} \le \frac{1}{s}\int_{\mathbf{K}_s\cap\partial E} (\mathbf{q}x-c)^2\,\mathrm{d}\mathcal{H}^{n-2}. \tag{24.35}$$

Let us now denote by I the set of those $r \in (2/3, 3/4)$ such that

$$\mathcal{H}^{n-2}\big(\partial E\cap\partial\mathbf{K}_{rs}\big) - \mathcal{H}^{n-2}(\partial\mathbf{D}_{rs}) \le \frac{48}{s}\Big(P(E;\mathbf{K}_s) - \mathcal{H}^{n-1}(\mathbf{D}_s)\Big), \tag{24.36}$$

$$\int_{\partial E\cap\partial\mathbf{K}_{rs}} (\mathbf{q}x-c)^2\,\mathrm{d}\mathcal{H}^{n-2} \le \frac{48}{s}\int_{\mathbf{K}_s\cap\partial^* E} (\mathbf{q}x-c)^2\,\mathrm{d}\mathcal{H}^{n-2}. \tag{24.37}$$

By (24.34) and (24.35) we find that $|I| > 1/24$; see Exercise 1.14. Finally, if $r \in I$, then by (24.33), (24.36), and (24.37) we find (24.28). □

24.3 Weak reverse Poincaré inequality

We now show how to combine the construction of the previous section with the perimeter minimizing property in order to prove a weak form of the reverse Poincaré inequality. We shall set, for $z \in \mathbb{R}^{n-1}$ and $s > 0$,

$$\mathbf{K}(z,s) = (z,0) + \mathbf{K}_s = \big\{x \in \mathbb{R}^n : |z - \mathbf{p}x| < s\,, |\mathbf{q}x| < 1\big\}. \tag{24.38}$$

Lemma 24.9 (Weak reverse Poincaré inequality) *If E is a $(\Lambda, 4)$-perimeter minimizer in $\mathbf{C}_4$ such that*

$$|\mathbf{q}x| < \frac{1}{8}, \qquad \forall x \in \mathbf{C}_2 \cap \partial E, \tag{24.39}$$

$$\Big|\Big\{x \in \mathbf{C}_2 \setminus E : \mathbf{q}x < -\frac{1}{8}\Big\}\Big| = \Big|\Big\{x \in \mathbf{C}_2 \cap E : \mathbf{q}x > \frac{1}{8}\Big\}\Big| = 0, \tag{24.40}$$

and if $z \in \mathbb{R}^{n-1}$ and $s > 0$ are such that

$$\mathbf{K}(z,s) \subset \mathbf{C}_2, \qquad \mathcal{H}^{n-1}\big(\partial E \cap \partial\mathbf{K}(z,s)\big) = 0, \tag{24.41}$$

then, for every $|c| < 1/4$,

$$\begin{aligned} &P(E;\mathbf{K}(z,s/2)) - \mathcal{H}^{n-1}(\mathbf{D}(z,s/2)) \\ &\le C(n)\bigg(\Big(P(E;\mathbf{K}(z,s)) - \mathcal{H}^{n-1}(\mathbf{D}(z,s))\Big)\int_{\mathbf{K}(z,s)\cap\partial^* E} \frac{(\mathbf{q}x-c)^2}{s^2}\,\mathrm{d}\mathcal{H}^{n-1}\bigg)^{1/2} \\ &+ C(n)\Lambda\, s^{n-1}. \end{aligned} \tag{24.42}$$

Proof In order to work with a lighter notation, we shall focus directly on the case $z = 0$. It is easily understood that this is not a real restriction.

Step one: By (24.39), (24.40), and the divergence theorem, we see as usual that the set function

$$\zeta(G) = P\big(E; \mathbf{C}_2 \cap \mathbf{p}^{-1}(G)\big) - \mathcal{H}^{n-1}(G), \qquad G \subset \mathbf{D}_2, \tag{24.43}$$

defines a Radon measure on $\mathbb{R}^{n-1}$, concentrated on $\mathbf{D}_2$.

Step two: By Theorem 13.8, if $\varepsilon_h \to 0^+$, then there exists a sequence $\{E_h\}_{h\in\mathbb{N}}$ of open sets with smooth boundary, such that

$$E_h \overset{\text{loc}}{\to} E, \qquad \mathcal{H}^{n-1} \llcorner \partial E_h \overset{*}{\rightharpoonup} \mathcal{H}^{n-1} \llcorner \partial E, \qquad \partial E_h \subset I_{\varepsilon_h}(\partial E).$$

As a consequence of $E_h \overset{\text{loc}}{\to} E$ and of the coarea formula (13.1),

$$\lim_{h\to\infty} \mathcal{H}^{n-1}\big(\partial \mathbf{K}_{rs} \cap (E^{(1)} \Delta E_h)\big) = 0, \tag{24.44}$$

for a.e. $r \in (2/3, 3/4)$. Moreover, provided h is large enough, by $\partial E_h \subset I_{\varepsilon_h}(\partial E)$,

$$|\mathbf{q}x| \le \frac{1}{4}, \quad \forall x \in \mathbf{C}_2 \cap \partial E_h,$$

$$\Big\{x \in \mathbf{C}_2 : \mathbf{q}x < -\frac{1}{4}\Big\} \subset \mathbf{C}_2 \cap E_h \subset \Big\{x \in \mathbf{C}_2 : \mathbf{q}x < \frac{1}{4}\Big\}.$$

Thus, given $\lambda \in (0, 1/4)$ and $|c| < 1/4$, we can apply Lemma 24.6 to every E_h with respect to s, λ, and c, and find the corresponding set $I_h \subset (2/3, 3/4)$ of "good values" for r. Since $|I_h| > 1/24$ for every $h \in \mathbb{N}$, we clearly have

$$\Big|\bigcap_{h\in\mathbb{N}} \bigcup_{k\ge h} I_k\Big| \ge \frac{1}{24} > 0.$$

Taking (24.44) into account, we conclude that there exist $h(k) \to \infty$ as $k \to \infty$ and $r \in (2/3, 3/4)$ such that

$$r \in \bigcap_{k\in\mathbb{N}} I_{h(k)}, \qquad \lim_{k\to\infty} \mathcal{H}^{n-1}\Big(\partial \mathbf{K}_{rs} \cap \big(E^{(1)} \Delta E_{h(k)}\big)\Big) = 0. \tag{24.45}$$

Hence by Lemma 24.6 there exists a sequence of open sets F_k of locally finite perimeter in $\mathbb{R}^n$ such that

$$F_k \cap \partial \mathbf{K}_{rs} = E_{h(k)} \cap \partial \mathbf{K}_{rs}, \tag{24.46}$$

and, moreover,

$$P(F_k; \mathbf{K}_{rs}) - \mathcal{H}^{n-1}(\mathbf{D}_{rs}) \le C(n)\Big\{\lambda\Big(P(E_{h(k)}; \mathbf{K}_s) - \mathcal{H}^{n-1}(\mathbf{D}_s)\Big) + \frac{1}{\lambda}\int_{\mathbf{K}_s \cap \partial E_{h(k)}} \frac{(\mathbf{q}x - c)^2}{s^2}\, d\mathcal{H}^{n-1}\Big\}. \tag{24.47}$$

We can test the $(\Lambda, 4)$-minimality of E in $\mathbf{C}_4$ by means of the comparison sets

$$G_k = \Big(F_k \cap \mathbf{K}_{rs}\Big) \cup \Big(E \setminus \mathbf{K}_{rs}\Big),$$

as $E\Delta G_k \subset\subset \mathbf{K}_s \subset\subset B(0,4)$. By (16.32), we infer that, for every $k \in \mathbb{N}$,

$$P(E;\mathbf{K}_{rs}) \le P(F_k;\mathbf{K}_{rs}) + \sigma_k + \Lambda\Big|\Big(E\Delta F_k\Big) \cap \mathbf{K}_{rs}\Big|, \tag{24.48}$$

where, by (24.45) and (24.46), $\sigma_k = \mathcal{H}^{n-1}(\partial\mathbf{K}_{rs} \cap (E^{(1)}\Delta F_k)) \to 0$ as $k \to \infty$. By step one and (24.48), we find that

$$\begin{aligned} &P(E;\mathbf{K}_{s/2}) - \mathcal{H}^{n-1}(\mathbf{D}_{s/2}) \le P(E;\mathbf{K}_{rs}) - \mathcal{H}^{n-1}(\mathbf{D}_{rs}) \\ &\le P(F_k;\mathbf{K}_{rs}) - \mathcal{H}^{n-1}(\mathbf{D}_{rs}) + \sigma_k + \Lambda\,|\,(E\Delta F_k) \cap \mathbf{K}_{rs}| \\ &\le C(n)\Big\{\lambda\Big(P(E_{h(k)};\mathbf{K}_s) - \mathcal{H}^{n-1}(\mathbf{D}_s)\Big) + \frac{1}{\lambda}\int_{\mathbf{K}_s\cap\partial E_{h(k)}} \frac{(\mathbf{q}x - c)^2}{s^2}\,d\mathcal{H}^{n-1}\Big\} \\ &\quad + \sigma_k + \Lambda\Big|\Big(E\Delta F_k\Big) \cap \mathbf{K}_{rs}\Big|. \end{aligned}$$

We let $k \to \infty$; by (24.41) and $|\,(E\Delta F_k) \cap \mathbf{K}_{rs}| \le |K_{rs}| \le C(n)\,s^{n-1}$, we have

$$\begin{aligned} &P(E;\mathbf{K}_{s/2}) - \mathcal{H}^{n-1}(\mathbf{D}_{s/2}) \\ &\le C(n)\Big\{\lambda\Big(P(E;\mathbf{K}_s) - \mathcal{H}^{n-1}(\mathbf{D}_s)\Big) + \frac{1}{\lambda}\int_{\mathbf{K}_s\cap\partial E} \frac{(\mathbf{q}x - c)^2}{s^2}\,d\mathcal{H}^{n-1} + \Lambda\,s^{n-1}\Big\}, \end{aligned} \tag{24.49}$$

whenever $\lambda \in (0, 1/4)$, $|c| < 1/4$. By step one,

$$P(E;\mathbf{K}_{s/2}) - \mathcal{H}^{n-1}(\mathbf{D}_{s/2}) \le P(E;\mathbf{K}_s) - \mathcal{H}^{n-1}(\mathbf{D}_s),$$

thus (24.49) is trivially true for $\lambda > 1/4$ provided we choose $C(n) > 4$. We have thus proved (24.49) for every $\lambda > 0$. A minimization over $\lambda > 0$ in (24.49) leads to (24.42). □

24.4 Proof of the reverse Poincaré inequality

We now prove Theorem 24.1. We shall set for the sake of brevity

$$\mathbf{e}_n(0, s) = \mathbf{e}(E, 0, s, e_n), \qquad s > 0.$$

Proof of Theorem 24.1 *Step one:* Up to replacing E with $E_{x_0,r}$ we may directly assume that E is a $(\Lambda r, 4)$-perimeter minimizer in $\mathbf{C}_4$, with

$$4\Lambda\,r \le 1, \qquad 0 \in \partial E, \qquad \mathbf{e}_n(0,4) \le \omega\Big(n, \frac{1}{8}\Big).$$

In particular, by Lemma 22.10, we have that

$$|\mathbf{q}x| < \frac{1}{8}, \qquad \forall x \in \mathbf{C}_2 \cap \partial E, \tag{24.50}$$

$$\left|\left\{x \in \mathbf{C}_2 \cap E : \mathbf{q}x > \frac{1}{8}\right\}\right| = \left|\left\{x \in \mathbf{C}_2 \setminus E : \mathbf{q}x < -\frac{1}{8}\right\}\right| = 0, \tag{24.51}$$

$$\mathcal{H}^{n-1}(G) = \int_{\mathbf{C}_2 \cap \partial^* E \cap \mathbf{p}^{-1}(G)} (\nu_E \cdot e_n) d\mathcal{H}^{n-1}, \quad \forall G \subset \mathbf{D}_2. \tag{24.52}$$

Thanks to (24.52), (24.2) follows by showing that, for every $c \in \mathbb{R}$,

$$P(E; \mathbf{C}) - \mathcal{H}^{n-1}(\mathbf{D}) \le C(n) \left\{ \int_{\mathbf{C}_2 \cap \partial E} |\mathbf{q}x - c|^2 \, d\mathcal{H}^{n-1}(x) + \Lambda r \right\}. \tag{24.53}$$

If $|c| \ge 1/4$ then by (24.50) we have

$$\int_{\mathbf{C}_2 \cap \partial E} |\mathbf{q}x - c|^2 \, d\mathcal{H}^{n-1}(x) \ge \frac{P(E; \mathbf{C})}{8^2},$$

and (24.53) follows immediately (provided we set $C(n) \ge 64$). We thus focus on the proof of (24.53) under the assumption that $|c| < 1/4$.

Step two: As usual, by (24.52), we see that

$$\zeta(G) = P\big(E; \mathbf{C}_2 \cap \mathbf{p}^{-1}(G)\big) - \mathcal{H}^{n-1}(G), \qquad G \subset \mathbf{D}_2,$$

defines a Radon measure on $\mathbb{R}^{n-1}$, concentrated on $\mathbf{D}_2$. By (24.50) and (24.51) we can apply Lemma 24.9 to E in every cylinder $\mathbf{K}(z, s)$ with $z \in \mathbb{R}^{n-1}$ and $s > 0$ such that

$$\mathbf{D}(z, 2s) \subset \mathbf{D}_2, \qquad \mathcal{H}^{n-1}\big(\partial E \cap \partial \mathbf{K}(z, 2s)\big) = 0, \tag{24.54}$$

to find that, under (24.54),

$$\zeta(\mathbf{D}(z, s)) \le C(n) \left\{ \left(\zeta(\mathbf{D}(z, 2s)) \inf_{|c|<1/4} \int_{\mathbf{K}(z,s) \cap \partial E} \frac{|\mathbf{q}x - c|^2}{s^2} \, d\mathcal{H}^{n-1}(x) \right)^{1/2} + \Lambda r s^{n-1} \right\}.$$

By an approximation argument, setting for brevity

$$h = \inf_{|c|<1/4} \int_{\mathbf{C}_2 \cap \partial^* E} |\mathbf{q}x - c|^2 \, d\mathcal{H}^{n-1}(x),$$

we conclude that, whenever $\mathbf{D}(z, 2s) \subset \mathbf{D}_2$ (and thus $s < 1$), we have

$$s^2 \zeta(\mathbf{D}(z, s)) \le C(n) \Big(\sqrt{s^2 \zeta(\mathbf{D}(z, 2s)) \, h} + \Lambda r \Big). \tag{24.55}$$

Notice that (24.55) applied with $z = 0$ and $s = 1$ implies a non-sharp form of (24.53), namely

$$P(E; \mathbf{C}) - \mathcal{H}^{n-1}(\mathbf{D}) \le C(n) \big(\sqrt{h} + \Lambda r \big).$$

In fact, the validity of (24.55) on every $\mathbf{D}(z, 2s) \subset \mathbf{D}_2$ allows us to prove (24.53) through the covering argument that we now describe. To this end, let

$$Q = \sup\{s^2\zeta(\mathbf{D}(z, s)) : \mathbf{D}(z, 2s) \subset \mathbf{D}_2\},$$

and notice that $Q < \infty$ as, whenever $\mathbf{D}(z, 2s) \subset \mathbf{D}_2$,

$$s^2\zeta(\mathbf{D}(z, s)) \le \zeta(\mathbf{D}_2) \le P(E; \mathbf{C}_2) \le C(n).$$

Given $\mathbf{D}(z, 2s) \subset \mathbf{D}_2$ we cover $\mathbf{D}(z, s)$ by finitely many balls $\{\mathbf{D}(z_k, s/4)\}_{k=1}^N$ with centers $z_k \in \mathbf{D}(z, s)$. Of course, this can be done with $N \le N(n)$. Hence, by the sub-additivity of ζ, (24.55), definition of Q, and since $\mathbf{D}(z_k, s) \subset \mathbf{D}_2$,

$$\begin{aligned} s^2\zeta(\mathbf{D}(z, s)) &\le 16 \sum_{k=1}^N \Big(\frac{s}{4}\Big)^2 \zeta\Big(\mathbf{D}\Big(z_k, \frac{s}{4}\Big)\Big) \\ &\le C(n)\sum_{k=1}^N \bigg\{\sqrt{\Big(\frac{s}{2}\Big)^2 \zeta\Big(\mathbf{D}\Big(z_k, \frac{s}{2}\Big)\Big) h} + \Lambda\, r\bigg\} \le C(n)N(n)\Big(\sqrt{Q\, h} + \Lambda\, r\Big). \end{aligned}$$

By the arbitrariness of $\mathbf{D}(z, 2s) \subset \mathbf{D}_2$, we have thus proved that

$$Q \le C(n)\Big(\sqrt{Q\, h} + \Lambda\, r\Big).$$

If now $\sqrt{Q\, h} \le \Lambda\, r$, then $Q \le C(n)\,\Lambda\, r$. If, instead, $\sqrt{Q\, h} \ge \Lambda\, r$, then $Q \le C(n)\, h$. In both cases, we may conclude that $Q \le C(n)(h + \Lambda\, r)$, which implies (24.55) for every $|c| < 1/4$. This concludes the proof of Theorem 24.1. □

25

Harmonic approximation and excess improvement

Let us now briefly sketch a picture of the situation after the results proved in the previous chapters. For the sake of clarity, in this overview, we shall directly refer to the case of a perimeter minimizer E in an open cylinder $\mathbf{C}(x_0, r_0)$, with $x_0 \in \partial E$ (thus, $\Lambda = 0$). In the Lipschitz approximation theorem, Theorem 23.7, we have shown the existence of two constants $\varepsilon_1(n)$ and $\delta_0(n)$ such that if $\mathbf{e}(E, x_0, 9\,r, \nu) \le \varepsilon_1(n)$ and $9r < r_0$, then there exists a Lipschitz function $u\colon \mathbb{R}^{n-1} \to \mathbb{R}$, which is "almost harmonic" on $\mathbf{D}(\mathbf{p}x_0, r)$, and whose graph covers the "good" part M_0 of $\mathbf{C}(x_0, r, \nu) \cap \partial E$,

$$M_0 = \left\{ x \in \mathbf{C}(x_0, r, \nu) \cap \partial E : \sup_{0<s<8\,r} \mathbf{e}(E, x, s, \nu) \le \delta_0(n) \right\}.$$

In turn, we were able to show that M_0 covers a portion of $\mathbf{C}(x_0, r, \nu)\cap\partial E$ which is large in proportion to the smallness of $\mathbf{e}(E, x_0, 9\,r, \nu)$. However, our final goal is showing that the two sets *coincide;* that is to say, we would like to show that, provided $\mathbf{e}(E, x_0, 9\,r, \nu)$ is small enough, then one actually has

$$\mathbf{e}(E, x, s, \nu) \le \delta_0(n)\,,$$

for *every* $x \in \mathbf{C}(x_0, r, \nu) \cap \partial E$ and $s \in (0, 8\,r)$. Clearly, this would imply that $\mathbf{C}(x_0, r, \nu) \cap \partial E$ coincides with the graph of u over $\mathbf{D}(\mathbf{p}x_0, r)$.

As we shall see in Chapter 26, this kind of "uniform smallness of the excess" property descends through a suitable iteration procedure from the following, more focused, "excess improvement by tilting" property (Theorem 25.3): for every sufficiently small $\alpha \in (0, 1)$, provided $\mathbf{e}(E, x_0, r, \nu)$ is small enough depending on α and n, there exists $\nu_0 \in S^{n-1}$ such that

$$\mathbf{e}(E, x_0, \alpha\, r, \nu_0) \le C(n)\, \alpha^2 \mathbf{e}(E, x_0, r, \nu)\,. \tag{25.1}$$

That is, up to "tilting" the direction ν into a suitable, new direction ν_0, to a reduction by a factor α of the scale at which we measure the excess, it corresponds to a reduction by a quadratic factor in α of the excess itself. The proof of this "excess improvement by tilting" property is the main result of this chapter.

Finally, going back to the framework of the Lipschitz approximation theorem, we may give the following rough sketch of the proof of Theorem 25.3.

Step one: We exploit the "almost harmonicity" of u on $\mathbf{D}(\mathbf{p}x_0, r)$ in order to find a harmonic function $v\colon \mathbf{D}(\mathbf{p}x_0, r) \to \mathbb{R}$ which is L^2-close to u (Lemma 25.2) and whose Dirichlet energy $\int_{\mathbf{D}(\mathbf{p}x_0,r)} |\nabla' v|^2$ is controlled by $\mathbf{e}(E, x_0, r, \nu)$;

Step two: We exploit the harmonicity of v to control the uniform distance of the graph of v from its tangent hyperplane at $(\mathbf{p}x_0, v(\mathbf{p}x_0))$ over $(n-1)$-dimensional balls of increasing radius (Lemma 25.1); specifically, we prove that

$$\frac{1}{\omega_{n-1}(\alpha r)^{n-1}} \int_{\mathbf{D}(\mathbf{p}x_0,\alpha r)} \frac{|v-w|^2}{\alpha^2} \le C(n)\alpha^2 \int_{\mathbf{D}(\mathbf{p}x_0,r)} |\nabla' v|^2 , \qquad (25.2)$$

where w is the first-order Taylor's expansion of v at x_0.

Step three: We exploit the L^2-proximity of v to u, and the $\mathcal{H}^{n-1}$-proximity of the graph of u to $\mathbf{C}(x_0, r) \cap \partial E$, to prove the left-hand side of (25.2) to control $\mathbf{f}(E, x_0, \alpha r, \nu_0)$, where ν_0 denotes the outer normal direction to the subgraph of v at $(\mathbf{p}x_0, v(\mathbf{p}x_0))$. By step one we have thus proved that

$$\mathbf{f}(E, x_0, \alpha r, \nu_0) \le C(n)\alpha^2 \mathbf{e}(E, x_0, r, \nu) ,$$

and (25.1) follows by the reverse Poincaré inequality, Theorem 24.1.

We now proceed as follows: in Section 25.1 we prove the required properties of harmonic functions, while in Section 25.2 we state and prove Theorem 25.3 along the lines described above.

25.1 Two lemmas on harmonic functions

In this section we gather a few properties of harmonic functions that we are going to exploit, and whose usefulness was roughly introduced in the discussion above. First of all, let us recall that if v is harmonic in B, then, by an application of the divergence theorem, the *mean value property*,

$$v(x) = \fint_{\partial B(x,r)} v \, d\mathcal{H}^{n-1} = \fint_{B(x,r)} v , \qquad \forall B(x,r) \subset\subset B ,$$

holds true; see e.g. [Eva98, Section 2.2.2]. By the mean value property it is not hard to prove estimates on higher order derivatives of a harmonic function v in terms of lower order derivatives. In the following lemma we provide a bound on the $L^\infty(B_{1/2})$-norm of the Hessian of v in terms of the $L^2(B)$-norm of ∇v.

Lemma 25.1 *If v is harmonic in B and $w(x) = v(0) + \nabla v(0) \cdot x$ ($x \in B$), then*

$$\sup_{B(0,\alpha)} |\, v - w| \le C(n)\alpha^2 \|\nabla v\|_{L^2(B)}\,, \tag{25.3}$$

for every $\alpha \in (0, 1/2]$. In particular,

$$\frac{1}{\omega_n \alpha^n} \int_{B(0,\alpha)} \frac{|\, v - w|^2}{\alpha^2} \le C(n)\, \alpha^2 \int_B |\nabla v|^2 .$$

Proof Let $e \in S^{n-1}$ and let $|x| < 1/2$. Since $e \cdot \nabla v$ is harmonic in B, by the mean-value property we find that, whenever $r < 1/4$,

$$\begin{aligned} |e \cdot \nabla v(x)| &= \left| \fint_{B(x,r)} e \cdot \nabla v \right| = \frac{C(n)}{r^n} \left| \int_{\partial B(x,r)} v\,(e \cdot \nu_{B(x,r)})\, d\mathcal{H}^{n-1} \right| \\ &\le \frac{C(n)}{r^n} \int_{\partial B(x,r)} |\, v(y)|\, d\mathcal{H}^{n-1}(y) \\ &= \frac{C(n)}{r^n} \int_{\partial B(x,r)} \left| \fint_{B(y,r)} v(z)\, dz \right| d\mathcal{H}^{n-1}(y) \le \frac{C(n)}{r^{n+1}} \int_{B(x,2r)} |\, v| . \end{aligned}$$

In particular

$$\sup_{B_{1/2}} |\nabla v| \le C(n) \|\, v\|_{L^2(B)} . \tag{25.4}$$

This inequality, applied to $e \cdot \nabla v$ in place of v, leads to

$$\sup_{B_{1/2}} |\nabla^2 v| \le C(n) \|\nabla v\|_{L^2(B)} .$$

By Taylor's formula, for every $x \in B$ there exists $t \in (0, 1)$ such that $|\, v(x) - w(x)| \le C|\nabla^2 v(t\,x)||x|^2$, and we are done. □

The second fact about harmonic functions that we shall need is the following simple stability lemma for (23.18), stating that almost solutions of (23.18) can be well approximated in $L^2(B)$ by exact solutions (i.e., by harmonic functions). We shall use the classical Poincaré inequality on $W^{1,2}(B)$, stating the existence of a positive constant $C(n)$ such that

$$\inf_{c \in \mathbb{R}} \|u - c\|_{L^2(B)} \le C(n) \|\nabla u\|_{L^2(B)}\,, \qquad \forall u \in W^{1,2}(B)\,,$$

and the compactness theorem for the immersion of $W^{1,2}(B)$ into $L^2(B)$.

Lemma 25.2 (Harmonic approximation) *For every $\tau > 0$ there exists $\sigma > 0$ with the following property. If $u \in W^{1,2}(B)$ is such that*

$$\int_B |\nabla u|^2 \le 1\,, \qquad \left| \int_B \nabla u \cdot \nabla \varphi \right| \le \sup_B |\nabla \varphi|\, \sigma\,,$$

for every $\varphi \in C^\infty_c(B)$, then there exists a harmonic function v on B such that

$$\int_B |\nabla v|^2 \le 1\,, \qquad \int_B |v-u|^2 \le \tau\,.$$

Proof We argue by contradiction, and thus assume the existence of $\tau > 0$ and of a sequence $\{u_h\}_{h\in\mathbb{N}} \subset W^{1,2}(B)$, such that

$$\int_B |\nabla u_h|^2 \le 1\,, \qquad \Big|\int_B \nabla u_h \cdot \nabla\varphi\Big| \le \frac{\|\nabla\varphi\|_{L^\infty(B)}}{h}\,,$$

for every $\varphi \in C^\infty_c(B)$, but

$$\int_B |u_h - v|^2 \ge \tau > 0\,,$$

for every harmonic function v such that $\int_B |\nabla v|^2 \le 1$. Let $c_h = \fint_B u_h$, then by the Poincaré inequality we have $\|u_h - c_h\|_{L^2(B)} \le C(n)\|\nabla u_h\|_{L^2(B)} \le C(n)$. Thus the sequence $\{w_h\}_{h\in\mathbb{N}}$, $w_h = u_h - c_h$, is bounded in $W^{1,2}(B)$. In particular, up to extracting a subsequence, we may assume $w_h \to w$ in $L^2(B)$ for some $w \in W^{1,2}(B)$ such that $\|\nabla w\|_{L^2(B)} \le 1$. We find a contradiction by showing that w is harmonic. Indeed, whenever $\varphi \in C^\infty_c(B)$ we have that

$$\Big|\int_B \nabla w \cdot \nabla\varphi\Big| \le \Big|\int_B \nabla(w - w_h)\cdot\nabla\varphi\Big| + \frac{\sup_B |\nabla\varphi|}{h}\,,$$

where $\int_B \nabla(w-w_h)\cdot\nabla\varphi = -\int_B (w-w_h)\Delta\varphi$. By the Hölder inequality, and letting $h \to \infty$, we obtain a contradiction. □

25.2 The "excess improvement by tilting" estimate

We now prove the following crucial theorem. For the sake of brevity, in the statement we directly set

$$\mathbf{e}(E, x, r, \nu) = \mathbf{e}(x, r, \nu)\,, \qquad x \in \partial E\,, r > 0\,, \nu \in S^{n-1}\,.$$

Theorem 25.3 (Excess improvement by tilting) *Given $\alpha \in (0, 1/72)$, there exist positive constants $\varepsilon_2(n, \alpha)$ and $C_2(n)$ with the following property. If E is a (Λ, r_0)-perimeter minimizer in $\mathbf{C}(x_0, r, \nu)$ with*

$$\Lambda r_0 \le 1\,, \qquad r < r_0\,, \qquad x_0 \in \partial E\,,$$

and if we have

$$\mathbf{e}(x_0, r, \nu) + \Lambda\, r \le \varepsilon_2(n, \alpha)\,, \tag{25.5}$$

then there exists $\nu_0 \in S^{n-1}$ *such that*

$$\mathbf{e}(x_0, \alpha\, r, \nu_0) \le C_2(n)\Big(\alpha^2\, \mathbf{e}(x_0, r, \nu) + \alpha\, \Lambda r\Big). \tag{25.6}$$

Remark 25.4 Note that, in order to infer (25.6) from (25.5), tilting the direction ν into ν_0 is necessary even in the case that E is a half-space. Indeed, if $E = \{x \in \mathbb{R}^n : x \cdot \nu_* = 0\}$ for some $\nu_* \in S^{n-1}$, then E is a perimeter minimizer in $\mathbb{R}^n$ (Exercise 16.12), $0 \in \partial E$, and the Taylor's expansion in the limit $\nu \to \nu_*$,

$$\mathbf{e}(E, 0, s, \nu) = \omega_{n-1}\Big(1 - (\nu \cdot \nu_*)\Big) + \mathrm{o}\Big(1 - (\nu \cdot \nu_*)\Big),$$

holds uniformly on $s > 0$. In particular, we may choose $\nu \in S^{n-1}$ so that $\mathbf{e}(E, 0, r, \nu)$ is arbitrarily small, positive, and constant in r.

Proof of Theorem 25.3 Up to replacing E with $E_{x_0,r/9}$, and up to a rotation taking ν into e_n, we can assume E to be a (Λ', r_0')-perimeter minimizer in $\mathbf{C}_9$,

$$\Lambda' = \frac{\Lambda r}{9}, \qquad r_0' = \frac{9\, r_0}{r} > 9, \qquad \Lambda' r_0' \le 1, \qquad 0 \in \partial E,$$

and, setting $\mathbf{e}_n(s) = \mathbf{e}(0, s, e_n)$ for $s > 0$,

$$\mathbf{e}_n(9) + \Lambda\, r \le \varepsilon_2(n, \alpha). \tag{25.7}$$

Given $\alpha \in (0, 1/72)$, we thus aim to find positive constants $\varepsilon_2(n, \alpha)$ and $C_2(n)$ such that the validity of (25.7) implies the existence of $\nu_0 \in S^{n-1}$ with

$$\mathbf{e}(0, 9\alpha, \nu_0) \le C_2(n)\,(\alpha^2\, \mathbf{e}_n(9) + \alpha \Lambda r). \tag{25.8}$$

To this end, let $\varepsilon_0(n)$, $\varepsilon_1(n)$, and $C_1(n)$ denote the constants determined in the height bound and in the Lipschitz approximation theorem (Theorem 22.8 and Theorem 23.7). In this way, if $\varepsilon_2(n, \alpha) \le \min\{\varepsilon_0(n), \varepsilon_1(n)\}$, then there exists $u \colon \mathbb{R}^{n-1} \to \mathbb{R}$ with $\mathrm{Lip}(u) \le 1$ such that, setting $\Gamma = \{(z, u(z)) : z \in \mathbf{D}\}$,

$$\mathcal{H}^{n-1}(M \Delta \Gamma) \le C_1(n)\, \mathbf{e}_n(9), \tag{25.9}$$

$$\sup_{\mathbb{R}^{n-1}} |u| \le C_1(n)\, \mathbf{e}_n(9)^{1/2(n-1)}, \tag{25.10}$$

$$\sup\Big\{|\,\mathbf{q}y| : y \in M\Big\} \le C_1(n)\, \mathbf{e}_n(9)^{1/2(n-1)}, \tag{25.11}$$

$$\int_{\mathbf{D}} |\nabla' u|^2 \mathrm{d}\mathcal{H}^{n-1} \le C_1(n)\, \mathbf{e}_n(9), \tag{25.12}$$

$$\left|\int_{\mathbf{D}} \nabla' u \cdot \nabla' \varphi\right| \le C_1(n) \sup_{\mathbf{D}} |\nabla' \varphi| \Big\{\mathbf{e}_n(9) + \Lambda\, r\Big\}, \tag{25.13}$$

for every $\varphi \in C^1_c(\mathbf{D})$. If we set

$$\beta = C_1(n)\Big\{\mathbf{e}_n(9) + \Lambda\, r\Big\}, \qquad u_0 = \frac{u}{\sqrt{\beta}}, \tag{25.14}$$

then $u_0 \in W^{1,2}(\mathbf{D})$, with, by (25.12) and (25.13),

$$\int_{\mathbf{D}} |\nabla' u_0|^2 \le 1\,, \qquad \left|\int_{\mathbf{D}} \nabla' u_0 \cdot \nabla' \varphi\right| \le \|\nabla'\varphi\|_{L^\infty(\mathbf{D})}\sqrt{\beta}\,, \quad \forall \varphi \in C^1_c(\mathbf{D})\,.$$

By Lemma 25.2, for every $\tau > 0$ there exists $\sigma(\tau) > 0$ such that if

$$\sqrt{\beta} \le \sigma(\tau)\,, \tag{25.15}$$

then there exists a harmonic function v_0 on $\mathbf{D}$ with

$$\int_{\mathbf{D}} |\nabla' v_0|^2 \le 1\,, \qquad \int_{\mathbf{D}} |v_0 - u_0|^2 \le \tau\,.$$

Therefore the function $v = \sqrt{\beta}\, v_0$ is harmonic on $\mathbf{D}$ and such that

$$\int_{\mathbf{D}} |\nabla' v|^2 \le \beta\,, \qquad \int_{\mathbf{D}} |v - u|^2 \le \tau\beta\,. \tag{25.16}$$

As $36\alpha < 1/2$, by Lemma 25.1, if we set $w(z) = v(0) + \nabla v(0)\cdot z$ ($z \in \mathbf{D}$), then

$$\sup_{\mathbf{D}_{36\alpha}} |v - w| \le C(n)(36\alpha)^2 \|\nabla' v\|_{L^2(\mathbf{D})} \le C(n)\alpha^2\sqrt{\beta}\,. \tag{25.17}$$

By (25.16) and (25.17), we easily find that

$$\frac{1}{\alpha^{n+1}}\int_{\mathbf{D}_{36\alpha}} |u - w|^2 \le C(n)\Big(\frac{\tau}{\alpha^{n+1}} + \alpha^2\Big)\beta\,. \tag{25.18}$$

We apply this argument with $\tau = \alpha^{n+3}$: provided $\varepsilon_2(n,\alpha)$ is such that

$$\sqrt{C_1(n)\varepsilon_2(n,\alpha)} \le \sigma(\alpha^{n+3})\,, \tag{25.19}$$

then, by (25.7) and (25.14), (25.15) holds true with $\tau = \alpha^{n+3}$, and (25.18) takes the form

$$\frac{1}{\alpha^{n+1}}\int_{\mathbf{D}_{36\alpha}} |u - w|^2 \le C(n)\,\alpha^2\beta\,. \tag{25.20}$$

We now relate the left-hand side of (25.20) with the excess of E at scale $18\,\alpha$ with respect to the direction ν_0 given by

$$\nu_0 = \frac{(-\nabla' v(0), 1)}{\sqrt{1 + |\nabla' v(0)|^2}}\,.$$

In this way, we shall be able to deduce (25.8) from (25.20) and the reverse Poincaré inequality, Theorem 24.1. To this end, it is convenient to divide the rest of the proof into three steps.

Step one: We claim that if (25.19) and

$$\beta^{1/(n-1)} \le \alpha^{n+3} \tag{25.21}$$

are in force (recall that $\beta = C_1(n)(\mathbf{e}_n(9) + \Lambda r)$), then

$$\frac{1}{\alpha^{n+1}} \int_{M \cap \mathbf{C}_{36\alpha}} |\nu_0 \cdot y - c|^2 \, d\mathcal{H}^{n-1}(y) \le C(n)\alpha^2 \beta \,, \tag{25.22}$$

where we have set

$$c = \frac{v(0)}{\sqrt{1 + |\nabla' v(0)|^2}} \,.$$

In order to prove (25.22) we decompose M by the graph Γ. On the one hand, by (25.20) and thanks to the fact that $\mathrm{Lip}(u) \le 1$ we find that

$$\begin{aligned} &\frac{1}{\alpha^{n+1}} \int_{M \cap \Gamma \cap \mathbf{C}_{36\alpha}} |\nu_0 \cdot y - c|^2 \, d\mathcal{H}^{n-1}(y) \\ &\le \frac{1}{\alpha^{n+1}} \int_{\mathbf{D}_{36\alpha}} \frac{|u - w|^2}{1 + |\nabla' v(0)|^2} \sqrt{1 + |\nabla' u|^2} \le C(n)\alpha^2 \beta \,. \end{aligned} \tag{25.23}$$

On the other hand,

$$\begin{aligned} &\int_{(M \setminus \Gamma) \cap \mathbf{C}_{36\alpha}} |\nu_0 \cdot y - c|^2 \, d\mathcal{H}^{n-1}(y) \\ &\le 2 \int_{(M \setminus \Gamma) \cap \mathbf{C}_{36\alpha}} |\mathbf{q}y|^2 + |v(0) + \nabla' v(0) \cdot \mathbf{p}y|^2 d\mathcal{H}^{n-1}(y) \qquad (25.24) \\ &\le 4\mathcal{H}^{n-1}(M \setminus \Gamma)\Big(\sup_{y \in M} |\mathbf{q}y|^2 + |v(0)|^2 + |\nabla' v(0)|^2 \Big) \\ &\le C(n)\beta\Big(\beta^{1/(n-1)} + |v(0)|^2 + |\nabla' v(0)|^2\Big) , \qquad (25.25) \end{aligned}$$

where in the last inequality we have applied (25.9) and (25.11). By the mean value property of v we trivially find $|v(0)|^2 \le C(n) \int_{\mathbf{D}} |v|^2$, so that (25.4), (25.20), and (25.10) imply

$$\begin{aligned} |v(0)|^2 + |\nabla' v(0)|^2 &\le C(n) \int_{\mathbf{D}} |v|^2 \le C(n)\Big(\int_{\mathbf{D}} |u - v|^2 + \int_{\mathbf{D}} u^2 \Big) \\ &\le C(n)\Big(\alpha^{n+3}\beta + \beta^{1/(n-1)}\Big) . \end{aligned} \tag{25.26}$$

Combining (25.25), (25.26), and (25.21), we finally deduce

$$\int_{(M \setminus \Gamma) \cap \mathbf{C}_{36\alpha}} |\nu_0 \cdot y - c|^2 \, d\mathcal{H}^{n-1}(y) \le C(n)\Big(\beta^{1+[1/(n-1)]} + \alpha^{n+3}\beta^2\Big) \le C(n)\alpha^{n+3}\beta \,,$$

which, together with (25.23), gives us (25.22).

Step two: We show that, provided $\varepsilon_2(n, \alpha)$ is suitably small, then

$$\mathbf{e}(0, 36\,\alpha, \nu_0) \le \omega\Big(n, \frac{1}{8}\Big) , \tag{25.27}$$

where $\omega(n, t_0)$ denotes the constant of Lemma 22.10 and Theorem 24.1. Indeed, by Proposition 22.5

$$\mathbf{e}(0, 36\,\alpha, \nu_0) \le C(n)\Big(\mathbf{e}_n(36\,\sqrt{2}\,\alpha) + |\nu_0 - e_n|^2\Big).$$

By definition of ν_0 and (25.26), we may roughly estimate that

$$|\nu_0 - e_n|^2 \le C|\nabla' v(0)|^2 \le C(n)\beta^{1/(n-1)},$$

while $\mathbf{e}_n(36\,\sqrt{2}\,\alpha) \le (9/36\,\sqrt{2}\,\alpha)^{n-1}\mathbf{e}_n(9)$ by Proposition 22.1. Hence,

$$\mathbf{e}(0, 36\,\alpha, \nu_0) \le C_3(n, \alpha)\beta^{1/(n-1)},$$

and (25.27) follows provided $\varepsilon_2(n, \alpha)$ is small enough with respect to the constant $C_3(n, \alpha)$ appearing in this last inequality.

Step three: By (25.27) we are now in the position to apply the reverse Poincaré inequality (24.2). Indeed, E is a (Λ', r_0')-perimeter minimizer in $\mathbf{C}_9$, thus in $\mathbf{C}(0, 36\,\alpha, \nu_0)$, with $\Lambda' r_0' \le 1$, $0 \in \partial E$, $36\alpha < r_0'$, and with (25.27) in force. Therefore, by (24.2) and since $\Lambda' = \Lambda r_9$, we find

$$\mathbf{e}(0, 9\alpha, \nu_0) \le C(n)\Big(\mathbf{f}(0, 18\,\alpha, \nu_0) + \Lambda' 9\alpha\Big) = C(n)\Big(\mathbf{f}(0, 18\,\alpha, \nu_0) + \alpha\Lambda r\Big).$$

At the same time, by (25.22), we have

$$\mathbf{f}(0, 18\,\alpha, \nu_0) \le 2^{n+1}\mathbf{f}(0, 36\,\alpha, \nu_0) \le C(n)\,\alpha^2\Big(\mathbf{e}_n(9) + \Lambda r\Big),$$

and thus (25.8) is proved, as desired. □

26

Iteration, partial regularity, and singular sets

We finally prove the $C^{1,\gamma}$-regularity theorems for local perimeter minimizers (Theorem 26.1), as well as for (Λ, r_0)-perimeter minimizers (with $\Lambda > 0$; see Theorem 26.3). In the first case we prove $C^{1,\gamma}$-regularity for every $\gamma \in (0, 1)$, while in the second case the presence of the perturbation term Λr in the key estimate (25.6) will force the restriction $\gamma \in (0, 1/2)$. In view of the higher regularity theory, local perimeter minimizers will in turn prove to be smooth, and in fact analytic. Hence, the real advantage in treating the two cases separately is just pedagogical. In Theorem 26.5, we apply these regularity results to prove the $C^{1,\gamma}$-regularity of the reduced boundary, together with an important characterization of singular sets. This last result will be the starting point for the analysis of singularities carried on in Chapter 28. As a first application of the characterization result, in Theorem 26.6, Section 26.4, we shall refine conclusion (i) of Theorem 21.14, by showing a sort of "C^1-convergence at regular points"-theorem for sequences of (Λ, r_0)-perimeter minimizers.

26.1 The $C^{1,\gamma}$-regularity theorem in the case $\Lambda = 0$

We now prove the $C^{1,\gamma}$-regularity theorem for local perimeter minimizers. As usual, in order to simplify the notation we shall set

$$\mathbf{e}(E, x, r, e_n) = \mathbf{e}_n(x, r) .$$

Below, $C_1(n)$ is as in Theorem 23.7 (Lipschitz approximation theorem).

Theorem 26.1 ($C^{1,\gamma}$-regularity theorem for local minimizers) *For every $\gamma \in (0, 1)$ there exist positive constants $\varepsilon_4(n, \gamma)$, $C_5(n, \gamma)$ with the following property. If E is a local perimeter minimizer in $\mathbf{C}(x_0, 9\,r)$ at scale r_0, with*

$$9r < r_0 , \qquad x_0 \in \partial E ,$$

and if

$$\mathbf{e}_n(x_0, 9\,r) \le \varepsilon_4(n,\gamma)\,,$$

then there exists a Lipschitz function $u\colon \mathbb{R}^{n-1} \to \mathbb{R}$ *with*

$$\sup_{\mathbb{R}^{n-1}} \frac{|u|}{r} \le C_1(n)\,\mathbf{e}_n(x_0, 9\,r)^{1/2(n-1)}\,, \qquad \mathrm{Lip}(u) \le 1\,, \tag{26.1}$$

such that

$$\mathbf{C}(x_0, r) \cap \partial E = x_0 + \Big\{(z, u(z)) : z \in \mathbf{D}_r\Big\}\,, \tag{26.2}$$

$$\mathbf{C}(x_0, r) \cap E = x_0 + \Big\{(z, t) : z \in \mathbf{D}_r\,, -r < t < u(z)\Big\}\,. \tag{26.3}$$

In fact, $u \in C^{1,\gamma}(\mathbf{D}(\mathbf{p}x_0, r))$, *with*

$$|\nabla' u(z) - \nabla' u(z')| \le C_5(n,\gamma)\,\mathbf{e}_n(x_0, 9\,r)^{1/2}\left(\frac{|z - z'|}{r}\right)^{\gamma}\,, \tag{26.4}$$

$$|\nu_E(x) - \nu_E(y)| \le C_5(n,\gamma)\mathbf{e}_n(x_0, 9\,r)^{1/2}\left(\frac{|x - y|}{r}\right)^{\gamma}\,, \tag{26.5}$$

for every $z, z' \in \mathbf{D}(\mathbf{p}x_0, r)$ *and* $x, y \in \mathbf{C}(x_0, r) \cap \partial E$.

The proof of this theorem is based on the iteration of the "excess improvement by tilting" estimate (25.6) (with $\Lambda = 0$). In order for this iteration to converge we need to replace the constant $C_2(n)$ appearing in (25.6) with the constant 1. This can be done at the price of trading the α^2 factor on the right-hand side of (25.6) with the larger factor $\alpha^{2\gamma}$, $\gamma \in (0,1)$.

Lemma 26.2 *For every* $\gamma \in (0,1)$ *there exist positive constants* $\alpha_0 = \alpha_0(n,\gamma) \in (0,1)$, $\varepsilon_3(n,\gamma)$, *and* $C_3(n,\gamma)$ *with the following property. If* E *is a local perimeter minimizer in* $\mathbf{C}(x_0, r, \nu)$ *at scale* r_0, *with*

$$r < r_0\,, \qquad x_0 \in \partial E\,,$$

and if

$$\mathbf{e}(x_0, r, \nu) \le \varepsilon_3(n,\gamma)\,, \tag{26.6}$$

then there exists $\nu_0 \in S^{n-1}$ *such that*

$$\mathbf{e}(x_0, \alpha_0\, r, \nu_0) \le \alpha_0^{2\gamma}\mathbf{e}(x_0, r, \nu)\,, \tag{26.7}$$

$$|\nu_0 - \nu|^2 \le C_3(n,\gamma)\mathbf{e}(x_0, r, \nu)\,. \tag{26.8}$$

Proof of Lemma 26.2 Given $\alpha \in (0, 1/72)$, let $C_2(n)$ and $\varepsilon_2(n,\alpha)$ be the constants of Theorem 25.3. Since $\gamma \in (0,1)$, we have

$$C_2(n)\alpha^2 = C_2(n)\alpha^{2(1-\gamma)}\alpha^{2\gamma} \le \alpha^{2\gamma}\,,$$

provided $\alpha \le C_2(n)^{-1/2(1-\gamma)}$. Therefore, if we define

$$\alpha_0(n,\gamma) = \min\left\{\left(\frac{1}{C_2(n)}\right)^{1/2(1-\gamma)}, \frac{1}{72}\right\}, \qquad \varepsilon_3(n,\gamma) = \varepsilon_2(n, \alpha_0(n,\gamma)),$$

then by (26.6) and Theorem 25.3 we deduce the existence of $\nu_0 \in S^{n-1}$ such that (26.7) holds true. We then integrate the inequality $|\nu_0 - \nu|^2 \le 2(|\nu_0 - \nu_E|^2 + |\nu_E - \nu|^2)$ over $\mathbf{C}(x, \alpha_0\, r, \nu_0) \cap \partial^* E$ to find that

$$\frac{P(E; \mathbf{C}(x, \alpha_0\, r, \nu_0))}{(\alpha_0\, r)^{n-1}} |\nu_0 - \nu|^2 \le 4\, \mathbf{e}(x_0, \alpha_0\, r, \nu_0) + \frac{2}{(\alpha_0\, r)^{n-1}} \int_{\mathbf{C}(x_0,\alpha_0\, r,\nu_0)\cap\partial^* E} |\nu_E - \nu|^2 \, d\mathcal{H}^{n-1}.$$

In turn, by the lower density estimate (21.12), and by (26.7), we find that

$$|\nu - \nu_0|^2 \le C(n)\left(\mathbf{e}(x_0, r, \nu) + \frac{1}{(\alpha_0\, r)^{n-1}} \int_{\mathbf{C}(x_0,\alpha_0\, r,\nu_0)\cap\partial^* E} |\nu_E - \nu|^2 \, d\mathcal{H}^{n-1}\right). \tag{26.9}$$

Since $\alpha_0 \le (1/\sqrt{2})$, we have $\mathbf{C}(x_0, \alpha_0\, r, \nu_0) \subset B(x_0, r) \subset \mathbf{C}(x_0, r, \nu)$, and thus

$$\frac{1}{(\alpha_0\, r)^{n-1}} \int_{\mathbf{C}(x_0,\alpha_0\, r,\nu_0)\cap\partial^* E} |\nu_E - \nu|^2 \, d\mathcal{H}^{n-1} \le \frac{2}{\alpha_0^{n-1}} \mathbf{e}(x_0, r, \nu).$$

In conclusion, if $C(n)$ is as in (26.9), then (26.8) follows with

$$C_3(n,\gamma) = C(n)\left(1 + \frac{2}{\alpha_0^{n-1}}\right). \qquad \square$$

Proof of Theorem 26.3 *Step one:* Given $\gamma \in (0,1)$, we show the existence of a constant $C_4 = C_4(n,\gamma)$ with the following property: if $\varepsilon_3(n,\gamma)$ denotes the constant of Lemma 26.2 and if

$$\mathbf{e}_n(x_0, 9\, r) \le \left(\frac{8}{9}\right)^{n-1} \varepsilon_3(n,\gamma), \tag{26.10}$$

then for every $x \in \mathbf{C}(x_0, r) \cap \partial E$ there exists $\nu(x) \in S^{n-1}$ such that

$$\mathbf{e}(x, s, \nu(x)) \le C_4 \left(\frac{s}{r}\right)^{2\gamma} \mathbf{e}_n(x_0, 9\, r), \quad \forall s \in (0, 4\, r), \tag{26.11}$$

$$|\nu(x) - e_n|^2 \le C_4\, \mathbf{e}_n(x_0, 9\, r), \tag{26.12}$$

$$\mathbf{e}_n(x, s) \le C_4\, \mathbf{e}_n(x_0, 9\, r), \quad \forall s \in (0, 8\, r). \tag{26.13}$$

Indeed, let us fix $x \in \mathbf{C}(x_0, r) \cap \partial E$, and set $t = 8r$ for the sake of brevity. By (26.10) and Proposition 22.4 we immediately find

$$\mathbf{e}_n(x, t) \le \left(\frac{9}{8}\right)^{n-1} \mathbf{e}_n(x_0, 9\, r) \le \varepsilon_3(n,\gamma). \tag{26.14}$$

Thus we can apply Lemma 26.2 to E at the point x and at scale t, and correspondingly find a direction $\nu_1 = \nu_1(x) \in S^{n-1}$ such that

$$\mathbf{e}(x, \alpha t, \nu_1) \le \alpha^{2\gamma} \mathbf{e}_n(x, t), \tag{26.15}$$

$$|\nu_1 - e_n|^2 \le C_3 \mathbf{e}_n(x, t), \tag{26.16}$$

where $\alpha = \alpha_0(n, \gamma)$ and $C_3 = C_3(n, \gamma)$ are as in Lemma 26.2. Since $\alpha \le 1$, by (26.15) and (26.14) we see that $\mathbf{e}(x, \alpha t, \nu_1) \le \varepsilon_3(n, \gamma)$. In particular we can apply Lemma 26.2 again to E at the point x, but this time at the smaller scale αt. Iterating, we prove the existence of a sequence of vectors $\nu_h = \nu_h(x) \in S^{n-1}$ such that

$$\mathbf{e}(x, \alpha^h t, \nu_h) \le \alpha^{2\gamma h} \mathbf{e}_n(x, t), \tag{26.17}$$

$$|\nu_h - \nu_{h-1}|^2 \le C_3 \mathbf{e}(x, \alpha^{h-1} t, \nu_{h-1}) \le C_3 \alpha^{2\gamma(h-1)} \mathbf{e}_n(x, t), \tag{26.18}$$

for every $h \in \mathbb{N}$, where we have set $\nu_0 = e_n$. By (26.18), if $j \ge h \ge 1$ then

$$|\nu_j - \nu_{h-1}| \le \sum_{k=h}^{j} |\nu_k - \nu_{k-1}| \le \sqrt{C_3 \mathbf{e}_n(x, t)} \sum_{k=h}^{\infty} \alpha^{\gamma(k-1)} = \frac{\sqrt{C_3 \mathbf{e}_n(x, t)}}{1 - \alpha^\gamma} \alpha^{\gamma(h-1)} . \tag{26.19}$$

Hence there exists $\nu(x) = \lim_{j \to \infty} \nu_j(x)$. Moreover, if we set $h = 1$ and let $j \to \infty$ in (26.19), by (26.14) we find that, for a suitable constant $C(n, \gamma)$,

$$|\nu(x) - e_n|^2 \le C(n, \gamma) \mathbf{e}_n(x_0, 9\, r), \tag{26.20}$$

which implies (26.12). We now turn to the proof of (26.11). Since $s \in (0, t/2)$, there exists $h \ge 0$ such that

$$\alpha^{h+1} t \le 2\, s < \alpha^h t. \tag{26.21}$$

In particular, by Proposition 22.5, (26.21), and Proposition 22.4,

$$\begin{aligned} \mathbf{e}(x, s, \nu(x)) &\le C(n) \Big(\mathbf{e}(x, \sqrt{2} s, \nu_h) + |\nu(x) - \nu_h|^2 \Big) \\ &\le C(n) \Big\{ \Big(\frac{\alpha^h t}{s} \Big)^{n-1} \mathbf{e}(x, \alpha^h\, t, \nu_h) + |\nu(x) - \nu_h|^2 \Big\}. \end{aligned} \tag{26.22}$$

The first term in (26.22) is controlled by (26.17), (26.14), and (26.21),

$$\begin{aligned} \Big(\frac{\alpha^h t}{s} \Big)^{n-1} \mathbf{e}(x, \alpha^h t, \nu_h) &\le \frac{C(n)}{\alpha^{n-1}} \alpha^{2\gamma h} \mathbf{e}_n(x, t) \le \frac{C(n)}{\alpha^{n-1+2\gamma}} \Big(\frac{s}{t} \Big)^{2\gamma} \mathbf{e}_n(x, t) \\ &\le C(n, \gamma) \Big(\frac{s}{t} \Big)^{2\gamma} \mathbf{e}_n(x_0, 9\, r). \end{aligned} \tag{26.23}$$

Concerning the second term in (26.22), by (26.19), (26.14), and (26.21),

$$|\nu(x) - \nu_h|^2 \le C(n, \gamma) \alpha^{2\gamma h} \mathbf{e}_n(x_0, 9\, r) \le C(n, \gamma) \Big(\frac{s}{t} \Big)^{2\gamma} \mathbf{e}_n(x_0, 9\, r). \tag{26.24}$$

We combine (26.22), (26.23), and (26.24) to prove that

$$\mathbf{e}(x, s, \nu(x)) \le C(n, \gamma)\left(\frac{s}{t}\right)^{2\gamma} \mathbf{e}_n(x_0, 9\,r), \qquad \forall s \in \left(0, \frac{t}{2}\right), \tag{26.25}$$

which is (26.11). We finally prove (26.13); if $s \in (0, t/4)$, then by Proposition 22.5 we deduce that

$$\mathbf{e}_n(x, s) \le C(n)\left(\mathbf{e}(x, \sqrt{2}s, \nu(x)) + |\nu(x) - e_n|^2\right) \le C(n, \gamma)\mathbf{e}_n(x_0, 9\,r),$$

where the second terms is bounded through (26.20), while in dealing with the first term we have taken into account $\sqrt{2}s < t/2$, Proposition 22.4, and (26.25); if, otherwise, $s \in (t/4, t)$, then, by $\mathbf{C}(x, s) \subset \mathbf{C}(x_0, 9\,r)$,

$$\mathbf{e}_n(x, s) \le \left(\frac{9\,r}{s}\right)^{n-1} \mathbf{e}_n(x_0, 9r) \le \left(\frac{9}{2}\right)^{n-1} \mathbf{e}_n(x_0, 9r).$$

We have thus achieved the proof of (26.13).

Step two: With $C_4(n, \gamma)$ as in step one, let us now define

$$\varepsilon_4(n, \gamma) = \min\left\{\varepsilon_0(n), \varepsilon_1(n), \left(\frac{8}{9}\right)^{n-1} \varepsilon_3(n, \gamma), \frac{\delta_0(n)}{C_4(n, \gamma)}\right\}, \tag{26.26}$$

where $\varepsilon_0(n)$ is the constant introduced in the height bound (Theorem 22.8), $\varepsilon_1(n)$ and $\delta_0(n)$ come from the Lipschitz approximation theorem (Theorem 23.7), and $\varepsilon_3(n, \gamma)$ is related to the "excess improvement by tilting" estimate of Theorem 25.3 through Lemma 26.2. Since $\mathbf{e}_n(x_0, 9r) \le \varepsilon_4(n, \gamma)$, (26.10) holds true. In particular, for every $x \in \mathbf{C}(x_0, r) \cap \partial E$, we can find $\nu(x) \in S^{n-1}$ such that (26.11), (26.12), and (26.13) hold true. We now let

$$M_0 = \left\{x \in \mathbf{C}(x_0, r) \cap \partial E : \sup_{0<s<8r} \mathbf{e}_n(x, s) \le \delta_0(n)\right\}.$$

By (26.13), $M_0 = \mathbf{C}(x_0, r) \cap \partial E$. Since $\mathbf{e}_n(x_0, 9r) \le \varepsilon_4(n, \gamma) \le \varepsilon_1(n)$, by Theorem 23.7 there exists a Lipschitz function $u \colon \mathbb{R}^{n-1} \to \mathbb{R}$ such that

$$\mathbf{C}(x_0, r) \cap \partial E \subset \Gamma = x_0 + \left\{(z, u(z)) : z \in \mathbb{R}^{n-1}\right\},$$

and (26.1) holds true. By Theorem 23.1 we have that

$$\mathbf{C}(x_0, r) \cap \partial E = x_0 + \left\{(z, u(z)) : z \in \mathbf{D}_r\right\},$$

that is (26.2), and, from step three in the proof of Theorem 23.7 we see that

$$\nu_E(x) = \frac{(-\nabla' u(\mathbf{p}x), 1)}{\sqrt{1 + |\nabla' u(\mathbf{p}x)|^2}}, \tag{26.27}$$

for $\mathcal{H}^{n-1}$-a.e. $x \in \mathbf{C}(x_0, r) \cap \partial E$. Finally, (26.3) follows immediately from Lemma 22.10 and (26.2).

Step three: Having in mind to apply Campanato's criterion to the proof of (26.4), we now show that for every $z \in \mathbf{D}(\mathbf{p}x_0, r)$ and $s \in (0, r)$,

$$\frac{1}{s^{n-1}} \int_{\mathbf{D}(z,s)} |\nabla' u - (\nabla' u)_{z,s}|^2 \le C(n,\gamma) \left(\frac{s}{r}\right)^{2\gamma} \mathbf{e}_n(x_0, 9\,r), \tag{26.28}$$

where $(\nabla' u)_{z,s}$ denotes the mean value of $\nabla' u$ on $\mathbf{D}(z, s)$. To this end, let us first notice that by (26.12), up to further decreasing $\varepsilon_3(n, \gamma)$, we can assume that

$$\mathbf{q}\nu(x) \ge \frac{1}{\sqrt{2}}, \qquad \forall x \in \mathbf{C}(x_0, r) \cap \partial E. \tag{26.29}$$

In particular, the set inclusion

$$\mathbf{C}(x, s) \subset \mathbf{C}\big(x, 2\,s, \nu(x)\big), \tag{26.30}$$

will hold whenever $x \in \mathbf{C}(x_0, r) \cap \partial E$ and $s > 0$. By (26.29), we can also define a vector field $\tau\colon \mathbf{C}(x_0, r) \cap \partial E \to \mathbb{R}^{n-1}$, by setting

$$\tau(x) = -\frac{\mathbf{p}\nu(x)}{\mathbf{q}\nu(x)}, \qquad x \in \mathbf{C}(x_0, r) \cap \partial E,$$

so that, for every $x \in \mathbf{C}(x_0, r) \cap \partial E$,

$$\mathbf{p}\nu(x) = \frac{-\tau(x)}{\sqrt{1 + |\tau(x)|^2}}, \qquad \mathbf{q}\nu(x) = \frac{1}{\sqrt{1 + |\tau(x)|^2}}, \qquad |\tau(x)| \le 1.$$

If $z \in \mathbf{D}(\mathbf{p}x_0, r)$, $s < r$, $x = (z, u(z))$, then $x \in \mathbf{C}(x_0, r) \cap \partial E$, and by (26.30),

$$\begin{aligned}
(2s)^{n-1}\mathbf{e}(x, 2\,s, \nu(x)) &\ge \int_{\mathbf{C}(x,s)\cap\partial^* E} \frac{|\nu_E - \nu(x)|^2}{2}\, d\mathcal{H}^{n-1} \\
&= \int_{\mathbf{D}(z,s)} \left| \frac{\nabla' u}{\sqrt{1 + |\nabla' u|^2}} - \frac{\tau(\mathbf{p}x)}{\sqrt{1 + |\tau(\mathbf{p}x)|^2}} \right|^2 \sqrt{1 + |\nabla' u|^2} \\
&\quad + \int_{\mathbf{D}(z,s)} \left| \frac{1}{\sqrt{1 + |\nabla' u|^2}} - \frac{1}{\sqrt{1 + |\tau(\mathbf{p}x)|^2}} \right|^2 \sqrt{1 + |\nabla' u|^2}.
\end{aligned}$$

As $|\tau(\mathbf{p}x)| \le 1$, we infer from the above chain of inequalities that

$$\begin{aligned}
\int_{\mathbf{D}(z,s)} |\nabla' u - (\nabla' u)_{z,s}|^2 &= \inf_{\xi\in\mathbb{R}^{n-1}} \int_{\mathbf{D}(z,s)} |\nabla' u - \xi|^2 \le \int_{\mathbf{D}(z,s)} |\nabla' u - \tau(\mathbf{p}x)|^2 \\
&\le 2 \int_{\mathbf{D}(z,s)} \left| \frac{\nabla' u - \tau(\mathbf{p}x)}{\sqrt{1 + |\tau(\mathbf{p}x)|^2}} \right|^2 \sqrt{1 + |\nabla' u|^2} \\
&\le C(n)\, s^{n-1} \mathbf{e}(x, 2s, \nu(x)).
\end{aligned}$$

which, by (26.11), implies (26.28).

Step four: We prove (26.5). From (26.28) and by Campanato's criterion Theorem 6.1 we immediately deduce the validity of (26.4). Since

$$v \in \mathbb{R}^{n-1} \mapsto \frac{(-v,1)}{\sqrt{1+|v|^2}} \in \mathbb{R}^n$$

defines a Lipschitz map on $\mathbb{R}^{n-1}$, we easily deduce from (26.27) and (26.4) that, if $x, y \in \mathbf{C}(x_0, r) \cap \partial E$, then

$$\begin{aligned}|\nu_E(x) - \nu_E(y)| &\le C\,|\nabla' u(\mathbf{p}x) - \nabla' u(\mathbf{p}y)| \le C(n,\gamma)\,\mathbf{e}_n(x_0, 9\,r)^{1/2}\Big(\frac{|\mathbf{p}x - \mathbf{p}y|}{r}\Big)^{\gamma}\\ &\le C(n,\gamma)\,\mathbf{e}_n(x_0, 9\,r)^{1/2}\Big(\frac{|x-y|}{r}\Big)^{\gamma}.\end{aligned}$$

□

26.2 The $C^{1,\gamma}$-regularity theorem in the case $\Lambda > 0$

In the case $\Lambda > 0$, the presence of the term $\Lambda\,r$ in the "excess improvement by tilting" inequality (25.6) forces us to prove a weaker version of Lemma 26.2, which in turns to a $C^{1,\gamma}$-regularity result for $\gamma \in (0, 1/2)$ only.

Theorem 26.3 ($C^{1,\gamma}$-regularity theorem for (Λ, r_0)-perimeter minimizers) *For every $\gamma \in (0, 1/2)$ there exist positive constants $\varepsilon_6(n,\gamma)$ and $C_8(n,\gamma)$ with the following property. If E is a (Λ, r_0)-perimeter minimizer in $\mathbf{C}(x_0, 9\,r)$ with*

$$\Lambda r_0 \le 1\,, \qquad 9r < r_0\,, \qquad x_0 \in \partial E\,,$$

and such that

$$\mathbf{e}_n(x_0, 9\,r) + \Lambda r \le \varepsilon_6(n,\gamma)\,, \tag{26.31}$$

then there exists a Lipschitz function $u\colon \mathbb{R}^{n-1} \to \mathbb{R}$ with

$$\sup_{\mathbb{R}^{n-1}} \frac{|u|}{r} \le C_1(n)\,\mathbf{e}_n(x_0, 9\,r)^{1/2(n-1)}\,, \qquad \mathrm{Lip}(u) \le 1\,, \tag{26.32}$$

such that

$$\mathbf{C}(x_0, r) \cap \partial E = x_0 + \Big\{(z, u(z)) : z \in \mathbf{D}_r\Big\}\,, \tag{26.33}$$

$$\mathbf{C}(x_0, r) \cap E = x_0 + \Big\{(z, t) : z \in \mathbf{D}_r\,, -r < t < u(z)\Big\}\,. \tag{26.34}$$

In fact, $u \in C^{1,\gamma}(\mathbf{D}(\mathbf{p}x_0, r))$, with

$$|\nabla' u(z) - \nabla' u(z')| \le C_8(n,\gamma)\Big(\mathbf{e}_n(x_0, 9\,r) + \Lambda\,r\Big)^{1/2}\Big(\frac{|z-z'|}{r}\Big)^{\gamma}, \tag{26.35}$$

$$|\nu_E(x) - \nu_E(y)| \le C_8(n,\gamma)\Big(\mathbf{e}_n(x_0, 9\,r) + \Lambda\,r\Big)^{1/2}\Big(\frac{|x-y|}{r}\Big)^{\gamma}, \tag{26.36}$$

for every $z, z' \in \mathbf{D}(\mathbf{p}x_0, r)$ and $x, y \in \mathbf{C}(x_0, r) \cap \partial E$.

As in the case $\Lambda = 0$, the first thing to do is to reformulate (25.6) in a form which is suitable for iteration.

Lemma 26.4 *For every $\gamma \in (0,1/2)$ there exist positive constants $\alpha_1 = \alpha_1(n,\gamma) < 1$, $\varepsilon_5(n,\gamma)$ and $C_6(n,\gamma)$ with the following property. Let E be a (Λ, r_0)-perimeter minimizer in $\mathbf{C}(x_0, r, \nu)$ and set*

$$\mathbf{e}^*(x,s,\nu) = \max\left\{\mathbf{e}(x,s,\nu), \frac{\Lambda s}{\alpha_1^{n-1+2\gamma}}\right\}, \qquad x \in \mathbb{R}^n, s > 0. \tag{26.37}$$

If we have $\Lambda r_0 \le 1$, $r < r_0$, $x_0 \in \partial E$, and

$$\mathbf{e}^*(x_0, r, \nu) \le \varepsilon_5(n,\gamma), \tag{26.38}$$

then there exists $\nu_0 \in S^{n-1}$ such that

$$\mathbf{e}^*(x_0, \alpha_1 r, \nu_0) \le \alpha_1^{2\gamma}\, \mathbf{e}^*(x_0, r, \nu), \tag{26.39}$$

$$|\nu_0 - \nu|^2 \le C_6(n,\gamma)\, \mathbf{e}^*(x_0, r, \nu). \tag{26.40}$$

Proof of Lemma 26.4 We define $\alpha_1(n,\gamma)$ and $\varepsilon_5(n,\gamma)$:

$$\alpha_1(n,\gamma) = \min\left\{\frac{1}{72}, \left(\frac{1}{2C_2(n)}\right)^{1/(1-2\gamma)}\right\}, \qquad \varepsilon_5(n,\gamma) = \frac{\varepsilon_2(n,\alpha_1)\,\alpha_1^{n-1+2\gamma}}{2},$$

where ε_2 and C_2 are the constants introduced in Theorem 25.3. Let us now prove (26.39). Taking into account that, by $2\gamma < 1$,

$$\frac{\Lambda\, \alpha_1 r}{\alpha_1^{n-1+2\gamma}} \le \alpha_1\, \mathbf{e}^*(x_0, r, \nu) \le \alpha_1^{2\gamma}\, \mathbf{e}^*(x_0, r, \nu),$$

we only have to show the existence of $\nu_0 \in S^{n-1}$ such that

$$\mathbf{e}(x_0, \alpha_1 r, \nu_0) \le \alpha_1^{2\gamma}\, \mathbf{e}^*(x_0, r, \nu). \tag{26.41}$$

If $\Lambda r \ge \mathbf{e}(x_0, r, \nu)$, then this is a trivial consequence of Proposition 22.1,

$$\mathbf{e}(x_0, \alpha_1 r, \nu) \le \frac{\mathbf{e}(x_0, r, \nu)}{\alpha_1^{n-1}} \le \alpha_1^{2\gamma} \frac{\Lambda r}{\alpha_1^{n-1+2\gamma}} \le \alpha_1^{2\gamma}\, \mathbf{e}^*(x_0, r, \nu),$$

(26.39) holds with $\nu = \nu_0$, and (26.40) follows immediately. If, instead,

$$\Lambda r \le \mathbf{e}(x_0, r, \nu), \tag{26.42}$$

then we first notice that, by our choice of $\varepsilon_5(n,\gamma)$, (26.38) allows us to apply (25.6) in Theorem 25.3, to find $\nu_0 \in S^{n-1}$ such that

$$\begin{aligned} \mathbf{e}(x_0, \alpha_1 r, \nu_0) &\le C_2(n)\big(\alpha_1^2 \mathbf{e}(x_0, r, \nu) + \alpha_1 \Lambda r\big) \le C_2(n)(\alpha_1^2 + \alpha_1)\mathbf{e}(x_0, r, \nu) \\ &\le 2C_2(n)\alpha_1 \mathbf{e}(x_0, r, \nu) \le 2C_2(n)\alpha_1 \mathbf{e}^*(x_0, r, \nu) \\ &\le \alpha_1^{2\gamma} \mathbf{e}^*(x_0, r, \nu), \end{aligned} \tag{26.43}$$

where in the last inequality we have exploited the definition of α_1. This concludes the proof of (26.39). Concerning the proof of (26.40), as already noticed, we may directly assume that (26.42), and thus (26.43), hold true. Arguing as in the proof of Lemma 26.2, we easily find that

$$|\nu - \nu_0|^2 \le C(n)\left(\mathbf{e}(x_0, \alpha_1 r, \nu_0) + \frac{\mathbf{e}(x_0, r, \nu)}{\alpha_1^{n-1}}\right), \tag{26.44}$$

which, combined with (26.43), immediately implies (26.40), with

$$C_6(n, \gamma) = C(n)\left(\alpha_1^{2\gamma} + \frac{1}{\alpha_1^{n-1}}\right).$$

Here, of course, $C(n)$ is the constant appearing in (26.44). □

Proof of Theorem 26.3 The proof is almost identical to that of Theorem 26.1, and we limit ourselves to briefly reviewing the argument. The main difference is that, in the iteration argument, we exploit (26.39) and (26.40) in place of (26.7) and (26.8). Precisely, we define $\mathbf{e}^*(x, s, \nu)$ as in (26.37), and set $\mathbf{e}_n^*(x, s) = \mathbf{e}^*(x, s, e_n)$. If ε_6 is suitably small, then (26.31) implies

$$\mathbf{e}_n^*(x_0, 9r) \le \left(\frac{8}{9}\right)^{n-1} \varepsilon_5(n, \gamma),$$

where $\varepsilon_5(n, \gamma)$ is the constant defined by Lemma 26.4. With some minor modifications, the argument of step one in the proof of Theorem 26.1 shows that for every $x \in \mathbf{C}(x_0, r) \cap \partial E$ there exists $\nu(x) \in S^{n-1}$ such that

$$\mathbf{e}^*(x, s, \nu(x)) \le C_7 \left(\frac{s}{r}\right)^{2\gamma} \mathbf{e}_n^*(x_0, 9\,r), \quad \forall s \in (0, 4\,r), \tag{26.45}$$

$$|\nu(x) - e_n|^2 \le C_7\, \mathbf{e}_n^*(x_0, 9\,r), \tag{26.46}$$

$$\mathbf{e}_n^*(x, s) \le C_7\, \mathbf{e}_n^*(x_0, 9\,r), \quad \forall s \in (0, 8\,r), \tag{26.47}$$

where $C_7 = C_7(n, \gamma)$. Thanks to (26.47), and, possibly, by decreasing the value of ε_6, one obtains that

$$M_0 = \left\{x \in \mathbf{C}(x_0, r) \cap \partial E : \sup_{0<s<8\,r} \mathbf{e}_n(x, s) \le \delta_0(n)\right\} = \mathbf{C}(x_0, r) \cap \partial E\,.$$

Requiring $\varepsilon_6(n, \gamma) \le \varepsilon_1(n)$, the Lipschitz approximation theorem guarantees the existence of a Lipschitz function $u\colon \mathbb{R}^{n-1} \to \mathbb{R}$ with $\mathrm{Lip}(u) \le 1$ such that (26.32), (26.33), and (26.34) hold true. Finally, (26.35) and (26.36) are deduced by Campanato's criterion and starting from (26.45) and (26.45), by the same argument used in step three of the proof of Theorem 26.1. □

26.3 $C^{1,\gamma}$-regularity of the reduced boundary, and the characterization of the singular set

The following theorem provides a particularly useful characterization of the singular set of a (Λ, r_0)-perimeter minimizer.

Theorem 26.5 (Regularity of the reduced boundary and the singular set) *If A is an open set in $\mathbb{R}^n$, $n \geq 2$, and E is a (Λ, r_0)-perimeter minimizer in A, with $\Lambda r_0 \leq 1$, then $A \cap \partial^* E$ is a $C^{1,\gamma}$-hypersurface for every $\gamma \in (0, 1/2)$ that is relatively open in $A \cap \partial E$, and it is $\mathcal{H}^{n-1}$-equivalent to $A \cap \partial E$. Moreover, there exists a positive constant $\varepsilon(n)$, which depends on the dimension n only, such that the singular set $\Sigma(E; A)$ of E in A,*

$$\Sigma(E; A) = A \cap (\partial E \setminus \partial^* E),$$

is characterized in terms of the spherical excess as follows:

$$\Sigma(E; A) = \Big\{ x \in A \cap \partial E : \inf_{0<r<r_0, B(x,r) \subset\subset A} \Big(\mathbf{e}(E, x, r) + \Lambda r \Big) \geq \varepsilon(n) \Big\}. \tag{26.48}$$

Proof If $\{\varepsilon_6(n, \gamma)\}_{0<\gamma<1/2}$ are the constants appearing in Theorem 26.3, then

$$\varepsilon(n) = \sup_{\gamma \in (0,1/2)} \varepsilon_6(n, \gamma) \tag{26.49}$$

is positive. Let us now consider the subset of $A \cap \partial E$ defined as

$$\Sigma = \Big\{ x \in A \cap \partial E : \inf_{0<r<r_0, B(x,r) \subset\subset A} \Big(\mathbf{e}(E, x, r) + \Lambda r \Big) \geq \varepsilon(n) \Big\}.$$

By Proposition 22.3 we find that

$$\lim_{r \to 0^+} \mathbf{e}(E, x, r) + \Lambda r = 0, \qquad \forall x \in A \cap \partial^* E,$$

so that $A \cap \partial^* E \subset (A \cap \partial E) \setminus \Sigma$. If now $x \in (A \cap \partial E) \setminus \Sigma$, then, by construction, there exist $r \in (0, r_0/9)$ with $B(x, r) \subset\subset A$ and $\gamma \in (0, 1/2)$, such that

$$\mathbf{e}(E, x, 9r) + \Lambda r < \varepsilon_6(n, \gamma).$$

As a consequence, by Proposition 22.3, there exists $\nu \in S^{n-1}$ such that

$$\mathbf{e}(E, x, 9r, \nu) + \Lambda r < \varepsilon_6(n, \gamma).$$

By Theorem 26.3, $\mathbf{C}(x, r, \nu) \cap \partial E$ is the $(n-1)$-dimensional graph of a function u of class $C^{1,\gamma}$, with (26.34) in force: in particular, it must be that $x \in \partial^* E$, $\mathbf{C}(x, r, \nu) \cap \partial E = \mathbf{C}(x, r, \nu) \cap \partial^* E$, and

$$A \cap \partial^* E = (A \cap \partial E) \setminus \Sigma,$$

that is $\Sigma = \Sigma(E; A)$. We have thus completed the proof of the theorem. □

26.4 C^1-convergence for sequences of (Λ, r_0)-perimeter minimizers

As a further application of the estimates in Theorem 26.3, we now show that the convergence of regular points of perimeter minimizers to a regular point of the limit set forces the convergence of the corresponding outer unit normals.

Theorem 26.6 (Convergence of outer unit normals) *If $\{E_h\}_{h\in\mathbb{N}}$ and E are (Λ, r_0)-perimeter minimizers in the open set $A \subset \mathbb{R}^n$ with $\Lambda r_0 \le 1$ and*

$$E_h \overset{\mathrm{loc}}{\to} E\,, \qquad x_h \in A \cap \partial E_h\,, \qquad x \in A \cap \partial^* E\,, \qquad \lim_{h\to\infty} x_h = x\,,$$

then, for h large enough, $x_h \in A \cap \partial^* E_h$. Moreover,

$$\lim_{h\to\infty} \nu_{E_h}(x_h) = \nu_E(x)\,.$$

Proof Since $x_h \to x$, up considering h large enough, to replace E_h with $E_h + (x - x_h)$, and A with $\{x \in A : \mathrm{dist}(x, \partial A) > \delta\}$ for some suitably small $\delta > 0$, we may directly assume that $x_h = x$ for every $h \in \mathbb{N}$. Since $x \in A \cap \partial^* E$, given $\gamma \in (0, 1/2)$, by Proposition 22.3 there exists $r > 0$ and $\nu \in S^{n-1}$ such that $9r < r_0$, $\mathbf{C}(x, 9\,r, \nu) \subset\subset A$, and

$$\mathbf{e}(E, x, 9\,r, \nu) + \Lambda\, r < \varepsilon_6(n, \gamma)\,, \qquad \mathcal{H}^{n-1}\Big(\partial^* E \cap \partial\mathbf{C}(x, r, \nu)\Big) = 0\,,$$

where $\varepsilon_6(n, \gamma)$ is the constant appearing in Theorem 26.3. Up to a common rotation of E and of all the E_h, we may assume that $\nu = e_n$. Thus,

$$\mathbf{e}_n(E, x, 9\,r) + \Lambda\, r < \varepsilon_6(n, \gamma)\,, \qquad \mathcal{H}^{n-1}\Big(\partial^* E \cap \partial\mathbf{C}(x, r)\Big) = 0\,.$$

By the continuity of the cylindrical excess (Proposition 22.6),

$$\mathbf{e}_n(E_h, x, 9\,r) + \Lambda\, r < \varepsilon_6(n, \gamma)\,,$$

provided h is large enough. In particular, by the characterization of singular sets in Theorem 26.5 and definition (26.49) of $\varepsilon(n)$, it turns out that $x \in A \cap \partial^* E_h$ for every h large enough. By Theorem 26.3 there exist Lipschitz functions $u, u_h : \mathbf{D}(\mathbf{p}x, r) \to \mathbb{R}$ with $\mathrm{Lip}(u), \mathrm{Lip}(u_h) \le 1$, such that

$$\mathbf{C}(x, r) \cap E = \Big\{(z, t) : z \in \mathbf{D}(\mathbf{p}x, r)\,, -r < t < u(z)\Big\}\,, \tag{26.50}$$

$$\mathbf{C}(x, r) \cap E_h = \Big\{(z, t) : z \in \mathbf{D}(\mathbf{p}x, r)\,, -r < t < u_h(z)\Big\}\,, \tag{26.51}$$

and such that, for every $z, z' \in \mathbf{D}(\mathbf{p}x, r)$, and $h \in \mathbb{N}$,

$$|\nabla' u_h(z) - \nabla' u_h(z')| \le C(n, \gamma) \left(\frac{|z - z'|}{r}\right)^{\gamma}. \tag{26.52}$$

Thanks to (26.50) and (26.51), we have

$$\begin{aligned} &\int_{\mathbf{D}(\mathbf{p}x,r)} |u_h - u| = \Big|\Big(E_h \Delta E\Big) \cap \mathbf{C}(x,r)\Big| \to 0\,, \qquad \text{as } h \to \infty\,, \\ &\text{so that} \qquad \int_{\mathbf{D}} \varphi \nabla' u = \lim_{h\to\infty} \int_{\mathbf{D}} \varphi \nabla' u_h\,, \quad \forall \varphi \in C^0_c(\mathbf{D})\,. \end{aligned} \tag{26.53}$$

Now, $\{\nabla' u_h\}_{h\in\mathbb{N}}$ is equi-continuous by (26.52), and bounded by $\mathrm{Lip}(u_h) \le 1$; thus it is compact in the uniform convergence by the Ascoli–Arzelá theorem. By (26.53), $\nabla' u$ is its only possible limit point. Hence, $\nabla' u_h \to \nabla' u$ uniformly on $\mathbf{D}(\mathbf{p}x, r)$, and thus, by (26.27), $\nu_{E_h}(x) \to \nu_E(x)$. □

27

Higher regularity theorems

When dealing with local perimeter minimizers, volume-constrained perimeter minimizers, and minimizers in prescribed mean curvature problems, the $C^{1,\gamma}$-regularity theory from the previous chapters provides only preliminary information on the actual degree of regularity of reduced boundaries. In Section 27.2 we prove some higher regularity theorems, which are based on the fruitful connection between Euler–Lagrange equations for variational integrals and elliptic equations in divergence form presented in Section 27.1.

27.1 Elliptic equations for derivatives of Lipschitz minimizers

A convex function $f \in C^2(\mathbb{R}^n)$ is called **locally uniformly convex** if for every $R > 0$ there exists $\lambda(R) > 0$ such that

$$\left(\nabla^2 f(\xi)e\right) \cdot e \geq \lambda(R)|e|^2\,, \qquad \forall e, \xi \in \mathbb{R}^n\,, |\xi| \leq R\,. \tag{27.1}$$

This is the case of the area integrand $f(\xi) = \sqrt{1 + |\xi|^2}$, as $\nabla^2 f(\xi) = M(\xi)$ with

$$M(\xi) = \frac{1}{\sqrt{1 + |\xi|^2}}\left(\mathrm{Id} - \frac{\xi \otimes \xi}{1 + |\xi|^2}\right), \qquad \xi \in \mathbb{R}^n\,, \tag{27.2}$$

and $(M(\xi)e) \cdot e \geq (1 + R^2)^{-3/2}|e|^2$ for every $|\xi| \leq R$ and $e \in \mathbb{R}^n$. As turns out, the regularity of local $C^{1,\gamma}$ minimizers of an integral functional $\mathcal{F}(u; B) = \int_B f(\nabla u)$, defined by a locally uniformly convex integrand f, can be investigated through the classical Schauder theory for second order elliptic equations. The starting point here is the fact, proved in Theorem 23.4, that u is a solution to the weak Euler–Lagrange equation associated with f,

$$\int_B \nabla f(\nabla u) \cdot \nabla\varphi = 0\,, \qquad \forall \varphi \in C^\infty_c(B)\,. \tag{27.3}$$

Recall that, if u is twice differentiable, then (27.3) takes the form

$$-\mathrm{div}\Big(\nabla f(\nabla u)\Big) = 0 \qquad \text{on } B. \tag{27.4}$$

In turn, if both f and u are smooth, then we can differentiate in the x_i direction the non-linear PDE (27.4), commute div and ∂_i, and find

$$-\mathrm{div}\Big(\nabla^2 f(\nabla u)\nabla(\partial_i u)\Big) = 0 \qquad \text{on } B. \tag{27.5}$$

This apparently complicated PDE has in fact a nice structure. Indeed, if we set

$$v(x) = \partial_i u(x), \qquad A(x) = \nabla^2 f(\nabla u(x)), \qquad x \in B,$$

then (27.5) is a second order elliptic equation in divergence form,

$$-\mathrm{div}\,(A\nabla v) = 0 \qquad \text{on } B, \tag{27.6}$$

for the directional derivatives of u. In fact, by the so-called *difference quotients method*, we can perform this argument even if u is merely Lipschitz, showing that the validity of the Euler–Lagrange equation in weak form implies the existence of $v = \partial_i u$ as a distributional derivative in L^2 which solves the weak form of (27.6); see, e.g., [GM05, Proposition 8.6] for the simple proof.

Theorem 27.1 (Elliptic equations for directional derivatives) *Let $f \in C^2(\mathbb{R}^n)$ be locally uniformly convex, let $a \in C^2(B \times \mathbb{R})$, $a = a(x, s)$, $(x, s) \in B \times \mathbb{R}$, and consider the functional $\mathcal{F}_a\colon \mathrm{Lip}(B) \to \mathbb{R}$ defined as*

$$\mathcal{F}_a(u; B) = \int_B f(\nabla u) - a(x, u), \qquad u \in \mathrm{Lip}(B). \tag{27.7}$$

(i) If u is a Lipschitz local minimizer u of $\mathcal{F}_a$ in B, then u solves the weak Euler–Lagrange equation

$$\int_B \nabla f(\nabla u)\cdot\nabla\varphi = \int_B \varphi\,\frac{\partial a}{\partial s}(x, u), \qquad \forall \varphi \in C^\infty_c(B). \tag{27.8}$$

(ii) If u is a Lipschitz function solving (27.8), then $u \in W^{2,2}_{\mathrm{loc}}(B)$, and, for every $i = 1, \ldots, n$, the distributional directional derivative $v = \partial_i u$ of u satisfies the elliptic equation in divergence form:

$$\int_B A\nabla v\cdot\nabla\varphi = \int_B \varphi\,(b\,v + c), \qquad \forall\varphi \in C^\infty_c(B), \tag{27.9}$$

*associated with the elliptic tensor field $A\colon B \to$ **Sym**(n) and the continuous functions $b, c\colon B \to \mathbb{R}$ defined, at $x \in B$, as*

$$A(x) = \nabla^2 f(\nabla u(x)), \qquad b(x) = \frac{\partial^2 a}{\partial s^2}(x, u(x)), \qquad c(x) = \frac{\partial^2 a}{\partial x_i \partial s}(x, u(x)).$$

If we apply this theorem to a $C^{1,\gamma}$ local minimizer u of $\mathcal{F}_a$, then the elliptic tensor field A will be of class $C^{0,\gamma}$. In this situation, the following classical result from Schauder's theory for second order elliptic equations will suffice to obtain several higher regularity results; see, e.g., [GM05, Theorems 5.17 and 5.18].

Theorem 27.2 *If $k \in \mathbb{N}$, $\gamma \in (0,1)$, $A \in C^{k,\gamma}_{\rm loc}(B;\mathbf{Sym}(n))$ is an elliptic tensor field in B, $T \in C^{k,\gamma}_{\rm loc}(B;\mathbb{R}^n)$, and $v \in W^{1,2}_{\rm loc}(B)$ is a weak solution of the second order elliptic equation* $\operatorname{div}(A\nabla v) = \operatorname{div} T$ *in B, that is, if*

$$\int_B (A\nabla v)\cdot\nabla\varphi = \int_B T\cdot\nabla\varphi\,, \qquad \forall\varphi\in C^\infty_c(B)\,, \tag{27.10}$$

then $v \in C^{k+1,\gamma}_{\rm loc}(B)$.

27.2 Some higher regularity theorems

Theorems 27.1 and 27.2 lead to higher regularity results for minimizers in Plateau-type problems (Theorem 27.3), relative isoperimetric problems (Theorem 27.4), and prescribed mean curvature problems (Theorem 27.5).

Theorem 27.3 *If E is a local perimeter minimizer in the open set $A \subset \mathbb{R}^n$, then $A \cap \partial^* E$ is an analytic vanishing mean curvature hypersurface.*

Proof The proof is based on the so-called *bootstrap argument*. By Theorems 26.1 and 26.5, for every $x \in A\cap\partial^*E$ and $\gamma\in(0,1)$, there exist $r>0$ and $u\colon\mathbb{R}^{n-1}\to\mathbb{R}$, with $\operatorname{Lip}(u)\le 1$ and $u\in C^{1,\gamma}(\mathbf{D}_r)$, such that, up to rotation,

$$\mathbf{C}(x,r)\cap\partial E = x+\Big\{(z,u(z)) : z\in\mathbf{D}_r\Big\}. \tag{27.11}$$

By Proposition 23.3, u is a local minimizer of the area functional. By Theorem 27.1, $u \in W^{2,2}_{\rm loc}(\mathbf{D}_r)$ and, if $v = \partial_i u$ denotes the ith derivative of u, then v is a weak solution of $-\operatorname{div}(A\nabla' v) = 0$ in $\mathbf{D}_r$ with $A = M\circ\nabla' u$ and M as in (27.2). Since $u\in C^{1,\gamma}(\mathbf{D}_r)$ and M is smooth, it follows that $A\in C^{0,\gamma}(\mathbf{D}_r)$. By Theorem 27.2, $v\in C^{1,\gamma}_{\rm loc}(\mathbf{D}_r)$. By the arbitrariness of i, $u\in C^{2,\gamma}_{\rm loc}(\mathbf{D}_r)$, and thus $A\in C^{1,\gamma}_{\rm loc}(\mathbf{D}_r)$. Iterating the use of Theorem 27.2, $u\in C^\infty(\mathbf{D}_r)$. Thus u is a smooth solution of the minimal surfaces equation

$$-\operatorname{div}'\left(\frac{\nabla' u}{\sqrt{1+|\nabla' u|^2}}\right) = 0 \qquad \text{on } \mathbf{D}_r;$$

therefore, it is analytic; see, e.g., [Mor66, KS00]. □

Theorem 27.4 *If E is a volume-constrained perimeter minimizer in $A \subset \mathbb{R}^n$, then $A \cap \partial^* E$ is an analytic constant mean curvature hypersurface.*

Proof By Example 21.3, there exist Λ and r_0 positive such that

$$P(E; B(x,r)) \le P(F; B(x,r)) + \Lambda \Big| |E| - |F| \Big| , \qquad (27.12)$$

whenever $E \Delta F \subset\subset B(x,r) \cap A$ and $r < r_0$. In particular, E is a (Λ, r_0)-perimeter minimizer in A, and thus by Theorems 26.3 and 26.5, for every $\gamma \in (0, 1/2)$ and $x \in A \cap \partial^* E$, there exist $r > 0$ and $u \colon \mathbb{R}^{n-1} \to \mathbb{R}$ with $\mathrm{Lip}(u) \le 1$ and $u \in C^{1,\gamma}(\mathbf{D}_r)$, such that, up to rotation, (27.11) holds true. By a straightforward adaptation of the proof of Proposition 23.3, we deduce from (27.12) that, given a compact set $K \subset \mathbf{D}_r$, there exists $\varepsilon > 0$ such that

$$\int_{\mathbf{D}_r} \sqrt{1 + |\nabla' u|^2} \le \int_{\mathbf{D}_r} \sqrt{1 + |\nabla'(u + \varphi)|^2} ,$$

whenever $\varphi \in C^\infty_c(\mathbf{D}_r)$, $\mathrm{spt}\, \varphi \subset K$, $\sup_{\mathbf{D}_r} |\varphi| \le \varepsilon$, and $\int_{\mathbf{D}_r} \varphi = 0$. Arguing as in the proof of Theorem 23.4, we thus find

$$\int_{\mathbf{D}_r} \frac{\nabla' u}{\sqrt{1 + |\nabla' u|^2}} \cdot \nabla' \varphi = 0 , \qquad \forall \varphi \in C^\infty_c(\mathbf{D}_r) \quad \text{with} \quad \int_{\mathbf{D}_r} \varphi = 0 . \quad (27.13)$$

If $\varphi_1, \varphi_2 \in C^\infty_c(\mathbf{D}_r)$ with $\int_{\mathbf{D}_r} \varphi_i \ne 0$ $(i = 1, 2)$, then

$$\varphi = \frac{\varphi_1}{\int_{\mathbf{D}_r} \varphi_1} - \frac{\varphi_2}{\int_{\mathbf{D}_r} \varphi_2}$$

is an admissible test function in (27.13). Writing down the resulting identity, and by the arbitrariness of φ_1 and φ_2, we conclude that, for some $c \in \mathbb{R}$,

$$\int_{\mathbf{D}_r} \frac{\nabla' u \cdot \nabla' \varphi}{\sqrt{1 + |\nabla' u|^2}} = c \int_{\mathbf{D}_r} \varphi , \qquad \forall \varphi \in C^\infty_c(\mathbf{D}_r) . \qquad (27.14)$$

Thus, u is a $C^{1,\gamma}$ solution to (27.8) with f the area integrand on $\mathbb{R}^{n-1}$ and $a(z, s) = c\, s$, $(z, s) \in \mathbb{R}^{n-1} \times \mathbb{R}$. By Theorem 27.1, $u \in W^{2,2}_{\mathrm{loc}}(\mathbf{D}_r)$, and every directional derivative v of u solves $-\mathrm{div}\,(A \nabla' v) = 0$ in weak form on $\mathbf{D}_r$, with $A = M \circ \nabla' u$ and M as in (27.2). By the bootstrap argument, u is smooth in $\mathbf{D}_r$, and (27.14) holds in strong form, that is

$$-\,\mathrm{div} \left(\frac{\nabla' u}{\sqrt{1 + |\nabla' u|^2}} \right) = c \qquad \text{on } \mathbf{D}_r .$$

Again by [KS00, Mor66], u is analytic on $\mathbf{D}_r$. □

Theorem 27.5 *If $\alpha \in (0, 1)$ and E is a minimizer in the prescribed mean curvature problem (12.32) defined by an open set A and a potential $g \in C^{2,\alpha}(\mathbb{R}^n)$,* *then $A \cap \partial^* E$ is a $C^{2,\beta}$-hypersurface for some $\beta \in (0, 1/2)$ and, in particular, the mean curvature of $A \cap \partial^* E$ is defined in the classical sense and satisfies $H_E = -g$ on $A \cap \partial^* E$.*

Proof By Example 21.2, E is a (Λ, r_0)-perimeter minimizer in A. By Theorems 26.3 and 26.5, for every $\gamma \in (0, 1/2)$ and $x \in A \cap \partial^* E$ there exist $r > 0$ and a function $u\colon \mathbb{R}^{n-1} \to \mathbb{R}$ with $\mathrm{Lip}(u) \le 1$ and $u \in C^{1,\gamma}(\mathbf{D}_r)$ such that, up to rotation, (27.11) holds true. If we now define $a\colon \mathbb{R}^{n-1} \times \mathbb{R} \to \mathbb{R}$ by setting

$$a(z, s) = -g(z + s\, e_n)\,, \qquad (z, s) \in \mathbb{R}^{n-1} \times \mathbb{R}\,,$$

then the argument proving Proposition 23.3 shows that u is a $C^{1,\gamma}$ local minimizer in $\mathbf{D}_r$ of the functional

$$\mathcal{F}_a(u; \mathbf{D}_r) = \int_{\mathbf{D}_r} \sqrt{1 + |\nabla' u(z)|^2} - a(z, u(z))\, dz\,.$$

By Theorem 27.1, $u \in W^{2,2}_{loc}(\mathbf{D}_r)$ and $v = \partial_i u$ satisfies

$$\int_{\mathbf{D}_r} (A \nabla' v) \cdot \nabla' \varphi = \int_{\mathbf{D}_r} \varphi\, (b\, v + c)\,, \qquad \forall \varphi \in C^\infty_c(\mathbf{D}_r)\,, \qquad (27.15)$$

where $A = M \circ \nabla' u$, M as in (27.2), and

$$b(z) = -\partial^2_{n\,n} g(z + u(z)\, e_n)\,, \qquad c(z) = -\partial^2_{n\,i} g(z + u(z)\, e_n)\,, \qquad \forall z \in \mathbf{D}_r\,.$$

Since u is Lipschitz and $\nabla^2 g \in C^{0,\alpha}(\mathbb{R}^n; \mathbf{Sym}(n))$, $b\, v + c \in C^{0,\alpha}(\mathbf{D}_r)$. By Potential Theory for the Poisson equation we prove the existence of $w \in C^{2,\alpha}_{\mathrm{loc}}(\mathbf{D}_r)$ such that

$$-\Delta' w = b\, v + c \qquad \text{on } \mathbf{D}_r\,;$$

see, e.g., [GT98, Chapter 4]. Setting $T = \nabla' w$, (27.15) takes the form

$$\int_{\mathbf{D}_r} (A \nabla' v) \cdot \nabla' \varphi = \int_{\mathbf{D}_r} T \cdot \nabla' \varphi\,, \qquad \forall \varphi \in C^\infty_c(\mathbf{D}_r)\,,$$

where $T \in C^{1,\alpha}_{\mathrm{loc}}(\mathbf{D}_r)$ and $A = M \circ \nabla' u \in C^{0,\gamma}(\mathbf{D}_r)$. By Theorem 27.2, $v \in C^{1,\beta}_{\mathrm{loc}}(\mathbf{D}_r)$, where $\beta = \min\{\alpha, \gamma\}$. In conclusion, $A \cap \partial^* E$ is a hypersurface of class $C^{2,\gamma\alpha}$ and, by Exercise 17.10, $H_E = -g$ on $A \cap \partial^* E$. □

Exercise 27.6 Discuss the regularity of minimizers in the prescribed mean curvature problem correspondingly to increasingly regular potentials g, and that of p-Cheeger sets; see Exercises 14.7, 17.12, and 21.5.

28

Analysis of singularities

In Theorem 26.5, we have proved the $C^{1,\gamma}$-regularity of the reduced boundaries $A \cap \partial^* E$ of any (Λ, r_0)-perimeter minimizers E in some open set A. This chapter is devoted to the study of the singular set

$$\Sigma(E; A) = A \cap (\partial E \setminus \partial^* E),$$

aiming to provide estimates on its possible size. From the density estimates of Theorem 21.11, we already know that $\mathcal{H}^{n-1}(\Sigma(E; A)) = 0$. This result can be largely strengthened, as explained in the following theorem.

Theorem 28.1 (Dimensional estimates of singular sets of (Λ, r_0)-perimeter minimizers) *If E is a (Λ, r_0)-perimeter minimizer in the open set $A \subset \mathbb{R}^n$, $n \geq 2$, with $\Lambda r_0 \leq 1$, then the following statements hold true:*

(i) if $2 \leq n \leq 7$, then $\Sigma(E; A)$ is empty;
(ii) if $n = 8$, then $\Sigma(E; A)$ has no accumulation points in A;
(iii) if $n \geq 9$, then $\mathcal{H}^s(\Sigma(E; A)) = 0$ for every $s > n - 8$.

There exists a perimeter minimizer E in $\mathbb{R}^8$ with $\mathcal{H}^0(\Sigma(E; \mathbb{R}^8)) = 1$. If $n \geq 9$, then there exists a perimeter minimizer E in $\mathbb{R}^n$ with $\mathcal{H}^{n-8}(\Sigma(E; \mathbb{R}^n)) = \infty$.

Let us now sketch the proof of this deep theorem. We start by looking at the blow-ups $E_{x,r}$ of a (Λ, r_0)-minimizer E at a singular point x. In Theorem 28.6, we show that, loosely speaking, the $E_{x,r}$ converge to a cone K with vertex at 0. This "tangent" cone is not a half-space, as it is necessarily singular at 0 (by construction, every $E_{x,r}$ is singular at 0). Moreover, K is a perimeter minimizer in $\mathbb{R}^n$. Thus the existence of singular points for (Λ, r_0)-minimizers is equivalent to the existence of cones, which are not half-spaces, but are perimeter minimizers in $\mathbb{R}^n$. The existence of such *singular (perimeter) minimizing cones* is, evidently, questionable. One starts examining the possibility of singular minimizing cones possessing only a point of singularity, namely, the vertex. Under this simplifying assumption, through some direct computations, it is shown that no singular minimizing cone exists in $\mathbb{R}^n$ provided $n \le 7$. This is a celebrated theorem by Simons, whose proof is discussed in Section 28.3. Simons' theorem is sharp. Precisely, in Theorem 28.15, we exhibit a singular minimizing cone with a point vertex singularity in $\mathbb{R}^8$, known as the Simons cone. The proof of Theorem 28.1 is then closed by the so-called *Federer's dimension reduction argument*, presented in Theorem 28.11, Section 28.4: if a singular minimizing cone K exists in $\mathbb{R}^n$ and the singular set of K contains other points than its vertex, then there exists a singular minimizing cone K' in $\mathbb{R}^{n-1}$. Finally, in Section 28.5 we combine Federer's argument and Simons' theorem to prove Theorem 28.1. We close these introductory considerations by reformulating Theorem 26.6, see Theorem 28.2, and by noticing how the conclusions of Theorem 28.1 can usually be strengthened when combined with symmetry assumptions, see Remark 28.3.

Theorem 28.2 (Closure and local uniform convergence of singularities) *If* $\{E_h\}_{h\in\mathbb{N}}$, E *are* (Λ, r_0)*-perimeter minimizers in* $A \subset \mathbb{R}^n$, $\Lambda r_0 \le 1$ *and*

$$E_h \stackrel{\mathrm{loc}}{\to} E\,, \qquad x_h \in \Sigma(E_h; A)\,, \qquad x \in A \cap \partial E\,, \qquad \lim_{h\to\infty} x_h = x\,,$$

then $x \in \Sigma(E; A)$. *Moreover, given* $\varepsilon > 0$ *and* $H \subset A$ *compact, there exists* $h_0 \in \mathbb{N}$ *such that*

$$\Sigma(E_h; A) \cap H \subset I_\varepsilon\big(\Sigma(E; A) \cap H\big)\,, \qquad \forall h \ge h_0\,. \tag{28.1}$$

Proof We must have $x \in \Sigma(E; A)$, for otherwise we would obtain a contradiction thanks to Theorem 26.6. We also prove (28.1) by contradiction. Indeed, let us assume the existence of $\varepsilon > 0$, $H \subset A$ compact, $h(k) \to \infty$ as $k \to \infty$, $y_k \in \Sigma(E_{h(k)}; A) \cap H$ for every $k \in \mathbb{N}$, with the property that

$$\mathrm{dist}\big(y_k, \Sigma(E; A) \cap H\big) \ge \varepsilon\,.$$

Since $\{y_k\}_{k\in\mathbb{N}} \subset H$, up to extracting a not-relabeled subsequence, we may assume that $y_k \to y$, with $y \in H \subset A$. By the first part of the theorem, it must be that $y \in \Sigma(E; A) \cap H$. In particular, there exists $k_0 \in \mathbb{N}$ such that $y_k \in B(y, \varepsilon) \subset I_\varepsilon(\Sigma(E; A) \cap H)$ for every $k \ge k_0$, a contradiction. □

Remark 28.3 Generally speaking, Theorem 28.1 can be strengthened when dealing with (Λ, r_0)-perimeter minimizers which possess a certain degree of symmetry. For example, let E be a (Λ, r_0)-perimeter minimizer in the open set $A \subset \mathbb{R}^n$, and assume that E is axially symmetric with respect to a line π, meaning that $E = Q(E)$ for every $Q \in \mathbf{O}(n)$ such that $Q(\pi) = \pi$. We claim that, in this case, $\Sigma(E; A) \setminus \{\pi\} = \emptyset$. Indeed, the axial symmetry of E implies that ∂E and $\partial^* E$, and thus $\Sigma(E;A)$, are axially symmetric with respect to π. In particular, if $x \in \Sigma(E;A) \setminus \{\pi\}$, then $\Sigma(E;A)$ contains the set

$$C = \left\{Qx : Q \in \mathbf{O}(n)\,, Q(\pi) = \pi\right\}.$$

Since $x \notin \pi$, we have that $\mathcal{H}^{n-2}(C) > 0$. As a consequence, $\dim(\Sigma(E;A)) \geq n-2$, a contradiction to Theorem 28.1.

28.1 Existence of densities at singular points

As a consequence of the monotonicity inequality Theorem 17.16, in Corollary 17.18, it was shown that the $(n-1)$-dimensional density of a stationary set exists at every point of its topological boundary. In fact, by the slicing theory introduced in Chapter 18 (precisely, through a slicing by spheres argument) we now prove an analogous property for (Λ, r_0)-perimeter minimizers.

Theorem 28.4 (Existence of densities at singular points) *If E is a (Λ, r_0)-perimeter minimizer in the open set A and $x_0 \in A \cap \partial E$, then the function*

$$\frac{P(E;B(x_0,r))}{r^{n-1}} + \Lambda\,\omega_n\,r$$

is increasing on $r < \min\{r_0\,, \mathrm{dist}(x_0, \partial A)\}$. In particular,

$$\theta_{n-1}(\partial E)(x_0) = \lim_{r\to 0^+} \frac{P(E;B(x_0,r))}{\omega_{n-1}\,r^{n-1}} \quad \textit{exists and it is finite}.$$

Proof Let us set $d = \min\{r_0\,, \mathrm{dist}(x_0, \partial A)\}$. Without loss of generality, we let $x_0 = 0$. We introduce the increasing function $\Phi\colon (0,d) \to [0,\infty)$ defined by

$$\Phi(r) = P(E;B_r)\,, \qquad r \in (0,d)\,.$$

Our assertions will follow easily from the inequality

$$\Phi(r) \le \frac{r\,\Phi'(r)}{n-1} + \Lambda\omega_n\,r^n\,, \qquad \text{for a.e. } r \in (0,d)\,, \tag{28.2}$$

which we now prove by the construction of suitable comparison sets.

Step one: Let F be an open set with smooth boundary in $\mathbb{R}^n$. For every $r > 0$, the set $F \cap \partial B_r$ is relatively open in ∂B_r. By Sard's lemma, $\partial F \cap \partial B_r$ is a smooth $(n-2)$-dimensional surface for a.e. $r > 0$. We now consider the cone $F(r)$, with vertex at 0 and passing through $F \cap \partial B_r$, that is we set

$$F(r) = \left\{\lambda\,y\colon \lambda > 0\,, y \in F \cap \partial B_r\right\}; \tag{28.3}$$

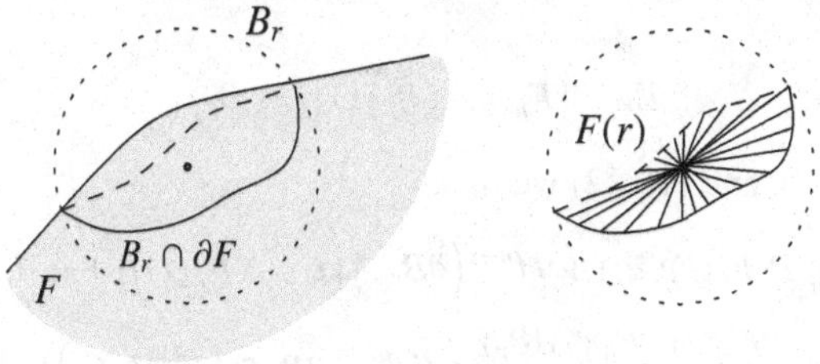

Figure 28.1 The set $F(r)$ is the cone with vertex at 0 spanned by $F \cap \partial B_r$. The relative perimeter of $F(r)$ in B_r is the $\mathcal{H}^{n-1}$-measure of the cone-like surface $\partial F(r) \cap \partial B_r$, that can be computed in terms of $\mathcal{H}^{n-2}(\partial F \cap \partial B_r)$ thanks to the coarea formula (18.35). When F is a perimeter minimizer, the inequality $\mathcal{H}^{n-1}(B_r \cap \partial F) \le \mathcal{H}^{n-1}(B_r \cap \partial F(r))$ leads us to establish the monotonicity of $r^{1-n}\,\mathcal{H}^{n-1}(B_r \cap \partial F)$.

see Figure 28.1. Arguing as in step one of the proof of Lemma 24.6, we see that for a.e. $r > 0$, the cone $F(r)$ is of locally finite perimeter in $\mathbb{R}^n$, with

$$\mu_{F(r)} = \nu_{F(r)}\,\mathcal{H}^{n-1} \llcorner \partial F(r)\,,$$
$$\nu_{F(r)}(x) \cdot x = 0\,, \qquad \forall x \in \partial F(r) \setminus \{0\}\,. \tag{28.4}$$

Let us apply (18.35) with $u(x) = |x|$ ($x \in \mathbb{R}^n$) and to the $\mathcal{H}^{n-1}$-rectifiable set $M = \partial F(r)$. By (28.4) we have $\nabla^M u = \nabla u$, so that $|\nabla^M u| = 1$ and

$$\int_0^r \mathcal{H}^{n-2}\big(\partial F(r) \cap \partial B_s\big)\,\mathrm{d}s = \int_{B_r \cap \partial F(r)} |\nabla^M u|\,\mathrm{d}\mathcal{H}^{n-1} = P(F(r); B_r)\,.$$

By construction,

$$\partial F(r) \cap \partial B_s = \Big\{\frac{t}{s}\,x : x \in \partial F \cap \partial B_r\Big\}, \qquad \forall t > 0\,,$$

and we thus find the useful formula

$$P(F(r); B_r) = \int_0^r \Big(\frac{s}{r}\Big)^{n-2} \mathcal{H}^{n-2}\big(\partial F \cap \partial B_r\big)\,\mathrm{d}s = \frac{r\,\mathcal{H}^{n-2}\big(\partial F \cap \partial B_r\big)}{n-1}\,. \tag{28.5}$$

Step two: Let $\{F_h\}_{h\in\mathbb{N}}$ be a sequence of open sets with smooth boundary, such that $F_h \overset{\mathrm{loc}}{\to} E$ and $|\mu_{F_h}| \overset{*}{\rightharpoonup} |\mu_E|$. Let us consider a radius $r \in (0,d)$ such that the following two properties hold true:

$$\mathcal{H}^{n-1}\big(\partial^* E \cap \partial B_r\big) = 0\,, \tag{28.6}$$

$\partial B_r \cap \partial F_h$ is an $(n-2)$-dimensional smooth surface for every $h \in \mathbb{N}$; (28.7)

notice that by Proposition 2.16 and by the Morse–Sard Lemma (Exercise 13.16), these properties indeed hold for a.e. $r \in (0,d)$. We now define $F_h(r)$ starting from F_h and accordingly to (28.3), and then introduce the comparison sets

$$E_h = \left(F_h(r) \cap B_r\right) \cup \left(E \setminus B_r\right).$$

By (28.6), (28.7), and (16.32) we find

$$\begin{aligned} P(E; B_r) &\le P(F_h(r); B_r) + \mathcal{H}^{n-1}\left(\partial B_r \cap \left(E^{(1)} \Delta F_h(r)\right)\right) + \Lambda\, |E_h \Delta E| \\ &\le \frac{r\,\mathcal{H}^{n-2}(\partial F_h \cap \partial B_r)}{n-1} + \mathcal{H}^{n-1}\left(\partial B_r \cap \left(E^{(1)} \Delta F_h\right)\right) + \omega_n \Lambda\, r^n , \end{aligned} \tag{28.8}$$

where we have also applied (28.5). Having proved (28.8) for a.e. $r \in (0, d)$, we may integrate it on an interval $(s, t) \subset (0, d)$ to find that

$$\begin{aligned} \int_s^t P(E; B_r)\,\mathrm{d}r \le \frac{1}{n-1} \int_s^t r\,\mathcal{H}^{n-2}\left(\partial F_h \cap \partial B_r\right) \mathrm{d}r \\ + \left| (E \Delta F_h) \cap B_d \right| + \omega_n \Lambda \frac{t^{n+1} - s^{n+1}}{n+1} . \end{aligned} \tag{28.9}$$

By (18.35), since $|\nabla^{\partial F_h} u| = |\nabla u| = 1$,

$$\int_s^t r\,\mathcal{H}^{n-2}\left(\partial F_h \cap \partial B_r\right) \mathrm{d}r = \int_{\partial F_h \cap (B_t \setminus \overline{B_s})} |x| |\nabla^{\partial F_h} u|\, \mathrm{d}\mathcal{H}^{n-1} \le t\, P(F_h; \overline{B_t} \setminus B_s) .$$

If we plug this estimate into (28.9) and then let $h \to \infty$, then by $F_h \overset{\mathrm{loc}}{\to} E$ and $|\mu_{F_h}| \overset{*}{\rightharpoonup} |\mu_E|$ we find that

$$\int_s^t P(E; B_r)\,\mathrm{d}r \le \frac{t\, P(E; \overline{B_t} \setminus B_s)}{n-1} + \omega_n \Lambda \frac{t^{n+1} - s^{n+1}}{n+1} . \tag{28.10}$$

Finally, if s is a differentiability point of Φ, then we easily find (28.2) by dividing (28.10) by $t - s$, and then by letting $t \to s^+$. □

28.2 Blow-ups at singularities and tangent minimal cones

We say that $K \subset \mathbb{R}^n$ is a **cone (with vertex at** $x \in \mathbb{R}^n$**)** if

$$K = K_{x,r} = \frac{K - x}{r}, \qquad \forall r > 0 .$$

In the following sections **we always tacitly assume we are dealing with cones with vertex at** 0. If a cone K is a perimeter minimizer in $\mathbb{R}^n$ and

$$\Sigma(K) = \Sigma(K; \mathbb{R}^n) = \partial K \setminus \partial^* K \neq \emptyset ,$$

then K is a **singular minimizing cone**.

Remark 28.5 If K is a singular minimizing cone, then $0 \in \Sigma(K)$, that is, the vertex of a singular minimizing cone is always a singular point. Indeed, if this were not the case, then by Theorem 15.5, $K = K_{0,r}$ would locally converge to a half-space, that is, K would be a half-space, forcing $\Sigma(K) = \emptyset$, a contradiction. Singular minimal cones having only the vertex singularity are studied in detail in Sections 28.3 and 28.6. Singular minimal cones with larger singular sets are the object of study of Sections 28.4 and 28.5.

Theorem 28.6 (Tangent singular minimizing cones) *If E is a (Λ, r_0)-perimeter minimizer in an open set $A \subset \mathbb{R}^n$ with $\Lambda r_0 \le 1$ and*

$$x \in \Sigma(E;A), \qquad \lim_{h\to\infty} r_h = 0, \qquad E_h = E_{x,r_h},$$

then there exist a singular minimizing cone $K \subset \mathbb{R}^n$ and a sequence $h(k) \to \infty$ as $k \to \infty$ such that

$$E_{h(k)} \overset{\text{loc}}{\to} K, \qquad \mu_{E_{h(k)}} \overset{*}{\rightharpoonup} \mu_K, \qquad |\mu_{E_{h(k)}}| \overset{*}{\rightharpoonup} |\mu_K|. \tag{28.11}$$

Remark 28.7 It is not known if the tangent singular minimizing cone K of Theorem 28.6 is uniquely determined by x only. In other words, it could be that for different vanishing sequences $\{r_h\}_{h\in\mathbb{N}}$ and $\{s_h\}_{h\in\mathbb{N}}$, the sequences of blow-ups $\{E_{x,r_h}\}_{h\in\mathbb{N}}$ and $\{E_{x,s_h}\}_{h\in\mathbb{N}}$ converge to different singular minimizing cones; see [Sim83, Remark 35.2 (2)]. A result of Leonardi [Leo00, Theorem 3.9] implies in particular that if E is a (Λ, r_0)-perimeter minimizer in A, $x \in \Sigma(E;A)$, and $r^{1-n}\, P(E; B(x,r))$ is Hölder continuous as a function of $r > 0$ in a neighborhood of 0, then there exists a unique tangent singular minimizing cone to E at x.

The first technical tool in the proof of Theorem 28.6 is the following characterization of cones among sets of locally finite perimeter.

Proposition 28.8 (Characterization of cones) *If E is a set of locally finite perimeter in $\mathbb{R}^n$, then $E^{(1)}$ is a cone if and only if, for $\mathcal{H}^{n-1}$-a.e. $x \in \partial^* E$,*

$$x \cdot \nu_E(x) = 0. \tag{28.12}$$

Proof Given $\nu \in S^{n-1}$ and $r \in (0,1)$ let us consider the open cone

$$H_{\nu,r} = \Big\{\lambda z \colon \lambda > 0, z \in B(\nu, r)\Big\},$$

see Figure 28.2. By Theorem 9.6 and Example 12.7 we know that $H_{\nu,r}$ is a set of locally finite perimeter and that $x \cdot \nu_{H_{\nu,r}}(x) = 0$ for every $x \in \partial H_{\nu,r} \setminus \{0\}$. If E is a set of locally finite perimeter, then, by Theorem 16.3, $E \cap H_{\nu,r} \cap B_s$ is a set

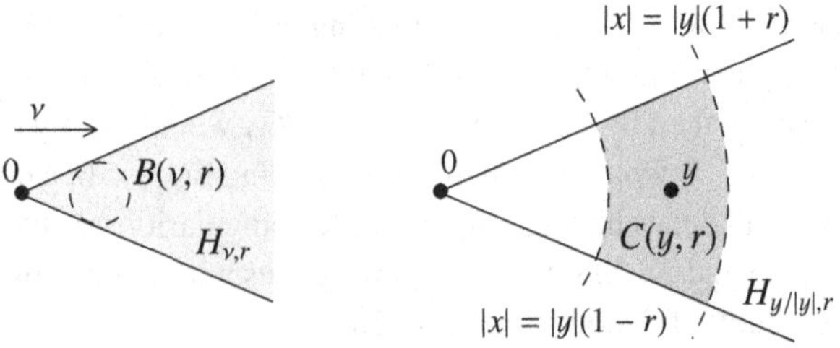

Figure 28.2 On the left, the cone $H_{\nu,r}$, which is defined starting from $\nu \in S^{n-1}$ and $r \in (0,1)$; on the right, the sector of annulus $C(y,r)$, which is defined starting from $y \in \mathbb{R}^n \setminus \{0\}$ and $r \in (0,1)$.

of finite perimeter, with

$$\mu_{E\cap H_{\nu,r}\cap B_s} = \nu_E\,\mathcal{H}^{n-1}\llcorner\Big(H_{\nu,r}\cap B_s\cap\partial^*E\Big) + \frac{x}{|x|}\,\mathcal{H}^{n-1}\llcorner\Big(E^{(1)}\cap H_{\nu,r}\cap\partial B_s\Big)$$
$$+\nu_{H_{\nu,r}}\,\mathcal{H}^{n-1}\llcorner\Big(E^{(1)}\cap B_s\cap\partial H_{\nu,r}\Big),\qquad \forall s>0\,.$$

By the divergence theorem, and since $x\cdot\nu_{H_{\nu,r}}(x)=0$ for $x\in\partial H_{\nu,r}\setminus\{0\}$,

$$n|E\cap H_{\nu,r}\cap B_s| = \int_{E\cap H_{\nu,r}\cap B_s}\operatorname{div}(x)\,\mathrm{d}x \tag{28.13}$$
$$= \int_{H_{\nu,r}\cap B_s\cap\partial^*E} x\cdot\nu_E(x)\,\mathrm{d}\mathcal{H}^{n-1}(x) + s\,\mathcal{H}^{n-1}\Big(E^{(1)}\cap H_{\nu,r}\cap\partial B_s\Big)\,.$$

At the same time, by the coarea formula (13.3), we have

$$u(s) = |E\cap H_{\nu,r}\cap B_s| = \int_0^s \mathcal{H}^{n-1}\Big(E\cap H_{\nu,r}\cap\partial B_t\Big)\,\mathrm{d}t\,,\qquad s>0\,.$$

Thus u is absolutely continuous on $(0,\infty)$ and by (28.13)

$$\frac{\mathrm{d}}{\mathrm{d}s}\frac{|E\cap H_{\nu,r}\cap B_s|}{s^n} = \frac{u'(s)}{s^n} - \frac{n\,u(s)}{s^{n+1}} = \frac{-\int_{H_{\nu,r}\cap B_s\cap\partial^*E}(x\cdot\nu_E(x))\,\mathrm{d}\mathcal{H}^{n-1}(x)}{s^{n+1}}\,, \tag{28.14}$$

for a.e. $s>0$. We now deal with the two implications separately. It is useful to introduce the notation (see Figure 28.2),

$$C(y,r) = H_{\nu,r}\cap B_{s(1+r)}\setminus B_{s(1-r)}\,,\qquad y\in\mathbb{R}^n\setminus\{0\}\,, r\in(0,1)\,,$$

where $\nu = y/|y|$ and $s = |y|$. We notice that for every $y\in\mathbb{R}^n\setminus\{0\}$ there exist positive constants $\alpha=\alpha(y)$, $\beta=\beta(y)$, and $r_0=r_0(y)$ such that

$$B(y,\alpha r)\subset C(y,r)\subset B(y,\beta r)\,,\quad \forall r\in(0,r_0)\,. \tag{28.15}$$

Step one: If $E^{(1)}$ is a cone, then $E^{(1)} \cap H_{\nu,r}$ is a cone. Hence, $s^{-n}\,u(s)$ is constant on $s \in (0,\infty)$,

$$\frac{|E \cap H_{\nu,r} \cap B_s|}{s^n} = \left|\frac{E \cap H_{\nu,r}}{s} \cap B\right| = |E \cap H_{\nu,r} \cap B|\,,$$

and thus by (28.14),

$$\int_{H_{\nu,r}\cap B_s \cap \partial^* E} x \cdot \nu_E(x)\,d\mathcal{H}^{n-1}(x) = 0\,, \quad \forall s > 0\,, r \in (0,1)\,, \nu \in S^{n-1}\,.$$

Since $C(y,r) = (H_{\nu,r} \cap B_{s(1+r)}) \setminus (H_{\nu,r} \cap B_{s(1-r)})$ (with $s = |y|$ and $\nu = y/|y|$),

$$\int_{C(y,r)\cap \partial^* E} x \cdot \nu_E(x)\,d\mathcal{H}^{n-1}(x) = 0\,, \quad \forall y \neq 0\,, r \in (0,1)\,. \tag{28.16}$$

On the other hand, if we apply the Lebesgue point theorem (Theorem 5.16) to the Radon measure $\mu = \mathcal{H}^{n-1} \llcorner \partial^* E$ and the locally bounded Borel function $f(x) = x \cdot \nu_E(x)$, then we find that for $\mathcal{H}^{n-1}$-a.e. $y \in \partial^* E$ it must be that

$$\lim_{r\to 0^+} \fint_{B(y,r)\cap\partial^* E} \Big|\big(x \cdot \nu_E(x)\big) - \big(y \cdot \nu_E(y)\big)\Big|\,d\mathcal{H}^{n-1}(x) = 0\,.$$

By (15.9) and thanks to the inclusion (28.15), we conclude that

$$\lim_{r\to 0^+} \frac{1}{r^{n-1}} \int_{C(y,r)\cap\partial^* E} \Big|\big(x \cdot \nu_E(x)\big) - \big(y \cdot \nu_E(y)\big)\Big|\,d\mathcal{H}^{n-1}(x) = 0\,,$$

for $\mathcal{H}^{n-1}$-a.e. $y \in \partial^* E \setminus \{0\}$, which, by (28.16), actually implies $y \cdot \nu_E(y) = 0$ for $\mathcal{H}^{n-1}$-a.e. $y \in \partial^* E$.

Step two: We now fix $y \in E^{(1)}$, $y \neq 0$, $\lambda > 0$, and show that $\lambda\,y \in E^{(1)}$. We start by noticing that, by (28.12) and (28.14) it must be that

$$\frac{|E \cap H_{\nu,r} \cap B_s|}{s^n} \quad \text{is constant for } s \in (0,\infty).$$

Therefore, for every $r \in (0,1)$,

$$\begin{aligned} |E \cap C(y,r)| &= |E \cap H_{\nu,r} \cap B_{s(1+r)}| - |E \cap H_{\nu,r} \cap B_{s(1-r)}| \\ &= \frac{|E \cap H_{\nu,r} \cap B_{\lambda s(1+r)}|}{\lambda^n} - \frac{|E \cap H_{\nu,r} \cap B_{\lambda s(1-r)}|}{\lambda^n} = \frac{|E \cap C(\lambda y, r)|}{\lambda^n}\,. \end{aligned}$$

Since $C(\lambda\,y,r) = \lambda\,C(y,r)$, we thus conclude that

$$\frac{|E \cap C(\lambda y, r)|}{|C(\lambda\,y,r)|} = \frac{|E \cap C(y,r)|}{|C(y,r)|}\,. \tag{28.17}$$

We now easily conclude that $\lambda\,y \in E^{(1)}$ thanks to the inclusions (28.15) and by Exercise 5.19. □

The second tool we shall need in the proof of Theorem 28.6 is a refinement of Theorem 17.16, known as the **monotonicity formula**.

Theorem 28.9 (Monotonicity formula) *If E is stationary for the perimeter in A and $x_0 \in A \cap \partial E$, then for a.e. $r \in (0, \mathrm{dist}(x_0, \partial A))$,*

$$\frac{\mathrm{d}}{\mathrm{d}r}\frac{P(E; B(x_0, r))}{r^{n-1}} = \frac{\mathrm{d}}{\mathrm{d}r}\int_{B(x_0,r)\cap\partial^* E} \frac{\big(\nu_E(x)\cdot(x-x_0)\big)^2}{|x-x_0|^{n+1}}\, \mathrm{d}\mathcal{H}^{n-1}(x). \qquad (28.18)$$

Proof We set for brevity $d = \mathrm{dist}(x_0, \partial A)$. Without loss of generality we assume $x_0 = 0$, so that $B_r = B(x_0, r)$. As in Theorem 17.16, we associate with every $\varphi \in C^\infty(\mathbb{R}; [0, 1])$ with $\varphi = 1$ on $(-\infty, 1/2)$, $\varphi = 0$ on $(1, \infty)$ and $\varphi' \le 0$ on $\mathbb{R}$, a function

$$\Phi(r) = \int_{\partial^* E} \varphi\Big(\frac{|x|}{r}\Big)\, \mathrm{d}\mathcal{H}^{n-1}(x), \qquad r \in (0, d). \qquad (28.19)$$

We now notice that, if we set

$$\Psi(r) = \int_{\partial^* E} \varphi\Big(\frac{|x|}{r}\Big)\frac{(x\cdot\nu_E(x))^2}{|x|^2}\, \mathrm{d}\mathcal{H}^{n-1}(x), \qquad r \in (0, d), \qquad (28.20)$$

then (17.39) in the proof of Theorem 17.16 takes the form

$$\frac{\Phi'(r)}{r^{n-1}} - (n-1)\frac{\Phi(r)}{r^n} = \frac{\Psi'(r)}{r^{n-1}}, \qquad \text{for a.e. } r \in (0, d). \qquad (28.21)$$

Let us now define the Lipschitz functions $\varphi_\varepsilon \colon \mathbb{R} \to [0, 1]$, $\varepsilon \in (0, 1)$, as

$$\varphi_\varepsilon(s) = 1_{(-\infty, 1-\varepsilon)}(s) + \frac{1-s}{\varepsilon} 1_{(1-\varepsilon, 1)}(s), \qquad s \in \mathbb{R}.$$

By an approximation argument, (28.21) implies that, if $\varepsilon \in (0, 1)$, then

$$\frac{\Phi_\varepsilon'(r)}{r^{n-1}} - (n-1)\frac{\Phi_\varepsilon(r)}{r^n} = \frac{\Psi_\varepsilon'(r)}{r^{n-1}}, \qquad \text{for a.e. } r \in (0, d). \qquad (28.22)$$

where Φ_ε and Ψ_ε are defined through (28.19) and (28.20) with the choice $\varphi = \varphi_\varepsilon$. Finally, let us set $\Phi_0(r) = P(E; B_r)$ and

$$\gamma(r) = \int_{B_r\cap\partial^* E} \frac{(\nu_E(x)\cdot x)^2}{|x|^{n+1}}\, \mathrm{d}\mathcal{H}^{n-1}(x), \quad r \in (0, d).$$

Evidently, $\Phi_\varepsilon \to \Phi_0$ on $(0, d)$ as $\varepsilon \to 0$. We claim that for a.e. $r \in (0, d)$ (precisely, at every $r \in (0, d)$ such that Φ_0 and γ are differentiable), we have

$$\Phi_\varepsilon'(r) \to \Phi_0'(r), \quad \Psi_\varepsilon'(r) \to r^{n-1}\gamma'(r), \qquad (28.23)$$

as $\varepsilon \to 0$. From (28.22) and (28.23), we shall then deduce

$$\frac{\Phi_0'(r)}{r^{n-1}} - (n-1)\frac{\Phi_0(r)}{r^n} = \gamma'(r),$$

which, in turn, is (28.18). We now prove (28.23). We easily compute that

$$\Phi_\varepsilon'(r) = \frac{1}{\varepsilon r}\int_{(B_r\setminus B_{r(1-\varepsilon)})\cap\partial^* E}\frac{|x|}{r}\,\mathrm{d}\mathcal{H}^{n-1}\,,$$

for every $r \in (0,d)$, so that

$$(1-\varepsilon)\,\frac{P(E;B_r)-P(E;B_{r-\varepsilon r})}{\varepsilon r} \le \Phi_\varepsilon'(r) \le \frac{P(E;B_r)-P(E;B_{r-\varepsilon r})}{\varepsilon r}\,.$$

If Φ_0 is differentiable at r, we thus find $\Phi_\varepsilon'(r) \to \Phi_0'(r)$ as $\varepsilon \to 0^+$. Similarly,

$$\begin{aligned}\frac{\Psi_\varepsilon'(r)}{r^{n-1}} &= -\int_{\partial^* E}\frac{|x|}{r^{n+1}}\,\varphi_\varepsilon'\Big(\frac{|x|}{r}\Big)\frac{(x\cdot\nu_E(x))^2}{|x|^2}\,\mathrm{d}\mathcal{H}^{n-1}\\ &= \frac{1}{\varepsilon r}\int_{(B_r\setminus B_{r(1-\varepsilon)})\cap\partial^* E}\Big(\frac{|x|}{r}\Big)^n\frac{(x\cdot\nu_E(x))^2}{|x|^{n+1}}\,\mathrm{d}\mathcal{H}^{n-1}\,,\end{aligned}$$

from which we deduce

$$\frac{\gamma(r)-\gamma(r-\varepsilon r)}{\varepsilon r} \ge \frac{\Psi_\varepsilon'(r)}{r^{n-1}} \ge (1-\varepsilon)^n\,\frac{\gamma(r)-\gamma(r-\varepsilon r)}{\varepsilon r}\,.$$

Again, if γ is differentiable at r, then, as $\varepsilon \to 0^+$, we find $\Psi_\varepsilon'(r) \to r^{n-1}\gamma'(r)$. We have thus completed the proof of (28.23), hence of the theorem. □

Proof of Theorem 28.6 Let $x \in \Sigma(E;A)$, $r_h \to 0^+$, $E_h = (E-x)/r_h$, $A_h = (A-x)/r_h$. Since E is a (Λ, r_0)-perimeter minimizer in A, then, by Remark 21.6, E_h is a $(\Lambda r_h, r_0/r_h)$-perimeter minimizer in A_h. Given $\varepsilon, R > 0$ with $\varepsilon R \le 1$, there there exists $h(\varepsilon,R) \in \mathbb{N}$ such that

$$B_R \subset\subset A_h\,,\qquad 2R < \frac{r_0}{r_h}\,,\qquad \Lambda r_h \le \varepsilon\,,\qquad \forall h \ge h(\varepsilon,R)\,.$$

In this way, $\{E_h\}_{h\ge h(\varepsilon,R)}$ is a sequence of (ε,R)-perimeter minimizers in B_R with $\varepsilon R \le 1$. By Proposition 21.13 and Theorem 21.14, given $R_0 < R$, we may find a set of finite perimeter $F_{R_0} \subset B_{R_0}$ and $h(k) \to \infty$ as $k \to \infty$, such that F_{R_0} is a (ε,R)-minimizer in B_{R_0}, and, as $k \to \infty$,

$$E_{h(k)} \cap B_{R_0} \to F_{R_0}\,,\qquad \mu_{E_{h(k)}} \overset{*}{\rightharpoonup} \mu_{F_{R_0}}\,,$$
$$\text{and}\quad |\mu_{E_{h(k)}}| \overset{*}{\rightharpoonup} |\mu_{F_{R_0}}| \qquad \text{in } B_{R_0}\,.$$

(Weak-star convergence in open sets was introduced in Exercise 4.32.) By the arbitrariness of R and ε, and by a simple diagonal argument, we can exploit this fact to show the existence of a set of finite perimeter F and (with a slight abuse of notation) of $h(k) \to \infty$ as $k \to \infty$, such that

$$E_{h(k)} \overset{\text{loc}}{\to} F\,,\qquad \mu_{E_{h(k)}} \overset{*}{\rightharpoonup} \mu_F\,,\qquad |\mu_{E_{h(k)}}| \overset{*}{\rightharpoonup} |\mu_F|\,, \tag{28.24}$$

as $k \to \infty$. Moreover, F is a (ε, R)-perimeter minimizer in $\mathbb{R}^n$ for every $\varepsilon, R > 0$, that is, F is a perimeter minimizer in $\mathbb{R}^n$. By Theorem 28.2, since 0 is a singular point for every E_h, we easily see that $0 \in \Sigma(F; \mathbb{R}^n)$. Let us now show that F is a cone. By (28.24), for a.e. $s > 0$ (precisely, for those s such that $\mathcal{H}^{n-1}(\partial^* F \cap \partial B_s) = 0$) we have that

$$P(F; B_s) = \lim_{k\to\infty} P(E_{h(k)}; B_s)$$
$$= \lim_{k\to\infty} \frac{P(E; B(x, s\, r_{h(k)}))}{r_{h(k)}^{n-1}} = \omega_{n-1} s^{n-1}\, \theta_{n-1}(\partial E)(x)\,,$$

where in the last identity we have used Theorem 28.4 to assert the existence of the $(n-1)$-dimensional density of ∂E at the singular point x. Thus,

$$\frac{P(F; B_s)}{s^{n-1}} = \omega_{n-1}\, \theta_{n-1}(\partial E)(x)\,, \qquad \text{for a.e. } s > 0\,,$$

that is, the $(n-1)$-dimensional density ratio of $\partial^* F$ at $0 \in \partial F$ is constant. By the monotonicity identity (28.18) we conclude that

$$\int_{(B_s \setminus B_t)\cap \partial^* F} \frac{(\nu_F(y)\cdot y)^2}{|y|^{n+1}}\, d\mathcal{H}^{n-1}(y) = 0\,,$$

whenever $0 < s < t$, which in turn implies $y \cdot \nu_F(y) = 0$ for $\mathcal{H}^{n-1}$-a.e. $y \in \partial^* F$. By Proposition 28.8, $F^{(1)}$ is a cone. □

28.3 Simons' theorem

Theorem 28.6 links the existence of singular points for (Λ, r_0)-perimeter minimizers to the existence of singular minimizing cones. The following important theorem by Simons [Sim68] excludes the existence (in low dimensions) of singular minimizing cones possessing only a vertex singularity.

Theorem 28.10 (Simons' theorem) *If $n \ge 2$ and there exists a singular minimizing cone $K \subset \mathbb{R}^n$ with $\Sigma(K) = \{0\}$, then $n \ge 8$.*

We have different proofs in the cases $n = 2$ and $3 \le n \le 7$. The first case is solved by an elementary comparison argument, while the second case rests on the choice of a suitable test function in the second variation formula for perimeter (17.60). In both cases, the starting remark is that, by Theorem 26.5, and since $\Sigma(K) = \{0\}$, $\partial K \setminus \{0\}$ is a smooth hypersurface in $\mathbb{R}^n$. Hence, the slice $K \cap \partial B = K \cap S^{n-1}$ is a smooth hypersurface in $\mathbb{R}^n$. In the case $n = 2$ this means that $K \cap S^1$ consists of finitely many open circular arcs, lying at mutually positive distance. We pick one of these intervals, with extreme points $x_1, x_2 \in S^1$ and notice that if $x_1 \neq -x_2$ then there exists a local variation of K

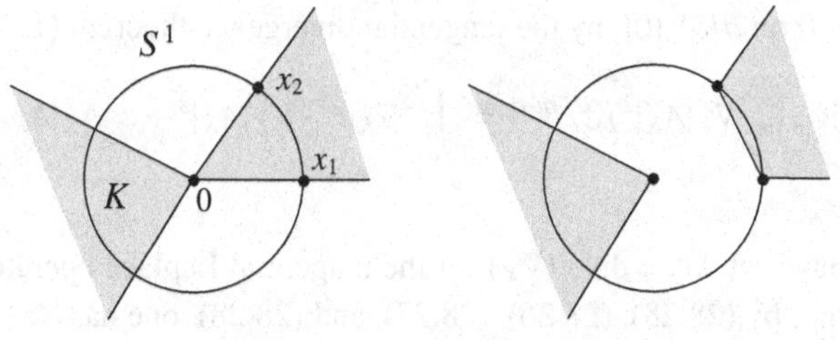

Figure 28.3 Simons' theorem in the plane. If $K \cap S^1$ contains two non-antipodal end-points x_1 and x_2, then a local variation of K that strictly decreases the perimeter is obtained by cutting the triangle of vertexes 0, x_1, and x_2 from K.

that is strictly decreasing its perimeter; see Figure 28.3. Thus $K \cap S^1$ consists of just one circular arc with antipodal end-points, that is, K is a half-plane, and $\Sigma(K) = \emptyset$, a contradiction. In fact, a similar argument can also be sketched in dimension $n = 3$. Since K is a cone with $\Sigma(K) = \{0\}$, at every $x \in \partial K \setminus \{0\}$, ∂K has a zero principal curvature. Since $n = 3$, the vanishing mean curvature condition reduces to $\kappa_1 + \kappa_2 = 0$ on $\partial K \setminus \{0\}$. Hence, $\kappa_1 = \kappa_2 = 0$ on $\partial K \setminus \{0\}$, and ∂K is, locally, and thus, by smoothness, globally, a plane, in contradiction to $\Sigma(K) \neq \emptyset$. In higher dimensions we have to abandon these simple geometric arguments, and rely on the stability condition (17.60). We now sketch this argument, referring readers to [Giu84, Theorem 10.10] and [Sim83, Appendix B] for fully detailed accounts. First, one exploits the smoothness of $\partial K \setminus \{0\}$ and an approximation argument to deduce from (17.60) that

$$\int_{\partial K} |\nabla_K \zeta|^2 d\mathcal{H}^{n-1} \geq \int_{\partial K} |A_K|^2 \zeta^2 \, d\mathcal{H}^{n-1}, \tag{28.25}$$

whenever $\zeta \in C^1_c(\mathbb{R}^n)$ with $\operatorname{spt}\zeta \cap \{0\} = \emptyset$ (here and in the following we write ∇_K in place of $\nabla^{\partial K}$). As our aim is to deduce from (28.25) that $|A_K| = 0$ on $\partial K \setminus \{0\}$, we are essentially forced to make $|A_K|$ appear on the left-hand side of (28.25). From this point of view it is natural to test (28.25) on a function ζ that, for x in a neighborhood of ∂K, satisfies

$$\zeta(x) = \varphi(x)\, |A_K(x)|, \tag{28.26}$$

where $\varphi \in C^\infty_c(\mathbb{R}^n)$, $\operatorname{spt}\varphi \cap \{0\} = \emptyset$. In order to settle possible smoothness issues, we should replace $|A_K|$ with $\sqrt{|A_K|^2 + \varepsilon}$ in (28.26), and then let $\varepsilon \to 0^+$ at the end of the argument. With this *caveat* in mind, we obtain

$$\begin{aligned}|\nabla_K(\varphi|A_K|)|^2 &= \varphi^2|\nabla_K |A_K||^2 + |A_K|^2|\nabla_K\varphi|^2 + 2\,\varphi|A_K|\,(\nabla_K\varphi)\cdot(\nabla_K|A_K|)\\ &= \varphi^2|\nabla_K |A_K||^2 + |A_K|^2|\nabla_K\varphi|^2 + \frac{1}{2}(\nabla_K\varphi^2)\cdot(\nabla_K|A_K|^2). \end{aligned} \tag{28.27}$$

Since $H_K = 0$ on $\partial K \setminus \{0\}$, by the tangential divergence theorem (17.24),

$$0 = \int_{\partial K} \mathrm{div}\,_K(\varphi^2 \nabla_K |A_K|^2)\, d\mathcal{H}^{n-1} = \int_{\partial K} \nabla_K \varphi^2 \cdot \nabla_K |A_K|^2 + \varphi^2 \Delta_K |A_K|^2\, d\mathcal{H}^{n-1}\,, \tag{28.28}$$

where we have set $\Delta_K = \mathrm{div}\,_K(\nabla_K)$ for the tangential Laplace operator on ∂K. Summarizing, by (28.25), (28.26), (28.27), and (28.28), one has

$$\int_{\partial K} \varphi^2 \Big(|A_K|^4 - |\nabla_K |A_K||^2 + \frac{1}{2} \Delta_K |A_K|^2 \Big)\, d\mathcal{H}^{n-1} \le \int_{\partial K} |A_K|^2 |\nabla_K \varphi|^2\, d\mathcal{H}^{n-1}\,, \tag{28.29}$$

whenever $\varphi \in C^\infty_c(\mathbb{R}^n)$, $\mathrm{spt}\,\varphi \cap \{0\} = \emptyset$. Next, one exploits the cone structure of K to show the lower bound

$$|A_K|^4 - |\nabla_K |A_K||^2 + \frac{1}{2} \Delta_K |A_K|^2 \ge \frac{2|A_K|^2}{|x|^2}\,;$$

see [Giu84, Lemma 10.9] or [Sim83, Lemma B.9]. Coming back to (28.29), we have proved the following necessary condition for perimeter minimality,

$$\int_{\partial K} |A_K|^2 \Big(|\nabla_K \varphi|^2 - 2\frac{\varphi^2}{|x|^2} \Big)\, d\mathcal{H}^{n-1} \ge 0\,, \tag{28.30}$$

for every $\varphi \in C^\infty_c(\mathbb{R}^n)$ with $\mathrm{spt}\,\varphi \cap \{0\} = \emptyset$. Since ν_K is a zero homogeneous vector field, and $A_K = \nabla \nu_K$, there must be a constant C such that

$$|A_K(x)| \le \frac{C}{|x|}\,, \qquad \forall x \in \partial K \setminus \{0\}\,.$$

Hence (28.30) actually holds by approximation on every Lipschitz function $\varphi \colon \mathbb{R}^n \to \mathbb{R}$ satisfying

$$\int_{\partial K} \frac{\varphi(x)^2}{|x|^4}\, d\mathcal{H}^{n-1}(x) < \infty\,.$$

It is therefore natural to restrict our attention to radially symmetric test functions $\varphi(x) = u(|x|)$. Since $x \cdot \nu_K(x) = 0$ we have $\nabla_K |x| = \nabla |x|$, that is the coarea factor of $|x|$ on ∂K is constant and equal to one. Thus,

$$\begin{aligned}\int_{\partial K} \frac{\varphi^2}{|x|^4}\, d\mathcal{H}^{n-1} &= \int_0^\infty dr \int_{\partial K \cap \partial B(x,r)} \frac{u(r)^2}{r^4}\, d\mathcal{H}^{n-2} \\ &= \mathcal{H}^{n-2}\big(\partial K \cap S^{n-1}\big) \int_0^\infty u(r)^2 r^{n-6}\, dr\,,\end{aligned}$$

and for the same reason $|\nabla_K \varphi| = |\nabla \varphi| = u(|x|)$. Hence (28.30) turns into

$$\int_{\partial K} |A_K(x)|^2 \Big(|u'(|x|)|^2 - 2\frac{u(|x|)^2}{|x|^2} \Big)\, d\mathcal{H}^{n-1}(x) \ge 0\,, \tag{28.31}$$

whenever $\int_0^\infty u(r)^2 r^{n-6} \mathrm{d}r < \infty$. We try the simplest possible choice, namely we define $u\colon (0,\infty) \to \mathbb{R}$ as

$$u(r) = \begin{cases} r^\alpha, & 0 < r < 1, \\ r^\beta, & r > 1, \end{cases}$$

for $\alpha, \beta \in \mathbb{R}$. Now, the admissibility of u requires $\int_0 r^{2\alpha+n-6}\mathrm{d}r$ and $\int^\infty r^{2\beta+n-6}\mathrm{d}r$ to be finite. Thus, we must impose $2\alpha+n-6 > -1$ and $2\beta+n-6 < -1$, that is,

$$\beta < \frac{5-n}{2} < \alpha. \tag{28.32}$$

By (28.31), if α and β satisfy (28.32), then we must have

$$0 \le (\alpha^2 - 2)\int_{B\cap\partial K} |A_K|^2 |x|^{2(\alpha-1)}\,\mathrm{d}\mathcal{H}^{n-1} + (\beta^2 - 2)\int_{(\partial K)\setminus B} |A_K|^2 |x|^{2(\beta-1)}\,\mathrm{d}\mathcal{H}^{n-1}. \tag{28.33}$$

Now, under the assumption $3 \le n \le 7$, there exist α and β such that

$$\beta < \frac{5-n}{2} < \alpha, \qquad \alpha^2 < 2, \qquad \beta^2 < 2. \tag{28.34}$$

Correspondingly, one infers from (28.33) that

$$\int_{B\cap\partial K} |A_K|^2|x|^{2(\alpha-1)}\mathrm{d}\mathcal{H}^{n-1}(x) = 0 = \int_{(\partial K)\setminus B} |A_K|^2|x|^{2(\beta-1)}\mathrm{d}\mathcal{H}^{n-1}(x),$$

i.e. that $|A_K| = 0$ on $\partial K \setminus \{0\}$. Hence ν_K is constant on $\partial K \setminus \{0\}$, so that K is a half-space, and $\Sigma(K) = \emptyset$, a contradiction.

28.4 Federer's dimension reduction argument

As seen in Section 28.2, the blow-ups at a singular point of a (Λ, r_0)-perimeter minimizer must converge to a singular minimizing cone. Simons' theorem allows us to exclude the existence in $\mathbb{R}^n$ of singular minimizing cones having just one singular point (the vertex), provided $2 \le n \le 7$. We now need to investigate the structure of singular minimizing cones with possibly larger singular sets. The main idea is to blow-up such sets in correspondence to their non-vertex singularities, to prove the existence of lower dimensional singular minimizing cones. This argument leads to the following theorem, which, together with Simons' theorem, will provide the key tool for proving Theorem 28.1 in Section 28.5.

Theorem 28.11 (Dimension reduction theorem) *If K is a singular minimizing cone in $\mathbb{R}^n$, $x_0 \in \Sigma(K)$, $x_0 \ne 0$, and if $r_h \to 0^+$, then, up to extracting a subsequence and up to rotation, the blow-ups K_{x_0,r_h} locally converge to a cylinder $F \times \mathbb{R}$, where F is a singular minimizing cone in $\mathbb{R}^{n-1}$.*

In turn, the proof of Theorem 28.11 is based on the two following lemmas.

Lemma 28.12 (Half-lines of singular points) *If K is a singular minimizing cone in $\mathbb{R}^n$, $x_0 \in \Sigma(K)$, and $x_0 \neq 0$, then $\{\lambda\, x_0 : \lambda > 0\} \subset \Sigma(K)$ and $n \geq 3$.*

Proof *Step one:* We show that, if $x \in \partial^* K$ and $\lambda > 0$, then

$$\lambda\, x \in \partial^* K\,, \qquad \nu_K(\lambda\, x) = \nu_K(x)\,.$$

The fact that $\lambda\, x \in \partial^* K$ is immediate from $\lambda\, K = K$ and the definition of reduced boundary. Moreover, if $\varphi \in C^1_c(\mathbb{R}^n)$, $\varphi_\lambda(y) = \varphi(y/\lambda)$, $y \in \mathbb{R}^n$, then

$$\int_{\partial^* K} \varphi\, \nu_K \, d\mathcal{H}^{n-1} = \int_K \nabla\varphi = \int_{\lambda K} \nabla\varphi = \lambda^{1-n} \int_K \nabla\varphi_\lambda = \lambda^{1-n} \int_{\partial^* K} \varphi_\lambda\, \nu_E \, d\mathcal{H}^{n-1}\,.$$

By approximation, we find that for every $r > 0$,

$$\frac{1}{\omega_{n-1} r^{n-1}} \int_{B(x,r)\cap\partial^* K} \nu_K \, d\mathcal{H}^{n-1} = \frac{1}{\omega_{n-1} (\lambda\, r)^{n-1}} \int_{B(\lambda\, x,\lambda\, r)\cap\partial^* K} \nu_K \, d\mathcal{H}^{n-1}\,.$$

As $r \to 0^+$, by Exercise 15.17, we find $\nu_K(x) = \nu_K(\lambda\, x)$.

Step two: By a simple change of variables and by step one, we see that

$$\frac{1}{r^{n-1}} \int_{B(x_0,r)\cap\partial^* K} \frac{|\nu_K - \nu|^2}{2} \, d\mathcal{H}^{n-1} = \frac{1}{(\lambda\, r)^{n-1}} \int_{B(\lambda\, x_0,\lambda\, r)\cap\partial^* K} \frac{|\nu_K - \nu|^2}{2} \, d\mathcal{H}^{n-1}\,,$$

for every $\lambda\,, r > 0$ and $\nu \in S^{n-1}$. If $x_0 \in \Sigma(K)$, then by the characterization of singular sets (26.48) and by definition of spherical excess, we find

$$\varepsilon(n) \leq \inf_{r>0} \mathbf{e}(K, x_0, r) = \inf_{r>0} \mathbf{e}(K, \lambda\, x_0, r)\,, \qquad \forall \lambda > 0\,.$$

Again by (26.48), we find $\{\lambda\, x_0 : \lambda > 0\} \subset \Sigma(K)$. In particular, $\mathcal{H}^1(\Sigma(K)) = \infty$. If $n = 2$, then, by (21.11), $\mathcal{H}^1(\Sigma(K)) = 0$. Thus, it is necessary that $n \geq 3$. □

Lemma 28.13 (Cylinders of locally finite perimeter)

(i) If F is of locally finite perimeter in $\mathbb{R}^{n-1}$, then $F \times \mathbb{R}$ is a of locally finite perimeter in $\mathbb{R}^n$, with

$$\mu_{F\times\mathbb{R}} = (\nu_F(\mathbf{p}x), 0)\, \mathcal{H}^{n-1} \llcorner \big((\partial^* F) \times \mathbb{R}\big)\,. \tag{28.35}$$

Moreover, if F is a perimeter minimizer in $\mathbb{R}^{n-1}$, then $F \times \mathbb{R}$ is a perimeter minimizer in $\mathbb{R}^n$.

(ii) If E is a set of locally finite perimeter in $\mathbb{R}^n$ such that

$$\nu_E(x) \cdot e_n = 0\,, \qquad \text{for } \mathcal{H}^{n-1}\text{-a.e. } x \in \partial^* E, \tag{28.36}$$

then there exists a set of locally finite perimeter F in $\mathbb{R}^{n-1}$ such that E is equivalent to $F \times \mathbb{R}$. If, moreover, E is a perimeter minimizer in $\mathbb{R}^n$, then F is a perimeter minimizer in $\mathbb{R}^{n-1}$.

Proof *Step one:* We prove the first part of statement (i). Let us first notice that if $\varphi \in C^1_c(\mathbb{R}^n)$, then $\varphi(z,\cdot) \in C^1_c(\mathbb{R})$ for every $z \in \mathbb{R}^{n-1}$, and, in particular

$$\int_{F\times\mathbb{R}} \partial_n \varphi = \int_F \mathrm{d}z \int_{\mathbb{R}} \partial_n \varphi(z,t)\,\mathrm{d}t = 0\,. \tag{28.37}$$

Let us now fix $R > 0$ and $T \in C^1_c(\mathbf{C}_R;\mathbb{R}^n)$, with $|T| \le 1$. For every $t \in \mathbb{R}$, define

$$S_t(z) = T(z,t) - (T(z,t)\cdot e_n)e_n\,, \qquad z \in \mathbb{R}^{n-1}\,,$$

so that $S_t \in C^1_c(\mathbf{D}_R;\mathbb{R}^{n-1})$, $|S_t| \le 1$ and $\operatorname{div} T = \partial_n(e_n \cdot T) + \operatorname{div}' S_t$. By (28.37), $\operatorname{spt} T \subset\subset \mathbf{C}_R$ and (12.1) we thus find

$$\int_{F\times\mathbb{R}} \operatorname{div} T = \int_{-R}^{R} \mathrm{d}t \int_F \operatorname{div}' S_t(z)\mathrm{d}z \le 2R\,P(F;\mathbf{D}_R)\,.$$

Since F is of locally finite perimeter in $\mathbb{R}^{n-1}$ and R is arbitrary, by (12.1) we conclude that $F\times\mathbb{R}$ is a set of locally finite perimeter in $\mathbb{R}^n$. Moreover, (28.35) follows by Exercise 18.10, as

$$\begin{aligned}\int_{\mathbb{R}^n} T\cdot \mathrm{d}\mu_{F\times\mathbb{R}} = \int_{F\times\mathbb{R}} \operatorname{div} T &= \int_{\mathbb{R}} \mathrm{d}t \int_F \operatorname{div}' S_t(z)\,\mathrm{d}z\\ &= \int_{\mathbb{R}} \mathrm{d}t \int_{\partial^* F} S_t(z)\cdot \nu_F(x)\,\mathrm{d}\mathcal{H}^{n-2}(z)\\ &= \int_{(\partial^* F)\times\mathbb{R}} T(z,t)\cdot(\nu_F(z),0)\,\mathrm{d}\mathcal{H}^{n-1}(z,t)\,.\end{aligned}$$

Step two: Now let F be a perimeter minimizer in $\mathbb{R}^{n-1}$. We want to show that, if $R > 0$ and $E\Delta G \subset\subset \mathbf{C}_R$, then $P(E;\mathbf{C}_R) \le P(G;\mathbf{C}_R)$. By (28.35) we have $|\mathbf{p}\nu_E(x)| = 1$ for $\mathcal{H}^{n-1}$-a.e. $x \in \partial^* E$. By (18.25),

$$\begin{aligned}P(E;\mathbf{C}_R) = \int_{\mathbf{C}_R\cap\partial^* E} |\mathbf{p}\nu_E|\,\mathrm{d}\mathcal{H}^{n-1} &= \int_{-R}^{R} \mathcal{H}^{n-2}\Big((\mathbf{C}_R)_t \cap \partial^* E_t\Big)\,\mathrm{d}z\\ &= 2R\,\mathcal{H}^{n-2}\Big(\mathbf{D}_R \cap \partial^* F\Big),\end{aligned} \tag{28.38}$$

where, as usual, $H_t = \{z \in \mathbb{R}^{n-1} : (z,t) \in H\}$ for $H \subset \mathbb{R}^n$, $t \in \mathbb{R}$. Since $G_t\Delta F \subset\subset \mathbf{D}_R$ for every $t \in \mathbb{R}$, and F is a perimeter minimizer in $\mathbb{R}^{n-1}$, by Theorem 18.11, for a.e. $t \in \mathbb{R}$ we find

$$\mathcal{H}^{n-2}\Big(\mathbf{D}_R \cap \partial^* F\Big) \le \mathcal{H}^{n-2}\Big(\mathbf{D}_R \cap \partial^* G_t\Big)\,.$$

By (28.38) and (18.25) we thus have

$$P(E;\mathbf{C}_R) \le \int_{-R}^{R} \mathcal{H}^{n-2}\Big(\mathbf{D}_R \cap \partial^* G_t\Big)\,\mathrm{d}t = \int_{\mathbf{C}_R\cap\partial^* G} |\mathbf{p}\nu_G|\,\mathrm{d}\mathcal{H}^{n-1} \le P(G;\mathbf{C}_R)\,.$$

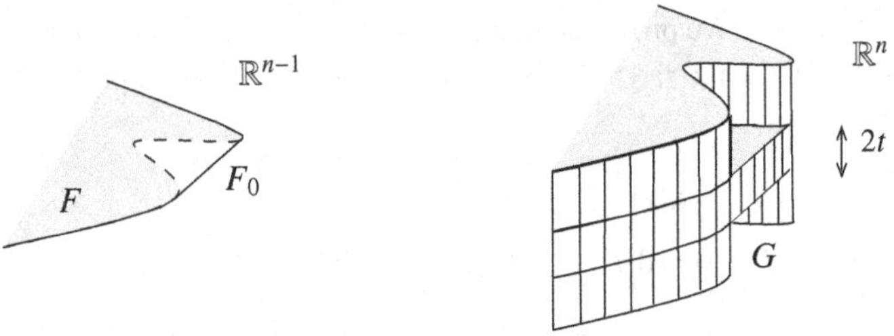

Figure 28.4 The comparison set G in the proof of Lemma 28.13(ii).

Step three: We prove (ii). By (28.36) we have that

$$\int_E \partial_n\varphi(x)\,dx = \int_{\partial^* E} \varphi\, e_n \cdot \nu_E \,d\mathcal{H}^{n-1} = 0\,, \qquad \forall \varphi \in C^1_c(\mathbb{R}^n)\,. \tag{28.39}$$

Given $x \in \mathbb{R}^n$, we apply (28.39) with $\varphi(y) = \rho_\varepsilon(y-x)$ ($y \in \mathbb{R}^n$) to find that $u_\varepsilon = 1_E \star \rho_\varepsilon$ satisfies $\partial_n u_\varepsilon(x) = 0$, that is, $u_\varepsilon(x) = u_\varepsilon(\mathbf{p}x, 0)$. Since $u_\varepsilon \to 1_E$ in $L^1_{\mathrm{loc}}(\mathbb{R}^n)$, if we set $F = \mathbf{p}(E) \subset \mathbb{R}^{n-1}$, then

$$1_E(x) = 1_E(\mathbf{p}x, 0) = 1_F(\mathbf{p}x)1_{\mathbb{R}}(\mathbf{q}x) = 1_{F\times\mathbb{R}}(x)\,,$$

that is, $E = F \times \mathbb{R}$. Since $E_t = F$ for every $t \in \mathbb{R}$, by Theorem 18.11, F is a set of locally finite perimeter. If now F *is not* a perimeter minimizer in $\mathbb{R}^{n-1}$, then there exist $\varepsilon > 0$ and $F_0 \subset \mathbb{R}^{n-1}$ such that, up to translation, $F\Delta F_0 \subset\subset \mathbf{D}_R$ and $P(F_0; \mathbf{D}_R) + \varepsilon \le P(F; \mathbf{D}_R)$ for some $R > 0$. Given $t > 0$,

$$I_t = \mathbb{R} \setminus (-t, t)\,, \qquad G = \big(F_0 \times (-t,t)\big) \cup \big(F \times I_t\big)\,;$$

see Figure 28.4. By Theorem 16.3 and (28.35), we find that

$$\begin{aligned}\mu_G = (\nu_{F_0}, 0)\,\mathcal{H}^{n-1} \llcorner \big(\partial^* F_0 \times (-t,t)\big) + (\nu_F, 0)\,\mathcal{H}^{n-1} \llcorner \big(\partial^* F \times I_t\big)\\ + e_n\,\mathcal{H}^{n-1} \llcorner \big((F_0\Delta F) \times \{t\}\big) - e_n\,\mathcal{H}^{n-1} \llcorner \big((F_0\Delta F) \times \{-t\}\big)\,.\end{aligned}$$

As $E\Delta G \subset\subset \mathbf{D}_R \times (-2t, 2t) = A$, by $P(F_0; \mathbf{D}_R) + \varepsilon \le P(F; \mathbf{D}_R)$ we thus find

$$\begin{aligned}P(G; A) - P(E; A) &= 2\,t\big(P(F_0; \mathbf{D}_R) - P(F; \mathbf{D}_R)\big) + 2\,\mathcal{H}^{n-1}(F_0\Delta F)\\ &\le -2\,t\,\varepsilon + 2\,\mathcal{H}^{n-1}(\mathbf{D}_R) < 0\,,\end{aligned}$$

provided t is large enough. Hence, if F is not a perimeter minimizer in $\mathbb{R}^{n-1}$, then E is not a perimeter minimizer in $\mathbb{R}^n$. □

Proof of Theorem 28.11 By Theorem 28.6, up to extracting a subsequence, the blow-ups $E_h = K_{x_0, r_h}$ locally converge to a singular minimizing cone E in

$\mathbb{R}^n$. By Lemma 28.12, $\Sigma(K)$ contains the half-line $\{\lambda x_0 : \lambda > 0\}$, so that

$$\Big\{ -x_0 + \lambda x_0 : \lambda > 0 \Big\} \subset \Sigma(E_h).$$

By Theorem 28.2, we thus find $\{-x_0 + \lambda x_0 : \lambda > 0\} \subset \Sigma(E)$ and, in particular,

$$\mathcal{H}^1(\Sigma(E)) = \infty. \tag{28.40}$$

We can clearly assume that $x_0 = e_n$. By Proposition 28.8, $\nu_K(y) \cdot y = 0$ for $\mathcal{H}^{n-1}$-a.e. $y \in \partial^* K$. Hence, $e_n \cdot \nu_K(y) = (e_n - y) \cdot \nu_K(y)$, for $\mathcal{H}^{n-1}$-a.e. $y \in \partial^* K$. By this last identity we conclude that, for every $R > 0$,

$$\begin{aligned}
\int_{B_R \cap \partial^* E_h} |e_n \cdot \nu_{E_h}| \, d\mathcal{H}^{n-1} &= \frac{1}{r_h^{n-1}} \int_{B(e_n, r_h R) \cap \partial^* K} |e_n \cdot \nu_K| \, d\mathcal{H}^{n-1} \\
&\le \frac{1}{r_h^{n-1}} \int_{B(e_n, r_h R) \cap \partial^* K} |y - e_n| \, d\mathcal{H}^{n-1}(y) \\
&\le \frac{r_h R}{r_h^{n-1}} P\Big(K; B(e_n, r_h R)\Big) \le n\omega_n \frac{(r_h R)^n}{r_h^{n-1}} \le C(n) R^n r_h,
\end{aligned} \tag{28.41}$$

where we have also applied (16.25). Since $\mu_{E_h} \stackrel{*}{\rightharpoonup} \mu_E$, by the Reshetnyak lower semicontinuity theorem (Theorem 20.11) and (28.41),

$$\int_{B_R \cap \partial^* E} |e_n \cdot \nu_E| \, d\mathcal{H}^{n-1} \le \liminf_{h \to \infty} \int_{B_R \cap \partial^* E_h} |e_n \cdot \nu_{E_h}| \, d\mathcal{H}^{n-1} = 0,$$

so that $e_n \cdot \nu_E(y) = 0$ for $\mathcal{H}^{n-1}$-a.e. $y \in \partial^* E$. By Lemma 28.13(ii), E is equivalent to $F \times \mathbb{R}$, where F is a perimeter minimizer in $\mathbb{R}^{n-1}$. Since E is a cone, F is a cone. By (28.35), $\partial^*(F \times \mathbb{R}) = (\partial^* F) \times \mathbb{R}$, thus $\Sigma(E) = \Sigma(F) \times \mathbb{R}$. If $\Sigma(F) = \emptyset$, then $\Sigma(E) = \emptyset$, against (28.40). Hence $\mathcal{H}^0(\Sigma(F)) > 0$, so that F is a singular minimizing cone in $\mathbb{R}^{n-1}$. □

28.5 Dimensional estimates for singular sets

Proof of Theorem 28.1 (i) Let $E \subset \mathbb{R}^n$ be a (Λ, r_0)-perimeter minimizer in the open set A, with $2 \le n \le 7$. If there exists $x \in \Sigma(E; A)$ then by Theorem 28.6 there exists a singular minimizing cone K in $\mathbb{R}^n$. By Simons' theorem it must be that $\mathcal{H}^0(\Sigma(K)) > 1$. By Lemma 28.12 we have $n \ge 3$, and by Theorem 28.11 there exists a singular minimizing cone K_1 in $\mathbb{R}^{n-1}$. By Simons' theorem and Lemma 28.12, it must be that $\mathcal{H}^0(\Sigma(K_1)) > 1$, hence $n - 1 \ge 3$. By repeating this argument four more times, we find $n \ge 8$. □

Proof of Theorem 28.1 (ii) Let $E \subset \mathbb{R}^8$ be a (Λ, r_0)-perimeter minimizer in some open set A, and assume by contradiction the existence of a sequence

$\{x_h\}_{h\in\mathbb{N}} \subset \Sigma(E;A)$ such that $x_h \to x$ for some $x \in A \cap \partial E$. By Theorem 28.2 we have that $x \in \Sigma(E;A)$. If we set $r_h = |x_h - x|$ and $E_h = E_{x,r_h}$, then by Theorem 28.6 there exists a singular minimizing cone K in $\mathbb{R}^n$ such that, up to possibly extracting a subsequence, $\{E_h\}_{h\in\mathbb{N}}$ locally converges to K. Up to extracting a further subsequence, we may also assume that $z_h = r^{-h}(x_h - x) \to z$ for some $z \in S^{n-1}$. Since $z_h \in \Sigma(E_h)$, by Theorem 28.2 we have $z \in \Sigma(K)$. As $z \neq 0$, it is true that $\mathcal{H}^0(\Sigma(K)) > 1$. Thus, by Theorem 28.11, there exists a singular minimizing cone in $\mathbb{R}^7$, in contradiction to Theorem 28.1 (i). □

The proof of part (iii) of Theorem 28.1 requires a last technical lemma. Statement (i) should be compared with (6.9) in Theorem 6.4.

Lemma 28.14

(i) If E is a Borel set such that $\mathcal{H}^s(E) < \infty$, $s > 0$, then

$$\limsup_{r\to 0^+} \frac{\mathcal{H}^s_\infty(E\cap B(x,r))}{\omega_s r^s} \geq \frac{1}{2^s}, \qquad \text{for } \mathcal{H}^s\text{-a.e. } x \in E.$$

(ii) If $\{E_h\}_{h\in\mathbb{N}}$ and E are (Λ, r_0)-perimeter minimizers in the open set $A \subset \mathbb{R}^n$ with $\Lambda r_0 \leq 1$, then for every compact set $H \subset A$ we have

$$\mathcal{H}^s_\infty\big(\Sigma(E;A)\cap H\big) \geq \limsup_{h\to\infty} \mathcal{H}^s_\infty\big(\Sigma(E_h;A)\cap H\big). \tag{28.42}$$

(iii) If $s \geq 0$, $F \subset \mathbb{R}^{n-1}$, and $\mathcal{H}^s_\infty(F) = 0$, then $\mathcal{H}^{s+1}_\infty(F\times\mathbb{R}) = 0$.

Proof *Step one:* If we set

$$F_\delta = \Big\{x \in E : \sup_{0<r<\delta} \frac{\mathcal{H}^s_\infty(E\cap B(x,r))}{\omega_s r^s} < \frac{1-\delta}{2^s}\Big\}, \qquad \delta \in (0,1),$$

then we have

$$\Big\{x \in E : \limsup_{r\to 0^+} \frac{\mathcal{H}^s_\infty(E\cap B(x,r))}{\omega_s r^s} < \frac{1}{2^s}\Big\} \subset \bigcup_{k\in\mathbb{N}} F_{1/k}\,.$$

We prove (i) by showing that $\mathcal{H}^s(F_\delta) = 0$ for every $\delta > 0$. Indeed, if $x \in F_\delta$ and G satisfies $x \in G$, $G \subset F_\delta$ and $\operatorname{diam} G < \delta$, then, by construction of F_δ,

$$\mathcal{H}^s_\delta(G) = \mathcal{H}^s_\infty(G) \leq \mathcal{H}^s_\infty\big(E\cap B\big(x, \operatorname{diam} G\big)\big) \leq (1-\delta)\,\omega_s\Big(\frac{\operatorname{diam} G}{2}\Big)^s.$$

Therefore, if $\mathcal{G}$ is a countable covering of F_δ by sets G having diameter at most δ, with $F_\delta \cap G \neq \emptyset$ and $G \subset F_\delta$, then we deduce

$$\mathcal{H}^s_\delta(F_\delta) \leq \sum_{G\in\mathcal{G}} \mathcal{H}^s_\delta(G) \leq (1-\delta)\,\omega_s \sum_{G\in\mathcal{G}} \Big(\frac{\operatorname{diam} G}{2}\Big)^s.$$

By the arbitrariness of $\mathcal{G}$, $\mathcal{H}^s_\delta(F_\delta) \leq (1-\delta)\mathcal{H}^s_\delta(F_\delta)$. Since $\mathcal{H}^s_\delta(F_\delta) \leq \mathcal{H}^s_\delta(E) \leq \mathcal{H}^s(E) < \infty$, it must be that $\mathcal{H}^s_\delta(F_\delta) = 0$, and thus $\mathcal{H}^s(F_\delta) = 0$.

Step two: If $\mathcal{F}$ is a finite covering by open sets of the compact set $\Sigma(E;A)\cap H$, then there exists $\varepsilon>0$ such that $I_\varepsilon(\Sigma(E;A)\cap H)\subset\bigcup_{F\in\mathcal{F}}F$. Then, by Theorem 28.2, there exists $h_0\in\mathbb{N}$ such that, for every $h\ge h_0$,

$$\Sigma(E_h;A)\cap H\subset I_\varepsilon\big(\Sigma(E;A)\cap H\big)\subset\bigcup_{F\in\mathcal{F}}F.$$

By definition of $\mathcal{H}^s_\infty$, we thus find

$$\limsup_{h\to\infty}\mathcal{H}^s_\infty(\Sigma(E_h;A)\cap H)\le\mathcal{H}^s_\infty\Big(\bigcup_{F\in\mathcal{F}}F\Big)\le\omega_s\sum_{F\in\mathcal{F}}\Big(\frac{\operatorname{diam}(F)}{2}\Big)^s,$$

and (28.42) follows by the arbitrariness of $\mathcal{F}$.

Step three: If $s=0$, then $F=\emptyset$, and there is nothing to prove. Let $s>0$. It is enough to prove that $\mathcal{H}^{s+1}_\infty(F\times(-t,t))=0$ for every $t>0$. For every $\varepsilon>0$ we can find a covering $\{F_h\}_{h\in\mathbb{N}}$ of F such that $\sum_{h\in\mathbb{N}}\delta_h^s\le\varepsilon$, where $\delta_h=\operatorname{diam}(F_h)$. For every $h\in\mathbb{N}$, let $\mathcal{G}_h$ be a finite covering of $(-t,t)$ with $\operatorname{diam}G\le\delta_h$ for every $G\in\mathcal{G}_h$ and $\#(\mathcal{G}_h)\le 2t\delta_h^{-1}$. Thus,

$$\bigcup_{h\in\mathbb{N}}\big\{F_h\times G: G\in\mathcal{G}_h\big\}$$

is a countable covering of $F\times(-t,t)$. Since $\operatorname{diam}(F_h\times G_h)\le\sqrt{2}\,\delta_h$,

$$\begin{aligned}\mathcal{H}^{s+1}_\infty(F\times(-t,t))&\le\omega_{s+1}\sum_{h\in\mathbb{N}}\sum_{G\in\mathcal{G}_h}\Big(\frac{\operatorname{diam}(F_h\times G)}{2}\Big)^{s+1}\\&\le C(s)\sum_{h\in\mathbb{N}}\#(\mathcal{G}_h)\,\delta_h^{s+1}\le C(s)\,t\sum_{h\in\mathbb{N}}\delta_h^s\le C(s)\,t\varepsilon,\end{aligned}$$

and we conclude the proof by letting $\varepsilon\to0^+$. □

As a final technical prerequisite, we recall that, if $E\subset\mathbb{R}^n$ and $\mathcal{H}^s(E)>0$ for some $s\ge0$, then there exists $F\subset E$ with $0<\mathcal{H}^s(F)<\infty$. For a proof of this fact, we refer the reader to [Mat95, Theorem 8.16].

Proof of Theorem 28.1 (iii) Let E be a (Λ,r_0)-perimeter minimizer in A with $\Lambda r_0\le1$ and $\mathcal{H}^s(\Sigma(E;A))>0$, $s>0$. By [Mat95, Theorem 8.16] and Lemma 28.14, there exist $x\in\Sigma(E;A)$ and a sequence $r_h\to0^+$, such that

$$\mathcal{H}^s_\infty\big(\Sigma(E;A)\cap B(x,r_h)\big)\ge\frac{\omega_s r_h^s}{2^{s+1}},\qquad\forall h\in\mathbb{N}.\tag{28.43}$$

By Theorem 28.6, the blow-ups $E_h=E_{x,r_h}$ are $(\Lambda r_h,r_0/r_h)$-perimeter minimizers in A_{x,r_h}, which locally converge (up to extracting subsequences) to a singular minimizing cone K in $\mathbb{R}^n$. Note that (28.43) is equivalent to

$$\mathcal{H}^s_\infty\big(\Sigma(E_h;A_{x,r_h})\cap B\big)\ge\frac{\omega_s}{2^{s+1}},\qquad\forall h\in\mathbb{N}.$$

If now $h_0 \in \mathbb{N}$ is such that $B_2 \subset A_{x,r_h}$ for every $h \geq h_0$, then

$$\mathcal{H}^s_\infty\Big(\Sigma(E_h; B_2) \cap \overline{B}\Big) \geq \frac{\omega_s}{2^{s+1}}, \qquad \forall h \geq h_0,$$

and, by (28.42), we conclude that

$$\mathcal{H}^s_\infty\Big(\Sigma(K) \cap \overline{B}\Big) \geq \frac{\omega_s}{2^{s+1}}. \tag{28.44}$$

By Proposition 3.4, $\mathcal{H}^s(\Sigma(K)) > 0$. We may thus repeat this argument with K and $\mathbb{R}^n$ in place of E and A respectively. In this way, also taking Theorem 28.11 into account, we prove the existence of a singular minimizing cone of the form $F \times \mathbb{R}$ in $\mathbb{R}^n$, with F a singular minimizing cone in $\mathbb{R}^{n-1}$, and with

$$\mathcal{H}^s_\infty\Big(\Sigma(F \times \mathbb{R}) \cap \overline{B}\Big) \geq \frac{\omega_s}{2^{s+1}}. \tag{28.45}$$

By Lemma 28.14, this implies $\mathcal{H}^{s-1}_\infty(\Sigma(F)) > 0$. If we now assume that $n \geq 9$ and $s > n-8$, then we may repeat this construction $n-8$ times, thus proving the existence of a singular minimizing cone K_* in $\mathbb{R}^8$ with $\mathcal{H}^{s-(n-8)}(\Sigma(K_*)) > 0$. This conclusion would contradict Theorem 28.1 (ii). Hence, $s \leq n-8$. □

28.6 Examples of singular minimizing cones

In this section we show examples of singular minimizing cones, proving in particular the following result, due to Bombieri, De Giorgi, and Giusti [BDGG69].

Theorem 28.15 *If $m \geq 4$, then the Simons cone*

$$K = \Big\{(x, y) \in \mathbb{R}^m \times \mathbb{R}^m : |x| < |y|\Big\} \tag{28.46}$$

is a singular minimizing cone in $\mathbb{R}^m \times \mathbb{R}^m$.

Corollary 28.16 *In $\mathbb{R}^n$, $n \geq 8$, there exists a singular minimizing cone K, with $\mathcal{H}^0(\Sigma(K)) = 1$ if $n = 8$, and with $\mathcal{H}^{n-8}(\Sigma(K)) = \infty$ if $n \geq 9$. This shows in particular the sharpness of the dimensional estimates in Theorem 28.1.*

Proof of Corollary 28.16 Theorem 28.15 with $m = 4$ provides a singular minimizing cone K in $\mathbb{R}^8$ with $\mathcal{H}^0(\Sigma(K)) = 1$. If $n \geq 9$ then $E = K \times \mathbb{R}^{n-8}$ is a perimeter minimizer by Lemma 28.13 (i). If $x \in \{0\} \times \mathbb{R}^{n-8}$, then $E_{x,r} = E$ for every $r > 0$. Since E is not a half-space, $\{0\} \times \mathbb{R}^{n-8} \subset \Sigma(E)$, and thus $\mathcal{H}^{n-8}(\Sigma(E)) = \infty$, as required. □

Over the years, many revisions of the original proof of Theorem 28.15 have been proposed, including contributions by Lawson [Law72], Federer [Fed75],

Massari and Miranda [MM83], and Davini [Dav04]. We follow here the approach developed in [DPP09] by De Philippis and Paolini, based on an effective variant of the classical **calibration method**. The rough idea is that, whenever E is a set of locally finite perimeter in $\mathbb{R}^n$ and $T\colon \mathbb{R}^n \to \mathbb{R}^n$ satisfies

$$|T| \le 1\,, \qquad \text{on } \mathbb{R}^n\,, \tag{28.47}$$

$$\operatorname{div} T \le 0\,, \qquad \text{on } E\,, \tag{28.48}$$

$$\operatorname{div} T \ge 0\,, \qquad \text{on } \mathbb{R}^n \setminus E\,, \tag{28.49}$$

$$T = \nu_E\,, \qquad \text{on } \partial^* E\,, \tag{28.50}$$

then E is a perimeter minimizer in $\mathbb{R}^n$. Indeed, let F with $E\Delta F \subset\subset B_R$ for some $R > 0$ be given. By applying the divergence theorem to T on $F \setminus E$,

$$
\begin{aligned}
\int_{F\setminus E} \operatorname{div} T &= \int_{\partial^*(F\setminus E)} T\cdot \nu_{F\setminus E}\, d\mathcal{H}^{n-1}\\
\text{by (16.5)} \qquad &= \int_{E^{(0)}\cap\partial^* F} T\cdot\nu_F\, d\mathcal{H}^{n-1} - \int_{F^{(1)}\cap\partial^* E} T\cdot\nu_E\, d\mathcal{H}^{n-1}\\
\text{by (28.50)} \qquad &= \int_{E^{(0)}\cap\partial^* F} T\cdot\nu_F\, d\mathcal{H}^{n-1} - P(E;F^{(1)})\\
\text{by (28.47)} \qquad &\le P(F;E^{(0)}) - P(E;F^{(1)})\,.
\end{aligned}
$$

By (28.49), we thus find $P(E;F^{(1)}) \le P(F;E^{(0)})$ (notice that $P(F;E^{(0)}) < \infty$ since $F^{(1/2)} \setminus B_R = E^{(1/2)} \setminus B_R$). Applying the divergence theorem to $E \setminus F$, we similarly prove $P(E;F^{(0)}) \le P(F;E^{(1)})$. Adding up these two inequalities, and noticing that, by Federer's theorem, $P(F;E^{(1/2)} \cap B_R) = P(E;F^{(1/2)} \cap B_R)$,

$$
\begin{aligned}
&P(F;B_R) = P(F;E^{(0)} \cup E^{(1)}) + P(F;E^{(1/2)} \cap B_R)\,,\\
\text{and} \qquad &P(E;B_R) = P(E;F^{(0)} \cup F^{(1)}) + P(E;F^{(1/2)} \cap B_R)\,,
\end{aligned}
$$

we find $P(E;B_R) \le P(F;B_R)$. We now prove the perimeter minimality of Simons' cones (28.46) by applying this argument with $T = \nabla f/|\nabla f|$, $f(x,y) = (|x|^4 - |y|^4)/4$. Some technicalities arise here, since T is not smooth at $(0,0)$.

Proof of Theorem 28.15 *Step one:* Let us us say that E is a **perimeter sub-minimizer in** $\mathbb{R}^n$, provided $P(E;B_R) \le P(F;B_R)$ whenever $R > 0$, $F \subset E$, $E \setminus F \subset\subset B_R$. We claim that, if $\{E_h\}_{h\in\mathbb{N}}$ are perimeter sub-minimizers in $\mathbb{R}^n$, which locally converge to E, and $E_h \subset E$ for every $h \in \mathbb{N}$, then E is a perimeter sub-minimizer in $\mathbb{R}^n$. Indeed, let $R > 0$, $F \subset E$, $E\setminus F \subset\subset B_R$. Since $E_h \cap F \subset E_h$ and $E_h \setminus (E_h \cap F) \subset\subset B_R$, by Lemma 12.22,

$$P(E_h;B_R) \le P(E_h \cap F;B_R) \le P(E_h;B_R) + P(F;B_R) - P(E_h \cup F;B_R)\,,$$

i.e. $P(E_h \cup F;B_R) \le P(F;B_R)$. Since $E_h \cup F$ locally converges to E, we conclude that $P(E;B_R) \le P(F;B_R)$ by Proposition 12.15, and the claim is proved.

Step two: Let $f\colon \mathbb{R}^m \times \mathbb{R}^m \to \mathbb{R}$ be defined as $f(x,y) = (|x|^4 - |y|^4)/4$, so that, according to (28.46), $K = \{f < 0\}$. The open sets with smooth boundary,

$$K_h = \left\{(x,y) \in \mathbb{R}^m \times \mathbb{R}^m : f(x,y) < -h^{-1}\right\}, \qquad h \in \mathbb{N},$$

are subsets of K, which locally converge to K as $h \to \infty$. We now claim that each K_h is a perimeter sub-minimizer in $\mathbb{R}^n$. Indeed, define $T\colon \mathbb{R}^n \to \mathbb{R}^n$ as

$$T = \frac{\nabla f(x,y)}{|\nabla f(x,y)|} = \frac{(|x|^2\,x, -|y|^2\,y)}{\sqrt{|x|^6 + |y|^6}}, \qquad (x,y) \in \mathbb{R}^n.$$

Since $T \in C^1(K;\mathbb{R}^n)$, we may prove the perimeter sub-minimality of K_h by applying the divergence theorem to T on $K_h \setminus F$, where $F \subset K_h$ and $K_h \setminus F \subset\subset \mathbb{R}^n$. All we need to check is that

$$|T| \le 1, \qquad \text{on } K, \tag{28.51}$$

$$\operatorname{div} T \le 0, \qquad \text{on } K, \tag{28.52}$$

$$T = \nu_{K_h}, \qquad \text{on } \partial K_h, \tag{28.53}$$

for every $h \in \mathbb{N}$. Since (28.51) and (28.53) are trivial consequences of the definitions of K_h and T, we are left to prove (28.52). To this end, let us first notice that if $A \subset \mathbb{R}^n$ is open and $F, S \in C^1(A;\mathbb{R}^n)$, then $F \circ S \in C^1(A;\mathbb{R}^n)$ with $\nabla(F \circ S) = \nabla F(S)\nabla S$. Since

$$\nabla\Big(\frac{z}{|z|}\Big) = \frac{1}{|z|}\Big(\mathrm{Id} - \frac{z \otimes z}{|z|^2}\Big), \qquad z \in \mathbb{R}^n,$$

and since $(z \otimes z)X = z \otimes (X^* z)$ for every $X \in \mathbb{R}^n \otimes \mathbb{R}^n$, we deduce that

$$\nabla\Big(\frac{S}{|S|}\Big) = \frac{1}{|S|}\Big(\nabla S - \frac{S \otimes [(\nabla S)^* S]}{|S|^2}\Big),$$

on the open set where $A \cap \{|S| > 0\}$. In particular we obtain

$$\operatorname{div}\Big(\frac{S}{|S|}\Big) = \frac{|S|^2 \operatorname{div} S - S \cdot [(\nabla S)^* S]}{|S|^3} = \frac{|S|^2 \operatorname{div} S - [(\nabla S) S] \cdot S}{|S|^3}, \tag{28.54}$$

on $A \cap \{|S| > 0\}$. By (28.54), since $|\nabla f| > 0$ on K, we have

$$\operatorname{div} T = \frac{|\nabla f|^2 \Delta f - [(\nabla^2 f)\nabla f] \cdot \nabla f}{|\nabla f|^3}, \qquad \text{on } K. \tag{28.55}$$

We now compute that

$$\nabla_x f = |x|^2 x, \quad \nabla_y f = -|y|^2 y, \quad |\nabla f|^2 = |x|^6 + |y|^6,$$

$$\nabla^2_{xx} f = |x|^2 \mathrm{Id}_{\mathbb{R}^m_x} + 2x \otimes x, \quad \nabla^2_{yy} f = -|y|^2 \mathrm{Id}_{\mathbb{R}^m_y} - 2y \otimes y, \quad \nabla^2_{xy} f = 0.$$

Therefore, taking (28.55) also into account,

$$\begin{aligned}
\Delta f &= \mathrm{trace}\,(\nabla^2 f) = (m+2)(|x|^2 - |y|^2)\,,\\
(\nabla^2 f)\nabla f &= ((\nabla^2_{xx} f)\nabla_x f, (\nabla^2_{yy} f)\nabla_y f) = (3|x|^4 x, 3|y|^4 y)\,,\\
((\nabla^2 f)\nabla f)\cdot\nabla f &= 3|x|^8 - 3|y|^8\,,\\
\mathrm{div}\,T &= \frac{(m+2)(|x|^6+|y|^6)(|x|^2-|y|^2) - 3|x|^8 + 3|y|^8}{|\nabla f|^3}\\
&= (|x|^4 - |y|^4)\frac{(m-1)(|x|^4+|y|^4) - (m+2)|x|^2|y|^2}{|\nabla f|^3}\,.
\end{aligned}$$

Now, $p(t) = (m-1)t^2 - (m+2)t + (m-1) \ge 0$ for every $t \in \mathbb{R}$ if and only if

$$(m+2)^2 \le 4(m-1)^2\,,$$

that is, if and only if $m \ge 4$. Hence, $\mathrm{div}\,T < 0$ on $K = \{f < 0\}$, as required.

Step three: By step one and step two, K is a perimeter sub-minimizer in $\mathbb{R}^n$. Moreover, the same computation as step two shows that $\mathrm{div}\,(-T) < 0$ on $\mathbb{R}^n \setminus \overline{K}$. Hence, by approximating $H = \mathbb{R}^n \setminus \overline{K} = \{f > 0\}$ with the open sets with smooth boundary $\{f > 1/h\}$, and then repeating the above argument, we see that H is a perimeter sub-minimizer in $\mathbb{R}^n$. Since K and H are perimeter sub-minimizer in $\mathbb{R}^n$, it turns out that, in fact, they are both perimeter minimizers in $\mathbb{R}^n$. Indeed, let F satisfy $K\Delta F \subset\subset B_R$ for some $R > 0$. By perimeter sub-minimality of K and H, and by carefully applying Federer's theorem and Theorem 16.3,

$$\begin{aligned}
0 &\le P(K\cap F; B_R) - P(K; B_R) = P(F; K) - P(K; F^{(0)})\,,\\
0 &\le P\big(H\cap(\mathbb{R}^n\setminus F); B_R\big) - P(H; B_R) = P(F; H) - P(H; F^{(1)})\\
&= P(F; K^{(0)}) - P(K; F^{(1)})\,.
\end{aligned}$$

Adding up these inequalities, we find $P(K; B_R) \le P(F; B_R)$. □

28.7 A Bernstein-type theorem

As a nice application of the ideas and tools developed in this chapter, following Fleming [Fle62], we prove the following Bernstein-type theorem; see Giusti [Giu84, Chapter 17] for more information on this subject.

Theorem 28.17 *If E is a perimeter minimizer in $\mathbb{R}^n$, where $2 \le n \le 7$, then E is a half-space.*

Proof We fix $x \in \partial E$. Since E is a perimeter minimizer in $\mathbb{R}^n$, by the monotonicity identity (Theorem 28.9), for a.e. $r > 0$, we have

$$\frac{\mathrm{d}}{\mathrm{d}r}\frac{P(E;B(x,r))}{r^{n-1}} = \frac{\mathrm{d}}{\mathrm{d}r}\int_{B(x,r)\cap\partial^*E}\frac{|\nu_E(y)\cdot(y-x)|^2}{|x-y|^{n+1}}\,\mathrm{d}\mathcal{H}^{n-1}(y)\,, \tag{28.56}$$

while the density estimates (17.40) and (16.25) imply

$$\omega_{n-1}s^{n-1} \le P(E_{x,r};B_s) \le n\omega_n s^{n-1}\,, \qquad \forall r,s>0\,.$$

Now let $r_h \to 0^+$ and $R_h \to \infty$ as $h \to \infty$. Up to extracting subsequences, by Corollary 12.27, there exist F_0 and F_∞ of locally finite perimeter such that

$$E_{x,r_h} \overset{\mathrm{loc}}{\to} F_0\,, \qquad E_{x,R_h} \overset{\mathrm{loc}}{\to} F_\infty\,, \qquad \text{as } h \to \infty.$$

Arguing as in Theorem 28.6, F_0 and F_∞ are perimeter minimizers, with

$$\frac{P(F_0;B_s)}{s^{n-1}} = \alpha_0, \qquad \frac{P(F_\infty;B_s)}{s^{n-1}} = \alpha_\infty\,, \qquad \forall s>0\,.$$

In particular, by (28.56) and Proposition 28.8, both F_0 and F_∞ are cones. Since $2 \le n \le 7$, by Theorem 28.1, F_0 and F_∞ are half-spaces. In particular,

$$\omega_{n-1} = \alpha_0 \le \frac{P(E;B_r)}{r^{n-1}} \le \alpha_\infty = \omega_{n-1}\,, \qquad \forall r>0\,.$$

By (28.56), E is a cone. As $2 \le n \le 7$, applying Theorem 28.1 again, conclude that E is a half-space, as claimed. □

Notes

We open these notes with some historical remarks about regularity theorems for variational problems in parametric form, with the intent of sketching a background for the results presented in this part of the book. After the pioneering work of Douglas and Radò on two-dimensional area minimizing surfaces of disk-type, the first partial regularity theorems for solutions to (suitable generalized formulations of) Plateau's problem in higher dimensions were obtained by De Giorgi [DG60] (in codimension one, using sets of finite perimeter), and by Besicovitch's student Reifenberg [Rei60, Rei64a, Rei64b] (in arbitrary codimension and in a different framework). Concerning De Giorgi's contribution, in [DG60] it is proved that if E is a local perimeter minimizer in some open set A, then $A \cap \partial^* E$ is an analytic hypersurface. Some key ideas are introduced in [DG60], like the use of approximation by harmonic functions in proving uniform excess decay estimates, which, in different disguises, will recur in all subsequent developments. A first estimate on the size of the singular set, namely $\mathcal{H}^{n-1}(A \cap (\partial E \setminus \partial^* E)) = 0$, was proved shortly after by Miranda [Mir65], who also studied, in [Mir67], the properties of sequences of local perimeter minimizers stated here in Theorems 21.14 and 26.6.

The first partial regularity results for rectifiable sets of arbitrary codimension minimizing sufficiently smooth elliptic functionals were obtained by Almgren [Alm68] (see also [Fed69, Chapter 5]) in the framework of the theories of currents and varifolds. In partcular, Almgren's regularity theorem, specialized to sets of finite perimeter, shows the following: if $\Phi\colon \mathbb{R}^n \times S^{n-1} \to (0,\infty)$ is of class C^k for some $k \ge 3$ (resp., analytic), if it is uniformly λ-elliptic for some $\lambda > 0$,

$$\nabla^2\Phi(x,\nu)(z,z) \ge \lambda\Big|z - (z\cdot\nu)\nu\Big|^2, \qquad \forall (x,\nu,z) \in \mathbb{R}^n \times S^{n-1} \times (\mathbb{R}^n \setminus \{0\}),$$

(here $\nabla^2\Phi$ is computed after extending Φ to $\mathbb{R}^n \times \mathbb{R}^n$ as $\Phi(x, t\nu) = t\,\Phi(x,\nu)$ for $t > 0$, $\nu \in S^{n-1}$), and if $E \subset \mathbb{R}^n$ is a local minimizer in A of the functional

$$\mathbf{\Phi}(E;A) = \int_{A\cap\partial^* E} \Phi(x,\nu_E(x))\,\mathrm{d}\mathcal{H}^{n-1}(x),$$

(same definition as in (16.22), with $\mathbf{\Phi}(E;A)$ in place of $P(E;A)$), then $A \cap \partial^* E$ is an hypersurface of class C^{k-1} (resp. analytic) in $\mathbb{R}^n$, with $\mathcal{H}^{n-1}(A \cap (\partial E \setminus \partial^* E)) = 0$.

The next breakthrough was obtained by Allard [All72], who proved partial regularity theorems for rectifiable sets of arbitrary codimension with sufficiently summable distributional mean curvature. Referring readers to [Sim83, Chapter 5] for an extremely clear presentation of Allard's regularity theorem, we limit ourselves to stating its following corollary: if A is open, and E is a set of locally finite perimeter with distributional mean curvature $H_E \in L^p_{\mathrm{loc}}(A \cap \partial^* E; \mathcal{H}^{n-1})$ for $p > n-1$, then $A \cap \partial^* E$ is a $C^{1,\alpha}$-hypersurface in $\mathbb{R}^n$ for $\alpha = 1 - (n-1)/p$, and $\mathcal{H}^{n-1}(A \cap (\partial E \setminus \partial^* E)) = 0$; see in particular [All72, 8.1(3)].

The analysis of singular sets for area minimizing integer currents of codimension one (local perimeter minimizers, in our setting) was completed through the contributions of several authors in the sixties: after previous partial results by Almgren, De Giorgi, and Fleming, Simons [Sim68] proved the non-existence of singular area minimizing hypercones with vertex singularity in $\mathbb{R}^n$ if $2 \le n \le 7$ (Theorem 28.10); Bombieri, De Giorgi and Giusti [BDGG69] showed that Simons' cone $\{(x,y) \in \mathbb{R}^4 \times \mathbb{R}^4 : |x| = |y|\}$ is area minimizing in $\mathbb{R}^8$ (Theorem 28.15), thus proving the sharpness of Simons' theorem; finally, Federer [Fed70], through the dimension reduction argument presented in Section 28.5, showed (in the framework of area minimizing currents) the dimensional estimates appearing in Theorem 28.1.

Dimensional estimates for singular sets of integer currents of codimension one minimizing smooth elliptic functionals were obtained by Schoen, Simon, and Almgren [SSA77], and shown to be optimal by Morgan [Mor90]. In our terminology, the results in [SSA77] imply that if E is a local minimizer of the functional $\mathbf{\Phi}(E;A)$ defined above, for Φ of class C^2 and uniformly λ-elliptic, then $\mathcal{H}^{n-3}(A \cap (\partial E \setminus \partial^* E)) = 0$.

These results leave open several questions about singular sets, even in the case of local perimeter minimizers: for example, is the dimension of the singular set always an integer? Are singularities stable under small perturbations of the boundary data? Are singular tangent cones unique? If this is the case at a point x, is the minimizer diffeomorphic to its singular tangent cone close to x? Let us also mention that the analysis of singularities becomes extremely more complex in higher codimension; see Almgren [Alm00], and De Lellis and Spadaro [DLS11b, DLS11a].

The deduction of partial regularity theorems from perturbed minimality inequalities (similar in spirit to the (Λ, r_0)-perimeter minimality condition considered here) was initiated by Almgren in [Alm76], a paper which, in turn, stimulated the revision and simplification of [Alm68] by Bombieri [Bom82], and, with different methods, by Schoen and Simon [SS82].

All these various inputs are combined in the beautiful paper by Duzaar and Steffen [DS02] to obtain sharp partial regularity results for integer currents of arbitrary codimension minimizing regular elliptic integrands; this last paper, together with Simon's presentation of Allard's regularity theorem, were particularly useful in the preparation of this part of the book. Readers willing to learn partial regularity theorems in higher codimension will find several points in common between the arguments presented here in the context of sets of finite perimeter and those used in the above mentioned papers in the case of currents and varifolds. For example, our proof of the height bound (Theorem 22.8) follows [Fed69, 5.3.4], [SS82, Lemma 2], and [DS02, Lemma 2.2]; the proof of the Lipschitz approximation theorem follows [Sim83, Section 20] and [DS02, Lemma 3.1]; the reverse Poincaré inequality is modeled after a comparison argument by Bombieri [Bom82, Section V] and a covering lemma by Simon [Sim96, 2.8], as presented in [DS02, Section 4]; finally, the approximation by harmonic functions and the excess decay improvement by tilting are derived as in [Sim83, Sections 21 and 22].

Regularity theorems for reduced boundaries can be proved under weaker minimality assumptions than the (Λ, r_0)-perimeter minimality considered here. For example, Massari [Mas74, Mas75] proves that if E has variational mean curvature equal to $-g$ in A for some $g \in L^p(A)$ (so that (21.2) holds with $\|g\|_{L^p(A)}|E\Delta F|^{(p-1)/p}$ in place of $\Lambda\,|E\Delta F|$) and if $p > n$, then $A \cap \partial^* E$ is a $C^{1,(p-n)/4p}$-hypersurface in $\mathbb{R}^n$. More generally, Tamanini [Tam84] shows that if E is an almost perimeter minimizer in A, in the sense that

$$P(E; B(x,r)) \le P(F; B(x,r)) + \omega(r)\, r^{n-1}\,,$$

whenever $x \in A$, $E\Delta F \subset\subset B(x,r)$ and $r < r_0$, then $A \cap \partial^* E$ is a C^1-hypersurface and the singular set has dimension at most $n-8$, provided $\omega\colon (0, r_0) \to [0,\infty)$ is such that $\omega(0^+) = 0$, $\omega(t)/t$ is increasing on $(0, r_0)$, and

$$\int_0^{r_0} \frac{\sqrt{\omega(r)}}{r}\, dr < \infty\,.$$

In fact, the result is more precise, and given $x \in A \cap \partial^* E$ we can find positive constants r and C such that

$$|\nu_E(x) - \nu_E(y)| \le C\Big(\int_0^{|x-y|} \frac{\sqrt{\omega(r)}}{r}\, dr + |x-y|^{1/2}\Big), \qquad \forall y \in B(x,r) \cap \partial^* E\,.$$

For example, if $\omega(r) = c\, r^{2\alpha}$ for some $c \ge 0$ and $\alpha \in (0, 1/2]$, then the above assumptions are satisfied, and Tamanini's theorem shows that $A \cap \partial^* E$ is a $C^{1,\alpha}$-hypersurface. Vice versa, if $A \cap \partial^* E$ is a $C^{1,\alpha}$-hypersurface for some $\alpha \in (0, 1/2]$, then E is an almost perimeter minimizer in A with $\omega(r) = c\, r^{2\alpha}$ for some $c \ge 0$. Tamanini's almost perimeter minimizers include of course (Λ, r_0)-minimizers, as well as further examples, like minimal boundaries with obstacles; see [Mir71, Tam82].

As shown in [GMT93], a planar set bounded by two bi-logarithmic spirals wrapping around the origin and having the origin as their common asymptotic end-point has variational mean curvature in $L^2(\mathbb{R}^2)$; at the same time, its boundary is not representable as the graph of a Lipschitz function in any neighborhood of the origin. This example is related to the following question by De Giorgi: is it true for a set $E \subset \mathbb{R}^n$ with variational mean curvature in $L^n(\mathbb{R}^n)$ that, for every $x \in \partial E$ outside a singular set of codimension at least 8, there exists $r > 0$ and a bi-Lipschitz map between $B(x,r) \cap \partial E$ and an $(n-1)$-dimensional disk? Paolini [Pao98], by combining ideas from both De Giorgi's and Reifenberg's approaches to the regularity of area minimizing surfaces, proved this statement with bi-$C^{0,\alpha}$ maps ($\alpha \in (0,1)$ arbitrary) in place of bi-Lipschitz maps. Ambrosio and Paolini [AP99, AP01] extended Paolini's theorem to any set E satisfying the following (extremely weak) local minimality condition,

$$P(E; B(x,r)) \le \big(1 + \omega(r)\big)\, P(F; B(x,r))\,,$$

whenever $E\Delta F \subset\subset B(x,r)$ and $r < r_0$, under the only assumption that $\omega\colon (0,r_0) \to [0,\infty)$ satisfies $\omega(0^+) = 0$; moreover, in the planar case, they are able to replace bi-$C^{0,\alpha}$ maps with bi-Lipschitz maps, thus answering affirmatively to De Giorgi's question in dimension $n = 2$.

In Section 28.6 we have proved the perimeter minimality of Simons' cones. Further examples of cones over products of spheres were provided by Lawson [Law72], Simoes [Sim73], and Bindschadler [Bin78], culminating with the complete characterization of singular area minimizing cones over products of spheres by Lawlor [Law88]. In particular, the results of Lawson and Simoes show that the cones

$$K = \left\{(x,y) \in \mathbb{R}^k \times \mathbb{R}^h : \sqrt{h-1}\,|x| < \sqrt{k-1}\,|y|\right\}$$

are perimeter minimizers in $\mathbb{R}^n$, $n = k + h$, if and only if $k, h \geq 2$ and either $k + h \geq 9$ or $(k,h) \in \{(3,5),(5,3),(4,4)\}$. An alternative proof of this result was provided by Davini [Dav04]. The calibration method of De Philippis and Paolini [DPP09] can also be used to prove the minimality of all the Lawson–Simoes cones, with the exception of six cases, and, in fact, to prove global stability inequalities for the associated Plateau problems see [DPM12].

Finally, our introductory exposition of the regularity theory for geometric variational problems was limited to interior regularity results; for the boundary situation, see [All75, Har77, HS79, HL86, Grü87, DS93, DS02].

PART FOUR

Minimizing clusters

Synopsis

A cluster $\mathcal{E}$ in $\mathbb{R}^n$ is a finite disjoint family of sets of finite perimeter $\mathcal{E} = \{\mathcal{E}(h)\}_{h=1}^N$ ($N \in \mathbb{N}$, $N \geq 2$) with finite and positive Lebesgue measure (note: the chambers $\mathcal{E}(h)$ of $\mathcal{E}$ are not assumed to be connected/indecomposable). By convention, $\mathcal{E}(0) = \mathbb{R}^n \setminus \bigcup_{h=1}^N \mathcal{E}(h)$ denotes the exterior chamber of $\mathcal{E}$. The perimeter $P(\mathcal{E})$ of $\mathcal{E}$ is defined as the total $(n-1)$-dimensional Hausdorff measure of the interfaces of the cluster,

$$P(\mathcal{E}) = \sum_{0 \leq h < k \leq N} \mathcal{H}^{n-1}\Big(\partial^*\mathcal{E}(h) \cap \partial^*\mathcal{E}(k)\Big).$$

Denoting by $\mathbf{m}(\mathcal{E})$ the vector in $\mathbb{R}^N_+$ whose hth entry agrees with $|\mathcal{E}(h)|$, we shall say that $\mathcal{E}$ is a **minimizing cluster in** $\mathbb{R}^n$ if $\operatorname{spt}\mu_{\mathcal{E}(h)} = \partial\mathcal{E}(h)$ for every $h = 1, \ldots, N$, and, moreover, $P(\mathcal{E}) \leq P(\mathcal{E}')$ whenever $\mathbf{m}(\mathcal{E}') = \mathbf{m}(E)$. By a **partitioning problem in** $\mathbb{R}^n$, we mean any variational problem of the form

$$\inf\Big\{P(\mathcal{E}) : \mathbf{m}(\mathcal{E}) = \mathbf{m}\Big\},$$

corresponding to the choice of some $\mathbf{m} \in \mathbb{R}^N_+$. Proving the following theorem will be the main aim of Part IV. The existence and regularity parts will be addressed, respectively, in Chapter 29 and Chapter 30.

Theorem (Almgren's theorem) *If $n, N \geq 2$ and $\mathbf{m} \in \mathbb{R}^N_+$, then there exist minimizers in the partitioning problem defined by* $\mathbf{m}$. *If $\mathcal{E}$ is an N-minimizing cluster in $\mathbb{R}^n$, then $\mathcal{E}$ is bounded. If $0 \leq h < k \leq N$, then $\partial^*\mathcal{E}(h) \cap \partial^*\mathcal{E}(k)$ is an analytic constant mean curvature hypersurface in $\mathbb{R}^n$, relatively open inside $\partial\mathcal{E}(h) \cap \partial\mathcal{E}(k)$. Finally,*

$$\sum_{h=0}^{N} \mathcal{H}^{n-1}\Big(\partial\mathcal{E}(h) \setminus \partial^*\mathcal{E}(h)\Big) = 0\,.$$

This existence and almost everywhere regularity theorem is one of the main results contained in the founding work for the theory of minimizing clusters and partitioning problems, that is Almgren's AMS Memoir [Alm76]. This theory, despite the various beautiful results which have been obtained since then, still presents many interesting open questions. We aim here to provide the reader with the necessary background to enter into these problems. Indeed, the techniques and ideas introduced in the proof of Almgren's theorem prove useful also in its subsequent developments, and are likely to play a role in possible further investigations in the subject.

Slightly rephrasing Almgren's words [Alm76, VI.1(6)], the aims of the theory are: (i) to show the existence of minimizing clusters; (ii) to prove the regularity of their interfaces outside singular closed sets; (iii) to describe the structure of these interfaces close to their singular sets, as well as the structure of the singular sets themselves; (iv) to construct examples of minimizing clusters; (v) to classify "in some reasonable way" the different minimizing clusters corresponding to different choices of $\mathbf{m} \in \mathbb{R}^N_+$; and (vi) to extend the analysis of these questions to multi-phase anisotropic partitioning problems (which are introduced below). We will deal with part (i) and (ii) of this programme in the case of mono-phase isotropic problems. We now provide a brief and partial review on the state of the art concerning Almgren's programme, and refer the reader to Morgan's book [Mor09, Chapters 13–16] for further references and information.

Planar minimizing clusters In the planar case, the constant mean curvature condition satisfied by the interfaces

$$\mathcal{E}(h,k) = \partial^*\mathcal{E}(h) \cap \partial^*\mathcal{E}(k)\,, \qquad 0 \le h < k \le N\,,$$

implies that each $\mathcal{E}(h,k)$ is a countable union of circular arcs, all with the same curvature $\kappa_{hk} \in \mathbb{R}$ (here, a straight segment is a circular arc with zero curvature); moreover, the blow-up clusters $\mathcal{E}_{x,r} = (\mathcal{E}-x)/r$ of $\mathcal{E}$ at a point x belonging to the singular set $\Sigma(\mathcal{E})$ of $\mathcal{E}$,

$$\Sigma(\mathcal{E}) = \bigcup_{0\le h<k\le N} \Big(\partial\mathcal{E}(h) \cap \partial\mathcal{E}(k)\Big) \setminus \mathcal{E}(h,k) = \bigcup_{h=0}^{N} \Big(\partial\mathcal{E}(h) \setminus \partial^*\mathcal{E}(h)\Big),$$

have the planar Steiner partition (see Figure 30.2) as their unique (up to rotations) possible limit in local convergence. Exploiting these two facts, in Theorem 30.7 we shall prove that the singular set is discrete, that every point $x \in \Sigma(E)$ is the junction of exactly three different interfaces, that the three circular arcs meeting at x form three 120-degree angles, and that each interface is made up of finitely many circular arcs (all with the same curvature); see Figure 1 and Section 30.3.

These general rules which planar minimizing clusters have to obey provide the starting point for attempting their characterization, at least in some special cases. For example, planar double bubbles have been characterized as the only planar minimizing 2-clusters in [FAB+93], and as the only *stable* (vanishing first variation and non-negative second variation) 2-clusters in [MW02]. There also exists a characterization of planar 3-clusters,

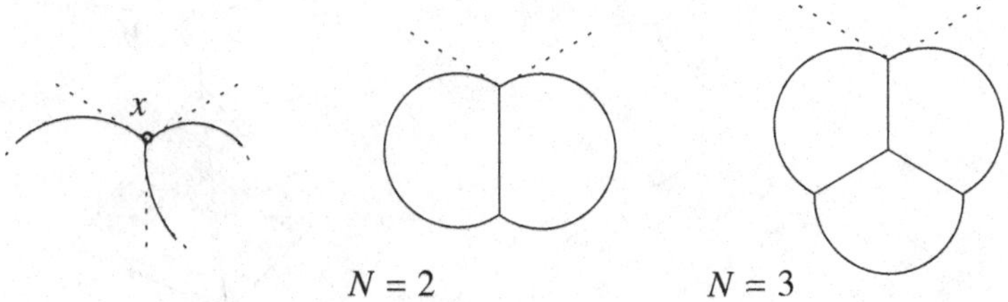

Figure 1 For planar clusters, the singular set is discrete, and the interfaces are circular arcs meeting in threes at singular points forming 120 degree angles. Starting from this information it is possible to characterize planar minimizing clusters with two and three chambers (the picture refers to the case in which the various chambers have equal areas).

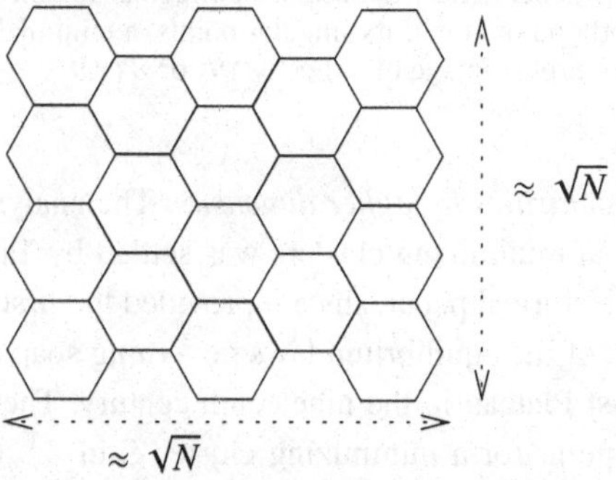

Figure 2 The honeycomb inequality: symmetric honeycombs (with unit area cells) provide the sharp asymptotic lower bound on the ratio perimeter over number of chambers for planar clusters with unit area chambers.

obtained in [Wic04]. Interestingly, no example of planar minimizing N-clusters is known if $N \geq 4$, although a list of possible candidates has been proposed in [CF10].

Describing the asymptotic properties of planar minimizing N-clusters as $N \to \infty$ provides another source of interesting questions. In this way, an interesting result is the so-called *honeycomb theorem* [Hal01]. A possible formulation of this result is as follows: if $\mathcal{E}$ is an N-cluster in $\mathbb{R}^2$ with $|\mathcal{E}(h)| = 1$ for every $h = 1, \dots, N$, then

$$\frac{P(\mathcal{E})}{N} > 2(12)^{1/4}. \tag{1}$$

This lower bound is sharp: if $\{\mathcal{E}_N\}_{N\in\mathbb{N}}$ denotes a sequence of planar N-clusters obtained by piling up approximately $\sqrt{N}$ rows consisting of approximately $\sqrt{N}$ many regular hexagons of unit area, then $P(\mathcal{E}_N)/N \to 2(12)^{1/4}$ as $N \to \infty$; see Figure 2. Moreover, in a sense that can be made precise, this is essentially the *unique* type of sequence $\{\mathcal{E}_N\}_{N\in\mathbb{N}}$ which asymptotically saturates (1).

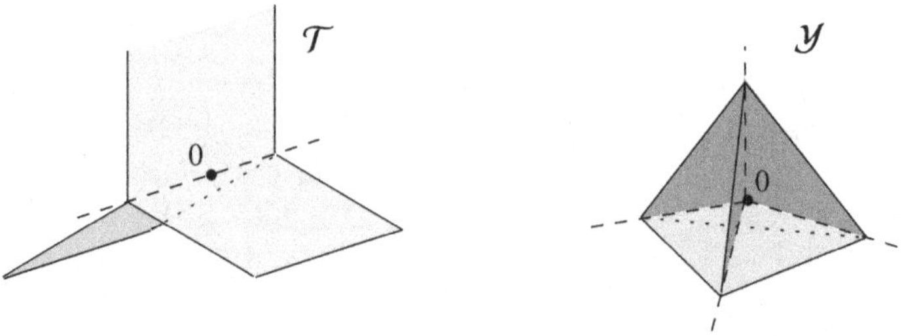

Figure 3 The only two possible tangent clusters at a singular point of a minimizing cluster in $\mathbb{R}^3$. The cone-like cluster $\mathcal{T}$ has as its interfaces three half-planes meeting along a line at 120 degree angles. The cone-like cluster $\mathcal{Y}$ has as its interfaces six planar angles of about 109 degrees of amplitude, which form the cone generated by the center of a regular tetrahedron and its edges. In a neighborhood of any of its singular points, a minimizing cluster in $\mathbb{R}^3$ is a $C^{1,\alpha}$-diffeomorphic image of either $B \cap \mathcal{T}$ or $B \cap \mathcal{Y}$.

Structure of singularities in higher dimension The analysis of singular sets of three-dimensional minimizing clusters was settled by Taylor in [Tay76]. This is considered a historical paper, since it provided the first complete mathematical justification of the equilibrium laws governing soap bubbles stated by the Belgian physicist Plateau in the nineteenth century. There it is proved that if x is a singular point for a minimizing cluster $\mathcal{E}$ in $\mathbb{R}^3$, then, up to rotations, the blow-up clusters $\mathcal{E}_{x,r} = (\mathcal{E} - x)/r$ of $\mathcal{E}$ at x locally converge as $r \to 0^+$ either to the 3-cluster $\mathcal{T}$ or the 4-cluster $\mathcal{Y}$ depicted in Figure 3. (Of course, according to our terminology, $\mathcal{T}$ and $\mathcal{Y}$ are not properly "clusters" as their chambers have infinite volumes.) Moreover, there exist an open neighborhood A of x in $\mathbb{R}^3$, $\alpha \in (0, 1)$, and a $C^{1,\alpha}$-diffeomorphism $f\colon \mathbb{R}^3 \to \mathbb{R}^3$, such that either

$$A \cap \mathcal{E} = f\big(B \cap \mathcal{T}\big) \qquad \text{or} \qquad A \cap \mathcal{E} = f\big(B \cap \mathcal{Y}\big),$$

where, as usual, B is the Euclidean unit ball in $\mathbb{R}^3$ centered at the origin. It should also be noted that Taylor's theorem actually applies to describe the singularities of (roughly speaking) any $\mathcal{H}^2$-rectifiable set M in $\mathbb{R}^3$ satisfying a suitably perturbed area minimality condition. In this way, Taylor's result has been extended to two-dimensional almost minimal rectifiable sets in $\mathbb{R}^n$ ($n \geq 3$) by David [Dav09, Dav10]. The extension of Taylor's theorem to the case of minimizing clusters in higher dimensions has been announced by White [Whi]. We finally remark that not much is known about general qualitative properties of minimizing clusters in dimension $n \geq 3$. For example, Tamanini [Tam98] has proved the existence of a constant $k(n)$ bounding the number of chambers

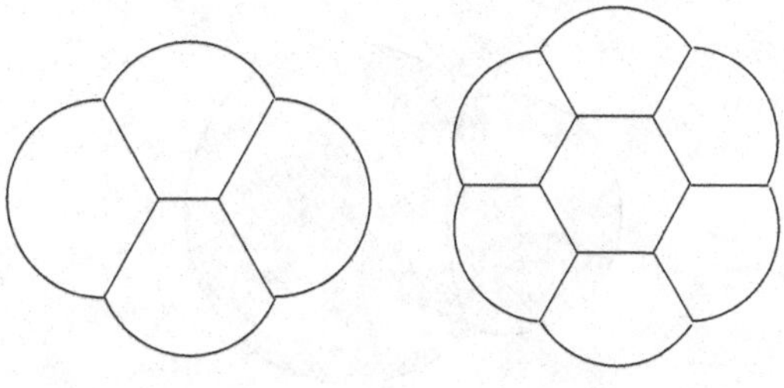

Figure 4 Conjectured minimizing clusters with 4 and 7 chambers which possess discrete groups of symmetries.

of a minimizing N-cluster which may meet at a given point of $\mathbb{R}^n$. However, no explicit bound on $k(n)$ is presently known.

Symmetry properties of minimizing clusters If $n \geq 2$ and $N \leq n-1$, then, given N sets in $\mathbb{R}^n$ with finite Lebesgue measure, we may find $n-(N-1)$ mutually orthogonal hyperplanes which cut each of the given sets into two halves of equal measure (this, by repeatedly applying a Borsuk–Ulam type argument). Notice that the intersection of $n-(N-1)$ mutually orthogonal hyperplanes in $\mathbb{R}^n$ defines a $(N-1)$-dimensional plane in $\mathbb{R}^n$. In this way, by standard reflection arguments (see, e.g., Section 19.5), we see that if $\mathbf{m} \in \mathbb{R}^N_+$ and $n-1 \geq N$, then there always exists a minimizing N-cluster $\mathcal{E}$ in $\mathbb{R}^n$ with $\mathbf{m}(\mathcal{E}) = \mathbf{m}$, which is symmetric with respect to an suitable $(N-1)$-dimensional plane of $\mathbb{R}^n$. Developing an idea due to White, Hutchings [Hut97] has actually proved that if $N \leq n-1$, then *every* minimizing N-cluster in $\mathbb{R}^n$ is symmetric with respect to an $(N-1)$-dimensional plane of $\mathbb{R}^n$ (a proof of this result in the language of sets of finite perimeter is presented in [Bon09]). The Hutchings–White theorem is the only general symmetry result for minimizing clusters known at present, although it is reasonable to expect that symmetries should appear also for special values of N and n outside the range $N \leq n-1$; see Figure 4.

The double bubble theorem In dimension $n \geq 3$, the only characterization result for minimizing N-clusters concerns the case $N = 2$. The starting point is the Hutchings–White theorem, which guarantees minimizing 2-clusters in $\mathbb{R}^n$ ($n \geq 2$) to be axially symmetric. In particular, the interfaces of a minimizing 2-cluster are constant mean curvature *surfaces of revolution*. This piece of information allows us to restrict the focus, so to say, on the mutual position of the various connected components of the chambers. Then, by careful first and second variations arguments, one comes to exclude all the alternative possibilities to the case of a double bubble, which is therefore the only minimizing

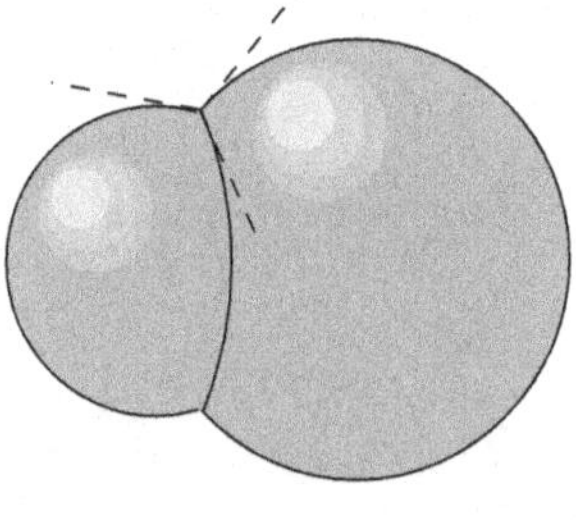

Figure 5 The topological boundary of a double bubble consists of three spherical caps which meet in a $(n-2)$-dimensional sphere forming three angles of 120 degrees.

2-cluster; see Figure 5. This beautiful result has been obtained by Hutchings, Morgan, Ritoré, and Ros in [HMRR02] in $\mathbb{R}^3$, and has been later extended to higher dimensions by Reichardt and collaborators [RHLS03, Rei08]. It should be noted that, at present, no characterization result for minimizing clusters is available in dimension $n \geq 3$ if $N \geq 3$. Another difficult problem is that of extending the honeycomb theorem (1) to higher dimensions. Following a conjecture by Lord Kelvin, it was believed for a long time that the asymptotic optimal tiling in $\mathbb{R}^3$ should be the one obtained by piling layers of relaxed truncated octahedra. Weaire and Phelan [WP94], however, disproved Kelvin's conjecture by showing a better competitor; see [Mor09, Chapter 15] for pictures and details.

Multi-phase anisotropic partition problems Finally, Almgren's existence and partial regularity theory applies to a wide class of partitioning problems, including the volume-constrained minimization of functionals of the type

$$\sum_{0\le h<k\le N} c_{hk} \int_{\mathcal{E}(h,k)} \Phi\left(x, \nu_{\mathcal{E}(h)}(x)\right) d\mathcal{H}^{n-1}(x),$$

under suitable assumptions on the coefficients $c_{hk} > 0$ and on the anisotropy $\Phi\colon \mathbb{R}^n \times S^{n-1} \to [0,\infty)$. Given $\mathbf{m} \in \mathbb{R}^N_+$, the volume-constrained minimization of this energy in the isotropic case leads to the *immiscible fluids problem*,

$$\inf\left\{ \sum_{0\le h<k\le N} c_{hk}\,\mathcal{H}^{n-1}\big(\mathcal{E}(h,k)\big) : \mathcal{E} \text{ is an } N\text{-cluster}, \mathbf{m}(\mathcal{E}) = \mathbf{m}\right\}.$$

We thus see the different chambers of the clusters as the regions occupied by possibly different fluids. The relative strengths of the mutual interactions between these different fluids are then weighted by the positive constants c_{hk}.

The lower semicontinuity of the multi-phase interaction energy is equivalent to the validity of the triangular inequality $c_{hk} \le c_{hi} + c_{ik}$ (Ambrosio and Braides [AB90], White [Whi96]). Assuming the strict triangular inequality $c_{hk} < c_{hi} + c_{ik}$, the regularity of the interfaces for minimizers is then addressed by reduction to the regularity theory for volume-constrained perimeter minimizers. The key tool to obtain this reduction is an *infiltration lemma* [Whi96, Leo01], which will be discussed (in the simple case when the c_{hk} are all equal) in Section 30.1.

29

Existence of minimizing clusters

This chapter will be entirely devoted to the proof of the existence of minimizing clusters. Precisely, we shall prove the following theorem.

Theorem 29.1 *For every* $\mathbf{m} \in \mathbb{R}^N_+$ *($N \in \mathbb{N}$, $N \geq 2$) and $n \geq 2$, there exists a minimizer in the variational problem*

$$\inf\Big\{P(\mathcal{E}) : \mathbf{m}(\mathcal{E}) = \mathbf{m}\,, \mathcal{E} \text{ is an } N\text{-cluster in } \mathbb{R}^n\Big\}. \tag{29.1}$$

If $\mathcal{E}$ *is a minimizer in (29.1), then* $\mathcal{E}$ *is bounded, that is, for some* $R > 0$,

$$\bigcup_{h=1}^{N} \mathcal{E}(h) \subset B_R\,.$$

This proof of this theorem, which presents several beautiful ideas, is rather long and technical. It could be advisable to limit a first reading to Section 29.1, where the basic definitions and remarks concerning clusters are introduced, Section 29.2, where an outline of the proof is presented, and Sections 29.5–29.6, where the technique of volume-fixing variations, of fundamental importance also in Chapter 30, is discussed. Sections 29.3–29.4 contain instead the other tools needed to prove Theorem 29.1, which are finally employed in Section 29.7 to prove the existence of minimizing clusters.

29.1 Definitions and basic remarks

An N-**cluster** $\mathcal{E}$ of $\mathbb{R}^n$ is a finite family of sets of finite perimeter $\mathcal{E} = \{\mathcal{E}(h)\}_{h=1}^N$, $N \in \mathbb{N}$, $N \geq 1$, with

$$0 < |\mathcal{E}(h)| < \infty\,, \qquad 1 \leq h \leq N\,, \tag{29.2}$$

$$|\mathcal{E}(h) \cap \mathcal{E}(k)| = 0\,, \qquad 1 \leq h < k \leq N\,. \tag{29.3}$$

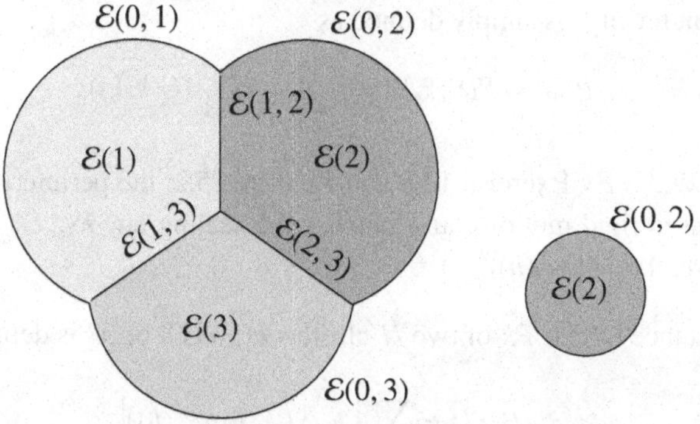

Figure 29.1 A 3-cluster in $\mathbb{R}^2$. In this example, the chamber $\mathcal{E}(2)$ is not indecomposable. The perimeter of the cluster simply amounts to the sum of the $\mathcal{H}^{n-1}$-dimensional measures of all its interfaces $\mathcal{E}(h,k)$; see (29.5).

The sets $\mathcal{E}(h)$ are called the **chambers** of $\mathcal{E}$. When the number N of the chambers of $\mathcal{E}$ is clear from the context, we shall use the term "cluster" in place of "N-cluster". The **exterior chamber** of $\mathcal{E}$ is defined as

$$\mathcal{E}(0) = \mathbb{R}^n \setminus \bigcup_{h=1}^{N} \mathcal{E}(h). \tag{29.4}$$

In particular, $\{\mathcal{E}(h)\}_{h=0}^N$ is a partition of $\mathbb{R}^n$ (up to a set of null Lebesgue measure) and, necessarily, $|\mathcal{E}(0)| = \infty$. The **volume vector** $\mathbf{m}(\mathcal{E})$ is defined as

$$\mathbf{m}(\mathcal{E}) = \Big(|\mathcal{E}(1)|, \dots, |\mathcal{E}(N)|\Big) \in \mathbb{R}^N.$$

We denote by bold letters the elements of $\mathbb{R}^N$, denoting by $\mathbf{m}(h)$ the hth component of $\mathbf{m} \in \mathbb{R}^N$. We let $\mathbb{R}^N_+$ be the cone of those $\mathbf{m} \in \mathbb{R}^N$ such that $\mathbf{m}(h) > 0$ for every $h = 1, \dots, N$. Notice that if $\mathcal{E}$ is an N-cluster, then $\mathbf{m}(\mathcal{E}) \in \mathbb{R}^N_+$ as $\mathbf{m}(\mathcal{E})(h) = |\mathcal{E}(h)| > 0$ for every $h = 1, \dots, N$.

Remark 29.2 It is important to notice that the chambers of a cluster are *not* assumed to be indecomposable (indecomposability, introduced in Exercise 16.9, is the commonly accepted notion of connectedness in the framework of sets of finite perimeter).

The **interfaces** of the N-cluster $\mathcal{E}$ in $\mathbb{R}^n$ are the $\mathcal{H}^{n-1}$-rectifiable sets

$$\mathcal{E}(h,k) = \partial^*\mathcal{E}(h) \cap \partial^*\mathcal{E}(k), \qquad 0 \le h, k \le N, h \neq k;$$

see Figure 29.1. We define the **relative perimeter** of $\mathcal{E}$ in $F \subset \mathbb{R}^n$ as

$$P(\mathcal{E}; F) = \sum_{0 \le h < k \le N} \mathcal{H}^{n-1}(F \cap \mathcal{E}(h,k)).$$

The **perimeter** of $\mathcal{E}$ is simply defined as

$$P(\mathcal{E}) = P(\mathcal{E};\mathbb{R}^n) = \sum_{0\le h<k\le N} \mathcal{H}^{n-1}(\mathcal{E}(h,k))\,. \tag{29.5}$$

Remark 29.3 By Exercise 12.8 and Remark 15.2, the perimeter of clusters is invariant to rigid motions, and satisfies the scaling law $P(\lambda\,\mathcal{E}) = \lambda^{n-1}P(\mathcal{E})$, $\lambda > 0$ (here, $\lambda\,\mathcal{E} = \{\lambda\,\mathcal{E}(h)\}_{h=1}^N$).

The distance in $F \subset \mathbb{R}^n$ of two N-clusters $\mathcal{E}$ and $\mathcal{E}'$ of $\mathbb{R}^n$ is defined as

$$d_F(\mathcal{E},\mathcal{E}') = \sum_{h=1}^N \Big|F \cap \Big(\mathcal{E}(h)\Delta\mathcal{E}'(h)\Big)\Big|\,.$$

We simply set $d(\mathcal{E},\mathcal{E}') = d_{\mathbb{R}^n}(\mathcal{E},\mathcal{E}')$. With this notation at hand, we say that a sequence of N-clusters $\{\mathcal{E}_k\}_{k\in\mathbb{N}}$ in $\mathbb{R}^n$ **locally converges to** $\mathcal{E}$, and write $\mathcal{E}_k \overset{\text{loc}}{\to} \mathcal{E}$, if for every compact set $K \subset \mathbb{R}^n$ we have $d_K(\mathcal{E}_k,\mathcal{E}) \to 0$ as $k \to \infty$. We say that $\{\mathcal{E}_k\}_{k\in\mathbb{N}}$ **converges to** $\mathcal{E}$, and write $\mathcal{E}_k \to \mathcal{E}$, if $d(\mathcal{E}_k,\mathcal{E}) \to 0$ as $k \to \infty$. We now prove a formula for the relative perimeter of clusters which, in turn, leads immediately to infer the lower semicontinuity of the relative perimeter in an open set with respect to local convergence, as well as a basic compactness criterion (Proposition 29.5).

Proposition 29.4 *If $\mathcal{E}$ is an N-cluster in $\mathbb{R}^n$, then for every $F \subset \mathbb{R}^n$ we have*

$$P(\mathcal{E};F) = \frac{1}{2}\sum_{h=0}^N P(\mathcal{E}(h);F)\,. \tag{29.6}$$

In particular, if A is open in $\mathbb{R}^n$ and $\mathcal{E}_k \overset{\text{loc}}{\to} \mathcal{E}$, then

$$P(\mathcal{E};A) \le \liminf_{k\to\infty} P(\mathcal{E}_k;A)\,. \tag{29.7}$$

Proof Clearly, (29.7) follows from (29.6) and Proposition 12.15. To deduce (29.6) from (29.5) it suffices to show that, for every $h = 1,\dots,N$,

$$\mathcal{H}^{n-1}\Big(\partial^*\mathcal{E}(h) \setminus \bigcup_{0\le k\le N,\,k\ne h} \mathcal{E}(h,k)\Big) = 0\,. \tag{29.8}$$

We now prove (29.8), using the notation $M_1 \approx M_2$ to indicate that M_1 and M_2 are $\mathcal{H}^{n-1}$-equivalent. By Federer's theorem (Theorem 16.2), if h, j, and k are three distinct indexes in $\{0,\dots,N\}$, then

$$\mathcal{E}(h)^{(1/2)} \cap \mathcal{E}(k)^{(1)} = \emptyset\,, \tag{29.9}$$

$$\mathcal{E}(h)^{(1/2)} \cap \mathcal{E}(k)^{(1/2)} \cap \mathcal{E}(j)^{(0)} \approx \mathcal{E}(h,k)\,, \tag{29.10}$$

where, as usual, $E^{(t)}$ denotes the set of points of density $t \in [0, 1]$ for $E \subset \mathbb{R}^n$. We now fix $h \in \{0, \dots, N\}$: by Federer's theorem and (29.9), if $k \neq h$, then

$$\begin{aligned}\partial^*\mathcal{E}(h) &\approx \left(\mathcal{E}(k)^{(0)} \cap \partial^*\mathcal{E}(h)\right) \cup \left(\mathcal{E}(k)^{(1)} \cap \partial^*\mathcal{E}(h)\right) \cup \left(\mathcal{E}(k)^{(1/2)} \cap \partial^*\mathcal{E}(h)\right)\\ &\approx \left(\mathcal{E}(k)^{(0)} \cap \mathcal{E}(h)^{(1/2)}\right) \cup \mathcal{E}(h, k);\end{aligned}$$

again by Federer's theorem, (29.9), and (29.10), if $j \neq h, k$, then

$$\begin{aligned}\mathcal{E}(k)^{(0)} \cap \mathcal{E}(h)^{(1/2)} &\approx \left(\mathcal{E}(j)^{(0)} \cap \mathcal{E}(k)^{(0)} \cap \mathcal{E}(h)^{(1/2)}\right)\\ &\quad \cup \left(\mathcal{E}(j)^{(1)} \cap \mathcal{E}(k)^{(0)} \cap \mathcal{E}(h)^{(1/2)}\right)\\ &\quad \cup \left(\mathcal{E}(j)^{(1/2)} \cap \mathcal{E}(k)^{(0)} \cap \mathcal{E}(h)^{(1/2)}\right)\\ &\approx \left(\mathcal{E}(j)^{(0)} \cap \mathcal{E}(k)^{(0)} \cap \mathcal{E}(h)^{(1/2)}\right) \cup \mathcal{E}(h, j);\end{aligned}$$

we may thus prove (29.8) by applying this argument finitely many times. □

Proposition 29.5 (Compactness criterion for clusters) *If $R > 0$, $\{\mathcal{E}_k\}_{k\in\mathbb{N}}$ is a sequence of N-clusters in $\mathbb{R}^n$, and*

$$\sup_{k\in\mathbb{N}} P(\mathcal{E}_k) < \infty\,, \tag{29.11}$$

$$\inf_{k\in\mathbb{N}} \min_{1\le h\le N} |\mathcal{E}_k(h)| > 0\,, \tag{29.12}$$

$$\mathcal{E}_k(h) \subset B_R\,, \qquad \forall k \in \mathbb{N}\,, h = 1, \dots, N\,, \tag{29.13}$$

then there exist an N-cluster $\mathcal{E}$ and a sequence $k(i) \to \infty$ as $i \to \infty$ such that $\mathcal{E}_{k(i)} \to \mathcal{E}$ as $i \to \infty$.

Proof By (29.6), for every $k \in \mathbb{N}$ and $h = 1, \dots, N$ we have $P(\mathcal{E}_k(h)) \le 2\,P(\mathcal{E}_k)$ and $\mathcal{E}_k(h) \subset B_R$. By (29.11), (29.13), and Theorem 12.26 we find sets of finite perimeter $\{\mathcal{E}(h)\}_{h=1}^N$ and a sequence $k(i) \to \infty$ as $i \to \infty$ such that, for every $h = 1, \dots, N$, we have $\mathcal{E}_{k(i)}(h) \to \mathcal{E}(h)$ as $i \to \infty$. As a consequence, $\{\mathcal{E}(h)\}_{h=1}^N$ is disjoint (up to modifications by sets of measure zero). Moreover, we have $|\mathcal{E}(h)| > 0$ for every $h = 1, \dots, N$ thanks to (29.12). Thus $\mathcal{E} = \{\mathcal{E}(h)\}_{h=1}^N$ defines a cluster with the desired properties. □

Exercise 29.6 (Density properties at interfaces) If $\mathcal{E}$ is an N-cluster in $\mathbb{R}^n$ and $x \in \mathcal{E}(h, k)$, $0 \le h < k \le N$, $j \neq h, k$, then

$$\nu_{\mathcal{E}(h)}(x) = -\nu_{\mathcal{E}(k)}(x)\,, \tag{29.14}$$

$$\theta_n(\mathcal{E}(j))(x) = 0\,, \tag{29.15}$$

$$\theta_{n-1}(\partial^*\mathcal{E}(j))(x) = 0\,. \tag{29.16}$$

Hint: Apply the results and the arguments of Chapter 16.

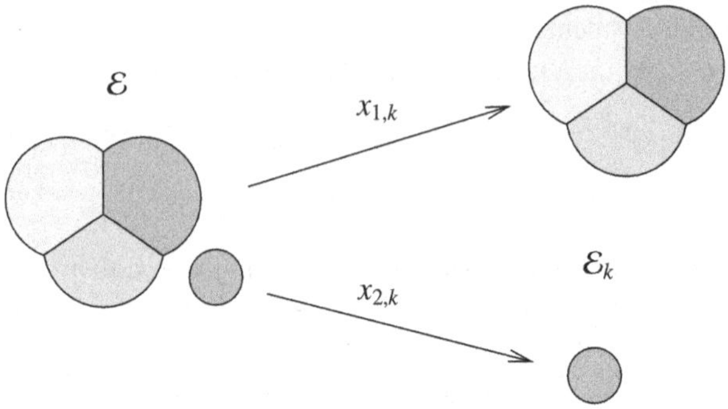

Figure 29.2 If the "disconnected" cluster $\mathcal{E}$ depicted on the left were minimizing, then we may define a minimizing sequence $\{\mathcal{E}_k\}_{k\in\mathbb{N}}$ for the partitioning problem associated with $\mathbf{m}(\mathcal{E})$ by applying two different sequences of translations $\{x_{1,k}\}_{k\in\mathbb{N}}$ and $\{x_{2,k}\}_{k\in\mathbb{N}}$ with $|x_{1,k}-x_{2,k}|\to\infty$ to the two "components" of $\mathcal{E}$. In this case, *any* sequence of the form $\{x_k+\mathcal{E}_k\}_{k\in\mathbb{N}}$ for $\{x_k\}_{k\in\mathbb{N}}\subset\mathbb{R}^n$ would not admit a convergent subsequence. Hence, in this example, the only way to gain compactness is to identify the two "diverging components", and then apply them to different sequences of translations.

Exercise 29.7 If $\mathcal{E}$ is an N-cluster and $\Lambda\subset\{0,\dots,N\}$, then

$$\mathcal{H}^{n-1}\Big(\partial^*\Big(\bigcup_{h\in\Lambda}\mathcal{E}(h)\Big)\setminus\bigcup_{h\in\Lambda,k\notin\Lambda}\mathcal{E}(h,k)\Big)=0\,. \tag{29.17}$$

Hint: If $h,k\in\Lambda$, then every $x\in\mathcal{E}(h,k)$ is of density 1 for $\bigcup_{i\in\Lambda}\mathcal{E}(i)$.

29.2 Strategy of proof

We now describe the strategy of proof of Theorem 29.1. Given $\mathbf{m}\in\mathbb{R}^N_+$, we want to prove the existence of a minimizer in the variational problem

$$\gamma=\inf\Big\{P(\mathcal{E}):\mathbf{m}(\mathcal{E})=\mathbf{m}\Big\}. \tag{29.18}$$

Taking into account Propositions 29.4 and 29.5, the only issue here is proving the existence of a minimizing sequence $\{\mathcal{E}_k\}_{k\in\mathbb{N}}$,

$$\lim_{k\to\infty}P(\mathcal{E}_k)=\gamma\,,\qquad \mathbf{m}(\mathcal{E}_k)=\mathbf{m}\,,\qquad \forall k\in\mathbb{N}\,, \tag{29.19}$$

satisfying the uniform confinement assumption (29.13) (indeed, (29.19) guarantees immediately the validity of (29.11) and (29.12)). Of course, this is not immediate due to the translation invariance of the problem (see also Remark 29.9). Precisely, if a minimizer $\mathcal{E}$ in (29.18) exists as we claim, then any sequence of the form $\{x_k+\mathcal{E}\}_{k\in\mathbb{N}}$ for $\{x_k\}_{k\in\mathbb{N}}\subset\mathbb{R}^n$ would be minimizing, even in case $|x_k|\to\infty$ as $k\to\infty$. If this were the only possible difficulty, then we may expect to establish the compactness of every minimizing sequence *up to translations*. In fact, an additional difficulty arises from the fact that, in principle, there could be "disconnected" minimizers. Then, by applying independent diverging sequences of translation to each "component" of the minimizer, we may construct a minimizing sequence which does not admit any compact subsequence *even up to translations* (see Figure 29.2). The only possibility of modifying such a minimizing sequence into a uniformly confined one would then be to isolate its various diverging components, to prove the uniform boundedness of their diameters, and then to apply a different sequence of translations to each sequence of distinct components. We thus aim to reduce to the situation described in the following proposition.

Proposition 29.8 *If $R > 0$, $L \in \mathbb{N}$, $\{\mathcal{E}_k\}_{k\in\mathbb{N}}$ are N-clusters in $\mathbb{R}^n$ satisfying (29.11) and (29.12), and $\{\Omega_k\}_{k\in\mathbb{N}}$ are finite sets of points in $\mathbb{R}^n$, with*

$$\mathcal{E}_k(h) \subset \bigcup_{x\in\Omega_k} B(x,R), \qquad \forall k \in \mathbb{N}, 1 \le h \le N, \tag{29.20}$$

$$\mathcal{H}^0(\Omega_k) \le L, \qquad \forall k \in \mathbb{N}, \tag{29.21}$$

then there exist N-clusters $\{\mathcal{E}'_k\}_{k\in\mathbb{N}}$ with $P(\mathcal{E}_k) = P(\mathcal{E}'_k)$, $\mathbf{m}(\mathcal{E}_k) = \mathbf{m}(\mathcal{E}'_k)$, and

$$\mathcal{E}'_k(h) \subset B_{13L^2R}, \qquad \forall k \in \mathbb{N}, 1 \le h \le N. \tag{29.22}$$

In particular, there exists an N-cluster $\mathcal{E}$ such that, up to extracting a subsequence, $\mathcal{E}'_k \to \mathcal{E}$. Finally, if $\{\mathcal{E}_k\}_{k\in\mathbb{N}}$ is further assumed to be a minimizing sequence for the partitioning problem (29.18), then $\{\mathcal{E}'_k\}_{k\in\mathbb{N}}$ is a minimizing sequence for (29.18) too, and $\mathcal{E}$ is a minimizer in (29.18).

Postponing the simple proof of this proposition until the end of the section, we now briefly describe the argument we shall adopt in order to construct a minimizing sequence satisfying the confinement assumptions (29.20)–(29.21); see also Figure 29.3. In Section 29.3, we prove a *nucleation lemma* which, applied to the chambers of the clusters of any minimizing sequence, will allow us to cover each $\mathcal{E}_k$ with a finite family $\mathcal{F}_k$ of balls of fixed radius S, up to a small error ε_0 in volume, and with $\sup_{k\in\mathbb{N}} \#\mathcal{F}_k < \infty$. Denoting by F_k the closed set defined as the union of the balls in $\mathcal{F}_k$, and applying the *truncation lemma* from Section 29.4, we shall then prove the existence of a sequence $\{r_k\}_{k\in\mathbb{N}}$, with $0 \le r_k \le C(n)\,\varepsilon_0^{1/n}$, such that, if $\mathcal{E}'_k$ denotes the cluster obtained by setting

$$\mathcal{E}'_k(h) = \mathcal{E}_k(h) \cap \left\{x \in \mathbb{R}^n : \mathrm{dist}(x, F_k) < r_k\right\}, \qquad 1 \le h \le N,$$

then

$$P(\mathcal{E}'_k) \le P(\mathcal{E}_k) - \frac{d(\mathcal{E}_k, \mathcal{E}'_k)}{C(n)\,\varepsilon_0^{1/n}}. \tag{29.23}$$

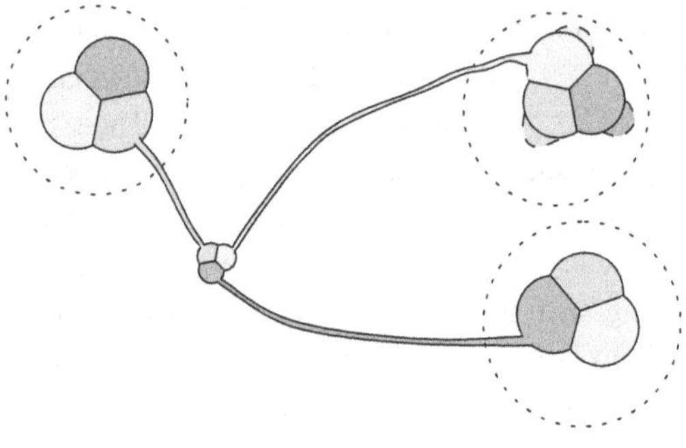

Figure 29.3 For each cluster $\mathcal{E}_k$, we find finitely many balls covering $\mathcal{E}_k$ up to a small error ε_0 (in the picture, which should be visualized in three-dimensions, ε_0 is the volume of the three cylindrical tubes, which may have arbitrarily small perimeter). Truncating the cluster with these balls, we decrease the perimeter. The volume lost in the truncation is recovered by slightly pushing the cluster inside the truncating balls, with a controlled increase of perimeters (in the picture, this modification corresponds to adding the regions with dashed boundary to the upper-right part of the cluster). Things can be tuned so that the modified truncated sequence is a minimizing sequence which satisfies the confinement assumptions of Proposition 29.8, and thus admits a convergent subsequence.

By construction, the sequence $\{\mathcal{E}'_k\}_{k\in\mathbb{N}}$ satisfies the confinement assumptions (29.20)–(29.21) with $R = S + C(n)\,\varepsilon_0^{1/n}$. Moreover, by (29.23),

$$\limsup_{k\to\infty} P(\mathcal{E}'_k) \le \gamma\,.$$

It may, however, happen that, in the truncation process, we have strictly decreased some volumes, that is, it could be that, for some $h = 1,\dots,N$ and for infinitely many ks, $\mathbf{m}(\mathcal{E}'_k)(h) < \mathbf{m}(\mathcal{E}_k)(h)$. We first notice that we control the size of any possible loss in volume, $d(\mathcal{E}_k,\mathcal{E}'_k)$ necessarily being bounded by $C(n)\,\varepsilon_0$. In Theorem 29.14 (Section 29.6) we show that, in the same vein of Lemma 17.21, any suitably small amount of volume can be recovered at the price of a proportional increase in the perimeter. Precisely, we show that if ε_0, and thus $d(\mathcal{E}_k,\mathcal{E}'_k)$, is small enough, then there exists a sequence of clusters $\{\mathcal{E}''_k\}_{k\in\mathbb{N}}$ (where each $\mathcal{E}''_k$ is a local variations of the corresponding $\mathcal{E}'_k$) such that

$$\mathbf{m}(\mathcal{E}''_k) = \mathbf{m}(\mathcal{E}_k)\,, \tag{29.24}$$

$$\mathcal{E}''_k(h)\Delta\mathcal{E}'_k(h) \subset\subset F_k\,, \tag{29.25}$$

$$P(\mathcal{E}''_k) \le P(\mathcal{E}'_k) + C\,d(\mathcal{E}_k,\mathcal{E}'_k)\,, \tag{29.26}$$

whenever $k\in\mathbb{N}$, $0\le h\le N$. A crucial point here is that the constant C in (29.26) will only depend on $\{\mathcal{E}_k\}_{k\in\mathbb{N}}$, but not on ε_0. In this way, selecting ε_0

small enough to entail $C(n)\,\varepsilon_0^{1/n} < C^{-1}$, and combining (29.23) and (29.26), we shall immediately obtain that

$$\limsup_{k\to\infty} P(\mathcal{E}_k'') \le \gamma\,.$$

By (29.24), $\{\mathcal{E}_k''\}_{k\in\mathbb{N}}$ will then be a minimizing sequence for (29.18). Moreover, by (29.25), $\{\mathcal{E}_k''\}_{k\in\mathbb{N}}$ will satisfy the confinement assumptions (29.20)–(29.21) of Proposition 29.8, which will then be applied to conclude the proof of Theorem 29.1.

Proof of Proposition 29.8 For every $k \in \mathbb{N}$ there exists a finite family of closed connected sets $\{F_{k,i}\}_{i=1}^{L(k)}$ with $\mathrm{dist}(F_{k,i}, F_{k,j}) > 0$ if $i \ne j$, and

$$\bigcup_{x\in\Omega_k} \overline{B}(x,R) = \bigcup_{i=1}^{L(k)} F_{k,i}\,.$$

By (29.21), for every $k \in \mathbb{N}$ and $1 \le i \le L(k)$, we have

$$L(k) \le L\,, \qquad \mathrm{diam}\, F_{k,i} \le 2LR\,.$$

In particular, there exists $\{x_{k,i} : k \in \mathbb{N}\,, 1 \le i \le L(k)\} \subset \mathbb{R}^n$ such that

$$x_{k,i} + F_{k,i} \subset B\big(9LR\,i\,e_n, 4LR\big)\,, \qquad \forall k \in \mathbb{N}\,, 1 \le i \le L(k)\,.$$

The constants have been chosen to ensure the balls $\{B(9LR\,i\,e_n, 4LR)\}_{i=1}^L$ lie at mutually positive distance, and their union is contained in the ball $\overline{B}_{13\,L^2 R}$. In this way, if we define, for every $k \in \mathbb{N}$,

$$f_k \colon \bigcup_{x\in\Omega_k} B(x,R) \to \mathbb{R}^n\,,$$

as $f_k(x) = x + x_{k,i}$ if $x \in F_{k,i}$, then we easily check that the cluster $\mathcal{E}_k'$,

$$\mathcal{E}_k'(h) = f_k(\mathcal{E}_k(h))\,, \qquad k \in \mathbb{N}\,, 1 \le h \le N\,,$$

satisfies $P(\mathcal{E}_k) = P(\mathcal{E}_k')$ (by Exercise 12.24), $\mathbf{m}(\mathcal{E}_k) = \mathbf{m}(\mathcal{E}_k')$ and (29.22). □

Remark 29.9 We have already met two variational problems with volume constraints in unbounded regions. The first one was of course the Euclidean isoperimetric problem (Chapter 14), corresponding to the case $N = 1$ in (29.18). The second one was the sessile liquid drop problem from Section 19.6. In both cases, we avoided the compactness issues described above by exploiting symmetrization principles. However, as arbitrary values of N and $\mathbf{m}$ minimizing clusters are not expected to possess any symmetry, we cannot rely on this kind of approach here.

29.3 Nucleation lemma

We prove here the *nucleation lemma* already described in Section 29.2. When applied to a sequence of sets $\{E_h\}_{h\in\mathbb{N}}$ with equibounded volumes and perimeters, this lemma provides a sequence $\{\mathcal{F}_h\}_{h\in\mathbb{N}}$ of finite families of balls of radius 2 with bounded cardinalities, such that $\mathcal{F}_h$ covers E_h up to an error ε, and $|E_h \cap B(x,1)|$ is bounded from below by ε^n (uniformly on $x \in \mathcal{F}_h$ and $h \in \mathbb{N}$).

Lemma 29.10 (Nucleation) *For every $n \geq 2$ there exists a positive constant $c(n)$ with the following property. If E is of finite perimeter, $0 < |E| < \infty$, and*

$$\varepsilon \leq \min\left\{|E|, \frac{P(E)}{2\,n\,c(n)}\right\}, \tag{29.27}$$

then there exists a finite family of points $I \subset \mathbb{R}^n$ such that

$$\Big|E \setminus \bigcup_{x\in I} B(x,2)\Big| < \varepsilon, \tag{29.28}$$

$$\Big|E \cap B(x,1)\Big| \geq \left(c(n)\,\frac{\varepsilon}{P(E)}\right)^n, \qquad \forall x \in I. \tag{29.29}$$

Moreover, $|x-y| > 2$ for every $x, y \in I$, $x \neq y$, and

$$\# I < |E| \left(\frac{P(E)}{c(n)\,\varepsilon}\right)^n. \tag{29.30}$$

Proof *Step one:* We reduce to proving the following *claim*: for every $n \geq 2$, there exists a positive constant $c(n)$ such that, if $F \subset \mathbb{R}^n$ is closed, and

$$\Big|\Big\{x \in E : \operatorname{dist}(x,F) > 1\Big\}\Big| \geq \varepsilon, \tag{29.31}$$

then there exists $x \in E^{(1)}$ with $\operatorname{dist}(x,F) > 1$ and

$$\Big|E \cap B(x,1)\Big| \geq \left(c(n)\,\frac{\varepsilon}{P(E)}\right)^n. \tag{29.32}$$

Indeed, assuming the claim, we define $I \subset \mathbb{R}^n$ as follows. First, we define x_1 by applying the claim to $F = \emptyset$. Having defined $\{x_i\}_{i=1}^s$ such that $|x_i - x_j| > 2$ for $1 \leq i < j \leq s$, and (29.32) is in force for $x = x_i$, we now have either that the lemma is proved by choosing $I = \{x_i\}_{i=1}^s$, or

$$\Big|E \setminus \bigcup_{i=1}^s B(x_i,2)\Big| \geq \varepsilon.$$

In the latter case, we apply the claim with $F = \bigcup_{i=1}^s \overline{B}(x_i,1)$ to define $x_{s+1} \in E^{(1)}$ such that (29.32) holds true for $x = x_{s+1}$ and $|x_{s+1} - x_i| > 2$ for $i = 1, \dots, s$.

Since $|E| < \infty$ and the balls $\{B(x_i, 1)\}_{i=1}^s$ are disjoint, the process ends up in finitely many steps, and (29.30) is immediately deduced from (29.29).

Step two: Preparing to prove the claim, we show that, if

$$\alpha > n\,, \qquad x \in E^{(1)}\,, \qquad \left|E \cap B(x,1)\right| \le \frac{1}{(2\alpha)^n}\,,$$

then there exists $r_x \in (0,1)$ such that

$$P(E; B(x, r_x)) > \alpha\, |E \cap B(x, r_x)|\,.$$

By contradiction, and setting as usual $m(r) = |E \cap B(x,r)|$, $r > 0$, let us assume that $\alpha\, m(r) \ge P(E; B(x,r))$ for every $r \in (0,1)$. Taking into account that for a.e. $r > 0$ we have $m'(r) = \mathcal{H}^{n-1}(E \cap \partial B(x,r))$, and by (15.15), we find that

$$\alpha\, m(r) + m'(r) \ge P(E; B(x,r)) + \mathcal{H}^{n-1}(E \cap \partial B(x,r)) = P(E \cap B(x,r))\,,$$

for a.e. $r \in (0,1)$. By the non-sharp isoperimetric inequality (12.38) (which is used in place of (14.2) just to work with simpler constants), we thus have

$$\alpha\, m(r) + m'(r) \ge m(r)^{(n-1)/n}\,, \qquad \text{for a.e. } r \in (0,1)\,.$$

Since m is increasing and, by assumption, $m(1) < (1/2\alpha)^n$, we have $\alpha\, m(r) \le m(r)^{(n-1)/n}/2$ for every $r \in (0,1)$, that is

$$m'(r) \ge \frac{m(r)^{(n-1)/n}}{2}\,, \qquad \text{for a.e. } r \in (0,1)\,.$$

By $x \in E^{(1)}$ we have $m(r) > 0$ for every $r > 0$, hence we may divide by $m(r)^{(n-1)/n}$ and reformulate the above inequality as

$$\left(m(r)^{1/n}\right)' \ge \frac{1}{2n} \quad \Rightarrow \quad m(r) \ge \left(\frac{r}{2n}\right)^n\,, \qquad \text{for a.e. } r \in (0,1)\,.$$

Letting $r \to 1^-$ we find $|E \cap B(x,1)| \ge (1/2n)^n$, in contradiction to $\alpha > n$.

Step three: We prove the claim of step one. Let F be a closed set satisfying (29.31). By contradiction, assume that if $x \in E^{(1)}$ and $\mathrm{dist}(x, F) > 1$, then

$$\left|E \cap B(x,1)\right| \le \left(c(n)\, \frac{\varepsilon}{P(E)}\right)^n\,, \tag{29.33}$$

with $c(n)$ to be determined in a moment. If α is such that the right-hand side of (29.33) equals $(1/2\alpha)^n$, that is, if

$$\alpha = \frac{1}{2\,c(n)}\,\frac{P(E)}{\varepsilon}\,,$$

then $\alpha > n$ by (29.27). By step two, for every $x \in E^{(1)}$ with $\mathrm{dist}(x, F) > 1$ there exists $r_x \in (0, 1)$ such that

$$P(E; B(x, r_x)) > \alpha |E \cap B(x, r_x)| . \tag{29.34}$$

Applying the Besicovitch covering theorem (Theorem 5.1) to the family of balls

$$\mathcal{F} = \left\{ \overline{B}(x, r_x) : x \in E^{(1)} , \mathrm{dist}(x, F) > 1 \right\} ,$$

we find a countably disjoint subfamily $\mathcal{F}'$ of $\mathcal{F}$ such that

$$\begin{aligned} \left| \left\{ x \in E : \mathrm{dist}(x, F) > 1 \right\} \right| &\le \xi(n) \sum_{\overline{B}(x,r_x) \in \mathcal{F}'} |E \cap B(x, r_x)| \\ &< \frac{\xi(n)}{\alpha} \sum_{\overline{B}(x,r_x) \in \mathcal{F}'} P(E; B(x, r_x)) , \end{aligned}$$

where the last inequality follows from (29.34). Since $\mathcal{F}'$ is countable and disjoint we have $\sum_{B \in \mathcal{F}'} P(E; B) = P(E; \bigcup_{B \in \mathcal{F}'} B)$, so that

$$\left| \left\{ x \in E : \mathrm{dist}(x, F) > 1 \right\} \right| < \frac{\xi(n)}{\alpha} P(E) .$$

Setting $c(n) = 1/2\xi(n)$ we contradict (29.31). □

Remark 29.11 The argument of step three in the above proof performed with $F = \emptyset$ and ε realizing (29.27) with equality shows that for every set of finite perimeter E with $0 < |E| < \infty$ there exists $x \in \mathbb{R}^n$ such that

$$\left| E \cap B(x, 1) \right| \ge \min \left\{ c(n) \frac{|E|}{P(E)} , \frac{1}{2n} \right\}^n .$$

29.4 Truncation lemma

We continue preparing the ground for the argument sketched in Section 29.2, by proving the following *truncation lemma.*

Lemma 29.12 (Truncation) *If F is a closed set in $\mathbb{R}^n$, $u(x) = \mathrm{dist}(x, F)$, $x \in \mathbb{R}^n$, $\alpha > 0$ and $\mathcal{E}$ is an N-cluster in $\mathbb{R}^n$ with*

$$\sum_{h=1}^{N} \left| \mathcal{E}(h) \setminus F \right| \le \alpha , \tag{29.35}$$

then there exists $r_0 \in [0, 7n\alpha^{1/n}]$ such that the N-cluster $\mathcal{E}'$ in $\mathbb{R}^n$ defined by

$$\mathcal{E}'(h) = \mathcal{E}(h) \cap \{u \le r_0\} , \qquad 1 \le h \le N , \tag{29.36}$$

satisfies

$$P(\mathcal{E}') \le P(\mathcal{E}) - \frac{d(\mathcal{E},\mathcal{E}')}{4\alpha^{1/n}}. \tag{29.37}$$

Proof If $\sum_{h=1}^N |\mathcal{E}(h)\setminus F| = 0$, then we set $r_0 = 0$. Otherwise, we prove that

(i) either $\mathcal{E}(h) \subset \{u \le 7n\alpha^{1/n}\}$ for $1 \le h \le N$ (and then set $r_0 = 7n\alpha^{1/n}$),
(ii) or there exists a positive $r_0 < 7n\alpha^{1/n}$ such that, if $\mathcal{E}'$ is defined as in (29.36), then (29.37) holds true.

To show this, let us consider the decreasing function $m : [0,\infty) \to [0,\alpha]$ defined by $m(r) = \sum_{h=1}^N |\mathcal{E}(h) \cap \{u > r\}|$, $r > 0$. Since u is a Lipschitz function with $|\nabla u| = 1$ a.e. on $\mathbb{R}^n$, by the coarea formula (18.3) for every $r > 0$ we have

$$m(r) = \sum_{h=1}^N \int_r^\infty \mathcal{H}^{n-1}\big(\{u = t\} \cap \mathcal{E}(h)\big)\,dt,$$

so that, for a.e $r > 0$,

$$m'(r) = -\sum_{h=1}^N \mathcal{H}^{n-1}\big(\{u = r\} \cap \mathcal{E}(h)\big).$$

Since $m(0) = \sum_{h=1}^N |\mathcal{E}(h)\setminus F| > 0$, if $\operatorname{spt} m = [0, r_1]$, then $r_1 > 0$. Arguing by contradiction, we now assume that $r_1 \ge 7n\alpha^{1/n}$ (so that $m(r) > 0$ for every $r < 7n\alpha^{1/n}$), and that, if $\mathcal{E}^r(h) = \mathcal{E}(h) \cap \{u \le r\}$ $(1 \le h \le N)$, then

$$P(\mathcal{E}) < P(\mathcal{E}^r) + \frac{m(r)}{4\alpha^{1/n}}, \qquad \forall r < 7n\alpha^{1/n}.$$

If this is the case, then, by (29.6), for a.e. $r < 7n\alpha^{1/n}$ we have

$$P\big(\mathcal{E};\{u > r\}\big) < \sum_{h=1}^N \mathcal{H}^{n-1}\big(\{u = r\} \cap \mathcal{E}(h)\big) + \frac{m(r)}{4\alpha^{1/n}}. \tag{29.38}$$

Since $P(\mathcal{E}(h) \cap \{u > r\}) = P(\mathcal{E}(h);\{u > r\}) + \mathcal{H}^{n-1}(\{u = r\} \cap \mathcal{E}(h))$ for a.e. r (see Exercise 18.3), adding $(1/2)\sum_{h=1}^N \mathcal{H}^{n-1}(\{u = r\} \cap \mathcal{E}(h))$ to both sides of (29.38) and taking (29.6) into account, we find that, for a.e. $r < 7n\alpha^{1/n}$,

$$\frac{1}{2}\sum_{h=1}^N P\big(\mathcal{E}(h) \cap \{u > r\}\big) < \frac{3}{2}\sum_{h=1}^N \mathcal{H}^{n-1}\big(\{u = r\} \cap \mathcal{E}(h)\big) + \frac{m(r)}{4\alpha^{1/n}}. \tag{29.39}$$

By the non-sharp isoperimetric inequality (12.38), we find

$$\frac{1}{2}\sum_{h=1}^N P\big(\mathcal{E}(h) \cap \{u > r\}\big) \ge \frac{1}{2} P\Big(\bigcup_{h=1}^N \mathcal{E}(h) \cap \{u > r\}\Big) \ge \frac{m(r)^{(n-1)/n}}{2},$$

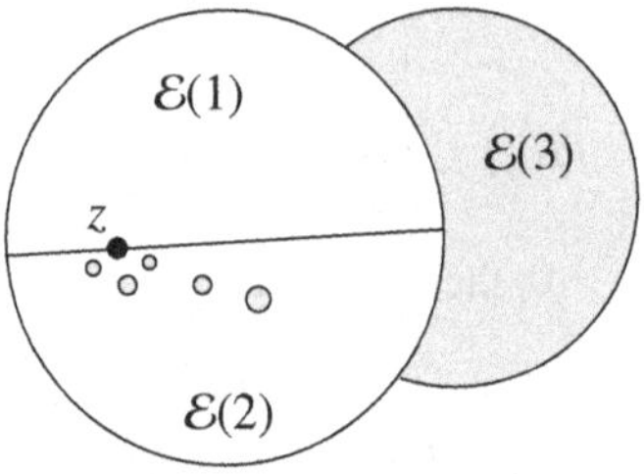

Figure 29.4 Given a ball $B(z, \varepsilon)$, $z \in \mathcal{E}(h, k)$, $0 \le h < k \le N$, it may happen that z is an accumulation point for some $\partial^*\mathcal{E}(i)$ with $i \ne h, k$. In particular, every local variation supported in $B(z, \varepsilon)$ could possibly deform *all* the chambers of $\mathcal{E}$. However, by the density estimates of Exercise 29.6, the only relevant volume variations are those of $\mathcal{E}(h)$ and $\mathcal{E}(k)$.

so that, by (29.39), and taking also into account that, by (29.35), $m(r) \le m(0)^{1/n} m(r)^{(n-1)/n} \le \alpha^{1/n} m(r)^{(n-1)/n}$, we have

$$\frac{m(r)^{(n-1)/n}}{2} < -\frac{3}{2}\, m'(r) + \frac{m(r)}{4\alpha^{1/n}} \le -\frac{3}{2}\, m'(r) + \frac{m(r)^{(n-1)/n}}{4},$$

for a.e. $r < 7n\alpha^{1/n}$. Since $m(r) > 0$ for every $r < 7n\alpha^{1/n}$, we may conclude that

$$\left(m(r)^{1/n}\right)' < -\frac{1}{6n}, \qquad \text{for a.e. } r < 7n\alpha^{1/n}.$$

Integrating over $r \in (0, 7n\alpha^{1/n})$, we find $m(0)^{1/n} \ge (7/6)\alpha^{1/n} > m(0)^{1/n}$. □

29.5 Infinitesimal volume exchanges

We now present a lemma which is the building block used to construct volume-fixing variations for clusters in Theorem 29.14. Precisely, we consider here the problem of deforming a given cluster $\mathcal{E}$ through a local variation to exchange volume between two given chambers $\mathcal{E}(h)$ and $\mathcal{E}(k)$ of $\mathcal{E}$, keeping at the same time the most explicit control we can obtain on the corresponding perimeter variation. In doing this, we have also to keep track of possible volume variations affecting the chambers $\mathcal{E}(i)$ with $i \ne h, k$; see Figure 29.4.

Lemma 29.13 (Infinitesimal volume exchange between two chambers) *If $n \ge 2$, $\mathcal{E}$ is a cluster in $\mathbb{R}^n$, $0 \le h < k \le N$, $z \in \mathcal{E}(h, k)$ and $\delta > 0$, then there exist positive constants ε_1 (depending on $\mathcal{E}$, z, and δ), ε_2 (depending on n, δ, and ε_1), and C_0 (depending on n and ε_1), and a one-parameter family of* *diffeomorphisms* $\{f_t\}_{|t|<\varepsilon_1}$ *with*

$$\Big\{x \in \mathbb{R}^n : x \neq f_t(x)\Big\} \subset\subset B(z,\varepsilon_1)\,, \qquad \forall |t| < \varepsilon_1\,, \tag{29.40}$$

which satisfies the following properties

(i) If $\mathcal{E}'$ *is a cluster,* $d(\mathcal{E},\mathcal{E}') < \varepsilon_2$ *(in particular, if* $\mathcal{E}' = \mathcal{E}$*), and* $|t| < \varepsilon_1$*, then*

$$\left|\frac{\mathrm{d}}{\mathrm{d}t}\left|f_t\Big(\mathcal{E}'(h)\Big) \cap B(z,\varepsilon_1)\right| - 1\right| < \delta\,, \tag{29.41}$$

$$\left|\frac{\mathrm{d}}{\mathrm{d}t}\left|f_t\Big(\mathcal{E}'(k)\Big) \cap B(z,\varepsilon_1)\right| + 1\right| < \delta\,, \tag{29.42}$$

$$\left|\frac{\mathrm{d}}{\mathrm{d}t}\left|f_t\Big(\mathcal{E}'(i)\Big) \cap B(z,\varepsilon_1)\right|\right| < \delta\,, \qquad \textit{for } i \neq h,k\,, \tag{29.43}$$

$$\left|\frac{\mathrm{d}^2}{\mathrm{d}t^2}\left|f_t\Big(\mathcal{E}'(i)\Big) \cap B(z,\varepsilon_1)\right|\right| < C_0\,, \qquad \textit{for } 0 \le i \le N\,, \tag{29.44}$$

(notice that $f_t(E) \cap B(z,\varepsilon_1) = f_t(E \cap B(z,\varepsilon_1))$ *for every* $E \subset \mathbb{R}^n$*).*

(ii) If Σ *is an* $\mathcal{H}^{n-1}$*-rectifiable set in* $\mathbb{R}^n$ *and* $|t| < \varepsilon_1$*, then*

$$\left|\mathcal{H}^{n-1}\Big(f_t(\Sigma)\Big) - \mathcal{H}^{n-1}(\Sigma)\right| \le C_0\,\mathcal{H}^{n-1}(\Sigma)\,|t|\,. \tag{29.45}$$

Proof *Step one:* Let $z \in \mathbb{R}^n$, $\varepsilon > 0$, $\nu \in S^{n-1}$, and let $u \in C^\infty_c((-n^{-1/2}, n^{-1/2}))$, $u \ge 0$, with $u \neq 0$. We define $v \in C^\infty_c(B)$ as

$$v(x) = c(n)\prod_{i=1}^n u(x_i)\,, \qquad x \in \mathbb{R}^n\,,$$

where $c(n)$ is chosen (depending on u) to have $\int_{\mathbb{R}^{n-1}} v(x',0)\mathrm{d}x' = 1$. Next, for every $\varepsilon > 0$, we define $v_\varepsilon \in C^\infty_c(B_\varepsilon)$ as $v_\varepsilon(x) = \varepsilon^{1-n}\,v(x/\varepsilon)$, $x \in \mathbb{R}^n$. In this way $\int_{\mathbb{R}^{n-1}} v_\varepsilon(x',0)\mathrm{d}x' = 1$, and, for a suitable constant $C(n)$,

$$|\nabla v_\varepsilon(x)| \le \frac{C(n)}{\varepsilon^n}\,, \qquad \forall x \in \mathbb{R}^n\,, \varepsilon > 0\,.$$

We now choose $Q_\nu \in \mathbf{O}(n)$ such that $Q_\nu(\nu) = e_n$, and, correspondingly, we define a smooth vector field $T[z,\varepsilon,\nu] \in C^\infty_c(B(z,\varepsilon);\mathbb{R}^n)$,

$$T[z,\varepsilon,\nu](x) = v_\varepsilon\Big(Q_\nu(x-z)\Big)\,\nu\,, \qquad x \in \mathbb{R}^n\,,$$

noticing that

$$\sup_{\mathbb{R}^n}\Big|\nabla T[z,\varepsilon,\nu]\Big| \le \frac{C(n)}{\varepsilon^n}\,. \tag{29.46}$$

Finally, we consider the smooth function $f\colon \mathbb{R}\times\mathbb{R}^n \to \mathbb{R}^n$ defined as

$$f(t,x) = f_t(x) = x + t\,T[z,\varepsilon,\nu](x)\,, \qquad (t,x) \in \mathbb{R}\times\mathbb{R}^n\,.$$

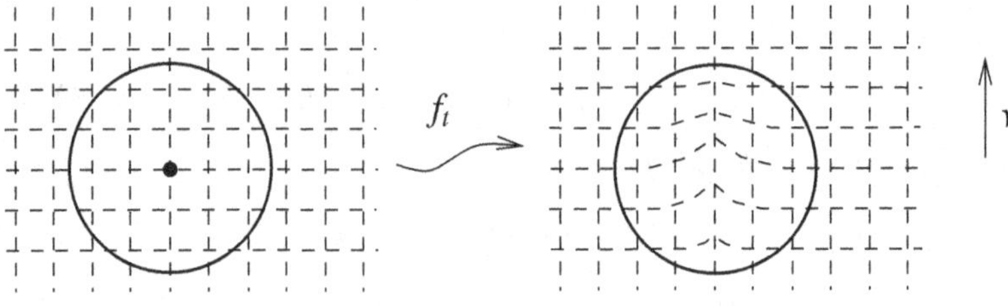

Figure 29.5 The local variation $\{f_t\}_{|t|<\varepsilon_1}$ pushes the space inside $B(z, \varepsilon_1)$ in the direction ν. The construction is performed so that the variation $\overline{f}_t$ defined by $\overline{z} \in \mathbb{R}^n$ and $\overline{\nu} \in S^{n-1}$ is obtained by composing f_t with a rigid motion taking z in $\overline{z}$ and ν in $\overline{\nu}$. Thus, both $|\nabla T[z, \varepsilon_1, \nu]|$ and $|\nabla T[\overline{z}, \varepsilon_1, \overline{\nu}]|$ are controlled by $C(n)/\varepsilon_1^n$.

Since $f_0(x) = x$ and $\operatorname{spt} T \subset B(z, \varepsilon)$ there exists $\varepsilon_1 > 0$ such that $\{f_t\}_{|t|<\varepsilon_1}$ is a one-parameter family of diffeomorphisms in $\mathbb{R}^n$ satisfying (29.40); see Figure 29.5.

Step two: Now let us fix $z \in \mathcal{E}(h, k)$, $0 \le h < k \le N$, and apply the construction of step one with $\varepsilon = \varepsilon_1$ and $\nu = \nu_{\mathcal{E}(h)}(z)$ in order to define $T = T[z, \varepsilon_1, \nu] \in C_c^\infty(B(z, \varepsilon_1); \mathbb{R}^n)$, and the corresponding one-parameter family of diffeomorphisms $\{f_t\}_{|t|<\varepsilon_1}$, $f_t(x) = x + t\,T(x)$, $x \in \mathbb{R}^n$. We now claim that, up to further decreasing the value of ε_1, and for a suitably chosen value of ε_2, properties (i) and (ii) hold true. To this end, let us first notice that, by the area formulae (8.1) and (11.6), if $0 \le i \le N$, $E \subset \mathbb{R}^n$ is of finite perimeter, and Σ is a $\mathcal{H}^{n-1}$-rectifiable set in $\mathbb{R}^n$, then

$$\Big|f_t(E) \cap B(z, \varepsilon_1)\Big| = \int_{E \cap B(z,\varepsilon_1)} J f_t(x)\, \mathrm{d}x\,, \tag{29.47}$$

$$\mathcal{H}^{n-1}\Big(f_t(\Sigma)\Big) = \int_\Sigma J^\Sigma f_t(x)\, \mathrm{d}\mathcal{H}^{n-1}(x)\,, \tag{29.48}$$

where, since $\nabla f_t = \mathrm{Id} + t\nabla T$, $J f_t$ and $J^\Sigma f_t(x)$ are smooth functions of the variable t. If V denotes a hyperplane of $\mathbb{R}^n$, then the map $S \in \mathbb{R}^n \otimes \mathbb{R}^n \mapsto J^V(\mathrm{Id} + S)$ is locally Lipschitz on $\mathbb{R}^n \otimes \mathbb{R}^n$, uniformly on V; taking (29.46) into account, we thus find that, for every $|t| < \varepsilon_1$ and for every $\mathcal{H}^{n-1}$-rectifiable set Σ in $\mathbb{R}^n$, and denoting by C a generic constant depending on ε_1 and n only,

$$|J^\Sigma f_t - 1| \le C\,|t| \qquad \mathcal{H}^{n-1}\text{-a.e. on } \Sigma\,;$$

in particular, by (29.48), we immediately deduce the validity of property (ii). We now focus on the proof of (i). Again by (29.46), and since T is compactly supported in $B(z,\varepsilon_1)$, we find that, if $|t|<\varepsilon_1$, then

$$\sup_{\mathbb{R}^n}\Big|\frac{d}{dt}Jf_t\Big|+\Big|\frac{d^2}{dt^2}Jf_t\Big|\le C\,,$$

$$\frac{d}{dt}Jf_t=\frac{d^2}{dt^2}Jf_t=0\,,\qquad\text{on }\mathbb{R}^n\setminus B(z,\varepsilon_1)\,.$$

By (29.47) we thus find, for every $|t|<\varepsilon_1$ and $E\subset\mathbb{R}^n$,

$$\Big|\frac{d^2}{dt^2}\big|f_t(E)\cap B(z,\varepsilon_1)\big|\Big|\le C\,,$$

from which (29.44) immediately follows; we similarly have, for $0\le i\le N$,

$$\Big|\frac{d}{dt}\big|f_t(\mathcal{E}'(i))\cap B(z,\varepsilon_1)\big|-\frac{d}{dt}\big|f_t(\mathcal{E}(i))\cap B(z,\varepsilon_1)\big|\Big|$$
$$\le C\big|\mathcal{E}'(i)\Delta\mathcal{E}(i)\big|\le C\,d(\mathcal{E},\mathcal{E}')\,;$$

hence, up to choosing ε_2 suitably small in terms of n, δ, and ε_1, it suffices to prove (29.41), (29.42), and (29.43) in the case that $\mathcal{E}'=\mathcal{E}$; similarly, taking (29.44) into account, up to decreasing the value of ε_1, we may directly prove (29.41), (29.42), and (29.43) on $\mathcal{E}'=\mathcal{E}$ and for $t=0$. To this end, let us now recall from Proposition 17.8 that, since $\operatorname{spt}T\subset\subset B(z,\varepsilon_1)$,

$$\frac{d}{dt}\Big|_{t=0}\big|f_t(E)\cap B(z,\varepsilon_1)\big|=\int_{B(z,\varepsilon_1)\cap\partial^*E}T\cdot\nu_E\,d\mathcal{H}^{n-1}\,. \tag{29.49}$$

If we thus set for brevity $Q_i=Q_{\nu_{\mathcal{E}(i)}}$, then by a change of variables and (29.49) we find

$$\frac{d}{dt}\Big|_{t=0}\Big|f_t\big(\mathcal{E}(i)\big)\cap B(z,\varepsilon_1)\Big|=\int_{B(z,\varepsilon_1)\cap\partial^*\mathcal{E}(i)}T\cdot\nu_{\mathcal{E}(i)}\,d\mathcal{H}^{n-1}$$
$$=\nu_{\mathcal{E}(h)}(z)\cdot\int_{B\cap(\partial^*\mathcal{E}(i)-z)/\varepsilon_1}v\big(Q_h(y)\big)\,\nu_{\mathcal{E}(i)}(z+\varepsilon_1 y)\,d\mathcal{H}^{n-1}(y)\,. \tag{29.50}$$

If $i\ne h,k$, then from (29.50) we find

$$\Big|\frac{d}{dt}\Big|_{t=0}\Big|f_t\big(\mathcal{E}(i)\big)\cap B(z,\varepsilon_1)\Big|\Big|\le\sup_{\mathbb{R}^n}|v|\,\frac{P(\mathcal{E}(i);B(z,\varepsilon_1))}{\varepsilon_1^{n-1}}\,;$$

since $z\in\mathcal{E}(h,k)$ and $i\ne h,k$, we have $\varepsilon^{1-n}P(\mathcal{E}(i);B(z,\varepsilon))\to0$ as $\varepsilon\to0^+$ (by (29.16)), and thus (29.43) follows (for $\mathcal{E}'=\mathcal{E}$ and $t=0$) if ε_1 is suitably small (depending on $\mathcal{E}$, z and δ). If $i=h$, then by (15.6) and (29.50), we find

$$\lim_{\varepsilon\to0}\nu_{\mathcal{E}(h)}(z)\cdot\int_{B\cap(\partial^*\mathcal{E}(h)-z)/\varepsilon}v\big(J_h(y)\big)\,\nu_{\mathcal{E}(h)}(z+\varepsilon y)\,d\mathcal{H}^{n-1}(y)$$
$$=\nu_{\mathcal{E}(h)}(z)\cdot\int_{B\cap\nu_{\mathcal{E}(h)}(z)^\perp}(v\circ J_h)\,\nu_{\mathcal{E}(h)}(z)\,d\mathcal{H}^{n-1}=\int_{\mathbb{R}^{n-1}}v(x',0)dx'=1\,,$$

which implies (29.41) for $\mathcal{E}' = \mathcal{E}$ and $t = 0$, provided ε_1 is small enough; similarly, if $i = k$, since $z \in \mathcal{E}(h,k)$ implies $\nu_{\mathcal{E}(k)}(z) = -\nu_{\mathcal{E}(h)}(z)$ and $\nu_{\mathcal{E}(k)}(z)^\perp = \nu_{\mathcal{E}(h)}(z)^\perp$, then, again by (15.6) and (29.50), we find

$$\lim_{\varepsilon\to 0} \nu_{\mathcal{E}(h)}(z)\cdot\int_{B\cap(\partial^*\mathcal{E}(k)-z)/\varepsilon} v\big(Q_h(y)\big)\,\nu_{\mathcal{E}(k)}(z+\varepsilon\,y)\,d\mathcal{H}^{n-1}(y)$$

$$= \nu_{\mathcal{E}(h)}(z)\cdot\int_{B\cap\nu_{\mathcal{E}(k)}(z)^\perp}(v\circ Q_h)\,\nu_{\mathcal{E}(k)}(z)\,d\mathcal{H}^{n-1} = -\int_{\mathbb{R}^{n-1}} v(x',0)dx' = -1\,,$$

which implies (29.42) for $\mathcal{E}' = \mathcal{E}$ and $t = 0$ if ε_1 is small enough. □

29.6 Volume-fixing variations

We prove here two results (Theorem 29.14 and Corollary 29.17) of fundamental technical importance in the proof of Almgren's theorem. These results carry into the context of clusters the volume-fixing variation lemma, Lemma 17.21, we met in the study of volume-constrained perimeter minimizers. We thus seek a way to modify (through local variations) a cluster $\mathcal{E}$ to achieve an arbitrary (but suitably small) change of the volumes of its chambers, with a controlled increase of perimeter. It is useful to introduce the space

$$V = \Big\{\mathbf{a}\in\mathbb{R}^{N+1} : \sum_{h=0}^{N}\mathbf{a}(h) = 0\Big\}. \tag{29.51}$$

Theorem 29.14 (Volume-fixing variations) *If $\mathcal{E}$ is an N-cluster in $\mathbb{R}^n$ then there exist positive constants η, ε_1, ε_2, C_1, and R with the following property: for every N-cluster $\mathcal{E}'$ with $d(\mathcal{E},\mathcal{E}') < \varepsilon_2$ there exists a function*

$$\Phi : \Big((-\eta,\eta)^{N+1}\cap V\Big)\times\mathbb{R}^n\to\mathbb{R}^n\,,$$

of class C^1 such that:

(i) if $\mathbf{a}\in(-\eta,\eta)^{N+1}\cap V$, then $\Phi(\mathbf{a},\cdot):\mathbb{R}^n\to\mathbb{R}^n$ is a diffeomorphism with $\{x\in\mathbb{R}^n : \Phi(\mathbf{a},x)\ne x\}\subset\subset B_R$;

(ii) if $\mathbf{a}\in(-\eta,\eta)^{N+1}\cap V$, then

$$\Big|\Phi\big(\mathbf{a},\mathcal{E}'(h)\big)\cap B_R\Big| = \Big|\mathcal{E}'(h)\cap B_R\Big| + \mathbf{a}(h)\,,\qquad \text{for } 0\le h\le N\,;$$

(iii) if $\mathbf{a}\in(-\eta,\eta)^{N+1}\cap V$ and Σ is a $\mathcal{H}^{n-1}$-rectifiable set in $\mathbb{R}^n$, then

$$\Big|\mathcal{H}^{n-1}\big(\Phi(\mathbf{a},\Sigma)\big)-\mathcal{H}^{n-1}(\Sigma)\Big| \le C_1\,\mathcal{H}^{n-1}(\Sigma)\sum_{h=0}^{N}|\mathbf{a}(h)|\,;$$

(iv) *there exist $M \in \mathbb{N}$, with $N \le M \le 2N^2$, and a finite family $\{z_\alpha\}_{\alpha=1}^M$ of interface points of $\mathcal{E}$, with $|z_\alpha - z_\beta| > 4\varepsilon_1$ for $1 \le \alpha < \beta \le M$, such that*

$$\left\{x \in \mathbb{R}^n : \Phi(\mathbf{a}, x) \neq x\right\} \subset \bigcup_{\alpha=1}^{M} B(z_\alpha, \varepsilon_1)\,; \tag{29.52}$$

(v) *if $\{y_\alpha\}_{\alpha=1}^M$ is another family of interface points of $\mathcal{E}$ with $|y_\alpha - y_\beta| \ge 4\varepsilon_1$ for $1 \le \alpha < \beta \le M$, and with the property that y_α and z_α belong to the same interface of $\mathcal{E}$ for $1 \le \alpha \le M$, then there exist positive constants η', ε_1', ε_2', C_1', and R' such that conclusions (i)–(iv) hold true with η', ε_1', ε_2', C_1' and R' in place of η, ε_1, ε_2, C_1, and R, and with $\{y_\alpha\}_{\alpha=1}^M$ in place of $\{z_\alpha\}_{\alpha=1}^M$.*

Proof *Step one:* Let $0 \le h, k \le N$, $h \neq k$. We say that $\mathcal{E}(h)$ and $\mathcal{E}(k)$ are: **neighboring chambers**, if $\mathcal{H}^{n-1}(\mathcal{E}(h,k)) > 0$; **linked chambers**, if there exists a (necessarily finite) sequence of pairwise neighboring chambers, starting with $\mathcal{E}(h)$ and ending up with $\mathcal{E}(k)$. If this is the case, the **order of link** between $\mathcal{E}(h)$ and $\mathcal{E}(k)$ is the minimum number of neighboring chambers which is needed to show $\mathcal{E}(h)$ and $\mathcal{E}(k)$ to be linked. Let us show that every chamber is linked to $\mathcal{E}(0)$ (and thus, to every other chamber). First, there exists at least one chamber of $\mathcal{E}$ which is linked to $\mathcal{E}(0)$: otherwise, by applying Exercise 29.7 with $\Lambda = \{1, \dots, N\}$, we would find

$$P\Big(\bigcup_{h=1}^{N} \mathcal{E}(h)\Big) = \sum_{h=1}^{N} \mathcal{H}^{n-1}(\mathcal{E}(h,0)) = 0\,;$$

in particular, it would be $\mathbf{m}(h) = 0$ for every $1 \le h \le N$, a contradiction. Next, now denoting by Λ the non-empty subset of $\{1, \dots, N\}$ corresponding to the chambers of $\mathcal{E}$ which are linked to $\mathcal{E}(0)$, assuming that Λ is a proper subset of $\{1, \dots, N\}$, and applying again Exercise 29.7, we see that

$$P\Big(\bigcup_{h\in\Lambda} \mathcal{E}(h)\Big) = \sum_{h\in\Lambda}\sum_{k\notin\Lambda} \mathcal{H}^{n-1}(\mathcal{E}(h,k)) = 0\,.$$

This implies that either $\Lambda = \emptyset$, or $\mathbf{m}(h) = 0$ for every $h \in \Lambda$, which are both contradictions. Thus, every chamber $\mathcal{E}(h)$ is linked to $\mathcal{E}(0)$ in a finite number of steps (with an order of link that is trivially bounded from above by N). If M_1 is the order of link between $\mathcal{E}(1)$ and $\mathcal{E}(0)$, then we may select M_1 points $\{z_\alpha\}_{\alpha=1}^{M_1}$ ($1 \le M_1 \le N$) by picking z_α on the αth interface of $\mathcal{E}$ we are crossing in moving from $\mathcal{E}(0)$ to $\mathcal{E}(1)$; we may then repeat the same operation, this time moving from $\mathcal{E}(1)$ to $\mathcal{E}(0)$, and never selecting the same point twice, thus choosing other points $\{z_\alpha\}_{\alpha=M_1+1}^{2M_1}$. We repeat this operation with the other

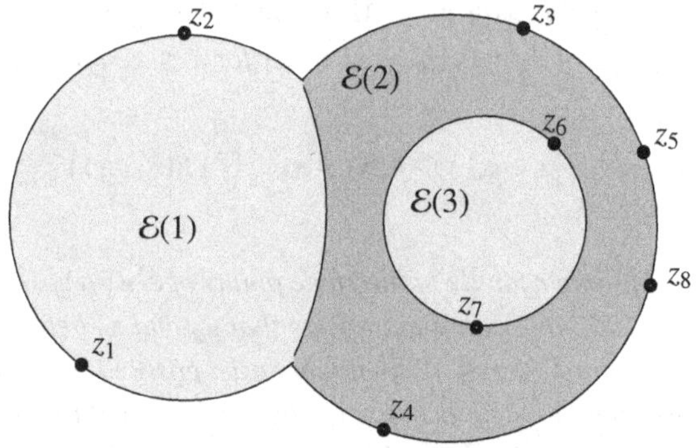

Figure 29.6 Proof of Theorem 29.14. In step one we construct interface points $\{z_\alpha\}_{\alpha=1}^M$ linking each chamber $\mathcal{E}(h)$ to the exterior chamber $\mathcal{E}(0)$. In the picture (where $N = 3$) z_1 and z_2 are used to move from $\mathcal{E}(0)$ to $\mathcal{E}(1)$ and then back to $\mathcal{E}(0)$; z_3 and z_4 are used to move from $\mathcal{E}(0)$ to $\mathcal{E}(2)$ and then back to $\mathcal{E}(0)$; finally, z_5 and z_6 are used to move from $\mathcal{E}(0)$ to $\mathcal{E}(3)$ passing though $\mathcal{E}(2)$, and then z_7 and z_8 are selected when moving back from $\mathcal{E}(3)$ to $\mathcal{E}(0)$, passing again through $\mathcal{E}(2)$. The chambers $\mathcal{E}(1)$ and $\mathcal{E}(2)$ have order of link 1 with $\mathcal{E}(0)$, the chamber $\mathcal{E}(3)$ has order of link 2 with $\mathcal{E}(0)$. The matrix $(L_{j\alpha})$ takes the following form (compare with the proof of (29.57)):

$$(L_{j\alpha}) = \begin{pmatrix} 1 & -1 & 1 & -1 & 1 & 0 & 0 & -1 \\ -1 & 1 & 0 & 0 & 0 & 0 & 0 & 0 \\ 0 & 0 & -1 & 1 & -1 & 1 & -1 & 1 \\ 0 & 0 & 0 & 0 & 0 & -1 & 1 & 0 \end{pmatrix}.$$

chambers $\mathcal{E}(2), \dots, \mathcal{E}(N)$. We have thus found distinct points $\{z_\alpha\}_{\alpha=1}^M$ in $\mathbb{R}^n$ with the following properties: each z_α is an interface point of $\mathcal{E}$, with

$$z_\alpha \in \mathcal{E}(h(\alpha), k(\alpha)), \qquad \mathcal{H}^{n-1}\big(\mathcal{E}(h(\alpha), k(\alpha))\big) > 0; \tag{29.53}$$

every two subsequent points z_α and $z_{\alpha+1}$ belong to the reduced boundary of the same chamber $\mathcal{E}(k(\alpha)) = \mathcal{E}(h(\alpha + 1))$,

$$k(\alpha) = h(\alpha + 1), \qquad 1 \le \alpha \le M - 1, \qquad k(M) = h(1) = 0; \tag{29.54}$$

each chamber of the cluster has at least two distinct interface points in the sequence, so that, in particular,

$$\{0, 1, ..., N\} = \{h(\alpha)\}_{\alpha=1}^M = \{k(\alpha)\}_{\alpha=1}^M; \tag{29.55}$$

and the total amount of considered points M satisfies

$$N \le M \le 2N^2, \tag{29.56}$$

as $M = \sum_{h=1}^{N}(2M_h)$ and $1 \le M_h \le N$ for each $h = 0, \ldots, N$. It is now convenient to associate a $(N+1) \times M$ matrix $(L_{j\alpha})$ with this construction, by setting

$$L_{j\alpha} = \begin{cases} 1, & \text{if } j = h(\alpha), \\ -1, & \text{if } j = k(\alpha), \\ 0, & \text{if } j \neq h(\alpha), k(\alpha). \end{cases} \qquad 1 \le \alpha \le M.$$

The values 1 and -1, corresponding to $j = h(\alpha)$ and $j = k(\alpha)$ respectively, suggest the idea of "inflating" the chamber $\mathcal{E}(h(\alpha))$ close to z_α, while "deflating" $\mathcal{E}(k(\alpha))$ by the same volume amount, thus leaving invariant the volume of all the other chambers ($L_{j\alpha} = 0$ if $j \neq h(\alpha), k(\alpha)$). We now notice a consequence of this construction that, in turn, is going to prove of crucial importance: namely,

$$\text{the rank of } (L_{j\alpha}) \text{ is } N. \tag{29.57}$$

To show this, we relabel the chambers of the cluster by increasing order of link, an operation which amounts to performing a suitable reordering of the rows of $(L_{j\alpha})$. After this reordering, if $1 \le h \le N$ and $\beta(h) \in \{1, \ldots, M\}$ is the first column index such that $L_{h\beta(h)} \neq 0$, then we have

$$L_{h\beta(h)} = -1, \tag{29.58}$$

$$L_{j\beta(h)} = 0, \qquad \forall\, h+1 \le j \le N \quad (\text{if } h \le N-1); \tag{29.59}$$

see also Figure 29.6. In particular, the column vectors $\{w_h\}_{h=1}^{N}$,

$$w_h = \sum_{j=0}^{N} L_{j\beta(h)} e_j\,,$$

define a system of N-linearly independent vectors in $\mathbb{R}^{N+1}$, which means that $(L_{j\alpha})$ has *at least* rank N. In fact, since every column of $(L_{j\alpha})$ just contains two non-zero entries, which are respectively equal to 1 and -1, it turns out that the linear map induced by $(L_{j\alpha})$ takes values in the N-dimensional vector space $V \subset \mathbb{R}^{N+1}$. Thus the rank of $(L_{j\alpha})$ is *at most* N, and (29.57) is proved.

Step two: We now state a uniform version of the inverse function theorem which we shall shortly need. (The symbols M, N, and V will have the same meaning as in the rest of the proof, and the generic points of $\mathbb{R}^M$ and $\mathbb{R}^{N+1}$ will be denoted, respectively, as $\mathbf{t} = (t_1, \ldots, t_M)$ and $\mathbf{a} = (\mathbf{a}(0), \mathbf{a}(1), \ldots, \mathbf{a}(N))$.) This is the statement we are interested in: if ε, κ, and C are positive constants, then there exists a positive constant η, depending on ε, κ, and C only, with the following property: for every class C^2 function $\psi\colon (-\varepsilon, \varepsilon)^M \to V$ with

$\psi(0) = 0$, $\nabla\psi(0)$ of rank N, and

$$|\nabla\psi(0)e| \geq \kappa\,|e|\,, \qquad \forall e \in W := \left(\ker \nabla\psi(0)\right)^{\perp}, \tag{29.60}$$

$$|\nabla^2\psi(\mathbf{t})| \leq C\,, \qquad \forall \mathbf{t} \in (-\varepsilon,\varepsilon)^M\,, \tag{29.61}$$

there exists a class C^2 function $\zeta\colon V \cap (-\eta,\eta)^{N+1} \to W$ with $\zeta(0) = 0$ and

$$\psi(\zeta(\mathbf{a})) = \mathbf{a}\,, \qquad |\,\zeta(\mathbf{a})| \leq \frac{2}{\kappa}\,|\,\mathbf{a}|\,, \tag{29.62}$$

for every $\mathbf{a} \in V \cap (-\eta,\eta)^{N+1}$. (The point here is that we want η to depend on ψ only through ε, κ, and C.) To show this, the reader may define $\bar\psi\colon W \cap (-\varepsilon,\varepsilon)^M \to V$ by setting $\bar\psi(\mathbf{t}) = \psi(\mathbf{t})$, notice that $\bar\psi$ is a class C^2 function between open sets in N-dimensional Euclidean spaces, and then repeat the proof [Spi65, Theorem 2-11] of the inverse function theorem. One has only to take care to use the Lipschitz continuity of $\nabla\bar\psi$ in place of its mere continuity in order to make explicit the range of validity of the various inequalities required in the argument.

Step three: Let $L \in \mathbb{R}^{N+1} \otimes \mathbb{R}^M$ denote the tensor associated with the $(N+1)\times M$-matrix $(L_{j\alpha})$ constructed in step one. Since L has rank N with $\mathrm{Im}\ L = V$, there exists an N-dimensional vector space W in $\mathbb{R}^M$ such that

$$L|_W\colon W \subset \mathbb{R}^M \to V \subset \mathbb{R}^{N+1} \quad \text{is an isomorphism.}$$

Thus, for every $\delta > 0$ there exists $\kappa > 0$ such that, if $L' \in \mathbb{R}^{N+1} \otimes \mathbb{R}^M$ satisfies

$$\mathrm{Im}\ L' = V\,, \qquad |L'_{j\alpha} - L_{j\alpha}| < \delta\,, \quad \forall 0 \leq j \leq N\,, 1 \leq \alpha \leq M\,, \tag{29.63}$$

then $W' = \left(\mathrm{Ker}\ L'\right)^{\perp}$ is an N-dimensional subspace of $\mathbb{R}^M$, L' is an isomorphism between W' and V, and

$$|L'\,e| \geq \kappa|\,e|\,, \qquad \forall e \in W'\,. \tag{29.64}$$

Step four: We now apply Lemma 29.13 at each $z = z_\alpha$ constructed in step one, and with δ defined as in step three. In this way we prove the existence of positive constants ε_1 (depending on $\mathcal{E}$, $\{z_\alpha\}_{\alpha=1}^M$, and δ), ε_2 (depending on δ and ε_1 – in particular in decreasing the value of ε_1, one needs to decrease the value of ε_2), and C_0 (depending on n and ε_1) and of one-parameter families of diffeomorphisms $\{f_t^\alpha\}_{|t|<\varepsilon_1}$, $1 \leq \alpha \leq M$, such that

$$\left\{x \in \mathbb{R}^n : x \neq f_t^\alpha(x)\right\} \subset\subset B(z_\alpha, \varepsilon_1)\,, \qquad \forall |\,t| < \varepsilon_1\,, \tag{29.65}$$

and, if $\mathcal{E}'$ is a cluster with $d(\mathcal{E},\mathcal{E}') < \varepsilon_2$, $1 \le \alpha \le M$, and $|t| < \varepsilon_1$, then

$$\left|\frac{\mathrm{d}}{\mathrm{d}t}\left|f_t^\alpha\Big(\mathcal{E}'(h(\alpha))\Big)\cap B(z_\alpha,\varepsilon_1)\right| - 1\right| < \delta\,, \tag{29.66}$$

$$\left|\frac{\mathrm{d}}{\mathrm{d}t}\left|f_t^\alpha\Big(\mathcal{E}'(k(\alpha))\Big)\cap B(z_\alpha,\varepsilon_1)\right| + 1\right| < \delta\,, \tag{29.67}$$

$$\left|\frac{\mathrm{d}}{\mathrm{d}t}\left|f_t^\alpha\Big(\mathcal{E}'(i)\Big)\cap B(z_\alpha,\varepsilon_1)\right|\right| < \delta\,, \qquad i \ne h(\alpha), k(\alpha)\,, \tag{29.68}$$

$$\left|\frac{\mathrm{d}^2}{\mathrm{d}t^2}\left|f_t^\alpha\Big(\mathcal{E}'(i)\Big)\cap B(z_\alpha,\varepsilon_1)\right|\right| < C_0\,, \qquad 0 \le i \le N\,, \tag{29.69}$$

and, if Σ is a $\mathcal{H}^{n-1}$-rectifiable set in $\mathbb{R}^n$ and $|t| < \varepsilon_1$, then

$$\left|\mathcal{H}^{n-1}\Big(f_t^\alpha(\Sigma)\Big) - \mathcal{H}^{n-1}(\Sigma)\right| \le C_0\,\mathcal{H}^{n-1}(\Sigma)\,|t|\,. \tag{29.70}$$

Moreover, we may choose $R > 0$ such that

$$\bigcup_{\alpha=1}^{M} B(z_\alpha,\varepsilon_1) \subset\subset B_R\,, \tag{29.71}$$

and, up to further decreasing the value of ε_1, we may assume that $|z_\alpha - z_\beta| > 4\varepsilon_1$ for $1 \le \alpha < \beta \le M$, so that, in particular,

$$\text{the balls } \{B(z_\alpha,\varepsilon_1)\}_{\alpha=1}^M \text{ lie at mutually positive distance.} \tag{29.72}$$

Finally, we define $\Psi\colon (-\varepsilon_1,\varepsilon_1)^M \times \mathbb{R}^n \to \mathbb{R}^n$ by setting

$$\Psi(\mathbf{t},x) = (f_{t_1}^1 \circ f_{t_2}^2 \circ \cdots \circ f_{t_M}^M)(x)\,, \qquad (\mathbf{t},x) \in (-\varepsilon_1,\varepsilon_1)^M \times \mathbb{R}^n\,.$$

By (29.65) and (29.72), for every $\mathbf{t} \in (-\varepsilon_1,\varepsilon_1)^M$ we have that $\Psi(\mathbf{t},\cdot)$ is a diffeomorphisms of $\mathbb{R}^n$, with

$$\Big\{x \in \mathbb{R}^n : \Psi(\mathbf{t},x) \ne x\Big\} \subset \bigcup_{\alpha=1}^{M} B(z_\alpha,\varepsilon_1)\,. \tag{29.73}$$

Let us now fix an N-cluster $\mathcal{E}'$ with $d(\mathcal{E},\mathcal{E}') < \varepsilon_2$, and construct the map Φ corresponding to $\mathcal{E}'$ and satisfying the conclusions of the theorem. To this end, let us define $\psi\colon (-\varepsilon_1,\varepsilon_1)^M \to V \subset \mathbb{R}^{N+1}$ by setting, for $0 \le h \le N$ and $\mathbf{t} \in (-\varepsilon_1,\varepsilon_1)^M$,

$$\begin{aligned}\psi_h(\mathbf{t}) &= \Big|\Psi(\mathbf{t},\mathcal{E}'(h)) \cap B_R\Big| - \Big|\mathcal{E}'(h) \cap B_R\Big| \\ &= \sum_{\alpha=1}^{M} \Big|f_{t_\alpha}^\alpha\Big(\mathcal{E}'(h)\Big) \cap B(z_\alpha,\varepsilon_1)\Big| - \Big|\mathcal{E}'(h) \cap B(z_\alpha,\varepsilon_1)\Big|\,.\end{aligned} \tag{29.74}$$

Combining (29.66), (29.67), (29.68), and (29.69) we see that $\psi(0) = 0$, $|\nabla^2\psi(\mathbf{t})| \le C_0$ for every $\mathbf{t} \in (-\varepsilon_1, \varepsilon_1)^M$, and $L' = \nabla\psi(0) \in \mathbb{R}^{N+1} \otimes \mathbb{R}^M$ satisfies (29.63); thus, by (29.64), if $W' = \ker \nabla\psi(0)^\perp$, then

$$|\nabla\psi(0)\,e| \ge \kappa |e|, \qquad \forall e \in W'. \tag{29.75}$$

(Here κ is independent of $\mathcal{E}'$: it just depends on δ and L as explained in step three.) We are thus in the position to apply step two to ψ to show the existence of η (depending on $\mathcal{E}$ and ε_2 but not on $\mathcal{E}'$) and of a class C^2 function $\zeta \colon (-\eta, \eta)^{N+1} \cap V \to \mathbb{R}^M$ such that

$$\psi(\zeta(\mathbf{a})) = \mathbf{a}, \qquad |\zeta(\mathbf{a})| \le \frac{2}{\kappa}|\mathbf{a}|, \tag{29.76}$$

for every $\mathbf{a} \in (-\eta, \eta)^{N+1} \cap V$. We finally set

$$\Phi(\mathbf{a}, x) = \Psi(\zeta(\mathbf{a}), x), \qquad (\mathbf{a}, x) \in \Big((-\eta, \eta)^{N+1} \cap V\Big) \times \mathbb{R}^n.$$

Assertion (i) is immediate, (iv) descends immediately from (29.73), and (v) is evident from the arguments leading to the construction of $\{z_\alpha\}_{\alpha=1}^M$ in step one. By (29.74), if $0 \le h \le N$, then

$$\Big|\Phi(\mathbf{a}, \mathcal{E}(h)) \cap B_R\Big| = \Big|\mathcal{E}(h) \cap B_R\Big| + \psi_h(\zeta(\mathbf{a})) = \Big|\mathcal{E}(h) \cap B_R\Big| + \mathbf{a}(h),$$

that is (ii). We are left to prove (iii): if Σ is a $\mathcal{H}^{n-1}$-rectifiable set in $\mathbb{R}^n$ and $\mathbf{a} \in (-\eta, \eta) \cap V$, then by (29.70) we find

$$\begin{aligned}
&\Big|\mathcal{H}^{n-1}(\Phi(\mathbf{a}, \Sigma)) - \mathcal{H}^{n-1}(\Sigma)\Big| \\
&= \Big|\sum_{\alpha=1}^M \mathcal{H}^{n-1}\Big(f^\alpha_{\zeta_\alpha(\mathbf{a})}(\Sigma) \cap B(z_\alpha, \varepsilon_1)\Big) - \mathcal{H}^{n-1}\Big(\Sigma \cap B(z_\alpha, \varepsilon_1)\Big)\Big| \\
&\le C_0 \sum_{\alpha=1}^M |\zeta_\alpha(\mathbf{a})|\,\mathcal{H}^{n-1}\Big(\Sigma \cap B(z_\alpha, \varepsilon_1)\Big) \le C_0 \sqrt{M}\,\mathcal{H}^{n-1}(\Sigma)\,|\zeta(\mathbf{a})|,
\end{aligned}$$

and (iii) follows from (29.76). □

Remark 29.15 As a consequence of Theorem 29.14, if $\mathcal{E}$ is an N-cluster in $\mathbb{R}^n$, A is an open set in $\mathbb{R}^n$, and for every $h = 1, \dots, N$ there exists a connected component A' of A such that

$$|\mathcal{E}(0) \cap A'| > 0, \qquad |\mathcal{E}(h) \cap A'| > 0, \tag{29.77}$$

then there exists positive constants ε_1, ε_2, C_1, and R such that the conclusions of Theorem 29.14 hold true with sequences of points $\{z_\alpha\}_{\alpha=1}^M$ satisfying

$$B(z_\alpha, \varepsilon_1) \subset\subset A, \qquad \forall \alpha = 1, \dots, M.$$

This is proved as follows: first, extending the terminology settled in the proof of Theorem 29.14, let us say that $\mathcal{E}(h)$ and $\mathcal{E}(k)$ ($0 \le h, k \le N$, $h \ne k$) are **neighboring chambers in** A if $\mathcal{H}^{n-1}(A \cap \mathcal{E}(h,k)) > 0$, and let us define accordingly the notion of linked chambers in A; next, a quick inspection of the proof of Theorem 29.14 shows that to prove our claim it suffices to check that the chambers $\mathcal{E}(h)$ and $\mathcal{E}(0)$ are linked in A for every $h = 1, \dots, N$; finally, this follows immediately by combining (29.77) with the following lemma. We also notice that, by construction, the values of the constants ε_1, ε_2, C_1, and R working for A will work as well for any open set A^* such that $A \subset A^*$.

Lemma 29.16 *If $\mathcal{E}$ is an N-cluster, A is a connected open set in $\mathbb{R}^n$, and*

$$|\mathcal{E}(h) \cap A| > 0, \qquad |\mathcal{E}(k) \cap A| > 0, \tag{29.78}$$

for some $0 \le h < k \le N$, then $\mathcal{E}(h)$ and $\mathcal{E}(k)$ are linked in A.

Proof Up to discarding those chambers which have null intersection with A, and thus replacing $\mathcal{E}$ with an N'-cluster $\mathcal{E}'$ with $N' \le N$, we can directly assume that $|\mathcal{E}(i) \cap A| > 0$ for every $i = 1, \dots, N$ (this choice will simplify notation). Moreover, without loss of generality, we can assume that either $h = 0$ and $k = 1$, or $h = 1$ and $k = 2$. We detail the argument in the case $h = 0$, $k = 1$ only, the other case being analogous. We define $\Lambda_0 \subset \{0, \dots, N\}$ so that $h \in \Lambda_0$ if and only if $\mathcal{E}(h)$ and $\mathcal{E}(0)$ are linked in A. Note that, in our terminology, $0 \notin \Lambda_0$. Similarly, define Λ_1. We now assume that $1 \notin \Lambda_0$, $0 \notin \Lambda_1$, so that Λ_0 and Λ_1 are disjoint subsets of $\{2, \dots, N\}$, and then derive a contradiction.

Step one: If Λ_0 is empty, then $\mathcal{H}^{n-1}(A \cap \mathcal{E}(h,0)) = 0$ for every $h = 1, \dots, N$. Hence, by Exercise 29.7,

$$P(\mathcal{E}(0); A) = \sum_{h=1}^{N} \mathcal{H}^{n-1}\big(A \cap \mathcal{E}(h,0)\big) = 0 .$$

By Exercise 12.17, either $|\mathcal{E}(0) \cap A| = 0$, or $0 = |A \setminus \mathcal{E}(0)| \ge |A \cap \mathcal{E}(1)|$, thus contradicting (29.78). Hence Λ_0 is non-empty. Similarly, Λ_1 is non-empty.

Step two: We now claim that $\Lambda_0 \cup \Lambda_1 = \{2, \dots, N\}$. Indeed, let Λ^* denote the complement of $\Lambda_0 \cup \Lambda_1$ in $\{2, \dots, N\}$. By construction, $\mathcal{H}^{n-1}(A \cap \mathcal{E}(h,k)) = 0$ whenever $h \in \Lambda^*$, $k \in \Lambda_0 \cup \Lambda_1 \cup \{0, 1\}$, $h \ne k$. By Exercise 29.7,

$$P\Big(\bigcup_{h \in \Lambda^*} \mathcal{E}(h); A\Big) = \sum_{h \in \Lambda^*} \sum_{0 \le k \le N, k \notin \Lambda^*} \mathcal{H}^{n-1}\big(A \cap \mathcal{E}(h,k)\big) = 0 .$$

Again by Exercise 12.17, either $|A \setminus \bigcup_{h \in \Lambda^*} \mathcal{E}(h)| = 0$ or $|A \cap \bigcup_{h \in \Lambda^*} \mathcal{E}(h)| = 0$. In the former case, Λ_0 and Λ_1 would be empty, contradicting step one; in the

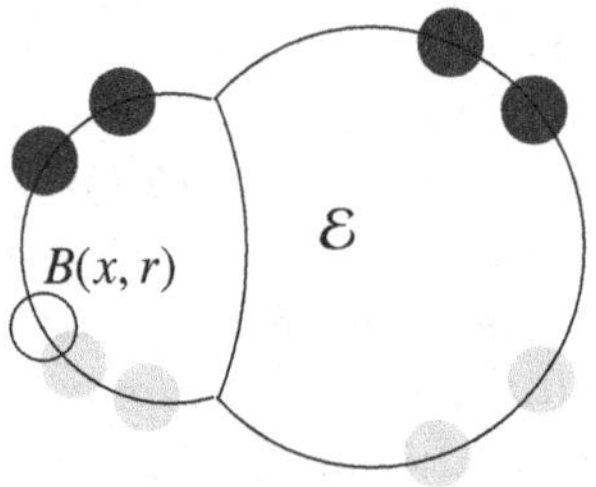

Figure 29.7 An illustration of Corollary 29.17 in the case $\mathcal{E}' = \mathcal{E}$. The dark grey balls represent the family $\{B(z_\alpha, \varepsilon_1)\}_{\alpha=1}^M$, the light grey ones the family $\{B(y_\alpha, \varepsilon_1)\}_{\alpha=1}^M$. If $\mathcal{F}$ is a variation of $\mathcal{E}$ supported in $B(x,r)$ as in the picture, and if $\mathrm{vol}\,(\mathcal{F}) \neq \mathrm{vol}\,(\mathcal{E})$, then, since $B(x,r)$ is disjoint from the family $\{B(z_\alpha, \varepsilon_1)\}_{\alpha=1}^M$, we can modify $\mathcal{F}$ inside the dark balls in order to construct a new variation $\mathcal{F}'$ of $\mathcal{E}$ with $\mathrm{vol}\,(\mathcal{F}') = \mathrm{vol}\,(\mathcal{E})$, at the price of a perimeter variation $|P(\mathcal{F}') - P(\mathcal{F})|$ proportional to the recovered volume $|\mathbf{m}(\mathcal{F}) - \mathbf{m}(\mathcal{E})|$. The construction is stable, in the sense that, with the same deformations, we can fix volumes of variations $\mathcal{F}$ of any cluster $\mathcal{E}'$ sufficiently close to $\mathcal{E}$.

latter case, since $|\mathcal{E}(h) \cap A| > 0$ for every $h = 1, \dots, N$, we conclude that $\Lambda^* = \emptyset$, which is exactly our claim.

Step three: By construction, if $h \in \{0\} \cup \Lambda_0$ and $k \in \{1\} \cup \Lambda_1$, then

$$\mathcal{H}^{n-1}\big(A \cap \mathcal{E}(h,k)\big) = 0\,.$$

By step two, $\{1\} \cup \Lambda_1$ is the complement of $\{0\} \cup \Lambda_0$ in $\{0, \dots, N\}$. Hence, if we set $E_0 = \bigcup_{h \in \{0\} \cup \Lambda_0} \mathcal{E}(h)$, then by Exercise 29.7

$$P(E_0; A) = \sum_{h \in \{0\} \cup \Lambda_0} \sum_{k \in \{1\} \cup \Lambda_1} \mathcal{H}^{n-1}\big(A \cap \mathcal{E}(h,k)\big) = 0\,.$$

By Exercise 12.17, either $0 = |E_0 \cap A| \geq |\mathcal{E}(0) \cap A|$ or $0 = |A \setminus E_0| \geq |\mathcal{E}(1) \cap A|$. Both options are in contradiction with (29.78). Therefore Λ_0 and Λ_1 cannot be disjoint, which means they have to coincide. The proof is complete. □

We now present a second variant of Theorem 29.14, which will be particularly useful in developing the regularity theory of Chapter 30 (for an illustration, see Figure 29.7).

Corollary 29.17 *If $\mathcal{E}$ is an N-cluster in $\mathbb{R}^n$, then there exist positive constants ε, r, and C such that, if $\mathcal{E}'$ and $\mathcal{F}$ are N-clusters, $x \in \mathbb{R}^n$, and, for $1 \leq h \leq N$,*

$$d(\mathcal{E}, \mathcal{E}') < \varepsilon\,, \qquad \mathcal{F}(h) \Delta \mathcal{E}'(h) \subset\subset B(x,r)\,, \tag{29.79}$$

then there exists an N-cluster $\mathcal{F}'$ *such that, for* $1 \le h \le N$,

$$\mathcal{F}'(h)\Delta\mathcal{F}(h) \subset\subset \mathbb{R}^n \setminus \overline{B}(x,r)\,, \tag{29.80}$$

$$\mathbf{m}(\mathcal{F}') = \mathbf{m}(\mathcal{E}')\,, \tag{29.81}$$

$$|P(\mathcal{F}') - P(\mathcal{F})| \le CP(\mathcal{E}')\,|\mathbf{m}(\mathcal{F}) - \mathbf{m}(\mathcal{E}')|\,. \tag{29.82}$$

In particular, if $\mathcal{E}$ *is a minimizing cluster and* $\mathcal{F}(h)\Delta\mathcal{E}(h) \subset\subset B(x,r)$ *for* $1 \le h \le N$, *then*

$$P(\mathcal{E}) \le P(\mathcal{F}) + C\,|\mathbf{m}(\mathcal{F}) - \mathbf{m}(\mathcal{E})|\,. \tag{29.83}$$

(Note that C depends on $\mathcal{E}$ *only, in particular, it is independent of x.)*

Remark 29.18 As a consequence of (29.83), if $\mathcal{E}$ is a minimizing cluster, then it is not possible to construct variations $\mathcal{F}_r$ of $\mathcal{E}$ supported in balls $B(x,r)$ of arbitrarily small radius, and such that $P(\mathcal{F}_r) \le P(\mathcal{E}) - c\,r^{n-1}$. Indeed, if this were the case, then by (29.83) we would find

$$P(\mathcal{E}) \le P(\mathcal{F}_r) + C\,|\mathbf{m}(\mathcal{F}_r) - \mathbf{m}(\mathcal{E})| \le P(\mathcal{F}_r) + C\,r^n \le P(\mathcal{E}) - c\,r^{n-1} + C\,r^n\,,$$

which leads to a contradiction for r sufficiently small.

Proof of Corollary 29.17 By Theorem 29.14, there exist positive constants η, ε_1, ε_2, and C (depending on $\mathcal{E}$), and two sequences $\{z_\alpha\}_{\alpha=1}^M$ and $\{y_\alpha\}_{\alpha=1}^M$ of interface points of $\mathcal{E}$ with $|z_\alpha - z_\beta| > 4\varepsilon_1$ and $|y_\alpha - y_\beta| > 4\varepsilon_1$ if $1 \le \alpha < \beta \le M$, and $|z_\alpha - y_\beta| > 4\varepsilon_1$ for $1 \le \alpha \le \beta \le M$, with the following properties: if $\mathcal{E}'$ is an N-cluster with $d(\mathcal{E},\mathcal{E}') < \varepsilon_2$, then there exist two class C^1 functions

$$\Phi_1, \Phi_2 \colon \Big((-\eta,\eta)^{N+1} \cap V\Big) \times \mathbb{R}^n \to \mathbb{R}^n\,,$$

such that property (ii) and (iii) of Theorem 29.14 hold true for Φ_1 and Φ_2, and, moreover,

$$\Big\{x \in \mathbb{R}^n : \Phi_1(\mathbf{a},x) \ne x\Big\} \subset\subset \bigcup_{\alpha=1}^M B(z_\alpha,\varepsilon_1)\,,$$

$$\Big\{x \in \mathbb{R}^n : \Phi_2(\mathbf{a},x) \ne x\Big\} \subset\subset \bigcup_{\alpha=1}^M B(y_\alpha,\varepsilon_1)\,,$$

for every $\mathbf{a} \in (-\eta,\eta)^{N+1} \cap V$. If $r < \varepsilon_1/2$, then for every $x \in \mathbb{R}^n$ we have either $B(x,r) \cap B(z_\alpha,\varepsilon_1) = \emptyset$ for every $\alpha = 1,\dots,M$, or $B(x,r) \cap B(y_\alpha,\varepsilon_1) = \emptyset$ for every $\alpha = 1,\dots,M$. Without loss of generality, let us assume that we are in the second case when considering x such that $\mathcal{F}(h)\Delta\mathcal{E}'(h) \subset\subset B(x,r)$ for every $h = 1,\dots,N$. Provided $\omega_n\,r^n < \eta$, if we set

$$\mathbf{a}(h) = |\mathcal{E}'(h) \cap B(x,r)| - |\mathcal{F}(h) \cap B(x,r)|\,, \qquad 0 \le h \le N\,,$$

then we have $\mathbf{a} \in (-\eta, \eta)^{N+1} \cap V$; in particular, we may define $\mathcal{F}'$ as

$$\mathcal{F}'(h) = \Phi_1(\mathbf{a}, \mathcal{F}(h)), \qquad 1 \le h \le N.$$

Clearly, we have (29.80); moreover, since $\Phi_1(\mathbf{a}, y) = y$ for $y \in B(x,r)$ and $\mathcal{F}(h)\Delta\mathcal{E}(h) \subset\subset B(x,r)$ for $1 \le h \le N$, we find that, if $1 \le h \le N$, then, taking also Theorem 29.14(ii) into account,

$$\begin{aligned}
|\mathcal{F}'(h)| = |\Phi_1(\mathbf{a}, \mathcal{F}(h))| &= \Big|\Phi_1\big(\mathbf{a}, \mathcal{F}(h) \cap B(x,r)\big)\Big| + \Big|\Phi_1\big(\mathbf{a}, \mathcal{F}(h) \setminus B(x,r)\big)\Big| \\
&= |\mathcal{F}(h) \cap B(x,r)| + \Big|\Phi_1\big(\mathbf{a}, \mathcal{E}'(h) \setminus B(x,r)\big)\Big| \\
&= |\mathcal{F}(h) \cap B(x,r)| + \Big|\Phi_1\big(\mathbf{a}, \mathcal{E}'(h)\big)\Big| - |\mathcal{E}'(h) \cap B(x,r)| \\
&= |\mathcal{F}(h) \cap B(x,r)| + \mathbf{a}(h) + |\mathcal{E}'(h)| - |\mathcal{E}'(h) \cap B(x,r)| \\
&= |\mathcal{E}'(h)|,
\end{aligned}$$

that is (29.81); moreover, by repeatedly applying Theorem 29.14(iii) with $\Sigma = \mathcal{E}'(h,k)$, we find

$$\begin{aligned}
|P(\mathcal{F}') - P(\mathcal{F})| &= \Bigg| \sum_{0 \le h < k \le N} \mathcal{H}^{n-1}\big(\Phi_1(\mathcal{E}'(h,k))\big) - \mathcal{H}^{n-1}\big(\mathcal{E}'(h,k)\big) \Bigg| \\
&\le C_0 \sum_{0 \le h < k \le N} \mathcal{H}^{n-1}\big(\mathcal{E}'(h,k)\big) \sum_{j=0}^{N} |\mathbf{a}(j)| = C_0\, P(\mathcal{E}') \sum_{j=0}^{N} |\mathbf{a}(j)|,
\end{aligned}$$

which gives (29.82) since

$$|\mathbf{a}(0)| \le \sum_{j=1}^{N} |\mathbf{a}(j)| = \sum_{j=1}^{N} \Big| |\mathcal{F}(j)| - |\mathcal{E}'(j)| \Big| \le \sqrt{N}\, |\mathbf{m}(\mathcal{F}) - \mathbf{m}(\mathcal{E}')|.$$

Finally, if $\mathcal{E}$ is a minimizing cluster, $\mathcal{F}$ is a variation of $\mathcal{E}$ supported in $B(x,r)$, and $\mathcal{F}'$ is constructed from $\mathcal{F}$ as above, then (29.81) gives $P(\mathcal{E}) \le P(\mathcal{F}')$, which, combined with (29.82), implies (29.83). □

29.7 Proof of the existence of minimizing clusters

We now combine the results of the previous four sections to achieve the proof of Theorem 29.1. Given $\mathbf{m} \in \mathbb{R}^N_+$ we set

$$m_{\min} = \min\Big\{\mathbf{m}(h) : 1 \le h \le N\Big\}, \qquad m_{\max} = \max\Big\{\mathbf{m}(h) : 1 \le h \le N\Big\},$$

we consider a minimizing sequence $\{\mathcal{E}_k\}_{k \in \mathbb{N}}$ for the partitioning problem defined by $\mathbf{m}$, that is, $\mathbf{m}(\mathcal{E}_k) = \mathbf{m}$ for every $k \in \mathbb{N}$ and

$$\lim_{k \to \infty} P(\mathcal{E}_k) = \gamma = \inf\Big\{P(\mathcal{E}) : \mathrm{vol}\,(\mathcal{E}) = \mathbf{m}\Big\}, \tag{29.84}$$

and, finally, we set

$$p_{\min} = \inf\left\{P(\mathcal{E}_k(h)) : 1 \le h \le N\,, k \in \mathbb{N}\right\},$$
$$p_{\max} = \sup\left\{P(\mathcal{E}_k(h)) : 1 \le h \le N\,, k \in \mathbb{N}\right\},$$

where $p_{\min} \ge n\omega_n^{1/n} m_{\min}^{(n-1)/n} > 0$ and $p_{\max} < \infty$ since $\gamma < \infty$. We now want to modify $\{\mathcal{E}_k\}_{k\in\mathbb{N}}$ to obtain a new minimizing sequence satisfying the assumptions of Proposition 29.8; see Section 29.2.

Step one: By Remark 29.11, there exist sequences $\{x_k(h)\}_{k\in\mathbb{N}}$ $(1 \le h \le N)$, such that, for every $k \in \mathbb{N}$ and $1 \le h \le N$,

$$\left|\mathcal{E}_k(h) \cap B(x_k(h), 1)\right| \ge \min\left\{c(n)\frac{m_{\min}}{p_{\max}}, \frac{1}{2n}\right\}^n . \tag{29.85}$$

If we now define $S > 0$ by $\omega_n S^n = 2\sum_{h=1}^N \mathbf{m}(h)$, then at least half of the room inside $B(x_k(h), S)$ is occupied by the exterior chamber of $\mathcal{E}_k$, that is

$$\left|\mathcal{E}_k(0) \cap B(x_k(h), S)\right| \ge \frac{\omega_n S^n}{2}, \tag{29.86}$$

for every $k \in \mathbb{N}$, $h = 1, \ldots, N$. We shall now prove the existence of positive constants ε_1 and C_1 and such that, up to extractly an un-relabeled subsequence in k, there exist class C^1 functions

$$\Psi_k\colon \left((-\varepsilon_1, \varepsilon_1)^{N+1} \cap V\right) \times \mathbb{R}^n \to \mathbb{R}^n,$$

such that $\Psi_k(\mathbf{a}, \cdot)$ is a C^1-diffeomorphism for every $\mathbf{a} \in (-\varepsilon_1, \varepsilon_1)^{N+1} \cap V$, with

$$\left\{x \in \mathbb{R}^n : \Psi_k(\mathbf{a}, x) \ne x\right\} \subset\subset \bigcup_{h=1}^N B(x_k(h), S), \tag{29.87}$$

$$\left|\Psi_k\big(\mathbf{a}, \mathcal{E}_k(h)\big)\right| = \left|\mathcal{E}_k(h)\right| + \mathbf{a}(h), \tag{29.88}$$

$$\left|\mathcal{H}^{n-1}\big(\Psi_k(\mathbf{a}, \Sigma)\big) - \mathcal{H}^{n-1}(\Sigma)\right| \le C_1\,\mathcal{H}^{n-1}(\Sigma)\sum_{h=0}^N |\mathbf{a}(h)|, \tag{29.89}$$

whenever $1 \le h \le N$, and Σ is a $\mathcal{H}^{n-1}$-rectifiable set in $\mathbb{R}^n$. In order to prove our claim (the proof of which may be safely skipped at a first reading), we start classifying the sequences $\{x_k(h)\}_{k\in\mathbb{N}}$ depending on their asymptotic behavior as $k \to \infty$. Precisely, given $h, h' \in \{1, \ldots, N\}$, let us say that $\{x_k(h)\}_{k\in\mathbb{N}}$ and $\{x_k(h')\}_{k\in\mathbb{N}}$ are *asymptotically close* provided,

$$\liminf_{k\to\infty} |x_k(h) - x_k(h')| < S\,.$$

Let us say that they *don't tear apart* if there exists $\{h_1, \ldots, h_\ell\} \subset \{1, \ldots, N\}$ such that for every $j = 1, ..., \ell - 1$, the sequences $\{x_k(h_j)\}_{k\in\mathbb{N}}$ and $\{x_k(h_{j+1})\}_{k\in\mathbb{N}}$

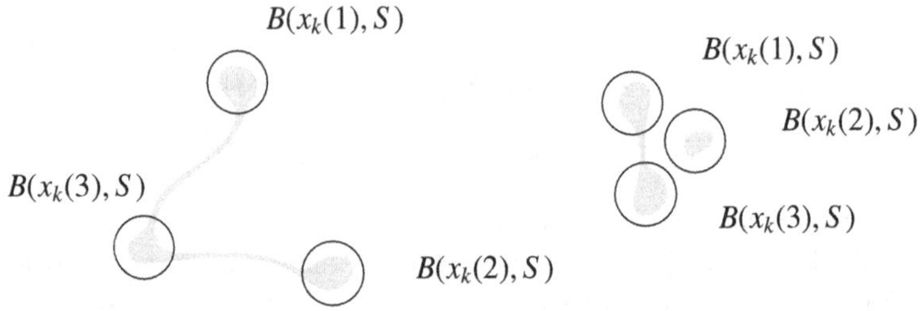

Figure 29.8 The partition $\{\Lambda_j\}_{j=1}^s$ in the two possible limit cases. The region occupied by the cluster $\mathcal{E}_k$ is depicted in grey. On the left, the case $s = N$ and $\Lambda_j = \{j\}$, in which all the various "nuclei" $\mathcal{E}_k(h) \cap B(x_k(h), S)$ are asymptotically tearing apart (note that these nuclei may be connected by filaments of small volume and perimeter). On the right, the case $s = 1$, $\Lambda_1 = \{1, \dots, N\}$, in which the various sequences $\{x_k(h)\}_{k\in\mathbb{N}}$ travel around at mutually bounded distances.

are asymptotically close, and $h_1 = h$, $h_\ell = h'$. Relying on these two concepts, and up to extracting un-relabeled subsequences, we may certainly define a partition $\{\Lambda_j\}_{j=1}^s$ of $\{1, \dots, N\}$ such that, if $1 \le j \le s$, then any two sequences $\{x_k(h)\}_{k\in\mathbb{N}}$ and $\{x_k(h')\}_{k\in\mathbb{N}}$ with $h, h' \in \Lambda_j$ don't tear apart (see Figure 29.8) and, moreover, since $\#\Lambda_j \le N$,

$$\lim_{k\to\infty} x_k(h) - x_k(h') \quad \text{exists and belongs to } B_{NS}\,. \tag{29.90}$$

We denote by h_j a distinguished index in Λ_j, and for every $h \in \{1, \dots, N\}$ we let $w(h) \in B_{NS}$ be defined as

$$w(h) = \lim_{k\to\infty} x_k(h) - x_k(h_j)\,, \qquad \text{if } h \in \Lambda_j\,. \tag{29.91}$$

In this way, provided k is large enough, we have

$$\bigcup_{h\in\Lambda_j} B(x_k(h), S) \subset\subset B\big(x_k(h_j), 2NS\big)\,, \qquad \forall k \in \mathbb{N}\,, 1 \le j \le s\,. \tag{29.92}$$

In other words, we have isolated s distinct nuclei into each cluster $\mathcal{E}_k$,

$$\bigcup_{h=1}^N \Big(\mathcal{E}_k(h) \cap \bigcup_{\overline{h}\in\Lambda_j} B(x_k(\overline{h}), S)\Big)\,, \qquad 1 \le j \le s\,,$$

each of them lying at distance at a most $2NS$ from the point $x_k(h_j)$. Hence, we may use the translation vectors

$$v_j = 4(N+1)S\,j\,e_n\,, \qquad j = 1, \dots, s\,, \tag{29.93}$$
$$y_{k,j} = v_j - x_k(h_j)\,, \qquad k \in \mathbb{N}\,, j = 1, \dots, s\,, \tag{29.94}$$

to put some order in the mutual positions of the various nuclei, and prevent any of them from disappearing to infinity. Precisely, for each $k \in \mathbb{N}$ we define a cluster $\mathcal{E}_k^*$ by setting, for $1 \le h \le N$,

$$\mathcal{E}_k^*(h) = \bigcup_{j=1}^{s} \Big(y_{k,j} + \Big(\mathcal{E}_k(h) \cap \bigcup_{\overline{h} \in \Lambda_j} B(x_k(\overline{h}), S) \Big) \Big).$$

Each cluster $\mathcal{E}_k^*$ is made of s disjoint pieces, contained in s balls of radius $2NS$ lying at mutual distance $4S$. Indeed, by (29.92), (29.93), and (29.94), if k is large enough, then

$$\mathcal{E}_k^*(h) \subset \bigcup_{j=1}^{s} B\big(v_j, 2N\,S\big),$$

where the balls $\{B(v_j, 2N\,S)\}_{j=1}^s$ lie at mutual distance $4S$. In particular, we may explicitly compute S' depending on S and N such that, for k large enough,

$$\mathcal{E}_k^*(h) \subset B_{S'}, \qquad \forall h = 1, \dots, N. \tag{29.95}$$

Note that we have indeed defined a sequence of N-clusters: given $h = 1, \dots, N$, if $j \in \{1, \dots, s\}$ is such that $h \in \Lambda_j$, then by (29.85) we have

$$\Big|\mathcal{E}_k^*(h) \cap B\big(x_k(h) + y_{k,j}, S\big)\Big| \ge \Big|\mathcal{E}_k(h) \cap B\big(x_k(h), S\big)\Big| \ge \min\Big\{c(n)\frac{m_{\min}}{p_{\max}}, \frac{1}{2n}\Big\}^n, \tag{29.96}$$

which, in particular, implies

$$\inf_{k \in \mathbb{N}} \min_{1 \le h \le N} |\mathcal{E}_k^*(h)| > 0. \tag{29.97}$$

Finally, by a repeated application of Lemma 12.22, we find

$$P(\mathcal{E}_k^*(h)) \le P(\mathcal{E}_k(h)) + \sum_{h=1}^{N} P(B(x_k(h), S)) \le p_{\max} + N\,n\,\omega_n\,S^{n-1},$$

so that

$$\sup_{k \in \mathbb{N}} \max_{1 \le h \le N} P(\mathcal{E}_k^*(h)) < \infty. \tag{29.98}$$

By virtue of (29.97), (29.95), and (29.98), we may apply Proposition 29.5 to show the existence of an N-cluster $\mathcal{E}^*$ such that, up to extracting an relabeled subsequence in k, we have $\mathcal{E}_k^* \to \mathcal{E}^*$. We now notice the following property of $\mathcal{E}^*$: by (29.91), (29.94), (29.86), (29.85), and by arguing as in (29.96), we see that, for $h \in \{1, \dots, N\}$ and $j \in \{1, \dots, s\}$ such that $h \in \Lambda_j$,

$$\Big|\mathcal{E}^*(h) \cap B\big(v_j + w(h), S\big)\Big| \ge \min\Big\{c(n)\frac{m_{\min}}{p_{\max}}, \frac{1}{2n}\Big\}^n,$$

$$\Big|\mathcal{E}^*(0) \cap B\big(v_j + w(h), S\big)\Big| \ge \frac{\omega_n\,S^n}{2}.$$

In particular, by Lemma 29.16, for every $h = 1, \dots, N$, $\mathcal{E}^*(h)$ is linked to $\mathcal{E}^*(0)$ inside $B(v_j + w(h), S)$. Thus, as explained in Remark 29.15, we may apply the constructions of Theorem 29.14 inside the open set

$$A = \bigcup_{j=1}^{s} \bigcup_{h\in\Lambda_j} B\Big(v_j + w(h), S\Big).$$

Precisely, since $d(\mathcal{E}_k^*, \mathcal{E}^*) \to 0$ as $k \to \infty$, there exist positive constants ε_1 and C_1 (depending on $\mathcal{E}^*$ and A) with the following property: for every k large enough (and for V defined as in (29.51)), there exists a class C^1-function

$$\Phi_k \colon \Big((-\varepsilon_1, \varepsilon_1)^{N+1} \cap V\Big) \times \mathbb{R}^n \to \mathbb{R}^n,$$

such that, if $\mathbf{a} \in (-\varepsilon_1, \varepsilon_1)^{N+1} \cap V$, $0 \le h \le N$ and Σ is $\mathcal{H}^{n-1}$-rectifiable, then

$$\Big|\Phi_k\Big(\mathbf{a}, \mathcal{E}_k^*(h)\Big) \cap A\Big| = \Big|\mathcal{E}_k^*(h) \cap A\Big| + \mathbf{a}(h), \tag{29.99}$$

$$\Big|\mathcal{H}^{n-1}\Big(\Phi_k(\mathbf{a}, \Sigma)\Big) - \mathcal{H}^{n-1}(\Sigma)\Big| \le C_1\, \mathcal{H}^{n-1}(\Sigma) \sum_{h=0}^{N} |\,\mathbf{a}(h)|. \tag{29.100}$$

$$\Big\{x \in \mathbb{R}^n : \Phi_k(\mathbf{a}, x) \ne x\Big\} \subset\subset A. \tag{29.101}$$

Since A is the union of the s disjoint sets

$$\bigcup_{h\in\Lambda_j} B\Big(v_j + w(h), S\Big), \qquad j = 1, \dots, s,$$

we conclude that, for every $x \in \mathbb{R}^n$ there exists at most one $j = 1, \dots, s$ such that $x \in y_{k,j} + \bigcup_{h\in\Lambda_j} B\Big(v_j + w(h), S\Big)$. Therefore, it makes sense to define

$$\Psi_k \colon \Big((-\varepsilon_1, \varepsilon_1)^{N+1} \cap V\Big) \times \mathbb{R}^n \to \mathbb{R}^n,$$

by setting, if $x \in y_{k,j} + \bigcup_{h\in\Lambda_j} B\Big(v_j + w(h), S\Big)$ for some $j = 1, \dots, s$,

$$\Psi_k(\mathbf{a}, x) = \Phi(\mathbf{a}\,, x - y_{k,j}) + y_{k,j}\,,$$

and $\Psi_k(\mathbf{a}, x) = x$ otherwise. The resulting functions are C^1-diffeomorphisms by (29.101), while the validity of (29.87), (29.88), and (29.89) follows at once from (29.99), (29.100), and (29.101).

Step two: We now apply the nucleation lemma again, this time in order to find, for each chamber $\mathcal{E}_k(h)$ of each cluster $\mathcal{E}_k$, a finite number (uniformly bounded in h and k) of balls of unit radius which identify the position of $\mathcal{E}_k(h)$ up to a small error. Precisely, if $\varepsilon_0 > 0$ satisfies

$$\varepsilon_0 \le \min\left\{m_{\min}\,, \frac{p_{\min}}{2nc(n)}\right\}, \tag{29.102}$$

then for every $k \in \mathbb{N}$ and $h = 1, \dots, N$ we may apply Lemma 29.10 to $\mathcal{E}_k(h)$ in order to find finitely many points $\{x_k(h,i)\}_{i=1}^{L(k,h)}$ with the property that

$$\left|\mathcal{E}_k(h) \setminus \bigcup_{i=1}^{L(k,h)} B(x_k(h,i),2)\right| < \varepsilon_0\,, \tag{29.103}$$

$$L(k,h) \le |\mathcal{E}_k(h)| \left(\frac{P(\mathcal{E}_k(h))}{c(n)\varepsilon_0}\right)^n \le \frac{m_{\max}\, p_{\max}^n}{(c(n)\,\varepsilon_0)^n}\,. \tag{29.104}$$

We next consider the closed sets $F_k \subset \mathbb{R}^n$,

$$F_k = \bigcup_{h=1}^N \left(\overline{B}(x_k(h), S) \cup \bigcup_{i=1}^{L(k,h)} \overline{B}(x_k(h,i),2)\right), \qquad k \in \mathbb{N}\,.$$

By (29.102), we may apply Lemma 29.12 with $\alpha = \varepsilon_0$ to each pair $(\mathcal{E}_k, F_k)$. Setting for brevity $u_k(x) = \mathrm{dist}(x, F_k)$ ($k \in \mathbb{N}$, $x \in \mathbb{R}^n$) we find a sequence

$$\{r_k\}_{k\in\mathbb{N}} \subset \left[0, 7n\varepsilon_0^{1/n}\right] \tag{29.105}$$

such that, if we define the cluster $\mathcal{E}_k'$ as

$$\mathcal{E}_k'(h) = \mathcal{E}_k(h) \cap \{u_k \le r_k\}\,, \qquad 1 \le h \le N\,,$$

then we have

$$P(\mathcal{E}_k') \le P(\mathcal{E}_k) - \frac{d(\mathcal{E}_k, \mathcal{E}_k')}{4\varepsilon_0^{1/n}}\,, \tag{29.106}$$

where, of course, by (29.102),

$$d(\mathcal{E}_k, \mathcal{E}_k') = \sum_{h=1}^N \left|\mathcal{E}_k(h) \cap \{u_k > r_k\}\right| \le \varepsilon_0\,. \tag{29.107}$$

Therefore, if we set

$$\mathbf{a}_k(h) = |\mathcal{E}_k(h)| - |\mathcal{E}_k'(h)| = \left|\mathcal{E}_k(h) \cap \{u_k > r_k\}\right|\,, \qquad 1 \le h \le N\,,$$

$$\mathbf{a}_k(0) = -\sum_{h=1}^N \mathbf{a}_k(h)\,,$$

then we have $\mathbf{a}_k \in (-\varepsilon_1, \varepsilon_1)^{N+1} \cap V$, provided ε_0 is small enough. By (29.87),

$$\left\{x \in \mathbb{R}^n : \Psi_k(\mathbf{a}_k, x) \ne x\right\} \subset\subset \bigcup_{h=1}^N B(x_k(h), S) \subset F_k \subset \{u \le r_k\}\,,$$

so that, for every k large enough and $h = 1, \dots, N$,

$$\Psi_k\Big(\mathbf{a}_k, \mathcal{E}_k(h)\Big) \cap \{u \le r_k\} = \Psi_k\Big(\mathbf{a}_k, \mathcal{E}_k'(h)\Big) \cap \{u \le r_k\} = \Psi_k(\mathbf{a}_k, \mathcal{E}_k'(h))\,.$$

By (29.88), we thus find

$$\begin{aligned}\mathbf{a}_k(h) &= \Big|\Psi_k\big(\mathbf{a}_k, \mathcal{E}_k(h)\big)\Big| - \Big|\mathcal{E}_k(h)\Big| \\ &= \Big|\Psi_k\big(\mathbf{a}_k, \mathcal{E}_k(h)\big) \cap \{u \le r_k\}\Big| - \Big|\mathcal{E}_k(h) \cap \{u \le r_k\}\Big| \\ &= \Big|\Psi_k\big(\mathbf{a}_k, \mathcal{E}'_k(h)\big) \cap \{u \le r_k\}\Big| - \Big|\mathcal{E}_k(h) \cap \{u \le r_k\}\Big| \\ &= \Big|\Psi_k\big(\mathbf{a}_k, \mathcal{E}'_k(h)\big)\Big| - \Big|\mathcal{E}_k(h) \cap \{u \le r_k\}\Big|,\end{aligned}$$

whenever $1 \le h \le N$. By definition of $\mathbf{a}_k(h)$, we conclude that the clusters $\{\mathcal{E}''_k\}_{k\in\mathbb{N}}$ defined as

$$\mathcal{E}''_k(h) = \Psi_k\big(\mathbf{a}_k, \mathcal{E}'_k(h)\big), \qquad 1 \le h \le N,$$

satisfy $\mathbf{m}(\mathcal{E}''_k) = \mathbf{m}(\mathcal{E}_k) = \mathbf{m}$. Moreover, by repeatedly applying (29.89) to the $\mathcal{H}^{n-1}$-rectifiable sets $\Sigma = \mathcal{E}'_k(h, \overline{h})$, $1 \le h < \overline{h} \le N$, we easily find that

$$\begin{aligned}P(\mathcal{E}''_k) &\le P(\mathcal{E}'_k) + C_1\, P(\mathcal{E}'_k) \sum_{h=0}^{N} |\mathbf{a}_k(h)| \\ &\le P(\mathcal{E}_k) - \frac{d(\mathcal{E}_k, \mathcal{E}'_k)}{4\varepsilon_0^{1/n}} + 2\gamma\, C_1 \sum_{h=0}^{N} |\mathbf{a}_k(h)|,\end{aligned}$$

where (29.106) was taken into account. Since $\sum_{h=0}^{N} |\mathbf{a}_k(h)| = 2\, d(\mathcal{E}_k, \mathcal{E}'_k)$, we finally conclude that, provided ε_0 is small enough,

$$P(\mathcal{E}''_k) \le P(\mathcal{E}_k) + \left(4\, C_1\, \gamma - \frac{1}{4\varepsilon_0^{1/n}}\right) d(\mathcal{E}_k, \mathcal{E}'_k) \le P(\mathcal{E}_k),$$

Hence $\{\mathcal{E}''_k\}_{k\in\mathbb{N}}$ is a minimizing sequence for (29.84). By construction, for every $h = 1, \dots, N$ and $k \in \mathbb{N}$, we have $\mathcal{E}''_k(h) \subset \{u_k \le r_k\}$. In particular, by (29.105), $\{\mathcal{E}''_k\}_{k\in\mathbb{N}}$ satisfies the confinement assumptions of Proposition 29.8. The existence of a minimizer in (29.84) thus follows by the Direct Method. Finally, if $\mathcal{E}$ is any such minimizer and $\mathcal{E}$ is not bounded, then by applying the above argument to the trivial minimizing sequence $\mathcal{E}_k = \mathcal{E}$, $k \in \mathbb{N}$, we find a way to strictly decrease the perimeter of $\mathcal{E}$, while keeping the volume constraint. The proof of Theorem 29.1 is thus completed.

30

Regularity of minimizing clusters

This chapter is devoted to the discussion of the regularity properties of minimizing clusters. We shall first prove the following $\mathcal{H}^{n-1}$-a.e. regularity result in arbitrary dimension (Theorem 30.1); then, in Section 30.3, we shall focus on the planar case, and achieve a more precise result, including a complete description of near-to-singularities behavior; see Theorem 30.7.

Theorem 30.1 *If $\mathcal{E}$ is a minimizing N-cluster in $\mathbb{R}^n$ and $0 \le h < k \le N$, then $\mathcal{E}(h,k)$ is an analytic hypersurface with constant mean curvature in $\mathbb{R}^n$, which is relatively open in $\partial\mathcal{E}(h) \cap \partial\mathcal{E}(k)$, and satisfies*

$$\sum_{h=0}^{N} \mathcal{H}^{n-1}\big(\partial\mathcal{E}(h) \setminus \partial^*\mathcal{E}(h)\big) = 0\,. \tag{30.1}$$

The key tool in the proof of Theorem 30.1 is the *infiltration lemma* of Section 30.1, which allows us, for example, to prove each chamber of the cluster to be a volume-constrained perimeter minimizer in a neighborhood of its interface points; see Corollary 30.3. In Section 30.2 we shall combine this result with some density estimates in order to complete the proof of (30.1).

30.1 Infiltration lemma

The following lemma shows that if some chambers of a minimizing cluster fill *most* of $B(x,2r)$, then they *completely* fill $B(x,r)$.

Lemma 30.2 *If $\mathcal{E}$ is a minimizing N-cluster in $\mathbb{R}^n$, then there exist positive constants $\varepsilon_0 < \omega_n$ (depending on n only) and r_0 (depending on $\mathcal{E}$) such that, if $x \in \mathbb{R}^n$, $r < r_0$, $\Lambda \subset \{0,\dots,N\}$, and*

$$\sum_{h\in\Lambda} \Big|\mathcal{E}(h) \cap B(x,r)\Big| \le \varepsilon_0\, r^n \tag{30.2}$$

(note that, necessarily, $\#\Lambda \le N$), then

$$\Big|\mathcal{E}(h)\cap B\Big(x,\frac{r}{2}\Big)\Big| = 0\,,\qquad \forall h\in\Lambda\,.$$

Corollary 30.3 *If $\mathcal{E}$ is a minimizing N-cluster in $\mathbb{R}^n$, $0\le h<k\le N$, $x\in \partial\mathcal{E}(h)\cap\partial\mathcal{E}(k)$, and*

$$\lim_{r\to 0}\frac{|\mathcal{E}(h)\cap B(x,r)|}{\omega_n\,r^n}+\frac{|\mathcal{E}(k)\cap B(x,r)|}{\omega_n\,r^n} = 1\,,\tag{30.3}$$

then there exists $r_x>0$ (depending on $\mathcal{E}$ and x) such that $|\mathcal{E}(i)\cap B(x,r_x)|=0$ if $i\ne h,k$, $0\le i\le N$, and $\mathcal{E}(h)$ and $\mathcal{E}(k)$ are volume-constrained perimeter minimizers in $B(x,r_x)$. In particular, if either $2\le n\le 7$, or $n\ge 8$ but $x\in\mathcal{E}(h,k)$, then

$$\partial\mathcal{E}(h)\cap\partial\mathcal{E}(k)\cap B(x,r_x) = \mathcal{E}(h,k)\cap B(x,r_x)$$

is a constant mean curvature analytic hypersurface in $\mathbb{R}^n$.

Remark 30.4 (Regularity of interfaces) Since, by Theorem 15.5, $\mathcal{E}(h,k)\subset \mathcal{E}(h)^{(1/2)}\cap\mathcal{E}(k)^{(1/2)}$, we see that (30.3) is satisfied at every interface point of the cluster. Thus, Corollary 30.3 implies that each interface $\mathcal{E}(h,k)$ is a constant mean curvature analytic hypersurface, which is relatively open inside $\partial\mathcal{E}(h)\cap\partial\mathcal{E}(k)$. Moreover, the mean curvature of $\mathcal{E}(h,k)$ takes the same constant value at possibly different connected components of $\mathcal{E}(h,k)$, as is easily seen by exploiting the local-volume constrained minimality of $\mathcal{E}(h)$ at points of $\mathcal{E}(h,k)$ through the same argument used in the proof of Theorem 17.20.

Proof of Corollary 30.3 By (30.3) and Lemma 30.2, for some $r_x>0$,

$$|\mathcal{E}(i)\cap B(x,r_x)| = 0\,,\qquad 0\le i\le N\,, i\ne hk\,.$$

Thus, if $\mathcal{E}(h)\Delta F\subset\subset B(x,r_x)$ and $|F|=|\mathcal{E}(h)|$, then the cluster $\mathcal{E}'$ defined by

$$\mathcal{E}'(h) = \Big(\mathcal{E}(h)\setminus B(x,r_x)\Big)\cup\Big(B(x,r_x)\cap F\Big),$$
$$\mathcal{E}'(k) = \Big(\mathcal{E}(k)\setminus B(x,r_x)\Big)\cup\Big(B(x,r_x)\setminus F\Big),$$

and by $\mathcal{E}'(i)=\mathcal{E}(i)$ if $i\ne h,k$, satisfies $\mathbf{m}(\mathcal{E}')=\mathbf{m}(\mathcal{E})$. Exploiting $P(\mathcal{E})\le P(\mathcal{E}')$, we find $P(\mathcal{E}(h);B(x,r_x))\le P(F;B(x,r_x))$. We have thus proved that $\mathcal{E}(h)$ is a volume-constrained perimeter minimizer in $B(x,r_x)$, and the proof is concluded by Theorem 27.4, Theorem 28.1, and Example 21.3. □

Proof of Lemma 30.2 We shall detail the case $\Lambda=\{h : 2\le h\le N\}$ only (the general case follows by minor modifications), and set $x=0$. By

Corollary 29.17 we find positive constants r_0 and C (depending on $\mathcal{E}$ but not on $x = 0$) such that, if $\mathcal{F}(h)\Delta\mathcal{E}(h) \subset\subset B_{r_0}$ for all $1 \le h \le N$, then

$$P(\mathcal{E}) \le P(\mathcal{F}) + C\,|\,\mathbf{m}(\mathcal{F}) - \mathbf{m}(\mathcal{E})|\,. \tag{30.4}$$

Let us now assume that, for some $r < r_0$, (30.2) holds true (with $\Lambda = \{h : 2 \le h \le N\}$), let us consider the sets of finite perimeter

$$E_s = B_s \cap \bigcup_{h=2}^{N} \mathcal{E}(h)\,, \qquad s > 0\,,$$

and let us define an increasing function $m\colon (0,\infty) \to (0,\infty)$ by $m(s) = |E_s|$, $s > 0$; in this way, for a.e. $s > 0$,

$$m'(s) = \sum_{h=2}^{N} \mathcal{H}^{n-1}\big(\mathcal{E}(h) \cap \partial B_s\big)\,, \tag{30.5}$$

$$0 = \sum_{h=0}^{N} \mathcal{H}^{n-1}\big(\partial^*\mathcal{E}(h) \cap \partial B_s\big)\,. \tag{30.6}$$

Finally, we define a cluster $\mathcal{F}$ depending on

$$\text{whether} \quad \sum_{h=2}^{N} \mathcal{H}^{n-1}\big(B_s \cap \mathcal{E}(h,1)\big) \ge \sum_{h=2}^{N} \mathcal{H}^{n-1}\big(B_s \cap \mathcal{E}(h,0)\big)\,, \tag{30.7}$$

$$\text{or} \quad \sum_{h=2}^{N} \mathcal{H}^{n-1}\big(B_s \cap \mathcal{E}(h,0)\big) > \sum_{h=2}^{N} \mathcal{H}^{n-1}\big(B_s \cap \mathcal{E}(h,1)\big)\,, \tag{30.8}$$

as follows (see Figure 30.1): if (30.7) holds true, then we set

$$\mathcal{F}(1) = \mathcal{E}(1) \cup E_s\,, \tag{30.9}$$

$$\mathcal{F}(h) = \mathcal{E}(h) \setminus B_s\,, \qquad h = 2, \dots, N\,, \tag{30.10}$$

$$\mathcal{F}(0) = \mathcal{E}(0)\,; \tag{30.11}$$

if, instead, (30.8) holds true, then we define $\mathcal{F}$ as

$$\mathcal{F}(1) = \mathcal{E}(1)\,,$$

$$\mathcal{F}(h) = \mathcal{E}(h) \setminus B_s\,, \qquad h = 2, \dots, N\,,$$

$$\mathcal{F}(0) = \mathcal{E}(0) \cup E_s\,. \tag{30.12}$$

In both cases, if $s < r_0$, then F is an N-cluster with

$$\mathcal{E}(h)\Delta\mathcal{F}(h) \subset\subset B_{r_0}\,, \qquad \forall h = 1, \dots, N\,, \tag{30.13}$$

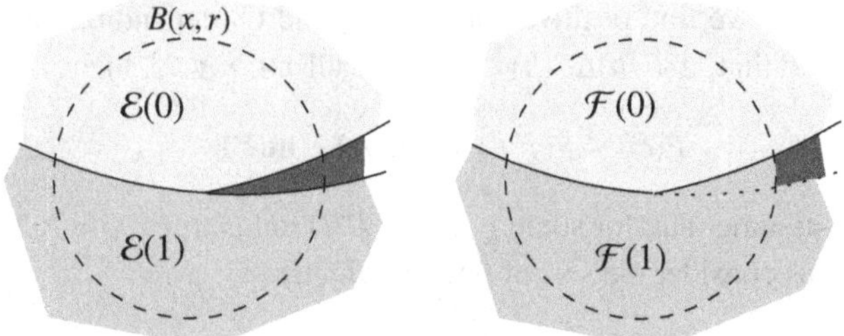

Figure 30.1 The cluster $\mathcal{F}$ constructed in the proof of the infiltration lemma. In both pictures, the dark region depicts the "infiltration", that is, the union of the chambers $\mathcal{E}(h)$ for $h = 2, \dots, N$. At scale r, we have just to decide whether we wish to annex the intersection E_s of the infiltration with $B(x, s)$ to $\mathcal{E}(1)$ or to $\mathcal{E}(0)$. We chose what saves more perimeter (this is the alternative between (30.7) and (30.8)). In the picture, $\mathcal{H}^{n-1}(\partial^*\mathcal{E}(1)\cap\partial^* E_s) \ge \mathcal{H}^{n-1}(\partial^*\mathcal{E}(0)\cap\partial^* E_s)$, therefore it is more convenient to cancel the interface contribution $\partial^*\mathcal{E}(1) \cap \partial^* E_s$ by annexing E_s to $\mathcal{E}(1)$.

and thus (30.4) holds true; since $P(\mathcal{E}; \mathbb{R}^n \setminus B_s) = P(\mathcal{F}; \mathbb{R}^n \setminus B_s)$, $P(\mathcal{E}; \partial B_s) = 0$ (by (30.6), for a.e. s), $P(\mathcal{F}; \partial B_s) = m'(s)$ (by (30.5), for a.e. s), and

$$|\mathbf{m}(\mathcal{F}) - \mathbf{m}(\mathcal{E})| \le \sum_{h=1}^{N} \Big| |\mathcal{F}(h)| - |\mathcal{E}(h)| \Big| \le 2\,|E_s|\,,$$

we find from (30.4) that, for a.e. $s < r_0$,

$$P(\mathcal{E}; B_s) \le P(\mathcal{F}; B_s) + m'(s) + 2C\, m(s)\,. \tag{30.14}$$

We now claim that, for a.e. $s < r_0$,

$$P(\mathcal{E}; B_s) - P(\mathcal{F}; B_s) \ge \frac{m(s)^{(n-1)/n} - m'(s)}{2}\,. \tag{30.15}$$

Indeed, by (12.17), (15.15), and (12.38),

$$\begin{aligned}\sum_{h=2}^{N} P(\mathcal{E}(h); B_s) \ge P(E_s; B_s) &= P(E_s \cap B_s) - m'(s)\\ &\ge m(s)^{(n-1)/n} - m'(s)\,,\end{aligned} \tag{30.16}$$

while, at the same time, by (29.6),

$$2\big(P(\mathcal{E}; B_s) - P(\mathcal{F}; B_s)\big) = \sum_{h=0}^{N} P(\mathcal{E}(h); B_s) - \sum_{h=0}^{N} P(\mathcal{F}(h); B_s)\,; \tag{30.17}$$

if we are in the case that (30.7) holds true, then $\mathcal{F}(1) = \mathcal{E}(1) \cup E_s$, $\mathcal{F}(0) = \mathcal{E}(0)$, and $P(\mathcal{F}(h); B_s) = 0$ by $|\mathcal{F}(h) \cap B_s| = 0$ for $2 \le h \le N$; hence,by (30.17) and (30.16),

$$2\Big(P(\mathcal{E};B_s)-P(\mathcal{F};B_s)\Big)\ge P(\mathcal{E}(1);B_s)-P\Big(\mathcal{E}(1)\cup E_s;B_s\Big)$$
$$+m(s)^{(n-1)/n}-m'(s)\,, \qquad (30.18)$$

where, by Exercise 29.7 and (30.7), we have that

$$\begin{aligned}
&P(\mathcal{E}(1);B_s)-P\Big(\mathcal{E}(1)\cup E_s;B_s\Big)\\
&=\sum_{0\le k\le N,k\ne 1}\mathcal{H}^{n-1}\Big(B_s\cap\mathcal{E}(k,1)\Big)-\sum_{1\le k\le N}\mathcal{H}^{n-1}\Big(B_s\cap\mathcal{E}(k,0)\Big)\\
&=\sum_{k=2}^{N}\mathcal{H}^{n-1}\Big(B_s\cap\mathcal{E}(k,1)\Big)-\sum_{k=2}^{N}\mathcal{H}^{n-1}\Big(B_s\cap\mathcal{E}(k,0)\Big)\ge 0\,.
\end{aligned}$$

Combining this inequality with (30.18) we conclude the proof of (30.15) in the case (30.7) holds true; since the complementary case follows by an identical argument, by (30.15) and (30.14) we find that

$$m(s)^{(n-1)/n}\le 3\,m'(s)+4\,C\,m(s)\,,\qquad \text{for a.e. } s\in(0,r_0).$$

We now use assumption (30.2), which gives, taking $\varepsilon_0<1$,

$$m(s)\le m(r)\le\varepsilon_0\,r^n\le r_0^n\,,\qquad \forall s<r\,;$$

hence, up to decreasing the value of r_0 to ensure $r_0<1/8C$, we find $4\,C\,m(s)\le m(s)^{(n-1)/n}/2$ for $s\in(0,r)$, which gives

$$m(s)^{(n-1)/n}\le 6m'(s)\,,\qquad \text{for a.e. } s\in(0,r). \qquad (30.19)$$

In particular, setting $\operatorname{spt} m=[r_*,\infty)$ for some $r_*\ge 0$, assuming without loss of generality that $r_*<r$, and by first dividing (30.19) by $m(s)^{(n-1)/n}>0$, and then integrating the resulting inequality $6n(m^{1/n})'\ge 1$ on (r_*,r), we find that

$$r-r_*\le 6n\Big(m(r)^{1/n}-m(r_*)^{1/n}\Big)\,,$$

which implies $r_*\ge r/2$ thanks to (30.2) as soon as $\varepsilon_0\le(1/12n)^n$. □

30.2 Density estimates

Upper $(n-1)$-dimensional density estimates (Lemma 30.5) follow immediately from Corollary 29.17; lower $(n-1)$-dimensional estimates and, more generally, upper and lower n-dimensional estimates (Lemma 30.6) are easily obtained from the infiltration lemma.

Lemma 30.5 *If $\mathcal{E}$ is a minimizing N-cluster in $\mathbb{R}^n$, then there exist positive constants r_0 and C_0 (depending on $\mathcal{E}$ only), such that*

$$\sum_{h=0}^{N} P\big(\mathcal{E}(h); B(x,r)\big) \le C_0\, r^{n-1}\,, \tag{30.20}$$

for every $r < r_0$ and $x \in \mathbb{R}^n$.

Proof By Corollary 29.17 there exist positive constants r_0 and C, depending on $\mathcal{E}$ only, such that whenever $\mathcal{F}$ is an N-cluster with $\mathcal{F}(h)\Delta\mathcal{E}(h) \subset\subset B(x, r_0)$ for some $x \in \mathbb{R}^n$, then $P(\mathcal{E}) \le P(\mathcal{F}) + C\,|\mathbf{m}(\mathcal{E}) - \mathbf{m}(\mathcal{F})|$; we now exploit this inequality setting, for $x \in \mathbb{R}^n$ and $r < r_0$ fixed,

$$\mathcal{F}(h) = \mathcal{E}(h) \setminus B(x,r)\,, \qquad 1 \le h \le N\,.$$

In this way, by (29.6) we easily find that, for a.e. $r > 0$,

$$P(\mathcal{E}) = \frac{1}{2}\sum_{h=0}^{N} P\Big(\mathcal{E}(h); B(x,r)\Big) + P\Big(\mathcal{E}; \mathbb{R}^n \setminus B(x,r)\Big),$$
$$P(\mathcal{F}) \le n\omega_n r^{n-1} + P\Big(\mathcal{E}; \mathbb{R}^n \setminus B(x,r)\Big);$$

taking into account that $|\,\mathbf{m}(\mathcal{E}) - \mathbf{m}(\mathcal{F})| \le \omega_n\, r^n$, we finally conclude that

$$\frac{1}{2}\sum_{h=0}^{N} P\Big(\mathcal{E}(h); B(x,r)\Big) \le n\omega_n r^{n-1} + C\,\omega_n\, r^n \le \Big(n\omega_n + C r_0\Big) r^{n-1}\,. \qquad \square$$

Lemma 30.6 *If $n \ge 2$, then there exist positive constants c_0, c_1, and c_2, depending on n only, and with $c_1 < 1$, with the following property: if $\mathcal{E}$ is a minimizing N-cluster in $\mathbb{R}^n$, then there exists $r_0 > 0$ (depending on $\mathcal{E}$ only) such that*

$$c_0\,\omega_n\, r^n \le |\,\mathcal{E}(h) \cap B(x,r)| \le c_1\,\omega_n\, r^n\,, \tag{30.21}$$
$$c_2\omega_{n-1}\, r^{n-1} \le P\big(\mathcal{E}(h); B(x,r)\big), \tag{30.22}$$

whenever $0 \le h \le N$, $x \in \partial\mathcal{E}(h)$, and $r < r_0$.

The n-dimensional density estimates (30.21), combined with the results of Section 30.1, lead immediately to the proof of Theorem 30.1.

Proof of Theorem 30.1 The regularity of the interfaces is proved in Remark 30.4. Now let $x \in \partial\mathcal{E}(h)$ for some $h = 0, \dots, N$. By (30.21), $x \in \mathbb{R}^n \setminus \Big(\mathcal{E}(h)^{(0)} \cup \mathcal{E}(h)^{(1)}\Big) = \partial^{\mathrm{e}}\mathcal{E}(h)$. Thus $\partial\mathcal{E}(h) \subset \partial^{\mathrm{e}}\mathcal{E}(h)$. By Federer's theorem, $\mathcal{H}^{n-1}(\partial\mathcal{E}(h) \setminus \partial^*\mathcal{E}(h)) = 0$. This proves (30.1). $\square$

Proof of Lemma 30.6 Since $0 < c_0 \le c_1 < 1$, (30.22) follows immediately by (30.21) and the relative isoperimetric inequality (12.45). We thus focus on the proof of (30.21). Let $\varepsilon_0 < \omega_n$ and r_0 be the constants appearing in Lemma 30.2. If $0 \le h \le N$ and $x \in \partial\mathcal{E}(h) = \operatorname{spt}\mu_{\mathcal{E}(h)}$, then

$$0 < |\mathcal{E}(h) \cap B(x,s)| < \omega_n\, s^n\,, \qquad \forall s > 0\,, \tag{30.23}$$

by Proposition 12.19; at the same time, if for some $r < r_0$ we have

$$|\mathcal{E}(h) \cap B(x,r)| < \varepsilon_0\, r^n\,,$$

then, by Lemma 30.2 (applied with $\Lambda = \{h\}$), we find

$$\Big|\mathcal{E}(h) \cap B\Big(x, \frac{r}{2}\Big)\Big| = 0\,,$$

in contradiction to (30.23); similarly, if for some $r < r_0$ we have

$$|\mathcal{E}(h) \cap B(x,r)| > (\omega_n - \varepsilon_0)\, r^n\,,$$

then, again by Lemma 30.2 (applied with $\Lambda = \{0,\dots,N\} \setminus \{h\}$), we find

$$\Big|\mathcal{E}(k) \cap B\Big(x, \frac{r}{2}\Big)\Big| = 0\,, \qquad \forall k \ne h\,,$$

and thus $|\mathcal{E}(h) \cap B(x, r/2)| = \omega_n\,(r/2)^n$, in contradiction to (30.23); in conclusion, for every $x \in \partial\mathcal{E}(h)$ and $r < r_0$ we have proved the validity of (30.21) with $c_0 = (\varepsilon_0/\omega_n)$ and $c_1 = 1 - c_0$. □

30.3 Regularity of planar clusters

We finally describe the singularities and the near-to-singularities behavior of planar minimizing clusters. In the following statement, a straight segment will be considered as a circular arc with zero curvature.

Theorem 30.7 (Structure of planar clusters) *If $\mathcal{E}$ is a minimizing cluster in the plane, then, for every $h = 0, \dots, N$, $\partial\mathcal{E}(h)$ is the union of finitely many circular arcs, and all the arcs belonging to $\mathcal{E}(h,k)$ for a given $k \ne h$ have the same curvature $\kappa_{kh} \in \mathbb{R}$. Moreover, the singular set of $\mathcal{E}$,*

$$\Sigma(\mathcal{E}) = \bigcup_{h=0}^{N} \partial\mathcal{E}(h) \setminus \partial^*\mathcal{E}(h) = \bigcup_{0 \le h < k \le N} \big(\partial\mathcal{E}(h) \cap \partial\mathcal{E}(k)\big) \setminus \mathcal{E}(h,k)$$

is discrete, and for each $x \in \Sigma(\mathcal{E})$, there exist exactly three circular arcs, belonging to three different interfaces, which share x as one of their endpoints, and meet at x forming three 120 degree angles. In particular, at a neighborhood of each of its boundary points, $\mathcal{E}$ is a diffeomorphic images of either *a pair of complementary half-spaces or a Steiner partition of the plane; see Figure 30.2.*

Exercise 30.8 Show that if $\mathcal{E}$ is a minimizing N-cluster in $\mathbb{R}^2$, then $E = \bigcup_{h=1}^N \mathcal{E}(h)$ is a connected set. *Hint:* Suitably translate two connected components of E until they touch at a point x, to construct a new minimizing cluster $\mathcal{E}'$ with $\mathrm{vol}\,(\mathcal{E}') = \mathrm{vol}\,(\mathcal{E})$ which either satisfies $P(\mathcal{E}') < P(\mathcal{E})$ or violates Theorem 30.7 at x.

We introduce the following terminology. An **improper M-cluster** in $\mathbb{R}^n$ is a family $\mathcal{F} = \{\mathcal{F}(i)\}_{i=1}^M$ of sets of locally finite perimeter in $\mathbb{R}^n$ such that $|\mathcal{F}(i)| = \infty$ for every $i = 1, \dots, M$, and, moreover,

$$\Big|\mathbb{R}^n \setminus \bigcup_{i=1}^M \mathcal{F}(i)\Big| = 0\,.$$

The perimeter of an improper cluster $\mathcal{F}$ in a bounded set A is then defined as

$$P(\mathcal{F}; A) = \frac{1}{2} \sum_{i=1}^M P(\mathcal{F}(i); A)\,.$$

We say that $\mathcal{F}$ is a **minimizing improper M-cluster** provided

$$P(\mathcal{F}; B_R) \le P(\mathcal{G}; B_R)\,,$$

whenever $R > 0$ and $\mathcal{G}$ is an improper M-cluster with $\mathcal{G}(i)\Delta\mathcal{F}(i) \subset\subset B_R$, $i = 1, \dots, M$. A **cone-like M-cluster** $\mathcal{F}$ is an improper M-cluster such that $r\mathcal{F}(i) = \mathcal{F}(i)$ for every $r > 0$, $i = 1, ..., M$. Finally, we denote by $\mathcal{H}$, the cone-like 2-cluster in $\mathbb{R}^2$ defined by the two complementary half-planes

$$\mathcal{H}(1) = \Big\{x \in \mathbb{R}^2 : x_1 > 0\Big\}, \qquad \mathcal{H}(2) = \Big\{x \in \mathbb{R}^2 : x_1 < 0\Big\},$$

and by $\mathcal{T}$, the cone-like 3-cluster in $\mathbb{R}^2$ given by the angular sectors

$$\mathcal{T}(1) = \Big\{x \in \mathbb{R}^2 : \frac{x_1}{|x|} > \frac{1}{2}\Big\}, \qquad \mathcal{T}(2) = \Big\{x \in \mathbb{R}^2 \setminus \mathcal{T}(1) : x_2 > 0\Big\},$$
$$\mathcal{T}(3) = \Big\{x \in \mathbb{R}^2 \setminus \mathcal{T}(1) : x_2 < 0\Big\};$$

see Figure 30.2. The following fact proves crucial for the validity of Theorem 30.7.

Proposition 30.9 *If $\mathcal{F}$ is a cone-like minimal cluster in $\mathbb{R}^2$, then, up to rotation, $\mathcal{F} \in \{\mathcal{H}, \mathcal{T}\}$.*

Proof Given three distinct points a, b, c in $\mathbb{R}^2$, let $[abc]$ denote the angle defined by these points with vertex in b and measured in radians. The basic geometric fact we need to point out is that, if $|a-c| = |b-c|$, $0 < [acb] < 2\pi/3$ and

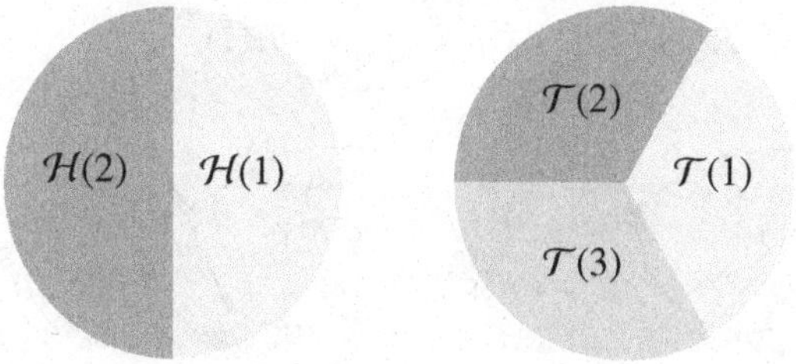

Figure 30.2 The minimizing cone-like improper clusters in $\mathbb{R}^2$. A **Steiner partition of** $\mathbb{R}^2$ is an improper 3-cluster of $\mathbb{R}^2$ which agrees with $\mathcal{T}$ up to rotation.

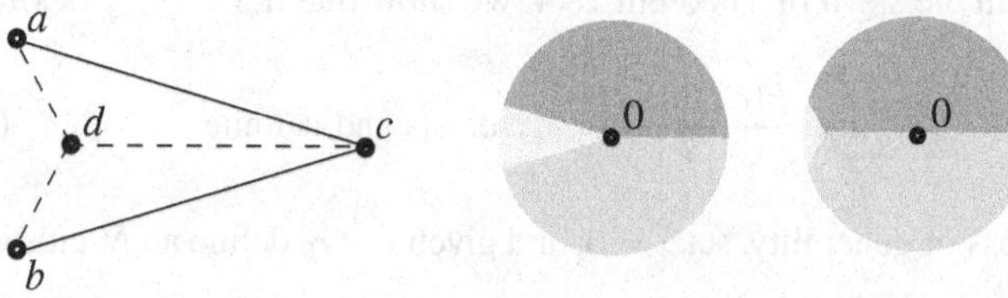

Figure 30.3 Since $[acb] < 2\pi/3$, it costs less length to first connect a and b to d, and then move straight to c, rather than directly connecting a and b to c. Correspondingly, no minimizing cone-like cluster in $\mathbb{R}^2$ can contain an angle smaller than $2\pi/3$.

d denotes the point in $\mathbb{R}^2$ lying in the interior of the triangle defined by a, b, and c, and such that $[adc] = [adb] = [bdc] = (2\pi/3)$, then

$$|a - d| + |b - d| + |c - d| < |a - c| + |b - c|, \tag{30.24}$$

see Figure 30.3. If $\mathcal{F}$ were a minimizing cone-like M-cluster in $\mathbb{R}^2$ with $M \geq 4$, then there would be at least four boundary points of $\mathcal{F}$ on S^1. At least two of them would define an angle smaller than $2\pi/3$ at the origin. We may thus exploit (30.24) to violate the minimality of $\mathcal{F}$. Thus, there are no minimizing cone-like M-clusters in $\mathbb{R}^2$ with $M \geq 4$. Similarly, the only possible minimizing cone-like 3-cluster in $\mathbb{R}^2$ is (up to rotations) $\mathcal{T}$. Since segments provide the least length way to connect two points in the plane, the only minimizing cone-like 2-cluster in $\mathbb{R}^2$, up to rotations, is given by $\mathcal{H}$. □

Proof of Theorem 30.7 By Corollary 29.17, there exist positive constants C and r_0, depending on $\mathcal{E}$ only, such that if $\mathcal{F}$ is an N-cluster with $\mathcal{F}(h)\Delta\mathcal{E}(h) \subset\subset B(x, r_0)$ for $1 \leq h \leq N$ and $x \in \mathbb{R}^n$, then

$$P(\mathcal{E}) \leq P(\mathcal{F}) + C\,|\,\mathbf{m}(\mathcal{E}) - \mathbf{m}(\mathcal{F})|. \tag{30.25}$$

We now divide the proof into several steps.

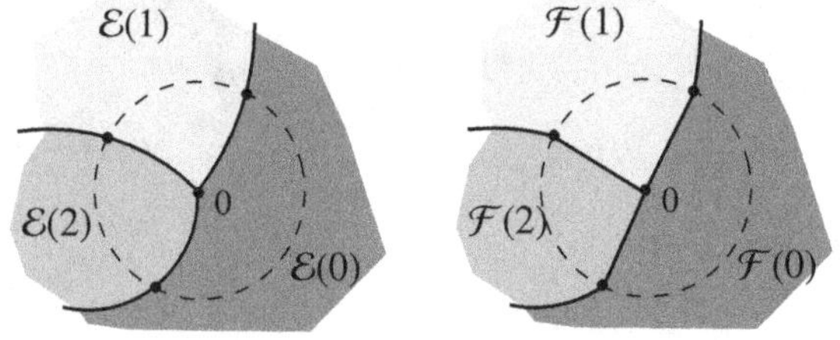

Figure 30.4 The comparison cluster used in the proof of (30.26).

Step one: In the spirit of Theorem 28.4, we show that if $x \in \bigcup_{h=1}^N \partial\mathcal{E}(h)$, then

$$\lim_{r\to 0^+} \frac{P(\mathcal{E}; B(x,r))}{r} \qquad \text{exists and is finite}. \tag{30.26}$$

Without loss of generality, set $x = 0$, and given $r < r_0$ define an N-cluster $\mathcal{F}$ as

$$\mathcal{F}(h) = \big(\mathcal{E}(h)\setminus B_r\big) \cup \big\{t\,x : t \in (0,1)\,, x \in \mathcal{E}(h)\cap\partial B_r\big\}, \qquad 1 \le h \le N\,;$$

see Figure 30.4. In this way, by (30.25), for a.e. $r \in (0, r_0)$ we find

$$P(\mathcal{E}; B_r) \le P(\mathcal{F}; B_r) + C\,\pi\,r^2\,. \tag{30.27}$$

In order to exploit this inequality, let us first notice that, by Theorem 30.1, there exists a finite set $I \subset \mathbb{R}$ such that $\bigcup_{h=0}^N \partial^*\mathcal{E}(h)$ is a countable union of circular arcs whose curvatures belong to I. In particular, if we set

$$\Gamma = \bigcup_{h=0}^N \left\{x \in \partial^*\mathcal{E}(h) : \nu_{\mathcal{E}(h)}(x) = \pm\frac{x}{|x|}\right\},$$

then, up to further decreasing the value of r_0, we may ensure that $\Gamma \cap B_{r_0} = \emptyset$; otherwise, there would be arcs of circle of arbitrarily large curvature among the interfaces of $\mathcal{E}$, thus contradicting the finiteness of I. Then, by applying the coarea formula (18.35) with $g = 1_{\partial^*\mathcal{E}(h)\setminus\Gamma}(1 - (\nu_{\mathcal{E}(h)}(x)\cdot x/|x|)^2)^{-1/2}$ and $u(x) = |x|$, we find that for $r \in (0, r_0)$,

$$\begin{aligned} P(\mathcal{E}; B_r) &= \frac{1}{2}\sum_{h=0}^N \mathcal{H}^1\Big((\partial^*\mathcal{E}(h)\setminus\Gamma)\cap B_r\Big) \\ &= \frac{1}{2}\sum_{h=0}^N \int_0^r \mathrm{d}s \int_{(\partial^*\mathcal{E}(h)\setminus\Gamma)\cap\partial B_s} \frac{\mathrm{d}\mathcal{H}^0(x)}{\sqrt{1 - (\nu_{\mathcal{E}(h)}(x)\cdot(x/|x|))^2}}\,. \end{aligned}$$

Thus $p(r) = P(\mathcal{E}; B_r)$ is absolutely continuous on $(0, r_0)$, with

$$p'(r) = \frac{1}{2}\sum_{h=0}^{N}\int_{\partial^*\mathcal{E}(h)\cap\partial B_r}\frac{d\mathcal{H}^0(x)}{\sqrt{1-(\nu_{\mathcal{E}(h)}(x)\cdot(x/|x|))^2}} \geq \frac{1}{2}\sum_{h=0}^{N}\mathcal{H}^0\big(\partial^*\mathcal{E}(h)\cap\partial B_r\big). \tag{30.28}$$

Thus, by taking into account that, for a.e. $r > 0$,

$$P(\mathcal{F}; B_r) = \frac{1}{2}\sum_{h=0}^{N} r\,\mathcal{H}^0\big(\partial^*\mathcal{E}(h)\cap\partial B_r\big),$$

we may infer from (30.27) and (30.28) that

$$p(r) \leq r\,p'(r) + C_1 r^2, \qquad \text{for a.e. } r < r_0. \tag{30.29}$$

where $C_1 = C\pi$. By (30.22), there exists $c_1 > 0$ such that $P(\mathcal{E}; B_r) \geq c_1 r$ for every $r > 0$. Therefore, setting $\Lambda = C_1/c_1$, by (30.29) we find

$$\left(\frac{e^{\Lambda r}\,p(r)}{r}\right)' = \frac{e^{\Lambda r}}{r^2}\Big(r\,p'(r) + \big(r\Lambda - 1\big)p(r)\Big) \geq \frac{e^{\Lambda r}}{r^2}\big(\Lambda\,r\,p(r) - C_1 r^2\big) \geq 0,$$

for a.e. $r \in (0, r_0)$, which immediately implies (30.26).

Step two: Let $x_0 \in \Sigma(\mathcal{E})$, let $r_k \to 0$ as $k \to \infty$ and define clusters $\{\mathcal{E}_k\}_{k\in\mathbb{N}}$ by

$$\mathcal{E}_k(h) = \frac{\mathcal{E}(h) - x_0}{r_k}, \qquad 1 \leq h \leq N. \tag{30.30}$$

By the upper density estimate (30.20), we see that

$$\sup_{k\in\mathbb{N}} P(\mathcal{E}_k; B_R) \leq C_0 R.$$

By Corollary 12.27, there exists a subsequence $\{r_k'\}_{k\in\mathbb{N}}$ of $\{r_k\}_{k\in\mathbb{N}}$ and a partition $\mathcal{F} = \{\mathcal{F}(h)\}_{i=0}^N$ of $\mathbb{R}^2$ into sets of locally finite perimeter such that

$$\mathcal{E}_k(h) \overset{\text{loc}}{\to} \mathcal{F}(h), \qquad \forall h = 0, \dots, N.$$

We notice that there must be at least three different indexes $h \in \{0, \dots, N\}$ such that $x_0 \in \partial\mathcal{E}(h)$, for otherwise, for some $0 \leq h < h' \leq N$, we would have

$$\limsup_{r\to 0^+}\frac{|\mathcal{E}(h)\cap B(x_0, r)|}{\pi r^2} + \frac{|\mathcal{E}(h')\cap B(x_0, r)|}{\pi r^2} = 1,$$

which, by Corollary 30.3 would imply x_0 was a regular point, against our assumption. Taking the lower density estimate (30.21) into account, this means that there are at least three distinct indexes $h, h', h'' \in \{0, \dots, N\}$ such that

$$|\mathcal{E}(j) \cap B(x_0, r)| \geq c_0 \pi r^2 , \qquad j = h, h', h'' .$$

As a consequence, if $j \in \{h, h', h''\}$ and $R > 0$, then

$$|\mathcal{F}(j) \cap B_R| = \lim_{k\to\infty} |\mathcal{E}_k(j) \cap B_R| = \lim_{k\to\infty} \frac{|\mathcal{E}(j) \cap B(x_0, r_k R)|}{r_k^2} \geq c_0 \pi R^2 ,$$

that is $|\mathcal{F}(j)| = \infty$ for $j = h, h', h''$. Thus we may relabel the sets in $\mathcal{F}$ so that $\mathcal{F} = \{\mathcal{F}(i)\}_{i=1}^M$ defines an improper M-cluster in $\mathbb{R}^2$, with $M \geq 3$, and

$$\mathcal{E}_k(\sigma(i)) \overset{\text{loc}}{\to} \mathcal{F}(i) , \qquad \text{as } k \to \infty ,$$

for some injective function $\sigma\colon \{1, \dots, M\} \to \{1, \dots, N\}$, and

$$\mathcal{E}_k(h) \overset{\text{loc}}{\to} \emptyset , \qquad \text{as } k \to \infty ,$$

if $h \neq \sigma(i)$ for every $i = 1, \dots, M$. The proof of Theorem 21.14 is easily adapted to the present context to show that, in fact, $\mathcal{F}$ is a minimizing improper M-cluster. Let us now prove that $\mathcal{F}$ is a cone-like improper M-cluster in $\mathbb{R}^2$. Now, for a.e. $s > 0$ (specifically, for those $s > 0$ such that $\mathcal{H}^1(\partial^*\mathcal{F}(i) \cap \partial B_s) = 0$ for $1 \leq i \leq M$), we have

$$\frac{P(\mathcal{F}; B_s)}{s} = \lim_{k\to\infty} \frac{P(\mathcal{E}_k; B_s)}{s} = \lim_{k\to\infty} \frac{P(\mathcal{E}; B(x_0, r_k\, s))}{r_k\, s} .$$

By step one the last limit exists, is finite, and independent of s. Therefore we conclude $P(\mathcal{F}; B_s)/s$ is a constant function on $(0, \infty)$. If we repeat the argument of step one with $\mathcal{F}$ in place of $\mathcal{E}$, then we see that, for a.e. $s > 0$,

$$\begin{aligned} P(\mathcal{F}; B_s) &\leq \frac{1}{2} \sum_{i=1}^M s\,\mathcal{H}^0\big(\partial^*\mathcal{F}(i) \cap \partial B_s\big) \\ &\leq \frac{1}{2} \sum_{i=1}^M s \int_{\partial^*\mathcal{F}(i)\cap\partial B_s} \frac{d\mathcal{H}^0(x)}{\sqrt{1 - (\nu_{\mathcal{F}(i)}(x) \cdot (x/|x|))^2}} = s \frac{d}{ds} P(\mathcal{F}; B_s) . \end{aligned}$$

Since $P(\mathcal{F}; B_s)/s$ is constant in s, we thus find that, if $i = 1, \dots, M$, then

$$\nu_{\mathcal{F}(i)}(x) \cdot x = 0 , \qquad \mathcal{H}^1\text{-a.e. on } \partial^*\mathcal{F}(i) .$$

By Proposition 28.8, each $\mathcal{F}(i)$ is a cone. Hence, $\mathcal{F}$ is a cone-like minimizing improper M-cluster in $\mathbb{R}^2$, with $M \geq 3$. Moreover, by Proposition 30.9, up to rotation, $\mathcal{F} = \mathcal{T}$. (Note that we have not proved yet that $(\mathcal{E} - x)/r$ locally converges, as $r \to 0^+$ and up to rotation, to $\mathcal{T}$: indeed, different vanishing sequences of radii could possibly converge to different rotations of $\mathcal{T}$.)

Step three: We prove $\Sigma(\mathcal{E})$ is discrete. Let $\{x_k\}_{k\in\mathbb{N}} \subset \Sigma(\mathcal{E})$. Since $\mathcal{E}$ is bounded, we may assume without loss of generality that, as $k \to \infty$,

$$x_k \to x_0 \in \bigcup_{h=0}^{N} \partial\mathcal{E}(h), \qquad \frac{x_k - x_0}{r_k} \to v \in S^1,$$

where $r_k = |x_k - x_0| > 0$ for $k \in \mathbb{N}$. We thus find that (30.30) defines a sequence $\{\mathcal{E}_k\}_{k\in\mathbb{N}}$ of minimizing N-clusters in $\mathbb{R}^2$, with $v \in \Sigma(\mathcal{E}_k)$ for every $k \in \mathbb{N}$. Up to extracting a subsequence, we may assume either that 0 is a regular point for $\mathcal{E}_k$ for every $k \in \mathbb{N}$, or that $0 \in \Sigma(\mathcal{E}_k)$ for every $k \in \mathbb{N}$. In the first case, by arguing as in step two, up to extracting a further subsequence, and up to a single rotation, we show the existence of $\sigma\colon \{1,2\} \to \{0,\dots,N\}$ such that

$$\mathcal{E}_k(\sigma(i)) \stackrel{\text{loc}}{\to} \mathcal{H}(i), \quad i = 1,2, \qquad \mathcal{E}_k(h) \stackrel{\text{loc}}{\to} \emptyset, \quad h \neq \sigma(1), \sigma(2).$$

In particular, for every $s < 1$ it would be

$$\lim_{k\to\infty} \frac{|\mathcal{E}_k(\sigma(1)) \cap B(v,s)|}{\pi s^2} + \frac{|\mathcal{E}_k(\sigma(2)) \cap B(v,s)|}{\pi s^2} = 1,$$

which, by Corollary 30.3, would imply v was a regular point of $\mathcal{E}_k$ for k large enough, a contradiction. In the second case, the argument of step two would now prove (again, up to extracting a further subsequence and up to a single rotation) the existence of $\sigma\colon \{1,2,3\} \to \{0,\dots,N\}$ such that

$$\mathcal{E}_k(\sigma(i)) \stackrel{\text{loc}}{\to} \mathcal{T}(i), \quad i = 1,2,3, \qquad \mathcal{E}_k(h) \stackrel{\text{loc}}{\to} \emptyset, \quad h \neq \sigma(1), \sigma(2), \sigma(3).$$

In this case, there would be $h, h' \in \{\sigma(1), \sigma(2), \sigma(3)\}$ such that, for every $s < 1$,

$$\lim_{k\to\infty} \frac{|\mathcal{E}_k(h) \cap B(v,s)|}{\pi s^2} + \frac{|\mathcal{E}_k(h') \cap B(v,s)|}{\pi s^2} = 1.$$

Once again, Corollary 30.3 would force v to be a regular point of $\mathcal{E}_k$ for k large enough. In conclusion, $\Sigma(\mathcal{E})$ is a discrete set.

Step four: By Theorem 30.1, we may decompose $\bigcup_{h=0}^{N} \partial^*\mathcal{E}(h)$ as a countable union of relatively open circular arcs $\{\gamma_i\}_{i\in\mathbb{N}}$ such that the end-points of each γ_i are singular points of $\mathcal{E}$. Now let x_0 be a singular point of $\mathcal{E}$. If there are M circular arcs γ_i that have x_0 as an end-point, then we easily find

$$\lim_{s\to 0^+} \frac{P(\mathcal{E}; B(x_0,s))}{s} \geq M.$$

Thus, by (30.20), only finitely many circular arcs γ_i can meet at a singular point x_0. Since, by step three, $\Sigma(\mathcal{E})$ is discrete, we conclude that $\{\gamma_i\}_{i\in\mathbb{N}}$ is in fact *a finite set.* It is now trivial to see that, in fact, for every $x_0 \in \Sigma(\mathcal{E})$ there exists $\sigma\colon \{1,2,3\} \to \{0,\dots,N\}$ such that, up to a single rotation, and as $r \to 0^+$,

$$\frac{\mathcal{E}(\sigma(i)) - x_0}{r} \to \mathcal{T}(i)\,, \qquad i = 1,2,3,$$
$$\frac{\mathcal{E}(h) - x_0}{r} \to \emptyset\,, \qquad h \neq \sigma(1),\sigma(2),\sigma(3)\,.$$

We may thus explicitly define a diffeomorphism $f\colon \mathbb{R}^2 \to \mathbb{R}^2$ and an open neighborhood A of x_0, such that (with an obvious abuse of notation),

$$A \cap \mathcal{E} = f(B \cap \mathcal{T})\,.$$

In particular, there are exactly three circular arcs γ_i which meet at x_0, and they have to form three 120 degrees angles in doing so. □

Notes

Chapter 29 is modeled after [Alm76]. In particular, Lemma 29.10 is Almgren's VI.13, Lemma 29.12 is (a particular case of) VI.14, Lemma 29.13 is VI.8-9, and Theorem 29.14 is VI.10-11-12. An alternative argument to Almgren's original existence proof is sketched by Morgan in [Mor09, Chapter 13].

The key result in Chapter 30, namely the infiltration lemma, is due to Leonardi [Leo01, Theorem 3.1]. An analogous lemma, under more restrictive assumptions, was announced by White [Whi96, Section 11, Property P]. We also mention [Mor94] for an alternative proof of Theorem 30.7.

References

[AB90] L. Ambrosio and A. Braides. Functionals defined on partitions in sets of finite perimeter II: Semicontinuity, relaxation and homogenization. *J. Math. Pures Appl.*, **69**:307–333, 1990.

[AC09] F. Alter and V. Caselles. Uniqueness of the Cheeger set of a convex body. *Nonlinear Anal.*, **70**(1):32–44, 2009.

[ACC05] F. Alter, V. Caselles, and A. Chambolle. A characterization of convex calibrable sets in $\mathbb{R}^n$. *Math. Ann.*, **332**(2):329–366, 2005.

[ACMM01] L. Ambrosio, V. Caselles, S. Masnou, and J. M. Morel. Connected components of sets of finite perimeter and applications to image processing. *J. Eur. Math. Soc. (JEMS)*, **3**(1):39–92, 2001.

[AFP00] L. Ambrosio, N. Fusco, and D. Pallara. *Functions of Bounded Variation and Free Discontinuity Problems*. Oxford Mathematical Monographs. New York, The Clarendon Press, Oxford University Press, 2000. xviii+434 pp.

[All72] W. K. Allard. On the first variation of a varifold. *Ann. Math.*, **95**:417–491, 1972.

[All75] W. K. Allard. On the first variation of a varifold: boundary behaviour. *Ann. Math.*, **101**:418–446, 1975.

[Alm66] F. J. Almgren Jr. *Plateau's Problem. An Invitation to Varifold Geometry*. Mathematics Monograph Series. New York, Amsterdam, W. A. Benjamin, Inc., 1966. xii+74 pp.

[Alm68] F. J. Almgren, Jr. Existence and regularity almost everywhere of solutions to elliptic variational problems among surfaces of varying topological type and singularity structure. *Ann. Math.*, **87**:321–391, 1968.

[Alm76] F. J. Almgren, Jr. Existence and regularity almost everywhere of solutions to elliptic variational problems with constraints. *Mem. Amer. Math. Soc.*, **4**(165):viii+199 pp, 1976.

[Alm00] F. J. Almgren, Jr. *Almgren's Big Regularity Paper. Q-valued Functions Minimizing Dirichlet's Integral and the Regularity of Area-minimizing Rectifiable Currents up to Codimension 2*, Volume 1 of World Scientific Monograph Series in Mathematics. River Edge, NJ, World Scientific Publishing Co., Inc., 2000. xvi+955 pp.

[AM98] L. Ambrosio and C. Mantegazza. Curvature and distance function from a manifold. *J. Geom. Anal.*, **8**:723–748, 1998.

[Amb97] L. Ambrosio. *Corso Introduttivo alla Teoria Geometrica della Misura ed alle Superfici Minime*. Pisa, Edizioni Scuola Normale Superiore di Pisa, 1997.

[AP99] L. Ambrosio and E. Paolini. Partial regularity for quasi minimizers of perimeter. *Ricerche Mat.*, **48**:167–186, 1999.

[AP01] L. Ambrosio and E. Paolini. Errata-corrige: "Partial regularity for quasi minimizers of perimeter". *Ricerche Mat.*, **50**:191–193, 2001.

[BCCN06] G. Bellettini, V. Caselles, A. Chambolle, and M. Novaga. Crystalline mean curvature flow of convex sets. *Arch. Rational Mech. Anal.*, **179**(1): 109–152, 2006.

[BdC84] J. L. Barbosa and M. do Carmo. Stability of hypersurfaces with constant mean curvature. *Math. Z.*, **185**(3):339–353, 1984.

[BDGG69] E. Bombieri, E. De Giorgi, and E. Giusti. Minimal cones and the Bernstein problem. *Invent. Math.*, **7**:243–268, 1969.

[BGM03] E. Barozzi, E. Gonzalez, and U. Massari. The mean curvature of a Lipschitz continuous manifold. *Rend. Mat. Acc. Lincei s. 9*, **14**:257–277, 2003.

[BGT87] E. Barozzi, E. Gonzalez, and I. Tamanini. The mean curvature of a set of finite perimeter. *Proc. Amer. Math. Soc.*, **99**(2):313–316, 1987.

[Bin78] D. Bindschadler. Absolutely area minimizing singular cones of arbitrary codimension. *Trans. Amer. Math. Soc.*, **243**:223–233, 1978.

[BM94] J. E. Brothers and F. Morgan. The isoperimetric theorem for general integrands. *Mich. Math. J.*, **41**(3):419–431, 1994.

[BNP01a] G. Bellettini, M. Novaga, and M. Paolini. On a crystalline variational problem I: First variation and global L^∞ regularity. *Arch. Ration. Mech. Anal.*, **157**(3):165–191, 2001.

[BNP01b] G. Bellettini, M. Novaga, and M. Paolini. On a crystalline variational problem II: *BV* regularity and structure of minimizers on facets. *Arch. Ration. Mech. Anal.*, **157**(3):193–217, 2001.

[Bom82] E. Bombieri. Regularity theory for almost minimal currents. *Arch. Ration. Mech. Anal.*, **7**(7):99–130, 1982.

[Bon09] M. Bonacini. *Struttura delle m-bolle di Perimetro Minimo e il Teorema della Doppia Bolla Standard*. Master's Degree thesis, Università degli Studi di Modena e Reggio Emilia, 2009.

[BNR03] M. Bellettini, M. Novaga, and G. Riey. First variation of anisotropic energies and crystalline mean curvature for partitions. *Interfaces Free Bound.*, **5**(3):331–356, 2003.

[CC06] V. Caselles and A. Chambolle. Anisotropic curvature-driven flow of convex sets. *Nonlinear Anal.*, **65**(8):1547–1577, 2006.

[CCF05] M. Chlebik, A. Cianchi, and N. Fusco. The perimeter inequality under Steiner symmetrization: cases of equality. *Ann. Math.*, **162**:525–555, 2005.

[CCN07] V. Caselles, A. Chambolle, and M. Novaga. Uniqueness of the Cheeger set of a convex body. *Pacific J. Math.*, **232**(1):77–90, 2007.

[CCN10] V. Caselles, A. Chambolle, and M. Novaga. Some remarks on uniqueness and regularity of Cheeger sets. *Rend. Semin. Mat. Univ. Padova*, **123**: 191201, 2010.

[CF10] S. J. Cox and E. Flikkema. The minimal perimeter for n confined deformable bubbles of equal area. *Electron. J. Combin.*, **17**(1), 2010.

[Dav04] A. Davini. On calibrations for Lawson's cones. *Rend. Sem. Mat. Univ. Padova*, **111**:55–70, 2004.

[Dav09] G. David. Hölder regularity of two-dimensional almost-minimal sets in $\mathbb{R}^n$. *Ann. Fac. Sci. Toulouse Math. (6)*, **18**(1):65–246, 2009.

[Dav10] G. David. $C^{1+\alpha}$-regularity for two-dimensional almost-minimal sets in $\mathbb{R}^n$. *J. Geom. Anal.*, **20**(4):837–954, 2010.

[DG54] E. De Giorgi. Su una teoria generale della misura $(r-1)$-dimensionale in uno spazio ad r-dimensioni. *Ann. Mat. Pura Appl. (4)*, **36**:191–213, 1954.

[DG55] E. De Giorgi. Nuovi teoremi relativi alle misure $(r-1)$-dimensionali in uno spazio ad r-dimensioni. *Ricerche Mat.*, **4**:95–113, 1955.

[DG58] E. De Giorgi. Sulla proprietà isoperimetrica dell'ipersfera, nella classe degli insiemi aventi frontiera orientata di misura finita. *Atti Accad. Naz. Lincei. Mem. Cl. Sci. Fis. Mat. Nat. Sez. I (8)*, **5**:33–44, 1958.

[DG60] E. De Giorgi. *Frontiere Orientate di Misura Minima*. Seminario di Matematica della Scuola Normale Superiore di Pisa. Editrice Tecnico Scientifica, Pisa, 1960. 57 pp.

[Din44] A. Dinghas. Über einen geometrischen Satz von Wulff für die Gleichgewichtsform von Kristallen. *Z. Kristallogr. Mineral. Petrogr. Abt. A.*, **105**:304–314, 1944.

[DL08] C. De Lellis. *Rectifiable Sets, Densities and Tangent Measures*. Zürich Lectures in Advanced Mathematics. Zürich, European Mathematical Society (EMS), 2008. vi+127 pp.

[DLS11a] C. De Lellis and E. N. Spadaro. Center manifold: a study case. *Disc. Cont. Din. Syst. A*, **31**(4):1249–1272, 2011.

[DLS11b] C. De Lellis and E. N. Spadaro. q-valued functions revisited. *Mem. Amer. Math. Soc.*, **211**:vi+79 pp., 2011.

[DM95] G. Dolzmann and S. Müller. Microstructures with finite surface energy: the two-well problem. *Arch. Rational Mech. Anal.*, **132**(2):101–141, 1995.

[DPM12] G. De Philippis and F. Maggi. Sharp stability inequalities for the Plateau problem. Preprint cvgmt.sns.it/paper/1715/, 2012.

[DPP09] G. De Philippis and E. Paolini. A short proof of the minimality of Simons cone. *Rend. Sem. Mat. Univ. Padova*, **121**:233–241, 2009.

[DS93] F. Duzaar and K. Steffen. Boundary regularity for minimizing currents with prescribed mean curvature. *Calc. Var.*, **1**:355–406, 1993.

[DS02] F. Duzaar and K. Steffen. Optimal interior and boundary regularity for almost minimizers to elliptic variational integrals. *J. Reine Angew. Math.*, **564**:73–138, 2002.

[EG92] L. C. Evans and R. F. Gariepy. *Measure Theory and Fine Properties of Functions*. Studies in Advanced Mathematics. Boca Raton, FL, CRC Press, 1992. viii+268 pp.

[Eva98] L. C. Evans. *Partial Differential Equations*, volume 19 of Graduate Studies in Mathematics. Providence, RI, American Mathematical Society, 1998. xviii+662 pp.

[FAB+93] J. Foisy, M. Alfaro, J. Brock, N. Hodges, and J. Zimba. The standard double soap bubble in $\mathbb{R}^2$ uniquely minimizes perimeter. *Pacific J. Math.*, **159**(1), 1993.

[Fal86] K. J. Falconer. *The Geometry of Fractal Sets*, volume 85 of Cambridge Tracts in Mathematics. Cambridge, Cambridge University Press, 1986. xiv+162 pp.

[Fal90] K. J. Falconer. *Fractal Geometry. Mathematical Foundations and Applications*. Chichester, John Wiley and Sons, Ltd., 1990. xxii+288 pp.

[Fal97] K. J. Falconer. *Techniques in Fractal Geometry*. Chichester, John Wiley and Sons, Ltd., 1997. xviii+256 pp.

[Fed69] H. Federer. *Geometric Measure Theory*, volume 153 of Die Grundlehren der mathematischen Wissenschaften. New York, Springer-Verlag New York Inc., 1969. xiv+676 pp.

[Fed70] H. Federer. The singular sets of area minimizing rectifiable currents with codimension one and of area minimizing flat chains modulo two with arbitrary codimension. *Bull. Amer. Math. Soc.*, **76**:767–771, 1970.

[Fed75] H. Federer. Real flat chains, cochains and variational problems. *Indiana Univ. Math. J.*, **24**:351–407, 1974/75.

[FF60] H. Federer and W. H. Fleming. Normal and intregral currents. *Ann. Math.*, **72**:458–520, 1960.

[Fin86] R. Finn. *Equilibrium Capillary Surfaces*, volume 284 of Die Grundlehren der mathematischen Wissenschaften. New York, Springer-Verlag New York Inc., 1986. xi+245 pp.

[Fle62] W. H. Fleming. On the oriented plateau problem. *Rend. Circ. Mat. Palermo (2)*, **11**:69–90, 1962.

[FM91] I. Fonseca and S. Müller. A uniqueness proof for the Wulff theorem. *Proc. R. Soc. Edinb. Sect. A*, **119**(12):125–136, 1991.

[FM92] I. Fonseca and S. Müller. Quasi-convex integrands and lower semicontinuity in L^1. *SIAM J. Math. Anal.*, **23**(5):1081–1098, 1992.

[FM11] A. Figalli and F. Maggi. On the shape of liquid drops and crystals in the small mass regime. *Arch. Rat. Mech. Anal.*, **201**:143–207, 2011.

[FMP08] N. Fusco, F. Maggi, and A. Pratelli. The sharp quantitative isoperimetric inequality. *Ann. Math.*, **168**:941–980, 2008.

[FMP10] A Figalli, F. Maggi, and A. Pratelli. A mass transportation approach to quantitative isoperimetric inequalities. *Inv. Math.*, **182**(1):167–211, 2010.

[Fus04] N. Fusco. The classical isoperimetric theorem. *Rend. Accad. Sci. Fis. Mat. Napoli (4)*, **71**:63–107, 2004.

[Gar02] R. J. Gardner. The Brunn–Minkowski inequality. *Bull. Am. Math. Soc. (NS)*, **39**(3):355–405, 2002.

[Giu80] E. Giusti. The pendent water drop. A direct approach. *Bollettino UMI*, **17-A**:458–465, 1980.

[Giu81] E. Giusti. The equilibrium configuration of liquid drops. *J. Reine Angew. Math.*, **321**:53–63, 1981.

[Giu84] E. Giusti. *Minimal Surfaces and Functions of Bounded Variation*, volume 80 of Monographs in Mathematics. Basel, Birkhäuser Verlag, 1984. xii+240 pp.

[GM94] E. H. A. Gonzalez and U. Massari. Variational mean curvatures. *Rend. Sem. Mat. Univ. Politec. Torino*, **52**(1):128, 1994.

[GM05] M. Giaquinta and L. Martinazzi. *An Introduction to the Regularity Theory for Elliptic Systems, Harmonic Maps and Minimal Graphs*, volume 2 of Lecture Notes. Scuola Normale Superiore di Pisa (New Series). Pisa, Edizioni della Normale, 2005. xiv+676 pp.

[GMS98a] M. Giaquinta, G. Modica, and J. Soucek. *Cartesian Currents in the Calculus of Variations. I. Cartesian Currents*, volume 37 of Ergebnisse der Mathematik und ihrer Grenzgebiete. 3. Folge. A Series of Modern Surveys in Mathematics. Berlin, Springer-Verlag, 1998. xxiv+711 pp.

[GMS98b] M. Giaquinta, G. Modica, and J. Soucek. *Cartesian Currents in the Calculus of Variations. II. Variational Integrals*, volume 38 of Ergebnisse der Mathematik und ihrer Grenzgebiete. 3. Folge. A Series of Modern Surveys in Mathematics. Berlin, Springer-Verlag, 1998. xxiv+697 pp.

[GMT80] E. Gonzalez, U. Massari, and I. Tamanini. Existence and regularity for the problem of a pendent liquid drop. *Pacific Journal of Mathematics*, **88**(2): 399–420, 1980.

[GMT93] E. H. A. Gonzales, U. Massari, and I. Tamanini. Boundaries of prescribed mean curvature. *Atti Accad. Naz. Lincei Cl. Sci. Fis. Mat. Natur. Rend. Lincei (9) Mat. Appl.*, **4**(3):197–206, 1993.

[Gon76] E. Gonzalez. Sul problema della goccia appoggiata. *Rend. Sem. Mat. Univ. Padova*, **55**:289–302, 1976.

[Gon77] E. Gonzalez. Regolarità per il problema della goccia appoggiata. *Rend. Sem. Mat. Univ. Padova*, **58**:25–33, 1977.

[Grü87] M. Grüter. Boundary regularity for solutions of a partitioning problem. *Arch. Rat. Mech. Anal.*, **97**:261–270, 1987.

[GT77] E. Gonzalez and I. Tamanini. Convessità della goccia appoggiata. *Rend. Sem. Mat. Univ. Padova*, **58**:35–43, 1977.

[GT98] D. Gilbarg and N. S. Trudinger. *Elliptic Partial Differential Equations of Second Order*. Berlin; New York, Springer, 1998. xiii+517 pp.

[Hal01] T. C. Hales. The honeycomb conjecture. *Discrete Comput. Geom.*, **25**(1): 1–22, 2001.

[Har77] R. Hardt. On boundary regularity for integral currents or flat chains modulo two minimizing the integral of an elliptic integrand. *Comm. Part. Diff. Equ.*, **2**:1163–1232, 1977.

[HL86] R. Hardt and F. H. Lin. Tangential regularity near the C^1 boundary. *Proc. Symp. Pure Math.*, **44**:245–253, 1986.

[HMRR02] M. Hutchings, F. Morgan, M. Ritoré, and A. Ros. Proof of the double bubble conjecture. *Ann. of Math. (2)*, **155**(2):459–489, 2002.

[HS79] R. Hardt and L. Simon. Boundary regularity and embedded solutions for the oriented Plateau problem. *Ann. Math.*, **110**:439–486, 1979.

[Hut81] J. E. Hutchinson. Fractals and self-similarity. *Indiana Univ. Math. J.*, **30** (5):713–747, 1981.

[Hut97] M. Hutchings. The structure of area-minimizing double bubbles. *J. Geom. Anal.*, **7**:285–304, 1997.

[Kno57] H. Knothe. Contributions to the theory of convex bodies. *Mich. Math. J.*, **4**:39–52, 1957.

[KP08] S. G. Krantz and H. R. Parks. *Geometric Integration Theory*, volume 80 of Cornerstones. Boston, MA, Birkhäuser Boston, Inc., 2008. xvi+339 pp.

[KS00] D. Kinderlehrer and G. Stampacchia. *An Introduction to Variational Inequalities and their Applications*, volume 31 of Classics in Applied Mathematics. Philadelphia, PA, Society for Industrial and Applied Mathematics (SIAM), 2000.

[Law72] H. B. Lawson, Jr. The equivariant Plateau problem and interior regularity. *Trans. Amer. Math. Soc.*, **173**:231–249, 1972.

[Law88] G. R. Lawlor. *A Sufficient Criterion for a Cone to be Area Minimizing*. PhD thesis, Stanford University, 1988.

[Leo00] G. P. Leonardi. Blow-up of oriented boundaries. *Rend. Sem. Mat. Univ. Padova*, **103**:211–232, 2000.

[Leo01] G. P. Leonardi. Infiltrations in immiscible fluids systems. *Proc. Roy. Soc. Edinburgh Sect. A*, **131**(2):425–436, 2001.

[Mas74] U. Massari. Esistenza e regolarità delle ipersuperfice di curvatura media assegnata in $\mathbb{R}^n$. *Arch. Rational Mech. Anal.*, **55**:357–382, 1974.

[Mas75] U. Massari. Frontiere orientate di curvatura media assegnata in L^p. *Rend. Sem. Mat. Univ. Padova*, **53**:37–52, 1975.

[Mat95] P. Mattila. *Geometry of Sets and Measures in Euclidean Spaces. Fractals and Rectifiability*, volume 44 of Cambridge Studies in Advanced Mathematics. Cambridge, Cambridge University Press, 1995. xii+343 pp.

[McC98] R. J. McCann. Equilibrium shapes for planar crystals in an external field. *Comm. Math. Phys.*, **195**(3):699–723, 1998.

[Mir65] M. Miranda. Sul minimo dell'integrale del gradiente di una funzione. *Ann. Scuola Norm. Sup. Pisa (3)*, **19**:626–665, 1965.

[Mir67] M. Miranda. Comportamento delle successioni convergenti di frontiere minimali. *Rend. Sem. Mat. Univ. Padova*, **38**:238257, 1967.

[Mir71] M. Miranda. Frontiere minimali con ostacoli. *Ann. Univ. Ferrara Sez. VII (N.S.)*, **16**:29–37, 1971.

[MM83] U. Massari and M. Miranda. A remark on minimal cones. *Boll. Un. Mat. Ital. A (6)*, **2**(1):123–125, 1983.

[MM84] U. Massari and M. Miranda. *Minimal Surfaces of Codimension One*. North-Holland Mathematics Studies, 91. Notas de Matematica [Mathematical Notes], 95. Amsterdam, North-Holland Publishing Co., 1984. xiii+243 pp.

[Mor66] C. B. Morrey, Jr. *Multiple Integrals in the Calculus of Variations*. Berlin, Heidelberg, New York, Springer-Verlag, 1966.

[Mor90] F. Morgan. A sharp counterexample on the regularity of Φ-minimizing hypersurfaces. *Bull. Amer. Math. Soc. (N.S.)*, **22**(2):295–299, 1990.

[Mor94] F. Morgan. Soap bubbles in $\mathbb{R}^2$ and in surfaces. *Pacific J. Math.*, **165**(2): 347–361, 1994.

[Mor09] F. Morgan. *Geometric Measure Theory. A Beginner's Guide. Fourth edition.* Amsterdam, Elsevier/Academic Press, 2009. viii+249 pp.

[MS86] V. D. Milman and G. Schechtman. *Asymptotic Theory of Finite-dimensional Normed Spaces*. With an appendix by M. Gromov. Number 1200 in Lecture Notes in Mathematics. Berlin, Springer-Verlag, 1986. viii+156 pp.

[MW02] F. Morgan and W. Wichiramala. The standard double bubble is the unique stable double bubble in $\mathbb{R}^2$. *Proc. Amer. Math. Soc.*, **130**(9):2745–2751, 2002.

[Pao98] E. Paolini. Regularity for minimal boundaries in $\mathbb{R}^n$ with mean curvature in R^n. *Manuscripta Math.*, **97**(1):15–35, 1998.

[Rei60] E. R. Reifenberg. Solution of the Plateau problem for m-dimensional surfaces of varying topological type. *Acta Math.*, **104**:1–92, 1960.

[Rei64a] E. R. Reifenberg. An epiperimetric inequality related to the analyticity of minimal surfaces. *Ann. of Math.*, **80**(2):1–14, 1964.

[Rei64b] E. R. Reifenberg. On the analyticity of minimal surfaces. *Ann. of Math.*, **80**(2):15–21, 1964.

[Rei08] B. W. Reichardt. Proof of the double bubble conjecture in $\mathbb{R}^n$. *J. Geom. Anal.*, **18**(1):172–191, 2008.

[RHLS03] B. W. Reichardt, C. Heilmann, Y. Y. Lai, and A. Spielman. Proof of the double bubble conjecture in $\mathbb{R}^4$ and certain higher dimensional cases. *Pacific J. Math.*, **208**(2), 2003.

[Roc70] R. T. Rockafellar. *Convex Analysis*, volume 28 of Princeton Mathematical Series. Princeton, NJ, Princeton University Press, 1970. xviii+451 pp.

[Sim68] J. Simons. Minimal varieties in Riemannian manifolds. *Ann. Math.*, **88**(1): 62–105, 1968.

[Sim73] P. Simoes. *A Class of Minimal Cones in* $\mathbf{R}^n$*,* $n \geq 8$*, that Minimize Area*. PhD thesis, University of California, Berkeley, 1973.

[Sim83] L. Simon. *Lectures on Geometric Measure Theory*, volume 3 of Proceedings of the Centre for Mathematical Analysis. Canberra, Australian National University, Centre for Mathematical Analysis, 1983. vii+272 pp.

[Sim96] L. Simon. *Theorems on Regularity and Singularity of Energy Minimizing Maps*. Basel, Boston, Berlin, Birkhaüser-Verlag, 1996.

[Spa09] E. N. Spadaro. *Q-valued Functions and Approximation of Minimal Currents*. PhD thesis, Universität Zürich, 2009.

[Spe11] D. Spector. Simple proofs of some results of Reshetnyak. *Proc. Amer. Math. Soc.*, **139**:1681–1690, 2011.

[Spi65] M. Spivak. *Calculus on Manifolds. A Modern Approach to Classical Theorems of Advanced Calculus*. New York, Amsterdam, W. A. Benjamin, Inc., 1965. xii+144 pp.

[SS82] R. Schoen and L. Simon. A new proof of the regularity theorem for rectifiable currents which minimize parametric elliptic functionals. *Indiana Univ. Math. J.*, **31**(3):415–434, 1982.

[SSA77] R. Schoen, L. Simon, and F. J. Almgren Jr. Regularity and singularity estimates on hypersurfaces minimizing parametric elliptic variational integrals. i, ii. *Acta Math.*, **139**(3–4):217–265, 1977.

[Str79] K. Stromberg. The Banach–Tarski paradox. *Amer. Math. Monthly*, **86**(3): 151–161, 1979.

[SZ98] P. Sternberg and K. Zumbrun. A Poincaré inequality with applications to volume-constrained area-minimizing surfaces. *J. Reine Angew. Math.*, **503**:63–85, 1998.

[SZ99] P. Sternberg and K. Zumbrun. On the connectedness of boundaries of sets minimizing perimeter subject to a volume constraint. *Comm. Anal. Geom.*, **7**(1):199–220, 1999.

[Tam82] I. Tamanini. Boundaries of Caccioppoli sets with Hölder-continuous normal vector. *J. Reine Angew. Math.*, **334**:27–39, 1982.

[Tam84] I. Tamanini. *Regularity Results for Almost Minimal Oriented Hypersurfaces in* $\mathbb{R}^N$. Quaderni del Dipartimento di Matematica dell'Università di Lecce. Lecce, Università di Lecce, 1984.

[Tam98] I. Tamanini. Counting elements of least-area partitions. *Atti Sem. Mat. Fis. Univ. Modena*, **46**(suppl.):963–969, 1998.

[Tay76] J. E. Taylor. The structure of singularities in soap-bubble-like and soap-film-like minimal surfaces. *Ann. of Math. (2)*, **103**(3):489–539, 1976.

[Tay78] J. E. Taylor. Crystalline variational problems. *Bull. Am. Math. Soc.*, **84**(4): 568–588, 1978.

[Wen91] H. C. Wente. A note on the stability theorem of J. L. Barbosa and M. do Carmo for closed surfaces of constant mean curvature. *Pacific J. Math.*, **147**(2):375–379, 1991.

[Whi] B. White. Regularity of singular sets for Plateau-type problems. See the announcement on author's webpage, math.stanford.edu/~white/bibrev.htm.

[Whi96] B. White. Existence of least-energy configurations of immiscible fluids. *J. Geom. Analysis*, **6**:151–161, 1996.

[Wic04] W. Wichiramala. Proof of the planar triple bubble conjecture. *J. Reine Angew. Math.*, **567**:1–49, 2004.

[WP94] D. Weaire and R. Phelan. A counter-example to Kelvin's conjecture on minimal surfaces. *Phil. Mag. Lett.*, **69**:107–110, 1994.

[Wul01] G. Wulff. Zur Frage der Geschwindigkeit des Wachsturms und der auflösungder Kristall-Flächen. *Z. Kristallogr.*, **34**:449–530, 1901.

Index

Printed in the United States
By Bookmasters

Printed in the United States
By Bookmasters